HUYGENS, Christian (Dutch, 1629–1695).  optics, mathematical theory of waves); pendulum ... ted in 1673; astronomy.

KEPLER, Johannes (German, 1571–1630). Astronomy; geometry, especially finding areas that helped in formulating his 3 laws of planetary motion {453}.

LAGRANGE, Joseph Louis (French, 1736–1813). One of greatest 18th century mathematicians. Analytical mechanics, including 3 body problem {456}; acoustics; calculus of variations; number theory; method of variation of parameters (1774) {200}; adjoint equation (1762) {409}; algebra (group theory); partial differential equations.

LAGUERRE, Edmond (French, 1834–1886). Mathematical analysis; complex variables, analytic functions {307}; Laguerre polynomials {346}.

LAPLACE, Pierre Simon (French, 1749–1827). Mechanics; astronomy; partial differential equations; Laplace's equation {566} discovered about 1787; probability.

LEGENDRE, Adrien Marie (French, 1752–1833). Number theory; elliptic functions {254}; astronomy; geometry; Legendre functions {344}.

LEIBNIZ, Gottfried Wilhelm (German, 1646–1716). Co-discoverer with Newton of calculus. Mathematical analysis; logic; philosophy; Leibniz's rule {260}; first to solve (1691) separable {35}, homogeneous {38}, and linear {53} first-order differential equations.

LIOUVILLE, Joseph (French, 1809–1882). Number theory (transcendental numbers); complex variables; Sturm-Liouville problems {356}; integral equations.

NEWTON, Isaac (English, 1642–1727). Co-discoverer with Leibniz of calculus. Mechanics, 3 laws of motion {71} and law of universal gravitation {116}; heat flow {107}; optics; mathematical analysis; series methods for solving differential equations (1671) {299}.

PARSEVAL, Marc Antoine (French, 1755–1836). Mathematical analysis; Parseval's identity connected with the theory of Fourier series {377, 394}.

PICARD, Charles Émile (French, 1856–1941). Algebraic geometry; topology; complex variables; Picard's method {314} and existence-uniqueness theorems for differential equations.

POINCARÉ, Henri Jules (French, 1854–1912). Non-linear differential equations and stability {482, 489}; topology; celestial mechanics including 3 body problem {456}; non-Euclidean geometry; philosophy.

POISSON, Siméon Denis (French, 1781–1840). Mathematical physics; electricity and magnetism; Poisson's equation {567}; Poisson's formula {599}; probability; calculus of variations; astronomy.

RICCATI, Jacopo Francesco (Italian, 1676–1754). Mathematical analysis; Riccati's equation {68} was solved in 1723 by Daniel Bernoulli and other younger members of his family.

RIEMANN, Georg Friedrich Bernhard (German, 1826–1866). One of greatest 19th century mathematicians (student of Gauss, Jacobi, and Dirichlet); complex variables {48}; non-Euclidean geometry; elliptic functions {254}; partial differential equations.

RODRIGUES, Olinde (French, 1794–1851). Mathematical analysis; Rodrigues' formula {344}.

SCHWARZ, Hermann Amandus (German, 1843–1921). Calculus of variations; existence theorems for partial differential equations; Schwarz's inequality {355}.

STURM, Jacques Charles Francois (Swiss, 1803–1855). Algebra (number of real roots of algebraic equations); geometry; fluid mechanics; acoustics; Sturm-Liouville problems {356}.

TAYLOR, Brook (English, 1685–1731). Mathematical analysis; Taylor series method {312}; singular solutions (1715); vibrations of strings {558}; motion of projectiles {446}; optics.

WRONSKI, Hcëné (Polish, 1778–1853). Determinants; Wronskian {183}; philosophy.

Alex F. Marin

PRENTICE-HALL, INC.
ENGLEWOOD CLIFFS, NEW JERSEY 07632

third edition

applied
differential
equations

MURRAY R. SPIEGEL
Mathematical Consultant and
Former Professor and Chairman,
Mathematics Department
Rensselaer Polytechnic Institute
Hartford Graduate Center

Library of Congress Cataloging in Publication Data

SPIEGEL, MURRAY R. (date)
  applied differential equations.

  Bibliography: p.
  Includes index.
  1. Differential equations.   I. Title.
QA371.S82   1981   515'.35   79-22567
ISBN 0-13-040097-1

Printed in the United States of America

20 19 18 17 16 15 14 13 12 11

Editorial production/supervision by Leslie I. Nadell
Cover design by Edsal Enterprises
Manufacturing buyers: Edmund Leone and John Hall

Prentice-Hall International (UK) Limited, *London*
Prentice-Hall of Australia Pty. Limited, *Sydney*
Prentice-Hall Canada Inc., *Toronto*
Prentice-Hall Hispanoamericana, S.A., *Mexico*
Prentice-Hall of India Private Limited, *New Delhi*
Prentice-Hall of Japan, Inc., *Tokyo*
Simon & Schuster Asia Pte. Ltd., *Singapore*
Editora Prentice-Hall do Brasil, Ltda., *Rio de Janeiro*

To
my mother

contents

CHAPTER THREE

APPLICATIONS OF FIRST-ORDER AND SIMPLE HIGHER-ORDER DIFFERENTIAL EQUATIONS 70

CHAPTER FOUR

LINEAR DIFFERENTIAL EQUATIONS 166

CHAPTER FIVE
APPLICATIONS OF LINEAR DIFFERENTIAL EQUATIONS

CHAPTER SIX
SOLUTION OF LINEAR DIFFERENTIAL EQUATIONS BY LAPLACE TRANSFORMS

CHAPTER SEVEN
SOLUTION OF DIFFERENTIAL EQUATIONS BY USE OF SERIES

CHAPTER EIGHT
◆ ORTHOGONAL FUNCTIONS AND STURM-LIOUVILLE PROBLEMS

part II

systems of ordinary differential equations

CHAPTER ELEVEN
◆ MATRIX EIGENVALUE METHODS FOR SYSTEMS OF LINEAR DIFFERENTIAL EQUATIONS 501

part III

partial differential equations

CHAPTER TWELVE
PARTIAL DIFFERENTIAL EQUATIONS IN GENERAL 540

CHAPTER THIRTEEN
SOLUTIONS OF BOUNDARY VALUE PROBLEMS
USING FOURIER SERIES 570

CHAPTER FOURTEEN
◆ SOLUTIONS OF BOUNDARY VALUE PROBLEMS USING BESSEL
AND LEGENDRE FUNCTIONS 619

APPENDIX

preface

The purpose of this book is to provide an introduction to differential equations and their applications for students of engineering, science, and mathematics. To accomplish this purpose, the book has been written with the following objectives:

1. To demonstrate how differential equations can be useful in solving many types of problems—in particular, to show the student how to (a) *translate* problems into the language of differential equations, i.e., set up mathematical formulations of problems; (b) *solve* the resulting differential equations subject to given conditions; and (c) *interpret* the solutions obtained. Elementary problems from many different and important fields are explained with regard to their mathematical formulation, solution, and interpretation. The applications are arranged so that topics of major interest to the student or instructor may be chosen without difficulty.

2. To motivate the student so that an understanding of the topics is achieved and an interest is developed. This is done with such aids as Examples, Questions, and Problems for Discussion.

3. To provide relatively *few* methods of solving differential equations that can be applied to a *large* group of problems. A minimum number of basic methods that the student would normally meet in practice have been emphasized; other less utilized methods which are nevertheless of interest may be found in the Exercises.

4. To give the student who may wish to investigate more advanced methods and ideas, or more complicated problems and techniques, an opportunity to do so. This is done by providing over 2200 Exercises graded in difficulty. The A Exercises are mostly straight-forward, requiring little originality and designed for practice purposes. The B Exercises involve more complex algebraic computations or greater originality than the A Group. The C Exercises are intended mainly to supplement the text material; they generally require a high degree of originality and background, designed to challenge the student.

5. To unify the presentation through a logical and orderly approach, the emphasis being on general concepts rather than on isolated details. For example, after introducing the very simple method of separation of variables for solving first-order ordinary differential equations, the concepts of transformation of variables and of making an equation exact upon multiplication by a suitable integrating factor are introduced. These are then used to solve equations of other types.

6. To separate the theory of differential equations from its applications so as to give ample attention to each. This is accomplished by treating theory and applications in separate chapters, particularly in the early chapters of the book. This is done for two reasons. First, from a pedagogical viewpoint, it seems inadvisable to mix theory and applications at an early stage since the beginner usually finds applied problems difficult to formulate mathematically; when he is forced to do so, in addition to learning techniques for solution, neither subject is generally mastered. By treating theory without applications and then gradually broadening out to applications (at the same time reviewing theory), the student may better learn both since attention is thereby focused on only one thing at a time. A second reason for separating theory and applications is to enable instructors wishing to present a minimum of applications to do so easily without being in the awkward position of having to "skip around" in chapters.

The book is divided into three main parts. Part I deals with *ordinary differential equations*, Part II with *systems of ordinary differential equations*, and Part III with *partial differential equations*. It is useful to discuss the chapters in each part.

Part I, ordinary differential equations. Chapter One provides a general introduction to differential equations including motivation for initial and boundary value problems with related concepts. In Chapter Two methods are given for solving some first-order and simple higher-order equations. These methods are applied in Chapter Three to fields such as physics (including mechanics, electricity, heat flow, etc.), chemistry, biology, and economics. Chapter Four handles basic methods for solving linear differential equations while Chapter Five uses these methods in applied problems.

In Chapter Six the Laplace transform is introduced and application is made to differential and integral equations. Among the topics considered are the gamma function, impulse functions and the Dirac delta function, the tautochrone problem, and servomechanisms.

Chapter Seven deals with series methods for solving differential equations including the method of Frobenius. Application is made to important special functions such as those of Bessel, Legendre, Hermite, and Laguerre.

Chapter Eight, which is optional, introduces the ideas of orthogonal functions and Sturm-Liouville problems using generalizations from vectors in two and three dimensions. Some of the topics treated in this chapter are eigenvalues and eigenfunctions and orthogonal series including Fourier series and Bessel series.

In the final chapter of Part I, Chapter Nine, an introduction to various numerical methods for solving differential equations is provided. Included in this chapter is a discussion of computer diagrams and elements of error analysis.

Part II, systems of ordinary differential equations. This part consists of two chapters. The first of these, Chapter Ten, is intended to be a general introduction and provides various methods for solving simultaneous differential equations together with applications such as planetary and satellite motion, vibrations, electricity, and biology. Included in this chapter are elementary principles of phase plane analysis and stability motivated by the predator-prey problem in ecology.

The second chapter, Chapter Eleven, which is another optional chapter, deals with matrix methods for solving linear systems. This chapter shows how important theoretical concepts such as eigenvalues and orthogonality arise quite naturally in the process of solution.

Part III, partial differential equations. This part is composed of three chapters. The first of these, Chapter Twelve, is intended to provide a general introduction to some of the ideas concerning partial differential equations. These include derivations of important equations arising in various fields such as heat conduction, vibration, and potential theory. The second chapter, Chapter Thirteen, deals with Fourier series methods for solving partial differential equations. Finally, Chapter Fourteen, which is optional, explores methods for solving partial differential equations using Bessel and Legendre functions. An interesting feature of this chapter is the atomic bomb problem which is treated along with the more conventional and relatively harmless types given in Chapters Twelve and Thirteen.

The chapters have been written and arranged so as to provide a maximum of flexibility. For example, Chapters Six and Eleven may be omitted without any loss in continuity if the instructor should decide not to cover Laplace transforms or matrix methods. Similarly, in Chapter Ten the complementary-particular solution method for solving systems of linear differential equations is illustrated without the use of matrices while in Chapter Eleven it is treated with matrices. Thus, the instructor can use one or the other or even both to demonstrate their relationships. As another example, in Chapter Thirteen, which deals with Fourier series methods for solving partial differential equations, the Fourier series are introduced in an historical manner, i.e., as Fourier might have discovered them. As a result, this chapter is essentially independent of Chapter Eight, which deals with orthogonal functions and series, providing the instructor with the option to omit Chapter Eight entirely. In cases where there might be some doubt, chapters and chapter sections have been labeled with a diamond to indicate that they are optional. However, chapters and sections which could have been labeled as optional (such as those concerning Laplace transforms, numerical methods, and particular applications), have not been labeled as such because coverage or omission of the included topics would generally be clear to the instructor depending on the kind of course being offered, the topics to be considered, and so on.

Because of its high degree of flexibility, the book can be used in a variety of courses ranging from a one-to-two semester course and including only ordinary or both ordinary and partial differential equations. The diagram on page xvi, which indicates possible chapter sequences, may be of use to an instructor in planning a course. For example, in a one semester course covering both ordinary and partial differential equations, a possible sequence of chapters is 1, 2, 3, 4, 5, 7, 9, 10, 12, 13. A double arrow indicates that chapters can be interchanged. Thus, for example, Chapter Seven could precede Chapter Six if desired.

The author wishes to take this opportunity to express his thanks to Esther and Meyer Scher for their continued interest and encouragement; to the staff of Prentice-Hall, especially Leslie Nadell and Bob Sickles, for their excellent cooperation; and to the following professors of mathematics who reviewed the manuscript and provided many helpful suggestions: Ebon E. Betz, United States Naval Academy; E. E. Burniston, North Carolina State University; John Burns, Virginia Polytechnic Institute and State University; Ronald Hirschorn, Queen's University; James Hurley, University of Connecticut; R. N. Kesarwani, University of Ottawa; Anthony L. Peressini, University of Illinois; William L. Perry, Texas A & M University; Daniel Sweet, University of Maryland; Henry Zatzkis, New Jersey Institute of Technology.

Murray R. Spiegel

POSSIBLE CHAPTER SEQUENCES

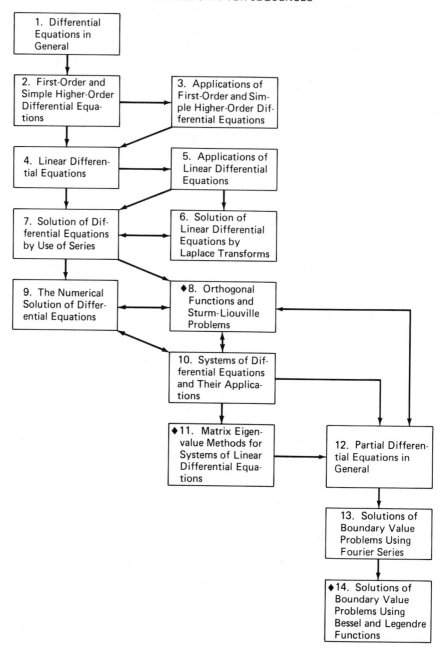

I

ordinary differential equations

one
differential equations in general

1 Concepts of Differential Equations

1.1 SOME DEFINITIONS AND REMARKS

The independent discovery of calculus by *Newton* and *Leibniz* in the 17th century provided impetus for the great advances in mathematics, science, and engineering which were to follow. One of the most important and fascinating branches of mathematics supplying the means for mathematical formulations and solutions of various problems in these areas is called *differential equations*, which we shall study in this book. In order to proceed, we first require some definitions.

Definition 1. A *differential equation* is an equation involving derivatives of an unknown function of one or more variables. If the unknown function depends on only one variable (so that the derivatives are ordinary derivatives) the equation is called an *ordinary differential equation*. However, if the unknown function depends on more than one variable (so that the derivatives are partial derivatives) the equation is called a *partial differential equation.**

Example 1. The equation $\quad \dfrac{dy}{dx} = 2x + y \quad$ or $\quad y' = 2x + y \qquad (1)$

in which y is an unknown function of the single variable x is an ordinary differential equation. We often write $y = f(x)$ and call x the *independent variable*, and y, which depends on x, the *dependent variable*. For brevity we can denote the value of y at x by $y(x)$, and its successive derivatives by $y'(x), y''(x), \ldots$, or simply $y', y'', \ldots$.

Example 2. The equation $\quad \dfrac{d^2 x}{dt^2} - 2\dfrac{dx}{dt} - 15x = 0 \qquad (2)$

in which x is an unknown function of the single variable t is an ordinary differential equation. We can write $x = g(t)$, where t is the independent variable and x the dependent variable. For brevity we can denote the value of x at t by $x(t)$, and can also denote the derivatives by $x'(t), x''(t), \ldots$, or simply $x', x'', \ldots$.

Example 3. The equation $\quad \dfrac{\partial^2 V}{\partial x^2} + 2\dfrac{\partial^2 V}{\partial y^2} = V \qquad (3)$

in which V is an unknown function of the two variables x and y is a partial differential equation. We can write $V = F(x, y)$, where x and y are independent variables and V the dependent variable. For brevity we can denote the value of V at x and y by $V(x, y)$.

---

* We exclude from the class of differential equations those which are identities such as

$$\frac{d}{dx}(xy) = x\frac{dy}{dx} + y$$

Definition 2. The *order* of a differential equation is the order of the highest derivative occurring in the equation.

Example 4. The highest derivative occurring in equation (1) is dy/dx, which is of first order, i.e., order 1. Thus, the differential equation is an equation of order 1, or a first-order ordinary differential equation.

Example 5. The highest derivative occurring in equation (2) is d^2x/dt^2, which is of second order, i.e., order 2. The differential equation is thus of order 2, or a second-order ordinary differential equation.

Example 6. The highest derivative occurring in equation (3) is $\partial^2 V/\partial x^2$ or $\partial^2 V/\partial y^2$, both of which are of second order. Thus, the differential equation is a second-order partial differential equation.

Remark 1. An nth-order ordinary differential equation can be expressed as

$$g(x, y, y', y'', \ldots, y^{(n)}) = 0 \tag{4}$$

If we can solve this for the highest derivative, we obtain one or more equations of nth order having the form

$$y^{(n)} = F(x, y, y', \ldots, y^{(n-1)}) \tag{5}$$

Example 7. The first-order equation $\qquad (y')^2 + xy' - y = 0 \tag{6}$

is equivalent to the two first-order equations

$$y' = \tfrac{1}{2}(\sqrt{x^2 + 4y} - x), \qquad y' = -\tfrac{1}{2}(\sqrt{x^2 + 4y} + x) \tag{7}$$

Remark 2. In addition to its order, it is useful to classify an ordinary differential equation as a *linear* or *non-linear* differential equation according to the following

Definition 3. *A linear ordinary differential equation* is an equation which can be written in the form

$$a_0(x)y^{(n)} + a_1(x)y^{(n-1)} + \cdots + a_{n-1}(x)y' + a_n(x)y = F(x) \tag{8}$$

where $F(x)$ and the coefficients $a_0(x), a_1(x), \ldots, a_n(x)$ are given functions of x, and $a_0(x)$ is not identically zero.* An ordinary differential equation which cannot be written in the form (8) is called a *non-linear differential equation*.

Example 8. Equations (1) and (2) are linear ordinary differential equations.

Example 9. Equation (6) or the two equivalent equations (7) are non-linear.

---

* In algebra $au + bv$ where a and b do not depend on u or v is often called a *linear* function of u and v. The terminology *linear* in Definition 3 is inspired by a generalization of this idea because the left side of (8) is a linear function of $y, y', \ldots, y^{(n)}$.

The ideas presented in Remarks 1 and 2 can also be extended to partial differential equations. As we shall often have occasion to observe throughout this book, linear differential equations are in general easier to handle than non-linear equations.

Definition 4. A *solution* of a differential equation is any function which satisfies the equation, i.e., reduces it to an identity.

Example 10. The functions defined by $x = e^{5t}$ and $x = e^{-3t}$ are two solutions of equation (2), since substitution of these leads respectively to

$$25e^{5t} - 2(5e^{5t}) - 15e^{5t} = 0, \qquad 9e^{-3t} - 2(-3e^{-3t}) - 15e^{-3t} = 0$$

which are identities. Another solution is $x = 0$, and there may be others. In fact $x = c_1 e^{5t} + c_2 e^{-3t}$ where c_1 and c_2 are arbitrary constants is a solution.

Example 11. The function defined by $V = e^{3x} \sin 2y$ is a solution of (3) since

$$\frac{\partial V}{\partial x} = 3e^{3x} \sin 2y, \quad \frac{\partial^2 V}{\partial x^2} = 9e^{3x} \sin 2y, \quad \frac{\partial V}{\partial y} = 2e^{3x} \cos 2y, \quad \frac{\partial^2 V}{\partial y^2} = -4e^{3x} \sin 2y$$

so that on substitution we find the identity $9e^{3x} \sin 2y + 2(-4e^{3x} \sin 2y) = e^{3x} \sin 2y$.

Remark 3. In Examples 10 and 11 solutions were given without any restrictions on the values assumed by the independent variables. Sometimes, however, we must restrict such values, as for instance when we want the functional values to be real or to have other properties. For example, if $f(x) = \sqrt{9 - x^2}$, then in order for $f(x)$ to be real we must have $-3 \leq x \leq 3$. Such values constitute what is called the *domain* of the function. When, as is often the case, the domain is not specified, we assume that the domain consists of all values for which the indicated operations produce meaningful results. Thus, for example, if a function is defined by $f(x) = 1/(x - 3)$, then the domain consists of all real values of x except 3; i.e., $x \neq 3$, since division by zero is not meaningful.

Example 12. The function defined by $y = \sqrt{9 - x^2}$ is a solution of

$$y' = -\frac{x}{y} \tag{9}$$

because

$$y' = \frac{dy}{dx} = \tfrac{1}{2}(9 - x^2)^{-1/2}(-2x) = \frac{-x}{\sqrt{9 - x^2}} \tag{10}$$

and substitution into the differential equation (9) yields

$$\frac{-x}{\sqrt{9 - x^2}} = -\frac{x}{\sqrt{9 - x^2}}$$

an identity. However, it is clear that if we want the function to be real and the derivative (10) to exist we must restrict x to the domain $-3 < x < 3$; i.e., we must exclude $x = -3$ and $x = 3$. We can thus say that $y = \sqrt{9 - x^2}$ is a solution of (9) over the interval $-3 < x < 3$. Other domains could be taken. For instance, the functions defined by $y = \sqrt{9 - x^2}$, $0 \leq x < 3$, or $y = \sqrt{9 - x^2}$, $1 < x < 2$, are also solutions of (9).

Similar remarks may be made for functions of two or more variables. For example, if $V = \sqrt{9 - (x^2 + y^2)}$, then in order for V and its partial derivatives with respect to x and y to exist and be real, we must restrict x and y so that $x^2 + y^2 < 9$, a domain which represents geometrically the interior of a circle of radius 3 in the xy plane having its center at the origin.

Remark 4. In all the above examples we dealt with solutions in which the dependent variable was solved *explicitly* in terms of the independent variables, and for this reason we often refer to the functions as *explicit functions*. As learned in calculus, however, we can have functions defined *implicitly* by equations involving the dependent and independent variables, in which case they are referred to as *implicit functions*.

Example 13. Given the relation $x^2 + y^2 = 9$ between x and y, we can think of y defined implicitly as a function of x. In fact, on noting that the relation is equivalent to $y = \pm\sqrt{9 - x^2}$, we see that one of these is the same as that given in Example 12. The situation points out the fact that in dealing with implicit functions further investigation may be required to determine the specific function which is wanted, since many functions may be included. Note that if we differentiate $x^2 + y^2 = 9$ implicitly, thinking of y as a function of x, we obtain

$$2x + 2yy' = 0 \quad \text{or} \quad y' = -\frac{x}{y} \tag{11}$$

which agrees with (9). The fact that we must be careful to know what we are doing can be illustrated by noting that $x^2 + y^2 = -9$ also satisfies (11) formally, but does not even define y as a real function of x.

Example 14. Assume that $\qquad y^3 - 3x + 3y = 5 \tag{12}$

defines y (implicitly) as a function of x. Then this function would be a solution of

$$y'' = -2y(y')^3 \tag{13}$$

because upon differentiation of (12) with respect to x we find

$$y' = \frac{1}{y^2 + 1}, \qquad y'' = \frac{-2y}{(y^2 + 1)^3} \tag{14}$$

so that substitution of the derivatives given by (14) into (13) yields the identity

$$\frac{-2y}{(y^2 + 1)^3} = -2y\left(\frac{1}{y^2 + 1}\right)^3 \tag{15}$$

The question of whether (12) actually does define y as a function of x requires further investigation, but until such decision is reached it can be referred to as a *formal solution*.

1.2 SIMPLE EXAMPLES OF INITIAL AND BOUNDARY VALUE PROBLEMS

Very often, especially in applied problems, a differential equation is to be solved subject to given conditions which must be satisfied by the unknown function. As a simple example, consider the following

PROBLEM FOR DISCUSSION

A particle P moves along the x axis (Fig. 1.1) in such a way that its acceleration at any time $t \geq 0$ is given by $a = 16 - 24t$. (a) Find the position x of the particle measured from origin O at any time $t > 0$, assuming that initially ($t = 0$) it is located at $x = 2$ and is traveling at a velocity $v = -5$. (b) Work part (a) if it is known only that the particle is located at $x = 2$ initially, and at $x = 7$ when $t = 1$.

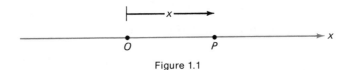

Figure 1.1

To formulate this problem mathematically, we first recall from calculus that the velocity and acceleration of a particle moving along the x axis are given by

$$v = \frac{dx}{dt} \quad \text{and} \quad a = \frac{d^2x}{dt^2} \tag{16}$$

respectively. Then from the first sentence in the problem we have at once

$$\frac{d^2x}{dt^2} = 16 - 24t \tag{17}$$

which is the required differential equation of the motion.

Solution to Part (a)   The conditions on the function x given in part (a) are

$$x = 2, \quad v = -5 \text{ at } t = 0 \quad \text{i.e.,} \quad x(0) = 2, x'(0) = -5 \tag{18}$$

It should be noted that the significance of the minus sign in $v = -5$ is that the particle is traveling toward the left initially. If we integrate (17) once, we find

$$\frac{dx}{dt} = 16t - 12t^2 + c_1 \tag{19}$$

where c_1 is an arbitrary constant. This constant can be determined from the second condition in (18) by using $t = 0$ in (19). We find $-5 = 0 + c_1$ i.e., $c_1 = -5$, so that

$$\frac{dx}{dt} = 16t - 12t^2 - 5 \tag{20}$$

Integration of (20) gives $x = 8t^2 - 4t^3 - 5t + c_2 \tag{21}$

where c_2 is another arbitrary constant which can be determined from the first condition in (18) by using $t = 0$ in (21). We find $2 = 0 + c_2$ or $c_2 = 2$. Thus

$$x = 8t^2 - 4t^3 - 5t + 2 \tag{22}$$

which is the required law of motion enabling us to determine the position at any time $t > 0$; for example, at time $t = 1$, $x = 1$, at time $t = 2$, $x = -8$, etc.

Solution to Part (b) In this part we still have the same differential equation (17) for the motion, but the conditions are changed to

$$x = 2 \text{ at } t = 0, \qquad x = 7 \text{ at } t = 1 \quad \text{or} \quad x(0) = 2, \qquad x(1) = 7 \tag{23}$$

In this case we integrate (17) as before to obtain (19). However, since we have no condition for dx/dt, we cannot yet determine c_1, and so must integrate (19) to obtain

$$x = 8t^2 - 4t^3 + c_1 t + c_2 \tag{24}$$

We can now use the two conditions in (23) to find the two arbitrary constants in (24). This leads to $2 = 0 + c_2$, $7 = 8(1)^2 - 4(1)^3 + c_1 + c_2$ or $c_1 = 1$, $c_2 = 2$

and so

$$x = 8t^2 - 4t^3 + t + 2 \tag{25}$$

The mathematical formulations of parts (a) and (b) in the above problem are, respectively,

(a) $$\frac{d^2x}{dt^2} = 16 - 24t, \qquad x(0) = 2, \, x'(0) = -5 \tag{26}$$

(b) $$\frac{d^2x}{dt^2} = 16 - 24t, \qquad x(0) = 2, \, x(1) = 7 \tag{27}$$

An important difference between these two is that in (a) the conditions on the unknown function x and its derivative x' or dx/dt are specified at *one value* of the independent variable (in this case $t = 0$), while in (b) the conditions on the unknown function x are specified at *two values* of the independent variable (in this case $t = 0$ and $t = 1$). The two types of problems presented in (a) and (b), respectively, are called *initial value problems* and *boundary value problems*. We thus make the following definitions.

Definition 5. An *initial value problem* is a problem which seeks to determine a solution to a differential equation subject to conditions on the unknown function and its derivatives specified at *one value* of the independent variable. Such conditions are called *initial conditions*.

Definition 6. A *boundary value problem* is a problem which seeks to determine a solution to a differential equation subject to conditions on the unknown function specified at *two or more values* of the independent variable. Such conditions are called *boundary conditions*.

As another example illustrating the above remarks, consider the following

ILLUSTRATIVE EXAMPLE 1

A curve in the xy plane has the property that its slope at any point (x, y) on it is equal to $2x$. Find the equation of the curve if it passes through the point $(2, 5)$.

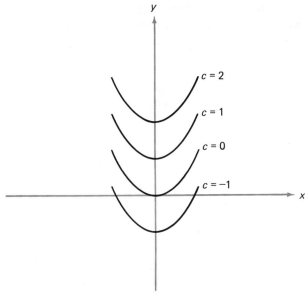

Figure 1.2

Solution Since the slope of a curve at any point (x, y) on it is given by dy/dx, the first sentence of the problem leads to

$$\frac{dy}{dx} = 2x \tag{28}$$

a first-order differential equation. Since the curve must pass through the point $(2, 5)$,

$$y = 5 \text{ when } x = 2 \quad \text{i.e.,} \quad y(2) = 5 \tag{29}$$

The problem of solving (28) subject to (29) is an initial value problem.

Integration of (28) gives $\qquad y = x^2 + c \tag{30}$

where c is an arbitrary constant. Using condition (29) in (30) yields $5 = (2)^2 + c$ so that $c = 1$. Thus the required curve is given by $\qquad y = x^2 + 1 \tag{31}$

Graphically, (30) represents a *family of curves* in the xy plane, each member of which is associated with a particular value of c. A few of these are shown in Fig. 1.2 for $c = 0, -1, 1, 2$. Since c can vary, it is often called a *parameter* to distinguish it from the main variables x and y. The differential equation (28) which is satisfied by all members of the family is often called the *differential equation of the family*.

Remark 5. The same terminology used in this example can also be used in the problem on page 7. Thus, (24) represents a family of curves in the tx plane, each member of which is associated with particular values of the two parameters c_1 and c_2, while (17) is the differential equation of the family. To specify the number of parameters involved, we sometimes speak of a *one-parameter family of curves*, a *two-parameter family of curves*, etc. The corresponding solutions of the differential equations can then be referred to as the *one-parameter solution* (or *one-parameter family of solutions*), the *two-parameter solution* (or *two-parameter family of solutions*), etc. We can also refer to these curves as *solution curves*.

In the process of mathematical formulation of applied problems, many kinds of differential equations can arise, as we shall see in later chapters. The following list shows a small sample of these.

$$\frac{d^2x}{dt^2} = -kx \tag{32}$$

$$x\frac{d^2y}{dx^2} + \frac{dy}{dx} + xy = 0 \tag{33}$$

$$v + M\frac{dv}{dM} = v^2 \tag{34}$$

$$EIy^{(IV)} = w(x) \tag{35}$$

$$\frac{d^2I}{dt^2} + 4\frac{dI}{dt} + 5I = 100\sin 20t \tag{36}$$

$$y'' = \frac{w}{H}\sqrt{1 + (y')^2} \tag{37}$$

$$\frac{\partial^2 V}{\partial x^2} + \frac{\partial^2 V}{\partial y^2} + \frac{\partial^2 V}{\partial z^2} = 0 \tag{38}$$

$$\frac{\partial U}{\partial t} = k\left(\frac{\partial^2 U}{\partial x^2} + \frac{\partial^2 U}{\partial y^2}\right) \tag{39}$$

$$\frac{\partial^2 Y}{\partial t^2} = a^2\frac{\partial^2 Y}{\partial x^2} \tag{40}$$

$$\frac{\partial^4 \phi}{\partial x^4} + 2\frac{\partial^4 \phi}{\partial x^2 \partial y^2} + \frac{\partial^4 \phi}{\partial y^4} = F(x, y) \tag{41}$$

Equation (32) is famous in the field of mechanics in connection with simple harmonic motion, as in small oscillations of a simple pendulum. It could, however, arise in many other connections.

Equation (33) arises in mechanics, heat, electricity, aerodynamics, stress analysis and many other fields.

Equation (34) arose in a problem in rocket flight.

Equation (35) is an important equation in civil engineering in the theory of deflection or bending of beams.

Equation (36) might arise in the determination of the current I as a function of the time t in an alternating current circuit, but it could also arise in mechanics, biology, and economics.

Equation (37) arises in connection with a problem on suspension of cables.

Equation (38) could arise in problems of electricity, heat, aerodynamics, potential theory, and many other fields.

Equation (39) arises in the theory of heat conduction, as well as in the diffusion of neutrons in an atomic pile for the production of nuclear energy. It also arises in the theory of Brownian motion.

Equation (40) arises in connection with the vibration of strings, as well as in the propagation of electric signals.

Equation (41) is famous in the theory of stress analysis.

These are but a few of the many equations which could arise and a few of the fields from which they are taken. Examinations of equations such as these by pure mathematicians, applied mathematicians, theoretical and applied physicists, chemists, engineers, and other scientists throughout the years have led to the conclusion that there are certain definite methods by which many of these equations can be solved. Such equations and methods together with the names of people associated with them will be given throughout the book.* In spite of all that is known, however, there remain many unsolved equations, some of them of great importance. Modern giant calculating machines are presently engaged in determining solutions of such equations vital for research involving national security, economic planning, and aerospace engineering as well as many other endeavors.

It is one of the aims of this book to provide an introduction to some of the important problems arising in science and engineering with which most scientists should be acquainted. To accomplish this aim, it will be necessary to demonstrate how one solves the equations which arise as a result of mathematical formulations of these problems. The student should keep in mind that there are three steps in the theoretical solution of scientific problems.

1. *Mathematical formulation of the scientific problem.* Scientific laws, which are of course based on experiment or observation, are translated into mathematical equations. In many instances a *mathematical model* is used to approximate the physical reality. Thus, for example, in dealing with the motion of a planet, such as the earth, around the sun, we may consider both the sun and earth as particles (or point masses). However, in a study of the rotation of the earth on its axis, such a model is clearly inappropriate, and so we may consider the earth to be a sphere or even more accurately an oblate spheroid.

2. *Solution of the equations.* The equations formulated in Step 1 need to be solved, subject to conditions arrived at from the problem, to determine the unknown, or unknowns, involved. The procedures used may yield an exact solution or, in cases where exact solutions cannot be obtained, approximate solutions. Often, recourse is made to the use of calculators in performing numerical computations. The process of obtaining solutions often leads to questions of a purely mathematical nature sometimes of more interest than the original scientific problem. In fact, many advances in mathematics were made as a result of the attempt to solve problems in science and engineering.

3. *Scientific interpretations of the solution.* By use of the known solutions, the scientist may be able to interpret what is going on from the applied point of view. He may make graphs or tables and compare theory with experiment. He may even base further research on such interpretations. Of course, if he finds that experiments

---

* A reference list of some of the important contributors to the theory and application of differential equations is provided on the front inside cover of the text.

or observations fail to agree with theory, he must revise the mathematical model and thus the mathematical formulation until reasonable agreement is achieved.

Every one of these steps is important in the final solution of an applied problem and, for this reason, we shall emphasize all three steps in this book.

Since, as one might expect, partial differential equations are much more complicated than ordinary differential equations, most of this book, that is, the eleven chapters in Parts I and II, is devoted to ordinary differential equations. Partial differential equations are treated in the three chapters of Part III. Thus, unless otherwise stated, whenever we refer to a differential equation we shall imply an ordinary differential equation.

A EXERCISES

1. Complete the following table.

	Differential equation	Ordinary or partial	Order	Inde-pendent variables	De-pendent variables
(a)	$y' = x^2 + 5y$				
(b)	$y'' - 4y' - 5y = e^{3x}$				
(c)	$\dfrac{\partial U}{\partial t} = 4\dfrac{\partial^2 U}{\partial x^2} + \dfrac{\partial U}{\partial y}$				
(d)	$\left(\dfrac{d^3 s}{dt^3}\right)^2 + \left(\dfrac{d^2 s}{dt^2}\right)^3 = s - 3t$				
(e)	$\dfrac{dr}{d\phi} = \sqrt{r\phi}$				
(f)	$\dfrac{d^2 x}{dy^2} - 3x = \sin y$				
(g)	$\dfrac{\partial^2 V}{\partial x^2} = \sqrt[3]{\dfrac{\partial V}{\partial y}}$				
(h)	$(2x + y)dx + (x - 3y)dy = 0$				
(i)	$y'' + xy = \sin y''$				
(j)	$\dfrac{\partial^2 T}{\partial x^2} + \dfrac{\partial^2 T}{\partial y^2} + \dfrac{\partial^2 T}{\partial z^2} = 0$				

2. Make a table similar to the one above for the differential equations (32)–(41) of page 10 and complete the table.

3. Which of the ordinary differential equations in the table of Exercise 1 are linear and which are non-linear?

4. Work Exercise 3 for the table constructed in Exercise 2.

5. Show that each of the functions defined in Column I, with one exception, is a solution of the corresponding differential equation in Column II, subject to the given conditions, if any.

I	II

I

(a) $y = e^{-x} + x - 1$.

(b) $y = Ae^{5x} + Be^{-2x} - \frac{1}{2}e^x$.

(c) $s = 8 \cos 3t + 6 \sin 3t$.

(d) $8x^3 - 27y^2 = 0$.

(e) $Y(x, t) = 4 \sin (2x - 3t)$.

(f) $y = c_1 e^{-2x} + c_2 e^x + c_3 e^{3x}$.

(g) $y = Ax^3 + Bx^{-4} - \dfrac{x^2}{3}$.

(h) $1 + x^2 y + 4y = 0$.

(i) $xy^2 - y^3 = c$.

(j) $V(x, y) = e^{2x-y} \cos (y - 2x)$.

II

$y' + y = x; \ y(0) = 0$.

$y'' - 3y' - 10y = 6e^x$.

$\dfrac{d^2 s}{dt^2} = -9s; s = 8, \dfrac{ds}{dt} = 18$ at $t = 0$.

$(y')^3 = y; \ y(0) = 0$.

$9 \dfrac{\partial^2 Y}{\partial x^2} = 4 \dfrac{\partial^2 Y}{\partial t^2}; \ Y(\pi, 0) = 0$.

$y''' - 2y'' - 5y' + 6y = 0$.

$x^2 y'' + 2xy' - 12y = 2x^2$.

$y'' = 2x(y')^2; \ y'(0) = 0, y''(0) = \frac{1}{8}$.

$y \, dx + (2x - 3y)dy = 0$.

$\dfrac{\partial^2 V}{\partial x^2} + 4 \dfrac{\partial^2 V}{\partial x \, \partial y} + 4 \dfrac{\partial^2 V}{\partial y^2} = 0$.

6. A particle moves along the x axis in such a way that its instantaneous velocity is given as a function of time t by $v = 12 - 3t^2$. At time $t = 1$, it is located at $x = -5$. (a) Set up an initial value problem describing the motion. (b) Solve the problem in (a). (c) Determine where the particle will be at times $t = 2$ and $t = 3$. (d) Determine the times when the particle is at the origin. In doing so, what assumptions are you making? (e) By using a graph or otherwise, describe the motion of the particle.

7. A particle moves along the x axis so that its instantaneous acceleration is given as a function of time t by $a = 10 - 12t^2$. At times $t = 2$ and $t = 3$, the particle is located at $x = 0$, and $x = -40$, respectively. (a) Set up a differential equation and associated conditions describing the motion. Is the problem an initial or boundary value problem? (b) Solve the problem in (a). (c) Determine the position and velocity of the particle at $t = 1$. (d) Sketch roughly the graph of x versus t and use it to describe the motion of the particle.

8. Work Exercise 7 if the particle is initially at $x = 3$ and has a velocity $v = -6$.

9. The slope of a family of curves at any point (x, y) of the xy plane is given by $4 - 2x$. (a) Set up a differential equation of the family. (b) Determine an equation for that particular member of the family which passes through the point $(0, 0)$. (c) Sketch several members of the family including the one found in (b).

10. Work Exercise 9 if the slope is given by $4e^{-2x}$.

11. Solve each of the following initial or boundary value problems. In each case give a possible physical or geometric interpretation.

(a) $\dfrac{dy}{dx} = 3 \sin x$, $y(\pi) = -1$.   (b) $\dfrac{dx}{dt} = 4e^{-t} - 2$, $x = 3$ when $t = 0$.

(c) $\dfrac{d^2x}{dt^2} = 8 - 4t + t^2$, $x = 1, \dfrac{dx}{dt} = -3$ when $t = 0$.

(d) $\dfrac{ds}{du} = 9\sqrt{u}$, $s(4) = 16$.   (e) $y'' = 12x(4 - x)$, $y(0) = 7, y(1) = 0$.

12. In each of the following a differential equation for a family of curves is given. Obtain the solution curves for each family and state the number of parameters involved. Find the particular members of each family which satisfy the conditions given,

(a) $y' = -4/x^2$, $y(1) = 2$.   (b) $y'' = 1 - \cos x$, $y(0) = 0, y'(0) = 2$.

(c) $y'' = \sqrt{2x + 1}$, $y(0) = 5, y(4) = -3$.

B EXERCISES

1. A particle moves along the x axis so that its velocity at any time $t \geq 0$ is given by $v = 1/(t^2 + 1)$. Assuming that it is at the origin initially, show that it will never get past $x = \pi/2$.

2. For what values of the constant m will $y = e^{mx}$ be a solution of each of the following differential equations. (a) $y' - 2y = 0$. (b) $y'' + 3y' - 4y = 0$. (c) $y''' - 6y'' + 11y' - 6y = 0$.

3. Would the method of Exercise 2 work in finding solutions to $x^2y'' - xy' + y = 0$? Explain. From your conclusions can you suggest a class of differential equations which always will have solutions of the form $y = e^{mx}$?

4. (a) If $y = Y_1(x)$ and $y = Y_2(x)$ are two different solutions of $y'' + 3y' - 4y = 0$ show that $y = c_1 Y_1(x) + c_2 Y_2(x)$ is also a solution, where c_1 and c_2 are arbitrary constants. (b) Use the result of (a) to find a solution of the differential equation which satisfies the conditions $y(0) = 3, y'(0) = 0$.

5. If $y = Y_1(x)$ and $y = Y_2(x)$ are solutions of $y'' + y^2 = 0$, is $y = Y_1(x) + Y_2(x)$ also a solution? Compare with Exercise 4 and discuss. Can you characterize the type of differential equation which has the property described in Exercise 4?

6. Show that a solution of $y' = 1 + 2xy$ subject to $y(1) = 0$ is $y = e^{x^2} \int_1^x e^{-t^2} \, dt$.

7. Is $x^2 + y^2 - 6x + 10y + 34 = 0$ a solution of the differential equation $\dfrac{dy}{dx} = \dfrac{3 - x}{y + 5}$?

8. The differential equation of a family of curves in the xy plane is given by $y''' = -24 \cos \dfrac{\pi x}{2}$.

(a) Find an equation for the family and give the number of parameters involved. (b) Find a member of this family which passes through the points $(0, -4)$, $(1, 0)$, and which has slope of 6 at the point where $x = 1$.

9. Is it possible for the differential equation of a three-parameter family to be of order 4? Explain.

C EXERCISES

1. In the equation $dy/dx + dx/dy = 1$, which variable is independent? Which variable is independent in the equation

$$\frac{d^2y}{dx^2} + \frac{d^2x}{dy^2} = 1$$

2. Show that the first-order equation $xy(y')^2 - (x^2 + y^2)y' + xy = 0$ is equivalent to two differential equations each of first order. Show that $y = cx$ and $x^2 - y^2 = c$, where c is any constant, are solutions of the equation.

3. Solve each of the following and interpret geometrically: (a) $\dfrac{dy}{dx} = y$, $y(0) = 1$; (b) $\dfrac{dy}{dx} = e^y$,

$y(1) = 0$; (c) $\dfrac{dy}{dx} = \sec y$, $y(0) = 0$. (*Hint:* Write each differential equation in terms of dx/dy rather than dy/dx.)

4. (a) Show that $\dfrac{d^2 y}{dx^2} = -\dfrac{d^2 x}{dy^2} \bigg/ \left(\dfrac{dx}{dy}\right)^3$ is an identity. (*Hint:* Differentiate both sides of $\dfrac{dy}{dx} =$

$1 \bigg/ \dfrac{dx}{dy}$ with respect to x.) (b) Use the result in (a) to transform the differential equation

$$\frac{d^2 x}{dy^2} + (\sin x)\left(\frac{dx}{dy}\right)^3 = 0$$

with independent variable y, into one with independent variable x. Can you obtain the solution of this equation?

5. Show that $x = a(\theta - \sin \theta)$, $y = a(1 - \cos \theta)$, where a is any constant other than zero, is a solution of $1 + (y')^2 + 2yy'' = 0$.

1.3 GENERAL AND PARTICULAR SOLUTIONS

On pages 8–9 we considered a differential equation having a solution involving one arbitrary constant (or parameter) which could be determined from a given condition. Similarly, on pages 7–8 we considered a second-order differential equation having a solution involving two arbitrary constants (parameters) which could be determined from two given conditions.

Suppose now that we are given an initial or boundary value problem which seeks to determine the solution of an nth-order differential equation satisfying n specified conditions. To accomplish this it would be nice if we could find a solution to the differential equation which contains n arbitrary constants, for hopefully we could then use the n conditions to solve for the n constants, and thus obtain the required solution. As an illustration, let us consider the following

PROBLEM FOR DISCUSSION

Solve the initial value problem $\dfrac{d^2 y}{dx^2} + y = 0$, $y(0) = 3$, $y'(0) = -4$. \hfill (42)

To satisfy the two initial conditions in (42), it is natural for us to look for a solution to the differential equation in (42), which we hope has two arbitrary constants, and then use the two conditions to determine these constants. At the present time, of course, we do not know how to determine such a solution, and methods for doing so must be left to a later chapter. Suppose, however, that by some means (such as being told or using trial and error) we arrive at

$$y = A \cos x + B \sin x \tag{43}$$

which contains the two needed arbitrary constants A and B and can be verified as a solution. Using the first condition of (42) in (43), i.e., $y(0) = 3$, or $y = 3$ when $x = 0$, we find $A = 3$, so that (43) becomes

$$y = 3 \cos x + B \sin x \qquad (44)$$

To find B we first take the derivative in (44) to obtain

$$y' = -3 \sin x + B \cos x \qquad (45)$$

Then using the second condition of (42) in (45), i.e., $y'(0) = -4$ or $y' = -4$ when $x = 0$, we find $B = -4$. The required solution is thus given by

$$y = 3 \cos x - 4 \sin x \qquad (46)$$

Now *in general* an nth-order differential equation will have a solution involving n arbitrary constants, and because of its special importance to us we give it the special name *general solution.** A solution obtained from this general solution by choosing particular values of the arbitrary constants (to satisfy given conditions for example) is then called a *particular solution*.

Example 15. In the problem (42) on page 15, $y = A \cos x + B \sin x$ is the general solution, while $y = 3 \cos x - 4 \sin x$ is a particular solution.

Example 16. In Illustrative Example 1, pages 8–9, $y = x^2 + c$ is the general solution of $y' = 2x$, while $y = x^2 + 1$ and $y = x^2 - 3$ are particular solutions.

Example 17. For the differential equation $\dfrac{dy}{dx} = 3y^{2/3}$ $\qquad (47)$

$y = (x + c)^3$ is the general solution, and $y = (x - 2)^3$ is a particular solution.

Example 18. For the differential equation $\quad y'' = 4y$ $\qquad (48)$

$y = Ae^{B + 2x}$ is *not* the general solution because we can write it as $y = (Ae^B)e^{2x}$ or $y = ce^{2x}$, which is a solution but has only *one* arbitrary constant, while the differential equation is of order 2.[†] However, the student can easily show by direct substitution that $y = c_1 e^{2x} + c_2 e^{-2x}$ containing two arbitrary constants c_1, c_2 is a solution of (48) and is thus its general solution.

The general solution of a differential equation may occur in implicit form. To examine this situation let us consider the following

PROBLEM FOR DISCUSSION

Solve $\qquad \dfrac{dy}{dx} = \dfrac{y}{2y - x}, \; y(1) = 2.$ $\qquad (49)$

---

* A theoretical justification for using the term *general solution* is supplied in reference [13] of the Bibliography. Henceforth numbers in square brackets will refer to the Bibliography.

[†] If a relationship involves a set of constants which cannot be replaced by a smaller set, the constants are sometimes said to be *essential*.

A solution is
$$y^2 - xy = c \qquad (50)$$

as can be verified by implicit differentiation of (50). Since (50) involves one arbitrary constant, we shall also refer to it as the general solution. To obtain that particular solution satisfying $y(1) = 2$, we substitute $x = 1$, $y = 2$ in (50) and find $c = 2$. Thus

$$y^2 - xy = 2 \qquad (51)$$

The required solution to the initial value problem is included in (51). To show this explicitly let us solve (51) for y in terms of x by the quadratic formula to obtain

$$y = \frac{x \pm \sqrt{x^2 + 8}}{2} \qquad (52)$$

Testing the condition $y = 2$ when $x = 1$ in (52) shows that we must exclude the minus sign in (52). The required solution is thus

$$y = \tfrac{1}{2}(x + \sqrt{x^2 + 8}) \qquad (53)$$

It is of interest to interpret the results graphically. The curves described by (52) are shown in Fig. 1.3. Although both curves represent solution curves for the differential equation (49), only one of them satisfies the given condition $y(1) = 2$, i.e., passes through the point $(1, 2)$. It is also of interest to note that for points on the line $y = x/2$ which is the line separating the two curves, the denominator on the right of the differential equation in (49) is zero.

The above remarks suggest the following. Given the initial value problem

$$y' = F(x, y), \qquad y(x_0) = y_0 \qquad (54)$$

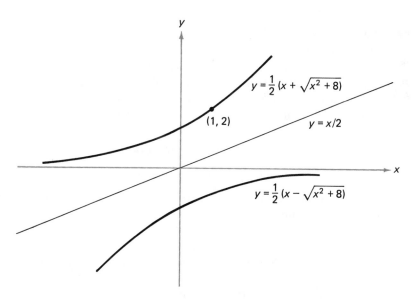

Figure 1.3

involving a first-order differential equation, we try to find a solution involving one arbitrary constant called the general solution. This can occur in any of the forms

$$y = f(x, c), \qquad U(x, y) = c, \qquad G(x, y, c) = 0 \tag{55}$$

the first representing an explicit function, and the last two implicit functions. The constant c is then determined so that the given condition in (54) is satisfied. Extensions to higher-order equations are easily made.

The problem of finding general solutions of differential equations is one which we shall be concerned with in later chapters. A much simpler problem is the reverse problem of finding the differential equation if we are given its general solution i.e., "given the answer, find the problem." To motivate the procedure, let us consider the following

PROBLEM FOR DISCUSSION

Find a differential equation having as its general solution

$$y = ce^{-2x} + 3x - 4 \tag{56}$$

Differentiating (56) yields $\qquad y' = -2ce^{-2x} + 3 \tag{57}$

Let us now eliminate c between the equations (56) and (57). We can do this either by solving for c in one equation and substituting in the other, or more simply in this case by multiplying equation (56) by 2 and adding to equation (57). The result is

$$y' + 2y = 6x - 5 \tag{58}$$

As a check we can substitute (56) into (58) to find the identity

$$y' + 2y = -2ce^{-2x} + 3 + 2(ce^{-2x} + 3x - 4) = 6x - 5$$

Thus, (58) is the required first-order differential equation having (56) as its general solution. As on page 9, we can interpret (56) as a *one-parameter family of curves*, and can call (58) the *differential equation of the family*.

The same idea can be used where general solutions contain more than one arbitrary constant, the procedure being simply to differentiate as many times as there are arbitrary constants and then to use these results to eliminate all the arbitrary constants. Let us illustrate the procedure by some examples.

ILLUSTRATIVE EXAMPLE 2

Find a differential equation whose general solution is $y = c_1 x + c_2 x^3$.

Solution   Since there are two arbitrary constants c_1 and c_2, we are led to differentiate twice, thus obtaining

$$y = c_1 x + c_2 x^3, \qquad y' = c_1 + 3c_2 x^2, \qquad y'' = 6c_2 x \tag{59}$$

We would now like to eliminate the arbitrary constants. To do this, let us solve for c_2 in the last of equations (59). We find $\quad c_2 = \dfrac{y''}{6x}. \tag{60}$

Using this in the second of equations (59) gives $y' = c_1 + xy''/2$, so that

$$c_1 = y' - \frac{xy''}{2} \tag{61}$$

Finally, using (60) and (61) in the first of equations (59), we have

$$y = \left(y' - \frac{xy''}{2} \right)x + (x^3)\left(\frac{y''}{6x} \right) \tag{62}$$

which on simplifying reduces to the required second-order differential equation

$$x^2 y'' - 3xy' + 3y = 0 \tag{63}$$

Check. $x^2 y'' - 3xy' + 3y = x^2(6c_2 x) - 3x(c_1 + 3c_2 x^2) + 3(c_1 x + c_2 x^3) = 0$

Note that $y = c_1 x + c_2 x^3$ represents graphically a *two-parameter family of curves* in the xy plane, and (63) is the *differential equation of this family.*

Remark 6. If we are given a solution having n arbitrary constants, it is often easy to obtain a differential equation of order greater than n having this solution. For instance, in Illustrative Example 2, $y = c_1 x + c_2 x^3$ would be a solution of the fourth order equation $y^{(IV)} = 0$. Of course it is not the general solution of this equation. When we look for the differential equation having a given general solution (e.g., $y = c_1 x + c_2 x^3$) we seek the one of *least order* i.e., of order equal to the number of arbitrary constants (in this case *two*).

ILLUSTRATIVE EXAMPLE 3

Find a differential equation for the family of all circles which have radius 1 and center anywhere in the xy plane.

Solution   The equation of a circle with center (A, B) and radius 1 is

$$(x - A)^2 + (y - B)^2 = 1 \tag{64}$$

Here we have two parameters or arbitrary constants A and B. What we are looking for is the differential equation whose general solution is given by (64), and we can use the same procedure given above. Differentiating (64) with respect to x,

$$2(x - A) + 2(y - B)y' = 0 \tag{65}$$

Solving for $(x - A)$ and substituting into (64), we have

$$(y - B)^2(y')^2 + (y - B)^2 = 1 \tag{66}$$

so that we have been successful in eliminating A. To eliminate B, solve for $(y - B)$

to obtain

$$y - B = \pm[1 + (y')^2]^{-1/2} \tag{67}$$

Differentiating and simplifying this last equation yields

$$\frac{y''}{[1 + (y')^2]^{3/2}} = \pm 1 \tag{68}$$

which is the required second-order differential equation. The student may remember that the left side of (68) is the curvature of a plane curve. Thus, (68) states that the curvature of a certain plane curve at any point on it is equal to 1 in absolute value. Only circles of radius 1 have this property.

1.4 SINGULAR SOLUTIONS

Whenever an initial or boundary value problem has been formulated, there are three questions concerning it which could and should be asked.

1. *Question of existence.* Does a solution of the differential equation satisfying the given conditions exist?

2. *Question of uniqueness.* If one solution satisfying the given conditions exists, can there be a different solution which also satisfies the conditions?

3. *Question of determination.* How do we find solutions satisfying the given conditions?

A natural tendency is to proceed directly to the third question and ignore the first two. However, suppose that we arrived at a mathematical formulation of some applied problem and could prove that there was no solution. Then clearly there would be no point in wasting time trying to find a solution. Again, even if we succeed in finding one solution, and have thus answered Question 1 in the affirmative, there is still the question of uniqueness. If two or more solutions could be found, this would violate the fundamental scientific principle that a system cannot behave in several different ways *under the same conditions.* In such case we would become suspicious of the validity of the mathematical formulation.

To show that there is some basis for asking the above questions, let us suppose that we are given the initial value problem

$$\frac{dy}{dx} = 3y^{2/3}, \qquad y(2) = 0 \tag{69}$$

As seen in Example 17, page 16, the differential equation in (69) has the general solution $y = (x + c)^3$. From the condition in (69) we have $(2 + c)^3 = 0$ so that $c = -2$. Thus we would arrive at the particular solution satisfying (69) given by

$$y = (x - 2)^3 \tag{70}$$

However another solution of (69) is given by $y = 0$. Still a third solution is given by

$$y = \begin{cases} (x - 2)^3 & x \geq 2 \\ 0 & x < 2 \end{cases} \tag{71}$$

Thus the solution is not unique.

The solution of (69) given by $y = 0$ cannot be obtained from the general solution $y = (x + c)^3$ by any choice of the constant c so that it is not a particular solution. It is customary to call any solution of a differential equation which cannot be obtained from the general solution by particular choice of the arbitrary constants a *singular solution.* As the name suggests, singular solutions are *unusual* or *odd* solutions. Occasionally, they turn up in applied problems whose mathematical formulations involve non-linear differential equations.

Remark 7. Some authors choose to use the terminology *general solution* to mean *all solutions* of a differential equation (i.e. including those which we have called singular solutions). We have not done this because, as has already been indicated, we are from a practical point of view generally not interested in all solutions, but only that solution of an *n*th-order differential equation containing *n* arbitrary constants which can then be determined hopefully from the *n* associated conditions.*

Because of situations such as presented above, mathematicians have become interested in theorems, called *existence and uniqueness theorems*, which serve to tell us where existence and uniqueness of solutions can be guaranteed instead of relying on chance.† We shall present an example of such a theorem in the next section. We shall also present a graphical procedure known as the method of *direction fields* which often provides insight into the kinds of solutions which can occur and their relationships. However, since we are mainly concerned in this book with applied problems which can be solved exactly and for which solutions exist and are unique, we have marked the next section as *optional*, indicating that the student may omit the section and proceed directly to Chapter Two without disturbing the continuity of presentation in any way.

A EXERCISES

1. Show that each differential equation in Column I has general solution given by the corresponding relation in Column II. Obtain particular solutions satisfying conditions in Column III.

I	II	III
(a) $y' + y \tan x = 0$.	$y = A \cos x$.	$y(\pi) = 4$.
(b) $y'' - y = 4x$.	$y = c_1 e^x + c_2 e^{-x} - 4x$.	$\begin{cases} y(0) = 2, \\ y'(0) = 0. \end{cases}$
(c) $y' = (y + x)/(y - x)$.	$x^2 - y^2 + 2xy = c$.	$y(-2) = 3$.
(d) $y''' = 0$.	$y = c_1 + c_2 x + c_3 x^2$.	$\begin{cases} y(0) = 1, \\ y(1) = 2, \\ y(2) = 9. \end{cases}$
(e) $x(y')^2 + 2yy' + xyy'' = 0$.	$xy^2 = Ax + B$.	$\begin{cases} y(3) = 1, \\ y'(3) = 2. \end{cases}$

2. Find a differential equation corresponding to each relation, with indicated arbitrary constants. Verify in each case that the differential equation has the relation as its general solution.

(a) $y = 3x^2 + ce^{-2x}$: c.
(b) $y = x + c \sin x$: c.
(c) $x^2 - ay^2 = 1$: a.
(d) $y \ln x = bx$: b.
(e) $y = c_1 \sin 4x + c_2 \cos 4x + x$: c_1, c_2.
(f) $I = \alpha t e^{-t} + \beta e^{-t} + 2 \sin 3t$: α, β.
(g) $x^2 + y^2 - cx = 0$: c.
(h) $y = ax^3 + bx^2 + cx$: a, b, c.
(i) $y = Ae^{-3x} + Be^{2x} + Ce^{4x}$: A, B, C.
(j) $r = a \ln \phi + b\phi + c$: a, b, c.

---

* The terminology *complete solution* is sometimes used to denote *all* solutions, i.e., the general solution together with singular solutions, if any.

† The author is aware of at least one scientist who, in desperation to solve a differential equation subject to given conditions, turned the problem over to a computer (at great cost), but still no solution was found. Finally, it was pointed out to him that by use of an advanced existence theorem it could be shown that the problem had no solution.

3. Find a differential equation for each of the following families of curves in the xy plane: (a) All circles with center at the origin and any radius. (b) All parabolas through the origin with x axis as common axis. (c) All circles with centers on $y = x$ and tangent to the y axis. (d) All ellipses with center at the origin and axes on the coordinate axes.

B EXERCISES

1. Find a differential equation of third order having $y = ax \sin x + bx \cos x$, where a and b are arbitrary constants, as a solution. Would you classify this as a general or particular solution of the third-order equation?

2. (a) How many arbitrary constants has $y = c_1 e^{5x + c_2} + c_3 e^{5x}$? (b) Find a differential equation having this as general solution.

3. Show that the general solution of $y'' + y = e^{-x^2}$ is

$$y = A \cos x + B \sin x + \sin x \int_0^x e^{-t^2} \cos t \, dt - \cos x \int_0^x e^{-t^2} \sin t \, dt$$

Find the particular solution satisfying $y(0) = y'(0) = 0$.

4. (a) By writing the differential equation of Example 17, page 16, in the form $\dfrac{dx}{dy} = \dfrac{1}{3y^{2/3}}$ and then integrating directly, verify the general solution obtained there. (b) From the form of the differential equation in (a), can you suggest how the singular solution $y = 0$ of the original equation might have been discovered? (See page 20.)

5. Find the general solution of $dy/dx = y^3$. Is $y = 0$ a singular solution? Explain.

6. (a) Show that $y' = y^p$, where p is a constant such that $0 < p < 1$ has solutions given by $y = 0$, and $y = [(1 - p)(x + c)]^{1/(1-p)}$. Examine the special cases $p = \frac{1}{2}, \frac{1}{3}, \frac{2}{3}$. (b) Discuss the relationships between these solutions. (c) Are there any solutions besides those given in (a)? (d) Discuss the cases where $p \le 0$, $p \ge 1$.

7. (a) Show that $y = A \cos \sqrt{\lambda} x + B \sin \sqrt{\lambda} x$ is the general solution of $y'' + \lambda y = 0$ where λ is a constant. (b) Show that there is no solution to the boundary value problem $y'' + \lambda y = 0$, $y(0) = 0, y(1) = 0$ besides $y = 0$ unless λ happens to be one of the values $\pi^2, 4\pi^2, 9\pi^2, 16\pi^2, \ldots$, and determine solutions in these cases. (c) From these observations, what can you conclude about the existence and uniqueness of solutions to the boundary value problem given in (b)?

C EXERCISES

1. (a) Show that the differential equation for $y = c_1 u_1(x) + c_2 u_2(x)$, where $u_1(x)$ and $u_2(x)$ are functions which are at least twice differentiable, may be written in determinant form as*

$$\begin{vmatrix} y & u_1(x) & u_2(x) \\ y' & u_1'(x) & u_2'(x) \\ y'' & u_1''(x) & u_2''(x) \end{vmatrix} = 0$$

(b) What happens in the case where

$$W = \begin{vmatrix} u_1(x) & u_2(x) \\ u_1'(x) & u_2'(x) \end{vmatrix} \equiv 0$$

The determinant W is called the *Wronskian* of $u_1(x)$ and $u_2(x)$.

* See Appendix for a brief review of determinants.

2. (a) Make use of the previous exercise to obtain the differential equation having as general solution $y = c_1 e^{3x} + c_2 e^{-2x}$. (b) Generalize the method of Exercise 1 and, thus, obtain a differential equation having general solution $y = c_1 x^4 + c_2 \sin x + c_3 \cos x$.

3. Discuss possible solutions of $(y' - 2x)(y' - 3x^2) = 0$.

4. Show that (a) $|y'| + 1 = 0$ has no solution. (b) $(y')^2 + 1 = 0$ has a solution but no real solution. (c) $|y'| + |y| = 0$ has a solution but not one involving an arbitrary constant.

5. (a) Find a differential equation for the family of tangents to $x^2 + y^2 = 1$. (*Hint:* Let $(\cos c, \sin c)$ be any point on the circle.) (b) Show that the general solution of the differential equation in (a) is defined by the family of tangent lines. (c) Show that $x^2 + y^2 = 1$ is a solution of the differential equation of (a). What kind of solution is it?

6. (a) Use B Exercise 6 to solve the initial value problem $y' = y^p$, $y(0) = 1$. (b) Determine the limit of the solution in (a) as $p \to 1$. (c) Is the function obtained in (b) a solution of $y' = y$, $y(0) = 1$? Discuss.

7. (a) Find the differential equation of the family $y = \left(1 + \dfrac{x}{n}\right)^n$ where $n \neq 0$ is the parameter.

(b) Show that $y = e^x$ is a solution of the equation found in (a) and explain why this is not too surprising.

◆ 2 Further Remarks Concerning Solutions

2.1 REMARKS ON EXISTENCE AND UNIQUENESS

Let us say at the outset that, for most of the differential equations with which we shall deal, there are unique solutions that satisfy certain prescribed conditions. However, lest the engineer or other scientist become too confident with this knowledge, we show, by means of an example, how important it is to be aware of existence and uniqueness problems. The student who feels that he will get by in 99 per cent of the cases is probably correct, but the author knows of a few who were "trapped" in the remaining 1 per cent category.

We consider the differential equation $\quad xy' - 3y = 0 \qquad$ (1)

which arose in a certain applied problem, the details of which we will omit. Suffice it to say that an experimental curve obtained appeared as in Fig. 1.4. From this curve it appears that $y = 0$ for $x \leq 0$ and that y increases (in some way) for $x \geq 0$. By simple methods which we discuss later, the scientist working the problem deduced that the general solution of (1) is $y = cx^3$, where c is an arbitrary constant. Theoretical considerations provided the condition $y = 1$ where $x = 1$ which agreed with experiment. Thus, the scientist decided that the required solution was given by $y = x^3$, the graph of which appears as Fig. 1.5. The theoretical and experimental graphs agreed for $x \geq 0$ but disagreed for $x < 0$. The scientist decided that the mathematics must be wrong. However, it turned out that the *handling* of the mathematics was wrong. The scientist had erroneously assumed, as he had previously always assumed, that a unique solution existed. It is not difficult to show that

$$y = \begin{cases} Ax^3, & x \geq 0 \\ Bx^3, & x \leq 0 \end{cases} \qquad (2)$$

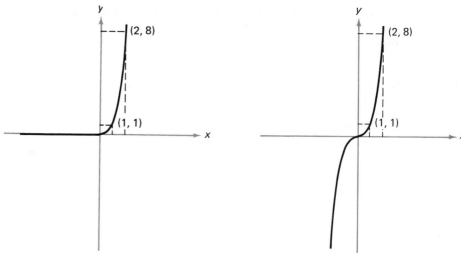

Figure 1.4   Figure 1.5

where A and B are constants, is also a solution. By choosing $A = 1$ to satisfy $y = 1$ where $x = 1$ and choosing $B = 0$ in the solution (2), we obtain

$$y = \begin{cases} x^3, & x \geq 0 \\ 0, & x \leq 0 \end{cases} \tag{3}$$

agreeing completely with experiment.

This example shows the need for knowing just when a unique solution does exist. Even though we cannot go into detail concerning proof, we shall not hide our heads from reality like the proverbial ostrich but, instead, shall satisfy our consciences by quoting the following

Existence–Uniqueness Theorem.   Given the first-order differential equation $y' = F(x, y)$, let $F(x, y)$ satisfy the following conditions:*

1. $F(x, y)$ is real, finite, single-valued, and continuous at all points (x, y) within a region R of the xy plane (which may comprise all points).

2. $\dfrac{\partial F(x, y)}{\partial y}$ is real, finite, single-valued, and continuous in R.

Then there exists one and only one solution $y = g(x)$ in R, such that $y = y_0$ when $x = x_0$, i.e. $y(x_0) = y_0.$†

---

* Some of the conditions given in the theorem are implied by others and are stated merely for emphasis.

† More precisely it can be concluded that $g(x)$ exists and is continuous in some interval $x_0 - h \leq x \leq x_0 + h$; i.e., $|x - x_0| \leq h$ where h is some positive number depending on $F(x, y)$.

Remark. This theorem provides *sufficient conditions* for the existence and unique-ness of a solution; i.e., if the conditions hold, the existence and uniqueness are guaranteed. However, the conditions are not *necessary conditions*; i.e., if the conditions are not all satisfied, there may *still* be a unique solution. It should be noted that the theorem does not tell us *how* to obtain this solution. Corresponding theorems for higher-order equations are available.*

A graphical interpretation of this theorem is that, if R is the region (shown shaded in Fig. 1.6) in which the specified conditions hold, then through any given point (x_0, y_0) in R there will pass one and only one curve C whose slope at any point of R is given by $y' = F(x, y)$. The solution $y = g(x)$ represents the equation of this curve in R. The theorem amounts to stating that there exists a unique solution to $y' = F(x, y)$ in R, which can be described by any of the equations given in (55), page 18, where c is to be determined from knowledge of the point (x_0, y_0), i.e., $y(x_0) = y_0$. This serves to support the use of the terminology *general solution*, since if the conditions of the theorem are satisfied there are *no other solutions in R*. Trouble can arise if we try to extend the solution beyond the region R. In fact singular solutions, if there are any, tend to occur on the boundary of the region.

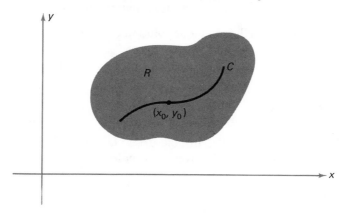

Figure 1.6

In a certain sense the Existence–Uniqueness Theorem may be more significant from the practical view where the conditions fail to hold than when they do hold. This is because when they fail to hold neither existence nor uniqueness can be guaranteed and this lack of guarantee should serve as a warning of potential trouble. For example, in the case of the scientist's differential equation (1), page 23, the conditions fail to hold in a rectangular region such as shown in Fig. 1.7 which contains points where $x \leqq 0$. Thus the scientist might not have felt so confident and perhaps even could have arrived at the correct theoretical solution (3) agreeing with experiment.

Let us consider some further examples illustrating use of the Existence–Uniqueness Theorem.

---

* Such a theorem is considered in C Exercise 2, page 33 and Chapter 4. For proofs of the theorems including the one stated above see [13].

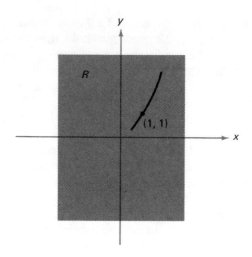

Figure 1.7

ILLUSTRATIVE EXAMPLE 1

Determine whether a unique solution exists for the initial value problem

$$\frac{dy}{dx} = \sqrt{9 - (x^2 + y^2)}, \qquad y(1) = 2 \tag{4}$$

Solution   We have $F(x, y) = \sqrt{9 - (x^2 + y^2)}, \qquad \dfrac{\partial F}{\partial y} = \dfrac{-y}{\sqrt{9 - (x^2 + y^2)}}$ (5)

and we see that potential trouble arises for points (x, y) for which $x^2 + y^2 = 9$. Suppose that we stay away from such points by choosing for example a region R

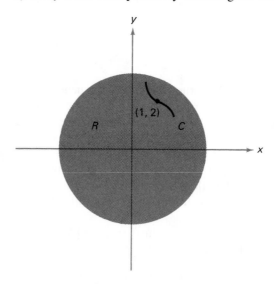

Figure 1.8

inside the circle $x^2 + y^2 = 8$ (see Fig. 1.8), which includes the required point $(1, 2)$ described by the initial condition. Then, since the conditions of the theorem are met, we can conclude that there does exist a unique solution to the initial value problem. In other words, there is a unique solution curve C contained in the region R which passes through the point $(1, 2)$ as indicated in Fig. 1.8.

ILLUSTRATIVE EXAMPLE 2

Use the Existence–Uniqueness Theorem to determine whether a unique solution exists for the initial value problem

$$\frac{dy}{dx} = 3y^{2/3}, \qquad y(2) = 0 \tag{6}$$

Solution   We have $\qquad F(x, y) = 3y^{2/3}, \quad \dfrac{\partial F}{\partial y} = \dfrac{2}{y^{1/3}} \tag{7}$

Thus we can expect trouble to arise in regions which include points where $y = 0$. From the Existence–Uniqueness Theorem we cannot guarantee the existence or uniqueness of a solution in such regions. By testing $y = 0$ we see that it is a solution showing that at least one exists, but we do not know if it is unique. Actually as seen on page 20, it is not unique.

ILLUSTRATIVE EXAMPLE 3

Given the first-order linear differential equation

$$\frac{dy}{dx} + P(x)y = Q(x) \tag{8}$$

where $P(x)$ and $Q(x)$ are supposed continuous in an interval $a \leq x \leq b$, prove that there exists a unique solution to the differential equation such that $y(x_0) = y_0$ where x_0 is inside the interval.

Solution   We have $\dfrac{dy}{dx} = F(x, y)$ where $F(x, y) = Q(x) - P(x)y, \quad \dfrac{\partial F}{\partial y} = -P(x) \tag{9}$

so that the conditions of the Existence–Uniqueness Theorem hold in a rectangular region R bounded by the lines $x = a$ and $x = b$. Thus the required result is proved.

ILLUSTRATIVE EXAMPLE 4

(a) Determine whether a unique solution exists for the initial value problem

$$\frac{dy}{dx} = y^2, \qquad y(0) = 1 \tag{10}$$

(b) Verify your conclusion in (a) by solving the problem.

Solution   (a) We have $\qquad F(x, y) = y^2, \dfrac{\partial F}{\partial y} = 2y \tag{11}$

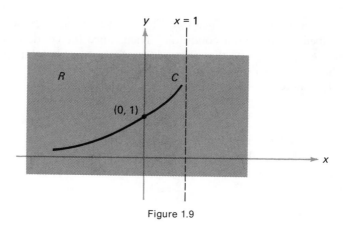

Figure 1.9

Thus the conditions of the Existence–Uniqueness Theorem are satisfied in any region R as shown in the rectangle of Fig. 1.9 and we can conclude that a unique solution exists in R.

(b) Write the given differential equation as $\dfrac{dx}{dy} = \dfrac{1}{y^2}$.

Then by integrating we have $x = -\dfrac{1}{y} + c$. Since $y = 1$ when $x = 0$, this gives $c = 1$

so that $x = -\dfrac{1}{y} + 1$ i.e., $y = \dfrac{1}{1 - x}$.

The unique solution curve C corresponding to this is also shown in Fig. 1.9. It should be noted that even though the region R can be chosen as large as we like and still satisfy the conditions of the existence theorem, the curve does not extend indefinitely to the right.* In fact, as seen, it does not go beyond $x = 1$ which represents an asymptote. The fact that $x = 1$ is a barrier is not at all evident from the given differential equation. These results indicate the complexities which can occur in non-linear equations.

2.2 DIRECTION FIELDS AND THE METHOD OF ISOCLINES

Suppose that we are given the differential equation

$$y' = F(x, y) \tag{12}$$

where $F(x, y)$ satisfies the conditions of the Existence–Uniqueness Theorem. At each point (a, b) of region R (see Fig. 1.10) we can construct a short line, called a *lineal element*, having slope $F(a, b)$. If we do this for a large number of points, we obtain a graph such as shown in Fig. 1.10 called the *direction field* of the differential equation. The lineal elements represent tangent lines to the solution curves at these points.

* The Existence–Uniqueness Theorem expresses this by not guaranteeing more than what is stated in the second footnote on page 24.

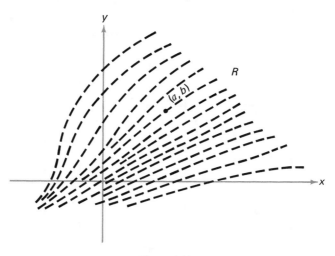

Figure 1.10

It is quite remarkable that by using this simple idea we can arrive at a pictorial representation of the general solution of a differential equation *without even solving the equation.* The technique is of course very useful, especially when an exact solution cannot be found. The graph indicates that the general solution of (12) is given by

$$y = f(x, c), \qquad U(x, y) = c \quad \text{or} \quad G(x, y, c) = 0 \tag{13}$$

where c is an arbitrary constant. Thus, each curve in Fig. 1.10 corresponds to a different value of c, or to put it another way, there will be one and only one curve passing through a given point in accordance with the Existence–Uniqueness Theorem. Let us illustrate the procedure involved in obtaining the direction field for a differential equation by considering the following

ILLUSTRATIVE EXAMPLE 5

Obtain the direction field for the differential equation

$$\frac{dy}{dx} = -\frac{x}{y} \tag{14}$$

Solution   It is convenient to choose points (x, y) for which x and y are integers, and to compute the corresponding slopes at these points. In more complicated equations use of the remarkable pocket calculator may serve to minimize laborious computations, and other points can be used to permit greater accuracy. The calculations are indicated in Fig. 1.11 for the case where x and y are between -4 and 4. Thus, for example, the slope corresponding to $x = 2$, $y = 3$, i.e., the point (2, 3), is $-\frac{2}{3}$. Since x/y does not exist for $y = 0$, entries for these cases are indicated by a dash.

The corresponding direction field is indicated in Fig. 1.12. The graph seems to indicate that the curves corresponding to the general solution are circles with center at the origin; i.e.,

$$x^2 + y^2 = c \tag{15}$$

x \ y	−4	−3	−2	−1	0	1	2	3	4
−4	−1	$-\dfrac{3}{4}$	$-\dfrac{1}{2}$	$-\dfrac{1}{4}$	0	$\dfrac{1}{4}$	$\dfrac{1}{2}$	$\dfrac{3}{4}$	1
−3	$-\dfrac{4}{3}$	−1	$-\dfrac{2}{3}$	$-\dfrac{1}{3}$	0	$\dfrac{1}{3}$	$\dfrac{2}{3}$	1	$\dfrac{4}{3}$
−2	−2	$-\dfrac{3}{2}$	−1	$-\dfrac{1}{2}$	0	$\dfrac{1}{2}$	1	$\dfrac{3}{2}$	2
−1	−4	−3	−2	−1	0	1	2	3	4
0	−	−	−	−	−	−	−	−	−
1	4	3	2	1	0	−1	−2	−3	−4
2	2	$\dfrac{3}{2}$	1	$\dfrac{1}{2}$	0	$-\dfrac{1}{2}$	−1	$-\dfrac{3}{2}$	−2
3	$\dfrac{4}{3}$	1	$\dfrac{2}{3}$	$\dfrac{1}{3}$	0	$-\dfrac{1}{3}$	$-\dfrac{2}{3}$	−1	$-\dfrac{4}{3}$
4	1	$\dfrac{3}{4}$	$\dfrac{1}{2}$	$\dfrac{1}{4}$	0	$-\dfrac{1}{4}$	$-\dfrac{1}{2}$	$-\dfrac{3}{4}$	−1

Figure 1.11

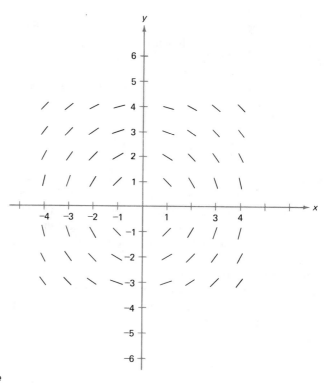

Figure 1.12

which becomes more evident as more points are chosen. The result (15) is actually correct since it is the general solution of (14); or to put it another way, the differential equation of the family of circles (15) is given by (14).

When finding the direction field for the differential equation

$$y' = F(x, y) \tag{16}$$

the labor involved can be somewhat reduced by setting

$$F(x, y) = m \tag{17}$$

where m is any constant, and realizing that any point on the curve represented by (17) has associated with it a lineal element of slope m. This is often called the *method of isoclines* (isocline meaning *constant slope*) and is illustrated in the following

ILLUSTRATIVE EXAMPLE 6

Use the method of isoclines to work Illustrative Example 5, page 29.

Solution   To obtain the required direction field, let us choose a particular value of m, say $m = 2$. Then on the corresponding line $y = -x/2$ we construct parallel lineal elements of slope $m = 2$, as shown in Fig. 1.13. We then do the same for other values of m and thus obtain the pattern indicated.

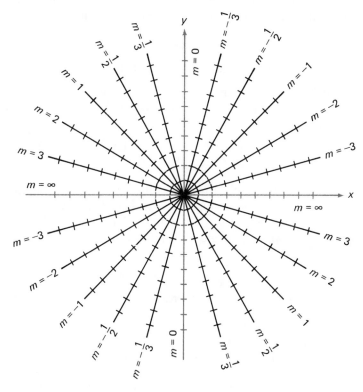

Figure 1.13

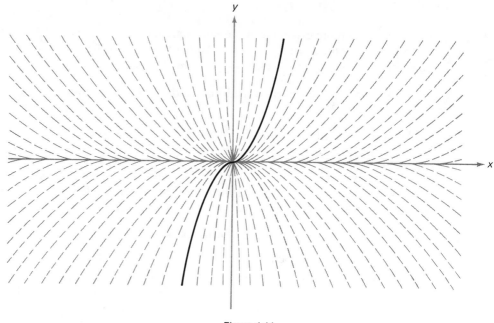

Figure 1.14

The dilemma of the scientist referred to on page 23 could also have been resolved by use of direction fields, as indicated in the following

ILLUSTRATIVE EXAMPLE 7

Obtain the direction field of the differential equation $xy' - 3y = 0$.

Solution   The direction field obtained by the method of isoclines, or as on pages 29–30, is shown in Fig. 1.14. It is interesting that this figure reveals the possibility of solutions such as

$$y = cx^3 \quad \text{or} \quad y = \begin{cases} Ax^3, & x \geq 0 \\ Bx^3, & x \leq 0 \end{cases}$$

the second of which, with $A = 1$, $B = 0$, gives the desired solution on page 24.

The direction field of Fig. 1.14 serves to illustrate very nicely the Existence–Uniqueness Theorem in view of the fact that there are infinitely many solutions through $(0, 0)$, while there are no solutions through $(0, y)$ where $y \neq 0$.

A EXERCISES

1. Use the Existence–Uniqueness Theorem to determine if unique solutions exist for each of the following initial value problems.
   (a) $y' = 3x + 2y$; $y(1) = 4$.
   (b) $y' = 1/(x^2 + y^2)$; $y(0) = 1$.
   (c) $y' = 1/(x^2 + y^2)$; $y(0) = 0$.
   (d) $y' + xy = x^2$; $y(0) = 2$.
   (e) $y' = (x - 2y)/(y - 2x)$; $y(1) = 2$.
   (f) $y' = 1/(x^2 - y^2)$; $y(1) = 2$.
   (g) $y' = x^2 + y^2$; $y(0) = 2$.
   (h) $y' = \sqrt{xy}$, $y(1) = 0$.
   (i) $y' = y \csc x$; $y(0) = 1$.
   (j) $y' = 1/\sqrt{x^2 + 4y^2 - 4}$; $y(3) = 2$.

2. (a) Show that the differential equation $y' = y^{1/2}$ has solutions $y = \frac{1}{4}(x + c)^2$ and $y = 0$.
(b) Discuss these solutions and their relation to the Existence–Uniqueness Theorem and to each other.

3. (a) Obtain the direction field for $y' = 2x - y$. (b) Explain the relationship existing among various solutions of the equation and the direction field. (c) Can you guess the approximate position of that curve passing through $(0, 0)$? (d) Show that the general solution of $y' = 2x - y$ is $y = 2x - 2 + ce^{-x}$, and thus compare with your guess in (c).

4. Obtain the direction fields for each of the following. (a) $y' = 2x$. (b) $y' = y/x$. (c) $y' = x + y$.
(d) $y' = 1/(x^2 + 4y^2)$.

5. (a) Obtain the direction field for $y' = y^{1/2}$. (b) Explain the connection between the direction field and the results of Exercise 2.

B EXERCISES

1. Discuss the existence and uniqueness of a solution to the initial value problem $y' = P(x)y + Q(x)y^n$, $y(x_0) = y_0$, for given constants n, x_0, y_0.

2. Find direction fields for the differential equations given in A Exercises 1(a)–(f), and relate the results obtained to the conclusions arrived at in those exercises.

3. Discuss the existence and uniqueness of a solution to the initial value problem $y' = 1 + xy + x^2y^2$, $y(0) = 1$. Generalize to the case $y' = a_0(x) + a_1(x)y + a_2(x)y^2$, $y(0) = y_0$.

C EXERCISES

1. Show that the differential equation $y' = \sqrt{y - x} + 1$ has solutions $y = x$, $y = x + \frac{1}{4}(x + c)^2$ and infinitely many other solutions, such as, for example,

$$y = \begin{cases} x, & x \leq -c \\ x + \frac{1}{4}(x + c)^2, & x \geq -c \end{cases}$$

How might you explain this?

2. An Existence–Uniqueness Theorem for the second-order differential equation $y'' = F(x, y, y')$, which can be written as two first-order equations $z' = F(x, y, z)$, $y' = z$, states that if $F(x, y, z)$ and its partial derivatives $\partial F/\partial y$, $\partial F/\partial z$ are real, single-valued, and continuous at all points (x, y, z) within a region R of three-dimensional space, then there exists one and only one solution $y = g(x)$ which passes through any given point of R. Note that if the point is (x_0, y_0, z_0) this is equivalent to $y = y_0$, $y' = z_0$ for $x = x_0$. (a) Using this theorem investigate the existence and uniqueness of solutions to $y'' + x(y')^2 = 1$. (b) Work part (a) for $xy'' + y' + xy = 0$, $y(1) = 2$, $y'(1) = 0$. (c) What difficulties occur in case conditions are prescribed at $x = 0$? (d) Can you think of a generalization of this theorem?

3. Do you think it would be possible to generalize the method of direction fields to second-order differential equations? Explain.

4. Show that $y = 0$, $y = c^2x + 2c$, $y = -1/x$ are solutions of

$$y' = \left(\frac{\sqrt{1 + xy} - 1}{x} \right)^2$$

Discuss these solutions and their relation to the Existence–Uniqueness Theorem and to each other.

two

first-order and simple higher-order ordinary differential equations

1 The Method of Separation of Variables

Suppose that we are given a first-order differential equation

$$\frac{dy}{dx} = F(x, y) \tag{1}$$

Then, regarding dy/dx as a quotient of differentials, it can also be written in the form

$$M(x, y)dx + N(x, y)dy = 0 \tag{2}$$

Thus, for example,

$$\frac{dy}{dx} = \frac{x - 3y}{2y - 5x} \tag{3}$$

can also be written

$$(x - 3y)dx + (5x - 2y)dy = 0 \tag{4}$$

where $M \equiv x - 3y$, $N \equiv 5x - 2y$. The problem of solving first-order differential equations hinges on the solution of equations (1) or (2). An especially simple type of equation occurring often in practice is one which can be written in the form

$$f(x)dx + g(y)dy = 0 \tag{5}$$

where one term involves only x while the other term involves only y. This equation can be solved immediately by integration. Thus, the general solution is

$$\int f(x)dx + \int g(y)dy = c \tag{6}$$

where c is the constant of integration. We can of course get back to (5) by taking the differential of both sides of (6), thus eliminating c; i.e.,

$$d\int f(x)dx + d\int g(y)dy = d(c) \quad \text{or} \quad f(x)dx + g(y)dy = 0$$

Because the method of solution depends on being able to write (1) or (2) in the form (5), where the variables are "separated" into two terms, it is called *the method of separation of variables*, and the variables are said to be *separable*. This fortunate state of affairs in which variables are separable does not, much to our dismay, occur all the time. For example, there is no way in which equation (4) can be written in the form (5). In such cases we are forced to look for other methods. The search for such methods will concern us in the remainder of this chapter.

ILLUSTRATIVE EXAMPLE 1

(a) Find the general solution of $\dfrac{dy}{dx} = \dfrac{x^2 + 1}{2 - y}$ and (b) determine the particular solution for which $y = 4$ when $x = -3$.

Solution (a) We can write the given equation in the form

$$(x^2 + 1)dx + (y - 2)dy = 0 \tag{7}$$

on separating the variables. Integration yields the required general solution

$$\int (x^2 + 1)dx + \int (y - 2)dy = c \quad \text{i.e.,} \quad \frac{x^3}{3} + x + \frac{y^2}{2} - 2y = c \qquad (8)$$

(b) Putting $x = -3$, $y = 4$ into (8) gives $c = -12$. The required particular solution is

$$\frac{x^3}{3} + x + \frac{y^2}{2} - 2y = -12 \qquad (9)$$

Remark 1. It should be noted that to get from the given differential equation to (7) we must multiply by $y - 2$ so that, strictly speaking we should assume $y \neq 2$. However, if we should forget this, or even choose to ignore it, we do not get something for nothing, because it shows up anyway in the fact that the curves (8) have vertical slopes at points where $y = 2$.

Remark 2. The solution can if desired be obtained explicitly in this case by solving for y in (9). The result is found by the quadratic formula to be

$$y = \frac{12 + \sqrt{-24x^3 - 72x - 720}}{6}$$

Occasionally, the fact that a differential equation may be "separable" is not too obvious, as in the following

ILLUSTRATIVE EXAMPLE 2

Solve $x\dfrac{dy}{dx} - y = 2x^2y$.

Solution   We can write the equation (on multiplying by dx) as

$$x\,dy - y\,dx = 2x^2y\,dx \quad \text{or} \quad (2x^2y + y)dx - x\,dy = 0$$

i.e., $$y(2x^2 + 1)dx - x\,dy = 0$$

Dividing by x and y, i.e., xy, gives

$$\left(\frac{2x^2 + 1}{x}\right)dx - \frac{dy}{y} = 0 \quad \text{or} \quad \int \left(\frac{2x^2 + 1}{x}\right)dx - \int \frac{dy}{y} = c$$

so that* $$x^2 + \ln |x| - \ln |y| = c \qquad (10)$$

This can also be written in a form free of logarithms by writing (10) successively as

$$x^2 + \ln \left|\frac{x}{y}\right| = c, \quad \ln \left|\frac{x}{y}\right| = c - x^2, \quad \left|\frac{x}{y}\right| = e^{c-x^2} = e^c e^{-x^2}$$

$$\frac{x}{y} = \pm e^c e^{-x^2}, \quad y = \pm e^{-c} x e^{x^2} \quad \text{or finally} \quad y = Axe^{x^2}$$

* It should be noted that $\int dx/x = \ln x$ only if $x > 0$. For $x > 0$ or $x < 0$, $\int dx/x = \ln |x|$. A table of some important integrals is given on the back inside cover of this book.

Check. Putting $y = Axe^{x^2}$ in the given differential equation, we have

$$x(2Ax^2e^{x^2} + Ae^{x^2}) - Axe^{x^2} = 2x^2(Axe^{x^2}) \quad \text{or} \quad 2Ax^3e^{x^2} = 2Ax^3e^{x^2}$$

A EXERCISES

1. Solve each of the following, subject to conditions where given.

(a) $\dfrac{dy}{dx} = -\dfrac{x}{y}$; $y = 2$ where $x = 1$.

(b) $\dfrac{dy}{dx} = -\dfrac{y}{x}$; $y(1) = 3$.

(c) $3x(y^2 + 1)dx + y(x^2 + 2)dy = 0$.

(d) $2y\,dx + e^{-3x}\,dy = 0$.

(e) $y' = \dfrac{x + xy^2}{4y}$; $y(1) = 0$.

(f) $r\dfrac{d\phi}{dr} = \phi^2 + 1$.

(g) $\sin^2 y\,dx + \cos^2 x\,dy = 0$; $y\left(\dfrac{\pi}{4}\right) = \dfrac{\pi}{4}$.

(h) $x\sqrt{1 + y^2}\,dx = y\sqrt{1 + x^2}\,dy$.

(i) $2y\cos x\,dx + 3\sin x\,dy = 0$; $y\left(\dfrac{\pi}{2}\right) = 2$.

(j) $y' = 8xy + 3y$.

(k) $\dfrac{dI}{dt} + 5I = 10$; $I(0) = 0$.

(l) $y\,dx + (x^3y^2 + x^3)dy = 0$.

2. The slope of a family of curves at any point (x, y) is given by $\dfrac{dy}{dx} = -\dfrac{3x + xy^2}{2y + x^2y}$.

Find the equation of that member of the family passing through $(2, 1)$.

B EXERCISES

Solve each of the following:

1. $\dfrac{dy}{dx} = \dfrac{(y - 1)(x - 2)(y + 3)}{(x - 1)(y - 2)(x + 3)}$.

2. $\dfrac{dr}{d\phi} = \dfrac{\sin \phi + e^{2r}\sin \phi}{3e^r + e^r \cos 2\phi}$; $r = 0$ where $\phi = \dfrac{\pi}{2}$.

3. $x^3e^{2x^2 + 3y^2}\,dx - y^3e^{-x^2 - 2y^2}\,dy = 0$.

4. $\dfrac{dU}{ds} = \dfrac{U + 1}{\sqrt{s} + \sqrt{sU}}$.

C EXERCISES

1. A particle moves along the x axis so that its velocity is proportional to the product of its instantaneous position x (measured from $x = 0$) and the time t (measured from $t = 0$). If the particle is located at $x = 54$ when $t = 0$ and $x = 36$ when $t = 1$, where will it be when $t = 2$?

2. Show that the differential equation $\dfrac{dy}{dx} = \dfrac{4y^2 - x^4}{4xy}$ is non-separable but becomes separable on changing the dependent variable from y to v according to the transformation $y = vx$. Use this to find the solution of the original equation.

3. (a) Show that the non-separable differential equation $[F(x) + yG(xy)]dx + xG(xy)dy = 0$ becomes separable on changing the dependent variable from y to v according to the transformation $v = xy$. (b) Use this to solve $(x^2 + y\sin xy)dx + x\sin xy\,dy = 0$.

2 The Method of Transformation of Variables

Since a differential equation whose variables are separable is so easy to solve, a rather obvious question which could be asked is the following.

Question. Are there any types of differential equations whose variables are not separable which can in some way be changed or transformed into ones whose variables are separable?

The answer to this question is "yes." In fact, one of the most important ways of solving a given differential equation is to make an appropriate change or *transformation of variables* so that the given equation reduces to some recognized type which can be solved. The situation is very much analogous to the "ingenious devices" often used in calculus for evaluating integrals by a change of variable. In some cases the particular transformation of variables to be used is suggested by the form of the equation. In other cases the transformation may be less obvious.

2.1 THE HOMOGENEOUS EQUATION

An equation which can always be transformed into one with separable variables is

$$\frac{dy}{dx} = f\left(\frac{y}{x}\right) \tag{1}$$

and any differential equation which is or can be put into this form is called a *homogeneous differential equation*. In order to change (1) into a separable equation, we use the transformation $y/x = v$ or $y = vx$, i.e., the change of the dependent variable from y to v keeping the same independent variable x. Then

$$\frac{dy}{dx} = v + x\frac{dv}{dx}$$

and equation (1) becomes

$$v + x\frac{dv}{dx} = f(v)$$

so that

$$\frac{dx}{x} = \frac{dv}{f(v) - v} \tag{2}$$

where the variables are separated. The solution is then obtained by integration.*

ILLUSTRATIVE EXAMPLE 1

Solve $\dfrac{dy}{dx} = \dfrac{x - y}{x + y}$.

Solution We may write the equation as $\qquad \dfrac{dy}{dx} = \dfrac{1 - y/x}{1 + y/x}$

---

* It should be noted that the method breaks down in case $f(v) = v$; i.e., $f(y/x) = y/x$. However, in this case the equation is already of separable type.

in which the right-hand side is a function of y/x, so that the equation is homogeneous. Letting $y = vx$, we have

$$v + x\frac{dv}{dx} = \frac{1 - v}{1 + v}, \qquad x\frac{dv}{dx} = \frac{1 - 2v - v^2}{1 + v}, \qquad \frac{dx}{x} = \frac{(1 + v)dv}{1 - 2v - v^2}$$

Thus, $\ln x = -\frac{1}{2}\ln(1 - 2v - v^2) + c_1$ or $\ln[x^2(1 - 2v - v^2)] = c_2$ so that $x^2(1 - 2v - v^2) = c$. Replacing v by y/x and simplifying, we find $x^2 - 2xy - y^2 = c$.

ILLUSTRATIVE EXAMPLE 2

Solve $\dfrac{dy}{dx} = \dfrac{ye^{y/x} + y}{x}$.

Solution The right-hand side can be written as $(y/x)e^{y/x} + (y/x)$, a function of y/x, so that the equation is homogeneous. Letting $y = vx$, we obtain

$$v + x\frac{dv}{dx} = ve^v + v \quad \text{or} \quad \frac{e^{-v}\,dv}{v} = \frac{dx}{x}$$

Hence

$$\int \frac{e^{-v}\,dv}{v} = \ln x + c$$

The integration cannot be performed in closed form.

Remark. The student should notice that an equation $y' = f(x, y)$ is homogeneous if upon placing $y = vx$ into the right-hand side of the equation it becomes a function of v alone. Thus, in Illustrative Example 1, for instance, $(x - y)/(x + y)$ becomes $(1 - v)/(1 + v)$.

2.2 OTHER SPECIAL TRANSFORMATIONS

As an example of a special transformation suggested by the form of a given differential equation, let us consider the following

ILLUSTRATIVE EXAMPLE 3

Solve the differential equation $y' = \sqrt{x + y}$.

Solution The equation is not separable. However, the presence of $x + y$ suggests that we might try a change of dependent variable from y to v given by

$$x + y = v^2 \tag{3}$$

where we have used v^2 in (3) rather than v to avoid square roots. From (3) we have

$$y' = \frac{dy}{dx} = \frac{d}{dx}(v^2 - x) = 2v\frac{dv}{dx} - 1$$

Thus the equation becomes $\qquad 2v\dfrac{dv}{dx} - 1 = v \tag{4}$

We can write this in separable form as

$$\frac{2v \, dv}{v + 1} = dx \quad \text{or} \quad \int \frac{2v \, dv}{v + 1} = \int dx$$

Performing the integration on the left, we have

$$\int \frac{2v \, dv}{v + 1} = 2 \int \frac{(v + 1) - 1}{v + 1} \, dv = 2 \int \left(1 - \frac{1}{v + 1} \right) dv = 2v - 2 \ln (v + 1)$$

so that the general solution of (4) is $2v - 2 \ln (v + 1) = x + c$.

Replacing v by $\sqrt{x + y}$ now yields the required general solution of the given equation

$$2\sqrt{x + y} - 2 \ln (\sqrt{x + y} + 1) = x + c$$

A EXERCISES

Solve each of the following:

1. $\dfrac{dy}{dx} = 1 + \dfrac{y}{x}$.

2. $\dfrac{dy}{dx} = \dfrac{y}{x} + \dfrac{y^2}{x^2}$; $y(1) = 1$.

3. $xy' = 2x + 3y$.

4. $(x^2 - y^2)dx - 2xy \, dy = 0$.

5. $(x + 2y)dx + (2x + y)dy = 0$.

6. $\dfrac{dy}{dx} = \dfrac{y + x \cos^2 (y/x)}{x}$; $y(1) = \dfrac{\pi}{4}$.

7. $xy' = y - \sqrt{x^2 + y^2}$.

8. $y \, dx = (2x + 3y)dy$.

9. $(x^3 + y^3)dx - xy^2 \, dy = 0$; $y(1) = 0$.

10. $\dfrac{dy}{dx} = \dfrac{1}{2} \left(\dfrac{x}{y} + \dfrac{y}{x} \right)$.

11. $y' = \dfrac{y}{x} + \sec^2 \dfrac{y}{x}$.

12. $(x - 4y)dx + (3x - 2y)dy = 0$.

B EXERCISES

Solve:

1. $\dfrac{dy}{dx} = \dfrac{\sqrt{x^2 + y^2}}{x}$.

2. $\dfrac{dy}{dx} = \dfrac{2x + 5y}{2x - y}$.

3. $\dfrac{dy}{dx} = \dfrac{6x^2 - 5xy - 2y^2}{6x^2 - 8xy + y^2}$.

4. $y' = (x + y)^2$.

5. $y' = \sqrt{2x + 3y}$.

6. Solve the equation $\dfrac{dy}{dx} = \dfrac{2x + 3y + 1}{3x - 2y - 5}$ by letting $x = X + h$ and $y = Y + k$, where X, Y

are new variables and h and k are constants, and then choosing h and k appropriately.

7. Solve $(3x - y - 9)y' = (10 - 2x + 2y)$.

8. Show that the method of Exercise 6 fails for the equation $(2x + 3y + 4)dx = (4x + 6y + 1)dy$. However, show that the substitution $2x + 3y = v$ leads to the solution.

9. Solve $(2x + 2y + 1)dx + (x + y - 1)dy = 0$.

10. Solve $\left[2x \sin \dfrac{y}{x} + 2x \tan \dfrac{y}{x} - y \cos \dfrac{y}{x} - y \sec^2 \dfrac{y}{x} \right] dx + \left[x \cos \dfrac{y}{x} + x \sec^2 \dfrac{y}{x} \right] dy = 0.$

C EXERCISES

1. Solve (a) $\dfrac{dy}{dx} = \dfrac{\sqrt{x+y} + \sqrt{x-y}}{\sqrt{x+y} - \sqrt{x-y}}$. (b) $\dfrac{dy}{dx} = \dfrac{1 + \sqrt{x-y}}{1 - \sqrt{x-y}}$.

2. Solve $\dfrac{dy}{dx} = \dfrac{2y}{x} + \dfrac{x^3}{y} + x \tan \dfrac{y}{x^2}$ by the transformation $y = vx^2$.

3. Solve $\dfrac{dy}{dx} = \dfrac{3x^5 + 3x^2y^2}{2x^3y - 2y^3}$ by letting $x = u^p$, $y = v^q$, and choosing the constants p and q appropriately. Could the equation be solved by letting $y = vx^n$ and choosing the constant n?

4. By letting $y = vx^n$ and choosing the constant n appropriately, solve

$$(2 + 3xy^2)dx - 4x^2y\, dy = 0$$

5. Solve (a) $\dfrac{dy}{dx} = \left(\dfrac{x - 3y - 5}{x + y - 1} \right)^2$. (b) $\sqrt{x+y+1}\, y' = \sqrt{x+y} - 1$.

6. Solve $\dfrac{dy}{dx} = \dfrac{y(1 + xy)}{x(1 - xy)}$.

7. Show that $x\, dy - y\, dx = \tan^{-1}(y/x)dx$ can be solved by the substitution $y = vx$ even though the equation is not homogeneous. Explain.

3

The Intuitive Idea of Exactness

Suppose that we are given a first-order differential equation

$$M\, dx + N\, dy = 0 \quad \text{or} \quad \dfrac{dy}{dx} = -\dfrac{M}{N} \tag{1}$$

Then, by analogy with equation (6), page 35, we may expect that the general solution is given by $\qquad U(x, y) = c \tag{2}$

where $U(x, y)$ must be determined and c is an arbitrary constant. In such case the differential equation corresponding to (2) is obtained by taking the differential of both sides of (2), i.e., $\qquad dU = \dfrac{\partial U}{\partial x} dx + \dfrac{\partial U}{\partial y} dy = 0 \tag{3}$

Thus the required differential equation is given by

$$\dfrac{\partial U}{\partial x} dx + \dfrac{\partial U}{\partial y} dy = 0 \quad \text{or} \quad \dfrac{dy}{dx} = -\dfrac{\partial U/\partial x}{\partial U/\partial y} \tag{4}$$

Now since equation (4) must be the same as equation (1), we must have

$$\frac{\partial U/\partial x}{\partial U/\partial y} = \frac{M}{N} \quad \text{or} \quad \frac{\partial U/\partial x}{M} = \frac{\partial U/\partial y}{N}$$

Calling each of these last ratios μ, which may be a function of x and y, we have

$$\frac{\partial U}{\partial x} = \mu M, \qquad \frac{\partial U}{\partial y} = \mu N \tag{5}$$

Substituting these values into (3) we find that

$$\mu(M \, dx + N \, dy) = dU = 0 \tag{6}$$

From (6) we see that $\mu(M \, dx + N \, dy)$ is the differential of a function U of x and y. We call this a *perfect* or *exact differential*. If we start with the differential equation

$$M \, dx + N \, dy = 0 \tag{7}$$

and multiply by μ to obtain $\quad \mu(M \, dx + N \, dy) = 0 \tag{8}$

where the left-hand side is an exact differential, we say that we have made equation (7) exact, and call (8) an *exact differential equation*. In such a case, (8) can be written

$$dU(x, y) = 0 \tag{9}$$

from which $\quad U(x, y) = c \tag{10}$

The function μ, which enables us to go from (7) to (9) and then by an integration to (10), is for obvious reasons called an *integrating factor*. Let us summarize these remarks in three definitions.

Definition 1. The differential of a function of one or more variables is called an *exact differential*.

Definition 2. If $M \, dx + N \, dy = 0$ is multiplied by $\mu(x, y)$ to obtain $\mu(M \, dx + N \, dy) = 0$, whose left-hand side is an exact differential, we say that we have made the differential equation exact.*

Definition 3. The multiplying function μ is called an *integrating factor* of the differential equation $M \, dx + N \, dy = 0$.

In the method of separation of variables we have, without realizing it, made use of the above ideas. For example, in Illustrative Example 2, page 36, we were given the differential equation

$$(2x^2y + y)dx - x \, dy = 0$$

---

* Naturally, we assume $\mu \neq 0$. Also, if $\mu \equiv 1$, the equation is already exact.

We then multiplied the equation by the "suitable" integrating factor $\mu = 1/xy$ to obtain

$$\left(2x + \frac{1}{x}\right)dx - \frac{dy}{y} = 0 \tag{11}$$

i.e., $\qquad d(x^2 + \ln|x| - \ln|y|) = 0 \quad\text{or}\quad x^2 + \ln|x| - \ln|y| = c$

For a more dramatic illustration of the above ideas, let us consider an equation whose variables do not happen to be separable as in the following

Motivation Example. Let it be required to solve the equation

$$(2y + 3x)dx + x\,dy = 0 \tag{12}$$

Suppose we "happen to know" that an integrating factor of this equation is x. Multiplication of (12) by x yields

$$(2xy + 3x^2)dx + x^2\,dy = 0 \tag{13}$$

which, by definition of integrating factor, should be exact. The student can verify that this is so, since we can write (13) as

$$d(x^2y + x^3) = 0 \tag{14}$$

and so, by integration, the general solution is $x^2y + x^3 = c$.

Theoretically, the example is perfectly clear, but students may be asking themselves the following

Questions. (1) How does one "know" that x is an integrating factor? (2) How does one "know" that we can write (13) as (14)?

Question 2 is answered in the next section. Question 1 will form a basis for later discussion in this chapter.

4 Exact Differential Equations

If the differential equation $M\,dx + N\,dy = 0$ is exact, then by definition there is a function $U(x, y)$ such that

$$M\,dx + N\,dy = dU \tag{1}$$

But, from elementary calculus, $\quad dU = \dfrac{\partial U}{\partial x}\,dx + \dfrac{\partial U}{\partial y}\,dy \tag{2}$

and so, upon comparison of (1) and (2), we see that

$$\frac{\partial U}{\partial x} = M, \qquad \frac{\partial U}{\partial y} = N \tag{3}$$

Differentiating the first of equations (3) with respect to y and the second with respect to x, we find*

$$\frac{\partial^2 U}{\partial y\, \partial x} = \frac{\partial M}{\partial y}, \qquad \frac{\partial^2 U}{\partial x\, \partial y} = \frac{\partial N}{\partial x} \tag{4}$$

Under suitable conditions, the order of differentiation is immaterial,† so that equations (4) lead to the condition

$$\frac{\partial M}{\partial y} = \frac{\partial N}{\partial x} \tag{5}$$

This is a *necessary* condition for exactness; i.e., if a differential equation is exact, then of necessity it follows that (5) is true. The converse theorem states that if (5) holds, then $M\, dx + N\, dy$ is an exact differential, i.e., we can find a function U such that $\partial U/\partial x = M$, $\partial U/\partial y = N$. This converse theorem can be proved‡ and shows that (5) is also a sufficient condition for exactness. We summarize these remarks in the following

Theorem. A necessary and sufficient condition for the exactness of the differential equation $M\, dx + N\, dy = 0$ is $\partial M/\partial y = \partial N/\partial x$. This means that (1) if $M\, dx + N\, dy = dU = 0$ (i.e., the equation is exact), then $\partial M/\partial y = \partial N/\partial x$ (necessity); (2) if $\partial M/\partial y = \partial N/\partial x$, then U exists such that $M\, dx + N\, dy = dU$ or, what is equivalent, U exists such that $\partial U/\partial x = M$ and $\partial U/\partial y = N$ (sufficiency).

To illustrate the theorem, consider equation (13) of page 43, i.e.,

$$(2xy + 3x^2)dx + x^2\, dy = 0$$

Here $\qquad M = 2xy + 3x^2, \qquad N = x^2, \qquad \dfrac{\partial M}{\partial y} = 2x, \qquad \dfrac{\partial N}{\partial x} = 2x$

Thus, by the sufficiency part of the theorem, we are guaranteed a function U so that

$$(2xy + 3x^2)dx + x^2\, dy = dU \tag{6}$$

or, what amounts to the same thing, there exists a function U so that

$$\frac{\partial U}{\partial x} = 2xy + 3x^2, \qquad \frac{\partial U}{\partial y} = x^2 \tag{7}$$

---

* We are naturally supposing that these derivatives exist; otherwise we have no right to take them.

† A sufficient condition under which the order of differentiation is immaterial is that U and its partial derivatives (of order two at least) be continuous in a region of the xy plane such as shown in Fig. 1.6, page 25. In what follows we shall assume that this condition is satisfied unless otherwise stated.

‡ See pages 45–46.

It remains for us to determine U. This is not difficult, for the first equation in (7) states merely that the partial derivative of U with respect to x is $2xy + 3x^2$. We should then be able to find U by the *reverse of differentiation* with respect to x, i.e., *integration* with respect to x, keeping y constant. The constant of integration which must be added is independent of x but *might* depend on y; i.e., the arbitrary constant may really be a function of y. Denote this by $f(y)$. Thus, we obtain

$$U = \int (2xy + 3x^2)\partial x + f(y) \tag{8}$$

the symbol ∂x emphasizing that integration is with respect to x, with y kept constant. Performing this integration, we find

$$U = x^2y + x^3 + f(y) \tag{9}$$

To find $f(y)$, substitute (9) into the second of equations (7). Then,

$$\frac{\partial}{\partial y}[x^2y + x^3 + f(y)] = x^2 \quad \text{or} \quad x^2 + f'(y) = x^2$$

so that $\qquad\qquad\qquad f'(y) = 0 \quad \text{or} \quad f(y) = \text{constant} = A$

Hence, $\qquad\qquad\qquad\qquad U = x^2y + x^3 + A$

Thus, the differential equation may be written $d(x^2y + x^3 + A) = 0$ and integration yields $x^2y + x^3 + A = B$ or $x^2y + x^3 = c$ where we have written $c = B - A$. It is observed that this is the same solution obtained in the last section. It is also observed that there was really no need to add the constant of integration, A, in finding $f(y)$.

To prove the sufficiency part of the theorem on page 44, it is enough to show that if $\partial M/\partial y = \partial N/\partial x$ then we can actually produce a function $U(x, y)$ such that

$$\frac{\partial U}{\partial x} = M, \qquad \frac{\partial U}{\partial y} = N \tag{10}$$

There certainly are functions U which will satisfy the first of equations (10); in fact all such functions are given by $\qquad U = \int M\,\partial x + f(y) \tag{11}$

All we have to do now is to show that there exists a function $f(y)$ such that (11) will also satisfy the second of equations (10), i.e., we must show that $f(y)$ exists so that

$$\frac{\partial}{\partial y}\left[\int M\,\partial x + f(y)\right] = N \quad \text{or} \quad f'(y) = N - \frac{\partial}{\partial y}\int M\,\partial x$$

To show this, we need only prove that

$$N - \frac{\partial}{\partial y}\int M\,\partial x \tag{12}$$

is a function of y alone. This will indeed be true if the partial derivative with respect to x of the expression (12) is zero. But this is easily shown as follows:

$$\frac{\partial}{\partial x}\left[N - \frac{\partial}{\partial y}\int M\,\partial x\right] = \frac{\partial N}{\partial x} - \frac{\partial}{\partial x}\frac{\partial}{\partial y}\int M\,\partial x$$

$$= \frac{\partial N}{\partial x} - \frac{\partial}{\partial y}\frac{\partial}{\partial x}\int M\,\partial x = \frac{\partial N}{\partial x} - \frac{\partial M}{\partial y} = 0$$

since $\partial M/\partial y = \partial N/\partial x$ by hypothesis. The sufficiency is therefore proved.

ILLUSTRATIVE EXAMPLE 1

Solve $2xy\,dx + (x^2 + \cos y)dy = 0$.

Solution   Here $M = 2xy$, $N = x^2 + \cos y$, $\dfrac{\partial M}{\partial y} = 2x = \dfrac{\partial N}{\partial x}$

and the equation is exact. Thus U exists such that

$$\frac{\partial U}{\partial x} = 2xy, \qquad \frac{\partial U}{\partial y} = x^2 + \cos y \tag{13}$$

Integrating the first equation with respect to x gives $U = x^2y + f(y)$. Substituting into the second equation of (13), we find

$$x^2 + f'(y) = x^2 + \cos y, \qquad f'(y) = \cos y, \qquad f(y) = \sin y$$

Hence, $U = x^2y + \sin y$ and the required general solution is $x^2y + \sin y = c$.

ILLUSTRATIVE EXAMPLE 2

Solve $y' = (xy^2 - 1)/(1 - x^2y)$, given that $y = 1$ where $x = 0$.

Solution   Writing the equation as $(xy^2 - 1)dx + (x^2y - 1)dy = 0$, we have

$$M = xy^2 - 1, \qquad N = x^2y - 1, \qquad \frac{\partial M}{\partial y} = \frac{\partial N}{\partial x} = 2xy$$

and the equation is exact. Thus, from $\partial U/\partial x = M$ and $\partial U/\partial y = N$ we find

$$U = \frac{x^2y^2}{2} - x - y = c$$

Using the condition that $y = 1$ where $x = 0$, we have finally $\frac{1}{2}x^2y^2 - x - y = -1$.

The student may find it easier to solve exact equations by a method of inspection known as "grouping of terms." It is based on ability to recognize certain exact differentials.

Solve $2xy\,dx + (x^2 + \cos y)dy = 0$ by "grouping of terms."

Solution   The equation is exact. If we group terms as follows:

$$(2xy\,dx + x^2\,dy) + \cos y\,dy = 0$$

then $\qquad d(x^2 y) + d(\sin y) = 0 \quad$ or $\quad d(x^2 y + \sin y) = 0$

Thus, the solution is $x^2 y + \sin y = c$, agreeing with Illustrative Example 1.

ILLUSTRATIVE EXAMPLE 4

Solve $y' = (xy^2 - 1)/(1 - x^2 y)$ by "grouping of terms."

Solution   The equation written $(xy^2 - 1)dx + (x^2 y - 1)dy = 0$ is exact. By grouping we obtain

$$(xy^2\,dx + x^2 y\,dy) - dx - dy = 0$$

or $\qquad d\left(\dfrac{x^2 y^2}{2}\right) - dx - dy = 0, \quad$ i.e., $\quad d\left(\dfrac{x^2 y^2}{2} - x - y\right) = 0$

Hence, $\qquad \dfrac{x^2 y^2}{2} - x - y = c$

In general, the grouping method yields results faster but may require more experience. The general method requires less ingenuity.

A EXERCISES

1. Write each equation in the form $M\,dx + N\,dy = 0$, test for exactness, and solve those equations which are exact.

(a) $3x\,dx + 4y\,dy = 0.$ (b) $y' = \dfrac{x - y}{x + y}.$ (c) $2xyy' = x^2 - y^2.$

(d) $y' = \dfrac{x}{x + y}.$ (e) $\dfrac{dy}{dx} = \dfrac{x - y\cos x}{\sin x + y}.$ (f) $\dfrac{dr}{d\phi} = \dfrac{r^2 \sin \phi}{2r\cos\phi - 1}.$

(g) $(ye^{-x} - \sin x)dx - (e^{-x} + 2y)dy = 0.$ (h) $\left(x^2 + \dfrac{y}{x}\right)dx + (\ln x + 2y)dy = 0.$

(i) $y' = \dfrac{y(y - e^x)}{e^x - 2xy}.$ (j) $(x^2 + x)dy + (2xy + 1 + 2\cos x)dx = 0.$

2. Solve each equation subject to the indicated conditions.

(a) $y' = \dfrac{y - 2x}{2y - x};\ y(1) = 2.$ (b) $2xy\,dx + (x^2 + 1)dy = 0;\ y(1) = -3.$

(c) $y' = \dfrac{2x - \sin y}{x\cos y};\ y(2) = 0.$ (d) $\dfrac{dx}{dy} = \dfrac{x\sec^2 y}{\sin 2x - \tan y};\ y(\pi) = \dfrac{\pi}{4}.$

(e) $(x^2 + 2ye^{2x})y' + 2xy + 2y^2 e^{2x} = 0;\ y(0) = 1.$

3. Show that each of the following equations is not exact but becomes exact upon multiplication by the indicated integrating factor. Thus, solve each equation.

(a) $(y^2 + 2x^2)dx + xy\ dy = 0$; x.   (b) $y\ dx + (4x - y^2)dy = 0$; y^3.

(c) $\cos x\ dy - (2y \sin x - 3)dx = 0$; $\cos x$.   (d) $(x - y)dx + (x + y)dy = 0$; $\dfrac{1}{(x^2 + y^2)}$.

B EXERCISES

1. Solve $\left[\dfrac{y}{(x + y)^2} - 1\right] dx + \left[1 - \dfrac{x}{(x + y)^2}\right] dy = 0$.

2. Prove that a necessary and sufficient condition for the equation $f(x)dx + g(x)h(y)dy = 0$ to be exact is that $g(x)$ be constant.

3. Prove that a necessary and sufficient condition for the equation $[f_1(x) + g_1(y)]dx + [f_2(x) + g_2(y)]dy = 0$ to be exact is that $g_1(y)dx + f_2(x)dy$ be an exact differential.

4. Determine the most general function $N(x, y)$ such that $(y \sin x + x^2 y - x \sec y)dx + N(x, y)dy = 0$ is exact, and obtain its solution.

5. Solve $\left(2x \sin \dfrac{y}{x} + 2x \tan \dfrac{y}{x} - y \cos \dfrac{y}{x} - y \sec^2 \dfrac{y}{x}\right) dx + \left(x \cos \dfrac{y}{x} + x \sec^2 \dfrac{y}{x}\right) dy = 0$.

Compare with B Exercise 10, page 41.

C EXERCISES

1. Show that $yf(xy)dx + xg(xy)dy = 0$ is not exact in general but becomes exact on multiplying by the integrating factor $\{xy[f(xy) - g(xy)]\}^{-1}$.

2. Use Exercise 1 to solve $(xy^2 + 2y)dx + (3x^2y - 4x)dy = 0$.

3. If $\partial P/\partial x = \partial Q/\partial y$ and $\partial P/\partial y = -\partial Q/\partial x$ show that the equation $P\ dx + Q\ dy = 0$ is not exact in general but becomes exact on multiplying by $1/(P^2 + Q^2)$. Illustrate by considering $P = x^2 - y^2, Q = 2xy$.

4. Let $f(z) = P(x, y) + iQ(x, y)$ where $f(z)$ is a polynomial in the complex variable $z = x + iy$ and P and Q are real. (a) Prove that

$$\frac{\partial P}{\partial x} = \frac{\partial Q}{\partial y}, \qquad \frac{\partial P}{\partial y} = -\frac{\partial Q}{\partial x}$$

(b) Discuss the relationship of (a) to Exercise 3. (c) Are the equations in (a), often called the *Cauchy–Riemann equations*, valid for other functions? Explain.

5   Equations Made Exact by a Suitable Integrating Factor

If the equation $M\ dx + N\ dy = 0$ is exact, i.e., if $\partial M/\partial y = \partial N/\partial x$, then the equation can be solved by the methods of the last section. In case the equation is not exact, we may be able to make the equation exact by multiplying it by a suitable integrating factor μ, so that the resulting equation

$$\mu M\ dx + \mu N\ dy = 0 \tag{1}$$

will be exact, i.e.,

$$\frac{\partial}{\partial y}(\mu M) = \frac{\partial}{\partial x}(\mu N) \tag{2}$$

Unfortunately there is no single method for obtaining integrating factors. If there were, our task would be greatly simplified. Fortunately, however, there are a few methods, which we shall discuss, which seem to arise often in practice. The first method which should always be looked for in practice is of course the method of separation of variables, where the integrating factor is usually apparent since M and N can each be written as a function of x times a function of y. Let us go through one such example using the ideas of integrating factor and exactness.

ILLUSTRATIVE EXAMPLE 1

Solve $\dfrac{dy}{dx} = \dfrac{3x + xy^2}{y + x^2y}$, if $y(1) = 3$.

Solution Writing the given equation as $(3x + xy^2)dx - (y + x^2y)dy = 0$

we have $\quad M = 3x + xy^2,\ N = -y - x^2y,\ \dfrac{\partial M}{\partial y} = 2xy,\ \dfrac{\partial N}{\partial x} = -2xy$

and so the equation is not exact. Noting that M and N can each be factored into a product of a function of x and a function of y, i.e.,

$$x(3 + y^2)dx - y(1 + x^2)dy = 0 \tag{3}$$

an integrating factor is $\quad \mu = \dfrac{1}{(3 + y^2)(1 + x^2)} \tag{4}$

Multiplying (3) by this integrating factor yields

$$\frac{x}{1 + x^2}\,dx - \frac{y}{3 + y^2}\,dy = 0 \tag{5}$$

which is separable and the equation is exact. Integration of (5) then yields

$$\tfrac{1}{2}\ln(1 + x^2) - \tfrac{1}{2}\ln(3 + y^2) = c \quad \text{or} \quad (1 + x^2) = A(3 + y^2)$$

Since $y = 3$ when $x = 1$, we find $A = \tfrac{1}{6}$. Thus, the required solution is

$$(1 + x^2) = \tfrac{1}{6}(3 + y^2) \quad \text{or} \quad y^2 - 6x^2 = 3 \tag{6}$$

5.1 EQUATIONS MADE EXACT BY INTEGRATING FACTORS INVOLVING ONE VARIABLE

Suppose it is desired to solve the differential equation

$$(2y^2x - y)dx + x\ dy = 0 \tag{7}$$

It is easy to show that the equation is not separable and not exact. By suitably grouping the terms in the form

$$(x\,dy - y\,dx) + 2y^2x\,dx = 0$$

and dividing by y^2 we may write the equation as

$$-d\left(\frac{x}{y}\right) + d(x^2) = 0 \quad \text{or} \quad -\frac{x}{y} + x^2 = c$$

which gives the general solution.

The method is commonly known as the *method of inspection* and is based on ingenuity in many cases. The student may observe from the method above that $1/y^2$ is an integrating factor of equation (7). Multiplication by this factor makes equation (7) exact, and we may then use our standard procedure. But how can we tell that $1/y^2$ is an integrating factor? Let us consider this problem.

Consider the case where $M\,dx + N\,dy = 0$ is not separable or exact. Let us multiply our equation by the integrating factor μ (as yet unknown). By definition of an integrating factor, the equation $\mu M\,dx + \mu N\,dy = 0$ is now exact, so that

$$\frac{\partial}{\partial y}(\mu M) = \frac{\partial}{\partial x}(\mu N) \tag{8}$$

We shall simplify our task by considering two cases:

Case 1, μ is a function of x alone. In this case we may write (8) as

$$\mu\frac{\partial M}{\partial y} = \mu\frac{\partial N}{\partial x} + N\frac{d\mu}{dx} \quad \text{or} \quad \frac{d\mu}{\mu} = \frac{1}{N}\left(\frac{\partial M}{\partial y} - \frac{\partial N}{\partial x}\right)dx \tag{9}$$

If the coefficient of dx on the right side of (9) is a function of x alone [say $f(x)$], then we have $d\mu/\mu = f(x)\,dx$ and so

$$\ln\mu = \int f(x)dx \quad \text{or} \quad \mu = e^{\int f(x)dx}$$

omitting the constant of integration. We may state this result as follows:

Theorem. If $\dfrac{1}{N}\left(\dfrac{\partial M}{\partial y} - \dfrac{\partial N}{\partial x}\right) = f(x)$, then $e^{\int f(x)dx}$ is an integrating factor.

Case 2, μ is a function of y alone. In this case, (8) may be written

$$\mu\frac{\partial M}{\partial y} + M\frac{d\mu}{dy} = \mu\frac{\partial N}{\partial x} \quad \text{or} \quad \frac{d\mu}{\mu} = \frac{1}{M}\left(\frac{\partial N}{\partial x} - \frac{\partial M}{\partial y}\right)dy$$

and we may prove the

Theorem. If $\dfrac{1}{M}\left(\dfrac{\partial N}{\partial x} - \dfrac{\partial M}{\partial y}\right) = g(y)$, then $e^{\int g(y)dy}$ is an integrating factor.

The following is a mnemonic scheme to summarize procedure. Consider

$$M\,dx + N\,dy = 0$$

Compute
$$\frac{\partial M}{\partial y} = (1), \qquad \frac{\partial N}{\partial x} = (2)$$

If (1) = (2), the equation is exact and can easily be solved.
If (1) ≠ (2), compute (1) minus (2), divided by N; call the result f.
If f is a function of x alone, then $e^{\int f\, dx}$ is an integrating factor.
If not, compute (2) minus (1), divided by M; call the result g.
If g is a function of y alone, then $e^{\int g\, dy}$ is an integrating factor.

ILLUSTRATIVE EXAMPLE 2

Solve $y\, dx + (3 + 3x - y)dy = 0$.

Solution   Here $M = y$, $N = 3 + 3x - y$, $\dfrac{\partial M}{\partial y} = 1$, $\dfrac{\partial N}{\partial x} = 3$

so the equation is not exact.

Now
$$\frac{1 - 3}{3 + 3x - y}$$

is not a function of x alone. But $\dfrac{3 - 1}{y} = \dfrac{2}{y}$

is a function of y alone. Hence $e^{\int (2/y)dy} = e^{2 \ln y} = e^{\ln y^2} = y^2$ is an integrating factor.
Multiplying the given equation by y^2, the student may show, indeed, that it becomes exact and the solution is

$$xy^3 + y^3 - \frac{y^4}{4} = c$$

ILLUSTRATIVE EXAMPLE 3

A curve having a slope given by $\dfrac{dy}{dx} = \dfrac{2xy}{x^2 - y^2}$ passes through the point (2, 1). Find its equation.

Solution   The differential equation may be written $2xy\, dx + (y^2 - x^2)dy = 0$.

Thus, $M = 2xy$, $N = y^2 - x^2$, $\dfrac{\partial M}{\partial y} = 2x$, $\dfrac{\partial N}{\partial x} = -2x$

so the equation is not exact. Now $\dfrac{2x - (-2x)}{y^2 - x^2} = \dfrac{4x}{y^2 - x^2}$

is not a function of x alone, but $\dfrac{-2x - 2x}{2xy} = \dfrac{-2}{y}$

is a function of y alone. Hence, an integrating factor is given by

$$e^{\int (-2/y)dy} = e^{-2 \ln y} = y^{-2}$$

Using this integrating factor, we find the general solution

$$x^2 + y^2 = cy$$

For the particular curve through (2, 1) we find $c = 5$, and the required equation is $x^2 + y^2 = 5y$, or in explicit form, $y = \frac{1}{2}(5 - \sqrt{25 - 4x^2})$.

Remark. The equation can also be solved as a homogeneous equation.

ILLUSTRATIVE EXAMPLE 4

Solve $y' = x - y$, given that $y = 2$ where $x = 0$.

Solution Writing the differential equation as $(x - y)dx - dy = 0$

we have $M = x - y$, $N = -1$, $\dfrac{\partial M}{\partial y} = -1$, $\dfrac{\partial N}{\partial x} = 0$

so the equation is not exact. Now $(-1 - 0)/-1 = 1$ is a function of x. Hence, $e^{\int 1\, dx} = e^x$ is integrating factor. The required solution is $(x - 1)e^x - ye^x = -3$ as the student may show.

A EXERCISES

1. Solve each of the following:
   (a) $(3x + 2y^2)dx + 2xy\, dy = 0$.
   (c) $(y^2 \cos x - y)dx + (x + y^2)dy = 0$.
   (b) $(2x^3 - y)dx + x\, dy = 0$; $y(1) = 1$.
   (d) $(x + x^3 \sin 2y)dy - 2y\, dx = 0$.

   (e) $\dfrac{dy}{dx} = \dfrac{\sin y}{x \cos y - \sin^2 y}$; $y(0) = \dfrac{\pi}{2}$.
   (f) $(2y \sin x - \cos^3 x)dx + \cos x\, dy = 0$.

   (g) $\dfrac{dy}{dx} + \dfrac{4y}{x} = x$.
   (h) $\dfrac{dx}{dy} = \dfrac{y^3 - 3x}{y}$.

   (i) $\dfrac{dI}{dt} = \dfrac{t - tI}{t^2 + 1}$; $I(0) = 0$.
   (j) $(y^3 + 2e^x y)dx + (e^x + 3y^2)dy = 0$.

2. The differential equation of a family is $y' = (x + y)/x$. Find the equation of a curve of this family passing through (3, 0).

3. Complete the solutions of the differential equations in Illustrative Examples 3 and 4.

B EXERCISES

Solve each of the following:

1. $\dfrac{dy}{dx} = \dfrac{3y^2 \cot x + \sin x \cos x}{2y}$.
2. $\dfrac{dy}{dx} = \dfrac{x}{x^2 y + y^3}$.

3. $(3x^2 + y + 3x^3 y)dx + x\, dy = 0$.
4. $(2x + 2xy^2)dx + (x^2 y + 2y + 3y^3)dy = 0$.

C EXERCISES

1. Show that if the equation $M\, dx + N\, dy = 0$ is such that $\dfrac{1}{xM - yN}\left(\dfrac{\partial N}{\partial x} - \dfrac{\partial M}{\partial y}\right) = F(xy)$

i.e., a function of the product xy, then an integrating factor is $e^{\int F(u)\, du}$ where $u = xy$.

2. Use the method of Exercise 1 to solve $(y^2 + xy + 1)dx + (x^2 + xy + 1)dy = 0$.

3. Solve $(2y^2 + 4x^2y)dx + (4xy + 3x^3)dy = 0$, given that there exists an integrating factor of the form $x^p y^q$, where p and q are constants.

4. In A Exercise 2, is there a member of the family passing through $(0, 0)$?

5.2 THE LINEAR FIRST-ORDER EQUATION

An equation which can be written in the form

$$\frac{dy}{dx} + P(x)y = Q(x) \tag{10}$$

where $P(x)$ and $Q(x)$ are given functions of x is called a *linear differential equation of first order.* It is easy to verify that the equation has $e^{\int P\,dx}$ as integrating factor, for upon multiplication of both sides of (10) by this factor we obtain

$$e^{\int P\,dx}\frac{dy}{dx} + Pye^{\int P\,dx} = Qe^{\int P\,dx} \tag{11}$$

which is equivalent to

$$\frac{d}{dx}(ye^{\int P\,dx}) = Qe^{\int P\,dx} \tag{12}$$

This is because if we use the rule of calculus for differentiating a product, the left side of (12) is

$$\frac{d}{dx}(ye^{\int P\,dx}) = y\frac{d}{dx}(e^{\int P\,dx}) + e^{\int P\,dx}\frac{dy}{dx}$$

$$= y(e^{\int P\,dx}P) + e^{\int P\,dx}\frac{dy}{dx} = e^{\int P\,dx}\frac{dy}{dx} + Pye^{\int P\,dx}$$

i.e., the left side of (11). From (12) we obtain by integration the solution[†]

$$ye^{\int P\,dx} = \int Qe^{\int P\,dx}\,dx + c \tag{13}$$

Remark. There is no need to memorize (13). It is much better to use the integrating factor $\mu = e^{\int P\,dx}$, multiply the given equation (10) by this factor and then write the left side as the derivative of μ times y as in (12).

ILLUSTRATIVE EXAMPLE 5

Solve $\dfrac{dy}{dx} + 5y = 50$.

* An equation which cannot be written in this form, as for example $\dfrac{dy}{dx} + xy^2 = \sin x$, is called a *non-linear first-order differential equation.*
 [†] It is interesting to note that (13) supplies the unique solution of a first-order linear differential equation predicted in Illustrative Example 3 on page 27.

Solution This is of the form (10) with $P = 5$, $Q = 50$. An integrating factor is $e^{\int 5\,dx} = e^{5x}$. Multiplying by e^{5x}, we may write the equation as

$$\frac{d}{dx}(ye^{5x}) = 50e^{5x} \quad \text{i.e.,} \quad ye^{5x} = 10e^{5x} + c \quad \text{or} \quad y = 10 + ce^{-5x}$$

The method of separation of variables also could have been used.

ILLUSTRATIVE EXAMPLE 6

Solve $\dfrac{dI}{dt} + \dfrac{10I}{2t + 5} = 10$, given that $I = 0$ where $t = 0$.

Solution The equation has the form (10), with I replacing y and t replacing x. An integrating factor is $e^{\int 10\,dt/(2t+5)} = e^{5\ln(2t+5)} = e^{\ln(2t+5)^5} = (2t + 5)^5$. Multiplying by $(2t + 5)^5$, we find

$$\frac{d}{dt}\left[(2t + 5)^5 I\right] = 10(2t + 5)^5 \quad \text{or} \quad I(2t + 5)^5 = \tfrac{5}{6}(2t + 5)^6 + c$$

Placing $I = 0$ and $t = 0$ in the equation, we have $c = -78{,}125/6$.

Thus, $$I = \tfrac{5}{6}(2t + 5) - \frac{78{,}125}{6(2t + 5)^5}$$

There is really no need to consider (10) as a new equation for it falls into a category already considered, namely, the category of equations having an integrating factor which is a function of one variable only. Nevertheless, the form in which (10) appears is of so frequent occurrence in practical applications, and the method of solution is so simple, that it is worthwhile to become acquainted with it. However, should the student fail to recognize that a particular equation has the form (10), he can rest assured that the method of integrating factors of one variable will work. For example, consider the equation

$$y\,dx + (3 + 3x - y)dy = 0$$

which we discussed in Illustrative Example 2, page 51. Should we happen to recognize that it can be written as

$$\frac{dx}{dy} + \frac{3x}{y} = \frac{y - 3}{y}$$

and is therefore of form (10) with x and y interchanged, we can solve it as a linear equation. (See A Exercise 3.) Otherwise we may look for an integrating factor involving one variable. To show that this method is applicable we write (10) as

$$[P(x)y - Q(x)]dx + dy = 0$$

Then $\quad M = P(x)y - Q(x), \quad N = 1, \quad \dfrac{\partial M}{\partial y} = P(x), \quad \dfrac{\partial N}{\partial x} = 0$

Now $[P(x) - 0]/1 = P(x)$ is a function of x alone, and so $e^{\int P(x)\,dx}$ is an integrating factor.

1. Solve each of the following:

(a) $\dfrac{dy}{dx} + \dfrac{y}{x} = 1.$

(b) $xy' + 3y = x^2.$

(c) $y^2 \dfrac{dx}{dy} + xy = 2y^2 + 1.$

(d) $\dfrac{dy}{dx} - \dfrac{2y}{x} = x^2 \sin 3x.$

(e) $I' + 3I = e^{-2t};\ I(0) = 5.$

(f) $y' + y \cot x = \cos x.$

(g) $y' = \dfrac{1}{x - 3y}.$

(h) $\dfrac{dr}{d\phi} = \phi - \dfrac{r}{3\phi};\ r = 1,\ \phi = 1.$

2. The current I, in amperes, in a certain electric circuit satisfies the differential equation

$$\frac{dI}{dt} + 2I = 10e^{-2t}$$

where t is the time. If $I = 0$ where $t = 0$, find I as a function of t.

3. Solve the differential equation of page 54 as a linear equation.

B EXERCISES

1. The equation $dy/dx + Py = Qy^n$, where P and Q are functions of x alone and n is a constant, is called *Bernoulli's differential equation* and arises in various applications. Show how to solve it where $n = 0$ or 1.

2. If $n \neq 0, 1$, none of the methods discussed so far applies. Show, however, that by changing the dependent variable from y to v according to the transformation $v = y^{1-n}$ the equation can be solved.

3. Solve $y' - y = xy^2$ by the method of Exercise 2.

4. Solve $y^2\, dx + (xy - x^3)dy = 0.$ 5. Solve $xy'' - 3y' = 4x^2.$ (*Hint:* Let $y' = v.$)

6. Solve Exercise 4 by looking for an integrating factor of the form $x^p y^q$, where p and q are suitably chosen constants.

7. Solve the equation $y' = \alpha y - \beta y^n$, where α, β, and $n \neq 0, 1$ are constants, as (a) a Bernoulli equation; (b) a separable equation.

C EXERCISES

1. Show that the differential equation $y' + Py = Qy \ln y$, where P and Q are functions of x, can be solved by letting $\ln y = v$.

2. Solve $xy' = 2x^2y + y \ln y.$

3. Show that a linear equation with independent variable x is transformed into another linear equation when x undergoes the transformation $x = f(u)$, where u is a new independent variable and f is any differentiable function.

4. Solve $xy' + 3 = 4xe^{-y};\ y(2) = 0.$

5.3 THE METHOD OF INSPECTION

On page 50, it was mentioned that an integrating factor of a differential equation could sometimes be found by inspection, a process based on ingenuity and experience. In that section we avoided the method of inspection for cases where the integrating factor involved only one variable. However in some instances integrating factors depend on both variables and "inspection" may be helpful. The inspection method usually applies when one notices certain special facts about the equation.

ILLUSTRATIVE EXAMPLE 7

Solve $(x^2 + y^2 + y)dx - x\,dy = 0$.

Solution All standard methods discussed so far fail for this equation. However if we write the equation as

$$(x^2 + y^2)dx + y\,dx - x\,dy = 0$$

and "happen to notice" that this can be written

$$dx + \frac{y\,dx - x\,dy}{x^2 + y^2} = 0 \quad \text{or} \quad dx - d\left[\tan^{-1}\frac{y}{x}\right] = 0$$

we immediately obtain by integration the solution $x - \tan^{-1}\frac{y}{x} = c$.

The student will observe that an integrating factor for this equation is $1/(x^2 + y^2)$.

ILLUSTRATIVE EXAMPLE 8

Solve $x\,dx + (y - \sqrt{x^2 + y^2})dy = 0$.

Solution Writing this as $\dfrac{x\,dx + y\,dy}{\sqrt{x^2 + y^2}} = dy$

we may notice that the left side can be written $d(\sqrt{x^2 + y^2})$, and so the equation may be written $d(\sqrt{x^2 + y^2}) = dy$. Integration leads to

$$\sqrt{x^2 + y^2} = y + c \quad \text{or} \quad x^2 = 2cy + c^2$$

The student might also have solved this problem by writing

$$\frac{dy}{dx} = \frac{x}{\sqrt{x^2 + y^2} - y}$$

and then using the transformation $y = vx$.

The following easily established results may help in the solution of differential equations by "inspection."

$$\frac{x\,dy - y\,dx}{x^2} = d\left(\frac{y}{x}\right), \qquad \frac{x\,dy - y\,dx}{y^2} = -d\left(\frac{x}{y}\right)$$

$$\frac{x\,dy - y\,dx}{x^2 + y^2} = d\left[\tan^{-1}\left(\frac{y}{x}\right)\right], \qquad \frac{x\,dx + y\,dy}{x^2 + y^2} = d\left[\frac{1}{2}\ln(x^2 + y^2)\right]$$

$$\frac{x\,dx + y\,dy}{\sqrt{x^2 + y^2}} = d(\sqrt{x^2 + y^2}), \qquad \frac{x\,dx - y\,dy}{\sqrt{x^2 - y^2}} = d(\sqrt{x^2 - y^2})$$

A EXERCISES

Solve each equation by the method of inspection or any other method.

1. $y\,dx + (2x^2y - x)dy = 0$.

2. $y\,dx + (y^3 - x)dy = 0$.

3. $(x^3 + xy^2 + y)dx - x\,dy = 0$.

4. $(x^3 + y)dx + (x^2y - x)dy = 0$.

5. $(x - \sqrt{x^2 + y^2})dx + (y - \sqrt{x^2 + y^2})dy = 0$.

6. $(x^2 + y^2 + y)dx + (x^2 + y^2 - x)dy = 0$.

7. $(x - x^2 - y^2)dx + (y + x^2 + y^2)dy = 0$.

8. $(x^2y + y^3 - x)dx + (x^3 + xy^2 - y)dy = 0$.

B EXERCISES

1. Show that $\sqrt{x^2 + y^2}(x\,dx + y\,dy) = \frac{1}{3}d[(x^2 + y^2)^{3/2}]$. Hence, solve

$$(y - x\sqrt{x^2 + y^2})dx + (x - y\sqrt{x^2 + y^2})dy = 0$$

2. Show that $\dfrac{x\,dy + y\,dx}{(xy)^4} = -\frac{1}{3}d[(xy)^{-3}]$. Hence, solve

$$(y - x^5y^4)dx + (x - x^4y^5)dy = 0$$

3. Show that $\dfrac{y\,dx - x\,dy}{x^2 - y^2} = \frac{1}{2}d\left[\ln\left|\dfrac{x - y}{x + y}\right|\right]$. Hence, solve

$$(x^3 - xy^2 + y)dx + (y^3 - x^2y - x)dy = 0$$

C EXERCISES

Solve:

1. $(x^3 + 2xy^2 - x)dx + (x^2y + 2y^3 - 2y)dy = 0$.

2. $\dfrac{dy}{dx} = \dfrac{x^3 + 2y}{x^3 + x}$.   **3.** $(xy^2 + x\sin^2 x - \sin 2x)dx - 2y\,dy = 0$.

4. $\left[x^2 + y(x - y)^2 \tan\left(\dfrac{y}{x}\right)\right]dx - \left[x^2 + x(x - y)^2 \tan\left(\dfrac{y}{x}\right)\right]dy = 0$.

6 Equations of Order Higher Than the First Which Are Easily Solved

We have now learned how to solve some first-order differential equations. To solve equations of higher order, it is natural to ask whether they can in some way be reduced to first-order equations, which can then be solved. There are actually two important types of higher-order equations which can be easily solved in this way.

As we have already found in Chapter One, the simplest type of differential equation which can arise is one which can be integrated directly. Let us review this briefly in the following

ILLUSTRATIVE EXAMPLE 1

Solve $y^{(IV)} = x$, given that $y = 0$, $y' = 1$, $y'' = y''' = 0$ where $x = 0$.

Solution   By one integration of the given equation we have $y''' = \dfrac{x^2}{2} + c_1$.

Since $y''' = 0$ where $x = 0$, this leads to $c_1 = 0$. Thus, $y''' = \dfrac{x^2}{2}$.

Integrating again, we get $y'' = \dfrac{x^3}{6} + c_2$.

Using $y'' = 0$ where $x = 0$, we have $c_2 = 0$. Hence, $y'' = \dfrac{x^3}{6}$.

Integrating again, we find $y' = \dfrac{x^4}{24} + c_3$.

Using $y' = 1$ where $x = 0$, we have $c_3 = 1$. Thus, $y' = \dfrac{x^4}{24} + 1$.

Integrating again, we obtain $y = \dfrac{x^5}{120} + x + c_4$.

Since $y = 0$ where $x = 0$, $c_4 = 0$. Thus, $y = \dfrac{x^5}{120} + x$.

Note that the given equation was considered as a first-order equation in y''' the second equation as a first-order equation in y'', etc. Note also that in the final result we have evaluated four arbitrary constants in agreement with the fact that we started out with a fourth-order equation.

6.2 EQUATIONS HAVING ONE VARIABLE MISSING

Our next method applies when one of the variables is missing in the equation. The method is often useful in applications.

ILLUSTRATIVE EXAMPLE 2

Solve $xy'' + y' = 4x$.

Solution   Here one of the variables, y, is absent from the equation. The method in such case is to let $y' = v$. Then $y'' = v'$, and the equation may be written

$$xv' + v = 4x \quad \text{or} \quad \frac{d}{dx}(xv) = 4x$$

Integration gives $xv = 2x^2 + c_1$, $v = 2x + (c_1/x)$. Replacing v by y', we have

$$y' = 2x + \frac{c_1}{x} \quad \text{or} \quad y = x^2 + c_1 \ln x + c_2$$

ILLUSTRATIVE EXAMPLE 3

Solve $2yy'' = 1 + (y')^2$.

Solution   In this case x is missing. Letting $y' = v$ as before, we find

$$2yv' = 1 + v^2 \quad \text{or} \quad 2y\frac{dv}{dx} = 1 + v^2 \tag{1}$$

Unfortunately we now have three variables x, v, and y. However, we may write

$$\frac{dv}{dx} = \frac{dv}{dy} \cdot \frac{dy}{dx} = \frac{dv}{dy} \cdot v$$

so that (1) becomes

$$2yv\frac{dv}{dy} = 1 + v^2$$

Separating the variables and integrating, we have

$$\int \frac{2v\,dv}{1 + v^2} = \int \frac{dy}{y}, \quad \ln(1 + v^2) = \ln y + c$$

Thus,

$$\frac{1 + v^2}{y} = c_1 \quad \text{or} \quad v = \pm\sqrt{c_1 y - 1}$$

i.e.,

$$\frac{dy}{dx} = \pm\sqrt{c_1 y - 1} \quad \text{or} \quad \int \frac{dy}{\sqrt{c_1 y - 1}} = \pm \int dx$$

Integration yields $2\sqrt{c_1 y - 1} = \pm c_1 x + c_2$, from which y can be obtained.

ILLUSTRATIVE EXAMPLE 4

Solve $y'' + y = 0$.

Solution   Letting $y' = v$, we can write the given equation as

$$\frac{dv}{dx} + y = 0, \quad \frac{dv}{dy} \cdot \frac{dy}{dx} + y = 0 \quad \text{or} \quad v\frac{dv}{dy} + y = 0$$

Separating the variables and integrating, we find

$$\int v\,dv + \int y\,dy = c \quad \text{or} \quad \tfrac{1}{2}v^2 + \tfrac{1}{2}y^2 = c$$

Then choosing $2c = c_1^2$, we have $v = \pm\sqrt{c_1^2 - y^2}$

i.e.,

$$\frac{dy}{dx} = \pm\sqrt{c_1^2 - y^2} \quad \text{or} \quad \int \frac{dy}{\sqrt{c_1^2 - y^2}} = \pm \int dx$$

Integration yields

$$\sin^{-1}(y/c_1) = \pm x + c_2$$

or $$y = c_1 \sin (\pm x + c_2) = c_1 \sin c_2 \cos x \pm c_1 \cos c_2 \sin x$$

which can be written $\quad y = A \sin x + B \cos x$

A EXERCISES

Solve each of the following subject to conditions where indicated.

1. $y'' = 2x$; $y(0) = 0$, $y'(0) = 10$. **2.** $y^{(IV)} = \dfrac{x}{3}$.

3. $y''' = 3 \sin x$; $y(0) = 1$, $y'(0) = 0$, $y''(0) = -2$.

4. $2y^{(IV)} = e^x - e^{-x}$; $y(0) = y''(0) = y'''(0) = 0$.

5. $I''(t) = t^2 + 1$; $I(0) = 2$, $I'(0) = 3$. **6.** $x^2 y'' = x^2 + 1$; $y(1) = 1$, $y'(1) = 0$.

7. $x^3 y''' = 1 + \sqrt{x}$. **8.** $y'' y' = 1$; $y(0) = 5$, $y'(0) = 1$.

9. $y'' + 4y = 0$; $y(0) = 3$, $y'(0) = 2$. **10.** $xy'' + 2y' = 0$.

11. $y'' - y = 0$. **12.** $yy'' = y'$.

13. $y'' + (y')^2 = 1$. **14.** $y'' = y'(1 + y)$. **15.** $y'' + xy' = x$.

B EXERCISES

Solve each of the following subject to conditions where indicated.

1. $y^{(IV)} = \ln x$; $y(1) = y'(1) = y''(1) = y'''(1) = 0$.

2. $y^{(V)} + 2y^{(IV)} = x$; $y(0) = y'(0) = y''(0) = y'''(0) = y^{(IV)}(0) = 0$.

3. $xy''' + y'' = 1$. **4.** $(y''')^2 = (y')^3$. **5.** $y''' - y' = 0$.

6. $1 + (y')^2 + yy'' = 0$. **7.** $x^2 y''' + 2xy'' = 1$.

C EXERCISES

1. If $y'' = -4/y^3$ and $y(2) = 4$, $y'(2) = 0$, find $y(4)$.

2. Solve $y'' = [1 + (y')^2]^{3/2}$ and interpret geometrically.

3. Solve $\dfrac{d^2y}{dx^2} \cdot \dfrac{d^2x}{dy^2} = 1$.

4. A curve in the xy plane has the property that its curvature at any point (x, y) is always equal to $\sin x$. If the curve has slope zero at the point $(0, 0)$ on it, what is its equation?

5. Work Exercise 4 if $\sin x$ is replaced by $2x$.

◆ 7 The Clairaut Equation

An equation of first order having many interesting properties is given by

$$y = xy' + f(y') \tag{1}$$

and is known as *Clairaut's differential equation* after the mathematician who first investigated these properties. We shall suppose that $f(y')$ defines a differentiable function of y'.

Example. $y = xy' + (y')^2$, $y = xy' + \tan y'$, $y = xy' + \sqrt{1 + (y')^2}$ are all Clairaut equations.

To solve (1), we first let $y' = v$ to obtain

$$y = xv + f(v) \tag{2}$$

We then use the device of differentiating both sides of (2) with respect to x to obtain

$$y' = xv' + v + f'(v)v' \quad \text{or} \quad v'[x + f'(v)] = 0$$

Two cases arise from the last equation.

Case 1, $v' = 0$. In this case we have, on integrating, $v = c$, where c is any constant. Then replacing v by c in (2) we obtain

$$y = cx + f(c) \tag{3}$$

It is remarkable that (3), which is obtained directly from the given differential equation (1) on simply replacing y' by c, yields the general solution of (1), as can be verified by substitution.

Case 2, $x + f'(v) = 0$. In this case we have, using (2),

$$x + f'(v) = 0, \qquad y = xv + f(v)$$

from which $\qquad\qquad x = -f'(v), \qquad y = -vf'(v) + f(v) \tag{4}$

Equations (4) are parametric equations for a curve where v is the parameter. This curve provides a solution of (1), as seen in B Exercise 4, page 64. However, since it is not a special case of the general solution (3), it is a singular solution.

To illustrate the procedure and to obtain some insight concerning the relationship between the singular and general solution, let us consider the following

ILLUSTRATIVE EXAMPLE

Solve $y = xy' + (y')^2$.

Solution As above, let $y' = v$ so that $y = xv + v^2$. Then differentiate with respect to x to obtain

$$y' = xv' + v + 2vv' \quad \text{or} \quad v'(x + 2v) = 0$$

Then there are two cases.

Case 1, $v' = 0$. In this case, $v = c$, which on substitution into $y = xv + v^2$ gives the general solution

$$y = cx + c^2 \tag{5}$$

Case 2, $x + 2v = 0$. In this case we obtain from $x + 2v = 0$ and $y = xv + v^2$ the parametric equations

$$x = -2v, \qquad y = -v^2 \quad \text{or} \quad y = -\frac{x^2}{4} \tag{6}$$

on eliminating v. Since this satisfies the given differential equation and is not a special case of (5), it is a singular solution.

The relationship between the singular solution and the general solution can be seen from Fig. 2.1. In this figure we have shown the graph of $y = -x^2/4$, which is a parabola, together with graphs of $y = cx + c^2$ for several values of c, which represent tangent lines to $y = -x^2/4$ (see B Exercise 3). The parabola $y = -x^2/4$, which "envelops" all the tangents $y = cx + c^2$, is for obvious reasons called the *envelope* of the family of tangent lines.

For the general Clairaut equation (1), the singular solution (4) represents the envelope of the family of straight lines (3), which in turn are tangent lines to the envelope. It is possible to obtain the envelope directly from this family, as indicated in B Exercise 1.

The fundamental Existence–Uniqueness Theorem of Chapter One also can provide clues to the presence of singular solutions and their connections with general solutions. Referring to the Clairaut equation in the Illustrative Example on page 61, for instance, we see on solving for y' that there are two values,

$$y' = \frac{-x + \sqrt{x^2 + 4y}}{2}, \qquad y' = \frac{-x - \sqrt{x^2 + 4y}}{2} \qquad (7)$$

Considering the first equation in (7), we note that the partial derivative with respect to y of the right-hand side is $1/\sqrt{x^2 + 4y}$, and this is real, single-valued and continuous if and only if $y > -x^2/4$, which describes geometrically the region above the parabola

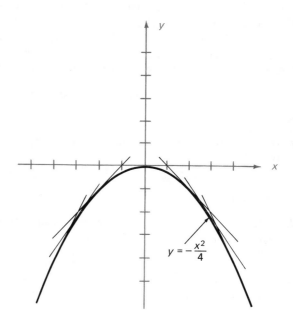

Figure 2.1

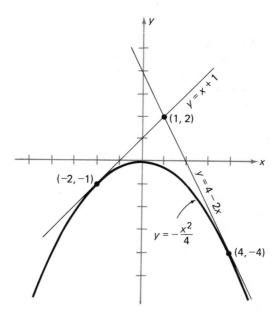

Figure 2.2

of Fig. 2.2. Given a point, say $(1, 2)$, in this region, we see from the general solution $y = cx + c^2$ that $c^2 + c - 2 = 0$ or $c = 1, -2$; i.e., $y = x + 1$, $y = 4 - 2x$. Of these, only $y = x + 1$, $x \geq -2$, satisfies the first equation of (7), while $y = x + 1$, $x \leq -2$, satisfies the second equation of (7) in accordance with the Existence-Uniqueness Theorem.

In a similar manner we can show that $y = 4 - 2x$, $x \leq 4$, is the only solution of the second equation in (7) passing through $(1, 2)$, while $y = 4 - 2x$, $x \geq 4$, is the only solution of the first equation. The situation is indicated in Fig. 2.2.

The concepts described above for the Clairaut equation serve to indicate some guiding principles concerning solutions for more general types of equations. For first-order ordinary differential equations, there will usually be a general solution in certain restricted regions as guaranteed by the Existence-Uniqueness Theorem. Singular solutions, if they occur, may manifest themselves at the boundaries of such regions. In some cases they can be seen from certain factors which may become zero or infinite.*

A EXERCISES

Obtain the general solution and singular solution for each of the following.

1. $y = xy' - (y')^2$.

2. $y = xy' + 1 + 4(y')^2$.

3. $y = xy' - \tan y'$.

4. $y = xy' + \sqrt{1 + (y')^2}$.

* For further discussion of envelopes and singular solutions, together with related topics, see the advanced exercises on page 64 and also reference [13].

1. (a) Show that the differential equation for the family of straight lines $y = cx - c^3$ is $y = xy' - (y')^3$. (b) Show that the envelope of the family $y = cx - c^3$ is also a solution of the differential equation in (a). What kind is it? [*Hint:* The envelope of a one-parameter family $F(x, y, c) = 0$, if it exists, may be found from the simultaneous solution of $F(x, y, c) = 0$ and $\partial F(x, y, c)/\partial c = 0$.] (c) Can the envelope be obtained directly from the differential equation?

2. Use the method of Exercise 1 to obtain the envelopes in (a) the Illustrative Example on page 61. (b) A Exercise 1. (c) A Exercise 2. (d) A Exercise 3. (e) A Exercise 4.

3. Show that $y = cx + c^2$ is tangent to $y = -x^2/4$.

4. Show that the curve defined by the parametric equations (4) on page 61 represents a solution of (1), page 60. Also show that the lines $y = cx + f(c)$ represent tangent lines to this curve; i.e., the curve is the envelope of the family of tangent lines.

5. The general solution of $y' = 3y^{2/3}$ is given by $y = (x + c)^3$ and the singular solution is $y = 0$. Examine the relationship between these solutions from the viewpoint of envelopes. Compare Chapter One pages 16, 20, and 27.

6. Find the general and singular solution of $y' = \sqrt{y}$.

1. Show that the equation $y = y' \tan x - y'^2 \sec^2 x$ can be reduced to a Clairaut equation by use of the transformation $z = \sin x$, and thus solve the equation.

2. Solve $y = \dfrac{x}{2}\left(\dfrac{dy}{dx} + \dfrac{dx}{dy}\right)$.

3. Work C Exercise 4, page 33, using the concept of an *envelope* (see B Exercise 1).

8 Review of Important Methods

We have in this chapter encountered various methods of solving first-order differential equations. It is natural for the student to wonder which of these methods can be expected to arise most often in practice so that appropriate emphasis may be given.

The following list, in which we present those methods of primary importance and those of secondary importance, may serve as a useful guide.

A. *Primary Importance*

1. *Separation of variables:* $f(x)dx + g(y)dy = 0$.
   Integrate to obtain the required solution $\int f(x)dx + \int g(y)dy = c$.

2. *Homogeneous equations:* $\dfrac{dy}{dx} = f\left(\dfrac{y}{x}\right)$.

   Let $y = vx$, where v is a new dependent variable depending on x, and thus reduce the equation to the separation of variables type.

3. *Linear equations:* $\dfrac{dy}{dx} + P(x)y = Q(x)$.

Multiply both sides by the integrating factor $\mu = e^{\int P\,dx}$ so that the equation can be written $\dfrac{d}{dx}(\mu y) = \mu Q$. Then integrate to obtain the required solution $\mu y = \int \mu Q\,dx + c$.

4. *Exact equations:* $M\,dx + N\,dy = 0$, where $\dfrac{\partial M}{\partial y} = \dfrac{\partial N}{\partial x}$.

Write the equation as $dU = 0$ and integrate to obtain the required solution $U(x, y) = c$.

B. Secondary Importance

1. *Equations which can be made exact:* See pages 48–54. This includes 1 and 3 above.
2. *Transformation of variables:* This involves special devices suggested by the particular form of the differential equation and often falls in the category of "ingenious devices." Method 2 given above is included here.
3. *Miscellaneous techniques,* such as the method of inspection and the Clairaut equation.

The following exercises are intended to serve as a review of the various methods.

MISCELLANEOUS EXERCISES ON CHAPTER 2

A EXERCISES

Solve each of the following differential equations subject to given conditions, if any.

1. $(x^2 + 1)(y^3 - 1)dx = x^2 y^2\,dy$.

2. $(y^2 + 2xy)dx + (x^2 + 2xy)dy = 0$.

3. $(x^2 + 2xy)dx + (y^2 + 2xy)dy = 0$.

4. $\dfrac{dy}{dx} + \dfrac{2y}{x} = x^2$.

5. $(3 - y)dx + 2x\,dy = 0;\ y(1) = 1$.

6. $\dfrac{dy}{dx} + 2x = 2$.

7. $s^2 t\,ds + (t^2 + 4)dt = 0$.

8. $2xyy' + x^2 + y^2 = 0$.

9. $\dfrac{dy}{dx} = \dfrac{2x^2 - ye^x}{e^x}$.

10. $x^2 y' + xy = x + 1$.

11. $\dfrac{dy}{dx} = \dfrac{y}{x} + \tan^{-1}\dfrac{y}{x}$.

12. $\dfrac{dy}{dx} = x + y$.

13. $y' + xy = x^3$.

14. $(3 - x^2 y)y' = xy^2 + 4$.

15. $r^2 \sin \phi\,d\phi = (2r \cos \phi + 10)dr$.

16. $y' = x^2 + 2y$.

17. $y' = \dfrac{2xy - y^4}{3x^2}$.

18. $(x^2 + y^2)dx + 2y\,dy = 0;\ y(0) = 2$.

19. $(x^2 + y^2)dx + (2xy - 3)dy = 0$.

20. $y'(2x + y^2) = y$.

21. $u^2v\,du - (u^3 + v^3)dv = 0.$

22. $(\tan y - \tan^2 y \cos x)dx - x \sec^2 y \, dy = 0.$

23. $\dfrac{dy}{dx} = \dfrac{x + 2y}{y - 2x}.$

24. $y' \sin x = y \cos x + \sin^2 x.$

25. $(x^2 - y^2)dx + 2xy\,dy = 0.$

26. $(2x^2 - ye^x)dx - e^x\,dy = 0.$

27. $(x + y)y' = 1.$

28. $(x + 2y)dx + x\,dy = 0.$

29. $\sin y\,dx + (x \cos y - y)dy = 0.$

30. $y' = e^{y/x} + \dfrac{y}{x}.$

31. $\sin x \cos y\,dx + \cos x \sin y\,dy = 0.$

32. $xy' = x^3 + 2y.$

33. $(3xy^2 + 2)dx + 2x^2y\,dy = 0.$

34. $(2y^2 - x)dy + y\,dx = 0.$

35. $y'' = y' + 2x.$

36. $(1 + y)y' = x\sqrt{y}.$

37. $\tan x \sin y\,dx + 3\,dy = 0.$

38. $x\,dy - y\,dx = x \cos\left(\dfrac{y}{x}\right)dx.$

39. $\dfrac{ds}{dt} = \sqrt{\dfrac{1 - t}{1 - s}}; s = 0$ where $t = 1.$

40. $(2y + 3x)dx + x\,dy = 0.$

41. $x^2y\,dx + (1 + x^3)dy = 0.$

42. $(\sin y - x)y' = 2x + y; y(1) = \dfrac{\pi}{2}.$

43. $\dfrac{dN}{dt} = -\alpha N; N = N_0$ at $t = 0.$

44. $\dfrac{dy}{dx} = \dfrac{y(x + y)}{x(x - y)}.$

45. $\dfrac{dI}{dt} + I = e^t.$

46. $xy' + y = x^2; y(1) = 2.$

47. $x\,dy - y\,dx = x^2y\,dy.$

48. $\dfrac{dq}{dp} = \dfrac{p}{q}e^{p^2 - q^2}.$

49. $(3y \cos x + 2)y' = y^2 \sin x; y(0) = -4.$

50. $(x + x \cos y)dy - (y + \sin y)dx = 0.$

51. $y' = 3x + 2y.$

52. $y^2\,dx = (2xy + x^2)dy.$

53. $\dfrac{dr}{d\phi} = \dfrac{r(1 + \ln \phi)}{\phi(1 + \ln r)}; \phi = e^2$ where $r = e.$

54. $\dfrac{dU}{dt} = -a(U - 100t); U(0) = 0.$

55. $(uv - 2v)du + (u - u^2)dv = 0.$

56. $\dfrac{dI}{dt} + 3I = 10 \sin t; I(0) = 0.$

57. $\dfrac{ds}{dt} = \dfrac{1}{s + t + 1}.$

58. $yy'' + (y')^2 = 0.$

59. $x\sqrt{1 - y^2} + yy'\sqrt{1 - x^2} = 0.$

60. $y' + (\cot x)y = \cos x.$

61. $y' = \left(\dfrac{y + 3}{2x}\right)^2.$

62. $xy' - 3y = x^4e^{-x}.$

63. $y' = \sin x \tan y.$

64. $y' = \dfrac{x}{y} + \dfrac{y}{x}.$

65. $x\,dy - y\,dx = 2x^2y^2\,dy.$

66. $xy' + y \ln x = y \ln y + y.$

67. $y' = 2 - \dfrac{y}{x}$.

68. $xy'' + y' = 1$.

69. $\dfrac{dI}{dt} = \dfrac{It^2}{t^3 - I^3}$.

70. $(e^y + x + 3)y' = 1$.

71. $\dfrac{dr}{d\phi} = e^\phi - 3r$; $r = 1$ at $\phi = 0$.

72. $yy'' = (y')^2$.

73. $x^4 y''' + 1 = 0$.

74. $\dfrac{dy}{dx} = \dfrac{x + 3y}{x - 3y}$.

75. $y' \cos x = y - \sin 2x$.

76. $e^{2x-y}\, dx + e^{y-2x}\, dy = 0$.

77. $r^3 \dfrac{dr}{d\phi} = \sqrt{a^8 - r^8}$.

78. $(2x^2 - ye^x)dx - e^x\, dy = 0$.

79. $x\, dy + 2y\, dx - x \cos x\, dx = 0$.

80. $\sqrt{1 + x^3}\, \dfrac{dy}{dx} = x^2 y + x^2$.

81. $(3y^2 + 4xy)dx + (2xy + x^2)dy = 0$.

82. $y' = y(x + y)$.

83. $y' = x(x + y)$.

84. $\dfrac{d^2 U}{dr^2} + \dfrac{1}{r}\dfrac{dU}{dr} = 4(1 - r)$; $U = 15, \dfrac{dU}{dr} = 0$ at $r = 1$.

85. $\dfrac{dy}{dx} = 1 - (x - y)^2$; $y(0) = 1$.

86. $\dfrac{dy}{dx} = \dfrac{e^{x-y}}{y}$.

B EXERCISES

1. Solve $xyy' + y^2 = \sin x$ by letting $y^2 = u$.

2. Show that $\sin^{-1} x + \sin^{-1} y = c_1$ and $x\sqrt{1 - y^2} + y\sqrt{1 - x^2} = c_2$ are general solutions of
$$\sqrt{1 - y^2}\, dx + \sqrt{1 - x^2}\, dy = 0$$
Can one of these solutions be obtained from the other?

3. (a) Solve $(1 + x^2)dy + (1 + y^2)dx = 0$, given that $y(0) = 1$. (b) Show that $y = (1 - x)/(1 + x)$ is a solution. Reconcile this with the solution obtained in (a).

4. Solve $y' = 2/(x + 2y - 3)$ by letting $x + 2y - 3 = v$.

5. Solve $y' = \sqrt{y + \sin x} - \cos x$. (*Hint*: Let $\sqrt{y + \sin x} = v$.)

6. Solve $y' = \tan (x + y)$.   **7.** Solve $y' = e^{x + 3y} + 1$.

8. Solve $y^{(IV)} = 2y''' + 24x$ subject to $y(0) = 1$, $y'(0) = y''(0) = y'''(0) = 0$.

9. Show that the equation $yF(xy)dx + xG(xy)dy = 0$ can be solved by the transformation $xy = u$. Thus, solve $(x^2 y^3 + 2xy^2 + y)dx + (x^3 y^2 - 2x^2 y + x)dy = 0$.

10. Solve $(y')^2 + (3y - 2x)y' - 6xy = 0$.   **11.** Solve $\dfrac{dy}{dx} = \dfrac{x + y^2}{2y}$; $y(0) = 1$.

12. Solve $y' = x + \sqrt{y}$.

1. Solve $\dfrac{dy}{dx} = \sqrt{\dfrac{5x - 6y}{5x + 6y}}$.

2. Show that if y_1 and y_2 are two different solutions of $y' + P(x)y = Q(x)$ then they must be related by $y_2 = y_1(1 + ce^{-\int Q\,dx/y_1})$. Hence, by observing that $y = x$ is a solution of $y' + xy = x^2 + 1$, find the general solution.

3. Show that $x^p y^q (\alpha y\,dx + \beta x\,dy) + x^r y^s (\gamma y\,dx + \delta x\,dy) = 0$ where p, q, r, s, α, β, γ, δ are given constants, has an integrating factor of the form $x^a y^b$, where a and b are suitable constants.

4. Solve $(x^2 y + 2y^4)dx + (x^3 + 3xy^3)dy = 0$, using the method of Exercise 3.

5. Show that $\dfrac{dy}{dx} = \dfrac{y(y^2 - x^2 - 1)}{x(y^2 - x^2 + 1)}$ can be solved by transforming to polar coordinates r, ϕ, where $x = r\cos\phi$, $y = r\sin\phi$. Hence, determine its solution.

6. Show that, if μ is an integrating factor of the differential equation $M\,dx + N\,dy = 0$ so that $\mu M\,dx + \mu N\,dy = dU(x, y)$, then $\mu\psi(U)$, where ψ is an arbitrary function, is also an integrating factor. Illustrate this by some examples.

7. Show that the differential equation $y' = P(x)F(y) + Q(x)G(y)$ can be reduced to a linear equation by the transformations

$$u = \frac{F(y)}{G(y)} \quad \text{or} \quad u = \frac{G(y)}{F(y)}$$

according as $\dfrac{FG' - GF'}{G}$ or $\dfrac{FG' - GF'}{F}$ is a constant.

8. Using the results of Exercise 7, obtain the general solutions of:
   (a) $y' = \sec y + x \tan y$.
   (b) $y' = P(x)y + Q(x)y^n$ (Bernoulli's equation).

9. Show that if the equation $M\,dx + N\,dy = 0$ is such that $\dfrac{x^2}{xM + yN}\left(\dfrac{\partial N}{\partial x} - \dfrac{\partial M}{\partial y}\right) = F\left(\dfrac{y}{x}\right)$ then an integrating factor is given by $e^{\int F(u)du}$, where $u = y/x$.

10. Prove that if a differential equation $M\,dx + N\,dy = 0$ is both exact and homogeneous, then its solution is $Mx + Ny = c$. Illustrate by using the differential equation

$$(x^2 + y^2)dx + 2xy\,dy = 0$$

11. (a) Prove that if μ and v are two different integrating factors of the equation $M\,dx + N\,dy = 0$ then its general solution is $\mu = cv$. (b) Illustrate part (a) by finding two different integrating factors of $x\,dy - y\,dx = 0$.

12. The *Riccati equation* is given by $y' = P(x)y^2 + Q(x)y + R(x)$.
   (a) Show that if one solution of this equation, say $y_1(x)$, is known, then the general solution can be found by using the transformation $y = y_1 + 1/u$ where u is a new dependent variable.
   (b) Show that if two solutions are known, say $y_1(x)$ and $y_2(x)$, the general solution is

$$\frac{y - y_1}{y - y_2} = ce^{\int P(y_1 - y_2)dx}$$

(c) Show that if three solutions are known, say $y_1(x)$, $y_2(x)$, and $y_3(x)$, then the general solution is

$$\frac{(y - y_1)(y_2 - y_3)}{(y - y_2)(y_1 - y_3)} = c$$

13. Solve the equation $y' = xy^2 - 2y + 4 - 4x$ by noting that $y = 2$ is a particular solution.

14. Solve $\dfrac{dy}{dx} + y^2 = 1 + x^2$.

15. Solve the equation $y' = \dfrac{y^2}{x - 1} - \dfrac{xy}{x - 1} + 1$ by noting that $y = 1$ and $y = x$ are solutions.

three

applications of first-order and simple higher-order differential equations

In this chapter we discuss applications of first-order and simple higher-order differential equations to problems of mechanics, electric circuits, geometry, rockets, biology, economics, chemistry, bending of beams, and others. The sections are arranged so that students may emphasize those topics which are particularly adapted to their interests or needs.

1 Applications to Mechanics

1.1 INTRODUCTION

The subject of physics deals with the investigation of the laws which govern the behavior of the physical universe. By physical universe we mean the totality of objects about us, not only things which we observe, but the things which we do not observe, such as atoms and molecules. The study of the motion of objects in our universe is a branch of mechanics called *dynamics*. Newton's three laws of motion, known to students of elementary physics, form the fundamental basis for its study. It turns out, however, that for objects moving very fast (for example, near the speed of light, 186,000 miles per second) we cannot use Newton's laws. Instead we must use a revision of these laws, evolved by Einstein and known as *relativistic mechanics*, or the mechanics of relativity. For objects of atomic dimensions, Newton's laws are not valid either. In fact, to obtain accurate pictures of the motion of objects of atomic dimensions, we need a set of laws studied in an advanced subject called *quantum mechanics*. Both relativistic and quantum mechanics are far too complicated to investigate in this book, since the student would need more extensive background in mathematics and physics to begin a study of these subjects.

Fortunately, to study motion of objects encountered in our everyday lives, neither objects which attain speeds near that of light nor objects which are of atomic dimensions, we do not need relativity or quantum mechanics. Newton's laws are accurate enough in these cases and we shall therefore embark on a discussion of these laws and their applications.

1.2 NEWTON'S LAWS OF MOTION

The three laws of motion first developed by Newton are:

     1. *A body at rest tends to remain at rest, while a body in motion tends to persist in motion in a straight line with constant velocity unless acted upon by external forces.*

     2. *The time rate of change in momentum of a body is proportional to the net force acting on the body and has the same direction as the force.*

     3. *To every action there is an equal and opposite reaction.*

The second law provides us with an important relation known to students of elementary physics and we shall often refer to it briefly as *Newton's law*.

The momentum of an object is defined to be its mass m multiplied by its velocity v. The time rate of change in momentum is thus $d/dt(mv)$. If we denote by F the net force which is acting on the body the second law states that

$$\frac{d}{dt}(mv) \propto F \tag{1}$$

where the symbol $\propto$ denotes proportionality. Introducing the constant of proportionality k, we obtain

$$\frac{d}{dt}(mv) = kF$$

If m is a constant, $\qquad m\frac{dv}{dt} = kF$ or $ma = kF$

where $a = dv/dt$ is the acceleration. Thus we see that

$$F = \frac{ma}{k} \tag{2}$$

The value of k depends on the units which we wish to use. At present two main systems are in use.

(a) **The CGS System or Centimeter, Gram, Second System.** In this system length is measured in centimeters (cm), mass in grams (g), and time in seconds (sec). The simplest value for k is $k = 1$, so that the law (2) is

$$F = ma \tag{3}$$

If a certain force produces an acceleration of one centimeter per second per second (1 cm/sec²) in a mass of 1 g, then from (3)

$$F = 1 \text{ g} \cdot 1 \text{ cm/sec}^2 = 1 \text{ g cm/sec}^2$$

We call such a force a *dyne*. The cgs system is also called the *metric system*.

(b) **The FPS System, or Foot, Pound, Second System.** In this system we may also use $k = 1$, so that the law is $F = ma$. If a certain force produces an acceleration of one foot per second per second (1 ft/sec²) in a mass of one pound (lb), we call this force a *poundal*. Thus, from $F = ma$ we have 1 poundal $= 1$ lb ft/sec².

Another way of expressing Newton's law involves use of weight rather than the mass of an object. Whereas the mass of an object is the same everywhere on the earth (or actually anywhere in the universe)* the weight changes from place to place.† It will be observed that for a body acted on only by its weight W, the corresponding acceleration is that due to gravity g. The force is W, and Newton's law is

$$W = mg \tag{4}$$

Dividing equation (3) by equation (4), we have

$$\frac{F}{W} = \frac{a}{g} \quad \text{or} \quad F = \frac{Wa}{g} \tag{5}$$

* Actually, we are talking here about "rest mass," because in relativity theory when an object is in motion its mass changes.
† On the earth's surface this change does not exceed 2 per cent.

We may use equation (5) with either cgs or fps units. In such case it is clear that F and W have the same units if a and g have.

With CGS Units: If W is in grams weight, a and g in cm/sec^2, then F is in grams weight. If W is in dynes, a and g in cm/sec^2, then F is in dynes. On the earth's surface $g = 980$ cm/sec^2, approximately.

With FPS Units: If W is in pounds weight, a and g in ft/sec^2, then F is in pounds weight. On the earth's surface $g = 32$ ft/sec^2, approximately.

In certain fields it is customary to use the cgs system in conjunction with the law $F = ma$, and to use the fps system in conjunction with the law $F = Wa/g$. Sometimes use is made of mass in terms of *slugs*.*

Note: It will be the custom in this book to use
1. $F = ma$, where F is in dynes, m in grams, a in cm/sec^2.
2. $F = Wa/g$, where F and W are in pounds, a and g in ft/sec^2.[†]

When other units are desired, appropriate changes can be made. If in a problem units are not specified, any system can be used so long as consistency prevails.[‡]

In the notation of the calculus we may write Newton's law in different forms by realizing that the acceleration can be expressed as a first derivative of a velocity v (i.e., dv/dt) or as a second derivative of a displacement s (i.e., d^2s/dt^2). Thus

$$F = m\frac{dv}{dt} = m\frac{d^2s}{dt^2} \qquad \text{(cgs)}$$

$$F = \frac{W}{g}\frac{dv}{dt} = \frac{W}{g}\frac{d^2s}{dt^2} \qquad \text{(fps)}$$

We now consider mathematical formulations of various mechanics problems involving concepts introduced, and the solution and interpretation of such problems.

ILLUSTRATIVE EXAMPLE 1

A mass of m grams falls vertically downward under the influence of gravity starting from rest. Assuming that air resistance is negligible set up a differential equation and associated conditions describing the motion and solve.

Mathematical Formulation. In the mathematical formulation of physical problems (or for that matter, any problems) it is useful to draw diagrams wherever possible. They help to fix ideas and consequently help us to translate physical ideas into mathematical equations. Let A (Fig. 3.1) be the position of mass m at time $t = 0$,

---

* The number of pounds weight divided by g (which equals 32 approximately) is known as the *number of slugs*. Thus, the mass of a 64 lb weight is 2 slugs.
  † Whenever we use the abbreviation lb we refer to pounds weight.
  ‡ For some purposes a variation of the cgs system known as the *meter, kilogram, second system* or *mks system* is sometimes used. Here length is in meters (100 cm), mass in kilograms (1000 g) and time in seconds. The force required to move a 1 kilogram mass at an acceleration of 1 meter/sec^2 is called a *newton*. The abbreviations for meter, kilogram, and newton are m, kg, and n respectively.

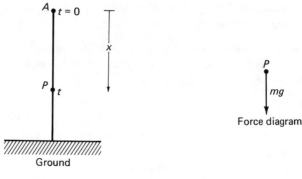

A $\bullet$ t = 0

x

P $\bullet$ t

Ground

Figure 3.1

P

mg

Force diagram

Figure 3.2

and let P be the position of m at any subsequent time t. In any physical problem involving vector quantities, such as force, displacement, velocity, and acceleration which necessarily require a knowledge of direction, it is convenient to set up a co-ordinate system, together with the assignment of positive and negative directions. In the present problem let A be the origin of our coordinate system and choose the x axis as vertical with "down" as the positive direction (and consequently "up" as the negative direction). The instantaneous velocity at P is $v = dx/dt$. The instan-taneous acceleration at P is $a = dv/dt$ or $a = d^2x/dt^2$. The net force acts vertically downward (considered positive as shown in the force diagram of Fig. 3.2). Its mag-nitude is mg. By Newton's law we have

$$m\frac{dv}{dt} = mg \quad \text{or} \quad \frac{dv}{dt} = g$$

Since we are told that the mass falls from rest we see that $v = 0$ when $t = 0$, or in other words $v(0) = 0$. Our mathematical formulation is the initial value problem

$$\frac{dv}{dt} = g, \qquad v(0) = 0 \tag{6}$$

Here we have a first-order equation and the required one condition.

Another way of formulating the problem is to write

$$m\frac{d^2x}{dt^2} = mg \quad \text{or} \quad \frac{d^2x}{dt^2} = g$$

In such case we have a second-order equation in the variables x and t, and we require two conditions for the determination of x. One of them is $v = 0$ or $dx/dt = 0$ at $t = 0$. The second may be arrived at by noting that $x = 0$ at $t = 0$ (since we chose the origin of our coordinate system at A). The mathematical formulation is

$$\frac{d^2x}{dt^2} = g, \qquad x = 0 \quad \text{and} \quad \frac{dx}{dt} = 0 \quad \text{at} \quad t = 0 \tag{7}$$

The procedure will be typical in the mathematical formulations of problems. When we set up differential equations to describe some phenomena or law, we shall always accompany them by enough conditions necessary to determine the arbitrary constants in the general solution.

Solution Starting with $dv/dt = g$, we obtain by integration $v = gt + c_1$. Since $v = 0$ when $t = 0$, $c_1 = 0$, or $v = gt$, i.e., $\dfrac{dx}{dt} = gt$.

Another integration yields $x = \frac{1}{2}gt^2 + c_2$. Since $x = 0$ at $t = 0$, $c_2 = 0$. Hence $x = \frac{1}{2}gt^2$. We could have arrived at the same result by starting with (7).

As an application, suppose we wish to know where the object is after 2 sec. Then, by the cgs system $x = \frac{1}{2} \times 980$ cm/sec$^2 \times (2$ sec$)^2 = 1960$ cm. By the fps system, $x = \frac{1}{2} \times 32$ ft/sec$^2 \times (2$ sec$)^2 = 64$ ft.

To find the velocity after 2 sec we write (in the fps system)

$$\frac{dx}{dt} = gt = 32 \text{ ft/sec}^2 \times 2 \text{ sec} = +64 \text{ ft/sec}$$

The plus sign indicates that the object is moving in the positive direction, i.e., downward. It should be noted that if we had taken the positive direction as upward the differential equation would have been $m(dv/dt) = -mg$, i.e.,

$$\frac{dv}{dt} = -g \quad \text{or} \quad \frac{d^2x}{dt^2} = -g$$

This would, of course, lead to results equivalent to those obtained.

ILLUSTRATIVE EXAMPLE 2

A ball is thrown vertically upward with an initial velocity of 128 ft/sec. What is its velocity after 2, 4, and 6 sec? When will it return to its starting position? What is the maximum height to which it will rise before returning?

Mathematical Formulation. Here we shall take the x axis as vertical, with the origin on the ground at A so that $x = 0$ when $t = 0$ (Fig. 3.3). We shall consider "up" as positive. The force acting on the ball (Fig. 3.4) is its weight and we must therefore consider it to be $-mg$ (the minus signifying down). The differential equation

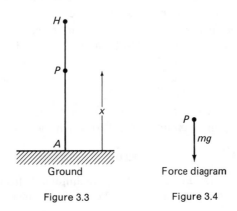

Figure 3.3

Figure 3.4

for the motion is

$$m\frac{d^2x}{dt^2} = -mg \quad \text{or} \quad \frac{d^2x}{dt^2} = -g$$

Two conditions are needed to determine x. One is supplied by the fact that $x = 0$ at $t = 0$. The other is supplied by the fact that the initial velocity is 128 ft/sec. This velocity is in the upward direction and we therefore consider it as positive. Thus

$$v = \frac{dx}{dt} = +128 \quad \text{at} \quad t = 0$$

The complete mathematical formulation is

$$\frac{d^2x}{dt^2} = -g, \quad x = 0 \quad \text{and} \quad \frac{dx}{dt} = 128 \quad \text{at} \quad t = 0 \qquad (8)$$

Solution Integration of the differential equation in (8) yields $\dfrac{dx}{dt} = -gt + c_1$

and since $dx/dt = 128$ where $t = 0$, $c_1 = 128$, so that $\dfrac{dx}{dt} = -gt + 128$.

Another integration yields $x = -\frac{1}{2}gt^2 + 128t + c_2$ and since $x = 0$ where $t = 0$, $c_2 = 0$. Hence,

$$x = -\tfrac{1}{2}gt^2 + 128t \quad \text{or} \quad x = 128t - 16t^2.$$

Velocity after 2, 4, 6 seconds. We have for the velocity at time t

$$v = \frac{dx}{dt} = 128 - 32t$$

Letting $t = 2$, we find $v = 64$, which means that the ball is rising at the rate of 64 ft/sec. Letting $t = 4$, we find $v = 0$, which means that the ball has stopped. Letting $t = 6$, we find $v = -64$, which means that the ball has turned and is coming down at the rate of 64 ft/sec.

Time for return. The ball is at position A, the starting position, when $x = 0$. This happens when $-16t^2 + 128t = 0$ or $-16t(t - 8) = 0$ i.e., $t = 0$ or $t = 8$. The value $t = 0$ is trivial, since we already know that $x = 0$ at $t = 0$. The other value $t = 8$ shows that the ball returns after 8 sec.

Maximum height of rise. The maximum value of x may be found by setting $dx/dt = 0$, which is equivalent to finding when $v = 0$. We have

$$v = \frac{dx}{dt} = 128 - 32t = 0 \qquad \text{where } t = 4$$

Since d^2x/dt^2 is negative, x is actually a maximum for $t = 4$. The value of x for $t = 4$ is 256. Hence, the maximum height to which the ball rises is 256 ft.

ILLUSTRATIVE EXAMPLE 3

A sky diver (and of course his parachute) falls from rest. The combined weight of sky diver and parachute is W. The parachute has a force acting on it (due to air resistance) which is proportional to the speed at any instant during the fall. Assuming that he falls vertically downward and that the parachute is already open when the jump takes place, describe the ensuing motion.

Mathematical Formulation. We draw, as usual, a physical and force diagram (Figs. 3.5 and 3.6). Assume A to be the origin and AB the direction of the positive x axis. The forces acting are : (a) the combined weight W acting downward; (b) the air resistance force R acting upward. The net force in the positive (downward) direction is $W - R$. Since the resistance is proportional to the speed we have

$$R \propto |v| \quad \text{or} \quad R = \beta |v|$$

where β is the constant of proportionality. Since v is always positive, we do not need absolute value signs, and may write simply $R = \beta v$. Hence the net force is $W - \beta v$, and we obtain by Newton's law

$$\frac{W}{g} \cdot \frac{dv}{dt} = W - \beta v$$

Since the sky diver starts from rest, $v = 0$ at $t = 0$. Thus the complete mathematical formulation is given by the initial value problem

$$\frac{W}{g} \cdot \frac{dv}{dt} = W - \beta v, \qquad v = 0 \quad \text{at} \quad t = 0$$

Solution   The differential equation has its variables separable. Thus,

$$\int \frac{W \, dv}{W - \beta v} = \int g \, dt \quad \text{or} \quad -\frac{W}{\beta} \ln (W - \beta v) = gt + c_1$$

Since $v = 0$ at $t = 0$, $c_1 = -\dfrac{W \ln W}{\beta}$ and, thus,

$$-\frac{W}{\beta} \ln (W - \beta v) = gt - \frac{W}{\beta} \ln W, \qquad \ln \left(\frac{W}{W - \beta v} \right) = \frac{\beta g t}{W}$$

Hence,

$$v = \frac{W}{\beta} (1 - e^{-\beta g t / W})$$

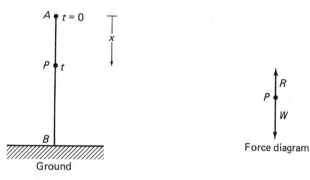

Figure 3.5

Figure 3.6

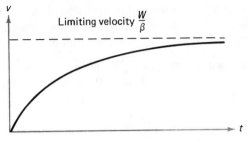

Figure 3.7

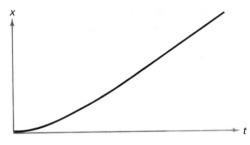

Figure 3.8

It will be noted that as $t \to \infty$, v approaches W/β, a constant limiting velocity. This accounts for the fact that we notice parachutes traveling at very nearly uniform speeds after a certain length of time has elapsed. We may also determine the distance traveled by the sky diver as a function of time.

From

$$\frac{dx}{dt} = \frac{W}{\beta}(1 - e^{-\beta g t/W})$$

we have

$$x = \frac{W}{\beta}\left(t + \frac{W}{\beta g}e^{-\beta g t/W}\right) + c_2$$

Using the fact that $x = 0$ at $t = 0$, we find $c_2 = -W^2/\beta^2 g$. Hence,

$$x = \frac{W}{\beta}\left(t + \frac{W}{\beta g}e^{-\beta g t/W} - \frac{W}{\beta g}\right)$$

The graphs of v and x as functions of t are shown in Figs. 3.7 and 3.8.

A EXERCISES

1. A mass of 25 g falls from rest under the influence of gravity. (a) Set up a differential equation and conditions for the motion. (b) Find the distance traveled and velocity attained 3 sec after the motion has begun. (c) How far does the mass travel between the 3rd and 4th seconds? between the 4th and 5th seconds?

2. A 200 g mass is thrown upward with initial velocity 2450 cm/sec. (a) Find the distances from the starting point and the velocities attained 2 and 4 sec after the motion has begun. (b) Find the highest point reached and the time required. (c) What are the total distances traveled after 2 sec? after 4 sec?

3. A 6 lb weight is dropped from a cliff $\frac{1}{4}$ mile high. Assuming no air resistance, at what time and with what velocity does it reach the ground?

4. An oil droplet, mass 0.2 g, falls from rest in air. For a velocity of 40 cm/sec, the force due to air resistance is 160 dynes. Assuming air resistance force proportional to velocity: (a) Find the velocity and distance traveled as a function of time. (b) Find the limiting velocity.

5. The force of water resistance acting on a boat is proportional to its instantaneous velocity, and is such that at 20 ft/sec the water resistance is 40 lb. If the boat weighs 320 lb and the only passenger weighs 160 lb, and if the motor can exert a steady force of 50 lb in the direction of motion: (a) Find the maximum velocity at which the boat will travel. (b) Find the distance traveled and velocity at any time, assuming the boat starts from rest.

6. A paratrooper and parachute weigh 200 lb. At the instant the parachute opens, he is traveling vertically downward at 40 ft/sec. If the air resistance varies directly as the instantaneous velocity and the air resistance is 80 lb when the velocity is 20 ft/sec: (a) Find the limiting velocity. (b) Determine the position and velocity at any time.

7. A 192 lb weight has limiting velocity 16 ft/sec when falling in air, which provides a resisting force proportional to the weight's instantaneous velocity. If the weight starts from rest: (a) Find the velocity of the weight after 1 sec. (b) How long is it before the velocity is 15 ft/sec?

8. Solve the previous problem if the force of air resistance varies as the square of the instantaneous velocity.

9. A particle moves along the x axis acted upon only by an opposing force proportional to its instantaneous velocity. It starts at the origin with a velocity of 10 ft/sec, which is reduced to 5 ft/sec after moving 2.5 ft. Find its velocity when it is 4 ft from the origin.

B EXERCISES

1. Show that a ball thrown vertically upward with initial velocity v_0 takes twice as much time to return as to reach the highest point. Find the velocity upon return. Air resistance is assumed negligible.

2. A body moves on a straight line with constant acceleration a. If v_0 is the initial velocity, v is the velocity, and s is the distance traveled after time t, show that:
(a) $v = v_0 + at$.   (b) $s = v_0 t + \frac{1}{2}at^2$.   (c) $v^2 = v_0^2 + 2as$.

3. A mass m is thrown upward with initial velocity v_0. Air resistance is proportional to its instantaneous velocity, the constant of proportionality being k. Show that the maximum height attained is

$$\frac{mv_0}{k} - \frac{m^2 g}{k^2} \ln\left(1 + \frac{kv_0}{mg}\right)$$

4. A paratrooper and his parachute weigh W pounds. When the parachute opens he is traveling vertically downward at v_0 feet per second. If the force of air resistance varies directly as the square of the instantaneous velocity and if air resistance is F pounds, where velocity is V feet per second: (a) Write differential equations for the velocity as a function of time and also of distance. (b) Find the velocity t sec after the parachute opens and the limiting velocity. What simplifications result if $v_0 = 0$? (c) Find the velocity as a function of distance traveled.

5. An object weighing 1000 lb sinks in water starting from rest. Two forces act on it, a buoyant force of 200 lb, and a force of water resistance which in pounds is numerically equal to $100v$, where v is in ft/sec. Find the distance traveled after 5 sec and the limiting velocity.

6. A 10 lb object is dropped vertically downward from a very high cliff. The law of resistance in the fps system is given by $0.001v^2$, where v is the instantaneous velocity. Determine (a) the velocity as a function of distance, (b) the velocity as a function of time, (c) the velocity of the object after having fallen 500 ft, (d) the limiting velocity, (e) the distance traveled after 10 sec.

7. A particle moves in a straight line toward a fixed point O on the line with an instantaneous speed which is proportional to the nth power of its instantaneous distance from O. (a) Show that if $n \geq 1$ the particle will never reach O. (b) Discuss the cases $n < 1$.

8. When a ball is thrown upward it reaches a particular height after time T_1 on the way up and time T_2 on the way down. (a) Assuming air resistance to be negligible, show that the height is given by $\frac{1}{2}g T_1 T_2$. (b) How can the result be used to find the height of a tree without climbing it?

9. Show that the ball in Exercise 8 was thrown upward with initial speed $\frac{1}{2}g(T_1 + T_2)$.

10. How long will it take for a chain of length L to just slide off a horizontal frictionless table if a portion of it having length a hangs over the side initially?

11. A particle of mass m moves along the x axis under the influence of a force given by $F(x)$. If v is the instantaneous velocity, show that

$$\tfrac{1}{2}mv^2 + V(x) = E$$

where E is a constant and

$$V(x) = \int_x^a F(u)du$$

assuming a is such that $V(a) = 0$. The result is known as the principle of the *conservation of energy*, with $\frac{1}{2}mv^2$ called the *kinetic energy*, $V(x)$ the *potential energy*, and E the *total energy*.

C EXERCISES

1. A 100 lb weight slides from rest down an inclined plane (Fig. 3.9) which makes an angle of 30° with the horizontal. Assuming no friction: (a) Set up a differential equation and conditions describing the motion. (b) What distance will the weight travel 5 sec after starting and what will be its velocity and acceleration at that instant? (*Hint*: Resolve the force due to the weight into two components, one parallel and one perpendicular to the plane. The component P parallel to the plane is the net force producing motion.)

2. Show that a weight W given an initial velocity v_0 slides a distance s down a frictionless inclined plane of inclination α in the time

$$\frac{\sqrt{v_0^2 + 2gs \sin \alpha} - v_0}{g \sin \alpha}$$

3. An object of mass m is thrown up an inclined plane of inclination α. Assuming no friction, show that the maximum distance reached is $v_0^2/(2g \sin \alpha)$.

4. If air resistance proportional to the instantaneous velocity (constant of proportionality k) is taken into account, show that the object in Exercise 3 reaches a maximum distance up the incline given by

$$\frac{mv_0}{k} - \frac{m^2g}{k^2} \sin \alpha \ln \left(1 + \frac{kv_0}{mg \sin \alpha} \right).$$

Verify that this distance reduces to that of Exercise 3 as $k \to 0$.

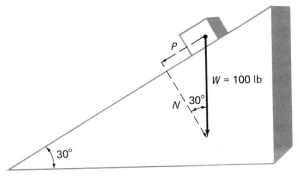

Figure 3.9

5. A 100 lb weight starts from rest down a 30° incline. If the coefficient of friction between weight and incline is 0.2, what distance will the weight travel after 5 sec? Find its velocity and acceleration at that instant (assume the weight does get started). (*Hint:* Referring to Fig. 3.9, the frictional force acting is given by the normal component N multiplied by the coefficient of friction.)

6. A weight W is given initial velocity v_0 down an incline of angle α. If the coefficient of friction between weight and plane is μ, show that after time T the weight travels a distance

$$v_0 T + \tfrac{1}{2}(g \sin \alpha - \mu g \cos \alpha)T^2 \quad \text{if } \tan \alpha > \mu$$

7. According to Einstein's special theory of relativity, the mass of a particle varies with its velocity v according to the formula

$$m = \frac{m_0}{\sqrt{1 - v^2/c^2}}$$

where m_0 is the rest mass and c is the velocity of light (186,000 miles/sec). The differential equation of motion is

$$F = m_0 \frac{d}{dt}\left(\frac{v}{\sqrt{1 - v^2/c^2}}\right)$$

If a particle starts from rest at $t = 0$ and moves in a straight line acted upon only by a constant force F, what distance will it cover and what will be its velocity in time t? Show that as time goes by, the velocity of the particle approaches the velocity of light.

8. An object having rest mass m_0 of 10,000 g moves on the x axis under a constant force of 50,000 dynes. If it starts from rest at $x = 0$ at time $t = 0$, determine where it will be at any time assuming: (a) The mass of the object is constant and equal to m_0. (b) The mass varies according to the law of special relativity.

9. Objects starting from rest fall without friction along chords of a vertical circle all of which end at the lowest point. Show that they reach the lowest point in the same time.

10. Is the result of Exercise 9 true if the circle is inclined at an angle with the vertical? Explain.

11. If an object is thrown vertically upward and if air resistance is present, will it take twice as much time for the object to return to the starting point as to reach the highest point? Explain. (See B Exercise 1.)

12. In Exercise 11 will the speed of the object on returning be the same as that with which it is thrown? Explain.

13. A satellite revolves in a circular orbit acted upon only by a force of resistance proportional to the square of its instantaneous speed. (a) If the speed is v_0 at time $t = 0$ and v_1 at time $t = T_1$ show that the speed at any time t is

$$v = \frac{v_0 v_1 T_1}{v_1 T_1 + (v_0 - v_1)t}$$

(b) Show that the number of revolutions made between times $t = 0$ and $t = T_1$ is

$$\frac{v_0 v_1 T_1}{\pi(v_0 - v_1)} \ln\left(\frac{v_0}{v_1}\right)$$

(c) Show that although the speed keeps decreasing the satellite revolves indefinitely.
(d) Do you believe that the problem represents a possible physical situation? Explain.

14. Work Exercise 13 if the force of resistance is proportional to the nth power of the instantaneous speed and examine the special case where $n = 3$.

15. Work B Exercise 10 if the table is inclined at angle α with the horizontal and the portion of length a hangs over the higher side initially? Is there any restriction on α? Explain.

2

Applications to Electric Circuits

2.1 INTRODUCTION

Just as mechanics has as a fundamental basis the laws of Newton, so the subject of electricity has a law describing the behavior of electric circuits, known as *Kirchhoff's law*, which we shall describe and use in this section. Actually, electrical theory is governed by a certain set of equations known in electromagnetic theory as Maxwell's equations. Just as we cannot enter into a discussion of relativity or quantum mechanics because of the student's insufficient background, so we cannot enter into discussion of Maxwell's equations. However, just as Newton's laws proved sufficient for motion of "everyday objects," so Kirchhoff's law is amply suited for study of the simple properties of electric circuits. For thorough study of electric circuits the student must, of course, take laboratory practice and watch demonstrations in lectures. In a mathematical book we can hope to present only a brief discussion.

The simplest electric circuit is a series circuit in which we have an emf (*electromotive force*), which acts as a source of energy such as a battery or generator, and a *resistor*, which uses energy, such as an electric light bulb, toaster, or other appliance.

In elementary physics we find that the emf is related to the current flow in the circuit. Simply stated, the law says that the instantaneous current I (in a circuit containing only an emf E and a resistor) is directly proportional to the emf. In symbols,

$$I \propto E \quad \text{or} \quad E \propto I$$

Hence, $$E = IR \tag{1}$$

where R is a constant of proportionality called the coefficient of resistance or, simply, resistance. The units, generally known as "practical units" are such that E is in *volts*, I is in *amperes* and R is in *ohms*. Equation (1) is familiar to the student of elementary physics under the name *Ohm's law*.

More complicated, but for many cases more practical, are circuits consisting of elements other than resistors. Two important elements are *inductors* and *capacitors*. An inductor opposes a change in current. It has an inertia effect in electricity in much the same way as mass has an inertia effect in mechanics. In fact the analogy is strong, and much could be said about it. A capacitor is an element which stores energy.

In physics we speak of a *voltage drop* across an element. In practice we can determine this voltage drop, or as it is sometimes called, *potential drop* or *potential difference*, by use of an instrument called a voltmeter. Experimentally the following laws are found to hold.

> 1. *The voltage drop across a resistor is proportional to the current passing through the resistor.*

If E_R is the voltage drop across the resistor and I is the current, then

$$E_R \propto I \quad \text{or} \quad E_R = RI$$

where R is the constant of proportionality called the coefficient of resistance or simply resistance.

> 2. *The voltage drop across an inductor is proportional to the instantaneous time rate of change of the current.*

If E_L is the voltage drop across the inductor, then

$$E_L \propto \frac{dI}{dt} \quad \text{or} \quad E_L = L\frac{dI}{dt}$$

where L is the constant of proportionality called the coefficient of inductance or simply the inductance.

> 3. *The voltage drop across a capacitor is proportional to the instantaneous electric charge on the capacitor.*

If E_C is the voltage drop across the capacitor and Q is the instantaneous charge,

then
$$E_C \propto Q \quad \text{or} \quad E_C = \frac{Q}{C}$$

where we have taken $1/C$ as the constant of proportionality, C being known as the coefficient of capacitance or simply capacitance.

2.2 UNITS

In electricity, as in mechanics, there exist more than one system of units. We consider and use only one such system in this book. Table 3.1 summarizes the important electrical quantities with their symbols and units. As in mechanics, time is in seconds.

Table 3.1

Quantity	Symbol	Unit
Voltage, emf, or potential	E or V	Volt
Resistance	R	Ohm
Inductance	L	Henry
Capacitance	C	Farad
Current	I	Ampere
Charge.	Q	Coulomb

The unit of current, the ampere (often abbreviated *amp*), corresponds to a coulomb of charge passing a given point in the circuit per second.

2.3 KIRCHHOFF'S LAW

The following is a statement of *Kirchhoff's law*:

> *The algebraic sum of all the voltage drops around an electric loop or circuit is zero.* [Another way of stating this is to say that the voltage supplied (emf) is equal to the sum of the voltage drops.]

It is customary to indicate the various circuit elements as shown:

As an example, consider an electric circuit consisting of a voltage supply E (battery or generator), a resistor R, and inductor L connected in series as shown in Fig. 3.10.* We adopt the following

> CONVENTION. The current flows from the positive ($+$) side of the battery or generator through the circuit to the negative ($-$) side, as shown in Fig. 3.10.

Since, by Kirchhoff's law, the emf supplied (E) is equal to the voltage drop across the inductor ($L\, dI/dt$) added to the voltage drop across the resistor (RI), we have as the required differential equation for the circuit

$$L\frac{dI}{dt} + RI = E$$

---

* Sometimes, for brevity, we shall speak of a battery E, resistor R, capacitor C, etc., instead of a battery having an emf of E volts, a resistor having a resistance of R ohms, a capacitor having a capacitance of C farads, etc.

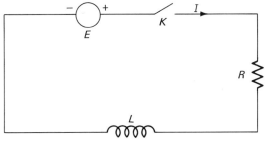

Figure 3.10

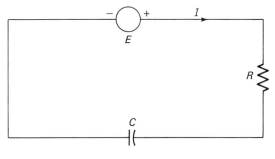

Figure 3.11

As another example, suppose we are given an electric circuit consisting of a battery or generator of E volts in series with a resistor of R ohms and a capacitor of C farads as in Fig. 3.11. Here the voltage drop across the resistor is RI and the voltage drop across the capacitor is Q/C, so that by Kirchhoff's law

$$RI + \frac{Q}{C} = E \tag{2}$$

As it stands this is not a differential equation. However by noting that the current is the time rate of change in charge, i.e., $I = dQ/dt$, (2) becomes

$$R\frac{dQ}{dt} + \frac{Q}{C} = E$$

which is the differential equation for the instantaneous charge. Accompanying the differential equations obtained are conditions which are derived, of course, from the specific problem considered.

ILLUSTRATIVE EXAMPLE 1

A generator having emf 100 volts is connected in series with a 10 ohm resistor and an inductor of 2 henries. If the switch K is closed at time $t = 0$, set up a differential equation for the current and determine the current at time t.

Mathematical Formulation. As usual we draw the physical diagram (Fig. 3.12). Calling I the current in amperes flowing as shown, we have: (1) voltage supplied = 100 volts, (2) voltage drop across resistance (RI) = $10I$, (3) voltage drop across

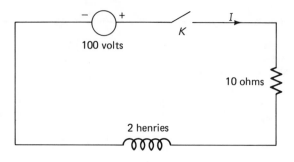

Figure 3.12

inductor $(L\,dI/dt) = 2\,dI/dt$. Hence, by Kirchhoff's law,

$$100 = 10I + 2\frac{dI}{dt} \quad \text{or} \quad \frac{dI}{dt} + 5I = 50 \tag{3}$$

Since the switch is closed at $t = 0$, we must have $I = 0$ at $t = 0$.

Solution   The differential equation (3) is a first-order linear equation with integrating factor e^{5t}. Multiplying by this factor gives

$$\frac{d}{dt}(e^{5t}I) = 50e^{5t} \quad \text{or} \quad e^{5t}I = 10e^{5t} + c \quad \text{i.e.,} \quad I = 10 + ce^{-5t}.$$

Since $I = 0$ at $t = 0$, $c = -10$. Thus, $I = 10(1 - e^{-5t})$.

Another Method.   Equation (3) may be also solved by separation of variables.

The graph of I versus t is shown in Fig. 3.13. Note that the current is zero at $t = 0$ and builds toward the maximum of 10 amp although theoretically it never reaches it. The student should note the similarity between this problem and the problem of the falling sky diver in Illustrative Example 3 of the last section.

ILLUSTRATIVE EXAMPLE 2

Set up and solve a differential equation for the circuit of Illustrative Example 1 if the 100 volt generator is replaced by one having an emf of 20 cos 5t volts.

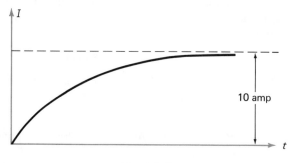

Figure 3.13

Mathematical Formulation. The only difference is that 20 cos 5t replaces 100 in equation (3). Hence, the required equation is

$$10I + 2\frac{dI}{dt} = 20 \cos 5t \quad \text{or} \quad \frac{dI}{dt} + 5I = 10 \cos 5t \tag{4}$$

Solution   Multiply the second equation in (4) by the integrating factor e^{5t}. Then

$$\frac{d}{dt}(e^{5t}I) = 10e^{5t} \cos 5t \quad \text{and} \quad e^{5t}I = 10 \int e^{5t} \cos 5t \, dt = e^{5t}(\cos 5t + \sin 5t) + c$$

or
$$I = \cos 5t + \sin 5t + ce^{-5t}$$

Since $I = 0$ at $t = 0$, we have $c = -1$. Thus, $I = \cos 5t + \sin 5t - e^{-5t}$.

ILLUSTRATIVE EXAMPLE 3

A decaying emf $E = 200e^{-5t}$ is connected in series with a 20 ohm resistor and 0.01 farad capacitor. Assuming $Q = 0$ at $t = 0$, find the charge and current at any time. Show that the charge reaches a maximum, calculate it, and find when it is reached.

Mathematical Formulation. Referring to Fig. 3.14 we have: (1) voltage supplied $(E) = 200e^{-5t}$, (2) voltage drop in resistor $(RI) = 20I$, (3) voltage drop across capacitor $(Q/C) = Q/0.01 = 100Q$. Hence, by Kirchhoff's law, $20I + 100Q = 200e^{-5t}$ and, since $I = dQ/dt$,

$$20\frac{dQ}{dt} + 100Q = 200e^{-5t} \quad \text{or} \quad \frac{dQ}{dt} + 5Q = 10e^{-5t}$$

Solution   An integrating factor is e^{5t}. Hence, $\frac{d}{dt}(Qe^{5t}) = 10$, $Qe^{5t} = 10t + c$.

Since $Q = 0$ at $t = 0$, $c = 0$. Hence, $Q = 10te^{-5t}$.

Current at Any Time.   Since $I = dQ/dt$, $I = \frac{d}{dt}(10te^{-5t}) = 10e^{-5t} - 50te^{-5t}$.

Maximum Charge. To find when Q is a maximum, set $dQ/dt = 0$, i.e., $I = 0$.

$$10e^{-5t} - 50te^{-5t} = 0, \quad \text{i.e.,} \quad t = \tfrac{1}{5} \text{ sec}$$

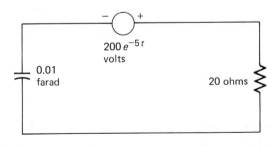

200 e^{-5t}
volts

0.01 farad

20 ohms

Figure 3.14

The student should show that this value of t actually gives a maximum. The value of the maximum charge is the value of Q when $t = \frac{1}{5}$, i.e.,

$$Q = 10 \times \tfrac{1}{5}e^{-1} = \frac{2}{e} \sim 0.74 \text{ coulomb}$$

A EXERCISES

1. At $t = 0$ an emf of 20 volts is applied to a circuit consisting of an inductor of 2 henries in series with a 40 ohm resistor. If the current is zero at $t = 0$, what is it at any time $t \geq 0$?

2. Work the previous exercise if the emf is 100 sin 10t.

3. A 20 ohm resistor and 5 henry inductor are in series in an electric circuit in which there is a current flow of 20 amp at time $t = 0$. Find the current for $t \geq 0$ if the emf is zero for $t > 0$.

4. A capacitor of 5×10^{-3} farads is in series with a 25 ohm resistor and an emf of 50 volts. The switch is closed at $t = 0$. Assuming that the charge on the capacitor is zero at $t = 0$, determine the charge and current at any time.

5. Work the previous exercise if the emf is 50 cos 6t, $t \geq 0$.

6. A circuit consists of a 10 ohm resistor and a 0.01 farad capacitor in series. The charge on the capacitor is 0.05 coulomb. Find the charge and the current at time t after the switch is closed.

7. A 4 ohm resistor and an inductor of 1 henry are connected in series with a voltage given by $100e^{-4t} \cos 50t$, $t \geq 0$. Find $I(t)$ if $I = 0$ at $t = 0$.

8. A 20 ohm resistor is connected in series with a capacitor of 0.01 farad and an emf in volts given by $40e^{-3t} + 20e^{-6t}$. If $Q = 0$ at $t = 0$, show that the maximum charge on the capacitor is 0.25 coulomb.

B EXERCISES

1. A circuit consists of a constant resistance R ohms in series with a constant emf E volts and a constant inductance L henries. If the initial current is zero, show that the current builds up to half its theoretical maximum in $(L \ln 2)/R$ sec.

2. An emf of $E_0 \cos \omega t$ volts, where E_0, ω are constants, is applied at $t = 0$ to a series circuit consisting of R ohms and C farads, where R and C are constants. If $Q = 0$ at $t = 0$, show that the charge at $t > 0$ is

$$Q = \frac{CE_0}{R^2 C^2 \omega^2 + 1}(\cos \omega t + \omega RC \sin \omega t - e^{-t/RC})$$

3. A resistance of R ohms varies with time t (seconds) according to $R = 1 + 0.01t, 0 \leq t \leq 1000$. It is connected in series with a 0.1 farad capacitor and 100 volt emf. The initial charge on the capacitor is 5 coulombs. Find (a) the charge and current as a function of time, (b) the theoretically maximum charge.

4. An inductor of L henries varies with time t (seconds) according to $L = 0.05 + 0.001t, 0 \leq t \leq 1000$. It is connected in series with a 40 volt emf and a 10 ohm resistor. If $I = 0$ at $t = 0$, find (a) $I(t)$, $t > 0$; (b) the theoretically maximum current.

5. A circuit has R ohms, C farads, and E volts in series with a switch, R, C, E being constants. The initial charge on the capacitor is zero. If the switch is closed until the charge is 99 per cent of its theoretical maximum and E is then suddenly reduced to zero, find Q thereafter.

1. An inductor of 0.1 henry, a resistor of 10 ohms, and an emf $E(t)$ volts, where

$$E(t) = \begin{cases} 10, & 0 < t \leqq 5 \\ 0, & t > 5 \end{cases}$$

are connected in series. Find the current $I(t)$, assuming $I(0) = 0$.

2. A periodic emf $E(t)$ shown graphically in Fig. 3.15 is applied at $t = 0$ to a circuit consisting of R ohms and C farads in series, where R and C are constants. Find the charge when $t = 4T$, assuming that at $t = 0$ the charge is zero.

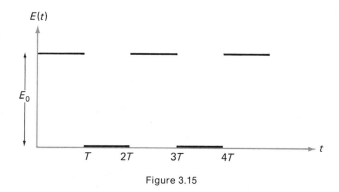

Figure 3.15

3. An emf $E_0 \cos \omega t$, $t \geq 0$ is applied to a circuit consisting of L henries and R ohms in series. The quantities E_0, ω, L, R are given constants. If the current is zero at $t = 0$; (a) Show that the current is composed of *transient terms* which become negligible after some time and *steady-state terms* which have the same period as the applied emf. (b) Show that the root-mean-square current (rms) defined by

$$\sqrt{\frac{\int_a^b I^2 \, dt}{b - a}}$$

where $b - a$ is the period $2\pi/\omega$ of the emf, approaches $\dfrac{\sqrt{2}}{2} \cdot \dfrac{E_0}{\sqrt{R^2 + L^2\omega^2}}$

3

Orthogonal Trajectories and Their Applications

Suppose we are given a family of curves as in Fig. 3.16 (heavy part). We may think of another family of curves (shown dashed) such that each member of this family cuts each member of the first family at right angles. For example, curve AB meets several members of the dashed family at right angles at the points L, M, N, O, and P. We say that the families are *mutually orthogonal*, or that either family forms a set of *orthogonal trajectories* of the other family. As an illustration, consider the family of all circles having center at the origin; a few such circles appear in Fig. 3.17. The orthogonal trajectories for this family of circles would be members of the family

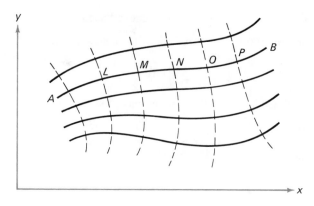

Figure 3.16

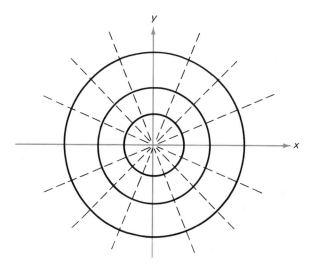

Figure 3.17

of straight lines (shown dashed). Similarly the orthogonal trajectories of the family of straight lines passing through the origin are the circles having center at the origin.

As a more complicated situation, consider the family of ellipses (Fig. 3.18) and the family of curves orthogonal to them. The curves of one family are the orthogonal trajectories of the other family. Applications of orthogonal trajectories in physics and engineering are numerous. As a very elementary application, consider Fig. 3.19. Here NS represents a bar magnet, N being its north pole, and S its south pole. If iron filings are sprinkled around the magnet we find that they arrange themselves like the dashed curves of Fig. 3.19. These curves are called *lines of force.** The curves perpendicular to these (shown heavy) are called *equipotential lines,* or *curves of equal*

---

* The student who has read the section on *direction fields* (page 28) should note the similarity between the iron filings and the lineal elements.

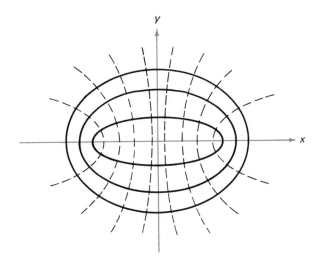

Figure 3.18

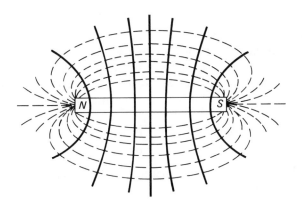

Figure 3.19

potential. Here, too, the members of one family constitute the orthogonal trajectories of the other family.

As another physical example consider Fig. 3.20, which represents a weather map so familiar in many of our daily newspapers. The curves represent *isobars*, which are curves connecting all cities which report the same barometric pressure to the weather bureau. The orthogonal trajectories of the family of isobars would indicate the general direction of the wind from high to low pressure areas. Instead of isobars, Fig. 3.20 could represent *isothermal curves* which are curves connecting points having the same temperature. In such case the orthogonal trajectories represent the general direction of heat flow.

Consider the example of isobars. Given any point (x, y), we may theoretically find the pressure at that point. Hence we may say that $P = f(x, y)$, i.e., pressure is a function of position. Letting P be a definite value, say P_1, we see that $f(x, y) = P_1$

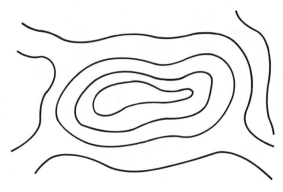

Figure 3.20

represents a curve, all points of which have pressure P_1, and thus is an isobar. By giving other values to P, other isobars are obtained. It is clear that these isobars could not intersect, for if they did, then the point of intersection would have two different pressures, and this would be impossible. If we use c instead of P we see that

$$f(x, y) = c \tag{1}$$

where c may take on values within a certain prescribed set, represents a family of isobars. In the first chapter we learned how to find the differential equation for a family of curves by differentiating so as to eliminate the arbitrary constants (c in our case). The problem which we face now is how to obtain the family of orthogonal trajectories. Actually this is simple because the differential equation of the family (1) is given by

$$df = \frac{\partial f}{\partial x}\, dx + \frac{\partial f}{\partial y}\, dy = 0 \quad \text{or} \quad \frac{dy}{dx} = -\frac{\partial f/\partial x}{\partial f/\partial y} \tag{2}$$

Now the slope of the orthogonal trajectories should be the negative reciprocal of the slope in (2), namely,

$$\frac{\partial f/\partial y}{\partial f/\partial x}$$

Thus, the differential equation for the family of orthogonal trajectories is

$$\frac{dy}{dx} = \frac{\partial f/\partial y}{\partial f/\partial x} \tag{3}$$

Upon solving this equation, the orthogonal trajectories are obtained. Illustrations of the procedure will now be given. A common mistake of students is to forget to eliminate the arbitrary constant in finding the differential equation of a family.

ILLUSTRATIVE EXAMPLE 1

Find the orthogonal trajectories of $x^2 + y^2 = cx$.

Mathematical Formulation. There are two ways for determining the differential equation of the family.

First way. Solve for c to obtain $c = (x^2 + y^2)/x$. Differentiating with respect to x, we have

$$\frac{x(2x + 2yy') - (x^2 + y^2)(1)}{x^2} = 0 \quad \text{or} \quad y' = \frac{dy}{dx} = \frac{y^2 - x^2}{2xy}$$

Second way. Differentiating $x^2 + y^2 = cx$ with respect to x, we find

$$2x + 2yy' = c$$

Eliminating c between this last equation and the given one, we find the same equation as before.

The family of orthogonal trajectories thus has the differential equation

$$\frac{dy}{dx} = \frac{2xy}{x^2 - y^2} \tag{4}$$

Solution To solve (4), note that it is a homogeneous equation. Letting $y = vx$, the student may show that $x^2 + y^2 = c_1 y$. The solution is also obtained by noting that (4), cleared of fractions, has an integrating factor depending on one variable $(1/y^2)$.

The two orthogonal families are shown in Fig. 3.21. The originally given family is shown dashed.

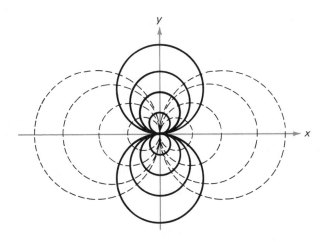

Figure 3.21

ILLUSTRATIVE EXAMPLE 2

Find the orthogonal trajectories of the family $y = x + ce^{-x}$ and determine that particular member of each family which passes through $(0, 3)$.

Mathematical Formulation. By differentiation of the given relation we have

$$y' = 1 - ce^{-x}$$

Elimination of c yields

$$y' = 1 + x - y$$

Thus, the differential equation for the family of orthogonal trajectories is

$$\frac{dy}{dx} = \frac{-1}{1 + x - y} \tag{5}$$

Solution   Write (5) in the form $dx + (1 + x - y)dy = 0$.

Here $\qquad\qquad M = 1, N = 1 + x - y, \dfrac{\partial M}{\partial y} = 0, \dfrac{\partial N}{\partial x} = 1$

so the equation is not exact. Now $\dfrac{0 - 1}{1 + x - y}$ is not a function of x alone. But

$\dfrac{1 - 0}{1} = 1$ is a function of y. Hence, $e^{\int 1 \, dy} = e^y$ is an integrating factor. Multiplying

by this factor and proceeding as usual for exact equations, we obtain

$$xe^y - e^y(y - 2) = c_1$$

The required curves passing through (0, 3) are found to be

$$y = x + 3e^{-x}, \qquad x - y + 2 + e^{3-y} = 0 \tag{6}$$

The student may find it instructive to obtain the graphs of equations (6). He may
also find it instructive to solve (5) by writing it as

$$\frac{dx}{dy} + x = y - 1$$

a linear equation with x as dependent variable.

A EXERCISES

1. The equation $y^2 = cx$ defines a family of parabolas. (a) Find a differential equation for the
family. (b) Find a differential equation for the orthogonal trajectories and solve. (c) Graph
several members of both families on the same set of axes.

2. Find the orthogonal trajectories of the family $y^3 = cx^2$ and draw a graph of the families.

3. Determine the orthogonal trajectories of each family and find particular members of each
passing through the indicated points.

   (a) $x^2 + cy^2 = 1$; (2, 1).   (b) $x^2 = cy + y^2$; (3, −1).   (c) $y = c \tan 2x + 1$; $\left(\dfrac{\pi}{8}, 0\right)$.

   (d) $y = ce^{-2x} + 3x$; (0, 3)   (e) $y^2 = c(1 + x^2)$; (−2, 5).

4. Show that the families $x^2 + 4y^2 = c_1$ and $y = c_2 x^4$ are orthogonal.

B EXERCISES

1. Find the constant a so that the families $y^3 = c_1 x$ and $x^2 + ay^2 = c_2$ are orthogonal.

2. Show that the family of parabolas $y^2 = 4cx + 4c^2$ is "self-orthogonal." Graph some members.

3. Show that the family $\qquad \dfrac{x^2}{a^2 + c} + \dfrac{y^2}{b^2 + c} = 1$

where a and b are given constants, is "self-orthogonal." This is called a family of "confocal conics." Graph some members of the family.

4. Determine the orthogonal trajectories of (a) $x^p + cy^p = 1$; $p = $ constant, (b) $x^2 + cxy + y^2 = 1$.

C EXERCISES

1. Determine the family of curves each member of which cuts each member of the family of straight lines $y = mx$ at an angle of $45°$.

2. Determine the curve passing through $(\frac{1}{2}, \sqrt{3/2})$ which cuts each member of the family $x^2 + y^2 = c^2$ at an angle of $60°$.

3. Find all curves cutting the family $y = ce^x$ at constant angle α.

4. Show that if a differential equation of a family of curves in polar coordinates (r, ϕ) is given by

$$\frac{dr}{d\phi} = F(r, \phi)$$

then a differential equation for the family of orthogonal trajectories is $\dfrac{dr}{d\phi} = -\dfrac{r^2}{F(r, \phi)}$.

(*Hint:* Use the fact from elementary calculus that in polar coordinates the tangent of the angle formed by the radius vector and the tangent line to a curve is $r\, d\phi/dr$.)

5. Find the orthogonal trajectories of $r = c \cos \phi$ and graph.

6. Determine the orthogonal trajectories of the spirals $r = e^{\alpha\phi}$.

7. Find the orthogonal trajectories of the cardioids $r = c(1 - \cos \phi)$.

8. Let $F(z) = u(x, y) + iv(x, y)$ where $z = x + iy$ and the functions $u(x, y)$, $v(x, y)$ are real. (a) Find $u(x, y)$ and $v(x, y)$ corresponding to $F(z) = z^2$ and show that the families $u(x, y) = c_1$ and $v(x, y) = c_2$ are orthogonal. (b) Work part (a) for the functions $F(z) = z^3$ and $F(z) = 2z^2 - iz - 3$. (c) Do you think the results indicated by these special cases hold in general? Explain (compare C Exercise 4, page 48).

4 Applications to Chemistry and Chemical Mixtures

There are many applications of differential equations to chemical processes. Some of these will be indicated in the following illustrative examples. Others are presented in the exercises.

ILLUSTRATIVE EXAMPLE 1

A tank is filled with 10 gallons (abbreviation *gal*) of brine in which is dissolved 5 lb of salt. Brine containing 3 lb of salt per gallon enters the tank at 2 gal per minute, and the well-stirred mixture leaves at the same rate.
    (a) Find the amount of salt in the tank at any time.
    (b) How much salt is present after 10 min?
    (c) How much salt is present after a long time?

Mathematical Formulation. Let A be the number of pounds of salt in the tank after t minutes. Then dA/dt is the time rate of change of this amount of salt given by

$$\frac{dA}{dt} = \text{rate of amount gained} - \text{rate of amount lost} \qquad (1)$$

Since 2 gal/min enter and there are 3 lb/gal salt we have as the amount of salt entering per minute

$$\frac{2 \text{ gal}}{\text{min}} \cdot \frac{3 \text{ lb}}{\text{gal}} = \frac{6 \text{ lb}}{\text{min}} \qquad (2)$$

which is the rate at which salt is gained. Since there are always 10 gal in the tank and since there are A pounds of salt at any time t, the concentration of salt at time t is A pounds per 10 gal. The amount of salt leaving per minute is, therefore,

$$\frac{A \text{ lb}}{10 \text{ gal}} \cdot \frac{2 \text{ gal}}{\text{min}} = \frac{2A \text{ lb}}{10 \text{ min}} = \frac{A \text{ lb}}{5 \text{ min}} \qquad (3)$$

From (1), (2), and (3) we have $\qquad \dfrac{dA}{dt} = 6 - \dfrac{A}{5}$

Since initially there are 5 lb of salt, we have $A = 5$ at $t = 0$. Thus, the complete mathematical formulation is

$$\frac{dA}{dt} = 6 - \frac{A}{5}, \qquad A = 5 \quad \text{at} \quad t = 0$$

Solution   Using the method of separation of variables, we have

$$\int \frac{dA}{30 - A} = \int \frac{dt}{5} \quad \text{or} \quad -\ln(30 - A) = \frac{t}{5} + c$$

Since $A = 5$ at $t = 0$, $c = -\ln 25$. Thus,

$$-\ln(30 - A) = \frac{t}{5} - \ln 25, \qquad \ln\frac{30 - A}{25} = -\frac{t}{5} \quad \text{or} \quad A = 30 - 25e^{-t/5}$$

which is the amount of salt in the tank at any time t.

At the end of 10 minutes the amount of salt is $A = 30 - 25e^{-2} = 26.6$ lb.

After a long time, i.e., as $t \to \infty$, we see that $A \to 30$ lb. This could also be seen from the differential equation by letting $dA/dt = 0$, since A is constant when equilibrium is reached.

ILLUSTRATIVE EXAMPLE 2

Two chemicals, A and B, react to form another chemical C. It is found that the rate at which C is formed varies as the product of the instantaneous amounts of chemicals A and B present. The formation requires 2 lb of A for each pound of B. If 10 lb of A and 20 lb of B are present initially, and if 6 lb of C are formed in 20 min, find the amount of chemical C at any time.

Mathematical Formulation. Let x pounds be the amount of C formed in time t hours. Then dx/dt is the rate of its formation. To form x pounds of C, we need

$2x/3$ lb of A and $x/3$ lb of B, since twice as much of chemical A as B is needed. Thus, the amount of A present at time t when x pounds of C is formed is $10 - 2x/3$, and the amount of B at this time is $20 - x/3$. Hence,

$$\frac{dx}{dt} \propto \left(10 - \frac{2x}{3}\right)\left(20 - \frac{x}{3}\right) \quad \text{or} \quad \frac{dx}{dt} = K\left(10 - \frac{2x}{3}\right)\left(20 - \frac{x}{3}\right)$$

where K is the constant of proportionality. This equation may be written

$$\frac{dx}{dt} = k(15 - x)(60 - x)$$

where k is another constant. There are two conditions. Since no chemical C is present initially, we have $x = 0$ at $t = 0$. Also $x = 6$ at $t = \frac{1}{3}$. We need two conditions, one to determine k, the other to determine the arbitrary constant from the solution of the differential equation. The complete formulation is

$$\frac{dx}{dt} = k(15 - x)(60 - x), \qquad x = 0 \quad \text{at} \quad t = 0, \qquad x = 6 \quad \text{at} \quad t = \frac{1}{3}$$

Solution   Separation of variables yields

$$\int \frac{dx}{(15 - x)(60 - x)} = \int k\, dt = kt + c_1$$

Now $\displaystyle \int \frac{dx}{(15 - x)(60 - x)} = \int \frac{1}{45}\left(\frac{1}{15 - x} - \frac{1}{60 - x}\right) dx = \frac{1}{45} \ln\left(\frac{60 - x}{15 - x}\right).$

Thus, we may show that $\displaystyle \frac{60 - x}{15 - x} = ce^{45kt}$

Since $x = 0$ at $t = 0$, we find $c = 4$. Thus, $\displaystyle \frac{60 - x}{15 - x} = 4e^{45kt}$

Since $x = 6$ at $t = \frac{1}{3}$, we have $e^{15k} = \frac{3}{2}$. Thus,

$$\frac{60 - x}{15 - x} = 4(e^{15k})^{3t} = 4\left(\frac{3}{2}\right)^{3t} \quad \text{or} \quad x = \frac{15[1 - (\frac{2}{3})^{3t}]}{1 - \frac{1}{4}(\frac{2}{3})^{3t}}$$

As $t \to \infty$, $x \to 15$ lb.

The preceding problem is a special case of the *law of mass action*, which is fundamental in the theory of rates of chemical reactions. For problems involving description and application of this law see the C Exercises.

A EXERCISES

1. A tank is filled with 8 gal of brine in which 2 lb of salt is dissolved. Brine having 3 lb of salt per gallon enters the tank at 4 gal per minute, and the well-stirred mixture leaves at the same rate. (a) Set up a differential equation for the amount of salt at time t. (b) Find the amount of salt as a function of time. (c) Find the concentration of salt after 8 min. (d) How much salt is there after a long time?

2. A tank has 40 gal of pure water. A salt water solution with 1 lb of salt per gallon enters at 2 gal per minute, and the well-stirred mixture leaves at the same rate. (a) How much salt is in the tank at any time? (b) When will the water leaving have $\frac{1}{2}$ lb of salt per gallon?

3. A tank has 60 gal of salt water with 2 lb of salt per gallon. A solution with 3 lb of salt per gallon enters at 2 gal per minute, and the mixture leaves at the same rate. When will 150 lb of salt be in the tank?

4. A tank has 100 gal brine with 40 lb of dissolved salt. Pure water enters at 2 gal per minute and leaves at the same rate. When will the salt concentration be 0.2 lb/gal? When will the concentration be less than 0.01 lb/gal?

5. Chemical A is transformed into chemical B. The rate at which B is formed varies directly as the amount of A present at any instant. If 10 lb of A is present initially and if 3 lb is transformed into B in 1 hr: (a) How much of A is transformed after 2, 3, and 4 hr? (b) In what time is 75% of chemical A transformed? (This type of reaction is called a *first-order reaction*.)

6. Chemical C is produced from a reaction involving chemicals A and B. The rate of production of C varies as the product of the instantaneous amounts of A and B present. The formation requires 3 lb of A for every 2 lb of B. If 60 lb each of A and B are present initially and 15 lb of C are formed in 1 hr find: (a) the amount of C at any time; (b) the amount of C after 2 hr; (c) the maximum quantity of C which can be formed.

B EXERCISES

1. A tank has 10 gal brine having 2 lb of dissolved salt. Brine with 1.5 lb of salt per gallon enters at 3 gal/min, and the well-stirred mixture leaves at 4 gal/min. (a) Find the amount of salt in the tank at any time. (b) Find the concentration of salt after 10 min. (c) Draw graphs of amount and concentration of salt versus time and give maxima in each case.

2. A tank has 60 gal of pure water. A salt solution with 3 lb of salt per gallon enters at 2 gal/min and leaves at 2.5 gal/min. (a) Find the concentration of salt in the tank at any time. (b) Find the salt concentration when the tank has 30 gal of salt water. (c) Find the amount of water in the tank when the concentration is greatest. (d) Determine the maximum amount of salt present at any time.

3. A chemical C is to be dissolved in water. Experimentally, the rate at which C enters into solution varies as the product of (i) the instantaneous amount of C which remains undissolved, (ii) the difference between the instantaneous concentration of the dissolved chemical and the maximum concentration possible at the given conditions of temperature and pressure (this maximum occurs when the solution is *saturated* and further increase of the chemical dissolved is not possible).

If 5 lb of C are placed in 2 gal of water, it is found that 1 lb dissolves in 1 hr. Assuming a saturated solution to have a concentration of 4 lb/gal: (a) How much of chemical C remains undissolved after 4 hr? (b) What is the concentration of the solution after 3 hr? (c) When will the concentration be 2 lb/gal?

4. When a pounds of a chemical are placed in b gallons of water, A pounds of the chemical dissolve in time T. If a saturated solution contains S pounds of the chemical per gallon (S is constant), show that the amount of the chemical undissolved at time t is

$$\frac{aR(bS - a)}{bS - aR} \quad \text{where} \quad R = \left[\frac{AbS}{a(A + bS - a)}\right]^{t/T}$$

Show that if $S \to \infty$ the amount of the chemical which is undissolved approaches $a[A/a]^{t/T}$.

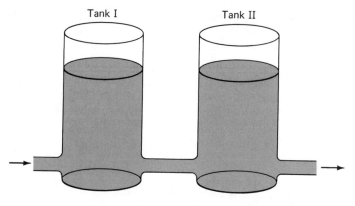

Tank I   Tank II

Figure 3.22

5. Two tanks (Fig. 3.22) each contain v gallons of water. Starting at time $t = 0$, a solution containing a lb/gal of a chemical solvent flows into tank I at the rate of b gal/min. The mixture then enters and leaves tank II at the same rate. Assuming thorough stirring in both tanks, show that the amount of the chemical in tank II after time $t > 0$ is $av(1 - e^{-bt/v}) - abte^{-bt/v}$.

C EXERCISES

Velocity of Chemical Reactions and the Law of Mass Action

A chemical equation describes how molecules of various substances combine to give other

substances. For example, $\qquad$ $2\,H_2 + O_2 \rightarrow 2\,H_2O$

states that 2 molecules of hydrogen combine with 1 molecule oxygen to yield 2 molecules water. In general, a chemical equation reads

$$aA + bB + cC + \cdots \rightarrow mM + nN + pP + \cdots$$

where A, B, C, ... represent molecules of the reacting substances, M, N, P, ... represent molecules of the substances formed by the reaction, and $a, b, c, \ldots, m, n, p$ are positive integers signifying the number of molecules taking part in the reaction. The rate at which a substance is formed is called the *velocity of the reaction*. Although no general rule applies in all cases, the law of mass action may apply in the determination of this rate.

> LAW OF MASS ACTION. *If the temperature is kept constant, the velocity of a chemical reaction is proportional to the product of the concentrations of the substances which are reacting.*

If we let $[A]$, $[B]$, $[C]$, ... denote the concentrations of A, B, C, ... at time t, these concentrations being expressed in moles per liter,* and if x is the number of moles per liter which have reacted after time t, the rate dx/dt of the reaction is given by

$$\frac{dx}{dt} = k[A]^a[B]^b[C]^c \cdots$$

* A mole is the molecular weight of a substance in grams, taking the atomic weight of oxygen as 16 g; hydrogen, H, as 1.008 g; carbon, C, as 12.01 g, etc. Thus, 1 mole $O_2 = 2(16) = 32$ g; 1 mole $H_2O = 16 + 2(1.008) = 18.016$ g; 1 mole $C_2H_6 = 2(12.01) + 6(1.008) = 30.068$ g.

where the constant of proportionality k is called the *velocity constant*. The *order* of the reaction is the sum of the exponents. The reactions are called unimolecular, bimolecular, trimolecular, etc., according as the order is one, two, three, etc.

1. In the bimolecular reaction $A + B \rightarrow M$, α moles per liter of A and β moles per liter of B are combined. If x denotes the number of moles per liter which have reacted after time t, the rate of reaction is given by

$$\frac{dx}{dt} = k(\alpha - x)(\beta - x)$$

(a) Show that if $\alpha \neq \beta$,

$$x = \frac{\alpha\beta[1 - e^{(\beta-\alpha)kt}]}{\alpha - \beta e^{(\beta-\alpha)kt}}$$

and find $\lim_{t \to \infty} x$, considering the two cases $\alpha > \beta$, $\beta > \alpha$.

(b) If $\alpha = \beta$, show that $x = \alpha^2 kt/(1 + \alpha kt)$ and find $\lim_{t \to \infty} x$.

2. In the third-order or trimolecular reaction $A + B + C \rightarrow M + N$, α, β, and γ moles per liter of A, B, and C are combined. If x denotes the number of moles per liter of A, B, or C which have reacted after time t (or the number of moles per liter of M or N which have been formed), then the rate of reaction is given by

$$\frac{dx}{dt} = k(\alpha - x)(\beta - x)(\gamma - x)$$

Solve the equation, considering that $x = 0$ at $t = 0$ and treating the cases: (a) α, β, γ are equal; (b) α, β, γ are not equal; (c) only two of α, β, γ are equal.

3. Substances A and B react chemically to yield P and Q according to the second-order reaction $A + B \rightarrow P + Q$. At time $t = 0$, M_A grams of A and M_B grams of B are combined. If x grams react in time t and if the law of mass action applies: (a) Show that

$$\frac{dx}{dt} = K\left(M_A - \frac{m_A}{m_A + m_B}x\right)\left(M_B - \frac{m_B}{m_A + m_B}x\right)$$

where m_A and m_B are the molecular weights of A and B, respectively. (b) Show that the number of grams of P and Q formed after time t are $m_P s$ and $m_Q s$, where

$$s = \frac{1 - e^{rt}}{M_A m_B - M_B m_A e^{rt}}, \qquad r = \frac{K M_A M_B (M_B m_A - M_A m_B)(m_A + m_B)}{(M_A + m_B)^2 m_A m_B}$$

and where we assume that $M_B/m_B \neq M_A/m_A$. Note that $m_A + m_B = m_P + m_Q$. Derive a similar result for the case $M_B/m_B = M_A/m_A$.

4. The amount 260 g of $CH_3COOC_2H_5$ (ethyl acetate) is combined with 175 g of NaOH (sodium hydroxide) in a water solution to yield CH_3COONa (sodium acetate) and C_2H_5OH (ethyl alcohol), according to the equation

$$CH_3COOC_2H_5 + NaOH \rightarrow CH_3COONa + C_2H_5OH$$

At the end of 10 min, 60 g of sodium acetate have been formed. If the reaction is known to obey the law of mass action, (a) calculate the velocity constant; (b) find the number of grams of sodium acetate and ethyl alcohol present after $\frac{1}{2}$ hr. (*Hint:* use the atomic weights C = 12.01, H = 1.008, O = 16, Na = 22.997.)

5 Applications to Steady-State Heat Flow

Consider a slab of material of indefinite length bounded by two parallel planes *A* and *B*, as in Fig. 3.23. We assume the material uniform in all properties, e.g., specific heat, density, etc. Suppose planes *A* and *B* are kept at 50°C and 100°C, respectively. Every point in the region between *A* and *B* reaches some temperature and does not change thereafter. Thus, all points on plane *C* midway between *A* and *B* will be at 75°C; plane *E* at 90°C. When the temperature at each point of a body does not vary with time, we say that *steady-state conditions* prevail or that we have *steady-state heat flow*.

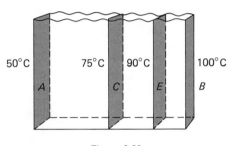

Figure 3.23

As another example consider a pipe of uniform material, the cross section of which is shown in Fig. 3.24. Suppose the outer surface kept at 80°C and the inner at 40°C. There will be a surface (shown dashed) each point of which will be at 60°C. This will not be midway between the inner and outer surfaces, however. Lines parallel to *A* and in a plane perpendicular to *A* (Fig. 3.23) are called *isothermal lines*. The dashed curve of Fig. 3.24 is an *isothermal curve*. The corresponding planes of Fig. 3.23 and cylinders of Fig. 3.24 are called *isothermal surfaces*. In the general case, isothermal curves will not be lines or circles, as in Fig. 3.23 or 3.24, but may be a

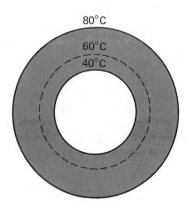

Figure 3.24

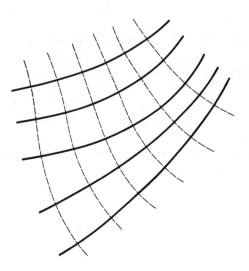

Figure 3.25

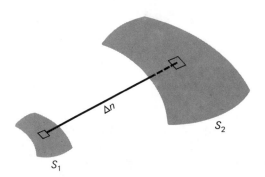

Figure 3.26

family of curves as shown in Fig. 3.25 (dashed curves). The orthogonal trajectories of the family are called *flow lines* (see Section 3 on orthogonal trajectories).

Consider small portions of two neighboring isothermal surfaces (Fig. 3.26) separated by a distance Δn. Assume that the temperature corresponding to surface S_1 is U_1, and that corresponding to S_2 is U_2. Call the temperature difference $U_2 - U_1 = \Delta U$. It is found experimentally that the amount of heat flowing from S_1 to S_2 per unit area per unit time is approximately proportional to $\Delta U/\Delta n$. The approximation becomes more accurate as Δn (and hence ΔU) gets smaller. In the limiting case as $\Delta n \to 0$, $\Delta U/\Delta n \to dU/dn^*$ which is called the *gradient* of U (rate of change of U in the direction normal to the isothermal surface or curve). If H is the amount of heat flow per unit area per unit time, we take as our physical law:

$$H \propto \frac{dU}{dn} \quad \text{or} \quad H = K \frac{dU}{dn} \tag{1}$$

---

* In case U depends on other factors besides n, then dU/dn is to be replaced by $\partial U/\partial n$.

If we wish to consider H a vector quantity (having direction as well as magnitude), we reason as follows. Consider as positive the direction from S_1 to S_2. If dU/dn is positive, then U is increasing and, hence, we must have $U_2 > U_1$. Thus, heat actually flows from S_2 to S_1 (higher to lower temperature); i.e., heat flow is in the negative direction. Similarly, if dU/dn is negative, U is decreasing, $U_2 < U_1$, and the flow is from S_1 to S_2; i.e., heat flow is in the positive direction. The direction of heat flow can be taken into account by a minus sign in (1), i.e.,

$$H \text{ (vector quantity)} = -K\frac{dU}{dn} \tag{2}$$

The amount of heat per unit time flowing across an area A is given by

$$q = -KA\frac{dU}{dn} \tag{3}$$

The constant of proportionality K, used above, depends on the material used and is called the *thermal conductivity*. The quantity of heat is expressed in *calories* in the cgs system, and in *British thermal units*, *Btu*, in the fps system. Consider now an illustration using the above principles.

ILLUSTRATIVE EXAMPLE

A long steel pipe, of thermal conductivity $K = 0.15$ cgs units, has an inner radius of 10 cm and an outer radius of 20 cm. The inner surface is kept at 200°C and the outer surface is kept at 50°C. (a) Find the temperature as a function of distance r from the common axis of the concentric cylinders. (b) Find the temperature when $r = 15$ cm. (c) How much heat is lost per minute in a portion of the pipe which is 20 m long?

Mathematical Formulation. It is clear that the isothermal surfaces are cylinders concentric with the given ones. The area of such a surface having radius r and length l is $2\pi rl$. The distance dn is dr in this case. Thus, equation (3) can be written

$$q = -K(2\pi rl)\frac{dU}{dr} \tag{4}$$

Since $K = 0.15$, $l = 20$ m $= 2000$ cm, we have

$$q = -600\pi r\frac{dU}{dr} \tag{5}$$

In this equation, q is of course a constant. The conditions are

$$U = 200°C \quad \text{at} \quad r = 10, \qquad U = 50°C \quad \text{at} \quad r = 20 \tag{6}$$

Solution Separating the variables in (5) and integrating yields

$$-600\pi U = q \ln r + c \tag{7}$$

Using the conditions (6), we have $-600\pi(200) = q \ln 10 + c$, $-600\pi(50) = q \ln 20 + c$ from which we obtain $q = 408,000$, $c = -1,317,000$. Hence, from (7) we find

$$U = 699 - 216 \ln r \tag{8}$$

If $r = 15$, we find by substitution that $U = 114°C$. From the value of q above, which is in calories per second, it is clear that the answer to part (c) is

$$q = 408,000 \times 60 \text{ cal/min} = 24,480,000 \text{ cal/min}$$

In the case of heat conduction, the heat flows from places of higher temperature to places of lower temperature. Physically, when one end of a bar is increased in temperature, the random motion of molecules at this end is increased with resultant increase in speed and number of collisions between molecules. The higher-speed molecules tend to move toward the other end of the bar resulting in further collisions and a consequent gradual increase in temperature of the remainder of the bar.

We can think of heat conduction as a spreading or *diffusion* of molecules. Such diffusion, however, is not limited to heat conduction. Thus, for example, if a bar is made of a porous material and we coat one end with a chemical, we find after some time that the chemical spreads or diffuses into the bar. Just as heat flows from places of higher temperature to lower temperature, so substances tend to diffuse from places of higher concentration or density to lower concentration. The analogy allows us to use the mathematical formulation provided above for heat conduction in problems of diffusion, so that U in such problems represents the concentration or density (in g/cm^3, lb/ft^3, etc.) rather than temperature. Problems of diffusion occur not only in chemistry but in biology, as in the transport of substances across cell membranes (a phenomenon often called *osmosis*), or in atomic physics, as in the diffusion of neutrons for atomic energy.

A EXERCISES

1. A slab of metal has a thickness of 200 cm, and its two other dimensions are very much larger. The plane faces of the slab are maintained at 75°C and 25°C, respectively. (a) Find the steady-state temperature in a plane at distance x from the 75°C face. (b) Construct a graph showing the steady-state temperature distribution. (c) How much heat is crossing a square centimeter of the plane each second, assuming that the thermal conductivity of the plane is 0.15 cgs units?

2. Work Exercise 1 if the slab is of thickness L, has a thermal conductivity K, and the temperatures of the plane faces are U_0 and U_1, respectively.

3. A long steel pipe, of thermal conductivity $K = 0.15$ cgs units, has an inner radius of 20 cm and outer radius of 30 cm. The outer surface is kept at 400°C, and the inner surface is kept at 100°C. (a) Find the temperature as a function of distance r from the common axis of the concentric cylinders. (b) Find the temperature where $r = 25$ cm. (c) How much heat is lost per minute in a portion of the pipe 10 m long?

4. A long hollow pipe has an inner diameter of 10 cm and outer diameter of 20 cm. The inner surface is kept at 200°C, and the outer surface is kept at 50°C. The thermal conductivity is 0.12 cgs units. (a) Find the temperature as a function of distance r from the common axis of the concentric cylinders. (b) Find the temperature where $r = 7.5$ cm. (c) How much heat is lost per minute in a portion of the pipe 20 m long?

5. A slab of metal 500 cm thick and with its other two dimensions very much larger has one face kept at 25°C and the opposite face at 65°C. If the thermal conductivity is 0.005 cgs units, (a) find the temperature in the slab as a function of distance x from the face having the lower temperature. (b) How much heat is transmitted per hour through a cross section parallel to the faces and having an area of 100 cm^2?

6. A cylinder of length L is composed of a porous material through which a certain chemical is capable of diffusing. One end of the cylinder is coated with the chemical at a concentration C_0 (g/cm³) and supply of the chemical at this concentration is maintained. The other end of the cylinder is placed so that it is just in contact with water, and as fast as the chemical diffuses to this end, it dissolves in the water. Assuming that the convex surface of the cylinder is coated with a substance through which the chemical cannot diffuse, (a) find the steady-state concentration of the chemical at any point x from the supply, and (b) construct its graph.

7. Describe the analogy of Exercise 6 to Exercises 1 and 2. Is there an analogy in the diffusion problem to part (c) involving thermal conductivity? Explain.

B EXERCISES

1. Show that the differential equation for steady-state temperature in a region bounded by two long coaxial cylinders is

$$\frac{d}{dr}\left(r\frac{dU}{dr}\right) = 0 \quad \text{or} \quad \frac{d^2U}{dr^2} + \frac{1}{r}\cdot\frac{dU}{dr} = 0$$

where U is the temperature at distance r from the common axis.

2. Show that the temperature U at distance r from the common axis of two coaxial cylinders of radii a and b $(a < b)$ kept at constant temperatures U_1 and U_2 respectively, is

$$U = \frac{U_1 \ln b/r + U_2 \ln r/a}{\ln b/a}$$

3. Show that the differential equation for steady-state temperature in a region bounded by two concentric spheres is

$$\frac{d}{dr}\left(r^2\frac{dU}{dr}\right) = 0 \quad \text{or} \quad \frac{d^2U}{dr^2} + \frac{2}{r}\cdot\frac{dU}{dr} = 0$$

where U is the temperature at distance r from the common center.

4. Show that the temperature U at distance r from the common center of two concentric spheres of radii a and b $(a < b)$ kept at constant temperatures U_1 and U_2, respectively, is

$$U = \frac{bU_2 - aU_1}{b - a} + \frac{ab(U_1 - U_2)}{(b - a)r}$$

5. Give an interpretation involving diffusion to (a) Exercises 1 and 2; (b) Exercises 3 and 4.

C EXERCISES

1. A long cylindrical pipe of length L containing steam at temperature U_1 has a radius equal to a. It is covered by an insulating material of thickness b and thermal conductivity K. Show that if the temperature outside of the insulation is U_2, then the heat loss is given by

$$\frac{2\pi LK(U_1 - U_2)}{\ln (1 + b/a)}$$

2. Suppose that a pipe has two layers of insulating material of equal thickness, but of different thermal conductivities. Prove that there is less heat loss when the outer layer has the larger thermal conductivity.

3. A slab of width L (whose other dimensions can be considered infinite) is composed of a radioactive material. Because of the reactions which take place in this material, heat is generated

which continuously flows outward from the faces of the slab. (a) Taking the x axis perpendicular to the faces so that the faces are located at $x = 0$ and $x = L$, respectively, show that the steady-state temperature in the slab is described by the equation

$$\frac{d^2 U}{dx^2} = -Q$$

where Q is a positive constant which depends on the rate at which heat is generated and the physical constants of the material. (b) Assuming that the temperatures of the faces are maintained at $0°C$, show that the temperature at the middle plane of the slab is $\frac{1}{8}QL^2$.

4. Work a problem similar to that of Exercise 3 involving radioactive material contained between two concentric cylinders of radii a and b, respectively.

6 Applications to Miscellaneous Problems of Growth and Decay

The differential equation $\qquad \dfrac{dy}{dt} = ay \qquad$ (1)

states that the time rate of change of a quantity y is proportional to y. If the constant of proportionality a is positive and y is positive, then dy/dt is positive and y is increasing. In this case we speak of y growing, and the problem is one of growth. On the other hand, if a is negative and y is positive, then dy/dt is negative and y is decreasing. Here the problem is one involving decay.

Since the solution of (1) is given by the exponential function $y = ce^{at}$ we often refer to (1) as the *law of exponential growth* if $a > 0$ and the *law of exponential decay* if $a < 0$. Equations very similar to (1) arise from many fields seemingly unrelated. In the following we consider two illustrative examples, one involving temperature, the other the phenomenon of radioactive disintegration.

ILLUSTRATIVE EXAMPLE 1

Water is heated to the boiling point temperature $100°C$. The water is then removed from the heat and kept in a room which is at a constant temperature of $60°C$. After 3 min the water temperature is $90°C$. (a) Find the water temperature after 6 min. (b) When will the water temperature be $75°C$? $61°C$?

Mathematical Formulation. Denote by U the temperature of the water t min after removal from the heat source. The temperature difference between water and room is $U - 60$. The time rate of change in U is dU/dt. On the basis of experience, one expects that temperature will change most rapidly when $(U - 60)$ is greatest and most slowly when $(U - 60)$ is small. Let us perform an experiment in which we take temperatures at various intervals of time, ΔU being the change in temperature and Δt the time to produce this change. By taking small Δt we expect that $\Delta U/\Delta t$ will be very close to dU/dt. If we plot $-\Delta U/\Delta t$ against $(U - 60)$, we could produce a graph similar to that in Fig. 3.27. The dots are points determined by experiment. Since the graph is a straight line, approximately, we assume that dU/dt is proportional

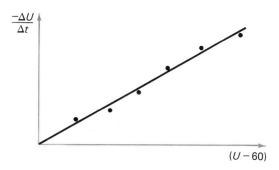

Figure 3.27

to $(U - 60)$, i.e.,
$$\frac{dU}{dt} = a(U - 60)$$

where a is a constant of proportionality. Now dU/dt is negative when $(U - 60)$ is positive, and so we shall write $a = -k$ where $k > 0$. The equation is

$$\frac{dU}{dt} = -k(U - 60)$$

This equation is known in physics as *Newton's law of cooling* and is of importance in many temperature problems. Actually, it is but an approximation of the true physical situation. The conditions which accompany this equation are provided from the facts that

$$U = 100°C \quad \text{where} \quad t = 0, \qquad U = 90°C \quad \text{where } t = 3 \text{ (minutes)}$$

Solution Solving the equation by separation of variables we have

$$\int \frac{dU}{U - 60} = \int -k \, dt, \qquad \ln (U - 60) = -kt + c_1 \quad \text{or} \quad U - 60 = ce^{-kt}$$

Where $t = 0$, $U = 100$, so that $c = 40$. Hence, $U - 60 = 40e^{-kt}$. Where $t = 3$, $U = 90$, so that $e^{-3k} = \frac{3}{4}$ or $e^{-k} = (\frac{3}{4})^{1/3}$. Hence,

$$U - 60 = 40(e^{-k})^t = 40(\tfrac{3}{4})^{t/3} \quad \text{i.e.,} \quad U = 60 + 40(\tfrac{3}{4})^{t/3} \tag{2}$$

Temperature after 6 minutes. Let $t = 6$ in (2), and obtain $U = 82.5°C$.

Times where temperature is 75°C, 61°C. If $U = 75°C$, then we find that $75 = 60 + 40(\tfrac{3}{4})^{t/3}$, $(\tfrac{3}{4})^{t/3} = \frac{3}{8}$, and $t = 10.2$. If $U = 61°C$, then $(\tfrac{3}{4})^{t/3} = \frac{1}{40}$ and $t = 38.5$.

Thus, it takes 10.2 min for the water at 100°C to drop in temperature to 75°C, and 38.5 min to drop in temperature from 100°C to 61°C.

By experimental methods similar to those indicated in the temperature problem we may arrive at the following:

> LAW OF RADIOACTIVE DISINTEGRATION. *The rate of disintegration of a radioactive substance is proportional, at any instant, to the amount of the substance which is present.*

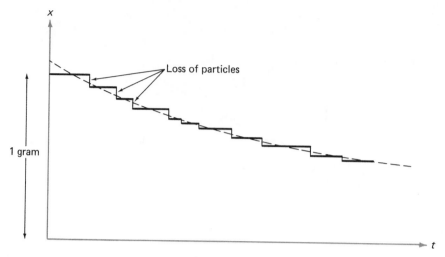

Figure 3.28

Before we formulate this law mathematically, let us consider the phenomenon of radioactivity in a little detail. When a radioactive element like radium or uranium disintegrates, it emits particles in a random fashion. Each of these particles has a definite mass, which is very small. If we start with a mass of 1 g of the radioactive material and consider what happens when particles are emitted, we find a situation similar to that shown in the graph of Fig. 3.28. Here, x is the amount of substance left after time t, assuming we start with 1 g at $t = 0$. Each time there is a drop in the value of x it means that particles have been emitted; the greater the drop, the larger the number of particles emitted. Thus, the quantity of radioactive substance is, in reality, a discontinuous function of t. What, then, is meant by dx/dt? To surmount this mathematical difficulty we approximate the actual graph by a smooth curve (dashed in Fig. 3.28). Thus, we are not much in error physically, and at the same time insure that we have a graph for which dx/dt will exist everywhere. Here we are forming a mathematical abstraction of a physical situation. The ideas presented here occur often in physics because of the finite size of even the smallest particle, in other words because of the atomic theory. Even in problems of electric circuits, the mathematical abstraction occurs (the student should try to see how). As a consequence one must always be alert in cases where these ideas are of importance. We now consider an example.

ILLUSTRATIVE EXAMPLE 2

It is found that 0.5 per cent of radium disappears in 12 years. (a) What percentage will disappear in 1000 years? (b) What is the half-life of radium?

Mathematical Formulation. Let A be the quantity of radium, in grams, present after t years. Then dA/dt (which exists by virtue of our mathematical abstraction) represents the rate of disintegration of radium. According to the law on page 107,

$$\frac{dA}{dt} \propto A \quad \text{or} \quad \frac{dA}{dt} = \alpha A$$

where α is a constant of proportionality. Since $A > 0$ and is decreasing, then $dA/dt < 0$ and we see that α must be negative. Writing $\alpha = -k$,

$$\frac{dA}{dt} = -kA$$

Let A_0 be the amount, in grams, of radium present initially. Then $0.005A_0$ gram disappears in 12 years, so that $0.995A_0$ gram remains. We thus have

$$A = A_0 \quad \text{at} \quad t = 0, \qquad A = 0.995A_0 \quad \text{at} \quad t = 12 \text{ (years)}$$

Solution Separating the variables, we have on integrating

$$\ln A = -kt + c_1 \quad \text{or} \quad A = ce^{-kt}$$

Since $A = A_0$ at $t = 0$, $c = A_0$. Hence, $A = A_0 e^{-kt}$. Since $A = 0.995A_0$ at $t = 12$,

$$0.995A_0 = A_0 e^{-12k}, \qquad e^{-12k} = 0.995, \qquad e^{-k} = (0.995)^{1/12} \tag{3}$$

Hence, $\qquad\qquad A = A_0 e^{-kt} = A_0 (e^{-k})^t = A_0 (0.995)^{t/12} \tag{4}$

or, if we solve for k in (3), we find $k = 0.000418$, so that

$$A = A_0 e^{-0.000418t} \tag{5}$$

Per cent disappearing in 1000 years. Where $t = 1000$ we have, from (4) or (5), $A = 0.658A_0$, so that 34.2 per cent disappears in 1000 years.

Half-life of radium. The half-life of a radioactive substance is defined as the time it takes for 50 per cent of the substance to disappear. Thus, for our problem, we wish to find the time when $A = \frac{1}{2}A_0$. Using (5) we find $e^{-0.000418t} = \frac{1}{2}$ or $t = 1660$ years, approximately.

A EXERCISES

1. Water at temperature 100°C cools in 10 min to 80°C in a room of temperature 25°C. (a) Find the temperature of the water after 20 min. (b) When is the temperature 40°C? 26°C?

2. Water at temperature 10°C takes 5 min to warm up to 20°C in a room at temperature 40°C. (a) Find the temperature after 20 min; after $\frac{1}{2}$ hr. (b) When will the temperature be 25°C?

3. If the half-life of radium is 1700 years, what percentage radium may be expected to remain after 50, 100, and 200 years?

4. Find the half-life of a radioactive substance if 25 per cent of it disappears in 10 years.

5. If 30 per cent of a radioactive substance disappears in 10 days, how long will it take for 90 per cent to disappear?

6. The electric charge, in coulombs, on a spherical surface leaks off at a rate proportional to the instantaneous charge. Initially, 5 coulombs are present, and one-third leaks off in 20 min. When will there be 1 coulomb remaining?

7. Bacteria in a certain culture increase at a rate proportional to the number present. If the original number increases by 50 per cent in $\frac{1}{2}$ hr, in how many hours can one expect three times the original number? Five times the original number?

8. A culture of "sick" bacteria grows at a rate which is inversely proportional to the square root of the number present. If there are 9 units present initially and 16 units present after 2 days, after how many days will there be 36 units?

9. In a certain solution there are 2 g of a chemical. After 1 hour there are 3 g of the chemical. If the rate of increase of the chemical is proportional to the square root of the time that it has been in solution, how many grams will there be after 4 hours?

10. Suppose that a substance decreases at a rate which is inversely proportional to the amount present. If 12 units of the substance are present initially and 8 units are present after 2 days, how long will it take the substance to disappear?

11. The maximum temperature which can be read on a certain thermometer is 110°F. When the thermometer reads 36°F, it is placed in an oven. After 1 and 2 minutes, respectively, it reads 60°F and 82°F. What is the temperature of the oven?

B EXERCISES

1. Neutrons in an atomic pile increase at a rate proportional to the number of neutrons present at any instant (due to nuclear fission). If N_0 neutrons are initially present, and N_1 and N_2 neutrons are present at times T_1 and T_2, respectively, show that

$$\left(\frac{N_2}{N_0}\right)^{T_1} = \left(\frac{N_1}{N_0}\right)^{T_2}$$

2. Uranium disintegrates at a rate proportional to the amount present at any instant. If M_1 and M_2 grams are present at times T_1 and T_2, respectively, show that the half-life is

$$\frac{(T_2 - T_1)\ln 2}{\ln (M_1/M_2)}$$

3. A man borrows $1000 subject to the condition that interest be compounded continuously. If the interest rate is 4 per cent per annum, how much will he have to pay back in 10 years? What is the equivalent simple interest rate?

4. Show that the time required, in years, for a sum of money to double itself if interest is compounded continuously at r per cent per annum is $(100 \ln 2)/r$.

5. In a certain culture, bacteria grow at a rate which is proportional to the pth power of the number present. If the number is N_0 initially, and if after time T_1 the number is N_1, show that the number is N_2 in a time given by

$$T_2 = \frac{N_2^{1-p} - N_0^{1-p}}{N_1^{1-p} - N_0^{1-p}}$$

Discuss the cases $p = 1$ and $p = 0$.

6. A radioactive isotope having a half-life of T minutes is produced in a nuclear reactor at the rate a grams per minute. Show that the number of grams of the isotope present after a long time is given by $aT/\ln 2$.

C EXERCISES

1. When light passes through a window glass some of it is absorbed. Experimentally, the amount of light absorbed by a small thickness of glass is proportional to the thickness of the glass and to the amount of incident light. Show that if r per cent of the light is absorbed by a thickness w, then the percentage of the light absorbed by a thickness nw is

$$100\left[1 - \left(1 - \frac{r}{100}\right)^n\right], \qquad 0 \leq r \leq 100$$

2. The amount of radioactive isotope C^{14} (carbon 14) present in all living organic matter bears a constant ratio to the amount of the stable isotope C^{12}. An analysis of fossil remains of a dinosaur shows that the ratio is only 6.24 per cent of that for living matter. Assuming that the half-life of C^{14} is approximately 5600 years, determine how long ago the dinosaur was alive. This method of *carbon dating* is often used by scientists, such as physicists, chemists, biologists, geologists and archaeologists who are interested in determining the age of various materials, such as rocks, ancient relics, paintings, etc.

7 The Hanging Cable

Let a cable or rope be hung from two points, A and B (Fig. 3.29), not necessarily at the same level. Assume that the cable is flexible so that due to loading (which may be due to the weight of the cable, to external forces acting, or to a combination of these) it takes a shape as in the figure. Let C be the lowest position on the cable, and choose x and y axes as in the figure, so that the y axis passes through C.

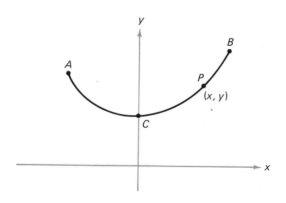

Figure 3.29

Consider that part of the cable between the minimum point C and any point P on the cable having coordinates (x, y). This part will be in equilibrium due to the tension T at P (Fig. 3.30), the horizontal force H at C, and the total vertical loading on the portion CP of the cable which we denote by $W(x)$ and which we assume acts at some point Q, not necessarily the center of arc CP. For equilibrium, the algebraic sum of the forces in the x (or horizontal) direction must equal zero, and the algebraic sum of the forces in the y (or vertical) direction must equal zero. Another way to say this is that the sum of forces to the right equals the sum of forces to the left, and the sum of forces in the upward direction equals the sum of forces in the downward direction.

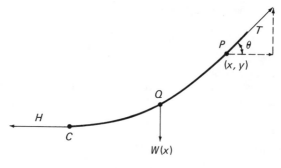

Figure 3.30

Let us resolve the tension T into two components (dashed in Fig. 3.30), the horizontal component having magnitude $T \cos \theta$, and the vertical component having magnitude $T \sin \theta$.

The forces in the x direction are H to the left and $T \cos \theta$ to the right, while the forces in the y direction are W down and $T \sin \theta$ up. Hence,

$$T \sin \theta = W, \qquad T \cos \theta = H \qquad (1)$$

Dividing these, using the fact that $\tan \theta = dy/dx = $ slope of tangent at P, we have

$$\frac{dy}{dx} = \frac{W}{H} \qquad (2)$$

In this equation, H is a constant, since it is the tension at the lowest point, but W may depend on x. By differentiation of (2) with respect to x,

$$\frac{d^2y}{dx^2} = \frac{1}{H} \cdot \frac{dW}{dx} \qquad (3)$$

Now dW/dx represents the increase in W per unit increase in x; i.e., it is the load per unit distance in the horizontal direction.

Equation (3) is fundamental. In the following illustrative examples and in the exercises we shall see that for different loads per unit horizontal distance we obtain various differential equations which yield various shapes of the cable.

ILLUSTRATIVE EXAMPLE 1

A flexible cable of small (negligible) weight supports a uniform bridge (Fig. 3.31). Determine the shape of the cable. (This is the problem of determining the shape of the cable in a *suspension bridge*, which is of great use in modern bridge construction.)

Mathematical Formulation. Equation (3) holds here and it remains only for us to determine dW/dx, the load per unit increase in the horizontal direction. In this case dW/dx is a constant, namely, the weight per unit length of the bridge. Calling this constant w, we have

$$\frac{d^2y}{dx^2} = \frac{w}{H} \qquad (4)$$

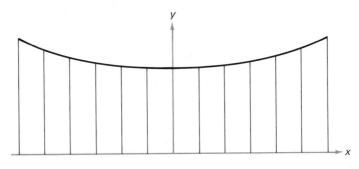

Figure 3.31

Denoting by b the distance of the lowest point of the cable from the bridge, we have

$$y = b \quad \text{where} \quad x = 0, \qquad \frac{dy}{dx} = 0 \quad \text{where} \quad x = 0$$

the second due to the fact that the point where $x = 0$ is a minimum point.

Integrating (4) twice, making use of the given conditions, we find that

$$y = \frac{wx^2}{2H} + b$$

The cable thus assumes the shape of a parabola.

A flexible rope having constant density hangs between two fixed points. Determine the shape of the rope.

Mathematical Formulation. Here, too, differential equation (3) holds and we have only to determine dW/dx. To do this, consider a portion of the rope (Fig. 3.32). This portion, supposedly infinitesimal, is magnified in the figure. The weight is distributed uniformly over the arc PQ in the figure. It is easy to see that if the density of the rope is w (a constant) then $dW/ds = w$. However, we want dW/dx. This is

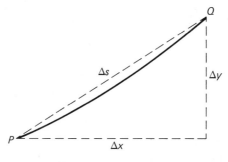

Figure 3.32

accomplished by writing

$$\frac{dW}{ds} = \frac{dW}{dx} \cdot \frac{dx}{ds} = w \quad \text{or} \quad \frac{dW}{dx} = w \frac{ds}{dx}$$

Since
$$\frac{ds}{dx} = \sqrt{1 + \left(\frac{dy}{dx}\right)^2}, \quad \frac{dW}{dx} = w \sqrt{1 + \left(\frac{dy}{dx}\right)^2}$$

Thus
$$\frac{d^2y}{dx^2} = \frac{w}{H} \sqrt{1 + \left(\frac{dy}{dx}\right)^2} \tag{5}$$

The same conditions prevail here as for Illustrative Example 1, namely, $y = b$ where $x = 0$, $dy/dx = 0$ where $x = 0$.

Solution   Equation (5) has both x and y missing. Letting $dy/dx = p$, we have

$$\frac{dp}{dx} = \frac{w}{H} \sqrt{1 + p^2}$$

and
$$\int \frac{dp}{\sqrt{p^2 + 1}} = \int \frac{w}{H} dx \quad \text{or} \quad \ln (p + \sqrt{p^2 + 1}) = \frac{wx}{H} + c_1$$

i.e.,
$$p + \sqrt{p^2 + 1} = c_2 e^{wx/H}$$

Since $p = dy/dx = 0$ where $x = 0$, we have $c_2 = 1$, so that

$$p + \sqrt{p^2 + 1} = e^{wx/H} \tag{6}$$

Solving for p by isolating the radical and squaring we have

$$p = \frac{dy}{dx} = \tfrac{1}{2}(e^{wx/H} - e^{-wx/H})$$

Hence, by integration,

$$y = \frac{H}{2w} (e^{wx/H} + e^{-wx/H}) + c_3 \tag{7}$$

By using $y = b$ where $x = 0$, we find $c_3 = b - H/w$. If we make the choice $b = H/w$, by suitably shifting the x axis, then $c_3 = 0$ and we have

$$y = \frac{H}{2w} (e^{wx/H} + e^{-wx/H}) \tag{8}$$

If we use the notation of hyperbolic functions, $\tfrac{1}{2}(e^{wx/H} + e^{-wx/H}) = \cosh wx/H$ and (8) may be written

$$y = \frac{H}{w} \cosh \frac{wx}{H} \tag{9}$$

The graph of (8) or (9) is called a *catenary* (from the Latin, meaning "chain").

A EXERCISES

1. A flexible cable of negligible weight supports a uniform bridge, as shown in Fig. 3.33. The dimensions are as indicated: P is the minimum point of curve APB. Using an appropriate set of axes, determine an equation for the curve APB.

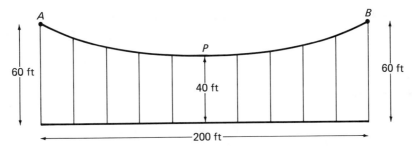

Figure 3.33

2. A cable of a suspension bridge has its supports at the same level, at a distance 500 ft apart. If the supports are 100 ft higher than the minimum point of the cable, use an appropriate set of axes to determine an equation for the curve in which the cable hangs, assuming the bridge is of uniform weight and that the weight of the cable is negligible. Find the slope of the cable at the supports.

3. The bridge of Fig. 3.33 has a variable weight of $400 + 0.001x^2$ pounds per foot of length, where x is the distance in feet measured from the center of the bridge. Find an equation for the curve APB of the cable.

B EXERCISES

1. A cable weighs 0.5 lb/ft. It is hung from two supports which are at the same level, 100 ft apart. If the slope of the cable at one support is $\frac{12}{5}$, (a) find the tension in the cable at its lowest point; (b) determine an equation for the curve in which the cable hangs.

2. A cable of a suspension bridge has its supports at the same level, at a distance L feet apart. The supports are a feet higher than the minimum point of the cable. If the weight of the cable is negligible but the bridge has a uniform weight of w pounds per foot show that, (a) the tension in the cable at its lowest point is $wL^2/8a$ pounds; (b) the tension at the supports is $(wL/8a)\sqrt{L^2 + 16a^2}$ pounds.

3. A cable has constant density w pounds per foot and is hung from two supports which are at the same level L feet apart. If the tension at the lowest point of the cable is H pounds, show that the tension in the cable at the supports is given, in pounds, by $H \cosh (wL/2H)$.

4. Show that the total weight of the cable in Exercise 3 is $2H \sinh (wL/2H)$.

C EXERCISES

1. A cable P feet long has a constant density w pounds per foot. It is hung from supports which are at the same level a distance L feet apart. The supports are a feet higher than the lowest point of the cable. Show that the tension H at the lowest point of the cable is given by

$$H = \frac{wL}{2 \ln \left[(P + 2a)/(P - 2a) \right]}$$

2. A cable of density 0.4 lb/ft is 250 ft long. It is hung from two supports which are at the same level 200 ft apart. Calculate (a) the distance of the supports above the lowest point of the cable, (b) the tension at that point.

3. A cable of density 0.5 lb/ft is hung from two supports which are at the same level 50 ft apart. The supports are 10 ft higher than the lowest point of the cable. Find (a) the length of the cable, (b) the tension at the lowest point of the cable, (c) the tension at the supports of the cable.

8 A Trip to the Moon

From time immemorial, mankind has gazed at the moon, wondering about many things, one of them being whether a day might come when men could travel there. Jules Verne wrote a fantastic novel about the possibilities. With the advent of modern technology this dream has at last become a reality because of the invention of rockets, which we discuss in the next section.

It was about the time of Newton's enunciation of his famous law of gravitation that speculations about possibilities of space travel were first given a mathematical basis.

> NEWTON'S LAW OF UNIVERSAL GRAVITATION. *Any two objects in the universe are attracted to each other with a force which varies directly as the product of their masses and inversely as the square of the distance between them.* In symbols
>
> $$F = \frac{GM_1M_2}{d^2} \tag{1}$$
>
> where M_1, M_2 are the masses of the objects; d is the distance between them; F is the force of attraction; and G is the constant of proportionality.

In this section we use this law to investigate the possibility of firing a projectile out of a huge cannon, for example, vertically upward toward the moon. While this means of transportation is practically useless, unless for some reason we have no desire to return, the problem is interesting for a variety of reasons, one of which is in estimating the muzzle velocity which this cannon ought to have. In attempting to solve this problem we make the following assumptions:

1. The earth and moon are perfect spheres having respective radii R_e and R_m, with masses M_e and M_m, and with the distance between their surfaces equal to a.

2. The projectile (or spaceship) of mass m is fired vertically upward toward the center of the moon with initial velocity v_0.

3. Rotations of the earth and moon are not taken into account.

4. Influence of the sun and other planets is not considered.

5. Air resistance is not taken into account.

Referring to Fig. 3.34, taking the direction from earth to moon as positive and r as the distance of m from the earth's surface at time t, we have by Newton's law and (1),

$$m\frac{d^2r}{dt^2} = -\frac{GM_em}{(r + R_e)^2} + \frac{GM_mm}{(a + R_m - r)^2}$$

or

$$\frac{d^2r}{dt^2} = -\frac{GM_e}{(r + R_e)^2} + \frac{GM_m}{(a + R_m - r)^2} \tag{2}$$

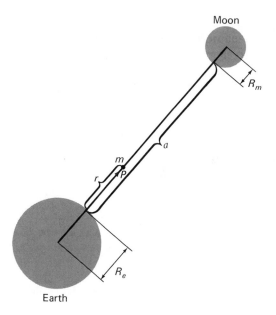

Figure 3.34

showing that the results are independent of the projectile's mass. It is possible to replace GM_e by more familiar quantities. To see this, note that the attraction of a mass m to the earth is its weight mg. Hence, from (1),

$$\frac{GM_e m}{R_e^2} = mg \quad \text{or} \quad GM_e = gR_e^2 \tag{3}$$

In a similar manner, denoting by g_m the acceleration due to gravity on the moon

$$GM_m = g_m R_m^2 \tag{4}$$

Using (3) and (4) in (2), then yields

$$\frac{d^2 r}{dt^2} = -\frac{gR_e^2}{(r + R_e)^2} + \frac{g_m R_m^2}{(a + R_m - r)^2} \tag{5}$$

The initial conditions are $r = 0$ and $dr/dt = v_0$, where $t = 0$. Since equation (5) does not involve t, let $dr/dt = v$. Then

$$v\frac{dv}{dr} = -\frac{gR_e^2}{(r + R_e)^2} + \frac{g_m R_m^2}{(a + R_m - r)^2}$$

Integrating and using the conditions at $t = 0$, we find

$$v^2 = \frac{2gR_e^2}{r + R_e} + \frac{2g_m R_m^2}{a + R_m - r} + v_0^2 - 2gR_e - \frac{2g_m R_m^2}{a + R_m} \tag{6}$$

which enables us to determine the instantaneous velocity v. Upon replacing v by dr/dt, we may determine r as a function of t. From this we may theoretically calculate the time taken to reach the moon. Actually the integrations which arise cannot be performed in closed form, and approximate techniques must be employed.*

Let us determine the muzzle velocity of our cannon needed to reach the *neutral point* (the place between earth and moon where there is zero gravitation) with velocity zero. The neutral position denoted by $r = r_n$ is determined from the equation

$$\frac{M_e}{(r_n + R_e)^2} = \frac{M_m}{(a + R_m - r_n)^2} \tag{7}$$

obtained by setting the right-hand side of (2) equal to zero. Since we want $v = 0$ when $r = r_n$, we have from (6),

$$v_0^2 = 2gR_e - \frac{2g_m R_m^2}{a + R_m - r_n} + \frac{2g_m R_m^2}{a + R_m} - \frac{2gR_e^2}{r_n + R_e} \tag{8}$$

To get some idea as to the magnitude of v_0, we employ the following approximate figures from astronomy:

$$a = 240{,}000 \text{ miles}, \qquad R_e = 4000 \text{ miles}, \qquad g = 32 \text{ ft/sec}^2$$

$$g_m = 5.3 \text{ ft/sec}^2 \text{ (about } \tfrac{1}{6}g\text{)}, \qquad M_e = 81 M_m$$

Since $M_e = 81 M_m$, approximately, equation (7) can be written

$$\frac{81 M_m}{(r_n + R_e)^2} = \frac{M_m}{(a + R_m - r_n)^2} \quad \text{or} \quad \frac{81}{(r_n + R_e)^2} = \frac{1}{(a + R_m - r_n)^2}$$

i.e.,

$$r_n = \frac{9a + 9R_m - R_e}{10}$$

Using this in (8), we have

$$v_0^2 = 2gR_e - \frac{20g_m R_m^2}{a + R_m + R_e} + \frac{2g_m R_m^2}{a + R_m} - \frac{20gR_e^2}{9(a + R_m + R_e)} \tag{9}$$

From (3) and (4) and the fact that $M_e = 81 M_m$, we have

$$gR_e^2 = 81 g_m R_m^2 \tag{10}$$

and since $g_m = g/6$, approximately, it follows from (10) that

$$R_m = \frac{\sqrt{6}}{9} R_e \tag{11}$$

approximately. Using (10) in (9), we have

$$v_0^2 = 2gR_e - \frac{200gR_e^2}{81(a + R_m + R_e)} + \frac{2gR_e^2}{81(a + R_m)} \tag{12}$$

---

* However, the integrals can be evaluated in terms of elliptic integrals.

Now a is approximately $60R_e$, and, from (11), R_m is approximately $\frac{1}{4}R_e$. Hence, (12) may be written

$$v_0^2 = 2gR_e - \frac{200gR_e}{81 \times 61.25} + \frac{2gR_e}{81 \times 60.25}$$

so that $v_0^2 = 2gR_e(1 - 0.02 + 0.0002)$, approximately, or $v_0 = \sqrt{2gR_e}(1 - 0.01) = 0.99\sqrt{2gR_e}$ by the binomial theorem. Using the values of g and R_e, we find that $v_0 = 6.9$ miles/sec, approximately. At present this speed is well out of attainment. The "Big Bertha" cannon of World War I had a muzzle velocity of 1 mile per second, so perhaps there may still be a chance.

From equation (9) we can determine the so-called *escape velocity*, which is the velocity that a projectile should have in order to leave the earth never to return, as-suming no other planets, suns, or anything else are taken into account. To find this velocity we have only to let $a \to \infty$ in (9). The result is $v_0 = \sqrt{2gR_e}$, which is ap-proximately 1 per cent greater than the velocity required to reach the neutral point. Actually, the escape velocity could have been found more easily by starting with (2) and removing the second term on the right (see A Exercise 4).

A EXERCISES

1. A projectile is fired vertically upward from the earth's surface with an initial velocity v_0 equal to the escape velocity. Neglecting the influence of the moon and other planets: (a) Show that the velocity of the projectile at distance r from the starting point is

$$v = \sqrt{\frac{2gR_e^2}{r + R_e}}$$

where R_e is the radius of the earth. (b) Calculate the velocity of the projectile after traveling 120,000 miles.

2. Show that the time for the projectile of Exercise 1 to travel the distance r is

$$\frac{2}{3R_e\sqrt{2g}}[(r + R_e)^{3/2} - R_e^{3/2}]$$

3. Assuming the moon and planets other than the earth to have no influence, how long would it take the projectile of Exercise 1 to cover the distances 120,000 miles and 240,000 miles?

4. Explain how the escape velocity of an object from the earth's surface can be found from equa-tion (2), page 116, by removing the second term on the right.

5. (a) Determine the approximate distance from the surface of the earth to the neutral point between the earth and the moon. (b) Assuming that a projectile is launched from the earth's surface with the escape velocity, what would its velocity be on reaching the neutral point?

B EXERCISES

1. An object is projected vertically upward from the earth's surface with an initial velocity v_0 of magnitude less than the escape velocity. If only the earth's influence is taken into account, show that the maximum height reached is $v_0^2 R_e/(2gR_e - v_0^2)$.

2. Assuming constant gravitational acceleration everywhere, show that an object projected from the earth's surface with velocity v_0 reaches a maximum height $v_0^2/2g$. Obtain this also by letting $R_e \to \infty$ in Exercise 1.

3. Compare the maximum heights reached according to Exercises 1 and 2 for an object projected upward with velocity 50 ft/sec, 6 miles/sec.

4. What is the velocity beyond which the heights in Exercises 1 and 2 differ by more than 5 per cent? 50 per cent?

C EXERCISES

1. Show that the time for the projectile of B Exercise 1 to reach the maximum height is

$$\frac{2gR_e^2}{(2gR_e - v_0^2)^{3/2}}\left(\sin^{-1}\frac{v_0}{\sqrt{2gR_e}} + \frac{v_0\sqrt{2gR_e - v_0^2}}{2gR_e}\right)$$

2. Show, by series or otherwise, that if $v_0^2 \ll 2gR_e$ (i.e., initial velocity much smaller than escape velocity), then the time to reach the maximum height is approximately v_0/g. This is the result obtained if the acceleration due to gravity is assumed constant everywhere.

3. An object falls a distance a from the center of the earth (radius R_e). Show that it (a) hits the earth's surface with a speed equal to $\sqrt{2gR_e(1 - R_e/a)}$; (b) reaches the earth's surface in a time given by

$$\sqrt{\frac{a}{2gR_e^2}}\left\{\sqrt{R_e(a - R_e)} + \tfrac{1}{2}a\,\cos^{-1}\left(\frac{2R_e - a}{a}\right)\right\}$$

assuming that air resistance is negligible.

4. Show that the velocity of escape from the surface of the moon is approximately 1.5 miles/sec.

9 Applications to Rockets

A rocket moves by the backward expulsion of a mass of gas formed by the burning of a fuel. This rejection of mass has the effect of increasing the forward velocity of the rocket, thus enabling it to continue onward. To consider the motion of rockets, we must treat the motion of an object whose mass is changing. In Section 1 of this chapter we pointed out that the net force acting on an object is equal to the time rate of change in momentum (Newton's second law). We will use this in finding the law of motion of a rocket.

Suppose that the total mass of the rocket at time t is M and that at the later time $t + \Delta t$ the mass is $M + \Delta M$, i.e., a mass $-\Delta M$ of gas has been expelled from the back of the rocket (note that the mass of gas expelled in the time Δt is $-\Delta M$, since ΔM is a negative quantity). Suppose that the velocity of the rocket relative to the earth at time t is V and at time $t + \Delta t$ is $V + \Delta V$, and let us take the upward direction of the rocket as positive. The expelled gas will have velocity $V + v$ relative to the earth, where v is a negative quantity, so that $-v$ represents the actual magnitude of the velocity of the gas relative to the rocket, which for our purposes will be considered constant. The total momentum of the rocket before the loss of gas is MV. After the loss of gas, the rocket has momentum $(M + \Delta M)(V + \Delta V)$, and the gas has momentum $-\Delta M(V + v)$, so that the total momentum after the loss is $(M + \Delta M)(V + \Delta V) - \Delta M(V + v)$. The change in momentum, i.e., the total momentum after the loss of gas minus the total momentum before the loss, is

$$(M + \Delta M)(V + \Delta V) - \Delta M(V + v) - MV = M\,\Delta V - v\,\Delta M + \Delta M\,\Delta V$$

The instantaneous time rate of change in momentum is the limit of the change in momentum divided by Δt as $\Delta t \to 0$, i.e.,

$$\lim_{\Delta t \to 0} \left(M\frac{\Delta V}{\Delta t} - v\frac{\Delta M}{\Delta t} + \frac{\Delta M}{\Delta t}\Delta V \right) \tag{1}$$

Remark. Since at a later time $t + \Delta t$ the mass of the rocket has *decreased*, one might be inclined to express it as $M - \Delta M$, where $\Delta M > 0$, rather than $M + \Delta M$, where $\Delta M < 0$. However, it is best to use the method of the text to conform with the idea familiar from calculus that, if a variable is decreasing, its derivative (approximated in this case by $\Delta M/\Delta t$) is negative.

Since $\Delta M \to 0$, $\Delta V \to 0$, $\Delta M/\Delta t \to dM/dt$, and $\Delta V/\Delta t \to dV/dt$ as $\Delta t \to 0$, (1) becomes

$$M\frac{dV}{dt} - v\frac{dM}{dt}$$

Now the time rate of change in momentum is the force F. Hence

$$F = M\frac{dV}{dt} - v\frac{dM}{dt} \tag{2}$$

is our basic equation for rocket motion.

ILLUSTRATIVE EXAMPLE

A rocket having initial mass M_0 grams starts radially from the earth's surface. It expels gas at the constant rate a grams per second, at a constant velocity b centimeters per second relative to the rocket, where $a > 0$ and $b > 0$. Assuming no external forces act on the rocket, find its velocity and distance traveled at any time.

Mathematical Formulation. Referring to the fundamental equation, we have $F = 0$, since there are no external forces. Since the rocket loses a grams per second, it will lose at grams in t seconds, and hence its mass after t seconds is given by $M = M_0 - at$. Also, the velocity of the gas relative to the rocket is given by $v = -b$.

Thus, (2) becomes $\quad (M_0 - at)\dfrac{dV}{dt} - ab = 0 \quad$ or $\quad \dfrac{dV}{dt} = \dfrac{ab}{M_0 - at} \tag{3}$

with the assumed initial condition $V = 0$ at $t = 0$.

Solution Integrating (3), we find $V = -b\ln(M_0 - at) + c_1$.

Since $V = 0$ at $t = 0$, $c_1 = b\ln M_0$, and

$$V = b\ln M_0 - b\ln(M_0 - at) \tag{4}$$

which is the required velocity of the rocket. Letting x be the distance which the rocket moves in time t measured from the earth's surface, we have $V = dx/dt$ so that

$$\frac{dx}{dt} = b \ln M_0 - b \ln (M_0 - at) = -b \ln \left(\frac{M_0 - at}{M_0} \right)$$

from which we obtain, upon integration, taking $x = 0$ at $t = 0$,

$$x = bt + \frac{b}{a}(M_0 - at) \ln \left(\frac{M_0 - at}{M_0} \right) \tag{5}$$

which is the required distance traveled. Note the equations (4) and (5) are valid only for $t < M_0/a$, which is the theoretical limit for the time of flight. The practical limit is much smaller than this.

A EXERCISES

1. If a constant gravitational field acts on the rocket in the illustrative example of the text, show that

$$(M_0 - at)\frac{dV}{dt} - ab = -g(M_0 - at)$$

Find the velocity of the rocket at any time $t < M_0/a$ after leaving the earth, assuming that its initial velocity is zero.

2. Determine the height of the rocket of Exercise 1 at time t.

B EXERCISES

1. A rocket has a mass of 25,000 kilograms (kg), which includes 20,000 kg of a fuel mixture. During the burning process the combustion products are discharged at a velocity relative to the rocket of 400 meters per second, involving a loss per second of 1000 kg of the fuel mixture. The rocket starts on the ground with zero velocity and travels vertically upward. If the only external force acting is that of gravitation (variation with distance neglected): (a) Find the velocity of the rocket after 15, 20, and 30 seconds. (b) Find the height reached when half the fuel mixture is burned.

2. A rocket has mass M, which includes a mass m of a fuel mixture. During the burning process the combustion products are discharged at a velocity $q > 0$ relative to the rocket. This burning involves a loss per second of a mass p of the fuel mixture. Neglecting all external forces except a constant gravitational force, show that the maximum theoretical height attained by the rocket is

$$\frac{qm}{p} + \frac{qM}{p} \ln \left(\frac{M - m}{M} \right) + \frac{q^2}{2g} \ln^2 \left(\frac{M - m}{M} \right)$$

if the rocket starts radially from the earth's surface with velocity zero.

3. In addition to the gravitational force acting on the rocket of Exercise 2, there is a force due to air resistance which is proportional to the instantaneous velocity of the rocket. (a) Find the velocity of the rocket at any time assuming that its initial velocity is zero. (b) Determine the height of the rocket at any time. (c) Find the maximum theoretical height attained.

4. An object of mass M_0, which is not acted upon by any external forces, moves in a straight line through space with velocity v_0. At $t = 0$ it begins to increase its mass at the constant rate of r grams per second. Show that the velocity at any time t is $V = M_0v_0/(M_0 + rt)$, and find the distance traveled.

C EXERCISES

1. A spherical mass grows at a rate proportional to its instantaneous surface area. Assuming that the sphere has an initial radius a and that it falls from rest under the influence of gravity (no variation with distance), show that its instantaneous acceleration is

$$\frac{g}{4}\left(1 + \frac{3a^4}{r^4}\right)$$

where r is its instantaneous radius. Thus show that a necessary and sufficient condition that the acceleration be constant is that the sphere have zero initial radius.

2. In a two-stage rocket the total mass initially, including fuel, is M. After some specified time T, the first stage of the rocket of mass M_0 falls away and the second stage of mass M_1 proceeds with the flight. Set up and solve the equations for the motion, assuming that initially the rocket is projected vertically from the earth's surface with speed v_0.

10 Physical Problems Involving Geometry

Many types of physical problems are dependent in some way upon geometry. For example, imagine a right circular cylinder, half full of water, rotating with constant angular velocity about its axis. The shape of the water's surface will be determined by the angular velocity of the cylinder. Here physics determines the geometrical shape of the water surface.

As another example, consider water emptying through a hole at the base of a conical tank. Here the geometrical shape of the container determines the physical behavior of the water.

In the illustrative examples which follow we consider three physical problems involving geometry; namely, the flow of water from a tank, the shape of a water surface in a rotating cylinder, and the shape of a reflector.

ILLUSTRATIVE EXAMPLE 1

A container having constant cross section A is filled with water to height H. The water flows out through an orifice, of cross section B, at the base of the container. Find the height of water at any time and find the time to empty the tank.

Mathematical Formulation. Let the container appear as in Fig. 3.35, where A is the constant cross-sectional area of the container, and B is the cross-sectional area of the orifice. Let h be the height of water in the tank at time t (level 1) and $h + \Delta h$ the height at time $t + \Delta t$ (level 2).

The basic principle which we use is the obvious one that the amount of water lost when the level drops from 1 to 2 is equal to the amount of water which escapes

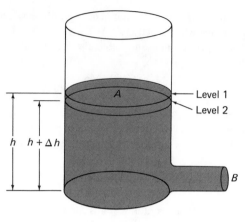

Figure 3.35

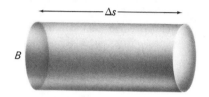

Figure 3.36

through the orifice. When the water level drops from 1 to 2, the volume lost is numerically equal to $A \, \Delta h$. However, we must be careful of signs. Since Δh is actually a negative quantity, we have, for the actual volume lost in time Δt, the amount $-A \, \Delta h$. The volume of water which escapes through the orifice is that volume which would be contained in a cylinder of cross section B and length Δs (Fig. 3.36), where Δs is the distance which the water would travel in time Δt if it were to keep traveling horizontally. We then have $-A \, \Delta h = B \, \Delta s$. Dividing by Δt and taking the limit as $\Delta t \to 0$, we find

$$-A \frac{dh}{dt} = B \frac{ds}{dt} = Bv \quad \text{or} \quad -A \, dh = Bv \, dt \tag{1}$$

where $v = ds/dt$ is the instantaneous velocity of efflux through the orifice.

We now need to have an expression for the velocity v of efflux of the water. It is clear that the greater the height of water, the greater is v. In fact, it is not hard to show that for ideal conditions* $v = \sqrt{2gh}$. Thus, (1) becomes

$$-A \, dh = B\sqrt{2gh} \, dt \tag{2}$$

Since the height is H initially, we have $h = H$ at $t = 0$.

---

* This follows since the potential energy mgh of a mass m of water equals the kinetic energy $\frac{1}{2}mv^2$. In practice, with losses, $v = c\sqrt{2gh}$, $0 < c < 1$, where c is the *discharge coefficient*.

Solution   Separation of variables in (2) yields

$$\int \frac{dh}{\sqrt{h}} = -\frac{B}{A} \sqrt{2g} \int dt, \qquad 2\sqrt{h} = -\frac{B}{A} \sqrt{2g}t + c$$

Using $h = H$ at $t = 0$, we find $c = 2\sqrt{H}$, so that

$$2\sqrt{h} = -\frac{B}{A} \sqrt{2g}t + 2\sqrt{H} \tag{3}$$

which expresses the height as a function of t.

The time for the tank to empty is found by finding t where $h = 0$. We obtain

$$t = \frac{A}{B} \sqrt{\frac{2H}{g}}$$

If $A = 4$ ft^2, $B = 1$ in.^{2}, $H = 16$ ft, $g = 32$ ft/sec^2, then $t = 576$ sec, or 9.6 min.

ILLUSTRATIVE EXAMPLE 2

A right circular cylinder having vertical axis is filled with water and is rotated about its axis with constant angular velocity ω. What shape does the water surface take?

Mathematical Formulation. When the cylinder rotates, the water surface assumes a shape as indicated in Fig. 3.37. Consider a particle of water P, of mass m, on the surface of the water. When steady-state conditions prevail, this particle will be moving in a circular path, the circle having center on the axis of rotation. For convenience we choose an xy coordinate system as shown in Fig. 3.38, where the y axis is the axis of rotation and where the x axis is perpendicular to the y axis and passes through the lowest point O of the surface. It will be clear that the surface is symmetric with

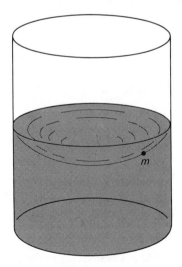

Figure 3.37

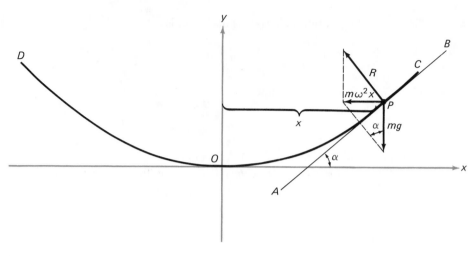

Figure 3.38

respect to the y axis. Let us investigate the forces on particle P when steady-state conditions are attained. First, there is the force due to the weight of the particle given by mg (Fig. 3.38). There is also a force on P due to the reaction of the other particles in the liquid. This reaction force is denoted by R and must be normal to the surface of the liquid.* The resultant of R and mg points to the center of the circle in which P rotates. This resultant force is the centripetal force acting on P and has magnitude $m\omega^2 x$, where x is the distance from P to the axis of rotation.

From the figure it is clear that

$$R \cos \alpha = mg, \qquad R \sin \alpha = m\omega^2 x$$

Dividing these equations and noting that the slope of the tangent APB is the same as the slope of the curve $DOPC$ at P and is therefore equal to $\tan \alpha$, or dy/dx, we have

$$\frac{dy}{dx} = \frac{\omega^2 x}{g} \tag{4}$$

which we must solve subject to $y = 0$ where $x = 0$.

Solution Integration of (4) using $x = 0$ where $y = 0$ we find

$$y = \frac{\omega^2 x^2}{2g} \tag{5}$$

Thus, in any plane through the y axis the water level assumes the shape of a parabola. In three dimensions it is a paraboloid of revolution.

ILLUSTRATIVE EXAMPLE 3

Find the shape of a reflector in order that light rays emitted by a point source be reflected parallel to a fixed line.

---

* If R were not normal to the surface, there would be a component of R tangential to the surface and the particle would move either toward or away from the axis of rotation.

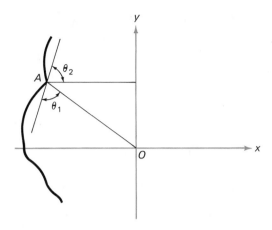

Figure 3.39

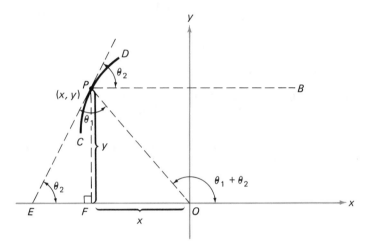

Figure 3.40

Mathematical Formulation. Let point O (origin of an xy coordinate system), Fig. 3.39, represent the point source of light. Light rays such as OA emerge from O, hit the reflector at A, and "bounce off" or are reflected, from then on traveling in a straight line. We wish to find the shape of the reflector so that all the rays emanating from O "bounce off" from the reflector parallel to the line Ox.

Let CD (Fig. 3.40) be a portion of the reflector and consider any point $P(x, y)$ on it. If θ_1 is the angle of incidence and θ_2 is the angle of reflection, then by an elementary principle of optics $\theta_1 = \theta_2$.* We wish to find a relation between the slope dy/dx of the curve (reflector) at P and the coordinates (x, y) of P. This may be obtained by use of elementary geometry.

* Actually $90 - \theta_1$, the angle which ray OP makes with the normal to arc CD at P, is the angle of incidence. Similarly $90 - \theta_2$, the angle between the reflected ray and the normal, is the angle of reflection. Clearly if $90 - \theta_1 = 90 - \theta_2$, then $\theta_1 = \theta_2$.

Since BP (Fig. 3.40) is parallel to Ox, we have $\angle OEP = \theta_2$. Hence $\angle xOP = \theta_1 + \theta_2 = 2\theta_1$ since $\theta_2 = \theta_1$. The slope of OP is $\tan 2\theta_1$ but by use of triangle OPF it is seen to be y/x. Hence $\tan 2\theta_1 = y/x$. But by elementary trigonometry,

$$\tan 2\theta_1 = \frac{2 \tan \theta_1}{1 - \tan^2 \theta_1} \tag{6}$$

Since $\tan \theta_1 = \tan \theta_2 = dy/dx = y'$, we have therefore

$$\frac{2y'}{1 - (y')^2} = \frac{y}{x} \quad \text{or} \quad \frac{dy}{dx} = \frac{-x \pm \sqrt{x^2 + y^2}}{y} \tag{7}$$

Solution   Equation (7) is homogeneous; hence, letting $y = vx$ we find

$$x \frac{dv}{dx} = \frac{-1 - v^2 \pm \sqrt{v^2 + 1}}{v} \quad \text{so that} \quad \int \frac{dx}{x} = -\int \frac{v \, dv}{v^2 + 1 \pm \sqrt{v^2 + 1}}$$

Letting $v^2 + 1 = u^2$ in the second integral so that $v \, dv = u \, du$, we find

$$\ln x + c_1 = -\int \frac{du}{u \pm 1} = -\ln (u \pm 1) = -\ln (\sqrt{v^2 + 1} \pm 1)$$

It follows that $x(\sqrt{v^2 + 1} \pm 1) = c$ or $\sqrt{x^2 + y^2} = c \pm x$

With squaring and simplification, this becomes

$$y^2 = \pm 2cx + c^2 \tag{8}$$

For a given value of c ($c \neq 0$) equation (8) represents two parabolas symmetric with respect to the x axis as shown in Fig. 3.41. The heavy curve has equation $y^2 =$

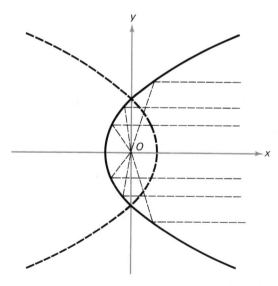

Figure 3.41

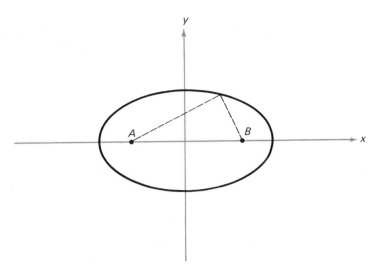

Figure 3.42

$2cx + c^2, c > 0$. The dashed curve has equation

$$y^2 = -2cx + c^2, \qquad c > 0 \quad \text{or} \quad y^2 = 2cx + c^2, \qquad c < 0$$

The focus for the family of parabolas is at the origin. In Fig. 3.41 we have also shown several light rays emanating from the focus and "bouncing off" the reflector parallel to the *x* axis.

If we revolve the parabola about the *x* axis, we obtain a paraboloid of revolution. An electric light bulb placed at the origin sends rays of light "bouncing off" the reflector to produce a direct beam of light rays which is, of course, the most efficient way of lighting. This important property accounts for the paraboloid shape of automobile headlights.

Remark. Another interesting problem is that of finding a curve in the *xy* plane having the property that sound waves (or light rays) emitted by a point source *A* will be reflected toward a point *B* by the curve (see Fig. 3.42). It turns out that the curve is an ellipse and the two points are its foci. If we revolve the ellipse about the major axis, we obtain a surface in the shape of an ellipsoid of revolution. Sound waves emitted from *A* will all be reflected by the surface toward point *B*. This has application to *acoustics*, since for example we can design an auditorium with this shape so that a lecturer or orchestra can be located at *A* and the audience located at *B*.

A EXERCISES

1. A right circular cylinder of radius 10 ft and height 20 ft is filled with water. A small circular hole at the bottom is 1 in. in diameter. When will all the water flow out if (a) $v = \sqrt{2gh}$; (b) $v = 0.6\sqrt{2gh}$?

2. A tank is in the form of a 12 ft cube. A leak at the bottom develops due to a small hole 2 in.^{2} in area. If the tank is initially three-quarters full, when will it be (a) half full; (b) empty? Assume $v = \sqrt{2gh}$.

3. A tank in the form of a right circular cone of height H, radius R, with its vertex below the base is filled with water. A hole, having cross section a at the vertex, causes the water to leak out. Assuming $v = c\sqrt{2gh}$, where c is the discharge coefficient, show that the time for the cone to empty is

$$T = \frac{2\pi R^2}{5ac}\sqrt{\frac{H}{2g}} = \frac{2A}{5ac}\sqrt{\frac{H}{2g}}$$

where $A = \pi R^2$ is the area of the base of the cone. If $H = 16$ ft, $a = 1$ in.^{2}, $R = 5$ ft, find T for the cases $c = 1$, $c = 0.6$.

4. The conical tank of Exercise 3 is inverted so that the vertex is above the base and a hole of area a is in the base. Find the time required to empty the tank. Compare with Exercise 3.

B EXERCISES

1. A cylindrical can is filled with a liquid of density ρ and is rotated about its axis with constant angular velocity ω. Show that the pressure at a distance r from the axis exceeds the pressure on the axis by $\frac{1}{2}\rho\omega^2 r^2$.

2. If the liquid in Exercise 1 is replaced by an ideal gas obeying Boyle's law (the pressure in a gas varies inversely as the volume if the temperature is constant), find the pressure as a function of distance from the axis.

3. If Boyle's law in Exercise 2 is replaced by the law $\rho = kP^\alpha$ where P and ρ are, respectively, the pressure and density of the gas, and α and k are constants, show that the pressure at distance r from the axis is

$$P = \left[\frac{k(1 - \alpha)\omega^2 r^2}{2} + P_0^{1-\alpha}\right]^{1/(1-\alpha)}, \qquad \alpha \neq 1$$

Obtain the result of Exercise 2 by letting $\alpha \to 1$.

4. Verify the remark on page 129 by setting up and solving an appropriate differential equation.

5. The pressure p and density ρ of the atmosphere above the earth's surface are related by the formula $p = k\rho^\gamma$ where k and γ are positive constants, Assuming that at sea level the pressure and density are given by p_0 and ρ_0 respectively, show that (a) the pressure variation with height h is given by $p^{1-1/\gamma} = p_0^{1-1/\gamma} - (1 - 1/\gamma)\rho_0 p_0^{-1/\gamma} h$ and (b) the height of the atmosphere can be considered as $\gamma p_0/(\gamma - 1)\rho_0$. (c) Discuss the cases $\gamma = 1$ and $\gamma > 1$.

6. A hemispherical bowl of radius R is filled with water. If there is a small hole of radius r at the bottom of the convex surface, show that the time taken for the bowl to empty is

$$T = \frac{14}{15c}\left(\frac{R}{r}\right)^2\sqrt{\frac{R}{2g}}$$

assuming that $v = c\sqrt{2gh}$, where v is the velocity of efflux of the water when the water level is at height h and c is the discharge coefficient.

C EXERCISES

1. A famous problem considered in an advanced part of mathematics called the *calculus of variations* is that of determining the shape of a wire in order that a bead placed on it will, under influence of gravity, travel from a given point to a given lower point in the shortest time, friction being neglected. Taking a rectangular coordinate system with y axis positive downward and x axis positive to the right and letting the bead start at $(0, 0)$ and travel to (a, b) it can be shown that the shape of the wire is given by the differential equation

$$1 + (y')^2 + 2yy'' = 0$$

Show that the curve is a portion of a cycloid. This problem is called the *brachistochrone* (shortest time) problem and was proposed by John Bernoulli in 1696.

2. Another problem of the calculus of variations is the determination of a curve joining points (a, b) and (c, d), having the property that the surface generated by revolving the curve about the x axis is a minimum. The required curve may be obtained from the differential equation

$$1 + (y')^2 = yy''$$

Show that the curve is a portion of a *catenary*.

The minimum surface property has an interesting physical significance. If two thin circular rings initially in contact are placed in a soap solution and then pulled carefully apart so that a soap film surface is formed, the surface has the property that its area is a minimum.

3. A man initially at O (see Fig. 3.43) walks along the straight shore Ox of a lake towing a rowboat, initially at A, by means of a rope of length a, which is always held taut. Show that the boat moves in a path (called a *tractrix*) with parametric equations

$$x = a \ln\left[\cot\frac{\theta}{2} - \cos\theta\right], \qquad y = a \sin\theta$$

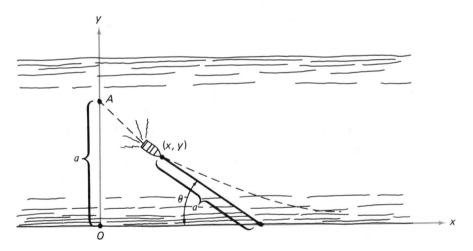

Figure 3.43

4. Towns A and B (Fig. 3.44) are directly opposite each other on the banks of a river of width D which flows east with constant speed U. A boat leaving town A travels with constant speed V always aimed toward town B. Show that (a) its path is given by

$$x = \tfrac{1}{2}D\left[t^{1-U/V} - t^{1+U/V}\right], \qquad y = Dt$$

and (b) it will not arrive at town B unless $V > U$.

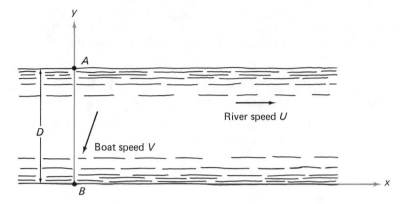

Figure 3.44

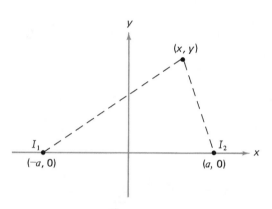

Figure 3.45

5. Two electric light bulbs having strengths I_1 and I_2 are situated at the points $(-a, 0)$ and $(a, 0)$ of a rectangular coordinate system (see Fig. 3.45). Show that the locus of all points in the plane at which the intensities of illumination from both lights are equal is given by

$$\frac{I_1(x + a)}{\sqrt{(x + a)^2 + y^2}} + \frac{I_2(x - a)}{\sqrt{(x - a)^2 + y^2}} = I_1 - I_2$$

(*Hint:* The intensity of illumination varies inversely as the square of the distance from the light source.)

11 Miscellaneous Problems in Geometry

Geometrical problems provide a fertile source of differential equations. We have already seen how differential equations arise in connection with orthogonal trajectories. In this section we consider various other geometrical problems.

The slope at any point of a curve is $2x + 3y$. If the curve passes through the origin, determine its equation.

Mathematical Formulation. The slope at (x, y) is dy/dx. Hence

$$\frac{dy}{dx} = 2x + 3y \tag{1}$$

is the required differential equation, which we solve subject to $y(0) = 0$.

Solution Equation (1) written as a first-order linear equation

$$\frac{dy}{dx} - 3y = 2x$$

has integrating factor e^{-3x}. Hence,

$$\frac{d}{dx}(ye^{-3x}) = 2xe^{-3x} \quad \text{or} \quad ye^{-3x} = \frac{-2xe^{-3x}}{3} - \frac{2e^{-3x}}{9} + c$$

Thus, since $y(0) = 0$, $c = \frac{2}{9}$ and we find $y = \frac{2}{9}e^{3x} - \frac{2x}{3} - \frac{2}{9}$.

The tangent line to a curve at any point (x, y) on it has its intercept on the x axis always equal to $\frac{1}{2}x$. If the curve passes through $(1, 2)$ find its equation.

Mathematical Formulation. To solve this problem we must find an expression for the x intercept OA of the tangent line AP to the curve QPR (Fig. 3.46). To accomplish this, let (X, Y) be any point on AP. Since y' is the slope of the line, its equation is

$$Y - y = y'(X - x)$$

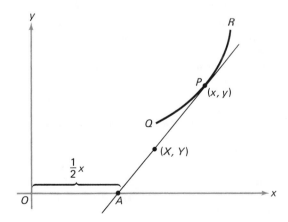

Figure 3.46

The required intercept is the value of X, where $Y = 0$. This is found to be

$$X = x - \frac{y}{y'}$$

Thus we must solve $\quad x - \dfrac{y}{y'} = \frac{1}{2}x, \quad y = 2 \quad$ where $\quad x = 1 \qquad$ (2)

Solution Equation (2) may be written $\dfrac{y}{y'} = \frac{1}{2}x$ or $\dfrac{dy}{dx} = \dfrac{2y}{x}$.

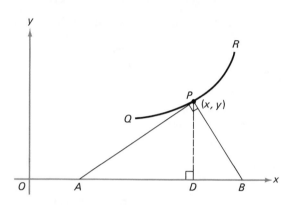

Figure 3.47

Table 3.2

Intercept of tangent line: On x axis:	$x - \dfrac{y}{y'}$
On y axis:	$y - xy'$
Intercept of normal line: On x axis:	$x + yy'$
On y axis:	$y + \dfrac{x}{y'}$
Length of tangent line from P: To x axis:	$\left\lvert \dfrac{y\sqrt{1 + (y')^2}}{y'} \right\rvert$
To y axis:	$\left\lvert x\sqrt{1 + (y')^2} \right\rvert$
Length of normal line from P: To x axis:	$\left\lvert y\sqrt{1 + (y')^2} \right\rvert$
To y axis:	$\left\lvert \dfrac{x\sqrt{1 + (y')^2}}{y'} \right\rvert$
Length of subtangent:	$\left\lvert \dfrac{y}{y'} \right\rvert$
Length of subnormal:	$\lvert yy' \rvert$

Separating the variables, integrating, and using the condition $y(1) = 2$, we find

$$y = 2x^2$$

In Illustrative Example 2 we were required to determine the x intercept of the tangent line to a curve. We could just as well have been required to determine the y intercept of the tangent line or the length of the tangent line included between point P and the x or y axes. Since many geometrical problems are based on such considerations, we discuss them briefly. In Fig. 3.47, P is any point (x, y) on curve QPR. It is customary to call the tangent line from P to point A on the x axis briefly the "tangent." Similarly PB is called the "normal." The projections of AP and PB on the x axis are called the "subtangent" and "subnormal," respectively. By procedures analogous to that used in Illustrative Example 2 the student may verify the entries in Table 3.2.

It should be observed that since length is a positive quantity we have used absolute value signs in the table. Also we have used rectangular coordinates. Similar considerations involving polar coordinates may be formulated. See the C Exercises.

A EXERCISES

1. The slope at any point (x, y) of a curve is $1 + y/x$. If the curve passes through $(1, 1)$ find its equation.

2. Find an equation for the family of curves such that the slope at any point is the sum of half the ordinate and twice the abscissa of the point.

3. The y intercept of the normal line to a curve at any point is 2. If the curve passes through $(3, 4)$ find its equation.

4. The y intercept of the tangent line to a curve at any point is always equal to the slope at that point. If the curve passes through $(2, 1)$ find its equation.

5. The length of the normal line from any point of a curve to the x axis is always equal to a constant $a > 0$. Show that the curve is a circle of radius a.

6. Find the equation of a curve passing through $(1, 1)$ having the property that the x intercept of its tangent line equals the y intercept of its normal line.

B EXERCISES

1. Find the equation of the curve through $(3, 4)$ such that the length of its subtangent at any point is equal to the distance of the point from the origin.

2. Show that the lengths of the tangent and normal lines from P (Fig. 3.47) to the x and y axes are given by the entries in Table 3.2.

3. The difference between the lengths of the "subtangent" and "subnormal" of a family of curves is 2. Find the equation of the family.

4. A curve in the first quadrant passes through $(0, 1)$. If the length of arc from $(0, 1)$ to any point (x, y) is numerically equal to the area bounded by the curve, x axis, y axis, and ordinate at (x, y) show that the curve is a portion of a catenary.

5. A point moves in the first quadrant of the xy plane so that the tangent to its path makes with the coordinate axes a triangle whose area is equal to the constant a^2. Find the path.

6. Find the path of the point in Exercise 5 if the tangent cut off by the coordinate axes is of length a.

7. A family of curves has the property that the tangent line to each curve at the point (x, y), the x axis, and the line joining (x, y) to the origin forms an isosceles triangle with the tangent line as base. (a) Determine an equation for the family and (b) that particular member which passes through the point $(2, 0)$.

C EXERCISES

1. Let (r, ϕ) be polar coordinates of any point on curve APB (Fig. 3.48). Point O is the pole of the coordinate system, Ox is the initial line, and OP is the radius vector. Let COD be a line perpendicular to OP at O. We define CP as the *polar tangent*, PD as the *polar normal*, CO as the *polar subtangent*, and OD as the *polar subnormal*. Demonstrate the validity of each of the following, where $r' = dr/d\phi$.

$$\text{Length of } subtangent = \left|\frac{r^2}{r'}\right|, \qquad \text{Length of } tangent = \left|\frac{r}{r'}\sqrt{r^2 + (r')^2}\right|$$

$$\text{Length of } subnormal = |r'|, \qquad \text{Length of } normal = \left|\sqrt{r^2 + (r')^2}\right|$$

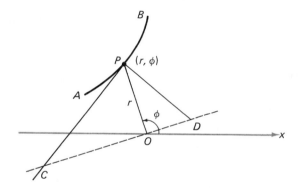

Figure 3.48

2. Find the family of curves whose polar subtangents are of constant length.

3. Find the family of curves whose polar normals are of constant length.

4. Find the family of curves whose polar subtangents and normals are equal in length.

5. Find the curve passing through the point with polar coordinates $(1, \pi/3)$ and such that the length of its polar subtangent at any point is equal to the distance of the point from the initial line.

6. (a) Show that the length of the perpendicular from the pole to the tangent line of any curve is $r^2/\sqrt{r^2 + (r')^2}$.

(b) Find all curves such that the length of the perpendicular from the pole to their tangent lines is constant and equal to $a > 0$.

12 The Deflection of Beams

Consider a horizontal beam AB of Fig. 3.49(a). We make the assumption that the beam is uniform in cross section and of homogeneous material. The axis of symmetry is indicated by the dashed line. When acted upon by forces which we assume are in a vertical plane containing the axis of symmetry, the beam, due to its elasticity, may become distorted in shape as shown in Fig. 3.49(b). These forces may be due to the weight of the beam, to externally applied loads, or a combination of both. The resulting distorted axis of symmetry dashed in Fig. 3.49(b) is called the *elastic curve*. The determination of this curve is of importance in the theory of elasticity and it will be part of the purpose of this section to show how this is done.

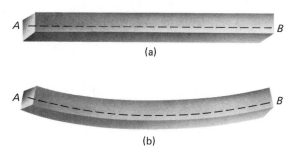

(a)

(b)

Figure 3.49

There are many ways in which beams can be supported. For example, Fig. 3.50(a) shows a beam in which the end A is rigidly fixed, while end B is free to move. This is called a *cantilever beam*. In Fig. 3.50(b) the beam is supported at ends A and B. This is called a *simply supported beam*. In such case the beam is hinged at the ends A and B so that although fixed at these ends, rotation can take place about the ends. Figure 3.50(c) shows still another way of supporting beams.

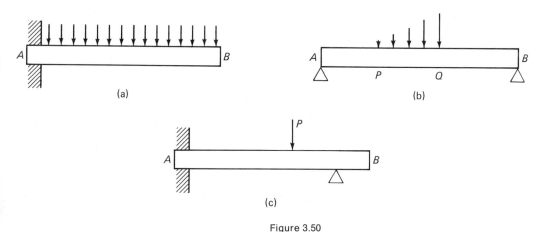

(a)

(b)

(c)

Figure 3.50

Just as there are different ways of supporting beams, there are different ways in which external loading forces may be applied. For example, in Fig. 3.50(a) there is a *uniformly distributed load* over the whole beam. There may be a *variable loading* over the whole beam or just part of it as in Fig. 3.50(b). On the other hand there may be a *concentrated load* as indicated in Fig. 3.50(c).

Consider the horizontal beam OB of Fig. 3.51(a). Let the axis of symmetry (shown dashed) lie on the x axis taken as positive to the right and having origin at O. Choose the y axis as positive downward. Due to the action of the external forces $F_1, F_2, \ldots$ (and the weight of the beam if appreciable) the axis of symmetry is distorted into the elastic curve shown dashed in Fig. 3.51(b), where we have taken the beam as fixed at O. The displacement y of the elastic curve from the x axis is called the *deflection* of the beam at position x. Thus if we determine the equation of the elastic curve, the deflection of the beam will be known. We now show how this can be accomplished.

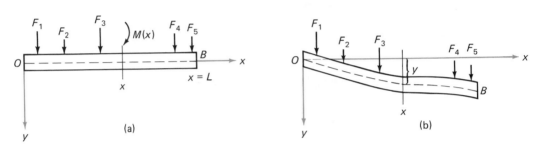

Figure 3.51

Let $M(x)$ denote the *bending moment* in a vertical cross section of the beam through x. This bending moment is defined as the algebraic sum of the moments of those forces which act *on one side of* x, the moments being taken about a horizontal line in the cross section at x. In computing moments we shall adopt the convention that *upward* forces produce *negative* moments and *downward* forces produce *positive* moments, assuming of course that the positive y axis is taken downward as stated above. As will be shown in Illustrative Example 1, it makes no difference which side of x we take since the bending moments computed from either side are equal.*

We will show later (see page 143) that the bending moment at x is related simply to the radius of curvature of the elastic curve at x, the relation being

$$EI \frac{y''}{[1 + (y')^2]^{3/2}} = M(x) \tag{1}$$

where E is Young's modulus of elasticity and depends on the material used in designing the beam, and I is the moment of inertia of the cross section of the beam at x with respect to a horizontal line passing through the center of gravity of this cross section. The product EI is called the *flexural rigidity*, and we shall take it as constant.

* This is true because the beam is in equilibrium.

If we assume that the beam bends only slightly, which is the case for many practical purposes, the slope y' of the elastic curve is so small that its square is negligible compared with unity, and equation (1) may be replaced by the good approximation

$$EIy'' = M(x) \tag{2}$$

Before giving a derivation of the result (1) from which the approximation (2) follows, let us see how equation (2) can be used.

ILLUSTRATIVE EXAMPLE 1

A horizontal, simply supported, uniform beam of length L bends under its own weight, which is w per unit length. Find the equation of its elastic curve.

Mathematical Formulation. In Fig. 3.52 the elastic curve of the beam (dashed) is shown relative to a rectangular set of axes having origin at O and indicated positive directions. Since the beam is simply supported at O and B, each of these supports carries half the weight of the beam, or $wL/2$. The bending moment $M(x)$ is the algebraic sum of the moments of these forces acting at one side of point P. Let us first choose the side to the *left* of P. In this case two forces act:

1. Upward force $wL/2$, distance x from P, producing a negative moment.
2. Downward force wx, distance $x/2$ (center of gravity of OP) from P, producing a positive moment.

The total bending moment at P is thus

$$M(x) = -\frac{wL}{2}x + wx\left(\frac{x}{2}\right) = \frac{wx^2}{2} - \frac{wLx}{2} \tag{3}$$

If we had chosen the side to the *right* of P, two forces would act:

1. Downward force $w(L - x)$, distance $(L - x)/2$ from P, producing a positive moment.
2. Upward force $wL/2$, distance $L - x$ from P, producing a negative moment.

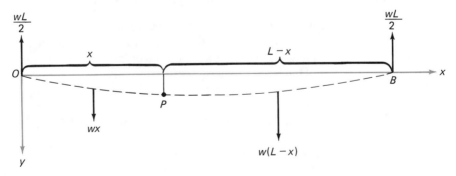

Figure 3.52

In this case the bending moment is

$$M(x) = w(L-x)\left(\frac{L-x}{2}\right) - \frac{wL}{2}(L-x) = \frac{wx^2}{2} - \frac{wLx}{2} \qquad (4)$$

which agrees with (3) and shows that in computing bending moments it makes no difference which side of P is used.

With the value of $M(x)$ obtained, the fundamental equation (2) is

$$EIy'' = \frac{wx^2}{2} - \frac{wLx}{2} \qquad (5)$$

Two conditions are necessary for determination of y. These are

$$y = 0 \qquad \text{where } x = 0 \quad \text{and} \quad \text{where } x = L$$

since the beam is not deflected at the ends.

Solution   Integrating (5) twice yields $EIy = \dfrac{wx^4}{24} - \dfrac{wLx^3}{12} + c_1 x + c_2.$

Since $y = 0$ when $x = 0$, we have $c_2 = 0$. Hence, $EIy = \dfrac{wx^4}{24} - \dfrac{wLx^3}{12} + c_1 x.$

Since $y = 0$ when $x = L$, $c_1 = wL^3/24$ and we have, finally,

$$y = \frac{w}{24EI}(x^4 - 2Lx^3 + L^3 x) \qquad (6)$$

as the required equation of the elastic curve. It is of practical interest to use (6) to find the maximum deflection. From symmetry or by calculus, the maximum is found to occur at $x = L/2$. Hence,

$$\text{maximum deflection} = \frac{5wL^4}{384EI}$$

ILLUSTRATIVE EXAMPLE 2

Find the elastic curve of a uniform cantilever beam of length L having a constant weight w per unit length and determine the deflection of the free end.

Mathematical Formulation. The dashed curve in Fig. 3.53 is the elastic curve of the cantilever beam. The origin O of the coordinate system is taken at the fixed end, and the positive x and y axes are as shown. In computing $M(x)$ it is simpler to consider the portion of the beam to the right of P, since only one force acts here, namely, the downward force $w(L-x)$, producing a positive moment given by

$$M(x) = w(L-x)\left(\frac{L-x}{2}\right) = \frac{w(L-x)^2}{2} \quad \text{so that} \quad EIy'' = \frac{w(L-x)^2}{2}$$

which we must solve subject to the conditions $y = y' = 0$ where $x = 0$, since there is no deflection at $x = 0$, and since the slope of the tangent to the elastic curve at $x = 0$ is zero.

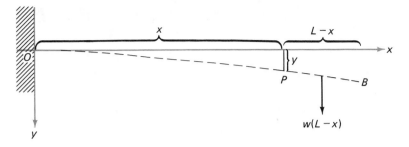

Figure 3.53

Solution Integrating twice and using the conditions it is easy to show that

$$y = \frac{w}{24EI}(x^4 - 4Lx^3 + 6L^2x^2)$$

is the equation of the elastic curve. By placing $x = L$ we find

$$\text{deflection of free end} = \frac{wL^4}{8EI}$$

ILLUSTRATIVE EXAMPLE 3

A horizontal, simply supported, uniform beam of length L and negligible weight bends under the influence of a concentrated load S at a distance $L/3$ from one end. Find the equation of the elastic curve.

Mathematical Formulation. The elastic curve is shown dashed in Fig. 3.54. The supports at O and B are at distances having a ratio 1:2 from the load S. Hence, they support loads having ratio 2:1, so that at O the amount $2S/3$ is supported,

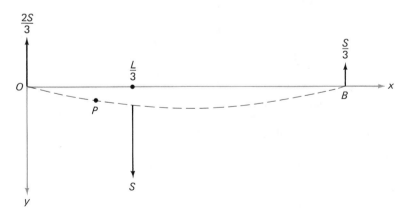

Figure 3.54

while at B the amount $S/3$ is supported. In determining the bending moment at P, three cases must be considered:

Case 1. P is to the left of S, i.e., $0 \leq x < L/3$. The portion OP has only one (upward) force acting at a distance x from P. The bending moment is $-2Sx/3$. Hence,

$$EIy'' = -\frac{2Sx}{3}, \qquad 0 \leq x < \frac{L}{3} \tag{7}$$

Case 2. P is to the right of S, i.e., $L/3 < x \leq L$. The portion OP has two forces, an upward force $2S/3$, distance x from P, and a downward force S, distance $x - L/3$ from P. The bending moment is $-2Sx/3 + S(x - L/3)$. Hence,

$$EIy'' = -\frac{2Sx}{3} + S\left(x - \frac{L}{3}\right), \qquad \frac{L}{3} < x \leq L \tag{8}$$

Case 3. P is at S. In this case the portion OP has two forces, an upward force $2S/3$, distance $L/3$ from P, and a downward force S, distance zero from P. The bending moment is

$$-\frac{2S}{3}\left(\frac{L}{3}\right) + S(0) = -\frac{2SL}{9}$$

Since this agrees with the bending moments in equations (7) and (8) we may combine Case 3 with those cases merely by rewriting the equations as

$$EIy'' = -\frac{2Sx}{3}, \qquad 0 \leq x \leq \frac{L}{3}$$

$$EIy'' = -\frac{2Sx}{3} + S\left(x - \frac{L}{3}\right), \qquad \frac{L}{3} \leq x \leq L \tag{9}$$

Since each equation is of second order, we expect to need four conditions. Two of these are clearly

$$y = 0 \quad \text{at} \quad x = 0, \qquad y = 0 \quad \text{at} \quad x = L \tag{10}$$

A third condition is obtained by realizing that the two values of y obtained from equations (9) must be equal at $x = L/3$. This is the *condition of continuity*. A fourth condition is obtained by realizing that there must be a tangent at $x = L/3$. This is the *condition for continuity in the derivative*.

Solution Integrating equations (9) each once, we have

$$EIy' = -\frac{Sx^2}{3} + c_1, \qquad 0 \leq x \leq \frac{L}{3}$$

$$EIy' = -\frac{Sx^2}{3} + \frac{S}{2}\left(x - \frac{L}{3}\right)^2 + c_2, \qquad \frac{L}{3} \leq x \leq L \tag{11}$$

Since these two values of y' must be equal at $x = L/3$, we have $c_1 = c_2$. Thus,

$$EIy' = -\frac{Sx^2}{3} + c_1, \qquad 0 \le x \le \frac{L}{3}$$

$$EIy' = -\frac{Sx^2}{3} + \frac{S}{2}\left(x - \frac{L}{3}\right)^2 + c_1, \qquad \frac{L}{3} \le x \le L \tag{12}$$

Integrating these each once, we have

$$EIy = -\frac{Sx^3}{9} + c_1 x + c_3, \qquad 0 \le x \le \frac{L}{3}$$

$$EIy = -\frac{Sx^3}{9} + \frac{S}{6}\left(x - \frac{L}{3}\right)^3 + c_1 x + c_4, \qquad \frac{L}{3} \le x \le L \tag{13}$$

Using the first and second conditions of (10) in the first and second equations of (13), respectively, and also using the condition for continuity, we find

$$c_3 = 0, \qquad c_4 = 0, \qquad c_1 = \frac{5SL^2}{81}$$

or

$$y = \begin{cases} \dfrac{S}{81EI}(5L^2 x - 9x^3), & 0 \le x \le \dfrac{L}{3} \\[2mm] \dfrac{S}{81EI}\left[5L^2 x - 9x^3 + \dfrac{27}{2}\left(x - \dfrac{L}{3}\right)^3\right], & \dfrac{L}{3} \le x \le L \end{cases} \tag{14}$$

Let us now derive equation (1). To do this we consider a small element of a curved beam between x and $x + dx$, as indicated, greatly enlarged in Fig. 3.55. In this figure, $ABCDA$ represents the midsection which contains the elastic curve. A fiber such as GH, shown below the midsection, is stretched due to the bending of the beam, while fibers above the midsection are contracted. Fibers in the midsection, however, will neither stretch nor contract. The beam element of Fig. 3.55 is shown in longitudinal cross section in Fig. 3.56. In this figure AB represents a part of the elastic curve having length l. The curve GH, which represents a stretched fiber, has length $l + dl$, where $dl = JH$ is the magnitude of the stretch. The distance between

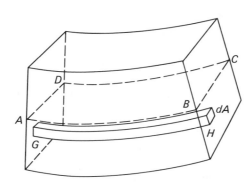

Figure 3.55

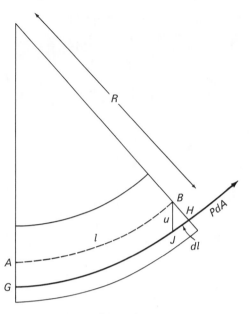

Figure 3.56

AB and *GH* is denoted by $u = AG = BJ = BH$. We now make use of a general law of elasticity due to Robert Hooke.

GENERALIZED HOOKE'S LAW. Suppose that in an elastic material a force acts perpendicular to an area and that the force per unit area, called the *stress*, is denoted by *P*. Suppose that because of this stress, a lineal element of length *l* in the direction of the force is stretched by an amount *dl*. The ratio *dl/l* is called the *strain*. Then, assuming that the elastic limit of the material is not exceeded, the *stress will be proportional to the strain*. In symbols this is expressed by

$$P = E\frac{dl}{l} \tag{15}$$

where the constant of proportionality, denoted by *E*, is called *Young's modulus of elasticity*.*

Now, if the fiber has cross section *dA* (see Fig. 3.55), then the force acting on the fiber is *P dA*. The moment *dM* of this force about the midsection is equal to this force multiplied by the distance *u* to the midsection, i.e.,

$$dM = uP \, dA \tag{16}$$

* Note that we have stretching if $dl > 0$ and contracting if $dl < 0$. An example of a case where the elastic limit is exceeded would be in a spring which is stretched so far that its properties (and as a result, Young's modulus) change. If too much force and resultant stretching occurs the spring will ultimately break.

If we let R denote the radius of curvature of the beam, we see from Fig. 3.56 that

$$\frac{dl}{u} = \frac{l}{R} \quad \text{or} \quad \frac{dl}{l} = \frac{u}{R} \tag{17}$$

so that (15) becomes

$$P = \frac{Eu}{R} \tag{18}$$

Using (18) in (16), we find

$$dM = \frac{Eu^2}{R} \, dA \tag{19}$$

Thus the total moment is

$$M = \frac{E}{R} \int u^2 \, dA \tag{20}$$

where the integral is taken over the entire transverse cross section of the beam, including fibers above and below the midsection.*

Now the integral in (20) is the moment of inertia of the beam about the midsection. With this denoted by I, (20) can be written

$$M = \frac{EI}{R} \tag{21}$$

But the radius of curvature is known from calculus to be

$$R = \frac{(1 + y'^2)^{3/2}}{y''} \tag{22}$$

Using this in (21) yields the required equation (1), and as we have already noted, (2) follows if we assume that the slope of the beam is small so that y'^2 is negligible compared with 1.

A EXERCISES

1. A uniform cantilever beam of length L and of negligible weight has a concentrated load S at the free end. Find (a) the equation of the elastic curve, (b) the maximum deflection.

2. A beam of length L and negligible weight is simply supported at both ends. A concentrated load S acts at its center. Find (a) the equation of the elastic curve, (b) the maximum deflection, (c) the numerical value of the slope at the ends.

3. Assume that, in addition to the concentrated load, the beam of Exercise 1 weighs w per unit length, where w is constant. (a) Find the equation of the elastic curve. (b) Find the maximum deflection. (c) Show that by letting $S = 0$, the results of Illustrative Example 2 are obtained.

4. Assume that, in addition to the concentrated load, the beam of Exercise 2 weighs w per unit length, where w is constant. (a) Find the equation of the elastic curve. (b) Find the maximum deflection. (c) Letting $S = 0$, obtain the results of Illustrative Example 1. (d) By letting $w = 0$, show that the results of Exercise 2 are obtained.

* Note that for fibers above the midsection, dl and u are negative. Since P is also negative here (i.e., it is a compressive force) everything works out so that (20) is correct.

5. A uniform cantilever beam of length L and of negligible weight has a concentrated load S at its center. Find the equation of the elastic curve, and determine the maximum deflection.

6. Determine the elastic curve for the beam of Exercise 5 if it is assumed that the beam has, in addition, a uniform weight w per unit length.

B EXERCISES

1. A beam of length L and uniform weight w per unit length has both ends horizontally fixed in masonry. Determine the equation of the elastic curve and find the maximum deflection. (*Hint:* Assume that the unknown moment at either of the ends is A, and determine A from the boundary conditions.)

2. Solve the previous problem if a concentrated load S acts at the center of the beam, in addition to the uniform weight.

3. One end of a beam of length L and uniform weight w per unit length is simply supported, while the other end is horizontally fixed. (a) Find the equation of the elastic curve. (b) Show that the maximum deflection occurs at a distance $(15 - \sqrt{33})L/16 = 0.578L$, approximately, from the fixed end and has an approximate magnitude given by $0.00542wL^4/EI$.

4. A beam of length L and negligible weight is simply supported at both ends and at the center. It carries a uniform load of w per unit length. Determine the equation of the elastic curve.

C EXERCISES

1. If $y_1(x)$ is the deflection at point x of a beam, due to a concentrated load S_1 at point P_1 of the beam, and $y_2(x)$ is the deflection at point x, due to a concentrated load S_2 at point P_2 of the beam, show that the deflection at point x, due to both concentrated loads, is given by $y_1(x) + y_2(x)$. This is the "principle of superposition" for beams.

2. Generalize Exercise 1 to n concentrated loads $S_1, \ldots, S_n$ at points $P_1, \ldots, P_n$.

3. Using the results of Exercises 1 and 2 and Illustrative Example 3, find the elastic curve formed by a beam of length L, which is simply supported at both ends and which has concentrated loads S at distances $L/3$ from each end.

4. A beam of length L is simply supported at both ends. It carries a variable load given by $w(x)$ per unit length, where x is the distance measured from one end of the beam. (a) Show that the differential equation for the determination of the elastic curve is

$$EIy'' = -\int_0^L \frac{u}{L} w(u)(L - x)du + \int_x^L w(u)(u - x)du \tag{23}$$

(*Hint:* Assume the interval $0 \leq x \leq L$, subdivided into n equal parts by the points

$$\frac{L}{n}, \quad \frac{2L}{n}, \quad \ldots, \quad \frac{(n-1)L}{n}$$

at which concentrated loads

$$\frac{L}{n} w\left(\frac{L}{n}\right), \quad \frac{L}{n} w\left(\frac{2L}{n}\right), \quad \ldots, \quad \frac{L}{n} w\left(\frac{(n-1)L}{n}\right)$$

act, respectively. Consider the limit as $n \to \infty$, applying the fundamental theorem of integral calculus. Note that the subdivisions need not be taken equal.) (b) By differentiations of (23) with respect to x, obtain

(A) $EIy''' = -\left[\int_x^L w(u)du - \int_0^L \dfrac{u}{L} w(u)du\right]$,   (B) $EIy^{(IV)} = w(x)$ (24)

assuming EI constant. Equations (24) are of great importance in beam theory. The quantity on the right of (24)A is the negative of the total loading to the right of x and is called the *vertical shear at* x. Equations (24) are valid for beams other than simply supported ones (see Exercise 8).

5. Using equation (24)B of Exercise 4(b) with $w(x)$ constant, arrive at the result of Illustrative Example 1.

6. Using equation (24)B of Exercise 4(b), determine the elastic curve of a beam of length L, which is simply supported at both ends and which has a loading proportional to (a) the distance from one end, (b) the square of the distance from one end.

7. A beam of length L and negligible weight is simply supported at both ends. A concentrated load S acts at distance p from one end. Show that the maximum deflection of the beam never exceeds the deflection of the center by more than 3 per cent.

8. Consider a portion of a beam between sections at x and $x + \Delta x$ (Fig. 3.57). Let M and $M + \Delta M$ denote the respective bending moments at these sections. Let $w(x)$ represent the loading per unit length at x, so that $w(x)\,\Delta x$ represents the loading on the portion between x and $x + \Delta x$, apart from infinitesimals of order higher than Δx. Let V be the vertical shear at x (i.e., the algebraic sum of the vertical forces to one side of x, say to the right) and $V + \Delta V$ the corresponding vertical shear at $x + \Delta x$. (a) From equilibrium considerations of this portion of the beam show that

$$\Delta V + w(x)\Delta x + \alpha(\Delta x)^2 = 0 \qquad \text{(sum of forces = 0)}$$

$$\Delta M + (V + \Delta V)\Delta x + w(x)\Delta x\theta\,\Delta x = 0, \qquad 0 \le \theta \le 1 \qquad \text{(sum of moments = 0)}$$

and thus dividing by Δx and letting $\Delta x \to 0$, show that

$$\frac{dV}{dx} = -w(x), \qquad \frac{dM}{dx} = -V$$

(b) Using the results in (a), show that the equation $EIy'' = M(x)$ becomes

$$EIy^{(IV)} = w(x)$$

assuming EI is constant. Compare with Exercise 4.

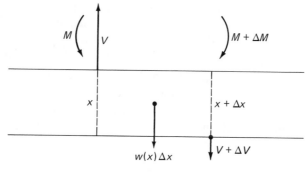

Figure 3.57

13 Applications to Biology

One of the most exciting fields of knowledge to which mathematical methods have been applied is that of *biology*. The possibility that mathematics could even be used successfully in dealing with the various natural processes of living beings from the lowliest microorganisms to mankind itself staggers the imagination. For what benefits might accrue if one could obtain the understanding provided by a mathematical formulation and solution of such problems as those pertaining to the brain and nervous system, the heart, the processes of digestion, cell production, and even the origin of life?

The rationale for attempting such a study is that if mathematics is capable of handling problems involving inanimate objects such as machines, why can it not also handle problems involving living beings, which can be thought of as very complex machines?

Although it is not possible in a book of this type to present many of the interesting mathematical applications to biology, we shall present a few and hope that the student may be motivated to pursue further study of this fascinating subject.

13.1 BIOLOGICAL GROWTH

A fundamental problem in biology involves that of *growth*, whether it be the growth of a cell, an organ, a human, a plant or a population. We have already dealt with some problems involving growth in Section 6, pages 106–111. In that section the fundamental differential equation was

$$\frac{dy}{dt} = \alpha y \tag{1}$$

having solution
$$y = ce^{\alpha t} \tag{2}$$

where c is an arbitrary constant. From this we see that growth occurs if $\alpha > 0$ while decay (or shrinkage) occurs if $\alpha < 0$.

One obvious defect of equation (1) and the corresponding solution (2) is that if $\alpha > 0$ then we have $y \to \infty$ as $t \to \infty$, so that as time passes the growth is unlimited. This conflicts with reality, for after a certain time elapses we know that a cell or individual stops growing, having attained a maximum size. The question naturally arises as to whether we can modify equation (1) so as to correspond to these biological facts. Let us see if we can formulate the problem mathematically.

Mathematical Formulation. To fix our ideas, let us suppose that y denotes the height of a human being (although as we have already mentioned it could refer to other things, such as cell size). It is natural to assume that the rate of change of height depends on the height in a more complicated manner than simple proportionality as in (1). Thus, we would have

$$\frac{dy}{dt} = F(y), \qquad y = y_0 \quad \text{for} \quad t = 0 \tag{3}$$

where y_0 represents the height at some specified time $t = 0$, and where F is some suitable but as yet unknown function. Since the linear function $F(y) = \alpha y$ is unsuitable, we are led to try a next order of approximation provided by a quadratic function $F(y) = \alpha y - \beta y^2$, where we choose the constant $\beta > 0$ in order to inhibit the growth of y as demanded by reality. The differential equation (3) thus becomes

$$\frac{dy}{dt} = \alpha y - \beta y^2, \qquad y = y_0 \quad \text{for} \quad t = 0 \tag{4}$$

It should be emphasized that this equation only provides a mathematical model which hopefully describes the biological facts of growth, and that if the model yields results in disagreement with reality it must be revised.

Solution Since equation (4) is one in which variables are separable, we have

$$\frac{dy}{\alpha y - \beta y^2} = dt \quad \text{or} \quad \int \frac{dy}{y(\alpha - \beta y)} = t + c$$

i.e., $\quad \int \frac{1}{\alpha} \left[\frac{1}{y} + \frac{\beta}{\alpha - \beta y} \right] dy = t + c \quad \text{or} \quad \frac{1}{\alpha} [\ln y - \ln (\alpha - \beta y)] = t + c \tag{5}$

Using the condition $y = y_0$ at $t = 0$, we see that $c = \frac{1}{\alpha} [\ln y_0 - \ln (\alpha - \beta y_0)]$. Thus

(5) becomes $\qquad \frac{1}{\alpha} [\ln y - \ln (\alpha - \beta y)] = t + \frac{1}{\alpha} [\ln y_0 - \ln (\alpha - \beta y_0)]$

Solving for y yields $\qquad y = \dfrac{\alpha/\beta}{1 + \left(\dfrac{\alpha/\beta}{y_0} - 1 \right) e^{-\alpha t}} \tag{6}$

If we take the limit of (6) as $t \to \infty$, we see, since $\alpha > 0$, that

$$y_{max} = \lim_{t \to \infty} y = \frac{\alpha}{\beta} \tag{7}$$

This shows that there is a limit to the growth of y as required by the biological facts, and tends to indicate the correctness of our mathematical model. As indicated in (7), this maximum is denoted by y_{max}.

To apply the result (6), let us suppose that the values of y corresponding to the times $t = 1$ and $t = 2$ (where we use some specified unit of time) are given respectively by y_1 and y_2. Then from (6) we see that

$$\frac{\alpha/\beta}{1 + \left(\dfrac{\alpha/\beta}{y_0} - 1 \right) e^{-\alpha}} = y_1, \qquad \frac{\alpha/\beta}{1 + \left(\dfrac{\alpha/\beta}{y_0} - 1 \right) e^{-2\alpha}} = y_2$$

or $\qquad \dfrac{\beta}{\alpha} (1 - e^{-\alpha}) = \dfrac{1}{y_1} - \dfrac{e^{-\alpha}}{y_0}, \quad \dfrac{\beta}{\alpha} (1 - e^{-2\alpha}) = \dfrac{1}{y_2} - \dfrac{e^{-2\alpha}}{y_0} \tag{8}$

To determine β/α and α in terms of y_0, y_1, and y_2, we can proceed as follows. Divide the members of the second equation in (8) by corresponding members of the first equation so as to eliminate β/α. This yields

$$1 + e^{-\alpha} = \frac{\dfrac{1}{y_2} - \dfrac{e^{-2\alpha}}{y_0}}{\dfrac{1}{y_1} - \dfrac{e^{-\alpha}}{y_0}} \tag{9}$$

Using some simple algebra $\qquad e^{-\alpha} = \dfrac{y_0(y_2 - y_1)}{y_2(y_1 - y_0)} \tag{10}$

If this is now substituted into the first of equations (8), we find

$$\frac{\beta}{\alpha} = \frac{y_1^2 - y_0 y_2}{y_1(y_0 y_1 - 2 y_0 y_2 + y_1 y_2)} \tag{11}$$

The values (10) and (11) can be used to write equation (6) in terms of suitably chosen values of y_0, y_1, and y_2. It is also of interest to note that the limiting value of y is

$$y_{max} = \lim_{t \to \infty} y = \frac{y_1(y_0 y_1 - 2 y_0 y_2 + y_1 y_2)}{y_1^2 - y_0 y_2} \tag{12}$$

Let us illustrate by some examples how these results can be applied.

ILLUSTRATIVE EXAMPLE 1

The mean heights in inches of male children at various ages are shown in Table 3.3. Use the data to predict the mean height of adult males at full growth.

Table 3.3 Mean Height of
Male Children
at Various Ages

Age	Height (in.)
Birth	19.4
1 year	31.3
2 years	34.5
3 years	37.2
4 years	40.3
5 years	43.9
6 years	48.1
7 years	52.5
8 years	56.8

Solution To cover the full set of data given in the table, let $t = 0, 1, 2$ correspond to the ages at birth, 4 years, and 8 years, respectively. Then we have

$$y_0 = 19.4, \qquad y_1 = 40.3, \qquad y_2 = 56.8$$

Substituting these values into equation (12) yields the value 66.9 inches, or 5 feet 7 inches as the required maximum mean height.

ILLUSTRATIVE EXAMPLE 2

According to the Bureau of the Census, the population of the United States for the years 1900–1960 is given in Table 3.4. Using the given data, determine (a) the theoretically maximum U.S. population, (b) the population expected in 1990, and (c) the population as of 1870.

Table 3.4 Population of
the United States

Year	Population (in millions)
1900	76.0
1910	92.0
1920	105.7
1930	122.8
1940	131.7
1950	151.1
1960	179.3

Solution To cover the full set of data given in the table, let $t = 0, 1, 2$ correspond to the years 1900, 1930, and 1960, respectively. Then we have

$$y_0 = 76.0, \qquad y_1 = 122.8, \qquad y_2 = 179.3 \qquad (13)$$

(a) Substituting these values into (12) yields $y_{max} = 346.3$; i.e., the maximum population of the United States will be about 346 million.

(b) Substituting the values (13) into (10) yields $e^{-\alpha} = 0.5117$. Using this and the result $\alpha/\beta = 346.3$ from part (a), we find from (6) that

$$y = \frac{346.3}{1 + (3.557)(0.5117)^t} \quad \text{or} \quad y = \frac{346.3}{1 + 3.557e^{-0.6678t}} \qquad (14)$$

Since the year 1990 corresponds to $t = 3$, we obtain on putting this value of t into (14) the result $y = 234.5$. Thus, the expected population in 1990 is about 235 million.

(c) The year 1870 corresponds to $t = -1$. Then putting this value of t into (14) we find $y = 43.6$. It is interesting that the actual population of the United States according to the Bureau of the Census was 39.8 million.

The above examples serve to show that the mathematical model (4) as expressed by equation (6) does have potential as a possible law of biological growth (or even growth in other fields, such as physics, chemistry, and economics). As we have already remarked, however, like any law intended for the mathematical description of nature (such as Newton's law or Kirchhoff's law), it must agree with experimental evidence or it will have to be changed. In this connection it is useful to have some criteria under which we can expect equation (6) to agree with reality. One of these is obtained by noting that the graph of (6) has the general appearance shown in Fig. 3.58. This

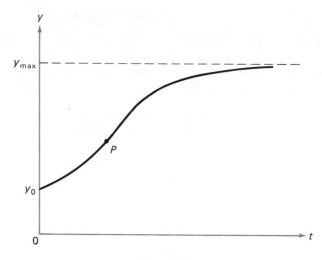

Figure 3.58

graph reveals that as t increases from 0, y increases from y_0 and gets closer and closer to y_{max}. The curve has increasing slope from $t = 0$ to a time corresponding to point P, and thereafter has decreasing slope. Thus, point P is a point of inflection and is obtained according to the usual methods of calculus by setting the second derivative $d^2y/dt^2 = 0$.

This S shaped curve is often called the *logistic curve* (from the Greek word *logistikos*, meaning *rational*) and is very widely used in the life sciences.

A EXERCISES

1. Write the equation of the logistic curve corresponding to the data of Table 3.3.

2. Check the accuracy of the results in (a) Illustrative Example 1 and (b) Illustrative Example 2 by comparing the calculated and actual data.

3. Table 3.5 shows the growth of a colony of bacteria over a number of days as measured by its size in square centimeters. (a) Find an equation for the area y in terms of time t. (b) Using the equation found in (a), compare computed values of the area with the actual ones. (c) What is the theoretical maximum size of the colony? Would you expect such a maximum to exist realistically? Explain.

Table 3.5 Growth of a Colony of Bacteria

Age t (days)	0	1	2	3	4	5	6
Area y (cm²)	1.20	3.43	9.64	19.8	27.2	33.8	37.4

4. The average height of a certain type of plant after 16, 32, and 48 days is given, respectively, by 21.6, 43.8, and 54.2 cm. Assuming that the pattern of growth follows the logistic curve, determine (a) the maximum theoretical height expected, (b) the equation of the logistic curve, and (c) the theoretical heights after 8, 24, and 60 days.

B EXERCISES

1. Derive the results (10) and (11) on page 150 and thus obtain (12).

2. Obtain the result (7), page 149, directly from the differential equation (4), page 149.

3. Show that the point of inflection of the logistic curve occurs where $y = \alpha/2\beta$ and that the corresponding value of t is

$$t_0 = \frac{1}{\alpha} \ln \left(\frac{\alpha}{\beta y_0} - 1 \right)$$

4. Show that the equation of the logistic curve can be written as $\qquad y = \dfrac{\alpha/\beta}{1 + e^{-\alpha(t - t_0)}}$

   where t_0 is the time corresponding to the inflection point.

5. Suppose that the "growth equation" in (4) is given by $\qquad \dfrac{dy}{dt} = \alpha y - \beta y^n$

   where α, β, and n are constants. Solve this and verify that the logistic curve is obtained in the special case where $n = 2$.

6. Show that the maximum value of y in Exercise 5 is given by $(\alpha/\beta)^{1/(1-n)}$.

7. Show how to modify (4), page 149, in the case where it represents population growth of a country if immigration into the country at a constant rate exists. Solve the resulting initial value problem.

8. Work Exercise 7 if there is immigration and emigration.

C EXERCISES

1. Suppose that the "growth equation" is given $\qquad \dfrac{dy}{dt} = \alpha y - \beta y^2 + \gamma y^3$

   where α, β, and γ are constants. (a) Solve this and verify that the logistic curve is obtained in the special case where $\gamma = 0$. (b) Determine whether a maximum value of y exists. (c) Do you believe that this equation would provide a better model for biological growth than that on page 149? Explain.

2. Show that if $\alpha/\beta y - 1$ is plotted on a logarithmic scale against t the resulting graph of the logistic curve is a straight line. Discuss how this observation can aid in obtaining values of the constants in the equation of the curve which may provide more accuracy. (*Hint:* Use B Exercise 4.)

3. Demonstrate the use of Exercise 2 by applying it to (a) Illustrative Example 1; (b) Illustrative Example 2; (c) A Exercise 3.

13.2 A PROBLEM IN EPIDEMIOLOGY

An important problem of biology and medicine deals with the occurrence, spreading, and control of a contagious disease, i.e., one which can be transmitted from one individual to another. The science which seeks to study such problems is called *epidemiology*, and if an unusually large percentage of a population gets the disease, we say that there is an *epidemic*.

Problems involving the spread of disease can be quite complicated. For example, it is known that some individuals may not actually get a disease even when exposed for long periods of time to others having the disease. We say in such a case that the individual has an *immunity* to the disease either because of having had the disease before and building up resistance to recurrence or by having initial resistance (*natural immunity*) so as not to get the disease at all. In some cases individuals are immune to the disease but are still capable of transmitting it to others; in such case they are called *carriers*. A well-known example of this is in the case of typhoid fever.

To present a simple mathematical model for the spread of a disease, let us assume that we have a large but finite population. To fix our ideas, let us suppose that we restrict ourselves to students in some large college or university who remain on campus for a relatively long period of time and do not have access to other communities. We shall suppose there are only two types of students, those who have the contagious disease, called *infected*, and those who do not have the disease, i.e., *uninfected*, but who are capable of developing it on first making contact with an infected student. We would like to obtain a formula for the number of infected students at any time, given that there is some specified number of infected students initially.

Mathematical Formulation. Suppose that at any time t there are N_i infected students and N_u uninfected students. Then if N is the total number of students, assumed constant, we have

$$N = N_i + N_u \tag{16}$$

The time rate of change in the number of infected students is then given by the derivative dN_i/dt, where in writing such a derivative we must suppose ideas similar to those considered in the problem of radioactivity on page 108. This derivative should depend in some way on N_i and thus N_u by virtue of (16). Assuming that dN_i/dt is a quadratic function of N_i as an approximation we have

$$\frac{dN_i}{dt} = a_0 + a_1 N_i + a_2 N_i^2 \tag{17}$$

where a_0, a_1, a_2 are constants. Now we would expect the time rate of change of N_i, i.e., dN_i/dt, to be zero where $N_i = 0$, i.e., there are no infected students, and where $N_i = N$, i.e., all students are infected. Then from (17) we have

$$a_0 = 0 \quad \text{and} \quad a_1 N + a_2 N^2 = 0 \quad \text{or} \quad a_2 = -\frac{a_1}{N}$$

so that (17) becomes $\dfrac{dN_i}{dt} = a_1 N_i - \dfrac{a_1 N_i^2}{N}$ or $\dfrac{dN_i}{dt} = \dfrac{a_1}{N} N_i(N - N_i)$

i.e., $$\frac{dN_i}{dt} = kN_i(N - N_i) \tag{18}$$

where we have written $k = a_1/N$ a constant. If we suppose that, initially, $t = 0$, there are N_0 infected students, then

$$N_i = N_0 \quad \text{at} \quad t = 0 \tag{19}$$

Our mathematical formulation thus consists of the initial value problem given by (18) and (19).

Solution The differential equation (18) is separable and is easily solved. In fact we may notice that the equation (18) and initial condition (19) are identical with the "growth equation" and initial condition given in (4), page 149, where

$$\alpha = kN, \qquad \beta = k, \qquad y_0 = N_0 \tag{20}$$

Thus from (6), page 149, $\qquad N_i = \dfrac{N}{1 + \left(\dfrac{N}{N_0} - 1\right)e^{-kNt}} \tag{21}$

Interpretation. The graph of (21) is the *logistic curve*, i.e., the S shaped curve already described on page 152. It follows that results already obtained in the discussion and exercises in connection with the "growth equation" apply as well to the "disease equation" (18) or (21). In particular, we see from the shape of the logistic curve that initially there is a gradual increase in the number of infected students, followed by a rather sharp rise in their number near the inflection point, and finally a tapering off. The limiting case occurs where all students become infected, as seen by noting from (21) that $N_i \to N$ as $t \to \infty$. From a realistic point, of course, this hopefully would not occur in practice since infected students once they are discovered would be isolated or *quarantined* so as to prevent contact with the others. The problem of epidemics where quarantine is imposed is more complicated and is considered in Chapter Ten.

In cases where large populations are involved, such as in a city or country, it is easy to see why disaster could prevail even with attempts to quarantine. Such was the case for example in the disease often called the "black plague," which ravaged much of Europe for several years in the middle of the 14th century.

A EXERCISES

1. Suppose that at time t the percentages of infected and uninfected students are given by $100p_i$ and $100p_u$, respectively, while the initial percentage of infected students is $100p_0$. Show that

$$p_i = \frac{p_0}{p_0 + (1 - p_0)e^{-\alpha t}}$$

2. If 5 per cent of the students at a university have a contagious disease and 1 week later a total of 15 per cent have developed the disease, what percentages will have developed it 2, 3, 4, and 5 weeks later assuming no quarantine? Graph the results.

3. Show that the logistic curve (21) has a point of inflection at the point where half the number of students have become infected.

4. Can Exercise 2 be worked if the assumption is made that the rate at which students become infected is proportional to the number infected? Justify your answer.

5. Show that the mathematical model for contagious diseases used on page 154 is the same as that in which it is assumed that the rate at which students become infected is proportional to the product of the numbers which are and are not infected. Can you justify the reasoning involved here?

B EXERCISES

1. An infected laboratory animal is introduced into a population of 24 uninfected animals. After 3 days another animal becomes infected. In how many more days can we expect all but one of the animals to be infected? What assumptions are you making?

2. Is it possible to determine the least number of days in which we can expect all the animals of Exercise 1 to become infected? Explain.

3. Suppose that the cumulative percentages of infected animals in some population at the ends of three equal intervals of time are given by $100p_0$, $100p_1$ and $100p_2$, respectively, where $0 < p_0 < p_1 < p_2 < 1$. (a) Show that

$$\frac{1 - 2p_1}{p_1^2} = \frac{1 - (p_0 + p_2)}{p_0 p_2}$$

(b) Use this to show that $p_1 = \frac{1}{2}$ if and only if $p_0 + p_2 = 1$, and explain the significance.
(c) Show that if $p_1^2 = p_0 p_2$ then we must have $p_0 = p_1 = p_2$ and give an interpretation of these results.

4. Suppose the percentage of infected animals at some time is $100p_0$, while at times t_1 and t_2 later the percentages are $100p_1$ and $100p_2$, respectively. Show that

$$\left[\frac{p_0(1 - p_1)}{p_1(1 - p_0)}\right]^{t_2} = \left[\frac{p_0(1 - p_2)}{p_2(1 - p_0)}\right]^{t_1}$$

C EXERCISES

1. In a certain community the number of individuals is increasing at a constant rate r. Suppose that at time $t = 0$ there is a number N_0 of infected individuals and that the population at this time is N. Using a mathematical model analogous to that given in the text, show that an initial value problem for the number of infected individuals N_i in the community at any time $t > 0$ is given by

$$\frac{dN_i}{dt} - k(N + rt)N_i = -kN_i^2, \qquad N_i(0) = N_0 \text{ where } k \text{ is a constant}$$

2. By solving the initial value problem in Exercise 1, show that

$$N_i = e^{kNt + (1/2)krt^2}\left[N_0^{-1} + k\int_0^t e^{kNv + (1/2)krv^2}\, dv\right]^{-1}$$

(*Hint*: The differential equation is a *Bernoulli equation* as discussed in B Exercises 1–7 on page 55.)

13.3 ABSORPTION OF DRUGS IN ORGANS OR CELLS

For purposes of mathematical analysis in biology, it is often convenient to consider an organism (such as a human, animal, or plant) as a collection of individual components called *compartments*. A compartment may be an organ (such as the stomach, pancreas, or liver) or a group of cells which together act as some unit. An important problem consists of determining the absorption of chemicals, such as drugs, by cells or organs. This has practical application in the field of medicine, since it may happen that certain harmful drugs could build up in an organ or group of cells leading ultimately to their destruction. The simplest type of problem deals with only one compartment. However, it may be important for some purposes to deal with systems involving two or more compartments which interact with one another. As would be expected, the difficulty of mathematical analysis tends to increase with the number of compartments. The following examples will serve to illustrate the kinds of problems which can arise.

ILLUSTRATIVE EXAMPLE 3

A liquid carries a drug into an organ of volume V cm^3 at a rate a cm^3/sec and leaves at a rate b cm^3/sec. The concentration of the drug in the entering liquid is c g/cm^3. (a) Write a differential equation for the concentration of the drug in the organ at any time together with suitable conditions, and (b) solve the equation.

Solution (a) The situation is described schematically by Fig. 3.59 which shows a single compartment of volume V together with inlet and outlet. If we let x be the concentration of drugs in the organ (i.e., the number of grams of the drug per cubic centimeter), the amount of drug in the organ at any time t is given by

$$(V \text{ cm}^3)(x \text{ g/cm}^3) = xV \text{ g} \tag{22}$$

The number of grams per second entering the organ at time t is given by

$$(a \text{ cm}^3/\text{sec})(c \text{ g/cm}^3) = ac \text{ g/sec} \tag{23}$$

The number of grams per second leaving the organ is given by

$$(b \text{ cm}^3/\text{sec})(x \text{ g/cm}^3) = bx \text{ g/sec} \tag{24}$$

Now the rate of change of the amount of drug in the organ is equal to the rate at which the drug enters minus the rate at which it leaves. Thus, from (22), (23), and (24)

$$\frac{d}{dt}(xV) = ac - bx \tag{25}$$

If we assume that the concentration of the drug in the organ at $t = 0$ is x_0, then

$$x = x_0 \quad \text{at} \quad t = 0 \tag{26}$$

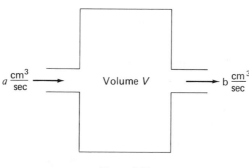

Figure 3.59

(b) If we assume that a, b, c, and V are constants, which is not required according to the formulation in part (a), equation (25) can be written

$$V \frac{dx}{dt} = ac - bx \tag{27}$$

Solving (27) by separation of variables using condition (26) gives

$$x = \frac{ac}{b} + \left(x_0 - \frac{ac}{b}\right)e^{-bt/V} \tag{28}$$

Two special cases are of interest.

Case 1. $a = b$. In this case the rate at which the drug enters is equal to the rate at which it leaves and (28) becomes

$$x = c + (x_0 - c)e^{-bt/V} \tag{29}$$

Case 2. $a = b$ and $x_0 = 0$. In this case the inflow and outflow rates are equal and the initial concentration of the drug in the organ is 0. Then we have

$$x = c(1 - e^{-bt/V}) \tag{30}$$

A EXERCISES

1. A liquid carries a drug into an organ of volume 500 cm³ at a rate of 10 cm³/sec and leaves at the same rate. The concentration of the drug in the entering liquid is 0.08 g/cm³. Assuming that the drug is not present in the organ initially, find (a) the concentration of the drug in the organ after 30 sec and 120 sec, respectively; (b) the steady-state concentration.

2. How long would it take for the concentration of the drug in the organ of Exercise 1 to reach (a) 0.04 g/cm³; (b) 0.06 g/cm³?

3. Work Exercise 1 if the initial concentration of the drug in the organ is 0.20 g/cm³.

4. Obtain the graph of (29) for the cases (a) $x_0 = 0$, (b) $x_0 = c$, (c) $x_0 > c$, (d) $x_0 < c$, and interpret.

B EXERCISES

1. Suppose that the maximum concentration of a drug present in a given organ of constant volume V must be c_{max}. Assuming that the organ does not contain the drug initially, that the liquid carrying the drug into the organ has constant concentration $c > c_{max}$, and that the inflow and outflow rates are both equal to b, show that the liquid must not be allowed to enter for a time longer than

$$\frac{V}{b} \ln\left(\frac{c}{c - c_{max}}\right)$$

2. Work Exercise 1 if the initial drug concentration in the organ is x_0.

3. Suppose that in the problem of the text the rate a at which the liquid of constant drug concentration c enters the organ is greater than the rate b at which it leaves. As a result, suppose that the volume of the organ expands at a constant rate r so that $V = V_0 + rt$. If the initial

concentration of the drug in the organ is x_0, show that the concentration at any later time t is

$$x = \frac{ac}{b+r} + \left(x_0 - \frac{ac}{b+r}\right)\left(\frac{V_0}{V_0 + rt}\right)^{(b+r)/r}$$

4. Show that in case $r \to 0$ the result of Exercise 3 reduces to (28), page 158.

C EXERCISES

1. Suppose that in the problem of the text the volume of the organ varies with time t according to $V = V_0 + r \sin \omega t$, where V_0, r, and ω are constants. Show that the drug concentration in the organ is given by the initial value problem

$$\frac{dx}{dt} + \left(\frac{b + r\omega \cos \omega t}{V_0 + r \sin \omega t}\right) x = \frac{ca}{V_0 + r \sin \omega t}, \quad x(0) = 0$$

where it is assumed that the drug is not present in the organ initially.

2. Show that the solution to Exercise 1 is given by

$$x = \frac{ace^{-f(t)}}{b(V_0 + r \sin \omega t)} \int_0^t f(u)du \quad \text{where} \quad f(t) = \frac{2b}{\omega\sqrt{V_0^2 - r^2}} \tan^{-1}\left(\frac{r + V_0 \tan \frac{1}{2}\omega t}{\sqrt{V_0^2 - r^2}}\right)$$

14 Applications to Economics

In recent years there has been an increasing interest in applications of mathematics to economics. However, since economics involves many unpredictable factors, such as psychology and political decisions, mathematical formulations of its problems are difficult. In spite of this, however, progress can be made as we shall attempt to illustrate in this section. It should be emphasized that, as in problems of science and engineering, any results arrived at theoretically must ultimately be tested in the light of reality.

14.1 SUPPLY AND DEMAND

Suppose that we have a commodity such as wheat or oil. Let p be the price of this commodity for some specified unit (e.g., bushel of wheat or barrel of oil) at any time t. Then we can think of p as a function of t so that $p(t)$ is the price at time t.

The number of units of the commodity which are desired per unit time by consumers at any time t is called the *demand* and is denoted by $D(t)$, or briefly D. This demand can depend not only on the price p at any time t, i.e., $p(t)$, but also on the direction in which consumers feel that prices may be going, i.e., the time rate of change of price or derivative $p'(t)$. For example, if prices are high at time t but consumers feel that they may move higher, the demand may tend to increase. In symbols this dependence of D on $p(t)$ and $p'(t)$ can be written

$$D = f(p(t), p'(t)) \tag{1}$$

We call f the *demand function*.

Similarly, the number of units of the commodity which are made available per unit of time by producers at any time t is called the *supply* and is denoted by $S(t)$, or briefly S. As in the case of demand, S also can depend on $p(t)$ and $p'(t)$. For example, if prices are high at time t but producers feel that they may move higher, the supply made available may tend to increase in anticipation of the higher prices. In symbols this dependence of S on $p(t)$ and $p'(t)$ can be written

$$S = g(p(t), p'(t)) \tag{2}$$

We call g the *supply function*.

In making the above remarks we have implicitly assumed the following.

(a) *A competitive or free economy*. The marketplace is one in which consumers and producers compete to determine prices. Because of this, producers have to worry about how far they can raise prices, since consumers may patronize others who offer lower prices or may reduce their demand.

(b) *No time lag in supply*. Equation (2) assumes that producers use the rate of change of price at time t, i.e., $p'(t)$, to decide on what supply is available. This is an approximation to reality, since in practice there is a time lag τ between actual production time and marketing to the consumer. In such case, (2) should be replaced by

$$S = g(p(t - \tau), p'(t - \tau)) \tag{3}$$

(c) *No consideration of other commodity prices or income*. Prices of other commodities in the market besides the one under consideration or the average income of consumers at various times may affect the supply or demand. These are not taken into account in the above economic model.

(d) *Prices, demand, and supply are continuous*. In practice, we cannot subdivide prices or numbers of a commodity indefinitely. For example, it makes no sense to speak of the number of bananas between 240 and 241, or that the price of a commodity will assume all values between two given ones. In spite of this, we adopt the assumption that such discrete variables can be approximated to a good degree of accuracy by continuous variables in much the same manner as indicated in the radioactivity problem on page 108.

Now if the supply S exceeds the demand D, there is a tendency for prices to adjust themselves in the marketplace until the supply is reduced to equal the demand, i.e., until $S = D$. This is especially true, for example, when there is the possibility that the commodity can spoil, such as bananas. Similarly, if the demand D exceeds the supply S, the prices will tend to adjust until the supply equals the demand, i.e., $S = D$. This leads us to adopt the following

ECONOMIC PRINCIPLE OF SUPPLY AND DEMAND. *The price of a commodity at any time t, i.e., $p(t)$, is determined by the condition that the demand at t is equal to the supply at t*, or using (1) and (2)

$$f(p(t), p'(t)) = g(p(t), p'(t)) \tag{4}$$

The equation (4) is a *first-order differential equation* for determining $p(t)$ if the forms of the functions f and g are known. If (3) is used instead of the right side of (4), the resulting equation is called a *first-order differential-difference equation*.

The question now naturally arises as to what form the functions f and g in (4) should take. The simplest ones are linear functions in $p(t)$ and $p'(t)$, i.e.,

$$D = f(p(t), p'(t)) = a_1 p(t) + a_2 p'(t) + a_3, \Big\}$$
$$S = g(p(t), p'(t)) = b_1 p(t) + b_2 p'(t) + b_3 \Big\} \tag{5}$$

where the a's and b's are constants. In such case (4) becomes

$$a_1 p(t) + a_2 p'(t) + a_3 = b_1 p(t) + b_2 p'(t) + b_3 \tag{6}$$

or

$$(a_2 - b_2)p'(t) + (a_1 - b_1)p(t) = b_3 - a_3 \tag{7}$$

Let us assume that $a_1 \neq b_1, a_2 \neq b_2, a_3 \neq b_3$. Then we can write (7) as

$$p'(t) + \left(\frac{a_1 - b_1}{a_2 - b_2}\right) p(t) = \frac{b_3 - a_3}{a_2 - b_2} \tag{8}$$

Solving this linear first-order equation subject to $p = p_0$ at $t = 0$ gives

$$p(t) = \frac{b_3 - a_3}{a_1 - b_1} + \left(p_0 - \frac{b_3 - a_3}{a_1 - b_1}\right) e^{-(a_1 - b_1)t/(a_2 - b_2)} \tag{9}$$

Case 1, $p_0 = (b_3 - a_3)/(a_1 - b_1)$. In this case we see from (9) that $p(t) = p_0$ and the price remains constant for all time.

Case 2, $(a_1 - b_1)/(a_2 - b_2) > 0$. In this case we see from (9) that the price $p(t)$ approaches $(b_3 - a_3)/(a_1 - b_1)$ as a limit when t increases, assuming of course that this limit is positive. For this case we have *price stability*, and the limit $(b_3 - a_3)/(a_1 - b_1)$ is called the *equilibrium price*.

Case 3, $(a_1 - b_1)/(a_2 - b_2) < 0$. In this case we see from (9) that the price $p(t)$ increases indefinitely as t increases, assuming that $p_0 > (b_3 - a_3)/(a_1 - b_1)$ i.e., we have continued inflation or *price instability*. This process may continue until economic factors change, which may result in a change in equation (7).

It is possible of course that from time to time the constants in (5) change so that over one interval of time we have one set of constants, over another interval a different set, etc. More generally, the a's and b's could themselves be functions of t. We shall consider these and other interesting cases in the exercises.

ILLUSTRATIVE EXAMPLE 1

The demand and supply of a certain commodity are given in thousands of units by $D = 48 - 2p(t) + 3p'(t)$, $S = 30 + p(t) + 4p'(t)$, respectively. If at $t = 0$ the price of the commodity is 10 units, find (a) the price at any later time t and (b) whether there is price stability or instability.

Solution   The price $p(t)$ is determined by equating demand and supply, i.e.,

$$48 - 2p(t) + 3p'(t) = 30 + p(t) + 4p'(t) \quad \text{or} \quad p'(t) + 3p(t) = 18. \tag{10}$$

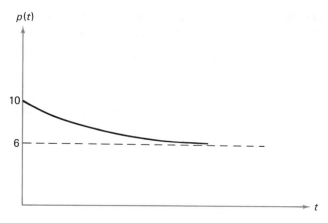

Figure 3.60

Solving this first-order linear equation subject to $p = 10$ at $t = 0$ gives

$$p(t) = 6 + 4e^{-3t} \tag{11}$$

(b) From (11) we see that, as $t \to \infty$, $p \to 6$. Thus, we have price stability, and the equilibrium price is 6 units.

The graph of price versus time is shown in Fig. 3.60.

14.2 INVENTORY

The above analysis made use of the situation where supply and demand are already equal and a price determined. However, it did not examine the dynamical situation where supply and demand are unequal but supply changes with time to meet demand. If, for example, supply is greater than demand, then producers have a certain quantity of the commodity in their possession, which is called their *inventory* of the commodity, which it is their hope to sell. On the other hand, if demand is greater than supply, the producers must acquire inventory. Our problem is to formulate mathematically how inventory changes with time as a result of supply and demand.

Mathematical Formulation. To accomplish this, let us denote by $q(t)$ the quantity or number of units of a commodity C available at time t. Then $q(t + \Delta t) = q(t) + \Delta q$ is the quantity available at time $t + \Delta t$. Thus, we have

$$\text{quantity accumulated in interval } t \text{ to } t + \Delta t = \Delta q = q(t + \Delta t) - q(t) \tag{12}$$

Now from the definitions on pages 159–160, we have

$S =$ number of units of C supplied per unit of time by producers at time t
$D =$ number of units of C demanded per unit of time by consumers at time t

Then in the time interval between t and $t + \Delta t$ the numbers of units supplied by producers and demanded by consumers are given very nearly by $S \, \Delta t$ and $D \, \Delta t$

respectively, the results being accurate apart from terms involving $(\Delta t)^2$ and higher. Thus, quantity accumulated in interval t to $t + \Delta t$

$$= S \, \Delta t - D \, \Delta t + \text{terms involving } (\Delta t)^2 \text{ or higher} \tag{13}$$

From (12) and (13) $\quad \Delta q = S \, \Delta t - D \, \Delta t + \text{terms involving } (\Delta t)^2 \text{ or higher} \tag{14}$

Thus, $\qquad \dfrac{\Delta q}{\Delta t} = S - D + \text{terms involving } \Delta t \text{ or higher} \tag{15}$

Taking the limit as $\Delta t \to 0$, $\qquad \dfrac{dq}{dt} = S - D \tag{16}$

In the special case where q is constant, we have $S = D$.

Equation (16) forms the basis for further analysis on prices. As an illustration, let us suppose that a producer wishes to protect his profits by a requirement that the rate at which he increases the price is proportional to the rate at which the inventory declines. In such case

$$\frac{dp}{dt} = -\alpha \frac{dq}{dt} \tag{17}$$

where $\alpha > 0$ is the proportionality constant assumed known, so that using (16),

$$\frac{dp}{dt} = -\alpha(S - D) \tag{18}$$

Since S and D can be expressed in terms of p, equation (18) is a differential equation for p. Consider the following

ILLUSTRATIVE EXAMPLE 2

Suppose that the supply and demand are given in terms of price p by $S = 60 + 2p$, $D = 120 - 3p$, respectively, and the proportionality constant in (17) is $\alpha = 4$. (a) Write the differential equation for p, and (b) determine the price at any time $t > 0$ assuming that $p = 8$ at $t = 0$.

Solution (a) From (18) the required differential equation for p is

$$\frac{dp}{dt} = -4(60 + 2p - 120 + 3p) \quad \text{or} \quad \frac{dp}{dt} + 20p = 240 \tag{19}$$

(b) Solving (19) as a first-order linear differential equation (or one with variables separable) yields

$$p = 12 + ce^{-20t}$$

Using $p = 8$ at $t = 0$ gives $c = -4$, and so

$$p = 12 - 4e^{-20t} \tag{20}$$

The graph of (20) is given in Fig. 3.61. Note that the price increases from 8 to the equilibrium price 12. This equilibrium price is also obtained by setting the supply equal to the demand, i.e., $60 + 2p = 120 - 3p$ or $p = 12$.

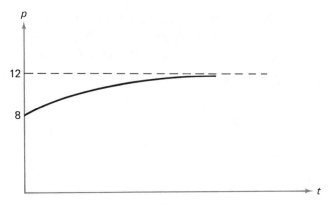

Figure 3.61

A EXERCISES

1. The demand and supply of a certain commodity are given in thousands of units, respectively, by $D = 120 + p(t) - 5p'(t)$, $S = 60 - 2p(t) - 3p'(t)$. At $t = 0$ the price of the commodity is 5 units. (a) Find the price at any later time and obtain its graph. (b) Determine whether there is price stability and the equilibrium price if one exists.

2. The demand and supply of a certain commodity are given in thousands of units, respectively, by $D = 40 + 3p(t) + p'(t)$, $S = 160 - 5p(t) - 3p'(t)$. At $t = 0$ the price of the commodity is 20 units. (a) Find the price at any later time and obtain its graph. (b) Determine whether there is price stability and the equilibrium price if one exists.

3. Work Exercise 2 if $D = 40 + p'(t)$, $S = 160 - 3p'(t)$.

4. To protect his profits, a producer decides that the rate at which he will increase prices should be numerically equal to three times the rate at which his inventory decreases. Assuming that the supply and demand are given in terms of the price p by $S = 80 + 3p$, $D = 150 - 2p$ and that $p = 20$ at $t = 0$, find the price at any time.

5. Determine the inventory of the producer in Exercise 4 at any time t if he has 2800 units of the commodity at time $t = 0$.

B EXERCISES

1. The supply and demand of a commodity are given in thousands of units by $S = 24(2 - e^{-2t}) + 16p(t) + 10p'(t)$, $D = 240 - 8p(t) - 2p'(t)$, respectively. At $t = 0$ the price of the commodity is 12 units. (a) Find the price at any later time and obtain its graph. (b) Determine whether there is price stability and the equilibrium price if one exists.

2. Suppose that the producer of Exercise 1 decides that to protect his profits the rate at which he will increase prices should be numerically equal to one fourth the rate at which his inventory decreases. Find the price at any time if the price at $t = 0$ is 12 units.

3. If the supply and demand of a commodity are given by (5) on page 161, discuss the cases (a) $a_2 = b_2$, (b) $a_1 = b_1$, $a_2 \neq b_2$, (c) $a_3 = b_3$, $a_2 \neq b_2$.

1. Solve completely the case where the supply and demand are given by $S = k_1 + k_2 p$, $D = k_3 + k_4 p$, and where $dp/dt = -\alpha \, dq/dt$. It is assumed that k_1, k_2, k_3, k_4, and α are constants. Discuss the economic implications treating all situations which arise.

2. The supply and demand of a commodity are given in thousands of units respectively by $D = 80 - 4p(t) \sin 4t + 5p'(t) \cos 4t$, $S = 120 + 8p(t) \sin 4t + 4p'(t) \cos 4t$. At $t = 0$ the price of the commodity is 25 units. Find the price at any time and discuss the economic implications.

four

linear differential equations

1 The General *n*th-Order Linear Differential Equation

In Chapter Two we found that the first-order linear differential equation

$$\frac{dy}{dx} + Py = Q \tag{1}$$

where P and Q are functions of x could be solved exactly by use of an integrating factor. In this chapter we shall concentrate on linear differential equations of order higher than one. Let us recall the following

Definition. A linear differential equation of order n has the form

$$a_0(x)\frac{d^ny}{dx^n} + a_1(x)\frac{d^{n-1}y}{dx^{n-1}} + a_2(x)\frac{d^{n-2}y}{dx^{n-2}} + \cdots + a_{n-1}(x)\frac{dy}{dx} + a_n(x)y = F(x) \tag{2}$$

where $a_0(x), a_1(x), \ldots, a_n(x)$, often denoted briefly by $a_0, a_1, \ldots, a_n$ and $F(x)$ depend only on x and not on y.*

If $n = 1$, equations (1) and (2) are equivalent. If $n = 2$, then (2) becomes

$$a_0(x)\frac{d^2y}{dx^2} + a_1(x)\frac{dy}{dx} + a_2(x)y = F(x) \tag{3}$$

which is a second-order linear differential equation. If all the coefficients $a_0, a_1, \ldots,$ a_n in (2) are constants, i.e., do not depend on x, we call the equation a *linear differential equation with constant coefficients*. However, if not all these coefficients are constants, we call the equation a *linear differential equation with variable coefficients*.

Example 1. The equations

$$\frac{d^2y}{dx^2} - 3\frac{dy}{dx} + 2y = 4\sin 3x \quad \text{and} \quad 3y''' + 2y'' - 4y' + 8y = 3e^{-x} + 5x^2$$

are linear differential equations with constant coefficients of orders 2 and 3.

Example 2. The equations $xy'' - 2y' + x^2y = x^3 - 2$ and $\dfrac{d^4y}{dx^4} + xy = 0$

are linear differential equations with variable coefficients of orders 2 and 4.

    Remark. We have already seen in Chapter Two that the first-order linear differential equation (1), or equivalently (2) with $n = 1$, can always be solved exactly by use of an appropriate integrating factor. Thus, the solution of (1) is given by

$$y = e^{-\int P\,dx}\int Qe^{\int P\,dx}\,dx + ce^{-\int P\,dx} \tag{4}$$

---

* An *n*th-order differential equation which cannot be written in this form is called *non-linear*.

and we say that this solution is exact even though we may not be able to actually express the indicated integrals in closed form. The question naturally arises as to whether (2) can be solved exactly, i.e., in terms of integrals such as in (4), for the case where $n \geq 2$. The answer is that no result such as (4) has ever been obtained for the equation (2) if $n \geq 2$. However, (2) with $n \geq 2$ can be solved exactly in various special cases, which we shall examine in this and later chapters. One class of linear differential equations which can always be solved and which fortunately also arises frequently in applications is that with *constant coefficients*. A large part of this chapter will be devoted to this case.

In discussing equation (2) it is convenient to use the symbols $D, D^2, D^3, \ldots$ to indicate the *operations* of taking the first, second, third, ... derivatives of whatever follows. Thus, Dy is the same as dy/dx, D^2y is the same as $d^2y/dx^2, \ldots$. We call the symbols $D, D^2, \ldots$ *operators* because they define an operation to be performed. Similarly xD^2 is an operator denoting the operation of taking a second derivative and then multiplying by x.

With this symbolic notation (2) may be written

$$a_0 D^n y + a_1 D^{n-1} y + a_2 D^{n-2} y + \cdots + a_{n-1} Dy + a_n y = F \tag{5}$$

or
$$(a_0 D^n + a_1 D^{n-1} + a_2 D^{n-2} + \cdots + a_{n-1} D + a_n)y = F \tag{6}$$

in which it is understood that each term in the parentheses is operating on y and the results are added. If we write for brevity

$$\phi(D) \equiv a_0 D^n + a_1 D^{n-1} + a_2 D^{n-2} + \cdots + a_{n-1} D + a_n \tag{7}$$

equation (6) may be written conveniently as $\phi(D)y = F$.

Example 3. The differential equation $\phi(D)y = F$, where $F = x^2 + e^x$, and $\phi(D) \equiv D^3 + 5D^2 + 6D - 8$ actually is a short-cut way of writing

$$\frac{d^3 y}{dx} + 5\frac{d^2 y}{dx^2} + 6\frac{dy}{dx} - 8y = x^2 + e^x$$

Example 4. The differential equation $\phi(D)y = F$, where $F = 4 \sin 2x$ and $\phi(D) \equiv xD^2 - 2D + 3x^2$ is a short-cut way of writing

$$x\frac{d^2 y}{dx^2} - 2\frac{dy}{dx} + 3x^2 y = 4 \sin 2x$$

On considering what the symbols mean, it should be clear to the student that

$$D^n(u + v) = D^n u + D^n v, \qquad D^n(au) = aD^n u \tag{8}$$

where u and v are differentiable functions, a is any constant, and n is any positive integer. Any operator such as D^n which has the properties shown in (8) is called a *linear operator.** It is easy to show that $\phi(D)$ given by (7) is a linear operator, i.e.,

$$\phi(D)(u + v) = \phi(D)u + \phi(D)v, \qquad \phi(D)(au) = a\phi(D)u \qquad (9)$$

The results follow directly by interpreting both sides and showing that they are equal.[†]

Let us consider the general nth-order linear equation (2). We are particularly interested in knowing how we may obtain its general solution. In order to do this we associate with the general equation

$$\phi(D)y = F(x) \qquad (10)$$

another equation $\qquad\qquad \phi(D)y = 0 \qquad (11)$

obtained by replacing $F(x)$ by zero. Of course, if $F(x)$ is already zero, then equations (10) and (11) are really the same. We assume $F(x) \not\equiv 0$.

Terminology. We refer to equation (10) with $F(x)$ not identically zero as the given equation, or the equation with right-hand member not zero, and shall refer to the associated equation (11) as the complementary equation, the reduced equation, or the equation with right-hand member zero.[‡]

Fundamental Theorem I. If $y = u(x)$ is any solution of the given equation (10), and $y = v(x)$ is any solution of the complementary equation (11), then $y = u(x) + v(x)$ is a solution of (10).

Proof. Since u is a solution of (10), we have

$$\phi(D)u = F(x) \qquad (12)$$

Since v is a solution of (11), $\qquad \phi(D)v = 0 \qquad (13)$

Adding (12) and (13), $\phi(D)(u + v) = F(x)$ which is another way of saying that $y = u(x) + v(x)$ is a solution.

In Chapter One we defined the general solution of an nth-order differential equation as one having n arbitrary constants. From Fundamental Theorem I above it follows that $y = u(x) + v(x)$ will be the general solution if $u(x)$ has no arbitrary constants, while $v(x)$ has n arbitrary constants. If $v(x)$ has n arbitrary constants, then it is the general solution of the complementary equation (11). If $u(x)$ has no arbitrary constants it is a particular solution of the given equation (10).

---

* It is possible to extend these ideas to cases where n is a negative integer or a fraction. For example, we may define a "half derivative" operator $D^{1/2} \equiv d^{1/2}/dx^{1/2}$ as one which, when operating twice on a function, yields the first derivative of the function. For an interesting account of such operators see [21]. See also C Exercises 10 and 11, page 285.

† Because $\phi(D)$ is a linear operator the symbol L is sometimes used in place of $\phi(D)$. We shall not use this however because L is used in many other connections, e.g., inductance, length, etc.

‡ The term *homogeneous* is used in many texts. However, this word is used in too many different and unrelated connections in mathematics. In this text we give the word a needed rest.

Terminology. We shall call the general solution of the complementary or re- duced equation (11) the *complementary solution** and will sometimes denote it by y_c. We shall refer to a selected solution of the given equation (no arbitrary constant) as a *particular solution†* of the given equation and will sometimes denote it by y_p. The above remarks are of such importance for the later work of this chapter that we summarize them in the following

Fundamental Theorem II. The general solution of $\phi(D)y = F(x)$ may be obtained by finding a particular solution y_p of this equation and adding it to the complementary solution y_c, which is the general solution of $\phi(D)y = 0$.

ILLUSTRATIVE EXAMPLE 1

Find the general solution of $\dfrac{d^2y}{dx^2} - 5\dfrac{dy}{dx} + 6y = 3x.$

Solution   In operator notation this may be written

$$(D^2 - 5D + 6)y = 3x \tag{14}$$

The complementary or reduced differential equation is

$$(D^2 - 5D + 6)y = 0 \tag{15}$$

It can be verified that $y_c = c_1 e^{3x} + c_2 e^{2x}$ is the complementary solution of (14) since it has two arbitrary constants and is thus the general solution of (15).

It can also be verified that $y_p = \frac{1}{2}x + \frac{5}{12}$ satisfies (14) and is thus a particular solution.

From the Fundamental Theorem II we see that the required general solution is

$$y = y_c + y_p = c_1 e^{3x} + c_2 e^{2x} + \tfrac{1}{2}x + \tfrac{5}{12}$$

The differential equation of Illustrative Example 1 is a linear differential equation with *constant* coefficients. As another example involving a differential equation with *variable* coefficients, let us consider the following

ILLUSTRATIVE EXAMPLE 2

Find the general solution of $x^2 y'' - 2xy' + 2y = \ln x.$

Solution   In operator notation this may be written

$$(x^2 D^2 - 2xD + 2)y = \ln x \tag{16}$$

The complementary or reduced differential equation is

$$(x^2 D^2 - 2xD + 2)y = 0 \tag{17}$$

Now we can verify that $y_c = c_1 x^2 + c_2 x$ is the complementary solution of (16) since it has the required two arbitrary constants and is thus the general solution of (17).

---

* The term *complementary function* or *complementary integral* is also used.
† The term *particular integral* is also used.

We can also verify that $y_p = \frac{1}{2}\ln x + \frac{3}{4}$ satisfies (16), and is thus a particular solution.

Thus, from Fundamental Theorem II, the required general solution is

$$y = y_c + y_p = c_1 x^2 + c_2 x + \tfrac{1}{2}\ln x + \tfrac{3}{4}$$

From these illustrative examples it is clear that to obtain the general solution of $\phi(D)y = F(x)$, the following questions must be answered.

Questions

1. How do we find the complementary solution?
2. How do we find a particular solution?

We shall answer these questions in Sections 3 and 4 for the very important case where the differential equations have constant coefficients, which fortunately turns out to be relatively easy. The variable coefficient case can be solved exactly only in special situations, and we consider some of these later.

2 Existence and Uniqueness of Solutions of Linear Equations

In Chapter One, page 24, an existence and uniqueness theorem for first-order differential equations was stated. We now state without proof a corresponding theorem for the general nth-order linear equation

$$[a_0(x)D^n + a_1(x)D^{n-1} + \cdots + a_n(x)]y = F(x) \tag{18}$$

Existence–Uniqueness Theorem. Let $a_0(x) \neq 0$, $a_1(x), \ldots, a_n(x)$ and $F(x)$ be functions which are continuous in the interval $a \leq x \leq b$, and suppose that $p_0, p_1, \ldots, p_{n-1}$ are given constants. Then there exists a unique solution $y(x)$ satisfying the differential equation (18) and also the conditions

$$y(c) = p_0, \qquad y'(c) = p_1, \ldots, y^{(n-1)}(c) = p_{n-1} \tag{19}$$

where $a \leq c \leq b$.

As in the Existence-Uniqueness Theorem on first-order equations, this theorem provides only sufficient conditions. That is, even if the conditions stated are not all satisfied, unique solutions may still exist.

A EXERCISES

1. Write each of the following in "operator notation."

(a) $\dfrac{d^2y}{dx^2} + 3\dfrac{dy}{dx} + 2y = x^3.$

(b) $3y^{(IV)} - 5y''' + y = e^{-x} + \sin x.$

(c) $\dfrac{d^2s}{dt^2} = -\beta\dfrac{ds}{dt} - \omega^2 s.$

(d) $x^2 y'' - 2xy' = y + 1.$

2. If $y = x^3 - 3x^2 + 2e^{-x}$ and $z = \sin 2x + 3 \cos 2x$ evaluate:

   (a) $(D^2 + 3D + 1)y$.
                              (b) $(2D^3 - D^2 - 4)z$.

   (c) $(D^2 + 2D)(y + z)$.
                     (d) $(x^2D^2 + 3xD - 2)(2y - 3z)$.

3. Complete the entries in the table. Verify that the general solutions in the fourth column satisfy the differential equations in the first column.

Differential equation	Complementary solution	Particular solution	General solution
$y'' - 3y' + 2y = x$	$y_c = c_1 e^x + c_2 e^{2x}$	$y_p = \dfrac{x}{2} + \dfrac{3}{4}$	
$(D^2 - 1)y = e^{-x}$		$y_p = -\dfrac{x}{2} e^{-x}$	$y = c_1 e^x + c_2 e^{-x} - \dfrac{x}{2} e^{-x}$
$(D^3 + D)y = \sin 2x$			$y = A \sin x + B \cos x$ $+ C + \frac{1}{6} \cos 2x$
			$y = c_1 e^{3x} + c_2 e^{2x} + e^x$
$(x^2D^2 + xD - 4)y = x^3$	$y_c = Ax^2 + \dfrac{B}{x^2}$	$y_p = \dfrac{x^3}{5}$	

B EXERCISES

1. Evaluate $F(x) \equiv (D - 1)(x^3 + 2x)$, where $D \equiv d/dx$. Then evaluate $(D - 2)F(x)$. The result may be written $(D - 2)(D - 1)(x^3 + 2x)$. Is this the same as $(D^2 - 3D + 2)(x^3 + 2x)$? Is the operator $(D - 2)(D - 1)$ the same as the operator $D^2 - 3D + 2$ when the operations are performed on any differentiable function? Prove your answer.

2. (a) Does $(D - a)(D - b) = D^2 - (a + b)D + ab$, where a and b are constants? (b) Two operators $\phi_1(D)$ and $\phi_2(D)$ are called *commutative* with respect to multiplication if

$$\phi_1(D)\phi_2(D)u = \phi_2(D)\phi_1(D)u.$$

Are the operators $(D - a)$ and $(D - b)$ commutative? (c) Operators $\phi_1(D)$, $\phi_2(D)$, and $\phi_3(D)$ are called *associative* with respect to multiplication if

$$\phi_1(D)[\phi_2(D)\phi_3(D)]u = [\phi_1(D)\phi_2(D)]\phi_3(D)u$$

Are the operators $(D - a)$, $(D - b)$, and $(D - c)$, where a, b, c are constants, associative?

C EXERCISES

1. Evaluate $F(x) \equiv (D - x)(2x^3 - 3x^2)$. Then evaluate $(D + x)F(x)$. The result may be written $(D + x)(D - x)(2x^3 - 3x^2)$. Is this result the same as $(D^2 - x^2)(2x^3 - 3x^2)$? Is $(D - x)(D + x)$ the same as $(D + x)(D - x)$?

2. Answer Questions 2 of the B Exercises when a, b, c are not all constants.

3. Represent $D^3 - 6D^2 + 11D - 6$ as a "product" of three factors if possible. Does the order make any difference? Prove your statements.

4. Discuss the existence and uniqueness of solutions of (a) $y'' - 3y' + 2y = x^2$; $y(0) = y'(0) = 0$. (b) $xy'' + y' + xy = 0$; $y(c) = p_1$, $y'(c) = p_2$.

5. Write the differential equation $(D^2 - 3D + 2)y = x$ as $(D - 1)(D - 2)y = x$. Letting $(D - 2)y = v$, write the equation as $(D - 1)v = x$ and thus find v. From this find y. Do you have the general solution? If so, write the complementary and particular solutions. This is called the method of *reduction of order*.

6. Use the method of Exercise 5 to obtain general solutions of each of the following and verify your answer in each case. (a) $(D^2 + D - 2)y = e^{-x}$. (b) $(D - 3)(D + 2)(D - 1)y = 3e^x - 2$. (c) $(D - 1)^2 y = 1$. (d) $(D - x)(D + 1)y = x$.

3 How Do We Obtain the Complementary Solution?

3.1 THE AUXILIARY EQUATION

In this section we concentrate our attention on methods of finding the general solution of the equation $\phi(D)y = 0$ with constant coefficients. Consider the following

PROBLEM FOR DISCUSSION

What is the general solution of the differential equation $y'' - 3y' + 2y = 0$? In operator notation this equation can be written $(D^2 - 3D + 2)y = 0$. Before we consider this further, let us return to a first-order linear differential equation with constant coefficients, for example, $(D - 2)y = 0$. We may write this equation

$$\frac{dy}{dx} - 2y = 0 \tag{1}$$

Solving this by one of many possible methods we find that $y = ce^{2x}$ is the general solution. This solution could be found by assuming a solution of the form $y = e^{mx}$, where m is an as yet undetermined constant. In order for this to be a solution, we have by substitution in (1) $(m - 2)e^{mx} = 0$, i.e., $m = 2$, since e^{mx} is never zero. From this it is a simple step to deduce that ce^{2x} is also a solution which gives the general solution. We now wonder whether the same technique will work as well on the equation $(D^2 - 3D + 2)y = 0$. Letting $y = e^{mx}$, the equation becomes

$$m^2 e^{mx} - 3me^{mx} + 2e^{mx} = 0 \quad \text{or} \quad (m^2 - 3m + 2)e^{mx} = 0$$

and it is clear that this will be satisfied if $m^2 - 3m + 2 = 0$ or $m = 1, 2$. Thus, it follows that $y = e^x$ and $y = e^{2x}$ are solutions. We now ask the question: What is the general solution? We know that the general solution has two arbitrary constants and the student may suspect that a good guess for this general solution would be

$y = c_1 e^x + c_2 e^{2x}$. This suspicion is confirmed by simply substituting into the equation and showing that it is satisfied. If the student likes he can strengthen his suspicions by the following

ILLUSTRATIVE EXAMPLE 1

Solve $y''' - 6y'' + 11y' - 6y = 0$.

Solution   Letting $y = e^{mx}$, where m is constant, we see that m must be such that $m^3 - 6m^2 + 11m - 6 = 0$. The three roots of this equation are $m = 1, 2, 3$. Hence, solutions are e^x, e^{2x}, and e^{3x}, and going along with our suspicions we guess that

$$y = c_1 e^x + c_2 e^{2x} + c_3 e^{3x}$$

is the required general solution. Our suspicions are again confirmed by substitution.

We may confirm permanently our suspicions by the following

Theorem 1.   If y_1 and y_2 are two solutions of $\phi(D)y = 0$, then $c_1 y_1 + c_2 y_2$, where c_1 and c_2 are arbitrary constants, is also a solution.

Proof. Since y_1 and y_2 are solutions, $\phi(D)y_1 = 0$, $\phi(D)y_2 = 0$. Multiplying these equations by c_1 and c_2, respectively, and remembering that $\phi(D)$ is a linear operator, we have $\phi(D)(c_1 y_1 + c_2 y_2) = 0$, which shows that $c_1 y_1 + c_2 y_2$ is a solution and proves the theorem.

The student should be able to extend this to the following

Theorem 2.   If $y_1, y_2, \ldots, y_k$ are k solutions of $\phi(D)y = 0$, then $c_1 y_1 + c_2 y_2 + \cdots + c_k y_k$ is also a solution.

In case the number of arbitrary constants in the solution is equal to the order of the differential equation, we have the general solution.* This theorem confirms the suspicions which we had above.

The student has probably noticed that the equation for determination of m has the same form as the differential equation written in terms of operators. Thus, for $(D^2 - 3D + 2)y = 0$, the "m equation" is $m^2 - 3m + 2 = 0$. In general for *constant coefficients* the differential equation $\phi(D)y = 0$ has the "m equation" $\phi(m) = 0$. This "m equation" is called the *auxiliary equation* and we shall refer to it by this name.

Remark. It is not difficult to show that Theorems 1 and 2 above hold for linear differential equations with variable coefficients, as well as those with constant coefficients. As a result we could solve such equations if we could find solutions $y_1, y_2, \ldots$ of $\phi(D)y = 0$. The difficulty which arises in the variable coefficient case is that solutions of the type $y = e^{mx}$ *rarely* exist, and so we cannot easily determine $y_1, y_2, \ldots$.

* The student should recall that when we speak of arbitrary constants we mean to imply *essential* arbitrary constants as indicated in the second footnote on page 16. This concept is intimately connected with that of linear independence of functions which is considered on page 181.

In the constant coefficient case, solutions of the type $y = e^{mx}$ *always* exist and so our task is much easier. See B Exercise 2.

See B Exercise 2.

A EXERCISES

1. Find general solutions of each of the following:
   (a) $y'' + 4y' - 5y = 0$.   (b) $(4D^2 - 25)y = 0$.   (c) $y'' = 4y$.
   (d) $2y''' - 5y'' + 2y' = 0$. (e) $I''(t) - 4I'(t) + 2I(t) = 0$. (f) $(D^3 + 2D^2 - 5D - 6)y = 0$.

2. Find solutions satisfying the given conditions: (a) $y'' - y = 0$, $y(0) = 2$, $y'(0) = -3$.
   (b) $(D^2 - 3D + 2)y = 0$, $y(0) = -1$, $y'(0) = 0$. (c) $(D^3 - 16D)y = 0$, $y(0) = 0$, $y'(0) = 0$, $y''(0) = 16$.

B EXERCISES

1. Solve $(D^3 + 5D^2 + 2D - 12)y = 0$.

2. Can $y = e^{mx}$ for constant m be used to solve $y'' - xy' + y = 0$? Explain.

3. Prove (a) Theorem 1 and (b) Theorem 2 for variable as well as for constant coefficients.

4. (a) Solve $[D^2 - (m_1 + m_2)D + m_1 m_2]y = 0$, m_1, m_2 constant, subject to $y(0) = 0$, $y'(0) = 1$ if $m_1 \neq m_2$. (b) What does the solution to (a) become if $m_2 \to m_1$? (c) Does the limiting solution of (b) satisfy the equation of (a) if $m_1 = m_2$? Can you devise a rule for repeated roots?

C EXERCISES

1. Solve $(D^4 - 20D^2 + 4)y = 0$.   2. Solve $(D^4 - 2D^3 - 16D^2 + 12D + 12)y = 0$.

3.2 THE CASE OF REPEATED ROOTS

Consider the following

PROBLEM FOR DISCUSSION

Solve $(D^2 - 6D + 9)y = 0$. Writing the auxiliary equation as usual, we have $m^2 - 6m + 9 = 0$. Solving this, we have $m = 3, 3$, i.e., two equal roots. Without thinking, we might write $y = c_1 e^{3x} + c_2 e^{3x}$ as the general solution. However, although this seems to have two arbitrary constants, in reality it does not since it can be written $y = (c_1 + c_2)e^{3x}$ or $y = ce^{3x}$, which has but one arbitrary constant. Hence we ask ourselves: What do we do in the case where the roots of the auxiliary equation are repeated?

The following method, which is of great generality and which is useful in later work, can come to our aid now. We state it as

Theorem 3.   If we know one solution, say $y = y_1$ of the nth-order equation $\phi(D)y = 0$, then the substitution $y = y_1 v$ will transform the given equation into an equation of the $(n - 1)$st order in v', i.e., dv/dx.

Remark. This theorem is valid for constant or variable coefficients and is also valid for the equation $\phi(D)y = F(x)$.

Proof. We shall prove the theorem for the case $n = 2$. Extensions to $n > 2$ follow the same pattern. For $n = 2$ the differential equation is

$$(a_0 D^2 + a_1 D + a_2)y = 0 \tag{2}$$

Now if $y = y_1 v$, we have

$$Dy = y_1 Dv + vDy_1, \qquad D^2 y = y_1 D^2 v + 2(Dy_1)(Dv) + vD^2 y_1$$

Substituting these into (2) and simplifying yields

$$a_0 y_1 D^2 v + (2a_0 Dy_1 + a_1 y_1)Dv + (a_0 D^2 y_1 + a_1 Dy_1 + a_2 y_1)v = 0 \tag{3}$$

Since y_1 satisfies (2), the last term on the left in (3) is zero so that the equation becomes

$$a_0 y_1 D^2 v + (2a_0 Dy_1 + a_1 y_1)Dv = 0 \tag{4}$$

which proves the theorem for $n = 2$, since (4) is of order $n - 1 = 2 - 1 = 1$ in $Dv = v'$.

In the special case $n = 2$, Theorem 3 can be used to obtain the general solution. For letting $Dv = u$ in (4) and dividing by $a_0 y_1 u$, we can write (4) as

$$\frac{du}{u} + \frac{2\, dy_1}{y_1} + \frac{a_1}{a_0}\, dx = 0$$

Integration then leads to

$$\ln u + 2 \ln y_1 + \int \frac{a_1}{a_0}\, dx = \ln c_1 \quad \text{or} \quad u = \frac{c_1 e^{-\int (a_1/a_0)dx}}{y_1^2}$$

Since $Dv = u$, we have by another integration

$$v = c_1 \int \frac{e^{-\int (a_1/a_0)dx}}{y_1^2}\, dx + c_2$$

Finally using $y = y_1 v$, we obtain the required general solution

$$y = c_1 y_1 \int \frac{e^{-\int (a_1/a_0)dx}}{y_1^2}\, dx + c_2 y_1 \tag{5}$$

Let us illustrate the theorem by application to $(D^2 - 6D + 9)y = 0$. We have seen that a solution is e^{3x}. According to the theorem we thus let $y = ve^{3x}$. Since

$$Dy = v'e^{3x} + 3ve^{3x}, \qquad D^2 y = v''e^{3x} + 6v'e^{3x} + 9ve^{3x}$$

we have on substituting into the equation and simplifying,

$$v''e^{3x} = 0$$

It follows that $v'' = 0$, or $v = c_1 + c_2 x$. Hence, we have $y = (c_1 + c_2 x)e^{3x}$. Thus, the general solution is $y = c_1 e^{3x} + c_2 x e^{3x}$. The dilemma raised in our problem for discussion is thus solved. It appears that to get another solution besides e^{3x} we simply multiply e^{3x} by x. We raise the rather obvious question now as to what happens when we have 3 roots which are the same. For example, suppose the roots are 2, 2, 2. This will occur when the auxiliary equation is $(m - 2)^3 = 0$, i.e.,

$m^3 - 6m^2 + 12m - 8 = 0$. The differential equation in such case would be

$$(D^3 - 6D^2 + 12D - 8)y = 0$$

Since e^{2x} is one solution, we substitute $y = ve^{2x}$ in accordance with the theorem and find that the given equation becomes, after simplification,

$$v'''e^{2x} = 0 \quad \text{or} \quad v''' = 0$$

so that $v = c_1 + c_2 x + c_3 x^2$ upon integration and the solution is given by

$$y = (c_1 + c_2 x + c_3 x^2)e^{2x}$$

The idea should now be apparent to the student. If we had 5 roots all equal to -1, the solution would be

$$(c_1 + c_2 x + c_3 x^2 + c_4 x^3 + c_5 x^4)e^{-x}$$

ILLUSTRATIVE EXAMPLE 2

Solve $(D^6 - 6D^5 + 12D^4 - 6D^3 - 9D^2 + 12D - 4)y = 0$.

Solution The auxiliary equation $m^6 - 6m^5 + 12m^4 - 6m^3 - 9m^2 + 12m - 4 = 0$ has roots $m = 1, 1, 1, 2, 2, -1$. Thus, there are (a) three roots equal to 1, (b) two roots equal to 2, (c) one root equal to -1, and the general solution (with the required six arbitrary constants) is

$$y = (c_1 + c_2 x + c_3 x^2)e^x + (c_4 + c_5 x)e^{2x} + c_6 e^{-x}$$

A EXERCISES

1. Find general solutions of the following:
   (a) $(D^2 - 4D + 4)y = 0$. (b) $16y'' - 8y' + y = 0$. (c) $4I''(t) - 12I'(t) + 9I(t) = 0$.
   (d) $(D^6 - 4D^4)y = 0$. (e) $(D^4 - 2D^3 + D^2)y = 0$. (f) $4y^{(IV)} - 20y'' + 25y = 0$.

2. Find solutions satisfying the given conditions: (a) $(D^2 - 2D + 1)y = 0$; $y(0) = 1$, $y'(0) = -2$.
   (b) $(D^3 - D^2)y = 0$; $y(0) = 1$, $y'(0) = y''(0) = 0$.

   (c) $\dfrac{d^2 s}{dt^2} = -16\dfrac{ds}{dt} - 64s$; $s = 0, \dfrac{ds}{dt} = -4$ where $t = 0$.

B EXERCISES

1. If a is constant and u is differentiable, prove the following: $(D - a)(e^{ax}u) = e^{ax}Du$, $(D - a)^2(e^{ax}u) = e^{ax}D^2u$, $(D - a)^3(e^{ax}u) = e^{ax}D^3u$. Can you prove by mathematical induction that $(D - a)^n(e^{ax}u) = e^{ax}D^nu$ where n is any positive integer? This is sometimes called the *exponential shift theorem*.

2. Use the results of Exercise 1 to show that the equation $(D - a)^n y = 0$ has general solution

$$y = (c_1 + c_2 x + c_3 x^2 + \cdots + c_n x^{n-1})e^{ax}$$

and thus establish the result of this section.

3. Prove the theorem on page 175 for (a) $n = 3$ and (b) $n = 4$. Can you prove it for $n > 4$?

C EXERCISES

1. Find the general solution of $y'' - xy' + y = 0$, given that $y = x$ is a solution.

2. Find a constant p such that $y = x^p$ satisfies $x^2 y'' + 3xy' + y = 0$. Write the general solution.

3. Given that $y = x$ is a solution of $(1 - x^2)y'' - 2xy' + 2y = 0$, find its general solution and also the general solution of $(1 - x^2)y'' - 2xy' + 2y = x$.

4. If $y = Y_1(x)$ satisfies $[a_0(x)D^2 + a_1(x)D + a_2(x)]y = 0$, solve $[a_0(x)D^2 + a_1(x)D + a_2(x)]y = R(x)$.

3.3 THE CASE OF IMAGINARY ROOTS

Consider the following

PROBLEM FOR DISCUSSION

Solve the equation $y'' + y = 0$. $\qquad$ (6)

If we let $y = e^{mx}$ as customary, we find $m^2 + 1 = 0$, and the roots are $m = \pm i$, i.e., the roots are *imaginary*. Formally the general solution is

$$y = c_1 e^{ix} + c_2 e^{-ix} \qquad (7)$$

but the question naturally arises: What do we mean by e^{ix} and e^{-ix}? A clue as to the possible meaning may be obtained by noticing that equation (6) has one of the variables missing. This equation has in fact been solved before (see pages 59–60 and the general solution is given by

$$y = A \sin x + B \cos x \qquad (8)$$

where A and B are arbitrary constants. Now it can be shown that the differential equation cannot have more than one general solution, so that we have to admit that although (7) and (8) look different they are actually the same. Let us therefore see what we can deduce from this. We have

$$c_1 e^{ix} + c_2 e^{-ix} \equiv A \sin x + B \cos x \qquad (9)$$

Putting $x = 0$ in this identity and supposing that $e^{\pm i \cdot 0} = 1$, we find

$$c_1 + c_2 = B \qquad (10)$$

Differentiating both sides of the identity (9), assuming that the usual rules hold when imaginaries are present, we have

$$c_1 i e^{ix} - c_2 i e^{-ix} \equiv A \cos x - B \sin x \qquad (11)$$

Placing $x = 0$, we have $\qquad (c_1 - c_2)i = A \qquad (12)$

With the values of A and B from (10) and (12), equation (9) becomes

$$c_1 e^{ix} + c_2 e^{-ix} \equiv c_1(\cos x + i \sin x) + c_2(\cos x - i \sin x) \qquad (13)$$

Since c_1 and c_2 are arbitrary, we may put $c_2 = 0$ and $c_1 = 1$. Then, from (13),

$$e^{ix} \equiv \cos x + i \sin x \qquad (14)$$

Similarly, $\qquad\qquad\qquad e^{-ix} \equiv \cos x - i \sin x \qquad (15)$

which also could have been obtained from (14) by replacing x by $-x$ or i by $-i$. We shall use (14) and (15), called *Euler's formulas*, as definitions of e^{ix} and e^{-ix}, since what we have done above makes plausible, but does not prove, these results.*

PROBLEM FOR DISCUSSION

Solve $(D^2 + 2D + 5)y = 0$. The auxiliary equation is $m^2 + 2m + 5 = 0$, and $m = -1 \pm 2i$. Then formally

$$y = c_1 e^{(-1+2i)x} + c_2 e^{(-1-2i)x} \tag{16}$$

i.e., $\quad y = c_1 e^{-x} e^{2ix} + c_2 e^{-x} e^{-2ix} \quad$ or $\quad y = e^{-x}(c_1 e^{2ix} + c_2 e^{-2ix})$

By analogy with the above results,

$$c_1 e^{2ix} + c_2 e^{-2ix} \equiv A \sin 2x + B \cos 2x$$

so that (16) becomes $\quad y = e^{-x}(A \sin 2x + B \cos 2x)$

This may, in fact, be verified as the required general solution. In general, if $a \pm bi$ are roots of the auxiliary equation, a corresponding solution is

$$e^{ax}(A \sin bx + B \cos bx)$$

ILLUSTRATIVE EXAMPLE 3

Solve $(D^4 - 5D^2 + 12D + 28)y = 0$.

Solution The auxiliary equation $m^4 - 5m^2 + 12m + 28 = 0$ has the roots $m = -2, -2, 2 \pm \sqrt{3}i$.

Since $m = -2$ is a double root, $(c_1 + c_2 x)e^{-2x}$ is a solution.
Since $2 \pm \sqrt{3}i$ are roots, $e^{2x}(c_3 \sin \sqrt{3}x + c_4 \cos \sqrt{3}x)$ is a solution. Hence,

$$y = (c_1 + c_2 x)e^{-2x} + e^{2x}(c_3 \sin \sqrt{3}x + c_4 \cos \sqrt{3}x)$$

PROBLEM FOR DISCUSSION

What is the general solution of a differential equation whose auxiliary equation has as roots $2, -1, 0, 0, 3 \pm 5i, 2, 0, 3 \pm 5i$? We have ten roots. These are:
(1) the triple root 0   (2) the root -1
(3) the double root 2   (4) the double roots $3 \pm 5i$
From (1) a solution is $(c_1 + c_2 x + c_3 x^2)e^{0x} = c_1 + c_2 x + c_3 x^2$.
From (2) a solution is $c_4 e^{-x}$.
From (3) a solution is $(c_5 + c_6 x)e^{2x}$.

* For those who are acquainted with series we may obtain these results by using the series

$$e^u = 1 + u + \frac{u^2}{2!} + \frac{u^3}{3!} + \frac{u^4}{4!} + \cdots$$

replacing u by ix and noting that

$$e^{ix} = 1 - \frac{x^2}{2!} + \frac{x^4}{4!} - \cdots + i\left(x - \frac{x^3}{3!} + \frac{x^5}{5!} - \cdots\right) = \cos x + i \sin x$$

However, this method needs justification, which we leave for a course in complex variables.

To obtain a solution corresponding to (4), note that a solution corresponding to the single pair $3 \pm 5i$ would be $e^{3x}(c_7 \sin 5x + c_8 \cos 5x)$. Since the pair is repeated, another solution is $xe^{3x}(c_9 \sin 5x + c_{10} \cos 5x)$. This is analogous to our results on real repeated roots. [If $3 \pm 5i$ were a triple root, another solution would be $x^2 e^{3x}(c_{11} \sin 5x + c_{12} \cos 5x).$] From the above it follows that the general solution with the ten arbitrary constants would be

$$y = c_1 + c_2 x + c_3 x^2 + c_4 e^{-x} + (c_5 + c_6 x)e^{2x}$$
$$+ e^{3x}(c_7 \sin 5x + c_8 \cos 5x) + xe^{3x}(c_9 \sin 5x + c_{10} \cos 5x)$$

We have now answered adequately the question: How do we obtain the complementary solution?

A EXERCISES

1. Find the general solution of each of the following:
   (a) $y'' + 4y = 0$. 　　　　(b) $(D^2 + 4D + 5)y = 0$. 　　(c) $4\dfrac{d^2 s}{dt^2} = -9s$.

   (d) $4y'' - 8y' + 7y = 0$. 　(e) $y^{IV} = -16y''$. 　　　　(f) $(D^3 + D^2 - 2)y = 0$.

2. Find solutions satisfying the given conditions.
   (a) $(D^2 + 1)y = 0$; $y(0) = 4$, $y'(0) = 0$. 　　(b) $U''(t) = -16U(t)$; $U(0) = 0$, $U'(0) = 4$.
   (c) $I''(t) + 2I'(t) + 5I(t) = 0$; $I(0) = 2$, $I'(0) = 0$.

B EXERCISES

1. Find the general solution of $(D^6 - 64)y = 0$; $D \equiv d/dx$.

2. (a) Find the solution of $(D^2 + a^2)(D^2 + b^2)y = 0$, subject to $y(0) = 1$, $y'(0) = y''(0) = y'''(0) = 0$, if $a \neq b$. (b) What does the solution in (a) become if $b \to a$? Is this a solution of the given equation when $b = a$? Does your result agree with the general results concerning repeated roots of the auxiliary equation?

3. Find the general solution of $(D^4 + 4D^2 + 4)y = 0$.

C EXERCISES

1. Solve $(D^4 + 4)y = 0$; $D \equiv d/dx$. 　　2. Solve $(D^4 + 6D^2 + 25)y = 0$; $D \equiv d/dx$.

MISCELLANEOUS REVIEW EXERCISES ON COMPLEMENTARY SOLUTIONS

A EXERCISES

1. Write the general solution of the differential equations whose auxiliary equations have the following roots:
   (a) $3, -1$. 　　　　　　　　　　　　(b) $4, 0, -2$.
   (c) $2, 2, 2, 0, 0$. 　　　　　　　　　(d) $-1, -i, i, -2$.
   (e) $2 \pm 3i, -1 \pm 2i, 5, -1$. 　　　　(f) $1 \pm \sqrt{3}i, -2, -2, 0, -1$.
   (g) $-1, 1, 0, -2, -1, 1, -1 \pm 2i$. 　　(h) $1 \pm i, 1 \pm i$.
   (i) $\pm\sqrt{3}, \pm 4, \frac{1}{2} \pm 2i, -1 \pm 3i$. 　　(j) $1, 1, 1, 0, 0, \pm i, \pm i$.

2. Find the general solution of each of the following:
   (a) $(D^2 + D + 1)y = 0$. 　　　　　(b) $(D^4 - 1)y = 0$.
   (c) $(D^6 + 2D^4 + D^2)y = 0$. 　　　(d) $(D^3 - 4D^2 + 4D)y = 0$.
   (e) $y''' = y''$. 　　　　　　　　　(f) $S^{(IV)}(t) + 2S''(t) - 8S(t) = 0$.

1. Determine constants a, b, c such that $(D^3 + aD^2 + bD + c)y = 0$ has solution

$$y = c_1 e^{-x} + e^{-2x}(c_2 \sin 4x + c_3 \cos 4x)$$

2. Find a differential equation whose auxiliary equation has roots $-1, -1, 1 \pm 2i, 1 \pm 2i$. Write the general solution.

1. Show that the general solution of $D^n y = y$, where n is a positive integer, is

$$y = \sum_{k=1}^{n} c_k e^{m_k x}$$

where $m_k = e^{2k\pi i/n}$, $k = 1, \ldots, n$. Can the solution be expressed in real form?

2. Find real solutions of (a) $D^3 y = y$; (b) $D^5 y = y$.

3. Solve $D^3 y = 4y$.

4. Let X_1 and X_2 be any two solutions of the differential equation $\ddot{X} + k^2 X = 0$ such that $X_1^2 + X_2^2 = 1$ where the dots denote derivatives with respect to t. Prove that $\dot{X}_1^2 + \dot{X}_2^2 = k^2$ and $\ddot{X}_1^2 + \ddot{X}_2^2 = k^4$. Generalize to higher-ordered derivatives.

3.4 LINEAR INDEPENDENCE AND WRONSKIANS

In Illustrative Example 1, page 174, we considered the differential equation

$$(D^3 - 6D^2 + 11D - 6)y = 0 \tag{17}$$

By direct substitution we found that e^x, e^{2x}, e^{3x} are solutions from which we obtained the general solution

$$y = c_1 e^x + c_2 e^{2x} + c_3 e^{3x}$$

Suppose, however, we somehow arrive at the three functions

$$e^{2x} + 2e^x, \qquad 5e^{2x} + 4e^x, \qquad e^x - e^{2x} \tag{18}$$

all of which are easily shown to be solutions. Could we then say that

$$y = A(e^{2x} + 2e^x) + B(5e^{2x} + 4e^x) + C(e^x - e^{2x}) \tag{19}$$

with the three constants A, B, C, is the general solution? The observant student on noting that (19) can be written

$$y = (2A + 4B + C)e^x + (A + 5B - C)e^{2x} \quad \text{or} \quad y = c_1 e^x + c_2 e^{2x}$$

would say that the solution does not really have three arbitrary constants and so cannot be the general solution. We now ask the question, "What is there about the functions (18) from which we could have foreseen this situation?" A clue is supplied by noting that there are constants $\alpha_1, \alpha_2, \alpha_3$, not all zero, such that

$$\alpha_1(e^{2x} + 2e^x) + \alpha_2(5e^{2x} + 4e^x) + \alpha_3(e^x - e^{2x}) \equiv 0$$

i.e., identically zero; for example $\alpha_1 = 3$, $\alpha_2 = -1$, $\alpha_3 = -2$. Thus we are led to the idea that although we have three solutions they are in a way dependent. This has led mathematicians to the following

Definition. A set of distinct functions $y_1(x)$, $y_2(x)$, ..., $y_n(x)$ denoted briefly by $y_1, y_2, \ldots, y_n$, is said to be *linearly dependent* in an interval if there exists a set of n constants, not all zero, such that in the interval

$$\alpha_1 y_1 + \alpha_2 y_2 + \cdots + \alpha_n y_n \equiv 0 \tag{20}$$

otherwise the set is said to be *linearly independent*. Equivalently, we can say that the set $y_1, y_2, \ldots, y_n$ is *linearly independent* in an interval if the existence of the identity (20) implies that all the constants must be zero, i.e., $\alpha_1 = \alpha_2 = \cdots = \alpha_n = 0$.

Remark 1. Note that if $y_1, y_2, \ldots, y_n$ are linearly dependent then we can express one of these functions in terms of the others. Thus for example if we suppose that $\alpha_n \neq 0$, then we can solve for y_n to obtain

$$y_n = \frac{-(\alpha_1 y_1 + \alpha_2 y_2 + \cdots + \alpha_{n-1} y_{n-1})}{\alpha_n} \tag{21}$$

showing that y_n is dependent on $y_1, \ldots, y_{n-1}$. When we cannot obtain any of the functions $y_1, y_2, \ldots, y_n$ in terms of the remaining ones we have linear independence. In the case (21) we often say that y_n is a *linear combination* of $y_1, y_2, \ldots, y_{n-1}$.

Remark 2. The concepts of linear dependence and independence of functions *always* refer to some given interval. To save repetition of the words "in an interval," we shall often refer simply to linearly dependent or independent functions without adding these words each time.

From the above ideas we feel, at least intuitively, that linearly independent functions must play an important role in solving linear differential equations [assumed to have the form (2), page 167] and this does in fact happen.

In solving linear differential equations up to now we have essentially *used* this concept of linear independence without actually stating it. For example, in dealing with the equation

$$(D^2 - 3D + 2)y = 0$$

we found the solutions e^x and e^{2x} and from this the general solution $y = c_1 e^x + c_2 e^{2x}$. Implicit in this is the assumption of the linear independence of these functions. To show this, let us assume that there are constants α_1 and α_2, not both zero, such that

$$\alpha_1 e^x + \alpha_2 e^{2x} \equiv 0$$

i.e., let us assume that the functions are linearly dependent. Then, on dividing both sides by e^x, we arrive at the result

$$\alpha_2 e^x \equiv -\alpha_1$$

which is clearly impossible unless both α_1 and α_2 are zero. Thus the functions are linearly independent.

We would like to have a condition for linear dependence or independence which does not involve the constants $\alpha_1, \ldots, \alpha_n$ required in the definition since using the definition can become tedious. Let us first examine the case for two functions y_1 and y_2. By definition, if y_1 and y_2 are linearly dependent, then we can find constants α_1 and α_2, not both zero, such that

$$\alpha_1 y_1 + \alpha_2 y_2 \equiv 0 \tag{22}$$

By differentiating this identity assuming that the derivatives y_1', y_2' exist, we find

$$\alpha_1 y_1' + \alpha_2 y_2' \equiv 0 \tag{23}$$

Multiplying (22) by y_2', (23) by y_2 and subtracting we find

$$\alpha_1(y_1 y_2' - y_2 y_1') \equiv 0 \tag{24}$$

Similarly, multiplying (22) by y_1', (23) by y_1 and subtracting, we find

$$\alpha_2(y_1 y_2' - y_2 y_1') \equiv 0 \tag{25}$$

From (24) and (25) we see that if α_1 and α_2 are not both zero then

$$W(y_1, y_2) = y_1 y_2' - y_2 y_1' = \begin{vmatrix} y_1 & y_2 \\ y_1' & y_2' \end{vmatrix} \equiv 0 \tag{26}$$

We call the determinant $W(y_1, y_2)$ in (26) the *Wronskian* of y_1 and y_2, and often denote it briefly by W.

Remark 3. We can also obtain (26) by considering (22) and (23) as simultaneous linear equations for the determination of α_1 and α_2. This is done by using Theorem 6 in the Appendix which states that equations (22) and (23) yield solutions α_1, α_2 which are not both zero if and only if the determinant of the coefficients of α_1, α_2 is zero, which leads to (26).

We have thus proved the following

Theorem 4. If y_1 and y_2 are linearly dependent in an interval, and if their derivatives y_1' and y_2' exist in the interval, then the Wronskian of y_1 and y_2 given by

$$W(y_1, y_2) = \begin{vmatrix} y_1 & y_2 \\ y_1' & y_2' \end{vmatrix} = y_1 y_2' - y_2 y_1'$$

is identically zero (i.e., $W \equiv 0$) in the interval.

This theorem can be stated in terms of linear independence as follows.

Theorem 5. If the Wronskian of y_1 and y_2 is not identically zero (i.e., $W \not\equiv 0$) in an interval, then y_1 and y_2 are linearly independent in the interval.

This is so because if y_1 and y_2 were linearly dependent in the interval their Wronskian would be identically zero in the interval by Theorem 4. This contradiction shows that they are not linearly independent, i.e., they are linearly dependent in the interval. The result also follows from (24) and (25) since if $W = y_1 y_2' - y_2 y_1' \neq 0$, we have $\alpha_1 = 0$, $\alpha_2 = 0$.

Example. Since the Wronskian of $y_1 = e^x$ and $y_2 = e^{2x}$ is

$$W = \begin{vmatrix} y_1 & y_2 \\ y_1' & y_2' \end{vmatrix} = \begin{vmatrix} e^x & e^{2x} \\ e^x & 2e^{2x} \end{vmatrix} = e^{3x}$$

which is not identically zero, the functions are linearly independent in any interval.

It is natural to ask the following

Question. If $W = 0$ identically, does it follow that y_1 and y_2 are linearly dependent, i.e., is the converse of Theorem 4 true?

Unfortunately, the answer to this is "no," as we show in the following

ILLUSTRATIVE EXAMPLE 4

Let $y_1 = x^2$ and $y_2 = x|x|$, where $|x|$ is the *absolute value of x*, be two functions defined in the interval $-1 \leq x \leq 1$. (a) Find the Wronskian of y_1 and y_2. (b) Show that the functions are not linearly dependent, i.e., are linearly independent, in the interval.

Solution (a) The derivative of $y_1 = x^2$ in the interval is clearly $y_1' = 2x$. Since

$$|x| = \begin{cases} x, & \text{if } x \geq 0 \\ -x, & \text{if } x \leq 0 \end{cases}$$

we have

$$y_2 = x|x| = \begin{cases} x^2, & \text{if } 0 \leq x \leq 1 \\ -x^2, & \text{if } -1 \leq x \leq 0 \end{cases}$$

Now for $0 \leq x \leq 1$ we have $y_2' = 2x$, while for $-1 \leq x \leq 0$, we have $y_2' = -2x$.

Thus

$$y_2' = \begin{cases} 2x, & \text{if } 0 \leq x \leq 1 \\ -2x, & \text{if } -1 \leq x \leq 0 \end{cases}$$

To determine the Wronskian, we must calculate it for $0 \leq x \leq 1$ and then for $-1 \leq x \leq 0$, i.e., there are two cases

Case 1, $0 \leq x \leq 1$. Here $W = \begin{vmatrix} x^2 & x^2 \\ 2x & 2x \end{vmatrix} = 0$.

Case 2, $-1 \leq x \leq 0$. Here $W = \begin{vmatrix} x^2 & -x^2 \\ 2x & -2x \end{vmatrix} = 0$.

From these two cases we have $W = 0$ identically in the interval $-1 \leq x \leq 1$.
(b) If $y_1 = x^2$ and $y_2 = x|x|$ are linearly dependent in $-1 \leq x \leq 1$, then we

should be able to find constants α_1, α_2 not both zero such that

$$\alpha_1 x^2 + \alpha_2 x|x| = 0$$

identically in $-1 \leq x \leq 1$. This means that we must have

$$\alpha_1 x^2 + \alpha_2 x^2 = 0 \quad \text{for} \quad 0 \leq x \leq 1, \qquad \alpha_1 x^2 - \alpha_2 x^2 = 0 \quad \text{for} \quad -1 \leq x \leq 0$$

for constants α_1, α_2 not both zero. However, this is clearly impossible. It follows that the functions are linearly independent in $-1 \leq x \leq 1$.

On seeing this example involving the function $y_2 = x|x|$, the suspicious student may very well ask, Why does the mathematician have to dream up a function like this in order to show that the converse of Theorem 4 is not true? The obvious answer of course is that, if a so-called "normal function" were used, it would not produce the desired result, and for such functions the converse would be true.

A clue is provided when one notes that, although the function $y_2 = x|x|$ has a first derivative in the interval $-1 \leq x \leq 1$ (just enough to enable one to calculate the Wronskian), it does not have a second derivative. Such a function could not possibly satisfy a second-order linear differential equation

$$(a_0 D^2 + a_1 D + a_2)y = 0 \tag{27}$$

as given in the Existence-Uniqueness Theorem on page 171 for $n = 2$. Because of this we are led to restricting functions considered to be solutions of (27).

Now if we take y_1 and y_2 to be solutions of (27), we have

$$a_0 y_1'' + a_1 y_1' + a_2 y_1 = 0, \qquad a_0 y_2'' + a_1 y_2' + a_2 y_2 = 0 \tag{28}$$

Multiplying the first equation of (28) by y_2, the second by y_1, and subtracting, we find

$$a_0(y_1 y_2'' - y_2 y_1'') + a_1(y_1 y_2' - y_2 y_1') = 0 \tag{29}$$

Since the Wronskian is given by $W = y_1 y_2' - y_2 y_1'$ and since

$$\frac{dW}{dx} = \frac{d}{dx}(y_1 y_2' - y_2 y_1') = y_1 y_2'' - y_2 y_1''$$

(29) becomes

$$a_0 \frac{dW}{dx} + a_1 W = 0$$

Solving, we obtain an important relation known as *Abel's identity*, given by

$$W = ce^{-\int (a_1/a_0)dx} \tag{30}$$

Since the exponential function in (30) is never zero, we see that the Wronskian W must either be *identically zero* in the given interval, in which case $c = 0$, or *never zero* in the interval, in which case $c \neq 0$. There cannot be anything in between. We summarize the results in the following

Theorem 6. Let y_1 and y_2 be two solutions of the differential equation

$$(a_0 D^2 + a_1 D + a_2)y = 0$$

where $a_0 \neq 0$, a_1, a_2 are continuous functions of x in some given interval. Then the Wronskian of y_1 and y_2 is given by *Abel's identity*

$$W = \begin{vmatrix} y_1 & y_2 \\ y_1' & y_2' \end{vmatrix} = ce^{-\int(a_1/a_0)dx} \tag{31}$$

and W is either identically zero in the interval or never zero in the interval.

Using this theorem, we can now prove the following theorem in case y_1 and y_2 are solutions of (27).

Theorem 7. Let y_1 and y_2 be solutions of the differential equation (27) in some given interval. Then

(a) y_1 and y_2 are linearly dependent if and only if $W = 0$ in the interval.
(b) y_1 and y_2 are linearly independent if and only if $W \neq 0$ in the interval.

We shall not present a proof of this theorem but shall instead leave it to C Exercise 9. It should be noted that there is no contradiction at all between Theorem 7 and Illustrative Example 4 on page 184, since the function $y_2 = x|x|$ cannot be a solution of any second-order linear differential equation because its second derivative does not exist in $-1 \leq x \leq 1$.

From Theorems 6 and 7 we can also obtain the following useful theorem.

Theorem 8. Let y_1 be a solution of the differential equation (27). Then a linearly independent solution is given by

$$y_2 = y_1 \int \frac{e^{-\int(a_1/a_0)dx}}{y_1^2} dx \tag{32}$$

A proof of this follows at once from (31) if we divide both sides by y_1^2 assumed different from zero. For in such case we obtain

$$\frac{d}{dx}\left(\frac{y_2}{y_1}\right) = \frac{ce^{-\int(a_1/a_0)dx}}{y_1^2} \quad \text{or} \quad \frac{y_2}{y_1} = \int \frac{ce^{-\int(a_1/a_0)dx}}{y_1^2} dx$$

on integrating and taking the constant of integration equal to zero. Then choosing the special case $c = 1$, we get (32).

Remark 4. The student will notice that Theorem 8 is closely related to Theorem 3 on page 175.

As a culmination of the above theory we have the following important theorem, which we have often made use of in the early part of this chapter but never actually stated in precise form.

Theorem 9. If y_1 and y_2 are linearly independent solutions of equation (27), then

$$y = c_1 y_1 + c_2 y_2 \tag{33}$$

is a solution of (27) for any constants c_1, c_2. Conversely, *every solution* of (27) has the form (33) for appropriate choices of the constants c_1, c_2.

We also have the following companion theorem to Theorem 9.

Theorem 10. If y_1 and y_2 are linearly independent solutions of equation (27) and y_p is a particular solution of

$$(a_0D^2 + a_1D + a_2)y = F(x) \tag{34}$$

then

$$y = c_1y_1 + c_2y_2 + y_p \tag{35}$$

is a solution of (34) for any constants c_1, c_2. Conversely, *every solution* of (34) has the form (35) for appropriate choices of the constants c_1, c_2.

Proofs of Theorems 9 and 10 are left to C Exercise 10, page 190.

Remark 5. It is important that the student understand the significance of Theorems 9 and 10. We have defined the *general solution* of an nth-order differential equation as that solution which involves n arbitrary constants. All solutions which could not be obtained from this general solution by any choice of these constants were then called *singular solutions*. As we have already mentioned, in most problems of an applied nature it is this general solution which provides the meaningful one after determination of the constants from the given conditions. If singular solutions do arise, they "generally" have little or no practical significance.

As far as we are concerned, the significance of Theorems 9 and 10 is thus *not* that we have found the solution involving two arbitrary constants (the same number as the order of the differential equation), which we have called the *general solution*, but that in actuality *all other solutions* which exist must be particular solutions, i.e., special cases of the general solution obtained by appropriate choices of the constants. Thus, Theorems 9 and 10 in fact guarantee that there are no singular solutions. This property of having no singular solutions is one peculiar to *linear* differential equations, but not all non-linear differential equations.

Some authors use the term general solution to mean *all solutions* of a differential equation. Still others use general solution only in cases of linear differential equations. Thus, the student who sees the term general solution in another book must be careful to determine which definition is being used.

Although the above theorems were given for two functions and second-order linear differential equations, they can be generalized to the case of n functions and nth-order linear differential equations. For example, the *Wronskian* in the case of n functions $y_1, y_2, \ldots, y_n$ is a natural generalization of (31) and is given by

$$W(y_1, y_2, \ldots, y_n) = \begin{vmatrix} y_1 & y_2 & \cdots & y_n \\ y_1' & y_2' & \cdots & y_n' \\ \vdots & \vdots & & \vdots \\ y_1^{(n-1)} & y_2^{(n-1)} & \cdots & y_n^{(n-1)} \end{vmatrix} \tag{36}$$

We have the following theorems corresponding to Theorems 6, 7, 9, and 10, respectively.

Theorem 11. Let $y_1, y_2, \ldots, y_n$ be n solutions of the differential equation

$$(a_0D^n + a_1D^{n-1} + \cdots + a_n)y = 0 \tag{37}$$

where $a_0 \neq 0, a_1, \ldots, a_n$ are continuous functions of x in some given interval. Then the Wronskian (36) denoted by W is given by *Abel's identity*

$$W = ce^{-\int (a_1/a_0)dx} \tag{38}$$

and W is either identically zero in the interval or never zero in the interval.

Theorem 12. Let $y_1, y_2, \ldots, y_n$ be solutions of the differential equation (37) where $a_0 \neq 0, a_1, \ldots, a_n$ are continuous functions of x in some given interval. Then

(a) $y_1, y_2, \ldots, y_n$ are linearly dependent if and only if $W = 0$ in the interval.
(b) $y_1, y_2, \ldots, y_n$ are linearly independent if and only if $W \neq 0$ in the interval.

Theorem 13. If $y_1, y_2, \ldots, y_n$ are linearly independent solutions of the equation (37), then

$$y = c_1 y_1 + c_2 y_2 + \cdots + c_n y_n \tag{39}$$

is a solution of (37) for any constants $c_1, c_2, \ldots, c_n$. Conversely, *every solution of* (37) has the form (39) for appropriate choices of the constants $c_1, c_2, \ldots, c_n$.

Theorem 14. If $y_1, y_2, \ldots, y_n$ are linearly independent solutions of the equation (37) and y_p is a particular solution of

$$(a_0 D^n + a_1 D^{n-1} + \cdots + a_n)y = F(x) \tag{40}$$

then

$$y = c_1 y_1 + c_2 y_2 + \cdots + c_n y_n + y_p \tag{41}$$

is a solution of (40) for any constants $c_1, c_2, \ldots, c_n$. Conversely, *every solution of* (40) has the form (41) for appropriate choices of the constants $c_1, c_2, \ldots, c_n$.

Let us now consider some examples which illustrate the above theorems.

ILLUSTRATIVE EXAMPLE 5

(a) Determine whether the solutions $y_1 = e^{2x} + 2e^x$, $y_2 = 5e^{2x} + 4e^x$, $y_3 = e^x - e^{2x}$ of the differential equation $(D^3 - 6D^2 + 11D - 6)y = 0$ are linearly dependent or linearly independent. (b) Can you obtain the general solution from the given solutions? Explain.

Solution (a) The Wronskian of the set of solutions is given by

$$W(y_1, y_2, y_3) = \begin{vmatrix} e^{2x} + 2e^x & 5e^{2x} + 4e^x & e^x - e^{2x} \\ 2e^{2x} + 2e^x & 10e^{2x} + 4e^x & e^x - 2e^{2x} \\ 4e^{2x} + 2e^x & 20e^{2x} + 4e^x & e^x - 4e^{2x} \end{vmatrix} = 0 \tag{42}$$

Since this is identically zero in any interval, we see by Theorem 12 that the solutions are linearly dependent (compare page 181).

(b) From Theorem 13 we see that three linearly independent solutions are needed to find the required solution. Although the three given solutions are linearly dependent, it is easy to show that any two are linearly independent. For example, in the case of the first and third solutions, the Wronskian is given by

$$W(y_1, y_3) = \begin{vmatrix} e^{2x} + 2e^x & e^x - e^{2x} \\ 2e^{2x} + 2e^x & e^x - 2e^{2x} \end{vmatrix} = -3e^{3x} \neq 0$$

showing that y_1, y_3 are linearly independent.

To obtain the required solution, we must find a solution Y which is linearly independent of both y_1 and y_3. One method for finding Y is to apply Theorem 3, page 175, twice (see B Exercise 7). In this way we obtain the required solution $y = c_1 e^x + c_2 e^{2x} + c_3 e^{3x}$.

ILLUSTRATIVE EXAMPLE 6

Verify Abel's identity for the differential equation $(D^3 - 6D^2 + 11D - 6)y = 0$.

Solution Comparing the given differential equation with (37), page 187, for $n = 3$, we have $a_0 = 1$, $a_1 = -6$. Then Abel's identity (38), page 188, becomes

$$W(y_1, y_2, y_3) = ce^{-\int (a_1/a_0)dx} = ce^{6x}$$

To verify this, we must obtain the Wronskian for three linearly independent solutions, and we can choose these to be $y_1 = e^x$, $y_2 = e^{2x}$, $y_3 = e^{3x}$. Thus,

$$W = \begin{vmatrix} e^x & e^{2x} & e^{3x} \\ e^x & 2e^{2x} & 3e^{3x} \\ e^x & 4e^{2x} & 9e^{3x} \end{vmatrix} = 2e^{6x} \tag{43}$$

which provides the required verification if $c = 2$. Of course, if we had chosen linearly dependent solutions, the verification would be immediate since $c = 0$ would agree with $W = 0$.

A EXERCISES

1. Determine which of the following sets of functions are linearly dependent and which are linearly independent. In each case use both the direct definition and the theorems involving the Wronskian.
   (a) e^{-4x}, e^{4x}.   (b) $2x^3, -3x^3$.   (c) $1, \cos x$.   (d) $x + 2, 2x - 3$.
   (e) $x^2, x^2 + 1, x^2 - 1$.   (f) $(x + 1)(x - 2), (2x - 1)(x + 3), (x + 2)(x - 1)$.
   (g) $\sin^2 x, \cos^2 x, 2$.   (h) $\sin x + \cos x, 3 \sin x - 2 \cos x, 4 \cos x$.

2. (a) Show that two linearly independent solutions of the equation $(D^2 - 6D + 9)y = 0$ are given by e^{3x} and xe^{3x}. (b) How can you write the general solution of the equation in (a)?

3. Write the general solutions of each of the following equations and justify your results.
   (a) $(D^2 + 2D - 3)y = 0$.   (b) $(D^2 - 2D + 5)y = 0$.
   (c) $(D^3 - 3D^2)y = 0$.   (d) $(D^4 - 8D^2 + 16)y = 0$.

B EXERCISES

1. (a) Prove that any three polynomials of first degree must be linearly dependent. (b) What is the greatest number of polynomials of second degree which will be linearly independent? Could some of these be of lower degree? Explain. (c) Generalize the results in (a) and (b).

2. Prove that if zero is added to any linearly independent set of functions, the resulting set is linearly dependent.

3. Let $P_n(x)$ be polynomials of degree n where $n = 1, 2, 3, \ldots$. Prove that any finite set of these polynomials is linearly independent.

4. Investigate the linear dependence of the set of functions $\tan^{-1} x$, $\tan^{-1}(2x)$, $\tan^{-1}\left(\dfrac{3x}{1 - 2x^2}\right)$.

5. (a) Prove without using Wronskians that the functions x^2 and x^3 are linearly independent in the interval $1 \leq x \leq 2$, and also in the interval $-1 \leq x \leq 1$. (b) Find the Wronskian of x^2 and x^3. In view of the fact that this Wronskian is zero at some point in the interval $-1 \leq x \leq 1$, does this conflict with the result in (a)? Explain.

6. Find n linearly independent solutions of the equation $D^n y = 0$ and write the general solution.

7. Complete part (b) of Illustrative Example 5, page 188. (*Hint:* First let $y = vy_1$ in the given differential equation to obtain a second-order equation in v''. Then let $v' = wy_2$ in this new equation to obtain a first-order equation in w'.)

8. Show that the function $y_2 = x|x|$, $-1 \leq x \leq 1$ of page 184 does not have a second derivative.

C EXERCISES

1. Suppose that in the differential equation (27), page 185, a_0, a_1 and a_2 are continuous functions of x in an interval and $a_0 \neq 0$ at any point of the interval. (a) Prove that if the Wronskian corresponding to two solutions y_1 and y_2 is zero at some point of the interval then it is identically zero in the interval and the solutions y_1 and y_2 are linearly dependent. (b) Show that the functions x^2 and x^3 of B Exercise 5 are linearly independent solutions of the equation $x^2 y'' - 4xy' + 6y = 0$. Is there any conflict with (a)? Explain.

2. Write Abel's identity for the equations (a) $y'' - 3y' + 2y = 0$; (b) $x^2 y'' + xy' + (x^2 - n^2)y = 0$; (c) $(1 - x^2)y'' - 2xy' + 2y = 0$.

3. Given one solution of a linear second-order differential equation, how could you get a linearly independent solution from Abel's identity? How could you then find the general solution? Illustrate by finding the general solution of the equations (a) $(D^2 - 2D + 1)y = 0$, and (b) $(1 - x^2)y'' - 2xy' + 2y = 0$. [*Hint:* In (b) note that x is a solution.]

4. (a) Show that the function $3x^2 - 1$ satisfies the differential equation $(1 - x^2)y'' - 2xy' + 6y = 0$ and has a minimum at $x = 0$. (b) Show that any linearly independent solution of the equation in (a) cannot have a minimum or maximum at $x = 0$.

5. Generalize the result of Exercise 4.

6. Prove *Abel's identity* for third- and higher-order linear differential equations.

7. If y_1 and y_2 are two linearly independent solutions of the equation $a_0 y'' + a_1 y' + a_2 y = 0$ where a_0, a_1, and a_2 are polynomials having no common factor other than a constant prove that the Wronskian is zero at only those points where $a_0 = 0$, and conversely.

8. Generalize the result of Exercise 7.

9. Prove Theorem 7, page 186.

10. Prove (a) Theorems 9 and 10, pages 186–187; (b) Theorems 11–14, pages 187–188.

4 How Do We Obtain a Particular Solution?

To find the general solution of $\phi(D)y = F(x)$ we must find a particular solution of this equation and add it to the general solution of the complementary or reduced equation $\phi(D)y = 0$. In the last section we discovered how to obtain the general solution of $\phi(D)y = 0$. In this section we shall discover how to obtain particular solutions of $\phi(D)y = F(x)$.

There are many methods whereby particular solutions can be obtained. A method often used in physics and engineering is the method of undetermined coefficients. This method is simple to understand, where it applies, but unfortunately it cannot be used in certain cases. However, such cases are rare in practice. When they do arise, other methods must be used. Because of its simplicity and widespread use, we discuss this method first.

4.1 METHOD OF UNDETERMINED COEFFICIENTS

The method of undetermined coefficients applies to the differential equation $\phi(D)y = F(x)$, where $F(x)$ contains a polynomial, terms of the form $\sin rx$, $\cos rx$, e^{rx} where r is constant, or combinations of sums and products of these. To gain some insight, consider the following

PROBLEM FOR DISCUSSION

Solve $y'' + 4y = 4e^{2x}$. $\qquad\qquad$ (1)

We are looking for a function whose second derivative added to four times the function yields $4e^{2x}$. Since the various derivatives of e^{2x} involve e^{2x}, we are led to consider $y = ae^{2x}$ where a is an undetermined constant, as a possible solution. Substituting in the given equation we have

$$4ae^{2x} + 4ae^{2x} = 4e^{2x}, \qquad 8ae^{2x} = 4e^{2x}, \qquad a = \tfrac{1}{2}$$

Thus, a particular solution is $y_p = \tfrac{1}{2}e^{2x}$. By methods of the last section, the general solution of $y'' + 4y = 0$ is

$$y = y_c = c_1 \cos 2x + c_2 \sin 2x$$

The general solution of the given equation is, therefore,

$$y = y_c + y_p = c_1 \cos 2x + c_2 \sin 2x + \tfrac{1}{2}e^{2x}$$

ILLUSTRATIVE EXAMPLE 1

Solve $(D^2 + 4D + 4)y = 6 \sin 3x$.

Solution The complementary solution is

$$y_c = (c_1 + c_2 x)e^{-2x}$$

To find a particular solution we ask ourselves: What functions differentiated once or twice yield $\sin 3x$ or constant multiples thereof? The answer is that terms like $\sin 3x$ or $\cos 3x$ will do. We therefore attempt as particular solution

$$y = a \sin 3x + b \cos 3x$$

where a and b are our undetermined coefficients. Substituting in the given equation, we find after simplifying

$$(D^2 + 4D + 4)y = (-5a - 12b) \sin 3x + (12a - 5b) \cos 3x = 6 \sin 3x$$

This will be an identity if and only if $-5a - 12b = 6$, $12a - 5b = 0$. Solving, $a = -\frac{30}{169}$, $b = -\frac{72}{169}$.

The particular solution is, therefore, $-\frac{30}{169} \sin 3x - \frac{72}{169} \cos 3x$. Hence, the general solution is $y = (c_1 + c_2 x)e^{-2x} - \frac{30}{169} \sin 3x - \frac{72}{169} \cos 3x$.

ILLUSTRATIVE EXAMPLE 2

Solve $(D^2 + 4D + 9)y = x^2 + 3x$.

Solution The complementary solution is $y_c = e^{-2x}(c_1 \cos \sqrt{5}x + c_2 \sin \sqrt{5}x)$.

To find a particular solution we ask ourselves: What function when differentiated will yield a polynomial? Clearly polynomials when differentiated yield polynomials. Let us therefore assume as particular solution $y = y_p = ax^3 + bx^2 + cx + d$, i.e., a polynomial of the third degree. Substituting this assumed solution in the given equation and simplifying, we find

$$9ax^3 + (12a + 9b)x^2 + (6a + 8b + 9c)x + 2b + 4c + 9d = x^2 + 3x$$

Hence, $9a = 0$, $12a + 9b = 1$, $6a + 8b + 9c = 3$, $2b + 4c + 9d = 0$. Solving, we have $a = 0$, $b = \frac{1}{9}$, $c = \frac{19}{81}$, $d = -\frac{94}{729}$. Hence, $y_p = \frac{1}{9}x^2 + \frac{19}{81}x - \frac{94}{729}$.

The general solution is, thus,

$$y = e^{-2x}(c_1 \cos \sqrt{5}x + c_2 \sin \sqrt{5}x) + \frac{x^2}{9} + \frac{19x}{81} - \frac{94}{729}$$

The fact that $a = 0$ means that we did not have to use a third-degree polynomial in our assumed solution; a second-degree polynomial would have done as well. In general, when a polynomial of nth degree occurs on the right of $\phi(D)y = F(x)$ we assume as particular solution a polynomial of nth degree.*

ILLUSTRATIVE EXAMPLE 3

Solve $(D^2 + 2D + 1)y = 2 \cos 2x + 3x + 2 + 3e^x$.

Solution The complementary solution is $y_c = (c_1 + c_2 x)e^{-x}$.

We have to decide what to assume as particular solution.
For the term $2 \cos 2x$, we assume $a \sin 2x + b \cos 2x$.

---

* On page 197 we shall see that there are some exceptions to this rule. The reasoning in those cases is, however, based on assumptions used here.

For the terms $3x + 2$ (first-degree polynomial) assume $cx + d$.

For the term $3e^x$, assume fe^x.

Hence, assume as particular solution $y_p = a \sin 2x + b \cos 2x + cx + d + fe^x$.

Substituting this particular solution in the given differential equation and simplifying, we have

$$(-3a - 4b) \sin 2x + (4a - 3b) \cos 2x + cx + d + 2c + 4fe^x$$
$$= 2 \cos 2x + 3x + 2 + 3e^x$$

It follows that $-3a - 4b = 0$, $4a - 3b = 2$, $c = 3$, $2c + d = 2$, $4f = 3$. Solving these, we find $a = \frac{8}{25}$, $b = -\frac{6}{25}$, $c = 3$, $d = -4$, $f = \frac{3}{4}$. Thus, the particular solution is $y_p = \frac{8}{25} \sin 2x - \frac{6}{25} \cos 2x + 3x - 4 + \frac{3}{4}e^x$ and the general solution of the given equation is

$$y = (c_1 + c_2 x)e^{-x} + \tfrac{8}{25} \sin 2x - \tfrac{6}{25} \cos 2x + 3x - 4 + \tfrac{3}{4}e^x$$

4.2 JUSTIFICATION FOR THE METHOD OF UNDETERMINED COEFFICIENTS. THE ANNIHILATOR METHOD

It is natural to seek reasons as to why the method of undetermined coefficients seems to work. To do so, let us return to the differential equation (1) considered on page 191, which we write in operator form as

$$(D^2 + 4)y = 4e^{2x} \tag{2}$$

Since we know how to solve differential equations with right sides replaced by zero, it is natural to ask whether such methods can be used directly to solve (2). Now we can change (2) into one with right-hand side zero as follows. First differentiate both sides of (2) or equivalently operate on both sides with the operator D to obtain

$$D(D^2 + 4)y = 8e^{2x} \tag{3}$$

We can now eliminate the term involving e^{2x} in the equations (2) and (3) by multiplying equation (2) by 2 and subtracting from equation (3). The result is

$$(D - 2)(D^2 + 4)y = 0 \tag{4}$$

Suppose now that we had been given (4) in the first place. Then using the methods of pages 173–180, we would obtain the solution

$$y = c_1 \cos 2x + c_2 \sin 2x + c_3 e^{2x}$$

Note that this contains the complementary solution of (2) and *also* the special particular solution $y_p = c_3 e^{2x}$ assumed in working the problem on page 191. Thus, we did not have to guess the form of the particular solution, but instead could have arrived at it as a natural consequence of solving (4). Using $y_p = c_3 e^{2x}$ in (2), we would find as before $c_3 = \frac{1}{2}$.

The process of operating on an equation such as (2) so as to obtain an equation with the right side zero is appropriately called the *method of annihilation* or *the annihilator method*. The operator needed to do this, such as $D - 2$ in (4), is called the *annihilation operator* or briefly the *annihilator*. Let us work another example.

Use the annihilator method to work Illustrative Example 1, page 191.

Solution   The equation to be solved is

$$(D^2 + 4D + 4)y = 6 \sin 3x \tag{5}$$

We must find an operator which when applied to (5) will make the right side zero. Since $D(\sin 3x) = 3 \cos 3x$, $D^2 \sin 3x = -9 \sin 3x$ so that

$$(D^2 + 9) \sin 3x = 0$$

we see that the required annihilator is $D^2 + 9$. Using this on (5), we have

$$(D^2 + 9)(D^2 + 4D + 4)y = 0 \tag{6}$$

Note that if we find the solution of (6) we get

$$y = (c_1 + c_2x)e^{-2x} + c_3 \sin 3x + c_4 \cos 3x$$

the last two terms being just what the assumed particular solution for (5) would be. From here the method would of course proceed as in Illustrative Example 1.

Work Illustrative Example 3, page 192, by the annihilator method.

Solution   To annihilate the term $2 \cos 2x$ on the right side of the given equation

$$(D^2 + 2D + 1)y = 2 \cos 2x + 3x + 2 + 3e^x \tag{7}$$

we must use the annihilator $D^2 + 4$. Similarly, to annihilate the terms $3x + 2$ and $3e^x$, respectively, we must use the annihilators D^2 and $(D - 1)$. The resulting annihilator which serves to eliminate all the terms is thus $D^2(D - 1)(D^2 + 4)$. Applying this to (7) gives

$$D^2(D - 1)(D^2 + 4)(D^2 + 2D + 1)y = 0 \tag{8}$$

The solution of (8) is

$$y = (c_1 + c_2x)e^{-x} + c_3 \sin 2x + c_4 \cos 2x + c_5e^x + c_6x + c_7$$

which is composed of the complementary and particular solutions of (7). From here the method proceeds as in Illustrative Example 3.

A EXERCISES

1. Find the general solution of each of the following:
   (a) $y'' + y = 2e^{3x}$.
   (b) $(D^2 + 2D + 1)y = 4 \sin 2x$.
   (c) $(D^2 - 4)y = 8x^2$.
   (d) $(D^2 + 4D + 5)y = e^{-x} + 15x$.
   (e) $4I''(t) + I(t) = t^2 + 2 \cos 3t$.
   (f) $(D^3 + 4D)y = e^x + \sin x$.

2. Find solutions satisfying the given conditions:
   (a) $y'' + 16y = 5 \sin x$; $y(0) = y'(0) = 0$.
   (b) $s''(t) - 3s'(t) + 2s(t) = 8t^2 + 12e^{-t}$; $s(0) = 0$, $s'(0) = 2$.

3. Demonstrate the use of the annihilator method by working (a) Exercises 1 (a)–(f); (b) Exercises 2 (a) and (b); (c) Illustrative Example 2, page 192.

B EXERCISES

1. Solve $y'' + y = 6 \cos^2 x$, given that $y(0) = 0$, $y(\pi/2) = 0$.

2. Solve the differential equation arising in an electric circuit problem:

$$\left(LD^2 + RD + \frac{1}{C} \right) Q = E_0 \sin \omega t$$

where $D \equiv d/dt$; L, R, C, E_0, and ω are given constants and $Q(0) = Q'(0) = 0$.

C EXERCISES

1. Find the general solution of $y'' - 3y' + 2y = 4 \sin^3 3x$.

2. Solve $y'' + y = F(x)$, where $\quad F(x) = \begin{cases} x, & 0 \le x \le \pi \\ 0, & x > \pi \end{cases}$

assuming that $y(0) = y'(0) = 0$ and that y and y' are continuous at $x = \pi$.

4.3 EXCEPTIONS IN THE METHOD OF UNDETERMINED COEFFICIENTS

Let us consider the following

PROBLEM FOR DISCUSSION

Solve $(D^2 + 3D + 2)y = 4e^{-2x}$. The complementary solution is $y_c = c_1 e^{-2x} + c_2 e^{-x}$. From the fact that $4e^{-2x}$ is on the right-hand side of the given equation, we would be led by the usual assumption to the particular solution $y = ae^{-2x}$. Substituting,

$$4ae^{-2x} - 6ae^{-2x} + 2ae^{-2x} = 4e^{-2x} \quad \text{or} \quad 0 = 4e^{-2x}$$

an *impossible* situation! If we think a little about what we have done, we should be able to see that this "catastrophe" could have been foreseen. We assumed the solution ae^{-2x}, which is no different essentially from the term $c_1 e^{-2x}$ of the complementary solution. It is therefore expected that ae^{-2x} would satisfy $(D^2 + 3D + 2)y = 0$.
A little experimentation shows that if we assume as particular solution axe^{-2x} instead of ae^{-2x} we get results. For if $y = axe^{-2x}$,

$$(D^2 + 3D + 2)(axe^{-2x}) = -ae^{-2x} = 4e^{-2x}, \quad \text{i.e.,} \quad a = -4$$

Hence, a particular solution is $-4xe^{-2x}$ and the general solution is

$$y = c_1 e^{-2x} + c_2 e^{-x} - 4xe^{-2x}$$

Remark. The reason why the assumption of a particular solution axe^{-2x} works in this problem is easily supplied by the annihilator method. To see this, we note that to annihilate the right side of the given differential equation $(D^2 + 3D + 2)y = 4e^{-2x}$, we must use the annihilator $D + 2$, thus obtaining

$$(D + 2)(D^2 + 3D + 2)y = 0$$

But because of repeated roots in the auxiliary equation for this case, the general solution is $y = c_1e^{-2x} + c_2e^{-x} + c_3xe^{-2x}$ which shows how the term axe^{-2x} arises.
Before jumping to conclusions concerning rules, let us look at more examples.

ILLUSTRATIVE EXAMPLE 6

Solve $(D^2 + 4)y = 6 \sin 2x + 3x^2$.

Solution We would normally assume as particular solution $a \sin 2x + b \cos 2x$ corresponding to the term $6 \sin 2x$, and $cx^2 + dx + f$ corresponding to the term $3x^2$. We will not, however, fall into the same trap as before because we see that the assumed particular solution $a \sin 2x + b \cos 2x$ is contained in the complementary solution. We are thus inclined, by virtue of previous experience, to write as our assumed particular solution $x(a \sin 2x + b \cos 2x)$, i.e., the result assumed previously, multiplied by x. Since $cx^2 + dx + f$ has no term appearing in the complementary solution, we use it as it is. Our assumed particular solution is therefore

$$y_p = x(a \sin 2x + b \cos 2x) + cx^2 + dx + f$$

Substitution into the differential equation yields

$$4a \cos 2x - 4b \sin 2x + 4cx^2 + 4 dx + (4f + 2c) = 6 \sin 2x + 3x^2$$

or $\qquad 4a = 0, \qquad -4b = 6, \qquad 4c = 3, \qquad 4d = 0, \qquad 4f + 2c = 0,$

i.e., $\qquad a = 0, \qquad b = -\frac{3}{2}, \qquad c = \frac{3}{4}, \qquad d = 0, \qquad f = -\frac{3}{8}$

The particular solution is, therefore, $y_p = -\frac{3}{2}x \cos 2x + \frac{3}{4}x^2 - \frac{3}{8}$ and so the general solution is $y = c_1 \sin 2x + c_2 \cos 2x - \frac{3}{2}x \cos 2x + \frac{3}{4}x^2 - \frac{3}{8}$.

ILLUSTRATIVE EXAMPLE 7

Solve $(D^3 - 3D^2 + 3D - 1)y = 2e^x$. $\qquad\qquad$ (9)

Solution From the auxiliary equation $m^3 - 3m^2 + 3m - 1 = (m-1)^3 = 0$ we have $m = 1, 1, 1$. The complementary solution is, thus,

$$(c_1 + c_2x + c_3x^2)e^x$$

Because of the term $2e^x$ on the right of (9) we would normally take as particular solution ae^x. This, however, is present in the complementary solution. Hence we are led from experience to assume as particular solution axe^x. Since this is also present, we are led to ax^2e^x. However, this too is present and we are led to ax^3e^x. At last, this is not in the complementary solution. Substituting ax^3e^x in the given equation, we have $(D^3 - 3D^2 + 3D - 1)y = 6ae^x = 3e^x$. Hence, $6a = 2$, $a = \frac{1}{3}$, and a particular solution is $\frac{1}{3}x^3e^x$. The general solution is $y = (c_1 + c_2x + c_3x^2)e^x + \frac{1}{3}x^3e^x$.

From these two examples we obtain the following rule, which is not hard to justify by using the method of annihilators.

Rule. To solve a linear differential equation with constant coefficients:

1. Write the complementary solution y_c.

2. Assume a particular solution corresponding to the terms on the right-hand side of the equation:

        (a) For a polynomial of degree n, assume a polynomial of degree n.
        (b) For terms $\sin rx$, $\cos rx$, or sums and differences of such terms, assume $a \sin rx + b \cos rx$.
        (c) For terms like e^{rx} assume ae^{rx}.

3. If any of the assumed terms in 2(a), (b), or (c) occur in the complementary solution, we must multiply these assumed terms by a power of x which is sufficiently high (but not higher) so that none of these assumed terms will occur in the complementary solution.

4. Write the assumed form for the particular solution and evaluate the coefficients, thus obtaining y_p.

5. Add y_p to y_c to obtain the required general solution.

A EXERCISES

1. Find the general solution of each of the following:

    (a) $(D^2 + 2D - 3)y = 2e^x$.
    (b) $(D^2 + 1)y = x^2 + \sin x$.
    (c) $(D^2 + D)y = x^2 + 3x + e^{3x}$.
    (d) $(D^2 - 2D + 1)y = e^x$.
    (e) $y'' + 4y = 8 \cos 2x - 4x$.
    (f) $(D^3 + D)y = x + \sin x + \cos x$.

2. Find solutions satisfying the given conditions:

(a) $\dfrac{d^2 I}{dt^2} + 9I = 12 \cos 3t$; $I(0) = 4$, $I'(0) = 0$.

(b) $\dfrac{d^2 s}{dt^2} + \dfrac{ds}{dt} = t + e^{-t}$; $s = 0$, $\dfrac{ds}{dt} = 0$ at $t = 0$.

B EXERCISES

1. Find the general solution of $(D^4 - 1)y = \cosh x$.   **2.** Solve $(D^2 + 1)y = x \sin x$.

3. Work A Exercises 1 (a)–(f) using the annihilator method.

C EXERCISES

1. (a) Solve $(D^2 + \omega^2)y = A \cos \lambda x$; ω, $\lambda > 0$, $\omega \neq \lambda$, where A, ω, λ are constants, subject to $y(0) = \alpha$, $y'(0) = \beta$. (b) What is the limit of the solution in (a) as $\lambda \to \omega$? (c) Is the limiting solution of (b) a solution of the given differential equation when $\lambda = \omega$?

2. Justify the method described in rule 3 above by using the annihilator method.

3. Solve $y'' + 4y = \sin^4 x$.

4. How can the results of Exercise 4 of the C Exercises, page 178, be used in solving the problems of this section? Give illustrations.

In case the right-hand side contains products of terms like e^{rx}, $\sin rx$, $\cos rx$, and polynomials, the method of undetermined coefficients can still be used, but the method may become unwieldy and lose its appeal. In such cases other methods discussed in sections which follow yield results more easily and faster. For the sake of completeness we shall now demonstrate a general procedure for determining what to use as assumed particular solution. To motivate this procedure consider the following

PROBLEM FOR DISCUSSION

Solve $(D^2 + D + 1)y = x^3 e^x$. The complementary solution is

$$e^{-x/2}\left(c_1 \cos \frac{\sqrt{3}}{2}x + c_2 \sin \frac{\sqrt{3}}{2}x\right)$$

To see what a particular solution might be we ask: What function when differentiated might yield $x^3 e^x$? If we assume the particular solution $ax^3 e^x$, as we very well might on first thought, we obtain on substitution in the differential equation the result

$$6axe^x + 9ax^2 e^x + 3ax^3 e^x = x^3 e^x \qquad (10)$$

and it is clear that we cannot determine a, and so $ax^3 e^x$ cannot be a solution. To get rid of the terms of (10) involving $x^2 e^x$, xe^x, and e^x, we might be led to try as particular solution $ax^3 e^x + bx^2 e^x + cxe^x + de^x$. Actually when this is substituted, we find

$$(D^2 + D + 1)y = 3ax^3 e^x + (9a + 3b)x^2 e^x + (6a + 6b + 3c)xe^x$$
$$+ (2b + 3c + 3d)e^x = x^3 e^x$$

or $3a = 1$, $9a + 3b = 0$, $6a + b + 3c = 0$, $2b + 3c + 3d = 0$

and $a = \frac{1}{3}$, $b = -1$, $c = \frac{4}{3}$, $d = -\frac{2}{3}$

Hence, the particular solution is $y_p = \frac{1}{3}x^3 e^x - x^2 e^x + \frac{4}{3}xe^x - \frac{2}{3}e^x$ and the general solution is

$$y = e^{-x/2}\left(c_1 \cos \frac{\sqrt{3}}{2}x + c_2 \sin \frac{\sqrt{3}}{2}x\right) + \frac{1}{3}x^3 e^x - x^2 e^x + \frac{4}{3}xe^x - \frac{2e^x}{3}$$

On the basis of observations of examples, such as the one just given, we are able to formulate a method for arriving at the particular solution which should be assumed. The method consists of differentiating the right-hand side of the equation indefinitely and keeping track of all essentially different terms which arise. If there are a finite number of these terms the method of undetermined coefficients is applicable and the assumed particular solution can be formed by taking each of these terms, multiplying it by an undetermined constant and then adding the results. After obtaining this particular solution, we must of course modify it in accordance with rule 3 on page 197 if any of its terms appear in the complementary solution. This method too can be justified by using the annihilator method, but we shall leave such justification to the exercises. Instead, we shall present the following illustrations of its use.

Solve $(D^2 + 1)y = x^2 \cos 5x$.

Solution The complementary solution is $y_c = c_1 \cos x + c_2 \sin x$. To find a particular solution consider the right-hand side $x^2 \cos 5x$. A single differentiation would give rise to terms (disregarding numerical coefficients) like

$$x^2 \sin 5x, \qquad x \cos 5x$$

Differentiation of each of these would produce terms like

$$x^2 \cos 5x, \qquad x \sin 5x, \qquad \cos 5x$$

Continuing in this manner, we find that no terms other than those in the following group arise:

$$x^2 \cos 5x, \qquad x^2 \sin 5x, \qquad x \cos 5x, \qquad x \sin 5x, \qquad \cos 5x, \qquad \sin 5x$$

Thus, we assume as particular solution,

$$y_p = ax^2 \cos 5x + bx^2 \sin 5x + cx \cos 5x + dx \sin 5x + f \cos 5x + g \sin 5x \quad (11)$$

since none of these terms appear in the complementary solution. The constants in (11) can be found by substituting y_p in the given differential equation and equating coefficients of corresponding terms.

If the right-hand side were $\ln x$, for example, successive differentiations yield the infinite set of functions $1/x, 1/x^2, \ldots$, and in this case the method is inapplicable. For such cases, other methods described later in this chapter may be employed.

Solve $(D^2 + 25)y = x^2 \cos 5x$.

Solution Proceeding as in Illustrative Example 8, we find the particular solution

$$y_p = ax^2 \cos 5x + bx^2 \sin 5x + cx \cos 5x + dx \sin 5x + f \cos 5x + g \sin 5x$$

However, the complementary solution is $y_c = c_1 \cos 5x + c_2 \sin 5x$, and we see that some of the terms in the above particular solution also appear in the complementary solution. Then in accordance with rule 3 on page 197, we must multiply the particular solution in this case by x to obtain

$$y_p = ax^3 \cos 5x + bx^3 \sin 5x + cx^2 \cos 5x + dx^2 \sin 5x + fx \cos 5x + gx \sin 5x$$

The constants $a, b, \ldots, f, g$ can now be obtained by substituting into the given equation and equating coefficients of like terms.

A EXERCISES

1. For each of the following, write the complementary solution and the expression in terms of undetermined coefficients which you would use in attempting to find a particular solution. You need not evaluate these coefficients.
   (a) $(D^2 + 1)y = xe^{-x} + 3 \sin x$.
   (b) $(D^2 - 2D - 3)y = x \sin 2x + x^3 e^{3x}$.
   (c) $(D^4 + D^2)y = 3x^2 - 4e^x$.
   (d) $(D^2 - 2D + 1)y = x^2 e^x$.
   (e) $(D^2 + 1)y = e^{-x} \cos x + 2x$.
   (f) $(D^2 - 4D + 3)y = 3e^x + 2e^{-x} + x^3 e^{-x}$.

2. Find the general solution of each of the following:

(a) $(D^2 - 1)y = xe^x$.

(b) $(D^2 + 4)y = x^2 + 3x \cos 2x$.

(c) $(D^2 + 2D + 1)y = \sin 3x + xe^{-x}$.

(d) $Q''(t) + Q(t) = t \sin t + \cos t$.

B EXERCISES

1. Find the general solution of $(D^3 - 5D^2 - 2D + 24)y = x^2 e^{3x}$.

2. Solve $(D^2 + \omega^2)y = t(\sin \omega t + \cos \omega t)$; $D \equiv d/dt$ subject to $y(0) = y'(0) = 0$.

3. Solve $y'' - 3y' + 2y = e^{-x}(1 + \cos 2x)$.

C EXERCISES

1. Solve $y'' + 4y = \cos x \cos 2x \cos 3x$. **2.** Solve $y''' + 4y'' - 6y' - 12y = (\sinh x)^4$.

3. Solve $(D^2 + 1)y = x^2 \cos 5x$ by using the results of Exercise 4 of the C Exercises, page 178.

4. Justify the method of finding complementary solutions described in Illustrative Example 8 by using the annihilator method.

4.5 THE METHOD OF VARIATION OF PARAMETERS

It has been mentioned that the method of undetermined coefficients is effective only when the right-hand side of the equation is of special type. It is natural for us to worry about what can be done in case the method of undetermined coefficients is inapplicable. Fortunately we have been saved from too much worry by the efforts of a famous mathematician named Lagrange, who discovered a very ingenious and powerful method which applies in cases where the method of undetermined coefficients does not work as well as to where it does. In fact by knowing this method alone one can do very well without the method of undetermined coefficients.

To show how the method of Lagrange works, we shall try to obtain the general solution of the innocent-looking differential equation

$$y'' + y = \tan x \tag{12}$$

The only disturbing thing about this is the presence of the term $\tan x$. The complementary solution is

$$A \cos x + B \sin x \tag{13}$$

where A and B are arbitrary constants.

The ingenuity of Lagrange lay in his assumption, which at first sight (like many nonobvious assumptions) looks ridiculous. He said:

> Assume that A and B in (13) are not constants but instead are functions of x, denoted by $A(x)$ and $B(x)$ respectively. The question is: What functions must they be so that $A(x) \cos x + B(x) \sin x$ will be a solution of (12)?

Since the method assumes that the quantities A and B vary, the method is generally called the *method of variation of parameters* or *variation of constants*. It is clear that since two functions $A(x)$ and $B(x)$ are to be determined, we expect that two conditions ought to be imposed. One of these arises from the fact that the assumed solution satisfies the differential equation. We are therefore at liberty to impose *one other condition*. With this in mind let us proceed. By differentiation of

$$y = A(x) \cos x + B(x) \sin x \tag{14}$$

we have $\quad y' = -A(x) \sin x + B(x) \cos x + A'(x) \cos x + B'(x) \sin x \tag{15}$

Upon realizing that a further differentiation would introduce a few more terms, we take advantage of our liberty to choose one condition on $A(x)$ and $B(x)$. We choose a condition which simplifies (15), i.e.,

$$A'(x) \cos x + B'(x) \sin x = 0 \tag{16}$$

identically. Thus, (15) becomes

$$y' = -A(x) \sin x + B(x) \cos x$$

One further differentiation yields

$$y'' = -A(x) \cos x - B(x) \sin x - A'(x) \sin x + B'(x) \cos x \tag{17}$$

Substituting (14) and (17) in the given differential equation, we find

$$-A'(x) \sin x + B'(x) \cos x = \tan x \tag{18}$$

From the two simultaneous equations (16) and (18), it is a simple exercise to obtain

$$A'(x) = -\frac{\sin^2 x}{\cos x}, \qquad B'(x) = \sin x$$

Hence,

$$A(x) = \int \frac{\cos^2 x - 1}{\cos x} \, dx = \int (\cos x - \sec x) dx = \sin x - \ln (\sec x + \tan x) + c_1$$

$$B(x) = -\cos x + c_2$$

It follows that the required solution (14) is

$$y = A(x) \cos x + B(x) \sin x = c_1 \cos x + c_2 \sin x - \cos x \ln (\sec x + \tan x)$$

Remark 1. It is not difficult to show that if $y = Au_1(x) + Bu_2(x)$ is the complementary solution of the equation $y'' + Py' + Qy = F(x)$, where P and Q may depend on x, the two principal equations (16) and (18) are

$$A'u_1(x) + B'u_2(x) = 0, \qquad A'u_1'(x) + B'u_2'(x) = F(x) \tag{19}$$

from which A and B may be obtained. Note that the determinant of the coefficients in (19) is the Wronskian given by

$$W(u_1, u_2) = \begin{vmatrix} u_1(x) & u_2(x) \\ u_1'(x) & u_2'(x) \end{vmatrix}$$

which is assumed different from zero.

It should be emphasized that the method of variation of parameters is a very useful one in that it is applicable for any differential equation (with constant or variable coefficients) where we can write the complementary solution. For this reason it can be used even in cases where we might apply the method of undetermined solutions. In fact, some ardent fans of the method advocate that it should be the only method used. To illustrate the method in the case where the method of undetermined coefficients can also be used, let us consider the following

ILLUSTRATIVE EXAMPLE 10

Solve $y'' - 3y' + 2y = xe^x + 2x$.

Solution The complementary solution in this case is

$$y = Ae^x + Be^{2x} \tag{20}$$

If we assume according to the method of variation of parameters that A and B are functions of x, then we have by differentiation $y' = Ae^x + 2Be^{2x} + A'e^x + B'e^{2x}$.

Taking

$$A'e^x + B'e^{2x} = 0 \tag{21}$$

so that

$$y' = Ae^x + 2Be^{2x} \tag{22}$$

we have by differentiating again

$$y'' = Ae^x + 4Be^{2x} + A'e^x + 2B'e^{2x} \tag{23}$$

Substitution of (20), (22), and (23) into the given differential equation yields

$$A'e^x + 2B'e^{2x} = xe^x + 2x \tag{24}$$

Solving the simultaneous differential equations (21) and (24) for A' and B' gives

$$A' = -x - 2xe^{-x}, \qquad B' = xe^{-x} + 2xe^{-2x}$$

Then by integration

$$A = -\frac{x^2}{2} + 2xe^{-x} + 2e^{-x} + c_1, \qquad B = -xe^{-x} - e^{-x} - xe^{-2x} - \tfrac{1}{2}e^{-2x} + c_2$$

Using these in (20) gives the required solution:

$$y = c_1e^x + c_2e^{2x} - \frac{x^2}{2}e^x - xe^{-x} - e^x + x + \frac{3}{2}$$

$$= C_1e^x + C_2e^{2x} - \frac{x^2}{2}e^x - xe^{-x} + x + \frac{3}{2}$$

The student should compare this with the method of undetermined coefficients.

Another Method. From the Remark on page 201, equations (21) and (24) can be written at once, and some time can be saved in using the method.

Remark 2. In practice it may be difficult or impossible to express the integrals obtained from A' and B' in closed form. For example, suppose that the right side of the equation to be solved in Illustrative Example 10 is $\sin x^3$ instead of $xe^x + 2x$. In this case we would arrive at the simultaneous equations

$$A'e^x + B'e^{2x} = 0, \qquad A'e^x + 2B'e^{2x} = \sin x^3$$

with solution $A' = -e^{-x} \sin x^3,$ $B' = e^{-2x} \sin x^3$

so that $A = -\int e^{-x} \sin x^3 \, dx + c_1,$ $B = \int e^{-2x} \sin x^3 \, dx + c_2$

The required general solution is

$$y = c_1 e^x + c_2 e^{2x} - e^x \int e^{-x} \sin x^3 \, dx + e^{2x} \int e^{-2x} \sin x^3 \, dx$$

even though we cannot carry out the integrations. This is no handicap however, since in an applied problem numerical integrations can be performed if necessary.

Lagrange's method of variation of parameters may be extended to higher-order linear equations. Such extension is considered in the exercises.

A EXERCISES

Solve each of the following by variation of parameters:

1. $y'' + y = \cot x.$

2. $y'' + y = \sec x.$

3. $y'' + 4y = \csc 2x.$

4. $y'' - y = e^x.$

5. $y'' + 3y' + 2y = 3e^{-2x} + x.$

6. $y'' + y' - 2y = \ln x.$

7. $2y'' + 3y' + y = e^{-3x}.$

8. $(D^2 - 1)y = x^2 e^x.$

9. $y'' - y = e^{-x^2}.$

10. $y'' - 4y' + 4y = \sqrt{x}.$

B EXERCISES

1. Use the method of variation of parameters to solve $y' + P(x)y = Q(x).$

2. Show that the solution of $y'' + a^2 y = F(x)$ subject to $y(0) = y'(0) = 0$ is

$$y = \frac{1}{a} \int_0^x F(u) \sin a(x - u) du$$

3. (a) Let $\phi(D)y = F(x)$ be a third-order linear differential equation. Assume that the complementary solution is known so that according to the method of variation of parameters $y = Au_1(x) + Bu_2(x) + Cu_3(x)$ where A, B, C are appropriate functions of x. Show that the equations for determining $A, B,$ and C are given by

$$\left. \begin{array}{l} A'u_1 + B'u_2 + C'u_3 = 0 \\ A'u_1' + B'u_2' + C'u_3' = 0 \\ A'u_1'' + B'u_2'' + C'u_3'' = F(x) \end{array} \right\}$$

(b) Solve $y''' - 2y'' - y' + 2y = e^x$ by using the method in (a).

4. Solve $y''' - 6y'' + 11y' - 6y = e^x + e^{-x}$ by variation of parameters.

5. Show that the general solution of $x^2 y'' - 2xy' + 2y = 0$ is $y = Ax^2 + Bx.$ Hence, find the general solution of $x^2 y'' - 2xy' + 2y = xe^{-x}.$

6. Prove the results in the Remark on page 201.

C EXERCISES

1. If the complementary solution of $y'' + P(x)y' + Q(x)y = R(x)$ is $Au_1(x) + Bu_2(x)$, show that its general solution is

$$y = c_1u_1(x) + c_2u_2(x) + u_2(x) \int \frac{R(x)u_1(x)dx}{W(u_1, u_2)} - u_1(x) \int \frac{R(x)u_2(x)dx}{W(u_1, u_2)}$$

where

$$W(u_1, u_2) = \begin{vmatrix} u_1(x) & u_2(x) \\ u_1'(x) & u_2'(x) \end{vmatrix} = u_1(x)u_2'(x) - u_2(x)u_1'(x)$$

is the *Wronskian* of $u_1(x)$ and $u_2(x)$ and is not identically zero. Discuss the case where $W \equiv 0$.

2. Use *Abel's identity* and Exercise 1 to show that the general solution of $y'' + P(x)y' + Q(x)y = R(x)$ is $y = c_1u_1(x) + c_2u_2(x) + \frac{u_2(x)}{c} \int R(x)u_1(x)e^{\int P(x)dx} \, dx - \frac{u_1(x)}{c} \int R(x)u_2(x)e^{\int P(x)dx} \, dx$.

3. Apply the results of Exercises 1 and 2 to Exercise 5 of the B Exercises.

4.6 SHORT-CUT METHODS INVOLVING OPERATORS

We have seen how to solve linear differential equations with constant coefficients by using the method of undetermined coefficients and the method of variation of parameters. The first method is applicable only when the right-hand side contains suitable functions (polynomials, exponentials, etc.). The second method however can be used not only for such suitable functions but for more complicated ones as well. In this case solutions may involve indefinite integrals (see Remark 2, page 202). In view of this, the so-called *operator* or *operational methods* for solving such equations to be described in this section may appear superfluous. However, before the student turns quickly to the next section, several reasons for including them should be pointed out. First, such methods often supply solutions to problems in a shorter, less tedious manner than other methods. Second, they illustrate the power of "symbolic methods" in general. Third, they serve as a forerunner of Laplace transform methods to be discussed in Chapter Six*, although they are of interest in themselves. With these comments we now present a brief discussion of operator methods.

Consider

$$\phi(D)y = F(x) \tag{25}$$

where $\phi(D)$ is the linear polynomial operator in $D \equiv d/dx$, i.e.,

$$\phi(D) \equiv a_0 D^n + a_1 D^{n-1} + \cdots + a_{n-1}D + a_n$$

where we shall here take $a_0, a_1, \ldots, a_n$ as given constants. If we "solve" formally for y in (25), treating the equation algebraically, we obtain

$$y = \frac{1}{\phi(D)} F(x) \tag{26}$$

---

* Historically, the electrical engineer *Heaviside* used such operational methods formally with great success. Laplace transform methods were developed to a large extent in an effort to place his methods on a sound mathematical basis.

Here $1/\phi(D)$ represents an operation to be performed on $F(x)$. A question which needs answering is: What is the nature of the operation? To gain insight, consider the simple equation $Dy = x$. Here we have symbolically*

$$y = \frac{1}{D} x$$

What shall we mean by this? A clue appears if we solve $Dy = x$. We obtain,

$$y = \int x \, dx = \frac{x^2}{2} + c$$

and hence it is natural to make the definition

$$\frac{1}{D} x \equiv \int x \, dx \tag{27}$$

the interpretation being that the operation of "multiplying" a function by $1/D$ corresponds to an integration of the function. It is natural to ask whether $1/D^2$ operating on a function corresponds to a double integration of the function, and in general whether $1/D^n$, n an integer, corresponds to an n-fold integration. The student can easily convince himself of this interpretation. Operators such as $1/D$, $1/D^2$, etc. are called *inverse operators*. Let us investigate other such operators. Consider

$$(D - p)y = f(x) \tag{28}$$

where p is a constant. Formally, we have

$$y = \frac{1}{D - p} f(x) \tag{29}$$

Since equation (28) can be solved exactly to give

$$y = e^{px} \int e^{-px} f(x) dx \tag{30}$$

it is natural to make the interpretation

$$\frac{1}{D - p} f(x) \equiv e^{px} \int e^{-px} f(x) dx \tag{31}$$

Note that this reduces to (27) if $p = 0$ and $f(x) = x$.

It is of interest to ask what one might mean by

$$(D - p_1)(D - p_2)y = f(x) \tag{32}$$

where p_1 and p_2 are constants. We know that

$$(D - p_2)y \equiv \frac{dy}{dx} - p_2 y$$

* Some may prefer to write D^{-1} (D *inverse*) instead of $1/D$. We shall use the two symbols interchangeably.

It thus seems plausible that

$$(D - p_1)(D - p_2)y \equiv \left(\frac{d}{dx} - p_1\right)\left(\frac{dy}{dx} - p_2 y\right)$$

$$\equiv \frac{d}{dx}\left(\frac{dy}{dx} - p_2 y\right) - p_1\left(\frac{dy}{dx} - p_2 y\right)$$

$$\equiv \frac{d^2 y}{dx^2} - (p_1 + p_2)\frac{dy}{dx} + p_1 p_2 y$$

$$\equiv [D^2 - (p_1 + p_2)D + p_1 p_2]y$$

Thus, the operator $(D - p_1)(D - p_2)$ is equivalent to $D^2 - (p_1 + p_2)D + p_1 p_2$. The converse may similarly be established. It follows that operators may be multiplied or factored like algebraic quantities. This is not, however, always possible if p_1 and p_2 are not constants. One may show that the operational factorization

$$a_0 D^n + a_1 D^{n-1} + \cdots + a_n \equiv a_0(D - p_1)(D - p_2) \cdots (D - p_n)$$

is always possible (and unique) when $a_0, \ldots, a_n$, and consequently $p_1, \ldots, p_n$ are constants. Furthermore the order of the factors is immaterial, i.e., the operators obey the commutative, associative, and distributive laws just as do algebraic quantities.* This important fact enables us to treat equation (32) in the same way that we treated the other equations; i.e., we may write (32) formally as

$$y = \frac{1}{(D - p_1)(D - p_2)} f(x)$$

By a double application of (31) we have

$$\frac{1}{(D - p_1)(D - p_2)} f(x) \equiv \frac{1}{D - p_1}\left[e^{p_2 x} \int e^{-p_2 x} f(x) dx\right]$$

$$\equiv e^{p_1 x} \int e^{-p_1 x}\left[e^{p_2 x} \int e^{-p_2 x} f(x) dx\right] dx$$

In a similar manner we may write

$$\frac{1}{(D - p_1) \cdots (D - p_n)} f(x) \equiv e^{p_1 x} \int e^{-p_1 x} e^{p_2 x} \int e^{-p_2 x} e^{p_3 x} \int \cdots e^{p_n x} \int e^{-p_n x} f(x) dx^n$$

(33)

The method works even when some or all of the constants $p_1, \ldots, p_n$ are equal.

Since the left side of (33) looks so much like an algebraic fraction, it is natural to ask whether one can, as in algebra, resolve it into partial fractions; i.e., can we write for example, in the case where $p_1, \ldots, p_n$ are distinct constants, the identity

$$\frac{1}{(D - p_1)(D - p_2) \cdots (D - p_n)} \equiv \frac{A_1}{D - p_1} + \frac{A_2}{D - p_2} + \cdots + \frac{A_n}{D - p_n}$$

(34)

* See the B and C Exercises on pages 172 and 173.

for suitably determined constants $A_1, A_2, \ldots, A_n$. If (34) is true, then

$$\frac{1}{(D - p_1) \ldots (D - p_n)} f(x) \equiv \frac{A_1}{D - p_1} f(x) + \frac{A_2}{D - p_2} f(x) + \cdots + \frac{A_n}{D - p_n} f(x)$$

and the right-hand side may be interpreted, by using (31) as

$$A_1 e^{p_1 x} \int e^{-p_1 x} f(x) dx + A_2 e^{p_2 x} \int e^{-p_2 x} f(x) dx + \cdots + A_n e^{p_n x} \int e^{-p_n x} f(x) dx$$

This turns out to be correct and is easier than the interpretation (33), since it involves only single integrations.

ILLUSTRATIVE EXAMPLE 11

Find the general solution of $(D^2 - 1)y = e^{-x}$.

Solution   We may write the equation as $(D - 1)(D + 1)y = e^{-x}$. Hence,

$$
\begin{aligned}
y &= \frac{1}{(D - 1)(D + 1)} e^{-x} = \left(\frac{\frac{1}{2}}{D - 1} - \frac{\frac{1}{2}}{D + 1} \right) e^{-x} \\
&= \tfrac{1}{2} e^{x} \int e^{-x}(e^{-x}) dx - \tfrac{1}{2} e^{-x} \int e^{x}(e^{-x}) dx \\
&= c_1 e^{x} + c_2 e^{-x} - \tfrac{1}{4} e^{-x} - \tfrac{1}{2} x e^{-x} \\
&= A e^{x} + B e^{-x} - \tfrac{1}{2} x e^{-x}
\end{aligned}
$$

It should be remarked that by omitting the constants of integration our methods can be used to obtain particular solutions.

ILLUSTRATIVE EXAMPLE 12

Find a particular solution of $(D^2 + 4D + 4)y = x^3 e^{-2x}$.

Solution   We write $(D + 2)^2 y = x^3 e^{-2x}$. Thus,

$$
\begin{aligned}
y &= \frac{1}{(D + 2)^2} (x^3 e^{-2x}) = \frac{1}{D + 2} \cdot \frac{1}{D + 2} (x^3 e^{-2x}) \\
&= \frac{1}{D + 2} \left[e^{-2x} \int e^{2x}(x^3 e^{-2x}) dx \right] = \frac{1}{D + 2} \left(e^{-2x} \int x^3 \, dx \right) \\
&= \frac{1}{D + 2} \left[e^{-2x} \left(\frac{x^4}{4} \right) \right] = e^{-2x} \int e^{2x} \cdot e^{-2x} \left(\frac{x^4}{4} \right) dx = \frac{x^5}{20} e^{-2x}
\end{aligned}
$$

In this example the work is much shorter than it would have been had we used the method of undetermined coefficients.

Operator methods are of great use in finding particular solutions. The methods employed may not always be easily justifiable, as is shown in the following

ILLUSTRATIVE EXAMPLE 13

Find a particular solution of $(D^2 - D + 1)y = x^3 - 3x^2 + 1$.

Solution   Let us write $y = \dfrac{1}{1 - D + D^2} (x^3 - 3x^2 + 1)$.

By ordinary long division in ascending powers of D we find

$$\frac{1}{1 - D + D^2} = 1 + D - D^3 - D^4 + \cdots$$

Hence, formally, $y = (1 + D - D^3 - D^4 + \cdots)(x^3 - 3x^2 + 1)$

$$= 1(x^3 - 3x^2 + 1) + D(x^3 - 3x^2 + 1) - D^3(x^3 - 3x^2 + 1)$$
$$\quad - D^4(x^3 - 3x^2 + 1) + \cdots$$
$$= x^3 - 6x - 5$$

The *remarkable* thing about this is that $y = x^3 - 6x - 5$ is actually a particular solution.

Because of the example just completed, the student must not get the impression that any manipulation of operators will lead to profitable results. While there are many ways in which they can be manipulated, there are also many ways in which they cannot. In the B and C Exercises we have presented a few of the more common short operator techniques and have indicated their proofs. The student may find them useful in getting results quickly in many cases.

A EXERCISES

Using operator methods, find the general solution of each equation.

1. $(D^2 - D)y = 1$.

2. $(D^2 - 2D + 1)y = e^x$.

3. $(D^2 + 3D + 2)y = e^x - e^{-x}$.

4. $(D^2 - 1)y = 2x^4 - 3x + 1$.

5. $(D^2 + D)y = 4x^3 - 2e^{2x}$.

6. $(D^2 + 2D + 1)y = x^2 e^{-x} + 1$.

7. $(D^2 - 4D + 4)y = e^{2x} \sin 3x$.

8. $(D^3 - D)y = 1 + x^5$.

B EXERCISES

1. If $\phi(D)$ is a polynomial operator in D with constant coefficients, and m is any constant, show that $\phi(D)e^{mx} = \phi(m)e^{mx}$. Hence, show that a solution of $\phi(D)y = 0$ is $y = e^{mx}$, where m may take on values which satisfy $\phi(m) = 0$. Note that $\phi(m) = 0$ is the auxiliary equation.

2. If a and b are constants and y_1, y_2 are suitable functions of x, show that

$$\frac{1}{\phi(D)}(ay_1 + by_2) = a\frac{1}{\phi(D)}y_1 + b\frac{1}{\phi(D)}y_2$$

This shows that $1/\phi(D)$ is a linear operator.

3. Show that if arbitrary constants are omitted

$$\frac{1}{\phi(D)}(e^{mx}) = \frac{1}{\phi(m)}(e^{mx}), \qquad \phi(m) \neq 0$$

Thus, evaluate $\dfrac{1}{D^2 - 2D - 3}e^{4x}$ and obtain the general solution of $(D^2 - 2D - 3)y = e^{4x}$.

4. Obtain a particular solution of $(D^3 + 3D^2 - 4D - 12)y = 2e^{3x} - 4e^{-5x}$. Also find the general solution.

5. (a) Prove that if m is a constant and F is differentiable,

$$D(e^{mx}F) = e^{mx}(D + m)F, \qquad D^2(e^{mx}F) = e^{mx}(D + m)^2 F$$

(b) By mathematical induction, extend the results of (a) to $D^n(e^{mx}F) = e^{mx}(D + m)^n F$.

6. Use Exercise 5 to show that $\phi(D)(e^{mx}F) = e^{mx}\phi(D + m)F$ where $\phi(D)$ is a polynomial in D with constant coefficients. The result is called the *operator-shift theorem*.

7. Use the operator-shift theorem of Exercise 6 to show that the equation $(D - m)^p y = 0$ has general solution $y = e^{mx}(c_1 + c_2 x + \cdots + c_p x^{p-1})$.

8. Use the operator-shift theorem to show that $\dfrac{1}{\phi(D)}(e^{mx}G) = e^{mx}\dfrac{1}{\phi(D + m)}G$.

[*Hint:* Let $\phi(D + m)F = G$ in the operator-shift theorem of Exercise 6.] The result is the *inverse operator-shift theorem*. Note that by this theorem, $1/\phi(D)$ has the same property as $\phi(D)$ in Exercise 6.

9. Using Exercise 8, evaluate

(a) $\dfrac{1}{D^2 - 4D + 3}(x^3 e^{2x})$ and obtain a particular solution of $(D^2 - 4D + 3)y = x^3 e^{2x}$.

(b) $\dfrac{1}{D^2 + 2D + 1}(2x^2 e^{-2x} + 3e^{2x})$ and obtain a particular solution of $(D^2 + 2D + 1)y = 2x^2 e^{-2x} + 3e^{2x}$.

10. Show that if ϕ is a polynomial with constant coefficients,

$$\phi(D^2)(\sin ax) = \phi(-a^2)(\sin ax), \qquad \phi(D^2)(\cos ax) = \phi(-a^2)(\cos ax)$$

Hence, derive the results,

$$\frac{1}{\phi(D^2)}(\sin ax) = \frac{1}{\phi(-a^2)}(\sin ax), \qquad \frac{1}{\phi(D^2)}(\cos ax) = \frac{1}{\phi(-a^2)}(\cos ax), \qquad \phi(-a^2) \neq 0$$

11. Using Exercise 10 evaluate **(a)** $\dfrac{1}{D^2 + 1}(\sin 3x)$. **(b)** $\dfrac{1}{D^4 - 3D^2 + 2}(2\cos 2x - 4\sin 2x)$.

12. The results of Exercise 10 can be used to find particular solutions of $\phi(D^2)y = \sin ax$ or $\phi(D^2)y = \cos ax$, but they fail for the equations $\phi(D)y = \sin ax$ or $\phi(D)y = \cos ax$.

(a) Show that $\phi(D)$ can always be written as $F_1(D^2) + DF_2(D^2)$. Consider

$$[F_1(D^2) + DF_2(D^2)]y = \sin ax$$

Operate on both sides by $F_1(D^2) - DF_2(D^2)$ to obtain

$$\{[F_1(D^2)]^2 - D^2[F_2(D^2)]^2\}y = F_1(-a^2)\sin ax - aF_2(-a^2)\cos ax$$

and hence show that a particular solution is

$$\frac{F_1(-a^2)\sin ax - aF_2(-a^2)\cos ax}{[F_1(-a^2)]^2 + a^2[F_2(-a^2)]^2}$$

(b) Obtain the result in (a) formally by writing

$$\frac{1}{\phi(D)} \sin ax = \frac{1}{F_1(D^2) + DF_2(D^2)} \sin ax$$

$$= \frac{F_1(D^2) - DF_2(D^2)}{[F_1(D^2) + DF_2(D^2)][F_1(D^2) - DF_2(D^2)]} \sin ax$$

$$= \frac{F_1(-a^2) - DF_2(-a^2)}{[F_1(-a^2)]^2 + a^2[F_2(-a^2)]^2} \sin ax$$

$$= \frac{F_1(-a^2) \sin ax - aF_2(-a^2) \cos ax}{[F_1(-a^2)]^2 + a^2[F_2(-a^2)]^2}$$

Arrive at a similar result for $\dfrac{1}{\phi(D)} \cos ax$.

13. Use the results of Exercise 12 to find particular solutions of: (a) $(D^2 - 3D + 2)y = \sin 3x$. (b) $(2D^3 + D^2 - 2D - 1)y = 3 \sin 2x + 4 \cos 2x$.

14. By combining the operator-shift theorem of Exercise 8 and the results of Exercise 12, evaluate

(a) $\dfrac{1}{D^2 + D - 2}(e^{2x} \sin 3x)$. $\qquad$ (b) $\dfrac{1}{D^3 + 2D^2 - 1}(e^{-x} \cos 2x)$.

15. Show how the identities $\qquad \sin \theta = \dfrac{e^{i\theta} - e^{-i\theta}}{2i}, \qquad \cos \theta = \dfrac{e^{i\theta} + e^{-i\theta}}{2}$

may be employed to obtain the results of Exercises 10–14.

16. Using the methods of Exercises 1–15 evaluate:

(a) $\dfrac{1}{D^2 + 2D - 5}(e^{3x})$. $\quad$ (b) $\dfrac{1}{D^3 - 1}(x^5 + 3x^4 - 2x^3)$. $\quad$ (c) $\dfrac{1}{D^2 + 1}(x^2 e^{2x})$.

(d) $\dfrac{1}{D^2 + D}(8 \sin 4x)$. $\quad$ (e) $\dfrac{1}{D^3 + 1}(\sin x + \cos x)$. $\quad$ (f) $\dfrac{1}{(D - 3)^3}(e^{3x} \cos 4x)$.

(g) $\dfrac{1}{D^2 - 4D + 3}[e^x(2 \sin x - 3 \cos x)]$. $\quad$ (h) $\dfrac{1}{(D + 2)^2(D + 1)^3}(x^3 e^{-x} + e^{-x} \sin x)$.

(i) $\dfrac{1}{D^3 + D^2 - D - 1}(x^2 + 3 \sin x)$. $\quad$ (j) $\dfrac{1}{D^3 - D^2 + D}(x^2 e^x - 4x^4)$.

C EXERCISES

1. (a) Use partial fractions to obtain $\qquad \dfrac{1}{(D - 2)^2(D + 3)}(x^2 e^{2x})$

and verify your result by other methods. (b) Justify the manipulations in (a) by proving that

$$\left(\frac{a}{\phi_1(D)} + \frac{b}{\phi_2(D)} \right) F = a \frac{1}{\phi_1(D)} F + b \frac{1}{\phi_2(D)} F$$

2. (a) Show how to modify the result of B Exercise 3 in case $\phi(m) = 0$ but $\phi'(m) \neq 0$ (i.e., m is a root but not a double root of $\phi(m) = 0$), obtaining

$$\frac{1}{\phi(D)} e^{mx} = \frac{1}{\phi'(m)} x e^{mx}$$

(b) In case $\phi(m) = 0$, $\phi'(m) = 0$, $\phi''(m) = 0, \ldots, \phi^{(p-1)}(m) = 0$, $\phi^{(p)}(m) \neq 0$, [i.e., m is a p-fold root of $\phi(r) = 0$], show that

$$\frac{1}{\phi(D)} e^{mx} = \frac{1}{\phi^{(p)}(m)} x^p e^{mx}$$

[*Hint:* If m is a p-fold root of $\phi(r) = 0$, then $(r - m)^p$ is a factor of $\phi(r)$; similarly $(D - m)^p$ is a factor of $\phi(D)$.]

3. Use the results of Exercise 2 to investigate the case of B Exercise 10 if $\phi(-a^2) = 0$.

4. Evaluate:

(a) $\dfrac{1}{(D - 1)^2} (e^x)$.   (b) $\dfrac{1}{D^2 - 3D + 2} (2e^x - e^{2x})$.   (c) $\dfrac{1}{(D + 2)^4} (e^{-2x})$.

(d) $\dfrac{1}{D^2 + 9} (\sin 3x)$.   (e) $\dfrac{1}{D^3 + D} (\cos x + \sin x)$.   (f) $\dfrac{1}{D^4 + 2D^2 + 1} (\sin x)$.

5. Solve $(D^2 + 1)y = x^2 \cos x$ by writing $\cos x$ as the real part of e^{ix} and factoring $D^2 + 1$ into $(D - i)(D + i)$.

6. By use of operators solve $(aD^2 + bD + c)y = F(x)$ where a, b, c are constants.

5

Remarks Concerning Equations with Variable Coefficients Which Can Be Transformed into Linear Equations with Constant Coefficients: The Euler Equation

The student has had occasion to see how some differential equations could be solved by the use of suitable and often ingenious transformations. It should come as no surprise that some linear differential equations with variable coefficients can be solved by transforming them into linear differential equations with constant coefficients. The particular transformation used may not always be obvious, and the student should not develop a sense of inferiority if he does not see it immediately. Consider the following

PROBLEM FOR DISCUSSION

Solve $x^2 y'' + xy' + 4y = 1$.   (1)

It is far from obvious how to proceed to solve this linear equation with variable coefficients. Mathematicians have found, however, that this equation belongs to a special type which can be solved by the transformation $x = e^z$. To see this, note that

$$y' = \frac{dy}{dx} = \frac{dy}{dz} \cdot \frac{dz}{dx} = \frac{dy/dz}{dx/dz} = e^{-z} \frac{dy}{dz}$$

so that
$$xy' = x\frac{dy}{dx} = \frac{dy}{dz} \tag{2}$$

Similarly, $y'' = \dfrac{d^2y}{dx^2} = \dfrac{d}{dx}\left(e^{-z}\dfrac{dy}{dz}\right) = \dfrac{d}{dz}\left(e^{-z}\dfrac{dy}{dz}\right) \bigg/ \dfrac{dx}{dz} = e^{-2z}\left(\dfrac{d^2y}{dz^2} - \dfrac{dy}{dz}\right)$

so that
$$x^2 y'' = x^2\frac{d^2y}{dx^2} = \frac{d^2y}{dz^2} - \frac{dy}{dz} \tag{3}$$

Because of (2) and (3), the differential equation becomes

$$\frac{d^2y}{dz^2} + 4y = 1 \tag{4}$$

so that
$$y = A\cos 2z + B\sin 2z + \tfrac{1}{4}$$

Since $z = \ln x$, the solution of the required equation is*

$$y = A\cos(2\ln x) + B\sin(2\ln x) + \tfrac{1}{4}$$

The transformation $x = e^z$ will transform

$$(a_0 x^n D^n + a_1 x^{n-1}D^{n-1} + \cdots + a_{n-1}xD + a_n)y = F(x) \tag{5}$$

where $a_0, a_1, \ldots, a_n$ are constants into a linear equation with constant coefficients. This was first discovered by Euler, and is known as *Euler's differential equation*, although it is sometimes attributed to Cauchy.

Remark. Suppose the operators D and $\mathscr{D}$ are defined by

$$D \equiv \frac{d}{dx}, \qquad \mathscr{D} \equiv \frac{d}{dz} \tag{6}$$

Then by (2) and (3), $\quad xDy = \mathscr{D}y, \quad x^2D^2y = \mathscr{D}(\mathscr{D} - 1)y \tag{7}$

which amounts to the operator equivalence

$$xD \equiv \mathscr{D}, \qquad x^2D^2 \equiv \mathscr{D}(\mathscr{D} - 1) \tag{8}$$

From these it may be conjectured that

$$x^3D^3 \equiv \mathscr{D}(\mathscr{D} - 1)(\mathscr{D} - 2), \qquad x^4D^4 \equiv \mathscr{D}(\mathscr{D} - 1)(\mathscr{D} - 2)(\mathscr{D} - 3), \ldots \tag{9}$$

This actually turns out to be the case, as seen in A Exercise 2 and B Exercise 2. These results enable us to transform the Euler equation easily.

---

* We have tacitly assumed here that $x > 0$. For $x < 0$ we can let $x = -e^z$ and obtain the solution $y = A\cos[2\ln(-x)] + B\sin[2\ln(-x)]$.

Example. Using (7) or (8), equation (1) can immediately be written as

$$\mathscr{D}(\mathscr{D} - 1)y + \mathscr{D}y + 4y = 1 \quad \text{or} \quad (\mathscr{D}^2 + 4)y = 1$$

which is the same as (4).

A EXERCISES

1. Solve each of the following Euler equations subject to any given conditions.
   (a) $x^2y'' - 2xy' + 2y = 0$.
   (b) $4x^2y'' + y = 0,\ y(1) = 1,\ y'(1) = 0$.
   (c) $x^2y'' = x + 2y$.
   (d) $(x^2D^2 -- xD + 2)y = \ln x$.
   (e) $x^2y'' + 5xy' + 4y = x^2 + 16(\ln x)^2$.
   (f) $x^2y'' + y = 16 \sin (\ln x)$.

   (g) $t^2 \dfrac{d^2I}{dt^2} + 3t \dfrac{dI}{dt} + I = t \ln t$.
   (h) $y'' = \dfrac{4}{25}\left(\dfrac{x - y}{x^2}\right),\ y(1) = 0,\ y'(1) = 2$.

   (i) $x^2y'' + xy' - 9y = x^{1/2} + x^{-1/2}$.
   (j) $x^2y''' - 2y' = 5 \ln x$.

2. If $x = e^z$, show that $x^3D^3y = \mathscr{D}(\mathscr{D} - 1)(\mathscr{D} - 2)y,\ x^4D^4y = \mathscr{D}(\mathscr{D} - 1)(\mathscr{D} - 2)(\mathscr{D} - 3)y$, where the operators D and $\mathscr{D}$ are as given in (6). Thus, demonstrate the results (9). Use the results to solve $x^3y''' + 3x^2y'' = 1 + x$.

3. Solve (a) $x^3y''' + xy' - y = x \ln x$; (b) $x^4y^{(IV)} + 6x^3y''' + 7x^2y'' + xy' - y = 1$.

4. Prove the result in the footnote on page 212.

B EXERCISES

1. Evaluate
$$I = \iint \frac{\ln x}{x^2}\, dx^2$$

   in two different ways. Can you generalize the result? (*Hint*: For one way, show that $x^2I'' = \ln x$; for the second way, use elementary calculus.)

2. If $x = e^z$, show that $x^ny^{(n)} = \mathscr{D}(\mathscr{D} - 1)(\mathscr{D} - 2)\cdots(\mathscr{D} - n + 1)y$ where $\mathscr{D} \equiv d/dz$. (*Hint*: Use mathematical induction.)

3. Determine the constant m so that $y = x^m$ is a solution of $x^2y'' + 3xy' - 3y = 0$. Hence, obtain the general solution. How can this be used to find the general solution of

$$x^2y'' + 3xy' - 3y = x^2 - 4x + 2$$

4. Can the method of Exercise 3 be used to determine general solutions of the equations (a) $x^2y'' - xy' + y = 0$, and (b) $x^2y'' - xy' + 4y = 0$? Use the method of the text to obtain the solutions. Discuss the advantages and the disadvantages to the method of Exercise 3.

5. Use the transformation $2x + 3 = e^z$ to solve $(2x + 3)^2y'' + (2x + 3)y' - 2y = 24x^2$.

6. Solve $(x + 2)^2y'' - y = 4$.

7. Solve $r \dfrac{d^2}{dr^2}(rR) - n(n + 1)R = 0$.

C EXERCISES

1. Use the transformation $z = \sin x$ to solve $y'' + (\tan x)y' + (\cos^2 x)y = 0$.

2. Let $x = z^m$ and choose the constant m appropriately to solve the differential equation $xy'' - y' - 4x^3y = 0$.

3. (a) Show that when the transformation $x = F(z)$ is made in the equation

$$y'' + P(x)y' + Q(x)y = R(x)$$

the resulting equation is

$$\frac{d^2y}{dz^2} + p(z)\frac{dy}{dz} + q(z)y = r(z) \quad \text{where} \quad p(z) = \frac{z'' + Pz'}{(z')^2}, \quad q(z) = \frac{Q}{(z')^2}, \quad r(z) = \frac{R}{(z')^2}$$

the primes denoting derivatives with respect to x. (b) Use the result of (a) to show that if z is chosen so that $q(z)$ is a constant (say unity), i.e., $z' = \sqrt{Q}$, $z = \int \sqrt{Q}\, dx$, and if this choice makes $p(z)$ also constant, the first equation of (a) can be solved. (c) Use the result of (a) to show that if z is chosen so that $z'' + Pz' = 0$ and if by this choice $q(z)$ is a constant, then the first equation of (a) can be solved.

4. Using Exercise 3, solve the equations of Exercises 1 and 2.

5. Use Exercise 3 to solve the Euler equation $x^2y'' - 2xy' + 2y = 3x - 2$.

6. Solve (a) $(\sin x)y'' + (3\sin^2 x - \cos x)y' + 2(\sin^3 x)y = 0$. (b) $x^4y'' + 2x^3y' + y = x^{-2}$.

6 Review of Important Methods

In this chapter we have been concerned with various methods of solving the linear differential equation

$$a_0\frac{d^ny}{dx^n} + a_1\frac{d^{n-1}y}{dx^{n-1}} + \cdots + a_ny = F(x) \tag{1}$$

or briefly

$$\phi(D)y = F(x) \tag{2}$$

where $D \equiv d/dx$ and

$$\phi(D) = a_0 D^n + a_1 D^{n-1} + \cdots + a_n \tag{3}$$

In (1) the coefficients $a_0, a_1, \ldots, a_n$, assumed real, may be functions of x, but in many important cases they are constants. The theory for solving (1) or (2) applies to either the constant coefficient case or the variable coefficient case and consists of the following steps.

STEP 1

Find the general solution of the complementary equation of (1) or (2), i.e.,

$$\phi(D)y = 0 \tag{4}$$

obtained on replacing the right-hand side by zero. To do this we must find n linearly independent solutions, say $y_1, y_2, \ldots, y_n$ (which can be checked as linearly independent by the Wronskian theory on pages 181–189). Once we have found these the required general solution of (4), also called the *complementary solution* of (2), is given by

$$y_c = c_1y_1 + c_2y_2 + \cdots + c_ny_n \tag{5}$$

where $c_1, c_2, \ldots, c_n$ are arbitrary constants. Of course, if $F(x) \equiv 0$, (5) would be the required solution. If $F(x) \neq 0$, we must proceed to Step 2.

There are two types of equations (2) which can arise according as the coefficients are all constants or not all constants.

Case (a). Coefficients All Constants. In this case we let $y = e^{mx}$ in (4) to obtain the auxiliary equation

$$a_0 m^n + a_1 m^{n-1} + \cdots + a_n = 0 \tag{6}$$

with n roots given by $m_1, m_2, \ldots, m_n$. The following three possibilities can occur

(i) *Roots real and distinct.* Here the complementary solution is

$$y_c = c_1 e^{m_1 x} + c_2 e^{m_2 x} + \cdots + c_n e^{m_n x}$$

(ii) *Some (or all) roots are repeated.* Here if root m_1, for example, occurs p_1 times then the terms of the complementary solution corresponding to these roots are given by

$$c_1 e^{m_1 x} + c_2 x e^{m_1 x} + c_3 x^2 e^{m_1 x} + \cdots + c_{p_1} x^{p_1 - 1} e^{m_1 x} = (c_1 + c_2 x + \cdots + c_{p_1} x^{p_1 - 1}) e^{m_1 x}$$

The sum of all such terms for all roots yields y_c.

(iii) *Some (or all) roots are imaginary.* Since the constants $a_0, a_1, \ldots, a_n$ in (6) are assumed real, any imaginary roots must occur in complex conjugate pairs. If $\alpha_1 \pm \beta_1 i$ is one such pair which is not repeated the term in y_c corresponding to the pair is given by $e^{\alpha_1 x}(x_1 \cos \beta_1 x + c_2 \sin \beta_1 x)$. If this pair occurs twice the corresponding term in y_c is given by

$$e^{\alpha_1 x}(c_1 \cos \beta_1 x + c_2 \sin \beta_1 x) + x e^{\alpha_1 x}(c_3 \cos \beta_1 x + c_4 \sin \beta_1 x)$$

etc.

Case (b). Coefficients Variable (Not All Constants). Here the method of letting $y = e^{mx}$ will not work (except in very special cases), and so specialized techniques must be used for finding $y_1, y_2, \ldots, y_n$, as for example in the case of *Euler's equation*, page 211, where the transformation $x = e^z$ is used to reduce the variable coefficient case to the constant coefficient case.

STEP 2

Find a particular solution y_p of the given equation (2). For the constant coefficient case there are three possible methods which can be used, as follows:

(a) Method of Undetermined Coefficients. This is applicable only when $F(x)$ consists of special types of functions such as polynomials of degree n, exponential functions of the form e^{rx} and trigonometric functions of the form $\sin rx$ and $\cos rx$. The procedure consists of assuming an appropriate form for the particular solution involving constant coefficients which must be determined by substitution into (2). This procedure is summarized on page 197 and we need not repeat it here.

The method is applicable also in cases where sums and products of the special functions occur in $F(x)$. In such case we use the procedure summarized on page 198.

As an alternative to the above procedures we can employ the annihilator method involving an appropriate operator to "wipe out" the right-hand side $F(x)$ of (2) resulting in an equation with right side zero. Then the procedure given in Step 1 can be used to arrive automatically at the appropriate form of the particular solution.

(b) **Method of Variation of Parameters.** This can be used for any functions $F(x)$ and in particular those used in method (a). Thus it is a "more powerful" method than (a). The method uses the complementary solution (5), assumes that the constants $c_1, c_2, c_3, \ldots$ are replaced by functions of x, denoted by $A, B, C, \ldots$, and then seeks to determine these functions so that (5) satisfies (2). In the case of a second-order equation with complementary solution $y_c = c_1 u_1(x) + c_2 u_2(x)$ the assumed solution of (2) is $y = Au_1(x) + Bu_2(x)$ involving the two functions A and B to be determined. Since one restriction is used up in the fact that y must satisfy the equation (2), we are at liberty to choose one condition relating A and B. This leads to the equations for determining A and B given by

$$A'u_1(x) + B'u_2(x) = 0, \qquad A'u_1'(x) + B'u_2'(x) = F(x)$$

where the second equation is found from the given differential equation. The method can also be used for variable coefficients and can be extended to higher order linear differential equations.

(c) **Short-Cut Operator Methods.** These methods (see pages 204–211) are sometimes useful in the sense that it may be possible in some cases to obtain solutions *faster* than with methods (a) or (b), provided that one develops the facility in dealing with such operational methods. Certainly the student should be acquainted with the first two methods above, and consider this operator method as optional.

STEP 3

Once we have obtained y_c and y_p, the general solution of (2) is given by

$$y = y_c + y_p \qquad \qquad (7)$$

Remark. The theorem on page 175 is often useful when one cannot find all the independent solutions needed in (5).

The following exercises are intended to serve as a review of the various methods.

MISCELLANEOUS EXERCISES ON CHAPTER FOUR

A EXERCISES

Solve each of the following differential equations subject to given conditions if any.

1. $y'' + 3y = x^2 + 1$; $y(0) = 0$, $y'(0) = 2$.

2. $\dfrac{d^2y}{dx^2} - 3\dfrac{dy}{dx} + 2y = \sin x$.

3. $(D^2 + 2D + 1)y = e^x + e^{-x}$.

4. $y''' - 4y = 4x + 2 + 3e^{-2x}$.

5. $\dfrac{d^2I}{dt^2} + 2\dfrac{dI}{dt} + 5I = 34\cos 2t$.

6. $\dfrac{d^4x}{dt^4} - x = 8e^{-t}$.

7. $y'' - 4y = xe^{2x}$; $y(0) = y'(0) = 0$.

8. $x^2 y'' - 6y = 0$; $y(1) = 2$, $y'(1) = 0$.

9. $\dfrac{d^3y}{dx^3} = 2\dfrac{d^2y}{dx^2} + 1.$

10. $y^{(\text{IV})} + 16y'' = 64 \cos 4x.$

11. $y'' + 4y = x(1 + \cos x).$

12. $\dfrac{d^2r}{d\phi^2} = 2r - e^{-2\phi}.$

13. $y''' - 4y'' + 4y' = 12e^{2x} + 24x^2.$

14. $y'' + y = \sec x,\ y(0) = 1,\ y'(0) = 2.$

15. $x^2y'' - 4xy' + 4y = 24(x + 1).$

16. $\dfrac{d^4s}{dt^4} - 2\dfrac{d^2s}{dt^2} + s = 100 \cos 3t.$

17. $4y'' - 4y' + y = \ln x.$

18. $D(D^2 - 1)(D^2 - 4)y = x^2 - x + e^x.$

19. $\dfrac{d^4I}{dt^4} + 9\dfrac{d^2I}{dt^2} = 20e^{-t};\ I(0) = I'(0) = 0.$

20. $x^2y''' - xy'' + y' = \dfrac{\ln x}{x}.$

B EXERCISES

1. Solve $y''' - 2y'' + 4y' - 8y = 64 \sin 2x.$

2. Find that solution of $x^2y'' + 2xy' - 6y = 0$ which is bounded in the interval $0 \leq x \leq 1$ and has the value 2 for $x = \frac{1}{2}$.

3. A particle moves along the x axis in such a way that its instantaneous acceleration is given by $a = 16e^{-t} - 20x - 8v$ where x is its instantaneous position measured from the origin, v is its instantaneous velocity, and t is the time of travel. If the particle starts from rest at the origin, find its position at any later time.

4. Solve $y'' + (\cos x)y' + (1 + \sin x)y = 0$ by first noting that $\cos x$ is a solution.

5. Solve Exercise 4 if the right-hand side is replaced by 1.

6. Solve $\dfrac{d^3y}{dx^3} = \dfrac{24(x + y)}{x^3}.$

7. Solve $xy''' + 2xy'' - xy' - 2xy = 1.$

C EXERCISES

1. Show that the equation $y'' + P(x)y' + Q(x)y = 0$ can be transformed into $u'' + f(x)u = 0$ by letting $y = u(x)v(x)$ and choosing $v(x)$ appropriately. Hence, solve

$$y'' + 4xy' + (3 + 4x^2)y = 0$$

2. Solve $xy'' + 2y' + xy = 0.$

3. Show that the equation $y'' + \lambda y = 0$, subject to the conditions $y(0) = y(\pi) = 0$, has nonzero solutions only for a certain set of values of the parameter λ. These values are called *eigenvalues*, or *characteristic values*, and the corresponding solutions are called *eigenfunctions*, or *characteristic functions*. Differential equations giving rise to eigenvalues and eigenfunctions are of importance in advanced work. We shall investigate such problems in later chapters.

4. (a) Show that by means of the substitution $y = -\dfrac{1}{Pu}\dfrac{du}{dx}$ the Riccati equation of Exercise 12, page 68, is transformed into the second-order linear differential equation

$$\dfrac{d^2u}{dx^2} - \left(\dfrac{P'}{P} + Q\right)\dfrac{du}{dx} + PRu = 0$$

(b) Use the method of (a) to solve $xy' = x^2y^2 - y + 1.$

5. Solve $y'' = (y')^2(2 + xy' - 4y^2 y')$. (*Hint:* Use C Exercise 4, page 15.)

6. (a) Solve the initial value problem $\dfrac{d^2 Q}{dt^2} + kQ = E(t)$, $\qquad Q(0) = Q_0,\ Q'(0) = 0$

where $E(t)$ is a given function of t and k is a positive constant.
   (b) Work part (a) if $E(t)$ is given by the graph in C Exercise 2, page 89.

7. (a) Show that the solution of $y'' = F(x)$, $y(0) = y(1) = 0$ is

$$y = \int_0^1 G(x, t) F(t)\, dt$$

where $\qquad\qquad G(x, t) = \begin{cases} t(x - 1), & 0 \le t \le x \\ x(t - 1), & x \le t \le 1 \end{cases}$

The function $G(x, t)$ is often called a *Green's function.*
   (b) Discuss how you might obtain $G(x, t)$ if it were not given. [*Hint:* One possibility is to write

$$y = \int_0^x G(x, t) F(t)\, dt + \int_x^1 G(x, t) F(t)\, dt$$

and substitute into the given equation and conditions to find suitable conditions on G in the two regions $0 \le t \le x$, $x \le t \le 1$.]
   (c) Apply your method in (b) to solve $y'' + y = F(x)$, $y(0) = y(1) = 0$.

five
applications of linear differential equations

In the preceding chapter, methods were given for the solution of

$$a \frac{d^2x}{dt^2} + b \frac{dx}{dt} + cx = F(t) \tag{1}$$

where a, b, c are given constants or functions of t and $F(t)$ is a given function of t. This equation occurs so frequently in applications of physics, engineering, and other sciences that it is worthy of special study. In this chapter we shall study applications of such differential equations to:

1. Vibratory or oscillatory motion of mechanical systems.
2. Electric circuit problems.
3. Miscellaneous problems.

1 Vibratory Motion of Mechanical Systems

1.1 THE VIBRATING SPRING. SIMPLE HARMONIC MOTION

Perhaps the simplest system available for study of vibratory motion consists of an ordinary spring of negligible weight [Fig. 5.1(a)] suspended vertically from a fixed support. Suppose that a weight W is hung on the spring [Fig. 5.1(b)]. When the weight is at rest we describe its position as the *equilibrium position*. If the weight is pulled down a certain distance and then released, it will undergo a vibratory motion about the equilibrium position [Fig. 5.1(c)]. It is our purpose in this section to discuss the motion of the weight in this and similar cases. In order to accomplish this purpose, we shall have to know the forces acting on the weight during its motion. It is clear from experience that there is some force tending to return or restore a displaced weight to its equilibrium position. This force is called the *restoring force*.

The law governing this force is a special case of the generalized Hooke's law on page 144. We shall refer to this special case as *Hooke's Law*, which is stated as

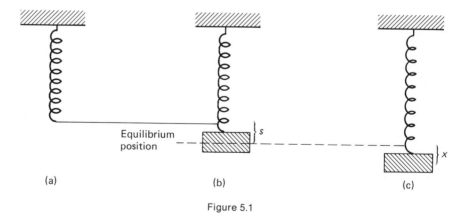

Equilibrium position

(a)   (b)   (c)

Figure 5.1

follows:

HOOKE'S LAW. *The force exerted by a spring, tending to restore the weight W to the equilibrium position, is proportional to the distance of W from the equilibrium position.* (Sometimes this is worded briefly as "force is proportional to stretch.")

Denote the magnitude of the restoring force by $|f|$, and let x denote the position of W measured from the equilibrium position. Assume the positive direction downward, so that x is positive when W is below the equilibrium position and negative when W is above this position. According to Hooke's law,

$$|f| \propto |x| \quad \text{i.e.,} \quad |f| = k|x|$$

where $k > 0$ is a constant of proportionality depending on the stiffness of the spring and called the *spring constant*. To determine the direction of the force, note that when $x > 0$ the force is directed upward and is thus negative. When $x < 0$ the force is directed downward and is thus positive. This can be satisfied only if the force is given both in magnitude and direction by $-kx$, so that Hooke's law reads

$$f = -kx \tag{2}$$

When weight W is put on the spring, it stretches a distance s as in Fig. 5.1(b). According to Hooke's law, the tension T_1 in the spring is proportional to the stretch, and so $T_1 = ks$. Since the spring and weight are in equilibrium it follows that

$$T_1 = ks = W \tag{3}$$

When the weight is pulled further and released, its position at any time is shown in Fig. 5.1(c). The tension T_2 in the spring at this time is, according to Hooke's law,

$$T_2 = k(s + x) \tag{4}$$

It follows that the net force in the positive direction is given by

$$W - T_2 = W - ks - kx = -kx$$

because of (3). Thus by Newton's law the equation of motion is

$$\frac{W}{g} \cdot \frac{d^2x}{dt^2} = -kx$$

Thus the net force is simply the restoring force and does not depend on the weight W.

ILLUSTRATIVE EXAMPLE 1

It is found experimentally that a 6 lb weight stretches a certain spring 6 in. If the weight is pulled 4 in. below the equilibrium position and released: (a) set up a differential equation and associated conditions describing the motion; (b) find the position of the weight as a function of time; and (c) determine the position, velocity, and acceleration of the weight $\frac{1}{2}$ sec after it has been released.

Mathematical Formulation. By Hooke's law (since 6 in. $= \frac{1}{2}$ ft), $|f| = k|x|$, or $6 = k \cdot \frac{1}{2}$; i.e., $k = 12$. The differential equation describing the motion is therefore

$$\frac{6}{32} \cdot \frac{d^2x}{dt^2} = -12x \quad \text{or} \quad \frac{d^2x}{dt^2} + 64x = 0 \tag{5}$$

Since initially ($t = 0$) the weight is 4 in. below the equilibrium position, we have

$$x = \tfrac{1}{3} \,(\text{ft}) \qquad \text{at } t = 0 \tag{6}$$

Also, since the weight is released (i.e., it has zero velocity) at $t = 0$,

$$\frac{dx}{dt} = 0 \qquad \text{at } t = 0 \tag{7}$$

The answer to (a) is provided by equation (5) with conditions (6) and (7).

Solution   The auxiliary equation for (5) is $m^2 + 64 = 0$ and has roots $m = \pm 8i$. Hence the differential equation has solution

$$x = A \cos 8t + B \sin 8t$$

From condition (6) we find $A = \frac{1}{3}$, so that

$$x = \tfrac{1}{3} \cos 8t + B \sin 8t \qquad \text{and} \qquad \frac{dx}{dt} = -\tfrac{8}{3} \sin 8t + 8B \cos 8t$$

Using condition (7), we now find $B = 0$. Hence, the required solution is

$$x = \tfrac{1}{3} \cos 8t \tag{8}$$

which provides the answer to part (b). Note that in equation (8), x is in feet. If it were desired to measure x in inches, the equation would be $x = 4 \cos 8t$.

Let us now turn to part (c). Differentiating (8) with respect to t, we see that

$$v = \frac{dx}{dt} = -\tfrac{8}{3} \sin 8t, \qquad a = \frac{d^2x}{dt^2} = -\frac{64}{3} \cos 8t$$

Placing $t = \frac{1}{2}$ and using the fact that 4 radians $= 4 \times (180/\pi)$ degrees $= 229$ degrees, approximately, we find

$$x = \tfrac{1}{3}(-0.656) = -0.219, \qquad v = -\tfrac{8}{3}(-0.755) = +2.01,$$

$$a = -\tfrac{64}{3}(-0.656) = +14.0$$

Thus after $\frac{1}{2}$ sec the weight is 0.219 ft *above* the equilibrium position and is traveling *downward* with velocity 2.01 ft/sec and acceleration 14.0 ft/sec^2.

The graph of (8) is shown in Fig. 5.2. It is seen from this graph that the weight starts at $x = \frac{1}{3}$ where $t = 0$, then proceeds through the equilibrium position given by $x = 0$ to the position $x = -\frac{1}{3}$ and then returns to the equilibrium position again, passes through, and goes back to $x = \frac{1}{3}$. This cycle then repeats over and over again. Physically the graph describes the periodic up and down motion of the spring which

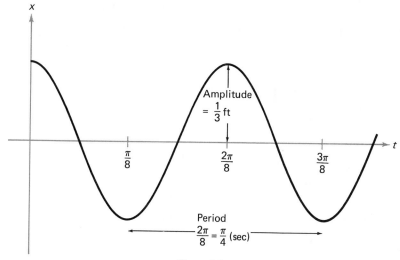

Figure 5.2

is called *simple harmonic motion*. In general, any motion described by

$$\frac{d^2x}{dt^2} = -ax \tag{9}$$

where $a > 0$ is a constant, will be simple harmonic motion. Physically (9) states that the acceleration is directly proportional to the displacement but in the opposite direction (as indicated by the minus sign).

We call the maximum displacement of the weight from the equilibrium position (i.e., $x = 0$) the *amplitude*. In the above example the amplitude is $\frac{1}{3}$ ft. The time for one complete cycle is called the *period*. From the graph it is seen that the period is $\pi/4$ seconds. Another way to see that the period is $\pi/4$ without the graph is to determine when the weight is at an extremity of its path (i.e., either the highest or the lowest point). Suppose for example, we take the lowest point given by $x = \frac{1}{3}$. From (8) we see that this will occur when

$$\cos 8t = 1, \text{ i.e., } 8t = 0, 2\pi, 4\pi, 6\pi, \ldots \text{ or } t = 0, \pi/4, 2\pi/4, 3\pi/4, \ldots$$

Hence the first time that $x = \frac{1}{3}$ is when $t = 0$, the second when $t = \pi/4$, the third time when $t = 2\pi/4$, etc. The difference between successive times is $\pi/4$, which is the period. The number of cycles per second is called the *frequency*. We have

$$\text{Period} = \text{number of seconds per cycle} = \pi/4$$

$$\text{Frequency} = \text{Number of cycles per second} = \frac{1}{\pi/4} = \frac{4}{\pi}$$

In general if T is the period, the frequency f is given by

$$f = \frac{1}{T}$$

In Illustrative Example 1, suppose the weight is pulled 4 in. below the equilibrium position and is then given a downward velocity of 2 ft/sec instead of being released. Determine the amplitude, period, and frequency of the motion.

Mathematical Formulation. The differential equation is

$$\frac{d^2x}{dt^2} + 64x = 0 \tag{10}$$

as in Illustrative Example 1. The initial conditions are

$$x = \frac{1}{3}, \quad \frac{dx}{dt} = 2 \quad \text{at } t = 0 \tag{11}$$

Solution The general solution of (10) is $x = A \cos 8t + B \sin 8t$. From the first of conditions (11), $A = \frac{1}{3}$. Hence, $x = \frac{1}{3} \cos 8t + B \sin 8t$. Differentiation gives

$$\frac{dx}{dt} = -\frac{8}{3} \sin 8t + 8B \cos 8t$$

and using the second of conditions (11), we find $B = \frac{1}{4}$. The required solution is

$$x = \frac{1}{3} \cos 8t + \frac{1}{4} \sin 8t \tag{12}$$

If x is measured in inches, the equation is

$$x = 4 \cos 8t + 3 \sin 8t \tag{13}$$

It is often useful to write (12) in an equivalent form, making use of the identity*

$$\left. \begin{array}{c} a \cos \omega t + b \sin \omega t = \sqrt{a^2 + b^2} \sin (\omega t + \phi) \\[2mm] \text{where} \quad \sin \phi = \dfrac{a}{\sqrt{a^2 + b^2}} \quad \text{and} \quad \cos \phi = \dfrac{b}{\sqrt{a^2 + b^2}} \end{array} \right\} \tag{14}$$

as indicated in Fig. 5.3. The angle ϕ is often called the *phase angle*. With the aid of this identity (14) becomes

$$x = \sqrt{(\tfrac{1}{3})^2 + (\tfrac{1}{4})^2} \sin (8t + \phi) = \tfrac{5}{12} \sin (8t + \phi) \tag{15}$$

where $\sin \phi = \frac{4}{5}$, $\cos \phi = \frac{3}{5}$. From tables, $\phi = 53°8'$ or 0.9274 radian so that

$$x = \tfrac{5}{12} \sin (8t + 0.9274) \tag{16}$$

if x is in feet, and

$$x = 5 \sin (8t + 0.9274)$$

---

* This is easy to prove, since

$$\sqrt{a^2 + b^2} \sin (\omega t + \phi) = \sqrt{a^2 + b^2} (\sin \omega t \cos \phi + \cos \omega t \sin \phi)$$

$$= \sqrt{a^2 + b^2} \left[(\sin \omega t) \left(\frac{b}{\sqrt{a^2 + b^2}} \right) + (\cos \omega t) \left(\frac{a}{\sqrt{a^2 + b^2}} \right) \right] = a \cos \omega t + b \sin \omega t$$

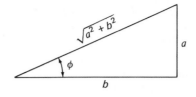

Figure 5.3

if x is in inches. The graph of (16) is shown in Fig. 5.4. The amplitude is 5 in., or $\frac{5}{12}$ ft, the period is $2\pi/8 = \pi/4$ seconds and the frequency is the reciprocal of the period, or $4/\pi$ cycles per second. In general if a motion can be described

$$x = A \sin(\omega t + \phi) \tag{17}$$

then
$$\left.\begin{array}{l} \text{amplitude} = A, \text{ period} = T = \dfrac{2\pi}{\omega} \\[2em] \text{frequency} = f = \dfrac{1}{T} = \dfrac{\omega}{2\pi} \end{array}\right\} \tag{18}$$

From the last statement, we have the relation $\omega = 2\pi f$, which is often useful.

Simple harmonic motion occurs in many other cases besides the vibrations of springs as in the motion of the pendulum of a grandfather clock, the rolling of a ship or plane, etc. We discuss some of these in the exercises and also in later sections.

A EXERCISES

1. A 2 lb weight suspended from a spring stretches it 1.5 in. If the weight is pulled 3 in. below the equilibrium position and released: (a) Set up a differential equation and conditions

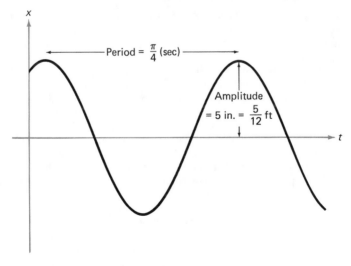

Figure 5.4

describing the motion. (b) Find the velocity and position of the weight as a function of time. (c) Find the amplitude, period, and frequency of the motion. (d) Determine the position, velocity, and acceleration $\pi/64$ sec after the weight is released.

2. A 3 lb weight on a spring stretches it 6 in. When equilibrium is reached the weight is struck with a downward velocity of 2 ft/sec. Find: (a) the velocity and position of the weight at time t sec after the impact; (b) the amplitude, period, and frequency; (c) the velocity and acceleration when the weight is 1 in. above the equilibrium position and moving upward.

3. A spring suspended from the ceiling has a constant of 12 lb/ft. An 8 lb weight is placed on the spring, and when equilibrium is reached, the weight is raised 5 in. above the equilibrium position and dropped. Describe the motion giving amplitude, period, and frequency.

4. Solve Exercise 3 if the weight is raised 5 in. and then thrust upward with velocity 5 ft/sec.

5. A spring is stretched 2 cm by a force of 40 dynes. A particle of mass 1 g is placed on the spring, and when equilibrium is reached, the mass is raised 5 cm above the equilibrium position and then released. (a) Find the amplitude, period, and frequency of the vibration. (b) At what times is the particle 2.5 cm above the equilibrium position?

6. A 256 lb weight is suspended from a vertical spring having spring constant of 200 lb/ft. If the weight is raised 3 in. above its equilibrium position and released: (a) Find the position of the weight at a time $\pi/3$ sec afterward and determine which way and how fast the weight is moving at this time. (b) Find the amplitude, period, and frequency of the vibration. (c) At what times is the weight 1.5 in. below the equilibrium position and moving downward?

7. A particle moves along the x axis toward the origin O under the influence of a force of attraction at O which varies directly as the distance of the particle from O. At $t = 0$ the particle is 4 cm from O and is moving toward O with velocity 6 cm/sec and acceleration 16 cm/sec². (a) Find the velocity and position as a function of time. (b) Find the amplitude, period, and frequency of the motion. (c) Find the maximum velocity and acceleration.

8. A particle of mass 2 g moves on the x axis attracted toward the origin O by a force which is directly proportional to its distance from O. At $t = 0$ the particle passes through O with velocity 20 cm/sec. The force on the particle is 100 dynes 2 cm from O. (a) Find the position, velocity, and acceleration as a function of time. (b) Find the amplitude, period, and frequency of the vibration. (c) Find the force on the particle at $t = \pi/4$.

9. A particle starts from rest, a distance 10 cm from a fixed point O. It moves along a horizontal straight line toward O under the influence of an attractive force at O. This force at any time varies as the distance of the particle from O. If the acceleration of the particle is 9 cm/sec² directed toward O when the particle is 1 cm from O, describe the motion.

10. A particle starts from rest 1 ft from a fixed point O. It moves along a horizontal line toward O subject to an attractive force at O which varies directly as its distance from O. The acceleration of the particle is 8 ft/sec² toward O when it is $\frac{1}{2}$ ft from O. (a) Find the velocity when the particle is $\frac{1}{2}$ ft from O. (b) Find the amplitude, period, and frequency of the motion. (c) Determine the position, velocity, and acceleration after $\pi/16$ sec.

11. A particle starts from rest 20 cm from a fixed point O. It moves along a horizontal line toward O under an attractive force at O which varies directly as its distance from O. At O its velocity is 40 cm/sec. (a) Find its velocity and acceleration 10 cm from O. (b) Determine the amplitude, period, and frequency of the motion. (c) Find its position, velocity, and acceleration after $\pi/3$ sec. (d) Find the times when the particle passes through O.

1. A weight W suspended from a vertical spring produces a stretch of magnitude a. When the weight is in equilibrium it is acted upon by a force which gives to it a velocity v_0 downward. Show that the weight travels a distance $v_0\sqrt{a/g}$ for a time $(\pi/2)\sqrt{a/g}$ before it starts to return.

2. A weight W on a vertical spring having constant k is oscillating with simple harmonic motion. When the weight reaches its lowest position it receives a blow which imparts to it a velocity v_0 downward. Assuming that this does not affect the properties of the spring, show that the weight oscillates with the same period as before but has a new amplitude given by $\sqrt{A_0^2 + (Wv_0^2/gk)}$, where A_0 is the original amplitude.

3. When a weight at the end of a vertical spring is set into vibration, the period is 1.5 sec. After adding 8 lb, the period becomes 2.5 sec. How much weight was originally on the spring?

4. A spring oscillates vertically. The maximum velocity and acceleration are given, respectively, by v_m and a_m. Show that the period of oscillation is $2\pi v_m/a_m$, and the amplitude is v_m^2/a_m.

5. A spring oscillates with amplitude A and period T. Show that the maximum velocity occurs at the center of the path and has magnitude $2\pi A/T$, while the maximum acceleration occurs at the ends of the path and has magnitude $4\pi^2 A/T^2$.

6. If a hole were bored through the earth's center, one would find that an object placed in it is acted upon by a force of attraction varying directly as the distance between the object and the earth's center. Assuming the earth is a sphere of 4000 mile radius: (a) Find the time for an object dropped in the hole to return. (b) Find its velocity on passing through the earth's center.

C EXERCISES

1. A spring rests taut but unstretched on a horizontal table. One end is attached to a point O on the table and the other to a weight W. The weight is displaced so that the spring is stretched a distance a and it is then released. If the coefficient of friction between weight and table is μ, and if the spring constant is k, show that when the spring returns to its unstretched position the magnitude of its velocity is

$$\sqrt{\frac{g}{W}(ka^2 - 2\mu Wa)}$$

and that this takes a time given by $\sqrt{\dfrac{W}{gk}}\left[\pi - \cos^{-1}\left(\dfrac{\mu W}{ka - \mu W}\right)\right]$.

2. When the spring of the previous exercise is in its unstretched position, the weight W is given a velocity v_0 away from O. Show that it travels a distance

$$\sqrt{\frac{\mu^2 W^2}{k^2} + \frac{Wv_0^2}{gk} - \frac{\mu W}{k}}$$

before it starts to return, and that this takes time given by $\sqrt{\dfrac{W}{gk}}\left(\cot^{-1}\dfrac{\mu}{v_0}\sqrt{\dfrac{gW}{k}}\right)$.

3. Compare Exercises 1 and 2 with B Exercises 1 and 2.

4. A particle moves with simple harmonic motion along the x axis described by the equation $x = a\sin \omega t$. (a) Show that the probability of finding the particle between positions x_1

and x_2, where $-a \leqq x_1 < x_2 \leqq a$, is given by

$$\frac{1}{\pi}\left\{\sin^{-1}\frac{x_2}{a} - \sin^{-1}\frac{x_1}{a}\right\}$$

(b) Where do you expect would be the greatest chance of finding the particle? Explain.

5. (a) Referring to Exercise 4, show that if $F(x)$ is the probability of finding the particle to the left of x then

$$F(x) = \int_{-\infty}^{x} P(v)dv \quad \text{where} \quad P(x) = \begin{cases} \dfrac{1}{\pi\sqrt{a^2 - x^2}}, & |x| \leqq a \\ 0, & |x| > a \end{cases}$$

We often call $F(x)$ the *distribution function* and $P(x)$ the *density function*. (b) Show that the root mean square displacement of the particle from its equilibrium position is $a/\sqrt{2}$.

6. Work B Exercise 6 for the case where the hole connects two points on the earth's surface but does not pass through the center of the earth.

1.2 THE VIBRATING SPRING WITH DAMPING. OVERDAMPED AND CRITICALLY DAMPED MOTION

The vibrating springs just considered were not very realistic, since the oscillations did not decrease, as one would expect from experience, but were instead forever maintained. In practice, frictional and other forces (such as air resistance) act to decrease the amplitude of the oscillations and ultimately to bring the system to rest. One way to get a better approximation to reality is to assume a *damping force*. The exact law for this force is not known, since it depends on so many variable factors, but it has been found from experiment that for small speeds, the magnitude of the damping force is approximately proportional to the instantaneous speed of the weight on the spring. The magnitude is therefore given by

$$\beta\left|\frac{dx}{dt}\right|$$

where β is the constant of proportionality called the *damping constant*. The damping force opposes the motion, so that when the weight is coming down the damping force acts up, while it acts downward when the weight is going up. Assuming downward as the positive direction, as we did before, we see that the damping force must be negative when dx/dt is positive, and must be positive when dx/dt is negative. Thus, with $\beta > 0$, it is clear that the damping force must be given both in magnitude and direction by $-\beta\, dx/dt$. When account is taken of the restoring force already considered, it follows by Newton's law that the differential equation of motion is

$$\frac{W}{g}\cdot\frac{d^2x}{dt^2} = -\beta\frac{dx}{dt} - kx \quad \text{or} \quad \frac{W}{g}\cdot\frac{d^2x}{dt^2} + \beta\frac{dx}{dt} + kx = 0$$

ILLUSTRATIVE EXAMPLE 3

Assume that a damping force, given in pounds numerically by 1.5 times the instantaneous velocity in feet per second, acts on the weight in Illustrative Example 1, page 221. (a) Set up the differential equation and associated conditions. (b) Find the position x of the weight as a function of time t.

Mathematical Formulation. Taking into account the damping force $-1.5\,dx/dt$ in Illustrative Example 1, we find for the equation of motion

$$\frac{6}{32}\cdot\frac{d^2x}{dt^2} = -12x - 1.5\frac{dx}{dt} \quad\text{or}\quad \frac{d^2x}{dt^2} + 8\frac{dx}{dt} + 64x = 0 \qquad (19)$$

The initial conditions are as in Illustrative Example 1:

$$x = \tfrac{1}{3} \quad\text{at}\quad t = 0 \qquad\text{and}\qquad \frac{dx}{dt} = 0 \quad\text{at}\quad t = 0 \qquad (20)$$

Solution The auxiliary equation corresponding to (19) has roots $m = -4 \pm 4\sqrt{3}\,i$, and so the general solution is

$$x = e^{-4t}(A\cos 4\sqrt{3}t + B\sin 4\sqrt{3}t)$$

Determining the constants A and B subject to conditions (20), we find

$$x = \tfrac{1}{9}e^{-4t}(3\cos 4\sqrt{3}t + \sqrt{3}\sin 4\sqrt{3}t) \qquad (21)$$

If we make use of the identity given in (14), page 224, (21) may be written

$$x = \frac{2\sqrt{3}}{9}\,e^{-4t}\sin\left(4\sqrt{3}t + \frac{\pi}{3}\right) \qquad (22)$$

The graph of (22), shown in Fig. 5.5, lies between the graphs of

$$x = \frac{2\sqrt{3}}{9}\,e^{-4t} \quad\text{and}\quad x = -\frac{2\sqrt{3}}{9}\,e^{-4t}$$

(shown dashed in Fig. 5.5), since the sine varies between -1 and $+1$.

The difference between the times of the successive maxima (or minima) of this graph can be shown to be constant and equal to $2\pi/4\sqrt{3}$ (see Exercise 2 of the B Exercises). It should be noted that the maxima (or minima) of the graph *do not* lie on the dashed curves as might be imagined (see Exercises 2 and 4 of the B Exercises). Thus, as shown in Fig. 5.5, point P represents a relative minimum, while point Q lies on the dashed curve. The constant difference in times between successive maxima (or minima) is called the *quasi* period, although we sometimes refer to it as the period. The adjective quasi is used, since the functional values do not repeat as they would if there were actual periodicity. The quasi period is also equal to twice the time taken between successive zeros, i.e., twice the time between successive passages of the weight through the equilibrium position (see Exercise 4 of the B Exercises).

The motion described in this example is called *damped oscillatory* or *damped vibratory* motion. It should be noted that (22) has the form

$$x = \mathscr{A}(t)\sin(\omega t + \phi) \qquad (23)$$

where
$$\mathscr{A}(t) = \frac{2\sqrt{3}}{9}\,e^{-4t}, \qquad \omega = 4\sqrt{3}, \qquad \phi = \frac{\pi}{3}$$

The quasi period is given by $2\pi/\omega = 2\pi/4\sqrt{3}$.

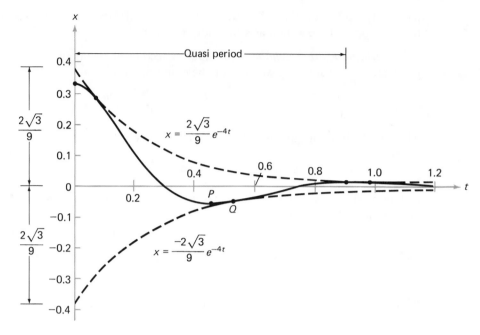

Figure 5.5

By analogy with the undamped case, $\mathscr{A}(t)$ is called the *amplitude*, or more exactly the *time-varying amplitude*. It is seen that the amplitude decreases with time, thus agreeing with our experience. One fact that should be noted is that the frequency with damping is less than that without damping. This is plausible since one would expect opposition to motion to increase the time for a complete cycle. The undamped frequency, i.e., with $\beta = 0$, is often called the *natural frequency*. It is of great importance in connection with the phenomenon of resonance to be discussed later.

The damping force may be too great compared with the restoring force to permit oscillatory motion. We consider this situation in the following

ILLUSTRATIVE EXAMPLE 4

In Illustrative Example 1, page 221, assume that a damping force in pounds numerically equal to 3.75 times the instantaneous velocity is taken into account. Find x as a function of t.

Mathematical Formulation. Taking into account the damping force $-3.75\, dx/dt$ in the differential equation of Illustrative Example 1, we find

$$\frac{6}{32}\cdot\frac{d^2x}{dt^2} = -12x - 3.75\frac{dx}{dt} \quad \text{or} \quad \frac{d^2x}{dt^2} + 20\frac{dx}{dt} + 64x = 0 \tag{24}$$

The initial conditions are as before given by (20).

Solution   The auxiliary equation has roots $m = -4, -16$. Hence,

$$x = Ae^{-4t} + Be^{-16t} \tag{25}$$

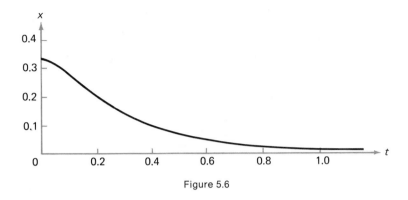

Figure 5.6

Using conditions (20), we find $x = \frac{4}{9}e^{-4t} - \frac{1}{9}e^{-16t}$. The graph appears in Fig. 5.6. It is seen that no oscillations occur; the weight has so much damping that it just gradually returns to the equilibrium position without passing through it. This type of motion is called *overdamped motion*.

An interesting case occurs when damping is such that any decrease in it produces oscillations. The motion is then *critically damped*.

ILLUSTRATIVE EXAMPLE 5

Instead of 3.75 in Illustrative Example 4, use 3, and find x as a function of t.

Mathematical Formulation. The equation becomes

$$\frac{6}{32} \cdot \frac{d^2x}{dt^2} = -12x - 3\frac{dx}{dt} \quad \text{or} \quad \frac{d^2x}{dt^2} + 16\frac{dx}{dt} + 64x = 0 \qquad (26)$$

and the conditions are still $x = \frac{1}{3}$ at $t = 0$ and $dx/dt = 0$ at $t = 0$.

Solution The roots of the auxiliary equation are $-8, -8$. Hence

$$x = Ae^{-8t} + Bte^{-8t}$$

and using the conditions at $t = 0$ we have

$$x = \frac{1}{3}e^{-8t} + \frac{8}{3}te^{-8t} \qquad (27)$$

The graph appears in Fig. 5.7 (heavy curve) and is to be compared with the curve of Fig. 5.6 (shown dashed in Fig. 5.7). A slight decrease in damping would produce oscillations such as is shown in Fig. 5.5.

It is interesting to inquire what would happen if the initial conditions in preceding illustrative examples were changed. It will be clear with but little thought that such a modification could not possibly change overdamped or critically damped motion into oscillatory motion. However some characteristics of the motion may be changed.

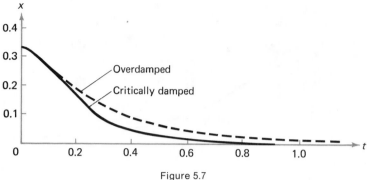

Figure 5.7

Assume the differential equation of Illustrative Example 5 but change the initial conditions to $x = 0$, $dx/dt = 5$ at $t = 0$.

Solution   The solution to the differential equation is, as in Illustrative Example 5, $x = Ae^{-8t} + Bte^{-8t}$. Using the given conditions, we find

$$x = 5te^{-8t}$$

The graph appears in Fig. 5.8. To interpret this motion, observe that initially the weight is at the equilibrium position and is given a velocity downward (positive direction) of 5 ft/sec. It travels until it reaches a maximum displacement (point P in figure) and then slowly returns to the equilibrium position, never passing it. The maximum displacement, occuring after $\frac{1}{8}$ sec, is found to be approximately 2.8 in.

Damped motion can of course arise in many other connections besides springs and is the basis for various important applications. Some examples are as follows:

Example 1.   Shock absorbers in automobiles provide damping which is needed to reduce vibrations and thus provide a smoother as well as safer ride.

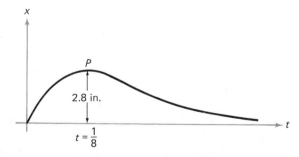

Figure 5.8

Example 2. On a screen or storm door there is an adjustment screw which provides sufficient damping for it to close without slamming shut or remaining open.

Equations analogous to that given above for a spring serve as models for such applications. As an illustration of such an equation, let us consider the screen or storm door of Example 2. Suppose that a top view of the open door is represented by OA in Fig. 5.9, while OB represents the top view of the door when it is shut. Let ϕ represent the angle which the door OA makes with OB at any time t. Then the differential equation for ϕ is given by

$$I\frac{d^2\phi}{dt^2} = -\beta\frac{d\phi}{dt} - k\phi \quad \text{or} \quad I\frac{d^2\phi}{dt^2} + \beta\frac{d\phi}{dt} + k\phi = 0 \tag{28}$$

We can arrive at (28) if we use the result from elementary physics that

$$\mathscr{T} = I\alpha \tag{29}$$

where

    1. $\mathscr{T}$ is the net *torque* or turning effect tending to produce rotation of an object about an axis,
    2. I is the *moment of inertia* of the object about the axis,
    3. α is the *angular acceleration* of the object about the axis.

Now there will be two torques acting on the door, a *restoring torque* tending to return the door to the equilibrium position OB, which is proportional to the angle ϕ and given by $k\phi$, and a *damping torque* proportional to the angular velocity $d\phi/dt$ and given by $\beta\,d\phi/dt$. Since these torques tend to oppose the outward motion of the door, the net torque is

$$-\beta\frac{d\phi}{dt} - k\phi \tag{30}$$

But since the net torque is also equal to the moment of inertia I of the door about its axis multiplied by the angular acceleration $d^2\phi/dt^2$ of the door, i.e.,

$$I\frac{d^2\phi}{dt^2} \tag{31}$$

we have on equating (30) and (31) the result (28).

    Equation (28) also can represent the vibrations of a ship or airplane about an axis through the center of gravity (see C Exercise 6).

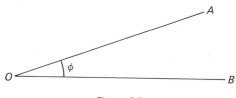

Figure 5.9

1. A 4 lb weight suspended from a spring stretches it 3 in. The weight is pulled 6 in. below the equilibrium position and released. Assume that the weight is acted upon by a damping force which in pounds is numerically equal to $2v$, where v is the instantaneous velocity in feet per second. (a) Set up a differential equation and conditions describing the motion. (b) Determine the position of the spring at any time after the weight is released. (c) Write the result of (b) in the form $\mathscr{A}(t) \sin(\omega t + \phi)$. Thus, determine the time-varying amplitude, quasi period, and phase angle.

2. A 2 lb weight suspended from a spring stretches it 6 in. A velocity of 5 ft/sec upward is imparted to the weight at its equilibrium position. Assume a damping force in pounds numerically equal to $0.6v$, where v is the instantaneous velocity in feet per second. (a) Find the position and velocity of the spring at any time. (b) Write the result of (a) in the form $\mathscr{A}(t) \sin(\omega t + \phi)$.

3. A 64 lb weight is suspended from a spring with constant 50 lb/ft. The weight is acted upon by a resisting force in pounds which is numerically equal to 12 times the instantaneous velocity in feet per second. If the weight is pulled 6 in. below the equilibrium position and released, describe the motion, giving the time-varying amplitude and quasi period of the motion.

4. A spring is stretched 10 cm by a force of 1250 dynes. A mass of 5 g is suspended from the spring and, after it has come to equilibrium, is pulled down 20 cm and released. Assuming that there is a damping force in dynes numerically equal to $30v$, where v is the instantaneous velocity in centimeters per second, find (a) the position, and (b) the velocity at any time.

5. Work Exercise 4 if the mass is pulled down 20 cm and then given a velocity of 120 cm/sec (a) downward; (b) upward.

6. A 2 lb weight on a spring stretches it 1.5 in. The weight is pulled 6 in. below its equilibrium position and released. Assume a damping force in pounds numerically equal to $2v$, where v is the instantaneous velocity in feet per second. (a) Find the position of the weight at any time. (b) Determine whether the motion is overdamped or critically damped.

7. In Exercise 6 assume the initial conditions are modified so that the weight is given a velocity downward of 10 ft/sec when it is at the equilibrium position. Find (a) the position and velocity at any time, (b) the maximum displacement of the weight from the equilibrium position.

8. A 3 lb weight on a spring stretches it 6 in. Assuming a damping force in pounds numerically equal to βv, where v is the instantaneous velocity in feet per second and $\beta > 0$, show that the motion is (a) critically damped if $\beta = 1.5$, (b) overdamped if $\beta > 1.5$, (c) oscillatory if $\beta < 1.5$.

B EXERCISES

1. The differential equation for the motion of a mass m suspended from a vertical spring of constant k, if damping proportional to the instantaneous velocity is taken into account, is $m\ddot{x} + \beta\dot{x} + kx = 0$, where the dots denote differentiation with respect to time t. Show that damped oscillations will take place provided that the damping constant is small enough so that $\beta < 2\sqrt{km}$ and that x is given by $x = Ce^{-\beta t/2m} \sin(\omega t + \phi)$ where $\omega = \sqrt{k/m - \beta^2/4m^2}$ and C and ϕ represent two arbitrary constants.

2. (a) Show that the times at which $x = Ce^{-\alpha t} \sin(\omega t + \phi)$ is a maximum (in absolute value) are given by $t_1, t_2, \ldots$, where

$$t_n = \frac{1}{\omega}\left[\tan^{-1}\frac{\omega}{\alpha} + (n-1)\pi - \phi\right]$$

Hence show that the quasi period is $2\pi/\omega$. (b) Show that the quasi period for the motion described in Exercise 1 is greater than the natural period (reciprocal of natural frequency).

3. By using the result of Exercise 2(a) show that the successive maximum distances from the equilibrium position are given by $x_n = Ce^{-\beta t_n/2m}\sqrt{1 - \beta^2/4mk}$ where t_n is given in Exercise 2(a).

Hence, show that $\dfrac{x_{n+1}}{x_n} = e^{-\beta\pi/2m\omega}$ i.e., the successive swings decrease in geometric progression.

In engineering, the quantity $\beta\pi/2m\omega$ is called the *logarithmic decrement*.

4. (a) Show that the times at which $x = Ce^{-\alpha t}\sin(\omega t + \phi)$ intersects the curves $x = Ce^{-\alpha t}$ and

$$x = -Ce^{-\alpha t} \text{ are given by} \qquad \tau_n = \frac{1}{\omega}\left[(2n - 1)\frac{\pi}{2} - \phi\right], \qquad n = 1, 2, 3, \ldots$$

Hence, show that the quasi period may also be obtained by considering the difference of the successive times where $\sin(\omega t + \phi) = 1$ (or -1). Compare with Exercise 2(a) above. (b) Let $X_1, X_2, \ldots$ denote the absolute values of the successive values of x corresponding to the times $\tau_1, \tau_2, \ldots$. Show that

$$\frac{X_{n+1}}{X_n} = e^{-\beta\pi/2m\omega}$$

and compare with the result of Exercise 3.

5. Compare the times t_n and τ_n of Exercises 2 and 4, respectively, and show that

$$\omega(\tau_n - t_n) = \frac{\pi}{2} - \tan^{-1}\frac{\omega}{\alpha} = \tan^{-1}\frac{\alpha}{\omega}$$

Also show that $\tau_n > t_n$ and that $\tau_n - t_n$ becomes smaller as the damping decreases.

C EXERCISES

1. A mass m is suspended vertically from a spring having constant k. At $t = 0$ the mass is struck so as to give it a velocity v_0 downward. A damping force βv, where v is the instantaneous velocity and β is a positive constant, acts on the mass. The damping is so large that $\beta > 2\sqrt{km}$. (a) Show that the instantaneous position of the mass at any time $t > 0$ is

$$x = \frac{v_0}{\gamma}e^{-\beta t/2m}\sinh\gamma t \qquad \text{where} \qquad \gamma = \frac{\sqrt{\beta^2 - 4km}}{2m}$$

measured from the equilibrium position. (b) Show that the mass travels downward for a time

$$\frac{1}{\gamma}\tanh^{-1}\frac{2m\gamma}{\beta}$$

and then gradually returns to the equilibrium position but never reaches it. Note that the time is independent of v_0. (c) Discuss the case $\gamma \to 0$ and compare with B Exercise 1.

2. Solve the previous problem in case the mass is pulled a distance x_0 below its equilibrium position and then given velocity v_0 downward.

3. A mass m is suspended vertically from a spring having constant k. The mass is pulled x_0 below its equilibrium position and given a velocity v_0 downward. A damping force βv, where v is the instantaneous velocity and β is a positive constant, acts on the mass. Show that if the

mass is chosen so that $m = \beta^2/4k$, then it will travel downward for a time given by

$$\frac{\beta^2 v_0}{2k(2kx_0 + \beta v_0)}$$

and then return gradually to the equilibrium position.

4. Interpret the results of Exercise 3 if $\beta v_0 = -2kx_0$.

5. Obtain the general solution of equation (28), page 233, and interpret physically.

6. Let I represent the moment of inertia of a ship (or airplane) about an axis through the center of gravity and in the direction from front to rear. Show how equation (28) can be used to describe the oscillations which take place.

7. Work Exercise 6 if I is the moment of inertia about an axis through the center of gravity but having direction from side to side.

1.3 THE SPRING WITH EXTERNAL FORCES

In the previous pages we discussed the problem of a spring where only restoring and damping forces were considered. We now consider cases where other external forces which depend on time may act. Such forces may occur, for example, when the support holding the spring is moved up and down in a prescribed manner such as in periodic motion, or when the weight is given a little push every time it reaches the lowest position. If we denote the external force by $F(t)$, the differential equation for motion of the spring is

$$\frac{W}{g} \cdot \frac{d^2x}{dt^2} = -kx - \beta\frac{dx}{dt} + F(t) \quad \text{or} \quad \frac{W}{g} \cdot \frac{d^2x}{dt^2} + \beta\frac{dx}{dt} + kx = F(t)$$

which may be written

$$a\frac{d^2x}{dt^2} + b\frac{dx}{dt} + cx = F(t) \tag{32}$$

(where $a = W/g$, $b = \beta$, $c = k$), often called the equation of *forced vibrations*.

ILLUSTRATIVE EXAMPLE 7

In Illustrative Example 3, page 228, assume that a periodic external force given by $F(t) = 24 \cos 8t$ is acting. Find x in terms of t, using the conditions given there.

Mathematical Formulation. The differential equation is

$$\frac{6}{32} \cdot \frac{d^2x}{dt^2} = -12x - 1.5\frac{dx}{dt} + 24 \cos 8t \quad \text{or} \quad \frac{d^2x}{dt^2} + 8\frac{dx}{dt} + 64x = 128 \cos 8t \tag{33}$$

and the initial conditions are $x = \frac{1}{3}$, $dx/dt = 0$ at $t = 0$.

Solution   The complementary solution of (33) is

$$x_c = e^{-4t}(A \cos 4\sqrt{3}t + B \sin 4\sqrt{3}t)$$

If we assume as particular solution $a \sin 8t + b \cos 8t$, we find $a = 2, b = 0$. Hence, the general solution of (33) is

$$x = e^{-4t}(A \cos 4\sqrt{3}t + B \sin 4\sqrt{3}t) + 2 \sin 8t$$

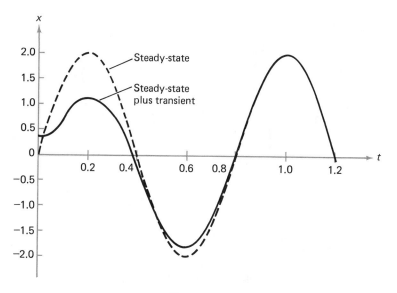

Figure 5.10

Using the initial conditions, we have $A = \frac{1}{3}$, $B = -11\sqrt{3}/9$ and thus,

$$x = \frac{e^{-4t}}{9}(3 \cos 4\sqrt{3}t - 11\sqrt{3} \sin 4\sqrt{3}t) + 2 \sin 8t \qquad (34)$$

The graph of (34) appears in Fig. 5.10. It will be observed that the terms in (34) involving e^{-4t} become negligible when t is large. These terms are called *transient terms* and are significant only when t is near zero. These transient terms in the solution, when they are significant, are sometimes called the *transient solution*. When the transient terms are negligible, the term $2 \sin 8t$ remains. This is called the *steady-state term* or *steady-state solution*, since it indicates the behavior of the system when conditions have become steady. It is seen that the steady-state solution (dashed curve in Fig. 5.10) is periodic and has the same period as that of the applied external force.

A EXERCISES

1. A vertical spring having constant 5 lb/ft has a 16 lb weight suspended from it. An external force given by $F(t) = 24 \sin 10t$, $t \geq 0$ is applied. A damping force given numerically in pounds by $4v$, where v is the instantaneous velocity of the weight in feet per second, is assumed to act. Initially the weight is at rest at its equilibrium position. (a) Determine the position of the weight at any time. (b) Indicate the transient and steady-state solutions. (c) Find the amplitude, period, and frequency of the steady-state solution.

2. A vertical spring having constant 8 lb/ft has a 64 lb weight suspended from it. A force given by $F(t) = 16 \cos 4t$, $t \geq 0$ is applied. Assuming that the weight, initially at the equilibrium position, is given an upward velocity of 10 ft/sec and that the damping force is negligible, determine the position and velocity of the weight at any time.

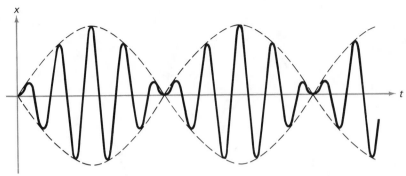

Figure 5.11

3. A spring is stretched 10 cm by a force of 500 dynes. A mass of 2 g is suspended from the spring and allowed to come to equilibrium. A force given in dynes by $F(t) = 200 \sin 5t$, $t \geq 0$, is then applied. Assuming that there is a damping force given numerically in dynes by $20v$, where v is the instantaneous velocity in centimeters per second, find the position of the mass (a) at any time; (b) after a long time.

B EXERCISES

1. The motion of a mass on a certain vertical spring is described by

$$\frac{d^2x}{dt^2} + 100x = 36 \cos 8t, \quad x = 0, \quad \frac{dx}{dt} = 0 \quad \text{at } t = 0$$

where x is the instantaneous distance of the mass from the equilibrium position, downward being taken as the positive direction. (a) Give a physical interpretation to the problem. (b) Show that the solution may be written $x = 2 \sin t \sin 9t$. (c) Show that the graph of x as a function of t is similar to that of Fig. 5.11.

The solution may be written $x = \mathscr{A}(t) \sin 9t$, where $\mathscr{A}(t) = \sin t$ is called the time-varying amplitude and is a slowly varying function (period $= 2\pi$) in comparison with the wave $\sin 9t$ (period $= 2\pi/9$). The wave $\sin 9t$ is said to be *amplitude modulated*. In the theory of acoustics these fluctuations of amplitude are called *beats*, the loud sounds corresponding to the large amplitudes. Beats may occur when two tuning forks having nearly equal frequencies are set into vibration simultaneously. A practical use of this is in tuning of pianos (or other instruments) where successful tuning is marked by adjusting the frequency of a note to that of a standard note until beats are eliminated. The phenomenon is also important in the theories of optics and electricity.

2. Work Exercise 1 if the initial conditions are changed so that
   (a) $x = 6$, $dx/dt = 0$ at $t = 0$.   (b) $x = 0$, $dx/dt = 10$ at $t = 0$.
   (c) $x = 6$, $dx/dt = 10$ at $t = 0$.

C EXERCISES

1. A spring of constant k with an attached mass m is suspended from a support which oscillates about line OP (Fig. 5.12) so that the instantaneous distance of the support from OP is $A \cos \omega t$, $t \geq 0$, where A is constant. Let x represent the instantaneous stretch of the spring.

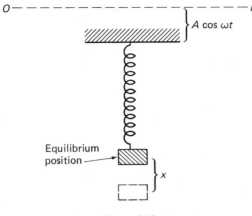

Figure 5.12

(a) Show that if damping is negligible the differential equation of motion of the mass is

$$m\frac{d^2x}{dt^2} + kx = mA\omega^2 \cos \omega t$$

(b) If at $t = 0$, $x = 0$ and $dx/dt = 0$, and if ω is nearly equal to $\sqrt{k/m}$, show that beats are produced with very large amplitudes which become larger the closer ω is to $\sqrt{k/m}$.

(c) Discuss the case $\omega = \sqrt{k/m}$.

2. Work Exercise 1 if the force is given by $At \cos \omega t$.

1.4 THE PHENOMENON OF MECHANICAL RESONANCE

When the frequency of a periodic external force applied to a mechanical system is related in a simple way (which will be described) to the natural frequency of the system, mechanical resonance may occur which builds up the oscillations to such tremendous magnitudes that the system may fall apart. A company of soldiers marching in step across a bridge may in this manner cause the bridge to collapse, (and in a famous disaster actually did) even though the bridge would have been strong enough to carry many more soldiers had they marched out of step. For this reason soldiers were required to "break step" on crossing a bridge. In an analogous manner, it may be possible for a musical note of proper characteristic frequency to shatter a glass. Because of the great damages which may thus occur, mechanical resonance is in general something which needs to be avoided, especially by the engineer in designing structures or vibrating mechanisms.

It should be mentioned, however, that mechanical resonance may also serve useful purposes. For example, if an automobile should get stuck in the snow (or mud), it can, by "rocking it," be set into vibration with its natural frequency. Then, on application of a force with this same frequency, the resulting mechanical resonance may often be enough to free it. In a more pleasurable vein, mechanical resonance also serves to produce wider and wider oscillations for a child or adult on a swing.

The following special case indicates the possible consequences of resonance.

Suppose an external force given by 3 cos 8*t* is applied to the spring of Illustrative Example 1, page 221. Describe the motion which ensues if it is assumed that initially the weight is at the equilibrium position ($x = 0$) and that its initial velocity is zero.

Mathematical Formulation. The differential equation is

$$\frac{6}{32} \cdot \frac{d^2x}{dt^2} = -12x + 3 \cos 8t \quad \text{or} \quad \frac{d^2x}{dt^2} + 64x = 16 \cos 8t \tag{35}$$

and the initial conditions are $x = 0$, $dx/dt = 0$ at $t = 0$.

Solution The complementary solution of (35) is $x_c = A \cos 8t + B \sin 8t$. For particular solution we must assume $x_p = t(a \cos 8t + b \sin 8t)$. Substituting in (35) we find $a = 0, b = 1$. Thus, the general solution is

$$x = A \cos 8t + B \sin 8t + t \sin 8t$$

From the initial conditions, it is readily found that $A = B = 0$. Hence

$$x = t \sin 8t \tag{36}$$

The graph of (36) lies between the graphs of $x = t$ and $x = -t$ as shown in Fig. 5.13. It is seen from the graph that the oscillations build up without limit. Naturally, the spring is bound to break within a short time.

It should be noted that in this example damping was neglected and resonance occurred because the *frequency of the applied external force was equal to the natural*

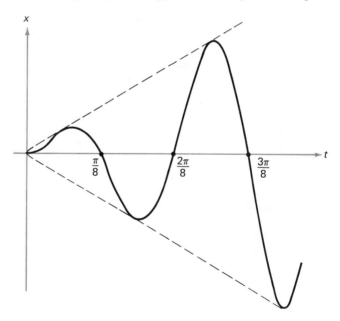

Figure 5.13

frequency of the undamped system. This is a general principle. In the case where damping occurs, the oscillations do not build up without limit but may nevertheless become very large. Resonance in this case occurs when the frequency of the applied external force is slightly smaller than the natural frequency of the system. For further discussion of this see B Exercise 1.

A EXERCISES

1. A vertical spring having constant 4 lb/ft has a 32 lb weight attached to it. A force given by $F(t) = 16 \sin 2t$, $t \geq 0$ is applied. Assuming that at $t = 0$ the weight is at rest at the equilibrium position and that the damping force is negligible, (a) set up a differential equation and conditions describing the motion; (b) determine the position and velocity of the weight at any time; (c) show that the motion is one of resonance.

2. In the previous problem suppose that at $t = 0$ the weight is 6 in. below the equilibrium position and is struck so as to give it a velocity of 4 ft/sec upward. Determine the position and velocity of the weight at any time. Is the motion still one of resonance?

3. A spring is stretched 20 cm by a force of 8000 dynes. A mass of 4 g is suspended from the spring and allowed to come to equilibrium. A force given by $F(t) = 60 \cos \omega t$, $t \geq 0$, is then applied. (a) Assuming that damping is negligible, find the position of the mass at any time for all $\omega > 0$. (b) For what values of ω will resonance occur?

B EXERCISES

1. The equation of forced vibration of a mass on a vertical spring is

$$m \frac{d^2x}{dt^2} + \beta \frac{dx}{dt} + kx = A \cos \omega t, \qquad t \geq 0$$

where x is the displacement of the mass from its equilibrium position and m, β, k, A, and ω are positive constants. (a) Show that a steady-state oscillation is given by

$$x = \frac{A}{\sqrt{(m\omega^2 - k)^2 + \beta^2\omega^2}} \cos(\omega t + \phi)$$

(b) Show that maximum oscillations (resonance) will occur if ω is so chosen that

$$\omega = \sqrt{\frac{k}{m} - \frac{\beta^2}{2m^2}}$$

provided $\beta^2 < 2km$. (c) Show that at resonance, the amplitude of the oscillation varies inversely as the damping constant β.

2. Discuss Exercise 1 if $\beta^2 \geq 2km$.

C EXERCISES

1. A mass on a spring undergoes a forced vibration given by

$$m \frac{d^2x}{dt^2} + kx = A \cos^3 \omega t, \qquad t \geq 0$$

Show that there are two values of ω at which resonance occurs and determine them.

2. A mass on a spring undergoes a forced vibration given by

$$m\frac{d^2x}{dt^2} + kx = \sum_{n=1}^{M} a_n \cos \frac{2\pi n t}{T}, \qquad t \geq 0, a_n \neq 0$$

(a) Show that the least period of the external force is T. (b) Show that resonance will occur if T has any one of the M values $2\pi n\sqrt{m/k}$, where $n = 1, 2, \ldots, M$.

It can be shown that a suitable function $F(t)$ defined in the interval $0 \leq t < T$ and which is such that $F(t + T) = F(t)$ outside the interval [i.e., $F(t)$ has period T] can be expanded in a series

$$A + \sum_{n=1}^{\infty} \left(a_n \cos \frac{2\pi n t}{T} + b_n \sin \frac{2\pi n t}{T} \right)$$

Such series are called *Fourier series*. Conditions under which such expansion is possible and determination of the constants A, a_n, b_n are presented in Chapter Eight.

3. If $m\, d^2x/dt^2 + kx = F(t)$, $t \geq 0$ such that $x = 0$, $dx/dt = 0$ at $t = 0$, show that

$$x = \frac{1}{\sqrt{km}} \int_0^t F(u) \sin\sqrt{\frac{k}{m}}(t - u)du$$

Find x and discuss a possible physical interpretation if $F(t) = \begin{cases} t, & 0 \leq t \leq T \\ 0, & t > T \end{cases}$.

4. Solve the previous problem if $F(t) = \begin{cases} F_0/\epsilon, & 0 \leq t \leq \epsilon \\ 0, & t > \epsilon \end{cases}$

where F_0 and ϵ are constants. Discuss the case $\epsilon \to 0$ and interpret the problem physically.

2 Electric Circuit Problems

In Chapter Three, Section 2, the student learned how to formulate differential equations arising from certain problems involving electric circuits. The case where a resistor, capacitor, and inductor were connected in series with a battery or generator was not considered. In this section we shall treat this case.

Consider the circuit of Fig. 5.14. When key K is closed, an instantaneous current flows. If Q is the instantaneous charge on capacitor C, then by Kirchhoff's law,

$$L\frac{dI}{dt} + RI + \frac{Q}{C} = E(t) \tag{1}$$

where $E(t)$, the emf, may depend on time, but where we assume L, R, C are constants. Since $I = dQ/dt$, (1) becomes

$$L\frac{d^2Q}{dt^2} + R\frac{dQ}{dt} + \frac{Q}{C} = E(t) \tag{2}$$

Comparison with the general equation of forced vibrations [equation (32), page 236], shows the striking analogy between mechanical and electrical quantities.

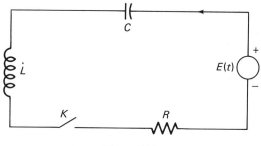

Figure 5.14

Charge Q corresponds to *position x.*
Inductance L corresponds to *mass m* or *W/g.*
Resistance R corresponds to *damping constant β.*
Inverse capacitance 1/C corresponds to *spring constant k.*
Electromotive force E(t) corresponds to *applied external force F(t).*
Current I = dQ/dt corresponds to *velocity v = dx/dt.*

Because of the remarkable analogy between these mechanical and electrical quantities, which holds in even more complicated cases, most statements made for mechanical systems apply to electric systems and vice versa. In fact, the analogy is often used in industry in studying a mechanical system which may be too complicated or too expensive to build, or when consequences may be too dangerous.

In particular, the phenomenon of resonance occurs in electric systems. However, contrary to the dangerous effects which may result in mechanical resonance, the effects of electrical resonance are mainly very useful. The fields of radio, television, radar, and communications would virtually be impossible were it not for electrical resonance. In such instances the current and consequently the power generated may build up to the large amounts needed in these fields. It is because of electrical resonance that we tune our radio to the frequency of the transmitting radio station in order to get reception.

ILLUSTRATIVE EXAMPLE

An inductor of 0.5 henry is connected in series with a resistor of 6 ohms, a capacitor of 0.02 farad, a generator having alternating voltage given by 24 sin 10t, t ≧ 0, and a switch K (Fig. 5.15).

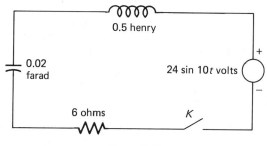

Figure 5.15

(a) Set up a differential equation for the instantaneous charge on the capacitor.

(b) Find the charge and current at time t if the charge on the capacitor is zero when the switch K is closed at $t = 0$.

Mathematical Formulation. Voltage drop across resistor is $6I$. Voltage drop across inductor is $0.5 \, dI/dt$. Voltage drop across capacitor is $Q/0.02 = 50Q$.

Hence, by Kirchhoff's law,

$$6I + 0.5 \frac{dI}{dt} + 50Q = 24 \sin 10t$$

or since $I = dQ/dt$,

$$0.5 \frac{d^2Q}{dt^2} + 6 \frac{dQ}{dt} + 50Q = 24 \sin 10t$$

or

$$\frac{d^2Q}{dt^2} + 12 \frac{dQ}{dt} + 100Q = 48 \sin 10t \tag{3}$$

The conditions are $Q = 0$ and $I = dQ/dt = 0$ at $t = 0$.

Solution The complementary solution of (3) is $e^{-6t}(A \cos 8t + B \sin 8t)$. Assuming the particular solution $a \sin 10t + b \cos 10t$, we find $a = 0$, $b = -\frac{2}{5}$. Hence, the general solution of (3) is

$$Q = e^{-6t}(A \cos 8t + B \sin 8t) - \tfrac{2}{5} \cos 10t$$

From the initial conditions we find $A = \frac{2}{5}$, $B = \frac{3}{10}$. Hence, the required solution is

$$Q = \tfrac{1}{10} e^{-6t}(4 \cos 8t + 3 \sin 8t) - \tfrac{2}{5} \cos 10t$$

It will be noted that the term with e^{-6t} is the *transient solution*; it soon becomes negligible. The term $-\frac{2}{5} \cos 10t$ is the *steady-state solution*; it remains after the transient term has virtually disappeared. The student should compare this with the example and graph on pages 236 and 237.

A EXERCISES

1. An emf of 500 volts is in series with a 20 ohm resistor, a 4 henry inductor, and a 0.008 farad capacitor. At $t = 0$, the charge Q and current I are zero. (a) Find Q and I for $t \geq 0$. (b) Indicate the transient and steady-state terms in Q and I. (c) Find Q and I after a long time.

2. A capacitor of 10^{-3} farads is in series with an emf of 20 volts and an inductor of 0.4 henries. At $t = 0$, $Q = 0$, and $I = 0$. (a) Find the natural frequency and period of the electric oscillations. (b) Find the maximum charge and current.

3. A 0.1 henry inductor, a 4×10^{-3} farad capacitor, and a generator having emf given by $180 \cos 40t$, $t \geq 0$, are connected in series. Find the instantaneous charge Q and current I if $I = Q = 0$ at $t = 0$.

4. A resistor of 50 ohms, an inductor of 2 henries and a 0.005 farad capacitor are in series with an emf of 40 volts and an open switch. Find the instantaneous charge and current after the switch is closed at $t = 0$, assuming that at that time the charge on the capacitor is 4 coulombs.

1. An inductor L, capacitor C, and resistor R are connected in series. At $t = 0$, the charge on the capacitor is Q_0, while the current is zero. Show that the charge Q and current I will be oscillatory if $R < 2\sqrt{L/C}$ and will be given by

$$Q = \frac{Q_0}{2\omega L}\, e^{-Rt/2L}\, \sqrt{R^2 + 4\omega^2 L^2}\, \sin(\omega t + \phi), \qquad I = -\frac{Q_0(R^2 + 4\omega^2 L^2)}{4\omega L^2}\, e^{-Rt/2L}\, \sin \omega t$$

where
$$\omega = \sqrt{\frac{1}{LC} - \frac{R^2}{4L^2}}, \qquad \phi = \tan^{-1}\frac{2\omega L}{R}$$

What is the quasi period of the oscillations (see page 229)?

2. If $R = 0$ in Exercise 1, show that the natural period of the oscillations is $2\pi\sqrt{LC}$. If $L = 0.5$ henries and $C = 4$ microfarads, find the natural frequency.

3. Discuss Exercise 1 if $R \geq 2\sqrt{L/C}$, showing the analogue of critically damped and over-damped motion in mechanical systems.

4. An inductor of 0.5 henries is connected in series with a resistor of 5 ohms, and a capacitor of 0.08 farads. At $t = 0$ the current is 10 amp, and the charge on the capacitor is zero. Show that the charge builds up to a maximum in 0.2 sec and determine the value of the maximum.

1. An inductor L, resistor R, and capacitor C are connected in series with an a-c generator having voltage given by $E_0 \cos \omega t$, $t \geq 0$. If L, R, C, E_0, and ω are given constants:
   (a) Show that the differential equation of the circuit is

$$L\frac{d^2 I}{dt^2} + R\frac{dI}{dt} + \frac{I}{C} = \frac{d}{dt}(\text{Re}\, E_0 e^{i\omega t}), \qquad t \geq 0$$

   where I is the instantaneous current and Re denotes "real part of."
   (b) Let $I = \text{Re}\,(Ae^{i\omega t})$, where A is a constant complex number and show that

$$A = \frac{E_0}{R + i(\omega L - 1/\omega C)} = \frac{E_0}{R + iX}$$

Write $R + iX = Ze^{i\phi}$, so that $Z = \sqrt{R^2 + X^2}$, $\phi = \tan^{-1} X/R$. Hence, show that

$$I = \frac{E_0}{Z}\cos(\omega t - \phi)$$

   is the steady-state current. Here Z is called the *impedance* and X the *reactance*.
   (c) Prove that the current becomes very large (electrical resonance) when the frequency of the applied voltage is given by $f = 1/(2\pi\sqrt{LC})$, whereas the charge on the capacitor is a maximum when

$$f = \frac{1}{2\pi}\sqrt{\frac{1}{LC} - \frac{R^2}{2L^2}}$$

2. Show that there are two frequencies, one below and one above the resonant frequency, at which the amplitudes are $1/n$th of the amplitude at resonance. Prove that the difference of these frequencies is independent of the capacitance and is given by $R\sqrt{n^2 - 1}/2\pi L$. The

ratio of this difference to the resonant frequency $f = \omega/2\pi$ is $R\sqrt{n^2 - 1}/\omega L$. The quantity $Q = \omega L/R$ is called the "Q of the circuit." If Q is large, resonance is "sharp," and we say that we have "sharp tuning." This is important in radio, television, and communications. For practical purposes n is usually taken as 2.

3 Miscellaneous Problems

3.1 THE SIMPLE PENDULUM

As a first illustration, let us consider the following

ILLUSTRATIVE EXAMPLE 1

A simple pendulum consists of a particle of mass m supported by a wire (or inelastic string) of length l and of negligible mass. If the wire is always straight and the system is free to vibrate in a vertical plane, find the period of vibration.

Mathematical Formulation. Let AB (Fig. 5.16) denote the wire, A being the fixed point of support, B the other end of the wire at which is attached mass m. Let θ be the angle of the wire with the vertical AO at any instant. While mass m is in motion two forces act on it, the tension τ in the string and the weight mg of the mass. Resolving weight mg into two components, one parallel to the path of motion and one perpendicular to it, we see that the component perpendicular to the path is balanced by the tension. The magnitude of the net force acting tangent to the path is $mg \sin \theta$.

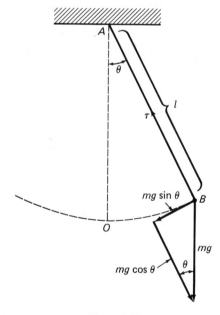

Figure 5.16

We choose signs so that $\theta > 0$ when the mass is on the right in the figure and $\theta < 0$ when it is on the left. This means essentially that we are choosing directions along the arc to the right as positive and to the left as negative. When $\theta > 0$, the resulting force is to the left, and when $\theta < 0$, the resulting force is to the right. The net force in magnitude and direction is thus given by $-mg \sin \theta$. Since the arc length is given by $s = l\theta$, we have by Newton's law,

$$m\frac{d^2s}{dt^2} = ml\frac{d^2\theta}{dt^2} = -mg \sin \theta \quad \text{or} \quad \frac{d^2\theta}{dt^2} = -\frac{g}{l}\sin \theta \qquad (1)$$

Equation (1), which is a non-linear equation, cannot be solved exactly in terms of elementary functions (however, see C Exercises 3, 4, 5). In order to proceed further we make an approximation. For small angles (roughly between $-5°$ and $+5°$) we may write $\sin \theta = \theta$, where θ is in radians. Thus, within the range of our approximation, we can replace (1) by the linear differential equation

$$\frac{d^2\theta}{dt^2} + \frac{g\theta}{l} = 0 \qquad (2)$$

Solution   The roots of the auxiliary equation of (2) are $\pm i\sqrt{g/l}$ so that

$$\theta = A \sin \sqrt{g/l}\,t + B \cos \sqrt{g/l}\,t \qquad (3)$$

From this it is clear that the period T is given by

$$T = \frac{2\pi}{\sqrt{g/l}} \quad \text{or} \quad T = 2\pi\sqrt{l/g} \qquad (4)$$

a formula familiar in elementary physics.

It should be noted that to find the period here, initial conditions are not needed. It should also be noted that our approximation of $\sin \theta$ by θ is equivalent to assuming that the motion is that of simple harmonic motion.

If the vibrations are not small, we cannot use the approximation θ for $\sin \theta$. However, in this case we can solve (1) in terms of *elliptic integrals*, and the motion is still periodic. Further discussion is given in the exercises.

3.2 VERTICAL OSCILLATIONS OF A BOX FLOATING IN A LIQUID

As a second illustration let us consider the following

ILLUSTRATIVE EXAMPLE 2

A cubical box 10 ft on a side floats in still water (density 62.5 lb/ft³). It is observed that the box oscillates up and down with period $\frac{1}{2}$ sec. What is its weight?

Mathematical Formulation. Figure 5.17 shows the cube in its equilibrium position, indicated by ABC. Figure 5.18 shows the cube nearly all submerged in water. In this position there is a force tending to push the box up again. To determine this

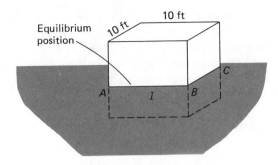

Figure 5.17

force we need the physical law known as:

> ARCHIMEDES' PRINCIPLE. *An object partially or totally submerged in a fluid is buoyed up by a force equal to the weight of the fluid displaced.*

From this principle it is clear that the weight of the cube equals the weight of the water occupied by that portion of the cube below the surface in Fig. 5.17, which is indicated by I. The region I needed to balance the weight of the cube is also shown in Fig. 5.18, from which it is evident that there is an additional unbalanced force equal to the weight of the water which would occupy the shaded region in that figure. Since the dimensions of the shaded region are x feet by 10 ft by 10 ft, and since the water weighs 62.5 lb/ft^3, the weight of the water normally occupying such a region would be $62.5 \times x \times 10 \times 10$ lb, or $6250x$ lb. This is numerically the net force acting to move the cube. It is analogous to the restoring force of the vibrating spring. If the weight of the box in pounds is W, Newton's law gives

$$\frac{W}{g} \cdot \frac{d^2x}{dt^2} = -6250x \quad \text{or} \quad \frac{d^2x}{dt^2} + \frac{200{,}000x}{W} = 0 \tag{5}$$

taking $g = 32$.

Solution   The general solution of (5) is

$$x = A \cos \sqrt{\frac{200{,}000}{W}}\, t + B \sin \sqrt{\frac{200{,}000}{W}}\, t \tag{6}$$

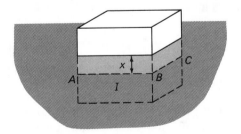

Figure 5.18

from which it is clear that the period is

$$T = \frac{2\pi}{\sqrt{200,000/W}} \quad \text{or} \quad T = \frac{2\pi\sqrt{W}}{200\sqrt{5}} \tag{7}$$

Equating this to $\frac{1}{2}$ sec, we find $W = 1270$ lb approximately. It is seen that the box vibrates with simple harmonic motion.

3.3 A PROBLEM IN CARDIOGRAPHY

The biological science which deals with the study of the heart is called *cardiology*. As might be expected, the nature and effects of vibrations of the heart as it pumps blood through the circulatory system of the body are a great source of mathematical applications. An important aspect involves the recording of such vibrations, which is known as *cardiography*. An instrument available for the recording of such vibrations is an *electrocardiograph*, which translates the vibrations into electrical impulses which are then recorded.

Instead of translating the heart vibrations into electrical impulses, it is of interest that these vibrations can be translated into mechanical vibrations. To see how this is possible, let us suppose that a person rests on a horizontal table which has springs so that it can vibrate horizontally but not vertically. Then, due to the pumping of the heart, the table will be found to undergo small vibrations the frequency and magnitude of which will depend on various parameters associated with the heart. Thus, by investigating the motion of the table some important conclusions may be drawn concerning the vibrations of the heart.

Mathematical Formulation. Let x denote the horizontal displacement of some specified point of the table (as for example one end) from some fixed location (such as a wall). Let M represent the combined mass of the person and the portion of the table which is set into motion. If we assume that there is a damping force proportional to the instantaneous velocity and a restoring force proportional to the instantaneous displacement, then the equation describing the motion of the table is

$$M\frac{d^2x}{dt^2} + \beta\frac{dx}{dt} + \gamma x = F \tag{8}$$

where β and γ are constants of proportionality and F is the force on the system due to the pumping action of the heart. Suppose that m is the mass of blood pumped out of the heart during each vibration and y is the instantaneous center of mass of this quantity of blood. Then we have by Newton's law

$$F = m\frac{d^2y}{dt^2} \tag{9}$$

It might be supposed as a first approximation that y can be expressed as a simple sinusoidal function of t given by

$$y = a \sin \omega t \tag{10}$$

where a, ω are constants. This, however, turns out to be an oversimplification, since (10) assumes that there is only one frequency associated with the vibrations of the heart, whereas evidence points out that there are many frequencies. This leads us to replace (10) by

$$y = a_1 \sin \omega t + a_2 \sin 2\omega t + a_3 \sin 3\omega t + \cdots \tag{11}$$

The series of terms on the right is often called a *Fourier series* after the man who first investigated it in connection with problems in heat flow, and we shall examine various properties of these series in later chapters. One important property is that a suitable periodic function with period $2\pi/\omega$ can be expressed in terms of such series by appropriate choices of the coefficients $a_1, a_2, a_3, \ldots$. The first term on the right of (11) represents a first approximation to the function, the sum of the first two terms a better approximation, etc. For our purposes here, we need only point out that substitution of (11) into (9) and then the result into (8) leads to a differential equation which can be solved subject to various possible initial conditions.

Solution Using only the first two terms of the series (11), we have

$$M \frac{d^2x}{dt^2} + \beta \frac{dx}{dt} + \gamma x = -m\omega^2 a_1 \sin \omega t - 4m\omega^2 a_2 \sin 2\omega t \tag{12}$$

The general solution of this equation consists of two parts, (1) the general solution of the equation with the right side replaced by zero, and (2) a particular solution. The first part will be the *transient* solution and will disappear rapidly provided that $\beta > 0$. The second part will be the *steady-state* solution in which we are interested. This steady-state solution is easily found to be

$$x = \frac{m\omega^2 a_1 [(M\omega^2 - \gamma) \sin \omega t + \beta\omega \cos \omega t]}{(M\omega^2 - \gamma)^2 + \beta^2\omega^2}$$

$$+ \frac{4m\omega^2 a_2 [(4M\omega^2 - \gamma) \sin 2\omega t + 2\beta\omega \cos 2\omega t]}{(4M\omega^2 - \gamma)^2 + 4\beta^2\omega^2} \tag{13}$$

A corresponding solution can be found assuming any number of terms in (11).

3.4 AN APPLICATION TO ECONOMICS

An important problem in economics concerns the behavior of the price P of a commodity at some time t. Obviously, the problem depends on many factors, some of which have already been described on page 160. Consequently, we must make some simplifying assumptions, realizing that we may incur the risk of over-simplifying and perhaps arriving at results in conflict with reality. As we have mentioned many times, we must in such case reject the assumptions and seek others.

Mathematical Formulation. It is not unrealistic to suppose that the commodity is such that increasing its price results in an increase of supply S, which as a consequence ultimately results in lowering of price. To seek a possible model of the price behavior of the commodity, let us look for factors which can lead to price changes. Clearly, prices change with time as a result of *inflation*. We shall assume that the inflation factor is given as a function of time by $F(t)$, which we do not specify at present. Let us also assume that apart from inflation the time rate of change in price is proportional to the difference between the supply S at time t and some equilibrium supply which we denote by S_0. If $S > S_0$, the supply is too large and the price tends to decrease, while if $S < S_0$, the supply is too small and the price tends to increase, so that the constant of proportionality must be negative and is denoted by $-k_1$. From these observations we are led to the equation

$$\frac{dP}{dt} = F(t) - k_1(S - S_0) \tag{14}$$

Let us also assume that the time rate of change of supply is proportional to the difference between the price and some equilibrium price, which we denote by P_0. Then we have, denoting the proportionality constant by k_2,

$$\frac{dS}{dt} = k_2(P - P_0) \tag{15}$$

If $P < P_0$, the price is too low, dS/dt is negative, and the supply decreases; if $P > P_0$, the price is too high, dS/dt is positive, and the supply increases. Thus, $k_2 > 0$.

Solution If we solve for S in (14), we find

$$S = S_0 + \frac{1}{k_1}\left[F(t) - \frac{dP}{dt}\right] \tag{16}$$

Substituting this into (14) and assuming that S_0 is constant, we have

$$\frac{d^2P}{dt^2} + k_1k_2P = F'(t) + k_1k_2P_0 \tag{17}$$

Various cases arise depending on the inflation factor $F(t)$.
    Let us consider the special case where

$$F(t) = \alpha \tag{18}$$

where α is a constant. In this case (17) has the general solution

$$P = P_0 + A\cos\sqrt{k_1k_2}\,t + B\sin\sqrt{k_1k_2}\,t \tag{19}$$

Substituting this into (14) then gives

$$S = S_0 + \frac{\alpha}{k_1} + A\sqrt{\frac{k_2}{k_1}}\sin\sqrt{k_1k_2}\,t - B\sqrt{\frac{k_2}{k_1}}\cos\sqrt{k_1k_2}\,t \tag{20}$$

Assuming that at some time $t = 0$, $P = P_0$, $S = S_0$, we find from (19) and (20) that

$$A = 0, \qquad B = \frac{\alpha}{\sqrt{k_1k_2}}$$

so that

$$P = P_0 + \frac{\alpha}{\sqrt{k_1k_2}}\sin\sqrt{k_1k_2}\,t \tag{21}$$

$$S = S_0 + \frac{\alpha}{k_1} - \frac{\alpha}{k_1}\cos\sqrt{k_1k_2}\,t \tag{22}$$

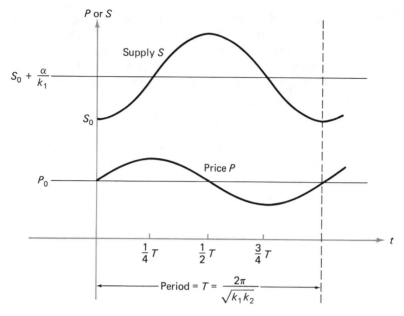

Figure 5.19

Interpretation. The equations (21) and (22) show that the price and supply oscillate sinusoidally about the values P_0 and $S_0 + \alpha/k_1$, respectively, with period

$$T = \frac{2\pi}{\sqrt{k_1 k_2}} \tag{23}$$

If we write (22) in the form

$$S = S_0 + \frac{\alpha}{k_1} + \frac{\alpha}{k_1} \sin \sqrt{k_1 k_2} \left(t - \frac{\pi}{2\sqrt{k_1 k_2}} \right) \tag{24}$$

we see that the price *leads* the supply by a time equal to one quarter of the period, i.e., $\frac{1}{4}T$, or that the supply *lags* the price by $\frac{1}{4}T$. This means that if the maximum price occurs at some particular time the maximum supply occurs at a time $\frac{1}{4}T$ later when the price has already fallen. The situation is shown graphically in Fig. 5.19, where for purposes of comparison the same vertical axis is used for both S and P although of course the units for these variables are not the same.

A EXERCISES

1. The small oscillations of a simple pendulum have period 2 sec. Determine the length of the pendulum. Find the corresponding length of a simple pendulum having twice this period.

2. The bob of a simple pendulum of length 2 ft is displaced so that the pendulum rod makes an angle of 5° with the vertical. If the pendulum is released from this position: (a) Find the angle θ which the rod makes with the vertical at any time. (b) Determine the frequency of the vibration. (c) Calculate the distance traveled by the pendulum bob during one period. (d) Find the velocity and acceleration of the bob at the center of its path.

3. A cube 5 ft on a side and weighing 500 lb floats in still water. It is pushed down slightly and released so that oscillations take place. Find the period and frequency of the vibrations.

4. A cylinder 4 ft in radius and 6 ft in altitude, weighing 1000 lb, vibrates in still water with its axis vertical. Find the frequency and period of the vibrations.

B EXERCISES

1. A cylinder of radius r and height h and having weight W floats with its axis vertical in a liquid of density ρ. If it is set into vibration, show that the period is $\dfrac{2}{r}\sqrt{\dfrac{\pi W}{\rho g}}$.

2. A cylinder vibrates with axis vertical in a liquid. It is found that the frequency of the vibrations in the liquid is half that in water. Determine the density of the liquid.

3. A simple pendulum vibrates in a medium in which the damping is proportional to the instantaneous velocity. If the pendulum bob passes through the equilibrium position $\theta = 0$ at $t = 0$ with velocity v_0, show that the angle θ which the pendulum rod makes with the vertical is

$$\theta = \frac{v_0}{\omega l} e^{-\beta t} \sin \omega t, \qquad \omega = \sqrt{\frac{g}{l} - \beta^2}$$

where β is the damping constant and l is the length of the pendulum. Find β if the distance traveled during one complete cycle is half the previous one. What is the quasi period and frequency?

4. By using three or more terms of (11) generalize the result (13), page 250.

5. Obtain the price and supply in the problem on page 250 for the cases
(a) $F(t) = \alpha + \beta t$; α, β constants.   (b) $F(t) = K \sin \omega t$; K, ω constants.

C EXERCISES

1. A sphere of radius R floating half submerged in a liquid is set into vibration. If x is the instantaneous displacement of its diametral plane from the equilibrium position, show that

$$\frac{d^2x}{dt^2} = -\frac{3g}{2}\left[\frac{x}{R} - \frac{1}{3}\left(\frac{x}{R}\right)^3\right]$$

Hence, show that for small vibrations the sphere vibrates with frequency equal to that of a simple pendulum of length $2R/3$.

2. If the sphere of the previous problem is pushed down until it is just barely under the liquid and is then released, show that the velocity of the sphere at the instant when the diametral plane coincides with the surface is $\frac{1}{2}\sqrt{5gR}$.

3. Using equation (1) for the vibrations of a simple pendulum show that

$$\left(\frac{d\theta}{dt}\right)^2 = \frac{2g}{l}(\cos \theta - \cos \theta_0)$$

Thus, show that the period T is given by

$$T = 2\sqrt{\frac{2l}{g}} \int_0^{\theta_0} \frac{d\theta}{\sqrt{\cos \theta - \cos \theta_0}}$$

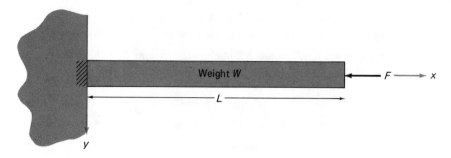

Figure 5.20

4. Use the identity $\cos u = 1 - 2 \sin^2 u/2$ to show that $T = 2 \sqrt{\dfrac{l}{g}} \displaystyle\int_0^{\theta_0} \dfrac{d\theta}{\sqrt{\sin^2 \dfrac{\theta_0}{2} - \sin^2 \dfrac{\theta}{2}}}$.

Then by letting $\sin \dfrac{\theta}{2} = \sin \dfrac{\theta_0}{2} \sin \phi$, where ϕ is a new variable, show that

$$T = 4 \sqrt{\dfrac{l}{g}} \int_0^{\pi/2} \dfrac{d\phi}{\sqrt{1 - k^2 \sin^2 \phi}}, \qquad k = \sin \dfrac{\theta_0}{2}$$

This integral is called an *elliptic integral of the first kind.*

5. (a) By using the result $\dfrac{1}{\sqrt{1-x}} = 1 + \dfrac{1}{2}x + \dfrac{1 \cdot 3}{2 \cdot 4} x^2 + \dfrac{1 \cdot 3 \cdot 5}{2 \cdot 4 \cdot 6} x^3 + \cdots, \qquad |x| < 1$

where $x = k^2 \sin^2 \phi$, show that

$$T = 2\pi \sqrt{\dfrac{l}{g}} \left[1 + \left(\dfrac{1}{2}\right)^2 k^2 + \left(\dfrac{1 \cdot 3}{2 \cdot 4}\right)^2 k^4 + \left(\dfrac{1 \cdot 3 \cdot 5}{2 \cdot 4 \cdot 6}\right)^2 k^6 + \cdots \right]$$

(b) Find the period of a pendulum when the rod makes a maximum angle of 30°, 60° with the vertical.

6. If the sphere of Exercise 1 is pushed down until the diametral plane is a distance $p(0 < p \le R)$ below the surface and is then released, show that the period T of the resulting vibrations is

$$T = 8R \sqrt{\dfrac{R}{g(6R^2 - p^2)}} \int_0^{\pi/2} \dfrac{d\phi}{\sqrt{1 - k^2 \sin^2 \phi}}, \qquad k^2 = \dfrac{p^2}{6R^2 - p^2}$$

Use the result of Exercise 5 to calculate T in the case where $p = R$. Determine T when p is near zero. Compare with Exercise 1. Does the result hold when $p = 0$?

7. A constant horizontal force F is applied to the free end of a uniform cantilever beam of length L and weight W as indicated in Fig. 5.20.

(a) Show that the deflection y is given by the equation $EIy'' = -Fy - \dfrac{W(L - x)^2}{2L}$.

(b) Find the maximum deflection of the beam.

six

solution of linear differential equations by Laplace transforms

1 Introduction to the Method of Laplace Transforms

1.1 MOTIVATION FOR LAPLACE TRANSFORMS

In preceding chapters the student learned how to solve linear differential equations with constant coefficients subject to given conditions called boundary or initial conditions. It will be recalled that the method consists of finding the general solution of the equations in terms of a number of arbitrary constants and then determining these constants from the given conditions.

During the 19th century it became fashionable for scientists and engineers, led and encouraged by the electrical engineer *Heaviside*, to use operator methods such as described on pages 204–211 to solve various problems involving differential equations. In these methods operators were treated as algebraic symbols and the resulting equations were manipulated according to rules of algebra (see, for instance, Illustrative Example 13, page 207). Remarkably, the methods led to correct answers. These successes encouraged scientists and engineers to use the methods even more, and incited rhetoric on the part of some mathematicians who did not enjoy seeing such blind mathematical manipulations rewarded by success. Comments degrading the unrigorous procedures were answered by remarks such as "Must one understand the processes of digestion in order to eat?" and "This series is divergent; hence it must have some practical use."

Some thoughtful mathematicians, seeing that the algebraic manipulations did lead to correct results, reasoned that there ought to be some way of placing the procedures on a rigorous mathematical foundation. Research toward this goal led to the powerful *method of Laplace transforms*, which we examine in this chapter. This method has various advantages over other methods. First, by using the method we can, as in Heaviside's approach, transform given differential equations into algebraic equations. Second, any given initial conditions are automatically incorporated into the algebraic problem so that no special consideration of them need be made. Finally, use of tables of Laplace transforms can reduce the labor of obtaining solutions just as tables of integrals reduce the labor of integration.

Laplace transforms have many other applications besides solving differential equations, such as evaluating integrals and solving integral equations, and we shall examine some of them later in the chapter.* With a view toward such applications, we now turn our attention to the definition and examples of the Laplace transform.

1.2 DEFINITION AND EXAMPLES OF THE LAPLACE TRANSFORM

Let $F(t)$, $t > 0$ be given. The *Laplace transform* of $F(t)$ is defined as[†]

$$f(s) = \mathscr{L}\{F(t)\} = \int_0^\infty e^{-st} F(t) dt \tag{1}$$

---

* Laplace transforms are also of theoretical interest in themselves. See [30] for example.

[†] In general if functions are denoted by capital letters such as F, G their Laplace transforms are denoted by corresponding lowercase letters f, g. Alternatively an overbar may be used to denote the Laplace transform. For example $\mathscr{L}\{f(t)\} = \bar{f}(s)$.

where s is a real parameter.* The symbol $\mathcal{L}$ is called the *Laplace transform operator.*

The improper integral in (1) is defined as

$$\lim_{M \to \infty} \int_0^M e^{-st} F(t)dt \tag{2}$$

and the Laplace transform is said to exist or not according as this limit exists or not. If (2) exists we say that the integral (1) converges. Conditions under which the Laplace transform exists are discussed on page 262.

Using definition (1) we can find the Laplace transform of various functions as indicated in the table on the inside back cover.[†]

ILLUSTRATIVE EXAMPLE 1

Find (a) $\mathcal{L}\{1\}$, (b) $\mathcal{L}\{e^{at}\}$.

Solution to (a) $\quad \mathcal{L}\{1\} = \int_0^\infty e^{-st}(1)dt = \lim_{M \to \infty} \int_0^M e^{-st}\, dt = \lim_{M \to \infty} \frac{e^{-st}}{-s} \Big|_0^M$

$$= \lim_{M \to \infty} \frac{1 - e^{-sM}}{s} = \frac{1}{s}, \qquad \text{if } s > 0$$

Solution to (b) $\quad \mathcal{L}\{e^{at}\} = \int_0^\infty e^{-st}(e^{at})dt = \lim_{M \to \infty} \int_0^M e^{-(s-a)t}\, dt$

$$= \lim_{M \to \infty} \frac{e^{-(s-a)t}}{-(s-a)} \Big|_0^M = \lim_{M \to \infty} \frac{1 - e^{-(s-a)M}}{s-a} = \frac{1}{s-a}, \qquad \text{if } s > a$$

These correspond to entries 1 and 5 of the table.

Note that from Illustrative Example 1, the existence of the Laplace transform of a function $F(t)$ depends on the values of s. Thus the Laplace transform of 1 exists if $s > 0$ but does not exist if $s \leq 0$. Similarly the Laplace transform of e^{at} exists if $s > a$ but does not exist if $s \leq a$. A similar situation arises in considering any Laplace transform.[‡] It is not difficult to show that if the Laplace transform of a function exists for $s = \alpha$ then it will also exist for all $s > \alpha$. There are functions whose Laplace transforms do not exist for any values of s. Thus, for example, since the integral

$$\int_0^\infty e^{-st} e^{t^2}\, dt$$

does not converge for any value of s, the Laplace transform of e^{t^2} does not exist.

---

* In the advanced theory of Laplace transforms it is convenient to assume that s is a complex variable so that $f(s)$ is a function of a complex variable. The letter p is sometimes used in place of s, especially by some engineers and physicists. In another definition sometimes used, the integral in (1) is multiplied by s (or p).

† Note that the Laplace transform of zero is zero, as is clear from the definition, and has not been included in the table.

‡ When writing Laplace transforms it may not always be convenient to indicate the actual range of values for which the Laplace transform exists, and we shall often omit it. One should of course be able to produce it on demand.

Find (a) $\mathscr{L}\{\sin \omega t\}$, (b) $\mathscr{L}\{\cos \omega t\}$.

Solutions Although these can be obtained directly from the definition (see B Exercise 1) we shall resort to the following device. Replace a by $i\omega$ in part (b) of Illustrative Example 1. Then using Euler's formula $e^{i\omega t} = \cos \omega t + i \sin \omega t$ we have

$$\mathscr{L}\{e^{i\omega t}\} = \int_0^\infty e^{-st}e^{i\omega t}\,dt = \int_0^\infty e^{-st}\cos \omega t\,dt + i\int_0^\infty e^{-st}\sin \omega t\,dt$$

$$= \mathscr{L}\{\cos \omega t\} + i\mathscr{L}\{\sin \omega t\} = \frac{1}{s - i\omega} = \frac{1}{s - i\omega}\cdot\frac{s + i\omega}{s + i\omega} = \frac{s + i\omega}{s^2 + \omega^2}$$

$$= \frac{s}{s^2 + \omega^2} + i\frac{\omega}{s^2 + \omega^2}$$

Equating real and imaginary parts we obtain

$$\mathscr{L}\{\cos \omega t\} = \frac{s}{s^2 + \omega^2}, \qquad \mathscr{L}\{\sin \omega t\} = \frac{\omega}{s^2 + \omega^2}$$

which correspond to entries 6 and 7, respectively, in the table.*

Since we have already hinted that Laplace transforms are useful in solving differential equations it should come as no surprise that we would be interested in finding Laplace transforms of derivatives. We can accomplish this directly from the definition. Thus we have

$$\mathscr{L}\{Y'(t)\} = \int_0^\infty e^{-st}Y'(t)dt = \lim_{M\to\infty}\int_0^M e^{-st}Y'(t)dt$$

$$= \lim_{M\to\infty}\left\{e^{-st}Y(t)\Big|_0^M + s\int_0^M e^{-st}Y(t)dt\right\}$$

$$= \lim_{M\to\infty}\left\{e^{-sM}Y(M) - Y(0) + s\int_0^M e^{-st}Y(t)dt\right\}$$

$$= s\int_0^\infty e^{-st}Y(t)dt - Y(0)$$

$$= sy(s) - Y(0)$$

where we have assumed that $\mathscr{L}\{Y(t)\} = y(s) = y$ and $\lim_{M\to\infty} e^{-sM}Y(M) = 0.$[†]

---

* It should be noted that the method is based on the equality

$$\int_0^\infty e^{-st}e^{i\omega t}\,dt = \frac{1}{s - i\omega}, \qquad s > 0$$

This can be justified using complex variable methods.

[†] It is also assumed that $Y(t)$ is continuous at $t = 0$. For the case where this is not so, see C Exercise 3.

To find Laplace transforms of higher-order derivatives we can use the definition and integration by parts. However it is easier to employ the result which we have just obtained. To do this let $G(t) = Y'(t)$. Then

$$\mathcal{L}\{Y''(t)\} = \mathcal{L}\{G'(t)\} = s\mathcal{L}\{G(t)\} - G(0)$$
$$= s\mathcal{L}\{Y'(t)\} - Y'(0) = s[sy(s) - Y(0)] - Y'(0)$$
$$= s^2 y(s) - sY(0) - Y'(0)$$

The results correspond to entries 15 and 16 of the table. Generalizations to higher derivatives can similarly be obtained and are indicated by entry 17 of the table.

Remark. Note that the derivatives of Y have Laplace transforms which are algebraic functions of s and contain the initial values of Y and its derivatives. It was this observation which provided mathematicians with an important clue as to the link between Laplace transforms and operational methods described on page 256, and the apparent connection between the operator D and the algebraic symbol s. It also served to show the advantages described on page 256. We shall illustrate the use of Laplace transform methods in differential equations on page 273 after we have examined some more properties of the transform.

It is of interest and importance to note that the Laplace transform operator $\mathcal{L}$ is a *linear operator* just as the operators $D, D^2, \ldots$, of Chapter Four, page 168. To prove this we need only show that

$$\mathcal{L}\{F(t) + G(t)\} = \mathcal{L}\{F(t)\} + \mathcal{L}\{G(t)\} = f(s) + g(s) \tag{3}$$

$$\mathcal{L}\{cF(t)\} = c\mathcal{L}\{F(t)\} = cf(s) \tag{4}$$

where $F(t)$ and $G(t)$ have Laplace transforms $f(s)$ and $g(s)$, respectively, and c is any constant. The proof follows directly from the properties of integrals. Since

$$\mathcal{L}\{F(t) + G(t)\} = \int_0^\infty e^{-st}\{F(t) + G(t)\}dt = \int_0^\infty e^{-st}F(t)dt + \int_0^\infty e^{-st}G(t)dt$$
$$= \mathcal{L}\{F(t)\} + \mathcal{L}\{G(t)\} = f(s) + g(s)$$

$$\mathcal{L}\{cF(t)\} = \int_0^\infty e^{-st}\{cF(t)\}dt = c\int_0^\infty e^{-st}F(t)dt = c\mathcal{L}\{F(t)\} = cf(s)$$

This *linear property* enables us to find Laplace transforms of sums.

ILLUSTRATIVE EXAMPLE 3

Find $\mathcal{L}\{3 - 5e^{2t} + 4\sin t - 7\cos 3t\}$.

Solution Using the linear property we have

$$\mathcal{L}\{3 - 5e^{2t} + 4\sin t - 7\cos 3t\}$$
$$= \mathcal{L}\{3\} + \mathcal{L}\{-5e^{2t}\} + \mathcal{L}\{4\sin t\} + \mathcal{L}\{-7\cos 3t\}$$
$$= 3\mathcal{L}\{1\} - 5\mathcal{L}\{e^{2t}\} + 4\mathcal{L}\{\sin t\} - 7\mathcal{L}\{\cos 3t\}$$
$$= \frac{3}{s} - \frac{5}{s-2} + \frac{4}{s^2+1} - \frac{7s}{s^2+9}, \qquad \text{if } s > 2$$

The results involving Laplace transforms of derivatives are often useful in finding Laplace transforms without direct use of the definition. Consider

ILLUSTRATIVE EXAMPLE 4

Find $\mathcal{L}\{t^n\}$, $n = 1, 2, 3, \ldots$.

Solution Let $Y(t) = t^n$ so that $Y'(t) = nt^{n-1}$, $Y(0) = 0$. Then we have

$$\mathcal{L}\{Y'(t)\} = s\mathcal{L}\{Y(t)\} - Y(0) \quad \text{or} \quad \mathcal{L}\{nt^{n-1}\} = s\mathcal{L}\{t^n\}$$

Thus

$$\mathcal{L}\{t^n\} = \frac{n}{s}\mathcal{L}\{t^{n-1}\}$$

Putting $n = 1, 2, \ldots$, we find for $s > 0$

$$\mathcal{L}\{t\} = \frac{1}{s}\mathcal{L}\{1\} = \frac{1}{s^2}, \quad \mathcal{L}\{t^2\} = \frac{2}{s}\mathcal{L}\{t\} = \frac{2}{s^3}, \quad \mathcal{L}\{t^3\} = \frac{3}{s}\mathcal{L}\{t^2\} = \frac{3 \cdot 2}{s^4} = \frac{3!}{s^4}$$

and in general

$$\mathcal{L}\{t^n\} = \frac{n(n-1)\cdots 1}{s^{n+1}} = \frac{n!}{s^{n+1}} \tag{5}$$

1.3 FURTHER PROPERTIES OF LAPLACE TRANSFORMS

In constructing tables of Laplace transforms certain properties prove to be useful. To develop one such property let us write the definition

$$f(s) = \mathcal{L}\{F(t)\} = \int_0^\infty e^{-st}F(t)dt \tag{6}$$

and formally replace s by $s - a$. Then we find

$$f(s - a) = \int_0^\infty e^{-(s-a)t}F(t)dt = \int_0^\infty e^{-st}\{e^{at}F(t)\}dt$$

and so

$$\mathcal{L}\{e^{at}F(t)\} = f(s - a) \tag{7}$$

Another important property arises by differentiating both sides of (6) with respect to s. We find

$$\frac{df}{ds} = f'(s) = \frac{d}{ds}\int_0^\infty e^{-st}F(t)dt = \int_0^\infty - te^{-st}F(t)dt = -\mathcal{L}\{tF(t)\}$$

assuming the differentiation under the integral sign justifiable.* Thus it follows that

$$\mathcal{L}\{tF(t)\} = -f'(s) \tag{8}$$

By further differentiation we have

$$\mathcal{L}\{t^2 F(t)\} = f''(s), \qquad \mathcal{L}\{t^3 F(t)\} = -f'''(s) \tag{9}$$

* If a and b are constants, the result
$$\frac{d}{ds}\int_a^b G(s, t)dt = \int_a^b \frac{\partial G}{\partial s}dt$$

is often called *Leibniz's rule* for differentiating an integral. For conditions under which the result holds, see any book on advanced calculus (e.g., [26] of the Bibliography).

or in general $$\mathcal{L}\{t^n F(t)\} = (-1)^n f^{(n)}(s) = (-1)^n \frac{d^n f}{ds^n} \qquad (10)$$

The above results are summarized in the following

Theorem. If $\mathcal{L}\{F(t)\} = f(s)$ then
1. $\mathcal{L}\{e^{at} F(t)\} = f(s - a)$.
2. $\mathcal{L}\{t^n F(t)\} = (-1)^n f^{(n)}(s), n = 1, 2, 3, \ldots$.

To illustrate these results let us consider some examples.

Find $\mathcal{L}\{e^{3t} \cos 4t\}$.

Solution Since $\mathcal{L}\{\cos 4t\} = \dfrac{s}{s^2 + 16}$ we have

$$\mathcal{L}\{e^{3t} \cos 4t\} = \frac{s - 3}{(s - 3)^2 + 16} = \frac{s - 3}{s^2 - 6s + 25}$$

Note that this is valid for $s > 3$. The result is a special case of entry 11 in the table.

Find (a) $\mathcal{L}\{t \sin t\}$, (b) $\mathcal{L}\{t^2 \sin t\}$.

Solution Since $\mathcal{L}\{\sin t\} = \dfrac{1}{s^2 + 1}$ we have

$$\mathcal{L}\{t \sin t\} = -\frac{d}{ds}\left(\frac{1}{s^2 + 1}\right) = \frac{2s}{(s^2 + 1)^2},$$

$$\mathcal{L}\{t^2 \sin t\} = (-1)^2 \frac{d^2}{ds^2}\left(\frac{1}{s^2 + 1}\right) = \frac{6s^2 - 2}{(s^2 + 1)^3}$$

which are valid for $s > 0$. Compare with entry 13 of the table.

1.4 THE GAMMA FUNCTION

We have already found (in Illustrative Example 4) that

$$\mathcal{L}\{t^n\} = \frac{n!}{s^{n+1}}, \qquad s > 0, n = 1, 2, 3, \ldots \qquad (11)$$

A natural question which arises is, How must (11) be modified if n is not a positive integer? To answer this let us first note that

$$\mathcal{L}\{t^n\} = \int_0^\infty e^{-st} t^n \, dt$$

Making the substitution $u = st, s > 0$, we find

$$\mathcal{L}\{t^n\} = \frac{1}{s^{n+1}} \int_0^\infty u^n e^{-u} \, du \qquad (12)$$

If we now use the notation $\quad \Gamma(n + 1) = \int_0^\infty u^n e^{-u} \, du$ (13)

then (12) becomes $\qquad\qquad \mathcal{L}\{t^n\} = \dfrac{\Gamma(n + 1)}{s^{n+1}}$ (14)

We call $\Gamma(n + 1)$ the *gamma function*. Let us consider this function a little more closely. On integrating by parts we find

$$\Gamma(n + 1) = \int_0^\infty u^n e^{-u} \, du = (u^n)(-e^{-u})\Big|_0^\infty - \int_0^\infty (nu^{n-1})(-e^{-u}) \, du$$

$$= n \int_0^\infty u^{n-1} e^{-u} \, du = n\Gamma(n)$$

The relation $\qquad\qquad \maltese \; \Gamma(n + 1) = n\Gamma(n)$ (15)

is called a *recurrence formula* for the gamma function. Since

$$\Gamma(1) = \int_0^\infty e^{-u} \, du = -e^{-u}\Big|_0^\infty = 1$$ (16)

we have $\quad \Gamma(2) = 1\Gamma(1) = 1, \; \Gamma(3) = 2\Gamma(2) = 2 \cdot 1 = 2!, \; \Gamma(4) = 3\Gamma(3) = 3!$

and in general when n is a positive integer

$$\Gamma(n + 1) = n!$$ (17)

Thus (14) agrees with (11) for this case.

It follows that the gamma function is actually a generalization of the factorial. One interesting result is that

$$\Gamma(\tfrac{1}{2}) = \sqrt{\pi}$$ (18)

To indicate a proof of this let us first note that on letting $u = x^2$

$$I = \Gamma(\tfrac{1}{2}) = \int_0^\infty u^{-1/2} e^{-u} \, du = 2 \int_0^\infty e^{-x^2} \, dx$$

Then $\qquad I^2 = \left(2 \int_0^\infty e^{-x^2} \, dx\right)\left(2 \int_0^\infty e^{-y^2} \, dy\right) = 4 \int_0^\infty \int_0^\infty e^{-(x^2+y^2)} \, dx \, dy$

By changing to polar coordinates (r, ϕ) this last integral can be transformed into

$$I^2 = 4 \int_{\phi=0}^{\pi/2} \int_{r=0}^\infty e^{-r^2} r \, dr \, d\phi = 4 \int_{\phi=0}^{\pi/2} \left(-\frac{1}{2} e^{-r^2}\right)\Big|_0^\infty d\phi = 4 \int_0^{\pi/2} \frac{1}{2} \, d\phi = \pi$$

from which $I = \Gamma(\tfrac{1}{2}) = \sqrt{\pi}$. Although this is a somewhat "hand-waving" approach, the method can be made mathematically rigorous by appropriate limiting procedures.

It is of interest to note that

$$\mathcal{L}\{t^{-1/2}\} = \frac{\Gamma(\tfrac{1}{2})}{s^{1/2}} = \sqrt{\frac{\pi}{s}}, \qquad s > 0$$ (19)

1.5 REMARKS CONCERNING EXISTENCE OF LAPLACE TRANSFORMS

In the definition (1) of the Laplace transform, the factor e^{-st} is a "damping factor" which for any fixed positive value of s tends to decrease as t increases. Intuitively speaking, we would expect the integral to converge, and thus the Laplace transform to exist, provided that $F(t)$ does not "grow too rapidly" as t increases. The mathematician makes this more precise by defining a class of functions which are such that there exist constants K and α for which

$$|F(t)| < Ke^{\alpha t}$$

Such functions are said to be of *exponential order* α, or briefly of *exponential order*. The function $F(t) = t$ is certainly of exponential order since we have (for example)

$$t < e^t$$

The function e^{t^2} on the other hand is not of exponential order, i.e., it grows too rapidly to be of this type.

Another class of functions which the mathematician finds important is the class of *piecewise* or *sectionally continuous functions*. We call a function piecewise continuous in an interval if it has only a finite number of discontinuities in the interval and if the right- and left-hand limits at each discontinuity exist.* For example,

$$F(t) = \begin{cases} 1, & 0 \leq t < 2 \\ t, & 2 \leq t \leq 4 \end{cases}$$

whose graph is shown in Fig. 6.1 is piecewise continuous in the interval $0 \leq t \leq 4$ since there is only one discontinuity, $t = 2$, in the interval and the right- and left-hand limits at this discontinuity exist (and are equal to 2 and 1, respectively).

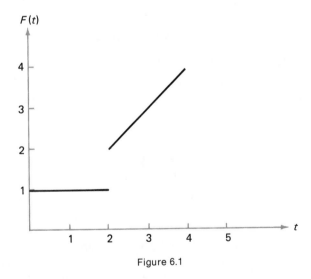

Figure 6.1

* The right-hand limit of a function $F(t)$ at the point t_0 is defined as $\lim_{\epsilon \to 0} F(t_0 + \epsilon)$ where ϵ approaches zero through positive values. Similarly, the left-hand limit of $F(t)$ at t_0 is $\lim_{\epsilon \to 0} F(t_0 - \epsilon)$ where ϵ approaches zero through positive values. To indicate that ϵ approaches zero through positive values we sometimes write $\lim_{\epsilon \to 0+} F(t_0 + \epsilon)$ and $\lim_{\epsilon \to 0+} F(t_0 - \epsilon)$ for the right- and left-hand limits, respectively, and we denote these limits, if they exist, by $F(t_0 + 0)$ and $F(t_0 - 0)$ respectively.

The following theorem is of fundamental importance. For a proof see C Exercise 5.

Theorem. If $F(t)$ is of exponential order α and is piecewise continuous in every finite interval $0 \leq t < T$ then the Laplace transform of $F(t)$ exists for all $s > \alpha$.

It should be emphasized that the hypotheses of this theorem guarantee the existence of the Laplace transform. However if these conditions are not satisfied it does not not follow that the Laplace transform does not exist. In fact it may or may not exist (see B Exercises 3 and 6). In situations such as these, the conditions are said to be *sufficient* but not *necessary* for the validity of the conclusions.

Let us illustrate the above theorem in the following examples.

ILLUSTRATIVE EXAMPLE 7

Find the Laplace transform of $F(t) = \begin{cases} 3, & 0 \leq t < 2 \\ 0, & t > 2 \end{cases}$.

Solution The function is piecewise continuous and of exponential order and so by the above theorem has a Laplace transform. This is given by

$$\mathcal{L}\{F(t)\} = \int_0^\infty e^{-st}F(t)dt = \int_0^2 e^{-st}(3)dt + \int_2^\infty e^{-st}(0)dt$$

$$= 3\left(\frac{e^{-st}}{-s}\right)\Big|_0^2 = 3\left(\frac{1 - e^{-2s}}{s}\right)$$

ILLUSTRATIVE EXAMPLE 8

Find the Laplace transform of $e^{\sqrt{t}}$.

Solution The function is piecewise continuous in any finite interval. We can also show that it is of exponential order. To do this we note that for $t > 1$, $\sqrt{t} < t$, and so

$$e^{\sqrt{t}} < e^t$$

from which we see that $F(t) = e^{\sqrt{t}}$ is of exponential order. It thus follows that $\mathcal{L}\{e^{\sqrt{t}}\}$ exists. However, even though this Laplace transform exists, we cannot determine it in closed form because the integral

$$\int_0^\infty e^{-st}e^{\sqrt{t}}\, dt = 2\int_0^\infty ue^{u - su^2}\, du \tag{20}$$

where $t = u^2$ cannot be evaluated exactly. It can however be evaluated approximately for any value of $s > 0$ by using numerical integration or the method of C Exercise 7(c).

This example serves to illustrate the fact, emphasized often in previous chapters, that there is a big difference between proving that something exists and finding it.

1.6 THE HEAVISIDE UNIT STEP FUNCTION

The function defined by $H(t - a) = \begin{cases} 1, & t > a \\ 0, & t < a \end{cases}$ $\tag{21}$

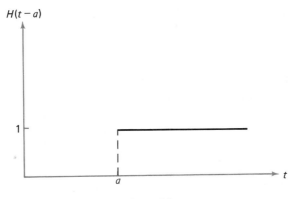

Figure 6.2

called the *Heaviside unit step function,* or more briefly the *Heaviside function* or the *unit step function,* is often useful in applications. The graph is shown in Fig. 6.2. The Laplace transform of this function is given by

$$\mathscr{L}\{H(t - a)\} = \int_0^\infty e^{-st}H(t - a)dt = \int_0^a e^{-st}(0)dt + \int_a^\infty e^{-st}(1)dt$$

$$= \int_a^\infty e^{-st} dt = \frac{e^{-as}}{s}$$

assuming that $s > 0$.

Various discontinuous functions can often be expressed in terms of the Heaviside function as indicated in the following

ILLUSTRATIVE EXAMPLE 9

Express the function $F(t) = \begin{cases} \sin t, & t > \pi \\ \cos t, & t < \pi \end{cases}$ in terms of the Heaviside unit

step function.

Solution The given function can be expressed as

$$F(t) = \cos t + \begin{cases} \sin t - \cos t, & t > \pi \\ 0, & t < \pi \end{cases} = \cos t + (\sin t - \cos t) \begin{cases} 1, & t > \pi \\ 0, & t < \pi \end{cases}$$

$$= \cos t + (\sin t - \cos t)H(t - \pi)$$

A EXERCISES

1. Using the definition, find the Laplace transform of each of the following functions. In each case specify the values of s for which the transform exists. Compare with results obtained from the table of Laplace transforms.
   (a) $3t - 2$.
   (b) $4 \sin 3t$.
   (c) $5 \cos 2t$.
   (d) $10e^{-5t}$.
   (e) $2e^t - 3e^{-t} + 4t^2$.
   (f) $3 \sin 5t - 4 \cos 5t$.
   (g) $6 \cosh 3t - 2 \sinh 5t$.
   (h) $t(e^{-3t} - t^2 + 1)$.

2. Given that $\mathcal{L}\{\sin \omega t\} = \dfrac{\omega}{s^2 + \omega^2}$ use entry 15 of the table to find $\mathcal{L}\{\cos \omega t\}$.

3. Find the Laplace transform of each of the following.
(a) $t^2 e^{3t}$.
(b) $e^{-2t}(5 \sin 2t - 2 \cos 2t)$.
(c) $t(\sin t + e^{-t})$.
(d) $(t^2 + 1)^2$.
(e) $t(\cosh 2t - 2t)$.
(f) $8 \sinh^2 3t$.

4. Use the gamma function to find (a) $\mathcal{L}\{t^{3/2}\}$, (b) $\mathcal{L}\{(t^{1/4} + t^{-1/4})^2\}$, (c) $\mathcal{L}\{t^{2/3}\}$, (d) $\mathcal{L}\{\sqrt{t}e^t\}$.

5. (a) Explain how you can be sure that the function $F(t) = \begin{cases} t, & 0 < t < 4 \\ 0, & t > 4 \end{cases}$ has a Laplace transform without actually finding it. (b) Find $\mathcal{L}\{F(t)\}$.

6. Verify entries 15 and 16 in the table for the functions (a) $Y(t) = te^t$, (b) $Y(t) = t^2 \sin 3t$.

7. Express each of the following in terms of Heaviside's function and obtain their graphs.

(a) $F(t) = \begin{cases} 2, & t > 1 \\ 1, & t < 1 \end{cases}$.
(b) $F(t) = \begin{cases} t^2, & t > 3 \\ 2t, & t < 3 \end{cases}$.
(c) $F(t) = \begin{cases} 0, & t > 2\pi \\ \cos t, & t < 2\pi \end{cases}$.

8. Find (a) $\mathcal{L}\{tH(t - 1)\}$. (b) $\mathcal{L}\{e^t H(t - 2) - e^{-t} H(t - 3)\}$.

B EXERCISES

1. Obtain $\mathcal{L}\{\sin \omega t\}$ by (a) direct evaluation; (b) using the fact that $\sin \omega t$ satisfies the differential equation $Y'' + \omega^2 Y = 0$. Do the same for $\mathcal{L}\{\cos \omega t\}$.

2. Find $\mathcal{L}\{te^{-t} \sin t\}$.

3. Prove that $\mathcal{L}\{e^{3t}\}$ exists but $\mathcal{L}\{e^{e^t}\}$ does not exist.

4. If $\mathcal{L}\{F(t)\}$ exists for $s = \alpha$, prove that it also exists for all $s > \alpha$.

5. Find the Laplace transform of the periodic function shown in Fig. 6.3.

6. Show that although the function $F(t) = t^{-1/2}$ does not satisfy the conditions of the theorem on page 264 it still has a Laplace transform. Is there any contradiction involved? Explain.

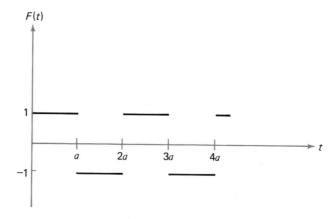

Figure 6.3

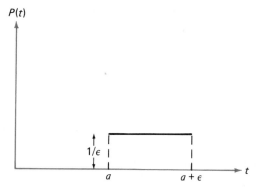

$P(t)$

$1/\epsilon$

a $\qquad$ $a + \epsilon$ $\qquad$ t

Figure 6.4

7. Express in terms of Heaviside's unit step function

$$F(t) = \begin{cases} 3 \sin t, & t \leqq \pi \\ t^2, & \pi < t \leqq 2\pi \\ t - \cos t, & t > 2\pi \end{cases}$$

8. The function

$$P(t) = \begin{cases} 0, & t < a \\ 1/\epsilon, & a \leqq t \leqq a + \epsilon \\ 0, & t > a + \epsilon \end{cases}$$

where $a \geqq 0$ and $\epsilon > 0$, whose graph is indicated in Fig. 6.4, is often called a *pulse function*.
(a) Express this function in terms of the Heaviside unit step function.
(b) Find the Laplace transform of this function.
(c) Show that the limit of the Laplace transform in (b) as $\epsilon \to 0$ exists and is equal to e^{-as}.

C EXERCISES

1. Show that $F(t) = 1$, $t > 0$ and $G(t) = \begin{cases} 5, & t = 3 \\ 1, & t \neq 3 \end{cases}$ have the same Laplace transforms, namely $1/s$, $s > 0$. Can you think of other functions with the same Laplace transform? Explain.

2. Let $F(t) = \begin{cases} 1, & t = 0 \\ t, & t > 0 \end{cases}$. (a) Find $\mathscr{L}\{F(t)\}$ and $\mathscr{L}\{F'(t)\}$. (b) Is it true for this case that $\mathscr{L}\{F'(t)\} = s\mathscr{L}\{F(t)\} - F(0)$?

3. Prove that if $Y(t)$ has a discontinuity at $t = 0$ then we must replace entry 15 of the table by $sy - Y(0+)$, where $Y(0+)$ means $\lim_{\epsilon \to 0+} Y(t)$, i.e., the limit as $\epsilon \to 0$ through positive values. Using this, explain the discrepancy in Exercise 2.

4. Let $F(t)$ be a periodic function of period P, starting at $t = 0$ (for example, see Fig. 6.3). (a) If $\mathscr{L}\{F(t)\}$ exists show that it is given by

$$\frac{\int_0^P e^{-st} F(t) dt}{1 - e^{-sP}}$$

(b) Use (a) to obtain the Laplace transform of $F(t) = |\sin t|$ where the period is π.

5. Prove the theorem on page 264. *Hint:* Write

$$\int_0^\infty e^{-st}F(t)\,dt = \int_0^T e^{-st}F(t)\,dt + \int_T^\infty e^{-st}F(t)\,dt$$

and then use the fact that

$$\left|\int_a^b e^{-st}F(t)\,dt\right| \le \int_a^b e^{-st}|F(t)|\,dt$$

6. (a) Prove that if $F(t)$ is piecewise continuous in every finite interval and of exponential order then its Laplace transform $f(s)$ approaches zero as $s \to \infty$. (b) Illustrate the result in (a) by giving several examples. (c) What would you conclude about $F(t)$ if $\lim_{s\to\infty} f(s) \ne 0$? (*Hint:* Use Exercise 5.)

7. Let $F(t) = e^t$. By using the series expansion $\quad e^t = 1 + t + \dfrac{t^2}{2!} + \dfrac{t^3}{3!} + \cdots$

and taking the Laplace transform term by term, verify formally that $\mathcal{L}\{e^t\} = 1/(s - 1), s > 1$. How can you use this method to find (a) $\mathcal{L}\{\sin t\}$; (b) $\mathcal{L}\{\cos t\}$; (c) $\mathcal{L}\{e^{\sqrt{t}}\}$? Can you justify the method?

8. Show that if $a > 0$ and $n > 1$ then (a) $\displaystyle\int_0^t H(u - a)\,du = (t - a)H(t - a),$

$$\text{(b)} \quad \int_0^t (u - a)^n H(u - a)\,du = \frac{(t - a)^{n+1}H(t - a)}{n + 1}.$$

9. Show that the function illustrated graphically in Fig. 6.3 can be represented by

$$H(t) + \sum_{n=1}^{\infty} (-1)^n H(t - na)$$

2   Impulse Functions and the Dirac Delta Function

Suppose that a force $F(t)$ depending on time t acts on a particle of mass m from time t_0 to time t_1. If v is the instantaneous velocity of the particle during this time interval, then by Newton's law we have

$$F = \frac{d}{dt}(mv) \tag{1}$$

so that

$$\int_{t_0}^{t_1} F\,dt = \int_{t_0}^{t_1} d(mv) = mv_1 - mv_0 \tag{2}$$

where v_0 and v_1 are the velocities of the particle at times t_0 and t_1, respectively.

Since mv is the momentum of the particle, the quantity on the right of (2) is the change in momentum over the interval. The integral on the left of (2) is often called the *impulse of the force* over the interval, or briefly the *impulse*. Thus, in words (2) states that   impulse $=$ change in momentum   (3)

Now suppose that the difference between t_1 and t_0 is given by $\epsilon > 0$, so that $t_1 = t_0 + \epsilon$. If we assume that the force is a constant, say A, in the interval, or if we assume that ϵ is taken small enough so that the force is approximately equal to A in the

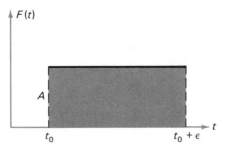

Figure 6.5

interval, then the force can be represented graphically as in Fig. 6.5. The impulse denoted by I is then given by

$$I = \int_{t_0}^{t_0+\epsilon} A\, dt = A\epsilon \tag{4}$$

and is represented geometrically by the area of the rectangle shown shaded in Fig. 6.5.

Suppose now that the impulse is a non-zero constant, say 1, so that as seen from (2) there is a unit change in momentum over the interval, i.e., $A\epsilon = 1$. Then as $\epsilon \to 0$ we have $A \to \infty$, so that geometrically the width of the rectangle in Fig. 6.5 gets smaller, while the height gets larger in such a way that the area stays equal to 1.

If the student prefers not to assume that $F(t)$ is as given in Fig. 6.5, but instead varies in some way as indicated in Fig. 6.6, then again assuming $I = 1$ we have

$$\int_{t_0}^{t_0+\epsilon} F(t)dt = 1 \tag{5}$$

In such case, even though the form of the function is not known, it is not difficult to see that if (5) is to be true then, as $\epsilon \to 0$, $F(t)$ must become infinite. The geometric interpretation is that as $\epsilon \to 0$ the peak of the curve of Fig. 6.6 must become infinite in such a way that the total area under it is equal to 1, agreeing essentially with the discussion given in connection with Fig. 6.5.

From these ideas we are led naturally to the mathematical question of attempting to describe functions, such as represented by Figs. 6.5 or 6.6, which become infinite as $\epsilon \to 0$ in such a way that their integrals (geometrically their associated areas and physically the impulse) remain constant. Such functions for obvious

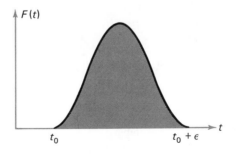

Figure 6.6

reasons are called *impulse functions* or, in the special case where the constant is equal to 1, *unit impulse functions*. Since we can consider $F(t)$ to be zero outside the interval from t_0 to $t_0 + \epsilon$, we can think more generally of a unit impulse function as characterized by the following two properties.

1. $F(t) = \begin{cases} \infty, & t = t_0 \\ 0, & t \neq t_0 \end{cases}$ (6)

2. $\int_{-\infty}^{\infty} F(t)dt = 1$ (7)

Such functions are certainly not the conventional kinds with which we have dealt, and in fact, as suggested by Property 1, are not really functions at all. To distinguish them from the familiar types of functions, they have been called *generalized functions*, and mathematicians have been successful in building up a theory concerning them called the *theory of distributions.**

In his research on quantum mechanics, Dirac found impulse functions of great use.† He introduced what was called the *delta function*, now often called the *Dirac delta function*, with the above properties, i.e.,

1. $\delta(t - t_0) = \begin{cases} \infty, & t = t_0 \\ 0, & t \neq t_0 \end{cases}$ (8)

2. $\int_{-\infty}^{\infty} \delta(t - t_0)dt = 1$ (9)

One important result concerning the delta function is that, if $f(t)$ is any function which is continuous at $t = t_0$, then

$$\int_{-\infty}^{\infty} \delta(t - t_0)f(t)dt = f(t_0)$$ (10)

This is sometimes called the *sifting* property of the delta function, since all values of $f(t)$ except those for which $t = t_0$ are sifted out and only the result $f(t_0)$ remains.

The sifting property can be made plausible if we realize that the integrand on the left of (10) is zero except where $t = t_0$. Then since $f(t) = f(t_0)$ at $t = t_0$, we can replace $f(t)$ by $f(t_0)$ in (10) to arrive at

$$\int_{-\infty}^{\infty} \delta(t - t_0)f(t_0)dt = f(t_0) \int_{-\infty}^{\infty} \delta(t - t_0)dt = f(t_0)$$

using the second property of the delta function.

If we take the special case $f(t) = e^{-st}$ in (10), we obtain the Laplace transform of the delta function, i.e.,

$$\mathscr{L}\{\delta(t - t_0)\} = \int_0^{\infty} e^{-st}\delta(t - t_0)dt = e^{-st_0}$$ (11)

If in particular $t_0 = 0$, we have $\qquad\qquad \mathscr{L}\{\delta(t)\} = 1$ (12)

---

* See reference [19] for example.
† See reference [8].

The fact that the Laplace transform of $\delta(t)$ does not approach zero as $s \to \infty$ serves as a reminder that the delta function is not a conventional function (see C Exercise 6, page 268). To gain facility in using the delta function, let us consider some examples.

ILLUSTRATIVE EXAMPLE 1

Evaluate $\int_{-\infty}^{\infty} e^{-t^2} \delta\left(t - \dfrac{\pi}{2}\right) dt$.

Solution Let $t_0 = \pi/2$ and $f(t) = e^{-t^2}$. Then from the sifting property (10) and $f(t_0) = f(\pi/2) = e^{-\pi^2/4}$, we have

$$\int_{-\infty}^{\infty} e^{-t^2} \delta\left(t - \frac{\pi}{2}\right) dt = e^{-\pi^2/4}$$

ILLUSTRATIVE EXAMPLE 2

Evaluate $\int_{-\infty}^{t} \delta(u - t_0) du$.

Solution Since $\delta(u - t_0) = 0$ for $u < t_0$, it follows that the integral is zero for $t < t_0$.

In case $t > t_0$, the integral can for all practical purposes be replaced by the integral in (9), since $\delta(u - t_0) = 0$ for $u > t_0$. Thus, the integral is 1 for $t > t_0$.

Then we have $\qquad \int_{-\infty}^{t} \delta(u - t_0) du = \begin{cases} 1, & t > t_0 \\ 0, & t < t_0 \end{cases}$

It should be noted that the function on the right is the Heaviside function so that

$$\int_{-\infty}^{t} \delta(u - t_0) du = H(t - t_0) \tag{13}$$

If we formally take the derivative of both sides of (13), we obtain

$$H'(t - t_0) = \delta(t - t_0) \tag{14}$$

i.e., the delta function is the formal derivative of the Heaviside unit step function. We say formal derivative because the Heaviside function does not have a derivative in the ordinary sense. In the theory of distributions mentioned before, such generalized derivatives are considered.

The Dirac delta function is useful in many applied problems arising in engineering, physics, and other sciences. For example, if we strike an object with a hammer or beat a drum with a stick, a rather large force acts for a short interval of time. This force can be approximated by the delta function multiplied by some appropriate constant of magnitude equal to the total impulse. The delta function can also arise in electricity, such as when there are large surges in voltage or current for short intervals of time.

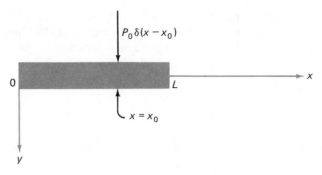

Figure 6.7

An interesting application of the delta function also arises in connection with beam problems. Suppose that the beam of Fig. 6.7 has a concentrated load P_0 acting on it at the location $x = x_0$ from the left end. If we assume that $w(x)$ is the force per unit length acting on the beam, then we will have

$$\int_{x_0}^{x_0 + \epsilon} w(x)dx = P_0 \tag{15}$$

where it is assumed that ϵ is small. The result (15) in the limit as $\epsilon \to 0$ suggests that we express $w(x)$ in terms of the delta function as

$$w(x) = P_0 \delta(x - x_0) \tag{16}$$

a result useful in finding the deflection of the beam due to a concentrated load. We shall present some applications of the delta function at the end of this chapter.

A EXERCISES

1. Evaluate each of the following integrals involving the delta function.

(a) $\int_{-\infty}^{\infty} \delta\left(t - \frac{3\pi}{2}\right) \cos 2t \, dt.$

(b) $\int_{-\infty}^{\infty} t^2 e^{-3t} \delta(t - \frac{1}{2}) dt.$

(c) $\int_0^1 t^3 \delta(t - \frac{1}{3}) dt.$

(d) $\int_0^1 t^3 \delta(t + \frac{1}{3}) dt.$

(e) $\int_0^{\infty} t^{3/2} \delta(t - \frac{9}{4}) dt.$

(f) $\int_0^{\pi} \delta\left(t - \frac{\pi}{2}\right) \sin t^2 \, dt.$

2. Evaluate each of the following Laplace transforms involving the delta function.
   (a) $\mathcal{L}\{e^t \delta(t - 2)\}.$ (b) $\mathcal{L}\{e^{-3t} \delta(t - \pi)\}.$ (c) $\mathcal{L}\{t \delta(t - 1)\}.$ (d) $\mathcal{L}\{te^{-t} \delta(t + 1)\}.$

B EXERCISES

1. (a) Find the Laplace transform of the function represented by the graph of Fig. 6.4. (b) Show that the limit of the Laplace transform found in (a) is e^{-st_0}. (c) Discuss the connection of the result found in (b) with the delta function.

2. Evaluate $\mathcal{L}\{\delta(t - \pi) \cos t^3\}.$

3. Discuss the relationship of B Exercise 8, page 267, with the delta function.

4. Show that
$$\int_{-\infty}^{\infty} \delta(at) f(t) dt = \begin{cases} \dfrac{f(0)}{a}, & \text{if } a > 0 \\[2mm] -\dfrac{f(0)}{a}, & \text{if } a < 0 \end{cases}$$

What assumptions must you make concerning $f(t)$?

5. Evaluate (a) $\displaystyle\int_{-\infty}^{\infty} \delta(2t + \pi) \sin 3t \, dt.$ **(b)** $\displaystyle\int_{-\infty}^{\infty} e^{(t-1)^2} \delta(3t) \, dt.$

C EXERCISES

1. Derive formally the following results concerning derivatives of the delta function.

(a) $\displaystyle\int_{-\infty}^{\infty} \delta'(t) f(t) dt = -f'(0).$ (b) $\displaystyle\int_{-\infty}^{\infty} \delta^{(n)}(t) f(t) dt = (-1)^n f^{(n)}(0).$

(c) $\mathscr{L}\{\delta^{(n)}(t)\} = s^n, \quad n = 1, 2, 3, \dots.$

2. Show that if $t_0 > 0$ and $f(t)$ is continuous at $t = \pm t_0$ then
$$\int_{-\infty}^{\infty} \delta(t^2 - t_0^2) f(t) dt = \frac{1}{2t_0} [f(t_0) + f(-t_0)]$$

3. Consider the sequence of functions $\phi_n(t) = n/\pi(1 + n^2 t^2), \, n = 1, 2, 3, \dots.$

(a) Show that $\displaystyle\int_{-\infty}^{\infty} \phi_n(t) dt = 1.$ (b) Assuming that $f(t)$ is continuous at $t = 0$, show that
$$\lim_{n \to \infty} \int_{-\infty}^{\infty} \phi_n(t) f(t) dt = f(0)$$

(c) Discuss a possible connection between the functions $\phi_n(t)$ and the delta function.

3 Application of Laplace Transforms to Differential Equations

3.1 SOLUTION OF SIMPLE DIFFERENTIAL EQUATIONS. INVERSE LAPLACE TRANSFORMS

In order to see how Laplace transforms can be used to find solutions to differential equations, let us consider the following

PROBLEM FOR DISCUSSION

Solve $Y'' + 4Y = 16t, \quad Y(0) = 3, \, Y'(0) = -6.$

If we take the Laplace transforms of both sides of the differential equation we find on using entry 16 in the table and denoting $\mathscr{L}\{Y\}$ by y,
$$\mathscr{L}\{Y''\} + \mathscr{L}\{4Y\} = \mathscr{L}\{16t\}, \, s^2 y - s Y(0) - Y'(0) + 4y = \frac{16}{s^2}$$

or
$$s^2 y - 3s + 6 + 4y = \frac{16}{s^2}$$

and
$$y = \frac{3s - 6}{s^2 + 4} + \frac{16}{s^2(s^2 + 4)} \tag{1}$$

It seems logical now that if we can find the function whose Laplace transform is the right side of (1) then we would have the solution. To do this we write (1) in the form

$$y = \frac{3s}{s^2 + 4} - \frac{6}{s^2 + 4} + \frac{16}{4}\left(\frac{1}{s^2} - \frac{1}{s^2 + 4}\right) = \frac{3s}{s^2 + 4} - \frac{10}{s^2 + 4} + \frac{4}{s^2}$$

Now we know that

The Laplace transform of $\cos 2t$ is $\dfrac{s}{s^2 + 4}$.

The Laplace transform of $\sin 2t$ is $\dfrac{2}{s^2 + 4}$.

The Laplace transform of t is $\dfrac{1}{s^2}$.

It would seem therefore that the function sought is

$$Y(t) = 3 \cos 2t - 5 \sin 2t + 4t$$

This does indeed satisfy the given differential equation and conditions and so is the required solution.

In the problem just considered we needed to find a function $F(t)$ whose Laplace transform $f(s)$ is known. Such a function is called an *inverse Laplace transform* and is denoted by $\mathcal{L}^{-1}\{f(s)\}$ where $\mathcal{L}^{-1}$ is called the *inverse Laplace transform operator*.

From the fact that $\mathcal{L}\{F(t) + G(t)\} = f(s) + g(s)$, $\mathcal{L}\{cF(t)\} = cf(s)$ we have

$$\mathcal{L}^{-1}\{f(s) + g(s)\} = F(t) + G(t) = \mathcal{L}^{-1}\{f(s)\} + \mathcal{L}^{-1}\{g(s)\} \tag{2}$$

$$\mathcal{L}^{-1}\{cf(s)\} = cF(t) = c\mathcal{L}^{-1}\{f(s)\} \tag{3}$$

which shows that $\mathcal{L}^{-1}$ is a linear operator.

In view of our previous experience regarding problems of existence and uniqueness (see page 20, for example) several questions arise.

1. **Existence.** Does the inverse Laplace transform of a given function $f(s)$ exist?
2. **Uniqueness.** If it does exist, is it unique?
3. **Determination.** How do we find it?

From the practical point of view, as we have already mentioned, number 3 seems most important. But the other two are also important. In what follows we shall investigate various methods by which inverse Laplace transforms can be found and at the same time shall show how to solve various differential equations using these methods. On page 282 we shall consider the questions of existence and uniqueness.

3.2 SOME METHODS FOR FINDING INVERSE LAPLACE TRANSFORMS

From the above discussion problem, it is at once clear that proficiency in solving differential equations using Laplace transforms is practically synonymous with proficiency in determining inverse Laplace transforms. Several methods are available for finding inverse Laplace transforms. These are

(a) Use of Laplace transform tables
(b) Use of theorems on inverse Laplace transforms
(c) The method of partial fractions
(d) The method of convolutions
(e) Miscellaneous methods

We shall now treat examples illustrating each of these.

(a) *Use of Laplace Transform Tables*

Suppose that we wish to find $\mathcal{L}^{-1}\{f(s)\}$ where $f(s)$ is known. Then we need only look in the Laplace transform table opposite $f(s)$. Consider for instance

ILLUSTRATIVE EXAMPLE 1

Solve $Y'' + Y = 16 \cos t,\qquad Y(0) = 0,\ Y'(0) = 0.$

Solution Taking Laplace transforms we find

$$\{s^2 y - sY(0) - Y'(0)\} + y = \frac{16s}{s^2 + 1} \quad \text{or} \quad y = \frac{16s}{(s^2 + 1)^2}$$

Hence, from entry 13, we have on letting $\omega = 1$ and dividing by 2 (justified by the linear property of Laplace transformation) the required solution

$$Y = 16\mathcal{L}^{-1}\left\{\frac{s}{(s^2 + 1)^2}\right\} = 8t \sin t$$

In some cases an inverse transform is not found directly from the table but may be obtained by combining transforms which are in the table. Consider for instance

ILLUSTRATIVE EXAMPLE 2

Find $\mathcal{L}^{-1}\left\{\dfrac{2s - 3}{s^2 + 1}\right\}.$

Solution The transform as given is not in the table. However by writing it as

$$2\mathcal{L}^{-1}\left\{\frac{s}{s^2 + 1}\right\} - 3\mathcal{L}^{-1}\left\{\frac{1}{s^2 + 1}\right\}$$

we see from entries 6 and 7 with $\omega = 1$ that the required result is $2 \cos t - 3 \sin t$.

It is quite evident that success in using tables depends on how extensive the tables are and also our ability to use the tables effectively.* In case we cannot find the required transform in the table other methods must be used.

* As in the case of differentiation and integration, the student should, for facility, become acquainted with certain basic Laplace transforms, for example entries 1–7.

(b) *Use of Theorems on Inverse Laplace Transforms*

Corresponding to each theorem developed for Laplace transforms there is a theorem on inverse Laplace transforms. For example corresponding to the theorem on page 261, we have

Theorem. If $\mathscr{L}^{-1}\{f(s)\} = F(t)$ then
1. $\mathscr{L}^{-1}\{f(s - a)\} = e^{at}F(t)$.
2. $\mathscr{L}^{-1}\{f^{(n)}(s)\} = (-1)^n t^n F(t)$.

ILLUSTRATIVE EXAMPLE 3

Solve $Y'' + 2Y' + 5Y = 0$, $Y(0) = 3, Y'(0) = -7$.

Solution   Taking Laplace transforms we find

$$\{s^2 y - sY(0) - Y'(0)\} + 2\{sy - Y(0)\} + 5y = 0, (s^2 + 2s + 5)y - 3s + 1 = 0$$

and so

$$y = \frac{3s - 1}{s^2 + 2s + 5}$$

To find the inverse, let us complete the square in the denominator and rewrite as

$$y = \frac{3(s + 1) - 4}{(s + 1)^2 + 4} = 3\left\{\frac{s + 1}{(s + 1)^2 + 4}\right\} - 4\left\{\frac{1}{(s + 1)^2 + 4}\right\}$$

From the first part of the theorem we see that

$$\mathscr{L}^{-1}\left\{\frac{s}{s^2 + 4}\right\} = \cos 2t, \qquad \mathscr{L}^{-1}\left\{\frac{s + 1}{(s + 1)^2 + 4}\right\} = e^{-t}\cos 2t$$

$$\mathscr{L}^{-1}\left\{\frac{1}{s^2 + 4}\right\} = \tfrac{1}{2}\sin 2t, \qquad \mathscr{L}^{-1}\left\{\frac{1}{(s + 1)^2 + 4}\right\} = \tfrac{1}{2}e^{-t}\sin 2t$$

Hence $y = 3e^{-t}\cos 2t - \tfrac{4}{2}e^{-t}\sin 2t = e^{-t}(3\cos 2t - 2\sin 2t)$

(c) *The Method of Partial Fractions*

In many problems we arrive at a transform which is a rational function of s [i.e., a function having the form $P(s)/Q(s)$, where $P(s)$ and $Q(s)$ are polynomials and the degree of $P(s)$ is less than that of $Q(s)$]. It is then often useful to express the given result as a sum of simpler fractions, called *partial fractions*. As an example consider

ILLUSTRATIVE EXAMPLE 4

Solve $Y'' - 3Y' + 2Y = 12e^{4t}$, $Y(0) = 1, Y'(0) = 0$.

Solution   Laplace transformation yields,

$$(s^2 - 3s + 2)y - s + 3 = \frac{12}{s - 4} \quad \text{or} \quad y = \frac{s^2 - 7s + 24}{(s - 1)(s - 2)(s - 4)} \tag{4}$$

To resolve this into partial fractions, two methods can be used.

First Method. If A, B, C are undetermined constants we have

$$\frac{s^2 - 7s + 24}{(s-1)(s-2)(s-4)} = \frac{A}{s-1} + \frac{B}{s-2} + \frac{C}{s-4} \tag{5}$$

Multiplying by $(s-1)(s-2)(s-4)$ we obtain

$$s^2 - 7s + 24 = A(s-2)(s-4) + B(s-1)(s-4) + C(s-1)(s-2)$$
$$= (A + B + C)s^2 + (-6A - 5B - 3C)s + (8A + 4B + 2C)$$

Since this is an identity we have on equating coefficients of like powers of s,

$$A + B + C = 1, \qquad -6A - 5B - 3C = -7, \qquad 8A + 4B + 2C = 24$$

Solving these we find $A = 6$, $B = -7$, $C = 2$. Thus

$$Y = \mathcal{L}^{-1}\left\{\frac{6}{s-1} - \frac{7}{s-2} + \frac{2}{s-4}\right\} = 6e^t - 7e^{2t} + 2e^{4t}$$

Second Method. Multiplying both sides of (5) by $s - 1$ we find

$$\frac{s^2 - 7s + 24}{(s-2)(s-4)} = A + \frac{B(s-1)}{s-2} + \frac{C(s-1)}{s-4}$$

Then letting $s \to 1$, $\qquad A = \dfrac{1 - 7 + 24}{(1-2)(1-4)} = 6$

Similarly, multiplying (5) by $s - 2$ and letting $s \to 2$ yields $B = -7$, and multiplying (5) by $s - 4$ and letting $s \to 4$ yields $C = 2$. The method then proceeds as before.

ILLUSTRATIVE EXAMPLE 5

Find $\mathcal{L}^{-1}\left\{\dfrac{5s^2 - 7s + 17}{(s-1)(s^2+4)}\right\}$.

Solution Assume that

$$\frac{5s^2 - 7s + 17}{(s-1)(s^2+4)} = \frac{A}{s-1} + \frac{Bs + C}{s^2 + 4} \tag{6}$$

First Method. Multiplying by $(s-1)(s^2+4)$ we have

$$5s^2 - 7s + 17 = (A + B)s^2 + (C - B)s + 4A - C$$

Then $\qquad A + B = 5, \qquad C - B = -7, \qquad 4A - C = 17$

Thus $A = 3$, $B = 2$, $C = -5$ and we have

$$\mathcal{L}^{-1}\left\{\frac{3}{s-1} + \frac{2s-5}{s^2+4}\right\} = 3\mathcal{L}^{-1}\left\{\frac{1}{s-1}\right\} + 2\mathcal{L}^{-1}\left\{\frac{s}{s^2+4}\right\} - 5\mathcal{L}^{-1}\left\{\frac{1}{s^2+4}\right\}$$

$$= 3e^t + 2\cos 2t - \tfrac{5}{2}\sin 2t$$

Second Method. Multiplying (6) by $s - 1$ and letting $s \to 1$ yields $A = 3$. Thus

$$\frac{5s^2 - 7s + 17}{(s-1)(s^2+4)} = \frac{3}{s-1} + \frac{Bs + C}{s^2 + 4}$$

To determine C it is convenient to put $s = 0$ and obtain $C = -5$. Finally by placing $s = -1$ for example we find $B = 2$. The method then proceeds as before.

Note that in both examples, the first method is general but solution of the simultaneous equations can be tedious. The second method, although shorter, is most effective when the denominator can be factored into real distinct linear factors. More complicated cases are considered in the exercises.

(d) *The Method of Convolutions*

We have already noted that if $f(s)$ and $g(s)$ are the Laplace transforms of $F(t)$ and $G(t)$, respectively, then

$$\mathcal{L}^{-1}\{f(s) + g(s)\} = F(t) + G(t), \qquad \mathcal{L}^{-1}\{f(s) - g(s)\} = F(t) - G(t)$$

It is of interest to ask whether there is some simple expression for the inverse Laplace transform of the product $f(s)g(s)$ in terms of $F(t)$ and $G(t)$. The answer is *yes* and the result is summarized in the following

Theorem. If $\mathcal{L}^{-1}\{f(s)\} = F(t)$ and $\mathcal{L}^{-1}\{g(s)\} = G(t)$ then

$$\mathcal{L}^{-1}\{f(s)g(s)\} = \int_0^t F(u)G(t-u)du = F*G$$

where we call $F*G$ the *convolution* of F and G. Equivalently, we have

$$\mathcal{L}\{F*G\} = \mathcal{L}\left\{\int_0^t F(u)G(t-u)du\right\} = f(s)g(s)$$

This theorem, called the *convolution theorem*, is often useful in obtaining inverse Laplace transforms. Before presenting a proof of the theorem, let us consider several examples of its use.

ILLUSTRATIVE EXAMPLE 6

Find $\mathcal{L}^{-1}\left\{\dfrac{s}{(s^2+1)^2}\right\}$.

Solution We have $\quad \mathcal{L}^{-1}\left\{\dfrac{s}{s^2+1}\right\} = \cos t, \qquad \mathcal{L}^{-1}\left\{\dfrac{1}{s^2+1}\right\} = \sin t$

Hence by the convolution theorem

$$\mathcal{L}^{-1}\left\{\dfrac{s}{(s^2+1)^2}\right\} = \mathcal{L}^{-1}\left\{\dfrac{s}{s^2+1}\cdot\dfrac{1}{s^2+1}\right\} = \cos t * \sin t$$

$$= \int_0^t \cos u \sin(t-u)du = \int_0^t \cos u[\sin t \cos u - \cos t \sin u]du$$

$$= \sin t \int_0^t \cos^2 u \, du - \cos t \int_0^t \sin u \cos u \, du$$

$$= \sin t[\tfrac{1}{2}t + \tfrac{1}{2}\sin t \cos t] - \cos t[\tfrac{1}{2}\sin^2 t] = \tfrac{1}{2}t \sin t$$

Note that this agrees with entry 13.

If we let $G(t) = 1$ in the convolution theorem we obtain

$$\mathcal{L}^{-1}\left\{\frac{f(s)}{s}\right\} = \int_0^t F(u)\,du \tag{7}$$

Thus multiplying $f(s)$ by $1/s$ corresponds to integrating $F(t)$ from 0 to t; multiplying by $1/s^2$ to integrating twice, etc. For a use of this consider

ILLUSTRATIVE EXAMPLE 7

Find (a) $\mathcal{L}^{-1}\left\{\dfrac{1}{s(s^2 + 1)}\right\}$,   (b) $\mathcal{L}^{-1}\left\{\dfrac{1}{s^2(s^2 + 1)}\right\}$.

Solutions (a) Since $\mathcal{L}^{-1}\left\{\dfrac{1}{s^2 + 1}\right\} = \sin t$ we have, using (7),

$$\mathcal{L}^{-1}\left\{\frac{1}{s(s^2 + 1)}\right\} = \int_0^t \sin u\,du = 1 - \cos t$$

(b) Since by (a) $\mathcal{L}^{-1}\left\{\dfrac{1}{s(s^2 + 1)}\right\} = 1 - \cos t$ we have, using (7),

$$\mathcal{L}^{-1}\left\{\frac{1}{s^2(s^2 + 1)}\right\} = \int_0^t (1 - \cos u)\,du = t - \sin t$$

By letting $G(t) = e^{at}$ in the convolution theorem we find

$$\mathcal{L}^{-1}\left\{\frac{f(s)}{s - a}\right\} = \int_0^t F(t)e^{a(t-u)}\,du = e^{at}\int_0^t e^{-au}F(u)\,du \tag{8}$$

Note that (7) is a special case of (8) with $a = 0$. From (8) we see that there is a correspondence between

$$\frac{1}{s - a} \quad \text{and} \quad e^{at}\int_0^t e^{-au}(\ \ )du \tag{9}$$

where the first can be considered as an operator acting on $f(s)$ while the second is considered as an operator acting on $F(t)$. The correspondence bears close resemblance to equation (31) in the section on operator methods on page 205. This provides a clue as to the connection between s and the operator D and thus the connection between Laplace transform methods and Heaviside's operational methods.

To present a proof of the convolution theorem, we first note that the Laplace transforms of $F(t)$ and $G(t)$ can be written, respectively, as

$$f(s) = \int_0^\infty e^{-sx}F(x)\,dx, \qquad g(s) = \int_0^\infty e^{-sy}G(y)\,dy \tag{10}$$

We can write these in terms of the Heaviside unit step function as

$$f(s) = \int_{-\infty}^\infty e^{-sx}F(x)H(x)\,dx, \qquad g(s) = \int_{-\infty}^\infty e^{-sy}G(y)H(y)\,dy \tag{11}$$

Then $I = f(s)g(s) = \left\{ \int_{-\infty}^{\infty} e^{-sx}F(x)H(x)dx \right\} \left\{ \int_{-\infty}^{\infty} e^{-sy}G(y)H(y)dy \right\}$

$$= \int_{-\infty}^{\infty} \int_{-\infty}^{\infty} e^{-s(x+y)}F(x)H(x)G(y)H(y)dx \, dy$$

using a procedure similar to that in the proof of $\Gamma(\tfrac{1}{2}) = \sqrt{\pi}$ on page 262.

Let us now write this double integral as the iterated integral

$$I = \int_{-\infty}^{\infty} F(x)H(x) \left[\int_{-\infty}^{\infty} e^{-s(x+y)}G(y)H(y)dy \right] dx \tag{12}$$

In the integral in brackets suppose we change the variable of integration from y to t, where $x + y = t$, i.e., $y = t - x$. Then (12) becomes

$$I = \int_{-\infty}^{\infty} F(x)H(x) \left[\int_{-\infty}^{\infty} e^{-st}G(t-x)H(t-x)dt \right] dx \tag{13}$$

or on changing the order of integration

$$I = \int_{-\infty}^{\infty} e^{-st} \left[\int_{-\infty}^{\infty} F(x)H(x)G(t-x)H(t-x)dx \right] dt \tag{14}$$

Now because of the definition of the Heaviside function, the integrand of the integral in brackets is zero except for $0 < x < t$, and for these values of x the Heaviside functions are equal to 1. Thus (14) becomes

$$I = \int_0^{\infty} e^{-st} \left[\int_0^t F(x)G(t-x)dx \right] dt$$

or

$$I = \int_0^{\infty} e^{-st} \left[\int_0^t F(u)G(t-u)du \right] dt = \mathscr{L} \left\{ \int_0^t F(u)G(t-u)du \right\} = \mathscr{L}\{F*G\}$$

on using $x = u$. This completes the proof. The interchanges of order of integrals in this proof can be justified if we assume that the functions F and G satisfy the conditions on page 264 for the existence of their Laplace transforms.

The convolution is often useful in solving *integral equations* where the unknown function to be determined is under the integral sign. As an illustration consider

ILLUSTRATIVE EXAMPLE 8

Solve the integral equation $Y(t) = 3t + \int_0^t Y(u) \sin(t-u)du$.

Solution The integral equation can be written in terms of convolutions as

$$Y(t) = 3t + Y(t)* \sin t$$

Then taking the Laplace transform and using the convolution theorem we have

$$y(s) = \frac{3}{s^2} + \frac{y(s)}{s^2+1} \quad \text{or} \quad y(s) = \frac{3(s^2+1)}{s^4} = \frac{3}{s^2} + \frac{3}{s^4}$$

Then taking the inverse Laplace transform we find $Y(t) = 3t + \tfrac{1}{2}t^3$.

The convolution theorem is often useful in obtaining solutions to differential equations in which there are functions whose Laplace transforms are difficult or even impossible to find. For instance let us consider

ILLUSTRATIVE EXAMPLE 9

Solve the initial value problem $\quad Y'' + Y = e^{-t^2}, \qquad Y(0) = 0, \; Y'(0) = 0.$ (15)

Solution Replace e^{-t^2} by $F(t)$ and denote its Laplace transform by $f(s)$. Then from the given differential equation and conditions we have

$$s^2 y - sY(0) - Y'(0) + y = f(s) \quad \text{or} \quad y = \frac{f(s)}{s^2 + 1}$$

Since

$$\mathscr{L}^{-1}\{f(s)\} = e^{-t^2}, \qquad \mathscr{L}^{-1}\left\{\frac{1}{s^2 + 1}\right\} = \sin t$$

we have

$$Y = \mathscr{L}^{-1}\left\{\frac{f(s)}{s^2 + 1}\right\} = \int_0^t e^{-u^2} \sin (t - u)\,du \tag{16}$$

The integral in (16) cannot be evaluated exactly.

It is interesting to note that if the right-hand side of the differential equation in (15) is e^{t^2} instead of e^{-t^2} the Laplace transform would not exist, but the solution without using Laplace transforms is in fact

$$Y = \int_0^t e^{u^2} \sin (t - u)\,du \tag{17}$$

Thus *formally* the method of Laplace transforms can be used to arrive at possible solutions which can then be checked.

The fact that Laplace transform techniques can lead to correct results even in cases where functions do not have Laplace transforms seems to indicate that there is something more basic than the Laplace transform, but still closely related to it, which can be used in place of it. From the above example it would seem that the *convolution itself* is the desired concept. This is indicated further by the fact that the convolution obeys many of the usual laws of algebra, such as the following

(a) $\qquad F*G = G*H$ $\qquad\qquad\qquad\qquad$ Commutative law
(b) $\qquad F*(G*H) = (F*G)*H$ $\qquad\qquad\quad$ Associative law
(c) $\quad F*(G + H) = F*G + F*H$ $\qquad\qquad$ Distributive law

This has led some authors to avoid the Laplace transform altogether and deal only with convolutions.* By use of this procedure it is possible to make rigorous impulse functions such as the *Dirac delta function.*

(e) *Miscellaneous Methods*

Various special methods may also be useful in finding inverse Laplace transforms.

---

* See for example, reference [19].

Find $\mathscr{L}^{-1}\{e^{-as}f(s)\}$ where $f(s) = \mathscr{L}\{F(t)\}$ and $a > 0$.

Solution We have by definition $$f(s) = \int_0^\infty e^{-st}F(t)dt$$

Then multiplying by e^{-as} we find $$e^{-as}f(s) = \int_0^\infty e^{-s(t+a)}F(t)dt$$

With $t + a = u$ this integral can be written as

$$e^{-as}f(s) = \int_a^\infty e^{-su}F(u-a)du = \int_0^a e^{-su}(0)du + \int_a^\infty e^{-su}F(u-a)du$$

$$= \mathscr{L}\{G(t)\}$$

where $$G(t) = \begin{cases} 0, & t < a \\ F(t-a), & t > a \end{cases}$$

We thus have on taking inverse Laplace transforms,

$$\mathscr{L}^{-1}\{e^{-as}f(s)\} = \begin{cases} 0, & t < a \\ F(t-a), & t > a \end{cases}$$

This result can also be expressed in terms of the Heaviside unit function as

$$\mathscr{L}^{-1}\{e^{-as}f(s)\} = F(t-a)H(t-a)$$

The result of this example is important and we state it for reference in the following

Theorem. If $\mathscr{L}^{-1}\{f(s)\} = F(t)$ then

$$\mathscr{L}^{-1}\{e^{-as}f(s)\} = \begin{cases} 0, & t < a \\ F(t-a), & t > a \end{cases} = F(t-a)H(t-a)$$

3.3 REMARKS CONCERNING EXISTENCE AND UNIQUENESS OF INVERSE LAPLACE TRANSFORMS

We have tacitly assumed above that there is only one function which has some given Laplace transform, i.e., we have assumed the inverse Laplace transform to be unique. That this is actually not so can be seen by noting that the function

$$G(t) = \begin{cases} 5, & t = 3 \\ 1, & t \neq 3 \end{cases} \tag{18}$$

differs from the function $F(t) = 1$, since the value of $G(t)$ at $t = 3$ is 5 while the value of $F(t)$ at $t = 3$ is 1. However the Laplace transform of *both* functions is given by $1/s$, $s > 0$. Thus the inverse Laplace transform of $1/s$ can be $F(t) = 1$ or the function $G(t)$ given in (18), or in fact any one of infinitely many functions.

A possible clue as to the reason why we do not get uniqueness is that the function $G(t)$ given in (18) is discontinuous at $t = 3$. As a matter of fact it can be shown that if we restrict ourselves to continuous functions then the inverse Laplace transform is unique. This theorem, which is rather difficult to prove, is called *Lerch's theorem.**

Now we know that if $F(t)$ is piecewise continuous in every finite interval and of exponential order then $\lim_{s \to \infty} f(s) = 0$ (see C Exercise 6, page 268. In case we should have $\lim_{s \to \infty} f(s) \neq 0$ it follows that the inverse transform cannot be piecewise continuous and of exponential order. Thus $\lim_{s \to \infty} f(s) = 0$ is a *necessary condition* for existence of an inverse Laplace transform which is piecewise continuous and of exponential order.†

A EXERCISES

1. Find inverse Laplace transforms of the following functions.

(a) $\dfrac{4}{s+2}$.

(b) $\dfrac{3s}{s^2+9}$.

(c) $\dfrac{15}{s^2+25}$.

(d) $\dfrac{6s-10}{s^2+4} - \dfrac{3}{s-4}$.

(e) $\dfrac{2-s}{5+s^2}$.

(f) $\dfrac{2+3s-s^2}{s^3}$.

2. Using theorems on inverse Laplace transforms find each of the following.

(a) $\mathscr{L}^{-1}\left\{\dfrac{1}{(s+2)^3}\right\}$.

(b) $\mathscr{L}^{-1}\left\{\dfrac{2s-10}{s^2-4s+20}\right\}$.

(c) $\mathscr{L}^{-1}\left\{\dfrac{s+1}{s^2+s+1}\right\}$.

(d) $\mathscr{L}^{-1}\left\{\dfrac{s}{(s-1)^4}\right\}$.

(e) $\mathscr{L}^{-1}\left\{\dfrac{2s-1}{4s^2+4s+5}\right\}$.

3. Use the method of partial fractions to find inverse Laplace transforms of the following.

(a) $\dfrac{s+17}{(s-1)(s+3)}$.

(b) $\dfrac{3s-8}{s^2-16}$.

(c) $\dfrac{s-11}{(s+1)(s-2)(s-3)}$.

(d) $\dfrac{2s^2+15s+7}{(s+1)^2(s-2)}$.

(e) $\dfrac{10}{s(s^2-2s+5)}$.

(f) $\dfrac{s+1}{(s^2+1)(s^2+4)}$.

4. Use the convolution method to find each of the following.

(a) $\mathscr{L}^{-1}\left\{\dfrac{1}{s^2-4}\right\}$.

(b) $\mathscr{L}^{-1}\left\{\dfrac{1}{s^2(s+1)^2}\right\}$.

(c) $\mathscr{L}^{-1}\left\{\dfrac{1}{(s^2+1)^2}\right\}$.

5. Solve each of the following and check solutions. (a) $Y'' - 4Y' + 3Y = 0$, $Y(0) = 3$, $Y'(0) = 5$.
(b) $Y'' + 2Y' = 4$, $Y(0) = 1$, $Y'(0) = -4$. (c) $Y'' + 9Y = 20e^{-t}$, $Y(0) = 0$, $Y'(0) = 1$.
(d) $Y'' - 2Y' + Y = 12t$, $Y(0) = 4$, $Y'(0) = 1$.
(e) $Y'' + 8Y' + 25Y = 100$, $Y(0) = 2$, $Y'(0) = 20$.

* More generally Lerch has shown that if two functions have the same Laplace transform then they differ at most by a *null function,* i.e., a function $N(t)$ such that for all $t > 0$

$$\int_0^t N(u)\,du = 0$$

A practical significance of this is that, in a certain sense, the inverse Laplace transform is "essentially unique." For further discussion see reference [6].

† The condition is not sufficient however. For sufficient conditions we must consider $f(s)$ as a function of the complex variable s. See [6] for example.

B EXERCISES

1. Use partial fractions to find $\mathcal{L}^{-1}\left\{\dfrac{11s^2 - 10s + 11}{(s^2 + 1)(s^2 - 2s + 5)}\right\}$.

$$\left[Hint: \text{ Assume } \frac{11s^2 - 10s + 11}{(s^2 + 1)(s^2 - 2s + 5)} = \frac{As + B}{s^2 + 1} + \frac{Cs + D}{s^2 - 2s + 5}.\right]$$

2. Find $\mathcal{L}^{-1}\left\{\dfrac{s^3 + 2s^2 + 4s + 5}{(s + 1)^2(s + 2)^2}\right\}$.

3. Solve $Y''' + 3Y'' + 3Y' + Y = 12e^{-t}$, $Y(0) = 1$, $Y'(0) = 0$, $Y''(0) = -3$.

4. Solve $Y^{(IV)} - Y = \cos t$ subject to $Y(0) = 1$, $Y'(0) = -1$, $Y''(0) = Y'''(0) = 0$.

5. Find (a) $\mathcal{L}^{-1}\{e^{-2s/}s^3\}$. (b) $\mathcal{L}^{-1}\{e^{-s}/(s + 1)^{3/2}\}$.

6. Solve $Y'' + Y = 0$, $Y(0) = 0$, $Y(\pi/2) = 4$. [$Hint$: Let $Y'(0) = C$ and find C.]

7. Prove (a) $F*G = G*F$. (b) $F*(G*H) = (F*G)*H$. Discuss.

8. Solve the following integral equations and check your answers.

   (a) $Y(t) = 1 + \int_0^t e^{2u} Y(t - u)du$. (b) $\int_0^t Y(u) \sin(t - u)du = Y(t) + \sin t - \cos t$.

9. Find $\mathcal{L}^{-1}\left\{\dfrac{s}{(s^2 - 1)^2}\right\}$.

10. Solve the integral equation $\int_0^t Y(u)Y(t - u)du = \frac{1}{2}(\sin t - t \cos t)$.

C EXERCISES

1. Show that $I = \displaystyle\int_0^\infty \frac{\sin tx}{x}\,dx = \frac{\pi}{2}$ if $t > 0$.

   ($Hint$: First show that $\mathcal{L}\{I\} = \displaystyle\int_0^\infty \frac{\mathcal{L}\{\sin tx\}}{x}\,dx$ and evaluate the last integral.)

2. Show that $\displaystyle\int_0^\infty \frac{\cos tx}{x^2 + 1}\,dx = \frac{\pi}{2}e^{-t}$ if $t \geq 0$.

3. Solve $tY'' - tY' + Y = 0$, $Y(0) = 0$, $Y'(0) = 1$.

4. Prove that $\mathcal{L}^{-1}\left\{\ln\left(1 + \dfrac{1}{s}\right)\right\} = \dfrac{1 - e^{-t}}{t}$.

5. Solve each of the following initial value problems involving the Dirac delta function.
   (a) $Y' + 2Y = 5\delta(t - 1)$, $Y(0) = 2$. (b) $Y'' + Y = 3\delta(t - \pi)$, $Y(0) = 6$, $Y'(0) = 0$.
   (c) $Y'' + 4Y' + 4Y = 6\delta(t - 2)$, $Y(0) = 0$, $Y'(0) = 0$.

6. Work C Exercise 4, page 242, by using the delta function.

7. Let $P(s)$ and $G(s)$ be polynomials in s where the degree of $P(s)$ is less than the degree of $Q(s)$ and where $Q(s) = 0$ has distinct roots $a_1, a_2, \ldots, a_n$. Prove *Heaviside's expansion formula*

$$\mathcal{L}^{-1}\left\{\frac{P(s)}{Q(s)}\right\} = \sum_{k=1}^{n} e^{a_k t}\frac{P(a_k)}{Q'(a_k)}$$

8. Use Exercise 7 to work (a) Illustrative Example 4, page 276; (b) Illustrative Example 5, page 277; (c) A Exercise 3(e).

9. Generalize Exercise 7 to the case where roots may not be distinct and illustrate by an example.

10. (a) Show that

$$\int_0^t \int_0^t \cdots \int_0^t F(x)dx^n = \int_0^t \frac{(t-x)^{n-1}}{(n-1)!} F(x)dx$$

where there are n integrals on the left. Does the result hold for $F(x) = e^{x^2}$? Explain.

(b) Show that the result in (a) is equivalent to the statement that

$$D^{-n}F(t) = \frac{1}{\Gamma(n)} \int_0^t (t-x)^{n-1}F(x)dx$$

11. Suppose that the result in Exercise 10(b) is taken as the definition of $D^{-n}F(t)$ for any positive number n. **(a)** Show that if in particular $n = \frac{1}{2}$ then

$$D^{-1/2}F(t) = \frac{1}{\sqrt{\pi}} \int_0^t \frac{F(x)}{\sqrt{t-x}} dx$$

(b) Operate with D on both sides of the result in (a) and assume that $D[D^{-1/2}] = D^{1/2}$ to arrive at the definition of the *half derivative* of $F(t)$ given by

$$\frac{d^{1/2}}{dt^{1/2}} F(t) = D^{1/2}F(t) = \frac{1}{\sqrt{\pi}} \frac{d}{dt} \int_0^t \frac{F(x)}{\sqrt{t-x}} dx$$

(c) Check the definition in (b) by finding the half derivative of t^2 twice to see if you get the whole derivative, i.e., $2t$.

4 Applications to Physical and Biological Problems

As we have already seen in preceding chapters a mathematical formulation of problems in mechanics, electricity, beams, etc., often leads to linear differential equations with constant coefficients. In this section we show how the Laplace transform is used to solve such problems.

4.1 APPLICATIONS TO ELECTRIC CIRCUITS

As a first example illustrating the use of Laplace transforms in the solution of applied problems, let us consider the following problem involving electric circuits.

ILLUSTRATIVE EXAMPLE 1

An electric circuit (see Fig. 6.8) consists of a resistor of resistance R ohms in series with a capacitor of capacitance C farads, a generator of E volts and a key. At time $t = 0$ the key is closed. Assuming that the charge on the capacitor is zero at $t = 0$, find the charge and current at any later time. Assume R, C, E are constants.

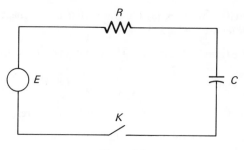

Figure 6.8

Mathematical Formulation. If Q and $I = dQ/dt$ are the charge and current at any time t then by Kirchhoff's law we have

$$RI + \frac{Q}{C} = E \quad \text{or} \quad R\frac{dQ}{dt} + \frac{Q}{C} = E \tag{1}$$

with initial condition $Q(0) = 0$.

Solution   Taking Laplace transforms of both sides of (1) and using the initial condition, we have, if q is the Laplace transform of Q,

$$R\{sq - Q(0)\} + \frac{q}{C} = \frac{E}{s}$$

or
$$q = \frac{CE}{s(RCs + 1)} = \frac{E/R}{s(s + 1/RC)}$$

$$= \frac{E/R}{1/RC}\left\{\frac{1}{s} - \frac{1}{s + 1/RC}\right\} = CE\left\{\frac{1}{s} - \frac{1}{s + 1/RC}\right\}$$

Then taking the inverse Laplace transform we find,

$$Q = CE(1 - e^{-t/RC}) \quad \text{and} \quad I = \frac{dQ}{dt} = \frac{E}{R}e^{-t/RC}$$

Laplace transform methods prove to be of great value in problems involving piecewise continuous functions. In such cases the properties of the Heaviside unit function (page 264) are useful. As an illustration of the procedure let us consider

ILLUSTRATIVE EXAMPLE 2

Work Illustrative Example 1 for the case where the generator of E volts is replaced by a generator with voltage given as a function of time by

$$E(t) = \begin{cases} E_0, & 0 \le t < T \\ 0, & t > T \end{cases}$$

Mathematical Formulation. Replacing E in Illustrative Example 1 by $E(t)$ we obtain the required differential equation

$$R\frac{dQ}{dt} + \frac{Q}{C} = E(t) \tag{2}$$

with initial condition $Q(0) = 0$. The equation (2) can also be expressed in terms of the Heaviside unit function as

$$R\frac{dQ}{dt} + \frac{Q}{C} = E_0[1 - H(t - T)] \tag{3}$$

Solution. *Method 1.* Taking Laplace transforms of both sides of (2) or (3) and using the initial condition we find

$$R\{sq - Q(0)\} + \frac{q}{C} = \frac{E_0(1 - e^{-sT})}{s}$$

or

$$q = \frac{E_0}{R}\frac{(1 - e^{-sT})}{s(s + 1/RC)} = \frac{E_0}{Rs(s + 1/RC)} - \frac{E_0}{Rs(s + 1/RC)}e^{-sT}$$

$$= CE_0\left\{\frac{1}{s} - \frac{1}{s + 1/RC}\right\} - CE_0\left\{\frac{1}{s} - \frac{1}{s + 1/RC}\right\}e^{-sT}$$

Taking the inverse Laplace transforms of both sides using the result stated in the Theorem on page 282, we find

$$Q = CE_0(1 - e^{-t/RC}) - CE_0(1 - e^{-(t - T)/RC})H(t - T)$$

$$= \begin{cases} CE_0(1 - e^{-t/RC}), & t < T \\ CE_0(e^{-(t - T)/RC} - e^{-t/RC}), & t > T \end{cases}$$

For $t = T$ we have $Q = CE_0(1 - e^{-T/RC})$.

Method 2, Using the Convolution Theorem. Let $e(s)$ be the Laplace transform of $E(t)$. Then as above we have

$$R\{sq - Q(0)\} + \frac{q}{C} = e(s)$$

or since $Q(0) = 0$,

$$q = \frac{e(s)}{R(s + 1/RC)}$$

Now

$$\mathscr{L}^{-1}\left\{\frac{1}{R(s + 1/RC)}\right\} = \frac{e^{-t/RC}}{R}, \qquad \mathscr{L}^{-1}\{e(s)\} = E(t)$$

Thus by the convolution theorem $\quad Q = \mathscr{L}^{-1}(q) = \frac{1}{R}\int_0^t E(u)e^{-(t - u)/RC}\,du$

For $0 < t < T$ we have $\qquad Q = \frac{1}{R}\int_0^t E_0 e^{-(t - u)/RC}\,du = CE_0(1 - e^{-t/RC})$

For $t > T$ we have
$$Q = \frac{1}{R} \int_0^T E_0 e^{-(t-u)/RC} \, du = CE_0 \{ e^{-(t-T)/RC} - e^{-t/RC} \}$$

which agrees with the result in Method 1.

4.2 AN APPLICATION TO BIOLOGY

As a biological application, in particular the absorption of drugs in an organ or cell, let us consider the following

ILLUSTRATIVE EXAMPLE 3

A liquid carries a drug into an organ of volume V cm^3 at a rate of a cm^3/sec and leaves at a rate of b cm^3/sec, where V, a, b are constants. At time $t = 0$ the concentration of the drug is zero and builds up linearly to a maximum of κ at $t = T$, at which time the process is stopped. What is the concentration of the drug in the organ at any time t?

Mathematical Formulation. The problem is the same as that on page 157, except that the concentration is a function of time denoted by $C(t)$ given by

$$C(t) = \begin{cases} \dfrac{\kappa t}{T}, & 0 \leqq t \leqq T \\ 0, & t > T \end{cases}$$

whose graph appears in Fig. 6.9. Denoting the instantaneous concentration of the drug in the organ by x, we thus have

$$\frac{d}{dt}(xV) = aC(t) - bx, \qquad x(0) = 0 \tag{4}$$

Solution We shall use the convolution method (Method 2 of Illustrative Example 2) to solve the initial value problem (4). Taking the Laplace transform of the differential equation in (4), calling $\mathscr{L}\{x\} = \bar{x}$ and $\mathscr{L}\{C(t)\} = c(s)$, we have

$$V\{s\bar{x} - x(0)\} = ac(s) - b\bar{x}$$

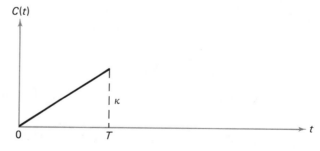

Figure 6.9

Then using $x(0) = 0$ yields

$$\bar{x} = \frac{ac(s)}{V(s + b/V)}$$

Now

$$\mathscr{L}^{-1}\left\{\frac{a}{V(s + b/V)}\right\} = \frac{a}{V}\,e^{-bt/V}, \qquad \mathscr{L}^{-1}\{c(s)\} = C(t)$$

Thus by the convolution theorem

$$x = \mathscr{L}^{-1}(\bar{x}) = \frac{a}{V}\int_0^t C(u)e^{-b(t-u)/V}\,du$$

For $0 \leq t < T$, we have

$$x = \frac{a}{V}\int_0^t \kappa u e^{-b(t-u)/V}\,du = \frac{\kappa a}{b}\,t - \frac{V\kappa a}{b^2}(1 - e^{-bt/V})$$

For $t > T$, we have $x = \dfrac{a}{V}\displaystyle\int_0^T \kappa u e^{-b(t-u)/V}\,du = \dfrac{V\kappa a}{b^2}\,e^{-bt/V} + \left(\dfrac{\kappa aT}{b} - \dfrac{V\kappa a}{b^2}\right)e^{-b(t-T)/V}$

The value of x for $t = T$ is found by letting $t = T$ in either of these.

Interpretation. From the last result we note that as t increases beyond T the drug gradually disappears. It follows that the drug concentration in the organ will reach a maximum at some time. The student may show (see B Exercise 9) that this time is given by $t = T$ and that this maximum which we shall call the *peak drug concentration* is given by

$$\frac{\kappa aT}{b} - \frac{V\kappa a}{b^2}(1 - e^{-bT/V})$$

In practice the peak drug concentration time will occur later than T due to the fact that the drug does not enter the organ instantaneously, as in the above model, but that instead there is a time lag.

4.3 THE TAUTOCHRONE PROBLEM—AN INTEGRAL EQUATION APPLICATION TO MECHANICS

As an example of a problem in mechanics which leads to an integral equation of convolution type let us consider the following

ILLUSTRATIVE EXAMPLE 4

A wire has the shape of a curve in a vertical xy plane with its lower end O located at the origin as indicated in Fig. 6.10. Assuming that there is no friction, find the shape which the curve must have so that a bead under the influence of gravity will slide from rest down to O in a specified constant time T regardless of where the bead is placed on the wire above O.

Mathematical Formulation. Before formulating the problem in mathematical terms, it may be instructive to examine whether the problem makes sense from a physical point of view. To do this suppose that we have two people with wires identical to that shown in Fig. 6.10. According to the problem, if the wires have just the right shape, then beads placed anywhere on the wires by the two people at a given instant should reach their lower ends simultaneously after time T. At first sight it

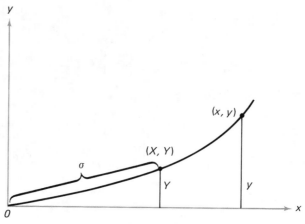

Figure 6.10

may be felt that this cannot happen, since it would seem that the higher a person places the bead on the wire, the longer it would take for the bead to reach the bottom because the distance to be traveled would be greater. However, the bead traveling the greater distance would also have greater speed near the bottom of the wire, so that presumably the race could result in a tie.

To formulate the problem mathematically, let (x, y) denote any starting point of the bead and (X, Y) any point between the starting point and point O. Let σ denote the length of the wire (i.e., the length of arc of the curve) as measured from O.

If we denote by P.E. and K.E. respectively the potential energy and kinetic energy of the bead, then according to the principle of conservation of energy of elementary mechanics we have

$$\text{P.E. at } (x, y) + \text{K.E. at } (x, y) = \text{P.E. at } (X, Y) + \text{K.E. at } (X, Y)$$

If the bead has mass m and t is the time of travel measured from the rest position this becomes

$$mgy + 0 = mgY + \tfrac{1}{2}m \left(\frac{d\sigma}{dt}\right)^2 \tag{5}$$

To get this we have used the fact that the potential energy is the weight mg multiplied by the height above the x axis, while the kinetic energy is $\tfrac{1}{2}mv^2$, where the speed $v = d\sigma/dt$ at (X, Y) but is zero at (x, y) since the bead is assumed to start from rest.

From (5) we obtain on solving for $d\sigma/dt$

$$\frac{d\sigma}{dt} = \pm\sqrt{2g(y - Y)} \tag{6}$$

However, since σ decreases as t increases so that $d\sigma/dt < 0$, we must choose the negative sign in (6) to obtain

$$\frac{d\sigma}{dt} = -\sqrt{2g(y - Y)} \tag{7}$$

Since $Y = y$ at $t = 0$ while $Y = 0$ at $t = T$, we have from (7) on separating variables and integrating

$$\int_{t=0}^{T} dt = -\frac{1}{\sqrt{2g}} \int_{Y=y}^{0} \frac{d\sigma}{\sqrt{y-Y}} \quad \text{or} \quad T = \frac{1}{\sqrt{2g}} \int_{Y=0}^{y} \frac{d\sigma}{\sqrt{y-Y}} \tag{8}$$

Now the arc length can be expressed as a function of y in the form $\sigma = F(y)$, so that (8) becomes

$$T = \frac{1}{\sqrt{2g}} \int_{Y=0}^{y} \frac{F'(Y)dY}{\sqrt{y-Y}} \tag{9}$$

Our problem is thus reduced to determining $F(y)$, i.e., solving the integral equation (9) and from this obtaining the required curve.

Solution The integral equation (9) is of convolution type and can be written as

$$T = \frac{1}{\sqrt{2g}} F'(y) * y^{-1/2}$$

Taking the Laplace transform and using the convolution theorem we find

$$\frac{T}{s} = \frac{1}{\sqrt{2g}} \mathscr{L}\{F'(y)\} \mathscr{L}\{y^{-1/2}\} \tag{10}$$

Now if we let $\mathscr{L}\{F(y)\} = f(s)$, then $\mathscr{L}\{F'(y)\} = sf(s) - F(0) = sf(s)$ since $F(0) = 0$. Also $\mathscr{L}\{y^{-1/2}\} = \Gamma(\tfrac{1}{2})/s^{1/2} = \sqrt{\pi}/\sqrt{s}$. Thus (10) becomes

$$\frac{T}{s} = \frac{1}{\sqrt{2g}} sf(s) \frac{\sqrt{\pi}}{\sqrt{s}} \quad \text{or} \quad sf(s) = T\sqrt{\frac{2g}{\pi}} s^{-1/2}$$

i.e.,

$$\mathscr{L}\{F'(y)\} = T\sqrt{\frac{2g}{\pi}} s^{-1/2}$$

Taking the inverse Laplace transform leads to

$$F'(y) = \frac{d\sigma}{dy} = \frac{T\sqrt{2g}}{\pi\sqrt{y}} \tag{11}$$

Now from the arc length formula of elementary calculus we have

$$d\sigma^2 = dx^2 + dy^2 \quad \text{or} \quad \left(\frac{d\sigma}{dy}\right)^2 = \left(\frac{dx}{dy}\right)^2 + 1$$

Using this together with (11) leads to

$$\left(\frac{dx}{dy}\right)^2 + 1 = \frac{a}{y}, \quad \text{where } a = \frac{2gT^2}{\pi^2}$$

From this we have

$$\frac{dx}{dy} = \sqrt{\frac{a-y}{y}} \tag{12}$$

on using the fact that the slope dy/dx and thus dx/dy cannot be negative. Separating variables in (12) and integrating we obtain

$$x = \int \sqrt{\frac{a - y}{y}}\, dy \tag{13}$$

To perform the integration in (13), it is convenient to let $y = a \sin^2 \phi$. Then

$$x = 2a \int \cos^2 \phi\, d\phi = a \int (1 + \cos 2\phi) d\phi = \frac{a}{2}(2\phi + \sin 2\phi) + c$$

so that $\qquad x = \frac{a}{2}(2\phi + \sin 2\phi) + c, \qquad y = a \sin^2 \phi = \frac{a}{2}(1 - \cos 2\phi)$

or on letting $2\phi = \theta$, $\qquad x = \frac{a}{2}(\theta + \sin \theta) + c, \qquad y = \frac{a}{2}(1 - \cos \theta)$

Using the fact that $x = 0$ when $y = 0$, we must have $c = 0$ so that the parametric equations of the required curve are given by

$$x = \frac{a}{2}(\theta + \sin \theta), \qquad y = \frac{a}{2}(1 - \cos \theta) \tag{14}$$

Interpretation. The curve described by (14) is a *cycloid* which is generated by a fixed point on a circle of diameter a as it rolls along the lower part of the dashed line $y = a$, as indicated in Fig. 6.11. In our case the required shape of the wire is represented by that part of the cycloid shown heavy in the figure. The size of the cycloid will naturally depend on the particular value T.

The curve obtained is often called a *tautochrone* from the Greek *tautos*, meaning *same* or *identical*, and *chronos*, meaning *time*. The problem of finding the required curve known as the *tautochrone problem** was proposed near the end of the 17th century and solved in various ways by some prominent mathematicians of that time. One of these was *Huygens*, who employed the principle to design a *cycloidal pendulum* for use in clocks. In this design (see Fig. 6.12) the pendulum bob B is at the end of a flexible string whose opposite end is fixed at O. The pendulum is constrained by the two neighboring arches of the cycloid OP and OQ so that it swings between them.

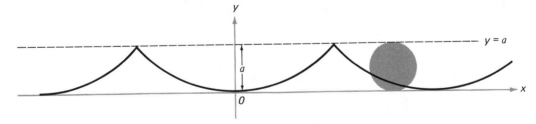

Figure 6.11

* The student should compare the tautochrone problem with the brachistochrone problem (C Exercise 1, page 130), which also involves the cycloid.

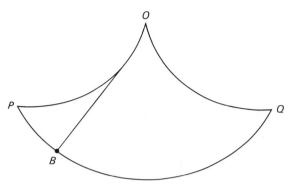

Figure 6.12

The period of oscillation is constant, and the path of the pendulum bob turns out to be a cycloid (see C Exercise 5, page 298).

The tautochrone problem, although seemingly only an interesting exercise in mechanics and differential equations, actually turned out to be of greater mathematical significance because it inspired *Abel* in 1823 to a study of *integral equations*. Research in this interesting and important branch of mathematics with numerous applications was done by many mathematicians of the 19th and early 20th centuries, but many unsolved problems still remain. The student wishing to study this subject should consult some of the references given in the Bibliography.

4.4 APPLICATIONS INVOLVING THE DELTA FUNCTION

As we mentioned on page 271, some important applied problems can be formulated in terms of the Dirac delta function. For instance let us consider

ILLUSTRATIVE EXAMPLE 5

A mass m is attached to the lower end of a vertical spring of constant k suspended from a fixed point. At time $t = t_0$ the mass is struck by applying a force in the upward direction lasting for a very short time. Describe the subsequent motion.

Mathematical Formulation. Assuming that the vertical axis of the spring is taken as the x axis and that the mass is initially at $x = 0$, we have

$$m\frac{d^2x}{dt^2} + kx = P_0\delta(t - t_0), \qquad x(0) = 0, x'(0) = 0 \tag{15}$$

Here we have assumed that the impulse of the force applied to the mass is constant and equal to P_0 so that the force can be taken as $P_0\delta(t - t_0)$.

Solution Take the Laplace transform of the differential equation in (15) using the initial conditions and the fact that $\mathcal{L}\{\delta(t - t_0)\} = e^{-st_0}$. Then if $\bar{x} = \mathcal{L}\{x\}$, we have

$$(ms^2 + k)\bar{x} = P_0e^{-st_0} \quad \text{or} \quad \bar{x} = \frac{P_0e^{-st_0}}{ms^2 + k}$$

Since

$$\mathcal{L}^{-1}\left\{\frac{P_0}{ms^2 + k}\right\} = \frac{P_0}{m}\mathcal{L}^{-1}\left\{\frac{1}{s^2 + k/m}\right\} = \frac{P_0}{m}\left(\frac{\sin\sqrt{k/m}\,t}{\sqrt{k/m}}\right) = \frac{P_0}{\sqrt{km}}\sin\sqrt{\frac{k}{m}}\,t$$

we have

$$x = \mathcal{L}^{-1}\left\{\frac{P_0 e^{-st_0}}{ms^2 + k}\right\} = \begin{cases} 0, & t < t_0 \\ \dfrac{P_0}{\sqrt{km}}\sin\sqrt{\dfrac{k}{m}}\,(t - t_0), & t > t_0 \end{cases} \tag{16}$$

Interpretation. From (16) we see that the mass remains at rest up to the time t_0, after which it oscillates sinusoidally with period $2\pi\sqrt{m/k}$ and amplitude $P_0/\sqrt{km}$. In this example we have not taken damping into account. This is left to B Exercise 12.

4.5 AN APPLICATION TO THE THEORY OF AUTOMATIC CONTROL AND SERVOMECHANISMS

Suppose that a missile M is tracking down an enemy aircraft or other missile E as indicated in Fig. 6.13. If at time t the enemy E turns through some angle, $\Psi(t)$, then M must also turn through this angle if it is to catch up with E and destroy it. If a man were aboard M, he could operate some steering mechanism to produce the required turn; but since the missile is unmanned for reasons of safety, such control must be accomplished automatically. To do this we need something to substitute for a man's eyes, such as a radar beam which will indicate or point to the direction which must be taken by M. We also need something providing a substitute for a man's hands which will turn a shaft through some angle in order to produce the desired turn. A mechanism, whether it involves electrical, mechanical, or other principles, designed to accomplish such automatic control is called a *servomechanism* or briefly a *servo*.

Mathematical Formulation. In the present application let us assume that the desired angle of turn as indicated by the radar beam is $\Psi(t)$. Also let $\Theta(t)$ denote the angle of turn of the shaft at time t. Ideally we would like to have $\Theta(t) = \Psi(t)$, but because things are happening so fast we must expect to have a discrepancy or error between the two given by

$$\text{error} = \Theta(t) - \Psi(t) \tag{17}$$

The existence of the error must be signaled back to the shaft, sometimes referred to as a *feedback signal*, so that a compensating turning effect or torque may be produced. If the error is large, the torque needed will be large. If the error is small, the torque

Figure 6.13

needed will be small. It is thus reasonable to design the servomechanism so that the required torque is proportional to the error (17). Now we know from mechanics that the torque is equal to the moment of inertia of the thing being turned (in this case the shaft together with whatever else is connected to it), denoted by I, multiplied by the angular acceleration given by $\Theta''(t)$. We thus have from (17)

$$I\Theta''(t) = -\kappa[\Theta(t) - \Psi(t)] \tag{18}$$

where $\kappa > 0$ is the constant of proportionality. The negative sign before κ is used because if the error is positive (i.e., the turn is too great) then the torque must oppose it (i.e., be negative), while if the error is negative the torque must be positive. Assuming that the initial angle and angular velocity are zero as possible conditions, we have

$$\Theta(0) = 0, \qquad \Theta'(0) = 0 \tag{19}$$

In arriving at (18) we have neglected damping. For this case, see B Exercise 15.

Solution Taking the Laplace transform of both sides of (18) using conditions (19), and assuming that $\mathscr{L}\{\Theta(t)\} = \theta(s)$, $\mathscr{L}\{\Psi(t)\} = \psi(s)$, we have

$$Is^2\theta(s) = -\kappa[\theta(s) - \psi(s)] \quad \text{or} \quad \theta(s) = \frac{\kappa\psi(s)}{Is^2 + \kappa} \tag{20}$$

Then by the convolution theorem

$$\Theta(t) = \sqrt{\frac{\kappa}{I}} \int_0^t \Psi(u) \sin\sqrt{\frac{\kappa}{I}}(t - u)du \tag{21}$$

Interpretation. The result (21) enables us to determine $\Theta(t)$ from $\Psi(t)$. In the theory of automatic control $\Psi(t)$ is often called the *input function* or briefly the *input*, $\Theta(t)$ is called the *output function* or briefly the *output*. The multiplying factor in (20),

$$\frac{\kappa}{Is^2 + \kappa}$$

which serves to characterize the behavior of the servomechanism in relating the input and output is called the *transfer function* or *response function*.

Servomechanisms arise in many connections in practice, as for example in the home where a thermostat is used to regulate the temperature and on ships or planes where an automatic pilot is needed. The basic idea of a servomechanism is portrayed schematically in the block diagram of Fig. 6.14. In the first block on the left we have the desired state (for example, the desired position and direction in the case of a missile or the desired temperature of a room). Since the desired state is not the same

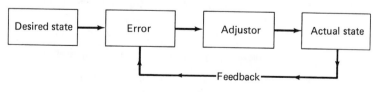

Figure 6.14

as the actual state, there is an error indicated by the second block. This error is fed into an adjustor, indicated by the third block, which attempts to rectify the error (such as an adjusting torque in the case of the missile) and leads to the actual state indicated by the block on the right. This actual state is then fed back (feedback signal) to reveal the new error of departure from the desired state, and the process is repeated over and over until the desired state is achieved.

A EXERCISES

1. An electric circuit consists of a resistor of resistance R ohms in series with an inductor of inductance L henries and a generator of E volts where R, L, and E are constants. If the current is zero at time $t = 0$ find it at any later time.

2. An object of mass m is thrown vertically upward with initial speed v_0. Assuming that the acceleration due to gravity is constant and equal to g, and neglecting air resistance, determine the position and velocity of the object at any later time.

3. Work Exercise 1 if the generator has voltage given by

   (a) $E(t) = \begin{cases} E_0, & 0 \le t \le T \\ 0, & t > T \end{cases}$

   (b) $E(t) = E_0 \sin \omega t$.

4. Work (a) Illustrative Example 3, page 76; (b) A Exercise 3, page 88; (c) Illustrative Example 1, page 221.

B EXERCISES

1. A mass m at the end of a vertical vibrating spring of constant k undergoes vibrations about its equilibrium position according to the equation

$$m \frac{d^2 X}{dt^2} + \beta \frac{dX}{dt} + kX = F_0 \cos \omega t$$

where β is a damping constant, and X is the displacement of the mass from its equilibrium position at any time t. (a) Solve this equation subject to the initial conditions $X(0) = X'(0) = 0$. (b) What is the steady-state solution? (c) Explain how the Laplace transform of the solution can be simplified so as to yield the steady-state solution.

2. Work Exercise 1 if the initial conditions are modified to read $X(0) = X_0$, $X'(0) = V_0$. Interpret the results physically.

3. Work (a) Illustrative Example 8, page 240; (b) the Illustrative Example on page 243.

4. Work (a) A Exercise 1, page 244; (b) B Exercise 1, page 245.

5. Work Exercise 1 if the external force on the mass m is given by

   (a) $F(t) = \begin{cases} F_0, & 0 < t < T \\ 0, & t > T \end{cases}$

   (b) $F(t) = F_0 t \cos \omega t$.

6. Work the problem on cardiography, page 249, by using Laplace transforms.

7. Work the problem on economics, page 250, by using Laplace transforms.

8. Use the parametric equations for the cycloid given by (14), page 289, to verify directly that the time taken for the bead to slide down the wire in Illustrative Example 4 is T.

9. Prove the result stated at the end of Illustrative Example 3, page 289.

10. Discuss how you would construct a wire in a vertical plane so that a bead placed anywhere on it would slide from rest to the lower end in 3 sec. What problems would you expect to arise from a physical point of view?

11. A particle of mass m is at rest at the origin O on the x axis. At $t = t_0$ it is acted upon by a force for a very short interval of time where the impulse of the force is a constant P_0. (a) Set up an initial value problem describing the motion. (b) Solve and interpret the results.

12. Work Illustrative Example 5, page 293, if damping is taken into account.

13. An electric circuit is made up of a resistance R and inductance L in series. At $t = t_0$ a very large voltage is introduced into the circuit but lasts for a very short time. Assuming that the initial current is zero, what is the current at any later time?

14. Work the servomechanism problem of page 294 if (a) $\Psi(t) = \alpha$, (b) $\Psi(t) = \alpha t$, (c) $\Psi(t) = \alpha \sin \omega t$, where α, ω are constants.

15. Work the servomechanism problem of page 294 if there is a damping force acting on the shaft which is proportional to the instantaneous angular velocity.

16. Discuss the characteristics of the servomechanism if the transfer function is given by

(a) $\dfrac{s+1}{s^2+1}$. (b) $\dfrac{s-2}{s^2-2s+1}$. (c) $\dfrac{s}{s^3-1}$.

C EXERCISES

1. The vibrations of a mass m at the end of a vertical spring of constant k are given by

$$m\frac{d^2X}{dt^2} + kX = F(t)$$

where $F(t)$ is the applied external force at any time t and X is the displacement of m from its equilibrium position at any time t. Suppose that the force is given by

$$F(t) = \begin{cases} F_0, & 0 < t < T \\ 2F_0, & T < t < 2T \\ 0, & t > 2T \end{cases}$$

(a) Find the displacement at any time t assuming that the initial displacement and velocity are zero. (b) Describe physically the vibrations of the mass.

2. Work Exercise 1 if a damping term $\beta\, dX/dt$ is taken into account.

3. Suppose that the force $F(t)$ in Exercise 1 is given by

$$F(t) = \begin{cases} F_0/\epsilon, & 0 < t < \epsilon \\ 0, & t > \epsilon \end{cases}$$

(a) Find the displacement at any time t assuming that the initial displacement and velocity are zero. (b) Discuss the result in (a) for the limiting case where $\epsilon \to 0$ and give a physical interpretation. (c) How is your result in (b) related to the Dirac delta function? (d) Could you obtain the result in (b) by letting $F(t) = F_0\delta(t)$ in the equation of Exercise 1 and then taking Laplace transforms using $\mathcal{L}\{\delta(t)\} = 1$? Explain.

4. Discuss Illustrative Example 4, page 289, in case the bead is given an initial speed v_0 at the starting point.

5. Prove the statements made about the cycloidal pendulum in the next to the last paragraph on page 293.

6. Work Illustrative Example 3, page 141, by using the delta function. [*Hint:* Use the differential equation $EIy^{(IV)} = w(x)$ obtained in C Exercise 8, page 147.]

7. Work A Exercise 1, page 145, by using the delta function.

8. Show how to find the solution of *Abel's integral equation* $\int_0^t \dfrac{Y(u)}{(t-u)^\alpha}\, du = F(t), \qquad 0 < \alpha < 1$

   where $F(t)$ is given and $Y(t)$ is to be determined.

9. (a) Discuss Exercise 8 for the special case $\alpha = \frac{1}{2}$ and explain the relationship of the solution with the *half derivative* of C Exercise 11 page 285. (b) Show that the solution of the tauto-chrone problem on page 289 depends on the solution of a "half-order" differential equation.

seven
solution of differential equations by use of series

1 Introduction to the Use of Series

1.1 MOTIVATION FOR SERIES SOLUTIONS

Until now we have been concerned with, and in fact have restricted ourselves to, differential equations which could be solved exactly and various applications which led to them. There are certain differential equations which are of great importance in higher mathematics and engineering or other scientific applications but which cannot be solved exactly in terms of elementary functions by any methods. For example, the innocent looking differential equations

$$y' = x^2 + y^2, \qquad xy'' + y' + xy = 0$$

cannot be solved exactly in terms of functions usually studied in elementary calculus, such as the rational algebraic, trigonometric and inverse trigonometric, exponential, and logarithmic functions.

It is rather frustrating to the engineer or scientist, and even the mathematician, to know that a solution exists and is unique but not be able to determine it exactly. Of course, it is a good thing to know that a solution exists and is unique because we can in such case usually employ numerical methods (which is the subject of Chapter Nine) in association with powerful computers which are available to provide us with desired answers. While this is wonderful from a practical point of view in our desperation to achieve answers, some of us may feel that we are in effect giving up and turning to this as a last resort.

Fortunately, to those among us who may feel this way there is another method which we can try when no other seems to yield an exact solution. To motivate a discussion of this method, let us consider the differential equation $y' = y$ subject to the condition that $y(0) = 1$. By separation of variables we easily discover that $y = e^x$ is the required solution. Now in calculus we learned that many functions such as e^x, $\sin x$, and $\cos x$ possess series expansions of the type

$$a_0 + a_1 x + a_2 x^2 + a_3 x^3 + \cdots \tag{1}$$

often called *power series*. We raise the following

Question. Let us assume that we could not solve $y' = y$, supposing, for instance, that we were not yet acquainted with properties of exponential functions. What possible way could we proceed to find the required solution (assuming one existed)?

One possible way in which we might begin would be to *assume* that the solution (if it exists) possesses a series expansion of the type (1), where $a_0, a_1, \ldots$ are constants at present undetermined. If (1) is to be a solution of $y' = y$, we must have

$$\frac{d}{dx}(a_0 + a_1 x + a_2 x^2 + a_3 x^3 + \cdots) = a_0 + a_1 x + a_2 x^2 + a_3 x^3 + \cdots$$

Now, *assuming* that term-by-term differentiation of infinite series is allowed, we have

$$a_1 + 2a_2 x + 3a_3 x^2 + 4a_4 x^3 + \cdots = a_0 + a_1 x + a_2 x^2 + a_3 x^3 + \cdots$$

Since this must be an identity, we must have coefficients of corresponding powers of x equal,* so that

$$a_1 = a_0, \quad 2a_2 = a_1, \quad 3a_3 = a_2, \quad 4a_4 = a_3, \ldots$$

From these we find

$$a_1 = a_0, \quad a_2 = \frac{a_1}{2} = \frac{a_0}{2}, \quad a_3 = \frac{a_2}{3} = \frac{a_0}{3 \cdot 2} = \frac{a_0}{3!}, \quad a_4 = \frac{a_3}{4} = \frac{a_0}{4!}, \ldots$$

the rule being apparent. Substituting these in the assumed solution, we have

$$y = a_0 \left(1 + x + \frac{x^2}{2!} + \frac{x^3}{3!} + \frac{x^4}{4!} + \cdots \right)$$

Using the condition that $y = 1$ when $x = 0$, we find $a_0 = 1$, so that

$$y = 1 + x + \frac{x^2}{2!} + \frac{x^3}{3!} + \frac{x^4}{4!} + \cdots \tag{2}$$

Since we have found the result (2), assuming so many things, it is natural for us to ask whether this really is the required solution. Anyone acquainted with series knows that (2) is the series expansion for e^x, so that we really have obtained the required solution. In cases where we have nothing to check with we may really be in doubt. The solution in the form (2) is just as good as $y = e^x$, and in fact, for many purposes it is better. For example, if one wished to know the value of y when $x = 0.6$, it is true that the answer $e^{0.6}$ may be found in the tables, but as a matter of fact the tabular value was probably computed by using (2) with x replaced by 0.6.

Although we have perhaps been unduly simple-minded in the above treatment, the general conclusions are applicable to many important cases. We must realize that we have not been rigorous because proofs of several steps have been omitted. We shall consider the question of rigor on page 306. In the meantime let us apply the method to another example.

ILLUSTRATIVE EXAMPLE 1

Solve $y'' + y = 0$ by use of series.

Solution   Let $y = a_0 + a_1 x + a_2 x^2 + a_3 x^3 + a_4 x^4 + a_5 x^5 + \cdots$ (3)

so that $y' = a_1 + 2a_2 x + 3a_3 x^2 + 4a_4 x^3 + 5a_5 x^4 + \cdots$ (4)

$y'' = 2a_2 + 6a_3 x + 12a_4 x^2 + 20a_5 x^3 + \cdots$ (5)

Substituting (3) and (5) into the given differential equation and combining like terms gives

$$(a_0 + 2a_2) + (a_1 + 6a_3)x + (a_2 + 12a_4)x^2 + (a_3 + 20a_5)x^3 + \cdots = 0$$

---

* This is a consequence of the fact that if we have $c_0 + c_1 x + c_2 x^2 + c_3 x^3 + \cdots \equiv 0$ where the series on the left is convergent in some interval, then we must have $c_0 = 0, c_1 = 0, c_2 = 0$, etc.

Since the right side is zero, this can be an identity if and only if each of the coefficients on the left is zero. Thus,

$$a_0 + 2a_2 = 0, \qquad a_1 + 6a_3 = 0, \qquad a_2 + 12a_4 = 0, \qquad a_3 + 20a_5 = 0, \ldots$$

This leads to

$$a_2 = -\frac{a_0}{2}, \qquad a_3 = -\frac{a_1}{6} = -\frac{a_1}{3!}, \qquad a_4 = -\frac{a_2}{12} = \frac{a_0}{4!}, \qquad a_5 = -\frac{a_3}{20} = \frac{a_1}{5!}, \ldots$$

Substituting these into (3), we have

$$y = a_0 + a_1 x - \frac{a_0}{2} x^2 - \frac{a_1}{3!} x^3 + \frac{a_0}{4!} x^4 + \frac{a_1}{5!} x^5 + \cdots$$

or

$$y = a_0 \left(1 - \frac{x^2}{2!} + \frac{x^4}{4!} - \cdots \right) + a_1 \left(x - \frac{x^3}{3!} + \frac{x^5}{5!} - \cdots \right) \tag{6}$$

Since (6) involves two arbitrary constants we suspect that it represents the general solution of the given differential equation.

We are fortunate in this example to have a check, since if we recall from calculus

$$\cos x = 1 - \frac{x^2}{2!} + \frac{x^4}{4!} - \cdots, \quad \sin x = x - \frac{x^3}{3!} + \frac{x^5}{5!} - \cdots$$

so that $y = a_0 \cos x + a_1 \sin x$ as we have already found by methods in Chapter Four. It is interesting to note that even if we had never heard of $\sin x$ or $\cos x$ we would be led to them by trying to solve the given differential equation. We could then obtain their properties, their graphs, etc. See C Exercise 3. The moral is of course that differential equations can lead to functions with which we do not happen to be familiar but that may have important or interesting properties. We shall see some examples of such functions later in this chapter.

Remark 1. Instead of using the series (3) we could also have used the power series about $x = a$, given by

$$a_0 + a_1(x - a) + a_2(x - a)^2 + a_3(x - a)^3 + \cdots \tag{7}$$

This is used for instance when the conditions are specified at $x = a$. Thus, for example, if we had to solve $y' = y$ given $y(1) = 2$, we could use the series (7) with $a = 1$. Alternatively, we could simply make the change of independent variable $v = x - a = x - 1$ and then use (3) with x replaced by v. This amounts to replacing x by $x - a = x - 1$. The series (7) is often called a *Taylor series* about $x = a$, and the special case where $a = 0$ is called a *Maclaurin series*.

1.2 USE OF THE SUMMATION NOTATION

In obtaining the solutions of the differential equations $y' = y$ and $y'' + y = 0$, we had to do a lot of writing and would have had to write even more if we used more terms. A short cut which not only reduces the labor involved but is often helpful in recognizing the general term of the series solution is provided by the *summation notation* with which the student is probably already familiar from a study of series or definite integrals in calculus.

According to the summation notation, a series such as

$$u_0 + u_1 + u_2 + u_3 + \cdots + u_n \tag{8}$$

with a finite number of terms is represented by

$$\sum_{j=0}^{n} u_j \tag{9}$$

which is read *the sum of all terms having the form* u_j, *where j goes from 0 to n.* The sign $\sum$ is a Greek capital letter *sigma*, j is called the *summation index* or briefly *index*, and we can read (9) briefly as *sigma, or sum, of* u_j *from* $j = 0$ *to n.* As for definite integrals $j = 0$ is referred to as the *lower limit*, while n is referred to as the *upper limit*.

In the case where we have an infinite series

$$u_0 + u_1 + u_2 + u_3 + \cdots \tag{10}$$

we represent it by

$$\sum_{j=0}^{\infty} u_j \tag{11}$$

where the upper limit n is replaced by ∞.

The following are some examples of the use of the summation index.

Example 1. $\displaystyle\sum_{j=0}^{n} j(j + 1) = 1 \cdot 2 + 2 \cdot 3 + 3 \cdot 4 + \cdots + n(n + 1).$

Example 2. $\displaystyle\sum_{j=0}^{\infty} a_j x^j = a_0 + a_1 x + a_2 x^2 + a_3 x^3 + \cdots.$

Example 3. $\displaystyle\sum_{j=0}^{\infty} \frac{x^j}{j!} = 1 + x + \frac{x^2}{2!} + \frac{x^3}{3!} + \cdots$ taking $0! = 1$.

We can of course use other limits besides 0 and n or ∞. For instance,

$$\sum_{j=2}^{6} u_j = u_2 + u_3 + u_4 + u_5 + u_6$$

i.e., the sum of u_j where j goes from 2 to 6. We can also use other indexes. Thus the sum just given can be represented in any of the forms

$$\sum_{k=2}^{6} u_k, \qquad \sum_{k=0}^{4} u_{k+2}, \qquad \sum_{j=0}^{4} u_{j+2}$$

The following are two important properties of the summation notation, whose proofs are easily verified by writing out the terms on each side.

1. $\displaystyle\sum_{j=0}^{n} u_j + \sum_{j=0}^{n} v_j = \sum_{j=0}^{n} (u_j + v_j)$

2. $\displaystyle\alpha \sum_{j=0}^{n} u_j = \sum_{j=0}^{n} \alpha u_j$ for any α independent of j

The results are also valid if other limits besides 0 and n are used provided they are the same in each sum. However, if any limit is infinite, the series must be convergent.

For purposes of finding power series solutions of differential equations, it is convenient to use the notation

$$\sum_{j=-\infty}^{\infty} a_j x^j \tag{12}$$

where we shall agree that $a_j = 0$ for all negative integer values of j, i.e., $j = -1, -2, -3, \ldots$. In such case the series (12) is equivalent to that in Example 2 above.

To illustrate the procedure involved in obtaining series solutions using the summation notation, let us consider the following

PROBLEM FOR DISCUSSION

Solve $y'' + y = 0$ (Illustrative Example 1).
We assume as before that

$$y = \sum_{j=-\infty}^{\infty} a_j x^j, \qquad a_j = 0, j = -1, -2, -3, \ldots \tag{13}$$

Then by differentiation
$$y' = \sum_{j=-\infty}^{\infty} j a_j x^{j-1} \tag{14}$$

$$y'' = \sum_{j=-\infty}^{\infty} j(j-1) a_j x^{j-2} \tag{15}$$

Substituting (13) and (15) in the given differential equation gives

$$\sum_{j=-\infty}^{\infty} j(j-1) a_j x^{j-2} + \sum_{j=-\infty}^{\infty} a_j x^j = 0 \tag{16}$$

We would now like to combine corresponding terms in the two series on the left of (16) so that the resulting series will reveal the coefficient of x^j. Since the coefficient of x^{j-2} in the first summation of (16) is $j(j-1)a_j$, it follows on replacing j by $j+2$ that the coefficient of x^j is $(j+2)(j+1)a_{j+2}$. Furthermore, there is no need to worry about the limits in the summation since the index runs from $-\infty$ to ∞ and changing j to $j+2$ has no effect on the limits, which incidentally is why we introduced them in the first place. It follows that (16) can be written as

$$\sum_{j=-\infty}^{\infty} (j+2)(j+1) a_{j+2} x^j + \sum_{j=-\infty}^{\infty} a_j x^j = 0 \tag{17}$$

or assuming the series to be convergent and using Property 1 above

$$\sum_{j=-\infty}^{\infty} [(j+2)(j+1) a_{j+2} + a_j] x^j = 0 \tag{18}$$

Now (18) is an identity for all values of x for which the series converges if and only if each coefficient is zero, i.e.,

$$(j+2)(j+1) a_{j+2} + a_j = 0 \tag{19}$$

If we put $j = -2$ in (19), we get $0 \cdot a_0 + 0 = 0$, showing that a_0 is arbitrary. Similarly, if we put $j = -1$ in (19), we get $0 \cdot a_1 + 0 = 0$, showing that a_1 is also arbitrary. Note that if we put $j = -3$ in (19) we get $0 \cdot a_{-1} + 0 = 0$, but we already know that $a_{-1} = 0$ from our agreement on page 304, so this value of j yields no information. For $j \geq 0$ we obtain from (19)

$$a_{j+2} = -\frac{a_j}{(j+2)(j+1)} \tag{20}$$

Putting $j = 0, 1, 2, \ldots$ in succession we find that

$$a_2 = -\frac{a_0}{2 \cdot 1}, \quad a_3 = -\frac{a_1}{3 \cdot 2}, \quad a_4 = -\frac{a_2}{4 \cdot 3} = \frac{a_0}{4!}, \quad a_5 = -\frac{a_3}{5 \cdot 4} = \frac{a_1}{5!}, \ldots \tag{21}$$

Using these in $\quad y = \sum_{j=-\infty}^{\infty} a_j x^j = a_0 + a_1 x + a_2 x^2 + \cdots$

yields as before $\quad y = a_0\left(1 - \frac{x^2}{2!} + \frac{x^4}{4!} - \cdots\right) + a_1\left(x - \frac{x^3}{3!} + \frac{x^5}{5!} - \cdots\right) \tag{22}$

The equations (19) or (20) are often called *recurrence formulas* because they enable us to find as many terms of the series as we like. Thus, for instance, if we wanted to find a_7, we would obtain from (20) on putting $j = 5$

$$a_7 = -\frac{a_5}{7 \cdot 6} = \frac{a_3}{7 \cdot 6 \cdot 5 \cdot 4} = -\frac{a_1}{7 \cdot 6 \cdot 5 \cdot 4 \cdot 3 \cdot 2} = -\frac{a_1}{7!} \tag{23}$$

Some further brevity in writing can be achieved in the above by omitting the summation limits in (13) and later sums, these limits being of course understood throughout.

In order to gain further practice in finding solutions using the summation technique, let us consider another example.

ILLUSTRATIVE EXAMPLE 2

Solve $y'' + 2xy' - y = 0$ subject to the conditions $y(0) = 0$, $y'(0) = 1$.

Solution   Let $\qquad\qquad y = \sum a_j x^j \tag{24}$

omitting summation limits $-\infty, \infty$. Then by differentiation

$$y' = \sum j a_j x^{j-1} \tag{25}$$

$$y'' = \sum j(j-1)a_j x^{j-2} \tag{26}$$

Using (24), (25), and (26) in the given differential equation and employing property 2 on page 303, we find

$$\sum j(j-1)a_j x^{j-2} + 2x \sum j a_j x^{j-1} - \sum a_j x^j = 0 \tag{27}$$

or $\qquad \sum j(j-1)a_j x^{j-2} + \sum 2j a_j x^j - \sum a_j x^j = 0 \tag{28}$

In order that all sums in (28) involve x^j, replace j by $j + 2$ in the first. Then (28) becomes

$$\sum (j+2)(j+1)a_{j+2} x^j + \sum 2j a_j x^j - \sum a_j x^j = 0 \tag{29}$$

Using the properties of sums on page 303 with the first extended to three sums instead of two, we can now write (29) as a single summation

$$\sum [(j + 2)(j + 1)a_{j+2} + (2j - 1)a_j]x^j = 0 \qquad (30)$$

Thus

$$(j + 2)(j + 1)a_{j+2} + (2j - 1)a_j = 0 \qquad (31)$$

or

$$a_{j+2} = \frac{-(2j - 1)a_j}{(j + 2)(j + 1)} \qquad (32)$$

Putting $j = 0, 1, 2, \ldots$, we obtain

$$a_2 = \frac{a_0}{2 \cdot 1}, \qquad a_3 = -\frac{a_1}{3 \cdot 2}, \qquad a_4 = \frac{-3a_2}{4 \cdot 3} = -\frac{3}{4!}a_0,$$

$$a_5 = \frac{-5a_3}{5 \cdot 4} = \frac{5}{5!}a_1, \qquad a_6 = \frac{-7a_4}{6 \cdot 5} = \frac{3 \cdot 7}{6!}a_0, \qquad a_7 = \frac{-9a_5}{7 \cdot 6} = -\frac{5 \cdot 9}{7!}a_1$$

etc. Using these values in $y = \sum a_j x^j = a_0 + a_1 x + a_2 x^2 + \cdots$, we find

$$y = a_0 \left(1 + \frac{x^2}{2!} - \frac{3x^4}{4!} + \frac{3 \cdot 7 x^6}{6!} - \frac{3 \cdot 7 \cdot 11 x^8}{8!} + \cdots \right)$$

$$+ a_1 \left(x - \frac{x^3}{3!} + \frac{5x^5}{5!} - \frac{5 \cdot 9 x^7}{7!} + \frac{5 \cdot 9 \cdot 13 x^9}{9!} - \cdots \right) \qquad (33)$$

From $y(0) = 0$ and $y'(0) = 1$, we obtain $a_0 = 0$, $a_1 = 1$ respectively. Thus (33) becomes

$$y = x - \frac{x^3}{3!} + \frac{5x^5}{5!} - \frac{5 \cdot 9 x^7}{7!} + \frac{5 \cdot 9 \cdot 13 x^9}{9!} - \cdots \qquad (34)$$

The series in (34) does not seem to be related in any obvious manner to any elementary functions with which we are familiar. Thus there is no immediate check on the correctness of our work. We can, however, show that (34) is the correct solution. Knowing this we can if we desire study its properties, obtain its graph, etc. We can if we like even give it a name, and should the equation prove to be important enough the discoverer could go down in history. As we shall see, there are many such cases of differential equations and associated functions named after their discoverers.

1.3 SOME QUESTIONS OF RIGOR

From the results obtained on preceding pages various questions arise.

Question 1. How do we know that the series which we obtained formally are actually solutions of their corresponding differential equations? This is a reasonable question to ask since in obtaining these series certain questionable operations were performed, such as, for example, the differentiation of a series term by term. Being scientists with a desire for truth, we do not want to be termed guilty of "blind manipulation" in producing a certain result. One way of proceeding is to attempt justification of each step as we proceed. Unfortunately this may be impossible. For example, how do we know that we can differentiate the series $a_0 + a_1x + a_2x^2 + \cdots$ if we do not know what the coefficients are? Clearly we have a vicious cycle. We cannot prove that we have a solution until we know $a_0, a_1, a_2, \ldots$ and we cannot honestly say that we have found the coefficients until we have justified the steps.

Question 2. How do we know whether a series of the form (13) can produce a solution of a given differential equation? Of course, we could try it and see, but this is not too scientific and it would be much nicer to know in advance whether it would work.

A very good way of avoiding the difficulties raised in both of the above questions is to try to find some kind of existence and uniqueness theorem which will tell us when a differential equation has power series solutions such as (13), or more generally (7). At first sight it would seem that the theorem on page 171 could be used. Unfortunately, however, the fact that a solution exists does not necessarily mean that we can find it in the form (7).

To simplify matters, let us restrict ourselves to linear second-order differential equations having the form

$$p(x)y'' + q(x)y' + r(x)y = 0 \tag{35}$$

where $p(x)$, $q(x)$, and $r(x)$ are polynomials. It turns out that such differential equations do arise a great deal in practice, and also once we have information regarding them it is easier to generalize to higher-order or more complicated equations.* On solving for y'' in (35) we get

$$y'' = -\frac{q(x)y' + r(x)y}{p(x)} \tag{36}$$

Now if there is to be a solution of the type (7) we would certainly want y'' to exist at $x = a$, and it could be catastrophic if the denominator $p(x)$ in (36) were zero for $x = a$. This leads us to introduce the following

Definition. A value of x which is such that $p(x) = 0$ is called a *singular point*, or *singularity*, of the differential equation (35). Any other value of x is then called an *ordinary point* or *non-singular point*.

Example 1. Given the differential equation $x(1 - x)y'' - (2x + 1)y' + 3y = 0$, $x = 0$ and $x = 1$ are both singular points, while any other values of x, such as $x = \frac{1}{2}, -3$ for example, are ordinary points.

Example 2. The differential equation $xy'' + y' + xy = 0$ has only one singular point $x = 0$. Any other values are ordinary points.

---

* The case where $p(x)$, $q(x)$, $r(x)$ are *analytic functions*, i.e., have power series expansions in some interval of convergence, involves one such generalization.

Example 3. The equations $y'' + y = 0$ and $y'' + 2xy' - y = 0$ have no singular points, or in other words every value of x represents an ordinary point.

Example 4. The equation $(x^2 + 1)y'' - 2y' + xy = 0$ has singular points given by $x^2 + 1 = 0$, i.e., $x = \pm i$. Thus singular points (and ordinary points) can be complex numbers.

We then have the following interesting and important theorem whose proof, which is somewhat long and tedious, will be omitted.*

Theorem 1. Let
$$p(x)y'' + q(x)y' + r(x)y = 0 \tag{37}$$
be a given differential equation where $p(x)$, $q(x)$, and $r(x)$ are polynomials. Suppose that a is any ordinary point of (37), i.e., $p(a) \neq 0$. Then we can draw the following two conclusions:

1. The general solution of (37) can be obtained by substituting the power series (or Taylor series) about $x = a$ given by
$$y = a_0 + a_1(x - a) + a_2(x - a)^2 + \cdots = \sum a_j(x - a)^j \tag{38}$$
   into the given differential equation. An equivalent result is that the general solution has the form $y = c_1 u_1(x) + c_2 u_2(x)$, where $u_1(x)$ and $u_2(x)$ are linearly independent solutions each having the form (38).
2. The series solutions obtained in part 1 converge for all values of x such that $|x - a| < R$, where R is the distance from point a to the *nearest* singularity. We often call R the *radius of convergence*. For other values of x the series solutions may or may not converge.

It is often easy to confirm the second conclusion of this theorem regarding convergence of the series solutions by use of the *ratio test* which the student learned in calculus. For purposes of review we summarize the results in the following

Theorem 2 (Ratio test). Given the series $\sum_{n=0}^{\infty} u_n$, suppose that
$$\lim_{n \to \infty} \left| \frac{u_{n+1}}{u_n} \right| = l \tag{39}$$
where l is a number which is greater than or equal to zero, i.e., $l \geq 0$. Then the series converges if $l < 1$ and diverges if $l > 1$. However, the test fails if $l = 1$; i.e., the series may or may not converge if $l = 1$, and other tests must be used in such case.

Before showing how Theorem 1 is used, let us consider the following

Remark 2. In calculus we obtained the series expansion
$$\frac{1}{1 + x^2} = 1 - x^2 + x^4 - x^6 + \cdots = \sum_{n=0}^{\infty} (-1)^n x^{2n}$$

---

* See reference [13].

308 *Chapter Seven*

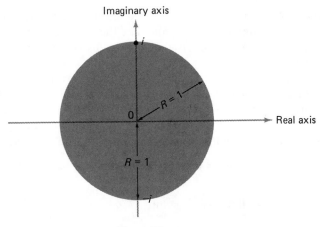

Imaginary axis

Real axis

$R = 1$

Figure 7.1

by using long division on the left or expansion into a Maclaurin series. Since the general term of the series is $u_n = (-1)^n x^{2n}$, we have by the ratio test

$$l = \lim_{n \to \infty} \left| \frac{(-1)^{n+1} x^{2(n+1)}}{(-1)^n x^{2n}} \right| = \lim_{n \to \infty} |x^2| = |x|^2$$

which shows that the series converges for $|x| < 1$ or $-1 < x < 1$ if x is real. Thus the series on the right is convergent for values of x in the interval $-1 < x < 1$, while there is *no such restriction* on x for the function on the left. It is natural to ask why the series on the right does not mean anything if $x = 2$, for example, when the function on the left has the value $\frac{1}{5}$.

Further investigation shows that the reason lies in the fact that the denominator on the left is zero for $x = \pm i$ or $|x| = 1$ on using the fact that the absolute value of a complex number $p + qi$ is $|p + qi| = \sqrt{p^2 + q^2}$. The convergence of the power series on the right is thus guaranteed only for those values of x which are less in absolute value than the distance R from $x = 0$ (which is an *ordinary point* for the function on the left) to $x = \pm i$ (which are *singular points* of the function on the left). In this case the distance is $R = 1$ so that the situation is described in Fig. 7.1, where the circle denotes the *circle of convergence* inside of which the series is guaranteed to converge and the radius $R = 1$ of this circle is the *radius of convergence*. It is this interpretation that we give to part 2 of Theorem 1. Note that Fig. 7.1 simply represents a rectangular coordinate system. Any complex number $p + qi$ can be plotted as the point (p, q) in this plane, p corresponding to the coordinate on the horizontal or *real axis* and q corresponding to the coordinate on the vertical or *imaginary axis*. For example, $i = 0 + 1i$ corresponds to the point $(0, 1)$.

Let us now give some examples illustrating the above theorems.

ILLUSTRATIVE EXAMPLE 3

(a) Use Theorem 1 to predict the convergence of the power series solution in Illustrative Example 2, page 305, and (b) verify by using the ratio test.

Solution (a) Since $y'' + 2xy' - y = 0$ has no singularity, the distance from $x = 0$ to the nearest singularity is infinite, i.e., the series converges for $|x| < \infty$ or all x. Another way of saying this is that the radius of convergence of the power series solution is infinite.

(b) We can test the series solution (34) for convergence by first finding the general term u_n and then using the ratio test given in Theorem 2. However, because we need only the ratio of successive terms which can be supplied by using the recurrence formula (32), it is simpler to proceed as follows. Let $j = 2n - 1$, where $n = 1, 2, 3, \ldots$, since (34) contains only terms $a_j x^j$ where j is odd. Thus we have

$$u_n = a_{2n-1} x^{2n-1}, \qquad u_{n+1} = a_{2n+1} x^{2n+1}$$

and

$$l = \lim_{n \to \infty} \left| \frac{u_{n+1}}{u_n} \right| = \lim_{n \to \infty} \left| \frac{a_{2n+1} x^{2n+1}}{a_{2n-1} x^{2n-1}} \right| = \lim_{n \to \infty} \left| \frac{a_{2n+1}}{a_{2n-1}} \right| |x|^2$$

$$= \lim_{n \to \infty} \frac{4n - 3}{(2n + 1)(2n)} |x|^2 = 0$$

Since $l = 0$ for all x it follows from the ratio test that the series converges for all x as predicted in part (a). The nice thing about Theorem 1 is that we do not even have to work out the power series and examine its convergence.

ILLUSTRATIVE EXAMPLE 4

Determine whether power series solutions about $x = a$, for the indicated values of a, exist for each of the following differential equations and predict the set of values for which each series is guaranteed to converge: (a) $y'' + y = 0$; $a = 0$. (b) $xy'' + y' + xy = 0$; $a = 2$, $a = 0$. (c) $(x^2 + 1)y'' - 2y' + 5xy = 0$; $a = 0$, $a = 1$.

Solution (a) The equation has no singularity. Thus the distance from $a = 0$ to the nearest singularity is infinite, i.e., the series converges for $|x| < \infty$ or all x. That this is in fact true is verified on page 302, where the series solutions for $\sin x$ and $\cos x$, convergent for all x, were obtained.

(b) The equation has a singularity at $x = 0$. Thus the distance from $a = 2$ to the nearest singularity is 2. It follows by Theorem 1 that we can find the general solution by using

$$y = a_0 + a_1(x - 2) + a_2(x - 2)^2 + a_3(x - 2)^3 + \cdots$$

The resulting series will be convergent for $|x - 2| < 2$, or $-2 < x - 2 < 2$, i.e., $0 < x < 4$, if x is real. At $x = 0$ and $x = 4$ the series may or may not converge.

For $a = 0$, which is a singularity, Theorem 1 does not apply since only power series about ordinary points are considered. We shall discuss power series solutions about singularities in the next section.

(c) The equation has a singularity at values of x for which $x^2 + 1 = 0$, i.e., $x = \pm i$. If $a = 0$ the distance to the nearest singularity is 1, as is indicated in Fig. 7.1 (note that both singularities i and $-i$ are equidistant from 0). The general solution can be found by using the series

$$y = a_0 + a_1 x + a_2 x^2 + a_3 x^3 + \cdots$$

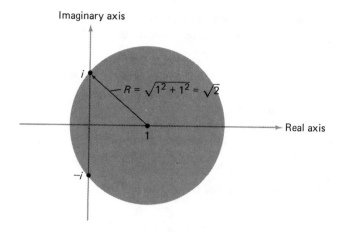

$$R = \sqrt{1^2 + 1^2} = \sqrt{2}$$

Real axis

Figure 7.2

and the resulting series will converge for $|x| < 1$, or if x is real, $-1 < x < 1$, but there is no guarantee of convergence at the endpoints -1 and 1.

If $a = 1$ the distance to the nearest singularity is seen from Fig. 7.2 to be $R = \sqrt{2}$ (note that both singularities i and $-i$ are equidistant from 1). The general solution can be found by using the series

$$y = a_0 + a_1(x - 1) + a_2(x - 1)^2 + a_3(x - 1)^3 + \cdots$$

and will be convergent for $|x - 1| < \sqrt{2}$, or if x is real, $1 - \sqrt{2} < x < 1 + \sqrt{2}$. However, it may or may not converge at the endpoints.

ILLUSTRATIVE EXAMPLE 5

Solve the initial value problem $xy'' - y = 0$, $y(2) = 0$, $y'(2) = 3$.

Solution The point $x = 0$ is a singularity of the differential equation. Since the initial conditions are specified at $x = 2$, we use the power series about $a = 2$, which is an ordinary point of the equation. It is convenient to make the transformation $v = x - 2$ so that $v = 0$ when $x = 2$. In such case the initial value problem becomes

$$(v + 2)\frac{d^2y}{dv^2} - y = 0, \qquad y = 0, \qquad \frac{dy}{dv} = 3 \quad \text{at} \quad v = 0 \tag{40}$$

To solve this assume $\qquad y = a_0 + a_1v + a_2v^2 + a_3v^3 + \cdots = \sum a_jv^j$

where as usual $a_j = 0$, $j < 0$. Then we have

$$(v + 2)\frac{d^2y}{dv^2} - y = (v + 2)\sum j(j - 1)a_jv^{j-2} - \sum a_jv^j$$

$$= \sum j(j - 1)a_jv^{j-1} + \sum 2j(j - 1)a_jv^{j-2} - \sum a_jv^j$$

$$= \sum [(j + 1)ja_{j+1} + 2(j + 2)(j + 1)a_{j+2} - a_j]v^j = 0$$

so that $\qquad (j + 1)ja_{j+1} + 2(j + 2)(j + 1)a_{j+2} - a_j = 0 \tag{41}$

Putting $j = -2$ and $j = -1$ in (41) shows that a_0 and a_1 are both arbitrary. How-ever, by noting that $y = a_0$, $dy/dv = a_1$ at $v = 0$, we see from the conditions in (40) that

$$a_0 = 0, \qquad a_1 = 3 \qquad (42)$$

We now have from (41) $\qquad a_{j+2} = \dfrac{a_j - (j + 1)ja_{j+1}}{2(j + 2)(j + 1)} \qquad (43)$

Putting $j = 0, 1, 2, \ldots$ in succession and using (42), we find

$$a_2 = 0, \qquad a_3 = \tfrac{1}{4}, \qquad a_4 = -\tfrac{1}{16}, \qquad a_5 = \tfrac{1}{40}, \qquad a_6 = -\tfrac{3}{320}, \ldots$$

Thus the required (unique) solution is

$$y = \sum a_j v^j = 3v + \frac{v^3}{4} - \frac{v^4}{16} + \frac{v^5}{40} - \frac{3v^6}{320} + \cdots$$

$$= 3(x - 2) + \frac{(x - 2)^3}{4} - \frac{(x - 2)^4}{16} + \frac{(x - 2)^5}{40} - \frac{3(x - 2)^6}{320} + \cdots \quad (44)$$

where by use of Theorem 1 the series is convergent for $|x - 2| < 2$ or, since x is real, $0 < x < 4$. The series may or may not converge at the endpoints.

Remark 3. It should be noted that the recurrence formula (43) is a *three-term recurrence formula* instead of the usual two-term result obtained before. This does not cause any difficulty although it tends to obscure the general term of the series. We can, however, obtain as many terms as we desire, and there is no cause for con-cern since we know the values of x for which the series converges from Theorem 1. The student may find it of interest to deduce the radius of convergence of the above series solution by use of (43) and the ratio test (see B Exercise 6).

1.4 THE TAYLOR SERIES METHOD

An alternative method for finding power series solutions of the differential equa-tion (35) about an ordinary point $x = a$ is available and is known as the *Taylor series method*. This method uses the values of the derivatives evaluated at the ordinary point, which are obtained from the differential equation by successive differentiation. When the derivatives are found, we then use the Taylor series expansion

$$y(x) = y(a) + y'(a)(x - a) + \frac{y''(a)(x - a)^2}{2!} + \frac{y'''(a)(x - a)^3}{3!} + \cdots \quad (45)$$

giving the required solution. Let us consider some examples of this method.

ILLUSTRATIVE EXAMPLE 6

Solve by the method of Taylor series $y' = x + y + 1$.

Solution From the differential equation we find

$$y' = x + y + 1, \qquad y'' = 1 + y', \qquad y''' = y'', \qquad y^{(IV)} = y''', \ldots \quad (46)$$

Assuming $y = c$ when $x = 0$, we find from (46)

$$y'(0) = c + 1, \qquad y''(0) = c + 2, \qquad y'''(0) = c + 2, \ldots$$

Substituting into (45) with $a = 0$, we have

$$y(x) = c + (c + 1)x + \frac{(c + 2)x^2}{2!} + \frac{(c + 2)x^3}{3!} + \cdots$$

or
$$y(x) = c + (c + 1)x + (c + 2)\left(\frac{x^2}{2!} + \frac{x^3}{3!} + \cdots\right)$$

$$= c + (c + 1)x + (c + 2)(e^x - 1 - x) = (c + 2)e^x - x - 2$$

This can be checked as the solution by substituting in the given differential equation.

ILLUSTRATIVE EXAMPLE 7

Work Illustrative Example 5, page 311, by the Taylor series method.

Solution By successive differentiation of $xy'' - y = 0$ we have

$$xy''' + y'' - y' = 0, \qquad xy^{(IV)} + 2y''' - y'' = 0, \qquad xy^{(V)} + 3y^{(IV)} - y''' = 0,$$

$$xy^{(VI)} + 4y^{(V)} - y^{(IV)} = 0, \ldots$$

For $x = 2$ these become

$$2y''(2) - y(2) = 0, \qquad 2y'''(2) + y''(2) - y'(2) = 0, \qquad 2y^{(IV)}(2) + 2y'''(2) - y''(2) = 0,$$

$$2y^{(V)}(2) + 3y^{(IV)}(2) - y'''(2) = 0, \qquad 2y^{(VI)}(2) + 4y^{(V)}(2) - y^{(IV)}(2) = 0, \ldots$$

Using the conditions $y(2) = 0$, $y'(2) = 3$, these yield

$$y''(2) = 0, \qquad y'''(2) = \tfrac{3}{2}, \qquad y^{(IV)}(2) = -\tfrac{3}{2}, \qquad y^{(V)}(2) = 3, \qquad y^{(VI)}(2) = -\tfrac{27}{2}, \ldots$$

Thus

$$y(x) = y(2) + y'(2)(x - 2) + \frac{y''(2)(x - 2)^2}{2!} + \frac{y'''(2)(x - 2)^3}{3!} + \cdots$$

$$= 3(x - 2) + \frac{3/2}{3!}(x - 2)^3 + \frac{-3/2}{4!}(x - 2)^4 + \frac{3}{5!}(x - 2)^5 + \frac{-27/2}{6!}(x - 2)^6 + \cdots$$

$$= 3(x - 2) + \frac{(x - 2)^3}{4} - \frac{(x - 2)^4}{16} + \frac{(x - 2)^5}{40} - \frac{3(x - 2)^6}{320} + \cdots$$

which is in agreement with the solution obtained on page 312.

The Taylor series method can also be used to obtain series solutions for non-linear differential equations. Unfortunately, however, there is no simple theorem such as Theorem 1 available for these non-linear equations, so that the convergence of the series obtained is in question and further investigations are required. As an example of the Taylor series method for non-linear equations consider

Solve $y' = x^2 + y^2$ given that $y(0) = 1$.

Solution   We have

$$y' = x^2 + y^2, \qquad y'' = 2x + 2yy', \qquad y''' = 2 + 2y'^2 + 2yy'',$$

$$y^{IV} = 6y'y'' + 2yy''', \qquad y^V = 6y''^2 + 8y'y''' + 2yy^{IV}, \dots$$

Putting $x = 0$ leads to

$$y(0) = 1, \qquad y'(0) = 1, \qquad y''(0) = 2, \qquad y'''(0) = 8, \qquad y^{IV}(0) = 28, \qquad y^V(0) = 144, \dots$$

Thus from the Taylor series expansion we obtain

$$y(x) = 1 + x + \frac{2x^2}{2!} + \frac{8x^3}{3!} + \frac{28x^4}{4!} + \frac{144x^5}{5!} + \cdots$$

Since the general rule of formation for the terms in this series is not apparent, we cannot conclude anything about the convergence.

Remark 4. As pointed out in the footnote on page 307, Theorem 1 can be generalized to the case where $p(x)$, $q(x)$, $r(x)$ are power series in $x - a$ rather than polynomials. The fundamental theorem in such case is as follows.

Theorem 3.   Let $\qquad\qquad p(x)y'' + q(x)y' + r(x)y = 0 \qquad\qquad$ (47)

be a given differential equation. Let $x = a$ be an ordinary point [i.e., $p(a) \neq 0$] and R the distance from a to the nearest singularity. Suppose also that the functions $p(x)$, $q(x)$, $r(x)$ have power series expansions of the form (7), page 302, which are convergent for $|x - a| < R$, i.e., inside a circle of radius R with center at a (the functions in such case are often said to be *analytic* inside the circle). Then we can draw the following two conclusions:

1. The general solution of (47) can be found by substituting the series

$$y = a_0 + a_1(x - a) + a_2(x - a)^2 + \cdots = \sum a_j(x - a)^j \qquad (48)$$

in (47) using the power series in $x - a$ for $p(x)$, $q(x)$, $r(x)$. Equivalently, the general solution has the form $y = c_1 u_1(x) + c_2 u_2(x)$, where $u_1(x)$ and $u_2(x)$ are linearly independent solutions each having the form (48).

2. The series solutions obtained in part 1 converge for all values of x such that $|x - a| < R$ (i.e., inside the circle). However, for other values of x the series solutions may or may not converge.

The solution in practice can also be obtained by use of the Taylor series method as in Illustrative Example 7. Further discussion is left to the advanced exercises.

1.5 PICARD'S METHOD OF ITERATION

There is a very interesting method due to the mathematician *Picard* which can be used to obtain series solutions of differential equations. Its value, however, arises more from the theoretical point of view rather than the practical. In fact, Picard used his method to prove existence and uniqueness theorems for solutions of various types of differential equations such as described on pages 24 and 171. In the following example we illustrate Picard's method for a first-order differential equation, although its use may be extended to differential equations of higher order.

ILLUSTRATIVE EXAMPLE 9

Solve $y' = x + y + 1$. (49)

Solution Assuming that $y = c$ when $x = 0$, we may integrate both sides of (49) with respect to x to obtain*

$$y = c + \int_0^x (x + y + 1)dx \qquad (50)$$

The integral on the right cannot be performed since we do not know how y depends on x. In fact, that is what we are looking for. The method of Picard consists in assuming that $y = c$ is a first approximation to y. We denote this first approximation by $y_1 = c$. When this value is substituted for y in the integrand of (50), we denote the resulting value of y by y_2, which is a second approximation to y, i.e.,

$$y_2 = c + \int_0^x (x + y_1 + 1)dx = c + \int_0^x (x + c + 1)dx = c + \frac{x^2}{2} + cx + x$$

Substituting this approximate value in the integrand of (50), we find the third approximation y_3, given by

$$y_3 = c + \int_0^x (x + y_2 + 1)dx = c + (c + 1)x + \frac{(c + 2)x^2}{2!} + \frac{x^3}{3!}$$

Similarly $\qquad y_4 = c + (c + 1)x + \frac{(c + 2)x^2}{2!} + \frac{(c + 2)x^3}{3!} + \frac{x^4}{4!}$

and, in general,

$$y_n = c + (c + 1)x + \frac{(c + 2)x^2}{2!} + \frac{(c + 2)x^3}{3!} + \cdots + \frac{(c + 2)x^{n-1}}{(n - 1)!} + \frac{x^n}{n!}$$

In the limiting case we see that

$$\lim_{n \to \infty} y_n = c + (c + 1)x + \frac{(c + 2)x^2}{2!} + \frac{(c + 2)x^3}{3!} + \cdots$$

That the series on the right is, in fact, the desired solution is seen by comparison with the results of Illustrative Example 6, page 312, where the same differential equation is solved. Thus we see that for this case $\lim_{n \to \infty} y_n = y$.

* In the integration we have used x in two senses, one for the dummy variable of integration and the other for the upper limit in the integral. There are various advantages to doing this and no confusion should result. However, the student might wish to use a different symbol for the dummy variable, such as t. In such case (50) would be written

$$y = c + \int_0^x (t + y + 1)dt$$

In general, given the differential equation

$$y' = f(x, y)$$

we can obtain by integration, as in the above example, a sequence of approximations $y_1, y_2, y_3, \ldots$. If $f(x, y)$ satisfies suitable conditions such as given in the theorem on page 24, this sequence converges to the required solution.* The method is not practical unless $f(x, y)$ is a simple function of x and y, such as in the above example, since otherwise successive integrations may be difficult or impossible. That this is so can be seen by considering for instance the differential equation

$$y' = \sin(x^2 + y^2)$$

in which even the first integration cannot be performed.

A EXERCISES

1. Find power series solutions for each of the following about a suitable point $x = a$ using the specified value of a if indicated. In each case determine the set of values of x for which the series converges and, if possible, sum the series in closed form.
   (a) $y' = -y; y(0) = 4$.
   (b) $y' = xy; y(0) = 5$.
   (c) $y' = 2x - y$.
   (d) $y'' - y = 0; y(0) = 1, y'(0) = 0$.
   (e) $xy'' + y' = 0; y(1) = 2, y'(1) = 3$.
   (f) $xy'' + y = 0; a = 1$.
   (g) $y'' + xy' + y = 0$.
   (h) $x^2y'' + xy' - y = 0; a = 2$.
   (i) $(1 - x^2)y'' + y = 0; y(0) = 1, y'(0) = 0$.
   (j) $(1 + x)y'' + 2y' = 0; a = 1$.

2. Determine whether power series solutions about $x = a$, for the indicated values of a, exist for each of the following differential equations, and predict the set of values for which these are guaranteed to converge.
   (a) $2y'' - 5y' + 3y = 0; a = 1$.
   (b) $x^2y'' - y = 0; a = 2$.
   (c) $(1 - x^2)y'' - 2xy' + 6y = 0; a = 0$.
   (d) $x(1 - x)y'' + y = 0; a = \frac{1}{3}$.
   (e) $(x^2 + 4)y'' - xy' + y = 0; a = 0$.
   (f) $(x^2 + x)y'' + (x - 2)y = 0; a = 1$.

3. Work Exercises 1(a)–(j) and 2(a)–(f) using the Taylor series method where applicable.

4. Find power series solutions for $y'' + xy = 0$ and discuss their convergence.

B EXERCISES

1. Use Picard's method to obtain solutions to each of the following. Find at least the fourth approximation to each solution. (a) $y' = -3y; y(0) = 1$. (b) $y' = x^2 - y; y(0) = 0$. (c) $y' = e^x + y; y(0) = 0$. (d) $2y' + xy - y = 0$.

2. Use Picard's method to obtain the fourth approximation to the solution of $y' = x + y$; $y(1) = 2$. (*Hint:* Take the lower limit in the integration equal to 1.)

3. Work A Exercises 1 (a)–(c) using Picard's method.

4. Solve using series $y'' + xy' - 2y = 0; y(0) = 1, y'(0) = 0$.

5. Find the general solution of $y'' - 2xy' + 4y = 0$ by using series methods.

* For a proof see [13].

6. Use (43), page 312, and the ratio test to find the radius of convergence of (44), page 312.

7. Show that the general solution in terms of power series about the singular point $x = 0$ can be obtained for the differential equation $x^2y'' - 2xy' + 2y = 0$, but not for $2x^2y'' + 2xy' + y = 0$. Does this have any connection with Theorem 1 on page 308? Explain.

8. How would you solve $y'' + xy = \sin x$ by use of series?

9. A Exercise 2(c) has solution $y = 3x^2 - 1$. How does this affect convergence prediction?

C EXERCISES

1. Solve using power series $y'' + e^x y = 0$, $y(0) = 1$, $y'(0) = 0$. For what values do you believe the series to be convergent? Explain.

2. Explain how you might solve $(\cos x)y'' + (\sin x)y = 0$ by use of series. For what values of x would you expect the series to converge?

3. Suppose that you had never heard of $\sin x$ and $\cos x$ but in some applied problem had arrived at the differential equation $y'' + y = 0$ and obtained the power series solutions as in Illustrative Example 1, page 301. Denoting these solutions by $C(x)$ and $S(x)$, i.e.,

$$C(x) = 1 - \frac{x^2}{2!} + \frac{x^4}{4!} - \frac{x^6}{6!} + \cdots, \qquad S(x) = x - \frac{x^3}{3!} + \frac{x^5}{5!} - \frac{x^7}{7!} + \cdots$$

show that these have the following properties: (a) $[C(x)]^2 + [S(x)]^2 = 1$. (b) $S(u + v) = S(u)C(v) + C(u)S(v)$. (c) $C(u + v) = C(u)C(v) - S(u)S(v)$.

4. Explain how you could find a series solution of $y' = e^{-xy}$, $y(0) = 1$.

5. Can you extend Picard's method to differential equations of order higher than the first? If so, illustrate by an example.

2

The Method of Frobenius

2.1 MOTIVATION FOR THE METHOD OF FROBENIUS

In the last section we showed how to obtain a power series solution to the differential equation

$$p(x)y'' + q(x)y' + r(x)y = 0 \tag{1}$$

about an ordinary point $x = a$, i.e., where $p(a) \neq 0$. A question naturally arises as to whether we can find power series solutions about $x = a$ if a is a singular point. This could of course be advantageous, when for example initial conditions are given at a singular point.

A mathematician by the name of *Frobenius* became interested in this question in the latter part of the 19th century and sought to investigate it. As a first step in the investigation, he decided to look at the Euler differential equation, which has a singular point and which can always be solved as we have already seen on page 211. He considered various examples of differential equations of this type together with their general solutions, such as in the following list:

1. $2x^2 y'' + 5xy' + y = 0$, $y = c_1 x^{-1} + c_2 x^{-1/2}$
2. $x^2 y'' - 2xy' + 2y = 0$, $y = c_1 x^2 + c_2 x$.
3. $x^2 y'' - xy' + y = 0$, $y = c_1 x + c_2 x \ln x$.

These examples showed that although the differential equations themselves did not look too different from each other their solutions were quite different. For instance, the second equation has solutions involving positive powers of x, the first has solutions involving negative powers of x, while the third has a solution involving $\ln x$. On assuming a power series solution of the form

$$y = a_0 + a_1 x + a_2 x^2 + \cdots = \sum a_j x^j \tag{2}$$

in each of these equations we would find the general solution in the second case, one solution in the third case, and no solution in the first case (aside from the trivial one $y = 0$). It became clear to Frobenius that the difficulty arises because of the type of series assumed in (2). Why not assume a more general type of series, such as

$$y = x^c(a_0 + a_1 x + a_2 x^2 + \cdots) = \sum a_j x^{j+c} \tag{3}$$

where c is an additional unknown constant, which reduces to (2) in the special case $c = 0$? This assumption would yield the general solution in both the first and second cases and one solution in the third.

In fact, it is not hard to show that any Euler equation of order 2 (or higher) has at least one solution of type (3). Frobenius was thus led to ask the following

Question. Under what conditions on $p(x)$, $q(x)$, and $r(x)$ where $x = 0$ is a singular point will the differential equation (1) have a solution of the type (3)?

Because a series of type (3) is associated with Frobenius, we shall often refer to it as a *Frobenius-type solution*.

Now it is of course impossible to know what went on in the mind of Frobenius in seeking the answer to this question, but the reasoning may have been something like the following.

Let us compare the general second-order Euler equation

$$k_0 x^2 y'' + k_1 xy' + k_2 y = 0 \tag{4}$$

where k_0, k_1, k_2 are constants with (1). For purposes of better comparison let us divide equation (1) by $p(x)$ and (4) by $k_0 x^2$ so that the coefficient of y'' in each case is 1. Thus we get

$$y'' + \frac{q(x)}{p(x)} y' + \frac{r(x)}{p(x)} y = 0, \qquad y'' + \frac{k_1}{k_0 x} y' + \frac{k_2}{k_0 x^2} y = 0$$

Now if we have

$$\frac{q(x)}{p(x)} = \frac{k_1}{k_0 x}, \qquad \frac{r(x)}{p(x)} = \frac{k_2}{k_0 x^2}$$

or equivalently if $\dfrac{xq(x)}{p(x)} = $ a constant, $\dfrac{x^2 r(x)}{p(x)} = $ a constant $\tag{5}$

then (1) is precisely an Euler equation.

If we do not have (5) but we do have

$$\lim_{x \to 0} \frac{xq(x)}{p(x)} = \text{a constant}, \qquad \lim_{x \to 0} \frac{x^2 r(x)}{p(x)} = \text{a constant} \qquad (6)$$

i.e., both limits in (6) exist, then (1) will not be an Euler equation but it ought to be *nearly* an Euler equation if we take x *close enough* to zero. In such case we would expect (1) to have a solution of Frobenius type (3).

Example 1. Suppose that we are given the differential equation

$$(2x^2 + 5x^3)y'' + (5x - x^2)y' + (1 + x)y = 0 \qquad (7)$$

then for sufficiently small values of x, i.e., for x close to zero, we can neglect $5x^3$ in comparison with $2x^2$, x^2 in comparison with $5x$, and x in comparison with 1, so that we can expect a good approximation to (7) near $x = 0$ to be given by

$$2x^2 y'' + 5xy' + y = 0 \qquad (8)$$

which is the first Euler equation given on page 318 with general solution $y = c_1 x^{-1} + c_2 x^{-1/2}$. In view of this we would expect (7) to have two solutions of Frobenius type (3) with $c = -1$ and $c = -\frac{1}{2}$, respectively. This is not too difficult to verify (see B Exercise 1).

While such a "hand-waving" approach is of course not a proof of anything, the kind of reasoning involved may help in leading us toward a *conjecture* of a possible theorem, which must then be proved. Before stating such a "possible theorem" let us first note that in imposing the requirement (6) we are placing a restriction on the singular point $x = 0$, since the requirement may or may not be met. This leads us to distinguish between *two kinds* of singular points as expressed in the following definition. We state this definition in a form applicable to any singular point $x = a$ rather than $x = 0$.

Definition. Let $x = a$ be a singular point of the differential equation (1). Then if

$$\lim_{x \to a} \frac{(x - a)q(x)}{p(x)} \quad \text{and} \quad \lim_{x \to a} \frac{(x - a)^2 r(x)}{p(x)} \qquad \text{both exist} \qquad (9)$$

we call $x = a$ a *regular singular point*; otherwise it is called an *irregular singular point*.

Example 2. Given the equation $(x - 2)y'' + 3y' - xy = 0$, we have $p(x) = x - 2$, $q(x) = 3$, $r(x) = -x$ and $x = 2$ is a singular point. Since

$$\lim_{x \to 2} \frac{(x - 2)(3)}{x - 2} = 3, \qquad \lim_{x \to 2} \frac{(x - 2)^2(-x)}{x - 2} = 0$$

i.e., both limits exist, $x = 2$ is a regular singular point.

Example 3. Given the equation $x^3(1 - x)y'' + (3x + 2)y' + x^4 y = 0$, we have $p(x) = x^3(1 - x)$, $q(x) = 3x + 2$, $r(x) = x^4$, and there are two singular points $x = 0$ and $x = 1$.

Since
$$\lim_{x \to 0} \frac{(x)(3x + 2)}{x^3(1 - x)}$$

does not exist, while
$$\lim_{x \to 0} \frac{(x^2)(x^4)}{x^3(1 - x)} = 0$$

the limits do not both exist and $x = 0$ is an irregular singular point. Since

$$\lim_{x \to 1} \frac{(x - 1)(3x + 2)}{x^3(1 - x)} = -5, \qquad \lim_{x \to 1} \frac{(x - 1)^2 (x^4)}{x^3(1 - x)} = 0$$

both limits exist and $x = 1$ is a regular singular point.

Based on the above remarks, we now conjecture the truth of the following

Theorem 1. Let $x = a$ be a regular singular point of the differential equation

$$p(x)y'' + q(x)y' + r(x)y = 0 \tag{10}$$

where $p(x)$, $q(x)$, $r(x)$ are polynomials, and suppose that $x = a$ is a regular singular point, i.e.,

$$\lim_{x \to a} \frac{(x - a)q(x)}{p(x)} \quad \text{and} \quad \lim_{x \to a} \frac{(x - a)^2 r(x)}{p(x)} \quad \text{both exist} \tag{11}$$

Then (10) has a Frobenius-type solution of the form

$$y = (x - a)^c [a_0 + a_1(x - a) + a_2(x - a)^2 + \cdots] = \sum a_j(x - a)^{j+c}$$

where the series apart from the factor $(x - a)^c$ converges for all x such that $|x - a| < R$ and where R is the distance from $x = a$ to the nearest singularity (other than a of course). However, for other values of x the series may or may not converge.*

Generalization to the case where $p(x)$, $q(x)$, $r(x)$ are analytic inside the circle of radius R with center at a (i.e., have power or Taylor series in $x - a$ convergent for $|x - a| < R$) can be made.

In stating this theorem we used the intuition provided by Theorem 1, page 308]. The remarkable thing about the above conjectured theorem is that it is *true*. However, as before we shall omit a proof and instead present examples of its use.[†]

Example 4. Since $x = 2$ is a regular singular point of $(x - 2)y'' + 3y' - xy = 0$. (see Example 2), the equation has a Frobenius-type solution of the form

$$y = (x - 2)^c [a_0 + a_1(x - 2) + a_2(x - 2)^2 + \cdots] = \sum a_j(x - 2)^{j+c}$$

---

* It should be noted that if we include the factor $(x - a)^c$ then the resulting series converges for $|x - a| < R$ if $c \geq 0$. However, if $c < 0$, the series does not converge for $x = a$.
† For a proof, see

where the series apart from the factor $(x - 2)^c$ is convergent for all x, i.e., for x such that $|x - 2| < \infty$, since there is no other singular point.

Example 5. Given the equation $x^3(1 - x)y'' + (3x + 2)y' + x^4y = 0$, we know from Example 3 that $x = 1$ is a regular singular point while $x = 0$ is an irregular singular point. Thus there exists a Frobenius-type solution of the form

$$y = (x - 1)^c[a_0 + a_1(x - 1) + a_2(x - 1)^2 + \cdots] = \sum a_j(x - 1)^{j+c}$$

where the series apart from the factor $(x - 1)^c$ is convergent for $|x - 1| < 1$ since the distance from 1 to the nearest singularity (i.e., 0) is 1. The series may or may not converge for other values of x.

Since $x = 0$ is an irregular singular point, we cannot conclude from the above theorem whether there are Frobenius-type solutions of the form

$$y = x^c[a_0 + a_1x + a_2x^2 + \cdots] = \sum a_j x^{j+c}$$

Remark 1. As indicated in Example 5, the theorem on page 320 does not enable us to draw any conclusion concerning Frobenius-type solutions in the case of an irregular singular point. Because the theory involving irregular singular points is rather difficult, we shall not concern ourselves with it. It should be pointed out, however, that we can always obtain a series solution about an ordinary point in such case and then use Theorem 1 on page 308.*

On comparing the above theorem with the one on page 308 the alert student may have noticed that the one above does not mention general solution, while the one on page 308 does. The reason is quite simple, because although the above theorem states conditions under which a Frobenius-type solution of a differential equation does exist, it makes no claim that *all solutions* must be of Frobenius type. We have seen this in fact in connection with the elementary Euler equation, for example the Case 3 on page 318, where ln x appears, which is certainly not of Frobenius type.

Fortunately, by making use of the theorem on page 175, if we know one solution of a second-order linear differential equation, we can find the general solution. Thus if we can find one Frobenius-type solution, we can find the general solution. As we shall see, this second solution involves a Frobenius-type series multiplied by ln x, as we might expect from observations of the third Euler equation on page 318.

2.2 EXAMPLES USING THE METHOD OF FROBENIUS

Let us now turn to some typical examples of the method of Frobenius.

ILLUSTRATIVE EXAMPLE 1

Find Frobenius-type solutions of $4xy'' + 2y' + y = 0$.

Solution We have $p(x) = 4x$, $q(x) = 2$, $r(x) = 1$, and $x = 0$ is a singular point.

Since
$$\lim_{x \to 0} \frac{xq(x)}{p(x)} = \lim_{x \to 0} \frac{(x)(2)}{4x} = \frac{1}{2} \quad \text{and} \quad \lim_{x \to 0} \frac{x^2 r(x)}{p(x)} = \lim_{x \to 0} \frac{(x^2)(1)}{4x} = 0$$

* See also C Exercises 3 and 4 on page 332.

both exist, we see that $x = 0$ is a regular singular point, so that there exists a Frobenius-type solution of the form

$$y = \sum a_j x^{j+c} \tag{12}$$

where the summation limits $-\infty$ and ∞ for j are omitted, and where $a_j = 0$ for negative integers j. Differentiation of (12) yields

$$y' = \sum (j + c)a_j x^{j+c-1}, \qquad y'' = \sum (j + c)(j + c - 1)a_j x^{j+c-2} \tag{13}$$

Substituting (12) and (13) in the given differential equation, we have

$$4xy'' + 2y' + y = \sum 4(j + c)(j + c - 1)a_j x^{j+c-1} + \sum 2(j + c)a_j x^{j+c-1} + \sum a_j x^{j+c}$$
$$= \sum 4(j + c + 1)(j + c)a_{j+1} x^{j+c} + \sum 2(j + c + 1)a_{j+1} x^{j+c} + \sum a_j x^{j+c}$$
$$= \sum [4(j + c + 1)(j + c)a_{j+1} + 2(j + c + 1)a_{j+1} + a_j]x^{j+c}$$
$$= \sum [(2j + 2c + 2)(2j + 2c + 1)a_{j+1} + a_j]x^{j+c} = 0$$

from which
$$(2j + 2c + 2)(2j + 2c + 1)a_{j+1} + a_j = 0 \tag{14}$$

Since $a_j = 0$ for j negative, the first value of j for which we get any information from (14) is $j = -1$ ($j = -2$ yields $0 = 0$). For $j = -1$ (14) becomes

$$(2c)(2c - 1)a_0 = 0 \tag{15}$$

or, since $a_0 \neq 0$,
$$(2c)(2c - 1) = 0 \tag{16}$$

from which $c = 0, \frac{1}{2}$. The equation (16) for determining c is often called the *indicial equation*, and the values of c are called the *indicial roots*, in this case $c = 0, \frac{1}{2}$. There are two cases which must be considered, corresponding to the two values of c.

Case 1, $c = 0$. In this case (14) becomes

$$(2j + 2)(2j + 1)a_{j+1} + a_j = 0 \quad \text{or} \quad a_{j+1} = -\frac{a_j}{(2j + 2)(2j + 1)}$$

Putting $j = 0, 1, 2, \ldots$, we find

$$a_1 = -\frac{a_0}{2!}, \qquad a_2 = -\frac{a_1}{4 \cdot 3} = \frac{a_0}{4!}, \qquad a_3 = -\frac{a_2}{6 \cdot 5} = -\frac{a_0}{6!}, \ldots$$

from which
$$y = \sum a_j x^j = a_0 \left(1 - \frac{x}{2!} + \frac{x^2}{4!} - \frac{x^3}{6!} + \cdots\right) \tag{17}$$

Case 2, $c = \frac{1}{2}$. In this case (14) becomes

$$(2j + 3)(2j + 2)a_{j+1} + a_j = 0, \qquad a_{j+1} = -\frac{a_j}{(2j + 3)(2j + 2)}$$

Putting $j = 0, 1, 2, \ldots$ yields

$$a_1 = -\frac{a_0}{3!}, \qquad a_2 = -\frac{a_1}{5 \cdot 4} = \frac{a_0}{5!}, \qquad a_3 = -\frac{a_2}{7 \cdot 6} = -\frac{a_0}{7!}, \ldots$$

from which
$$y = \sum a_j x^{j+1/2} = a_0 \left(x^{1/2} - \frac{x^{3/2}}{3!} + \frac{x^{5/2}}{5!} - \frac{x^{7/2}}{7!} + \cdots\right) \tag{18}$$

From the solutions (17) and (18) we obtain the general solution

$$y = A\left(1 - \frac{x}{2!} + \frac{x^2}{4!} - \frac{x^3}{6!} + \cdots\right) + B\left(x^{1/2} - \frac{x^{3/2}}{3!} + \frac{x^{5/2}}{5!} - \frac{x^{7/2}}{7!} + \cdots\right) \quad (19)$$

The observant student will note that the two series in (19) represent, respectively, $\cos \sqrt{x}$ and $\sin \sqrt{x}$, so that the general solution is

$$y = A \cos \sqrt{x} + B \sin \sqrt{x} \quad (20)$$

Note also that the series in (19) converge for all values of x. This follows at once from the theorem on page 320, since $x = 0$ is the only singular point.

ILLUSTRATIVE EXAMPLE 2

Find Frobenius-type solutions for $x^2 y'' + xy' + (x^2 - 1)y = 0$.

Solution Here $p(x) = x^2$, $q(x) = x$, $r(x) = x^2 - 1$ and $x = 0$ is a singular point. Since

$$\lim_{x \to 0} \frac{xq(x)}{p(x)} = \lim_{x \to 0} \frac{x(x)}{x^2} = 1, \qquad \lim_{x \to 0} \frac{x^2 r(x)}{p(x)} = \lim_{x \to 0} \frac{x^2(x^2 - 1)}{x^2} = -1$$

it follows that $x = 0$ is a regular singular point so that, by the theorem on page 320, there is a Frobenius-type solution of the form

$$y = \sum a_j x^{j+c}$$

Substituting this in the given differential equation yields

$$x^2 \sum (j + c)(j + c - 1)a_j x^{j+c-2} + x \sum (j + c)a_j x^{j+c-1} + (x^2 - 1) \sum a_j x^{j+c} = 0$$

or $\sum (j + c)(j + c - 1)a_j x^{j+c} + \sum (j + c)a_j x^{j+c} + \sum a_j x^{j+c+2} - \sum a_j x^{j+c} = 0$

which can be written as the single summation

$$\sum [(j + c)(j + c - 1)a_j + (j + c)a_j + a_{j-2} - a_j]x^{j+c} = 0$$

Setting the coefficient of x^{j+c} equal to zero and simplifying we get

$$(j + c + 1)(j + c - 1)a_j + a_{j-2} = 0 \quad (21)$$

Putting $j = 0$ in (21) yields $(c + 1)(c - 1)a_0 = 0$, which since $a_0 \neq 0$ yields the indicial equation

$$(c + 1)(c - 1) = 0 \quad \text{so that} \quad c = -1, 1$$

are the indicial roots. Thus there are two cases.

Case 1, $c = -1$. Putting $c = -1$ in (21), we have

$$j(j - 2)a_j + a_{j-2} = 0 \quad \text{or} \quad a_j = -\frac{a_{j-2}}{j(j - 2)}$$

Putting $j = 1$ leads to $a_1 = 0$, but putting $j = 2$ does not lead to a meaningful value for a_2 since we assumed $a_0 \neq 0$. Thus we cannot obtain any solution in this case.

Case 2, c = 1. Putting $c = 1$ in (21) yields

$$(j + 2)ja_j + a_{j-2} = 0 \quad \text{or} \quad a_j = -\frac{a_{j-2}}{j(j + 2)}$$

Putting $j = 1, 2, 3, \ldots$ leads to

$$a_1 = 0, \quad a_2 = -\frac{a_0}{2 \cdot 4}, \quad a_3 = 0, \quad a_4 = -\frac{a_2}{4 \cdot 6} = \frac{a_0}{2 \cdot 4^2 \cdot 6},$$

$$a_5 = 0, \quad a_6 = -\frac{a_4}{6 \cdot 8} = -\frac{a_0}{2 \cdot 4^2 \cdot 6^2 \cdot 8}, \ldots$$

Note that all a_j with odd subscripts are zero because they are all multiples of $a_1 = 0$. The required solution is thus given by

$$y = \sum a_j x^{j+1} = a_0 \left(x - \frac{x^3}{2 \cdot 4} + \frac{x^5}{2 \cdot 4^2 \cdot 6} - \frac{x^7}{2 \cdot 4^2 \cdot 6^2 \cdot 8} + \cdots \right) \qquad (22)$$

From Theorem 1 on page 320, it follows that this series converges for all x since $x = 0$ is the only singular point.

Unlike Illustrative Example 1 the method of Frobenius in this case leads to only one solution. To find the general solution, we must find another linearly independent solution. Fortunately, for this purpose Theorem 3, page 175, can come to our rescue since if we know one solution, say $Y_1(x)$ or briefly Y_1, of a second-order linear differential equation, we can find the general solution by letting $y = vY_1$. Let us carry this process through in the following

ILLUSTRATIVE EXAMPLE 3

Find the general solution of $x^2 y'' + xy' + (x^2 - 1)y = 0$.

Solution Letting $y = vY_1$, we have $y' = vY_1' + v'Y_1$, $y'' = vY_1'' + 2v'Y_1' + v''Y_1$

so that the given differential equation becomes

$$x^2(vY_1'' + 2v'Y_1' + v''Y_1) + x(vY_1' + v'Y_1) + (x^2 - 1)vY_1 = 0$$

or $\quad x^2 v'' Y_1 + (2x^2 Y_1' + xY_1)v' + [x^2 Y_1'' + xY_1' + (x^2 - 1)Y_1]v = 0 \qquad (23)$

But since Y_1 satisfies the given equation we have $x^2 Y_1'' + xY_1' + (x^2 - 1)Y_1 = 0$ so that the last term in (23) is zero and we have

$$x^2 v'' Y_1 + (2x^2 Y_1' + xY_1)v' = 0$$

a first-order equation in v'. Letting $v' = u$, this becomes

$$x^2 u' Y_1 + (2x^2 Y_1' + xY_1)u = 0$$

or on division by $x^2 Y_1 u$, $\qquad \dfrac{u'}{u} + \dfrac{2Y_1'}{Y_1} + \dfrac{1}{x} = 0$

Integrating and denoting the constant of integration by $\ln A$, we have

$$\ln u + 2 \ln Y_1 + \ln x = \ln A \quad \text{or} \quad \ln (u Y_1^2 x) = \ln A$$

i.e.,

$$u Y_1^2 x = A \quad \text{or} \quad v' = \frac{A}{x Y_1^2}$$

from which

$$v = A \int \frac{dx}{x Y_1^2} + B$$

where B is another constant of integration. From $y = v Y_1$ we now obtain

$$y = A Y_1 \int \frac{dx}{x Y_1^2} + B Y_1 \tag{24}$$

which is the general solution. The result (24) shows that in addition to Y_1 being a solution we have as a linearly independent solution

$$Y_2 = Y_1 \int \frac{dx}{x Y_1^2} \tag{25}$$

Let us see what (25) becomes if we use the known solution Y_1, which we take as the series in (22) apart from the arbitrary constant a_0, i.e.,

$$Y_1 = x - \frac{x^3}{2 \cdot 4} + \frac{x^5}{2 \cdot 4^2 \cdot 6} - \cdots \tag{26}$$

We have from (26)

$$\frac{1}{Y_1} = \frac{1}{x(1 - x^2/8 + x^4/192 - \cdots)} = \frac{1}{x}\left(1 + \frac{x^2}{8} + \alpha_1 x^4 + \cdots\right) \tag{27}$$

where we have used long division in the second factor and have denoted the coefficient of x^4 by α_1, which we need not determine explicitly for our present purposes. If we now square (27) and multiply by $1/x$ we have

$$\frac{1}{x Y_1^2} = \frac{1}{x^3}\left(1 + \frac{x^2}{4} + \alpha_2 x^4 + \cdots\right) \tag{28}$$

where the new coefficient of x^4 is denoted by α_2. Integration of (28) leads to

$$\int \frac{dx}{x Y_1^2} = -\frac{1}{2x^2} + \frac{1}{4} \ln x + \frac{\alpha_2 x^2}{2} + \cdots \tag{29}$$

Multiplication of (29) by Y_1 shows that

$$Y_2 = \tfrac{1}{4} Y_1 \ln x + F \tag{30}$$

where F is a Frobenius-type series. However, because of the term involving $\ln x$ we see that Y_2 cannot be represented by a Frobenius series and explains why we did not get the general solution in Illustrative Example 2. The result also serves to illustrate the remark on page 321 concerning the presence of $\ln x$ in the solution. It is of interest that if we were looking for a solution to the differential equation which is *bounded* at $x = 0$ we would have to discard Y_2, which amounts to choosing $A = 0$ in (24). This is often done in applied problems since physical variables cannot be unbounded, in general.

Remark 2. There are other ways of obtaining the second solution (i.e., the one which is not of Frobenius type) in a form which shows the general term. The method is indicated in C Exercise 1.

ILLUSTRATIVE EXAMPLE 4

Solve $xy'' + 2y' - xy = 0$.

Solution Here $p(x) = x$, $q(x) = 2$, $r(x) = -x$, and $x = 0$ is a singularity. Since

$$\lim_{x \to 0} \frac{xq(x)}{p(x)} = \lim_{x \to 0} \frac{x(2)}{x} = 2, \qquad \lim_{x \to 0} \frac{x^2 r(x)}{p(x)} = \lim_{x \to 0} \frac{x^2(-x)}{x} = 0$$

it follows that $x = 0$ is a regular singular point. Thus there exists a Frobenius-type solution of the form

$$y = \sum a_j x^{j+c}$$

Using this, the differential equation becomes

$$\sum (j + c)(j + c - 1)a_j x^{j+c-1} + \sum 2(j + c)a_j x^{j+c-1} - \sum a_j x^{j+c+1} = 0$$

or $$\sum [(j + c + 1)(j + c)a_{j+1} + 2(j + c + 1)a_{j+1} - a_{j-1}]x^{j+c} = 0$$

so that $$(j + c + 1)(j + c)a_{j+1} + 2(j + c + 1)a_{j+1} - a_{j-1} = 0$$

i.e., $$(j + c + 1)(j + c + 2)a_{j+1} - a_{j-1} = 0 \quad (31)$$

Putting $j = -1$ in (31) yields $c(c + 1)a_0 = 0$, and since $a_0 \neq 0$ we obtain $c = -1, 0$ for the indicial roots.

Case 1, $c = -1$. In this case from (31) we have

$$j(j + 1)a_{j+1} - a_{j-1} = 0 \quad \text{or} \quad a_{j+1} = \frac{a_{j-1}}{j(j + 1)} \quad (32)$$

Putting $j = 0$, the first equation in (32) gives $0 \cdot a_1 - 0 = 0$, so that a_1 is arbitrary. This can also be seen from the second equation in (32), which *formally* gives $a_1 = 0/0$, which is indeterminate. Now putting $j = 1, 2, 3, \ldots$ in (32) gives

$$a_2 = \frac{a_0}{1 \cdot 2} = \frac{a_0}{2!}, \qquad a_3 = \frac{a_1}{2 \cdot 3} = \frac{a_1}{3!}, \qquad a_4 = \frac{a_2}{3 \cdot 4} = \frac{a_0}{4!}, \qquad a_5 = \frac{a_3}{4 \cdot 5} = \frac{a_1}{5!}, \ldots$$

Thus $$y = \sum a_j x^{j+c} = \sum a_j x^{j-1}$$

$$= a_0 \left(x^{-1} + \frac{x}{2!} + \frac{x^3}{4!} + \frac{x^5}{5!} + \cdots \right) + a_1 \left(1 + \frac{x^2}{3!} + \frac{x^4}{5!} + \cdots \right) \quad (33)$$

which gives the general solution. Since $x = 0$ is the only singular point, the first series on the right of (33) converges for all $x \neq 0$ (because of the term x^{-1}) while the second series on the right converges for all x including $x = 0$. We can write the solution in terms of hyperbolic functions, i.e.

$$y = \frac{a_0 \cosh x + a_1 \sinh x}{x} \tag{34}$$

Case 2, $c = 0$. There is no need to carry out this case since we already have obtained the general solution from Case 1. If we do carry it out we get only the second series in (33). Because of this possibility, it is usually advisable to consider the smaller indicial root first.

ILLUSTRATIVE EXAMPLE 5

Obtain a Frobenius-type solution about $x = 1$ for the differential equation $x(x - 1)y'' + xy' + y = 0$.

Solution We have $p(x) = x(x - 1)$, $q(x) = x$, $r(x) = 1$, so that $x = 0$ and $x = 1$ are singular points. Since

$$\lim_{x \to 1} \frac{(x - 1)q(x)}{p(x)} = \lim_{x \to 1} \frac{(x - 1)(x)}{x(x - 1)} = 1 \quad \text{and} \quad \lim_{x \to 1} \frac{(x - 1)^2 r(x)}{p(x)} = \lim_{x \to 1} \frac{(x - 1)^2(1)}{x(x - 1)} = 0$$

it follows that $x = 1$ is a regular singular point, and so by the theorem on page 320 there is a Frobenius-type solution about $x = 1$ given by

$$y = \sum a_j (x - 1)^{j+c} \tag{35}$$

It is convenient to make the change of variable $v = x - 1$ so that the given differential equation becomes

$$v(v + 1)\frac{d^2 y}{dv^2} + (v + 1)\frac{dy}{dv} + y = 0 \tag{36}$$

and (35) can be written

$$y = \sum a_j v^{j+c} \tag{37}$$

Substituting (37) into (36) yields

$$(v^2 + v) \sum (j + c)(j + c - 1)a_j v^{j+c-2} + (v + 1) \sum (j + c)a_j v^{j+c-1} + \sum a_j v^{j+c} = 0$$

or

$$\sum \{[(j + c)^2 + 1]a_j + (j + c + 1)^2 a_{j+1}\} v^{j+c} = 0$$

Thus

$$[(j + c)^2 + 1]a_j + (j + c + 1)^2 a_{j+1} = 0 \tag{38}$$

Letting $j = -1$ yields $c^2 a_0 = 0$, i.e., $c = 0$ since $a_0 \neq 0$. Thus the roots of the indicial equation are both equal to zero. Putting $c = 0$ into (38) gives

$$(j^2 + 1)a_j + (j + 1)^2 a_{j+1} = 0 \quad \text{or} \quad a_{j+1} = \frac{-(j^2 + 1)}{(j + 1)^2} a_j \tag{39}$$

Now putting $j = 0, 1, 2, \ldots$, we have

$$a_1 = -a_0, \qquad a_2 = \frac{-(1^2 + 1)}{2^2} a_1 = \frac{(1^2 + 1)}{2^2} a_0,$$

$$a_3 = -\frac{(2^2 + 1)}{3^2} a_2 = -\frac{(1^2 + 1)(2^2 + 1)}{2^2 3^2} a_0, \ldots$$

Using these in (37) and also replacing v by $x - 1$, we obtain the required solution

$$y = a_0 \left[1 - (x - 1) + \frac{(1^2 + 1)}{2^2} (x - 1)^2 - \frac{(1^2 + 1)(2^2 + 1)}{2^2 3^2} (x - 1)^3 + \cdots \right] \quad (40)$$

From the theorem on page 320 it follows that the series solution (40) converges for $|x - 1| < 1$, since the distance from $x = 1$ to the nearest singularity $x = 0$ is 1. It is interesting to verify this by use of the ratio test. To do this we note that the jth term is $a_j(x - 1)^j$. Thus we have

$$\left| \frac{a_{j+1}(x - 1)^{j+1}}{a_j(x - 1)^j} \right| = \left| \frac{a_{j+1}}{a_j} \right| |x - 1| = \frac{j^2 + 1}{(j + 1)^2} |x - 1|$$

on using (39). Then

$$\lim_{j \to \infty} \left| \frac{a_{j+1}(x - 1)^{j+1}}{a_j(x - 1)^j} \right| = |x - 1|$$

so that by the ratio test the series converges if $|x - 1| < 1$. We can also conclude that it diverges for $|x - 1| > 1$, but it may or may not converge for $|x - 1| = 1$.

Note that we have not obtained the general solution in this case. To do so we proceed as in Illustrative Example 3 (see C Exercise 5).

From the above examples it is clear that the Frobenius-type solutions which we get depend on the indicial roots and that various cases may arise.

1. If the indicial roots differ by a constant which is not an integer, i.e., not $0, \pm 1, \pm 2, \ldots$, the general solution is *always* obtained. This is the situation in Illustrative Example 1, page 321.

2. If the indicial roots differ by an integer not equal to zero, there are *two possibilities*:

(a) No solution is obtained by using the smaller indicial root. However, in all cases a solution can be determined by using the *larger root*. This is the situation in Illustrative Example 2, page 323.

(b) The general solution is obtained by use of the smaller indicial root. For example, if the indicial roots are -2 and -1, then -2 is the smaller root. This is the situation in Illustrative Example 4, page 326.

3. If the indicial roots differ by zero, i.e., are equal, only one solution is obtained. This is the situation in Illustrative Example 5, page 327.

It is seen that cases 1 and 2(b) provide no difficulty since the general solution is obtained. In cases 2(a) and 3, only one solution (i.e., with one arbitrary constant) is obtained, and we must find another independent solution in order to determine the general solution. To do this we can always use the theorem on page 175, as in Illustrative Example 3, page 324, or use the method given in C Exercise 1, page 330.

Remark 3. In Theorem 1 on page 320 information is supplied concerning the convergence of Frobenius-type series solutions. However, as the student may have noted, no such information has been supplied regarding the convergence of a second series solution as in Illustrative Example 3 page 324, for instance. Such a theorem is available which we state here for reference purposes.

Theorem 2. Suppose that the conditions of Theorem 1, page 320 (or its generalization for analytic functions) are satisfied so that a Frobenius-type solution of the form

$$(x - a)^{c_1} \sum a_j(x - a)^j$$

exists corresponding to the indicial root c_1. Then if the second (linearly independent) solution is not a Frobenius-type solution, it must have the form

$$(x - a)^{c_2}[\sum b_j(x - a)^j + K \ln |x - a| \sum a_j(x - a)^j] \tag{41}$$

where K and c_2 are constants and where the power series $\sum a_j(x - a)^j$ and $\sum b_j(x - a)^j$ converge for $|x - a| < R$ where R, as before, is the distance from a to the nearest singular point.*

Example. The second solution Y_2 given by (30) in Illustrative Example 3, page 324, has the form (41).

Remark. The method of Frobenius can also be extended to linear differential equations with variable coefficients of order higher than two, but since the theory involved is analogous to that given above, and since such equations do not occur in most applied problems, we shall not consider this extension. See however C Exercise 7.

A EXERCISES

1. Find Frobenius-type series solutions about $x = 0$ for each of the following differential equations. Determine the general solution in each case and discuss the convergence.

(a) $xy'' + 2y' + xy = 0$.
(b) $2xy'' + y' - xy = 0$.
(c) $x(1 - x)y'' + 2y' + 2y = 0$.
(d) $(1 - x^2)y'' - 2xy' + 12y = 0$.
(e) $xy'' + y = 0$.
(f) $x^2y'' + xy' + (x^2 - \frac{1}{4})y = 0$.
(g) $xy'' + (1 - x)y' + y = 0$.
(h) $xy'' - y' + 4x^3y = 0$.
(i) $4x^2y'' - 4xy' + (3 - 4x^2)y = 0$.
(j) $xy'' + y' + y = 0$.
(k) $x(x^2 + 2)y'' - y' - 6xy = 0$.
(l) $x^2y'' + xy' + (x^2 - 4)y = 0$.

2. Show that the roots of the indicial equation may be found by letting $y = x^c$ in the given differential equation and determining c so that the coefficient of the lowest power of x is zero. Illustrate by working some of the above exercises.

3. Use the method of Frobenius to obtain the exact solution of $\dfrac{d^2U}{dr^2} + \dfrac{2}{r}\dfrac{dU}{dr} + \alpha U = 0$.

* Note that in the special case where $K = 0$, (41) corresponds to a linearly independent solution of Frobenius type corresponding to the indicial root $c_2 \neq c_1$.

B EXERCISES

1. Verify that the differential equation (7) on page 319 has the indicial roots $c = -1, -\frac{1}{2}$, and solve the equation.

2. How would you obtain the general solution of $y'' - xy' - y = 5\sqrt{x}$? [*Hint:* Assume a *particular solution* of the form $x^c(b_0 + b_1 x + b_2 x^2 + \cdots)$ and determine the constants so that the given equation is satisfied. Then solve $y'' - xy' - y = 0$ by assuming a Frobenius-type solution and obtain the complementary solution.)

3. Solve $xy'' + 2y' + xy = 2x$.

4. Solve (a) $(1 - x)y'' + (2 - 4x)y' - y = 0$, (b) $(1 - x)y'' + (2 - 4x)y' - y = 4x^2$.

5. Is $x = 0$ a regular singular point for the equation $x^2 y'' + y' + xy = 0$? How would you find a solution of this equation?

C EXERCISES

1. If the method of Frobenius yields a series solution with only one arbitrary constant, procedures are available for determining the general solution. Two cases may arise: (A) roots of indicial equation are equal; (B) roots of indicial equation differ by an integer. The procedure will be indicated in an example illustrating each case.

ILLUSTRATIVE EXAMPLE 6 (CASE A)

Find the general solution of $xy'' + y' + xy = 0$.

Solution We proceed as usual, using the summation notation. Substituting $y = \sum a_j x^{j+c}$ in the given equation, we find,

$$xy'' + y' + xy = \sum [a_{j-1} + (j + c + 1)^2 a_{j+1}]x^{j+c} \qquad (42)$$

The customary procedure now is to say that

$$a_{j-1} + (j + c + 1)^2 a_{j+1} = 0 \qquad (43)$$

and, upon letting $j = -1$, obtain the indicial roots $c = 0, 0$. This yields the Frobenius-type solution given, apart from a multiplicative constant, by

$$U(x) = 1 - \frac{x^2}{2^2} + \frac{x^4}{2^2 4^2} - \frac{x^6}{2^2 4^2 6^2} + \cdots \qquad (44)$$

which provides one solution of the given equation. To find a second solution, let us return once again to the equation (43), where $j \geq 0$ and where we suppose for a moment that $c \neq 0$, i.e.,

$$a_{j-1} + (j + c + 1)^2 a_{j+1} = 0, \qquad j \geq 0$$

Then

$$a_{j+1} = -\frac{a_{j-1}}{(j + c + 1)^2}, \qquad j \geq 0$$

Putting $j = 0, 1, 2, 3, \ldots$, we have

$$a_1 = 0, \qquad a_2 = \frac{-a_0}{(c + 2)^2}, \qquad a_3 = 0, \qquad a_4 = \frac{-a_2}{(c + 4)^2} = \frac{a_0}{(c + 2)^2(c + 4)^2}$$

and, in general,
$$a_{2j-1} = 0, \qquad a_{2j} = \frac{(-1)^j a_0}{(c+2)^2(c+4)^2 \cdots (c+2j)^2}$$

Consider now

$$Y = \sum a_j x^{j+c} = a_0 x^c \left[1 - \frac{x^2}{(c+2)^2} + \frac{x^4}{(c+2)^2(c+4)^2} - \cdots \right] \tag{45}$$

It is clear from (42) or by direct substitution that

$$xY'' + Y' + xY = c^2 a_0 x^{c-1} \tag{46}$$

We now resort to the following device; differentiate (46) with respect to c. Then

$$\frac{\partial}{\partial c}(xY'') + \frac{\partial}{\partial c}(Y') + \frac{\partial}{\partial c}(xY) = 2ca_0 x^{c-1} + c^2 a_0 x^{c-1} \ln x$$

or

$$x\left(\frac{\partial Y}{\partial c}\right)'' + \left(\frac{\partial Y}{\partial c}\right)' + x\left(\frac{\partial Y}{\partial c}\right) = 2ca_0 x^{c-1} + c^2 a_0 x^{c-1} \ln x$$

assuming the interchange of order of differentiation is valid. It follows that $(\partial Y/\partial c)|_{c=0}$ is a solution of the given differential equation. It is not difficult to show now that

$$\frac{\partial Y}{\partial c}\bigg|_{c=0} = \frac{x^2}{2^2} - \frac{x^4}{2^2 4^2}\left(1 + \frac{1}{2}\right) + \frac{x^6}{2^2 4^2 6^2}\left(1 + \frac{1}{2} + \frac{1}{3}\right) - \cdots$$
$$+ \ln x \left(1 - \frac{x^2}{2^2} + \frac{x^4}{2^2 4^2} - \cdots\right)$$

is a solution besides (44). The general solution is therefore

$$y = AU(x) + B\left[U(x) \ln x + \frac{x^2}{2^2} - \frac{x^4}{2^2 4^2}\left(1 + \frac{1}{2}\right) + \frac{x^6}{2^2 4^2 6^2}\left(1 + \frac{1}{2} + \frac{1}{3}\right) - \cdots\right] \tag{47}$$

The second solution can also be obtained by using the theorem on page 175.

ILLUSTRATIVE EXAMPLE 6 (CASE B)

Find the general solution of $xy'' + y = 0$.

Solution   Letting $y = \sum a_j x^{j+c}$ as usual, we find
$$xy'' + y = \sum [(j+c+1)(j+c)a_{j+1} + a_j]x^{j+c} \tag{48}$$

Normally we would write

$$(j+c+1)(j+c)a_{j+1} + a_j = 0$$

and place $j = -1$ to find $c(c-1) = 0$ or $c = 0, 1$. Considering these two cases, we find that the general solution is not obtained. Thus, we proceed as in the previous illustrative example. We assume that c is variable and that

$$(j+c+1)(j+c)a_{j+1} + a_j = 0, \qquad j \geq 0$$

Then $a_1 = \dfrac{-a_0}{c(c+1)}$, $a_2 = \dfrac{-a_1}{(c+1)(c+2)} = \dfrac{a_0}{c(c+1)^2(c+2)}$, $a_3 = \dfrac{-a_0}{c(c+1)^2(c+2)^2(c+3)}, \ldots$

Consider now

$$Y = \sum a_j x^{j+c}$$

$$= a_0 x^c \left(1 - \frac{x}{c(c+1)} + \frac{x^2}{c(c+1)^2(c+2)} - \frac{x^3}{c(c+1)^2(c+2)^2(c+3)} + \cdots\right) \quad (49)$$

It is clear from (48) or by actual substitution that

$$xY'' + Y = c(c-1)a_0 x^{c-1} \quad (50)$$

A direct differentiation with respect to c as in case A fails to produce promising results. We note that the series for Y is meaningless where $c = 0$. By writing $a_0 = cp_0$, where p_0 is a new arbitrary constant, the factor c disappears, and hence the resulting series does have meaning when $c = 0$. For this value of a_0, (50) becomes

$$xY'' + Y = c^2(c-1)p_0 x^{c-1} \quad (51)$$

Differentiation with respect to c now shows that

$$x\left(\frac{\partial Y}{\partial c}\right)'' + \frac{\partial Y}{\partial c} = (c^3 - c^2)p_0 x^{c-1} \ln x + (3c^2 - 2c)p_0 x^{c-1}$$

and thus $(\partial Y/\partial c)|_{c=0}$ is a solution. From (49), writing cp_0 in place of a_0, we have

$$Y = p_0 x^c \left(c - \frac{x}{c+1} + \frac{x^2}{(c+1)^2(c+2)} - \cdots\right)$$

The student may now show that the required general solution is

$$y = AZ(x) + B\left\{Z(x)\ln x - \left[1 + \frac{x}{1^2} - \left(\frac{1}{1^2 2^2} + \frac{2}{1^3 \cdot 2}\right)x^2 + \cdots\right]\right\} \quad (52)$$

where

$$Z(x) = \frac{x}{1} - \frac{x^2}{1^2 \cdot 2} + \frac{x^3}{1^2 \cdot 2^2 \cdot 3} - \frac{x^4}{1^2 \cdot 2^2 \cdot 3^2 \cdot 4} + \cdots \quad (53)$$

2. Find the general solution of each of the following using Exercise 1: (a) $xy'' + y' + y = 0$. (b) $xy'' + xy' + y = 0$. (c) $x^2 y'' + xy' + (x^2 - 4)y = 0$.

3. (a) Show that Frobenius-type solutions (about $x = 0$) do not exist for the equation $x^4 y'' + 2x^3 y' + y = 0$. (b) Letting $x = 1/u$ (often used for such cases), solve this equation and thus account for the result in (a).

4. Use the method of Exercise 3 to solve: (a) $x^4 y'' + 2x^2(x+1)y' + y = 0$.

$$\text{(b) } 4x^3 y'' + 10x^2 y' + (2x+1)y = 0.$$

5. Obtain the general solution of the differential equation in Illustrative Example 5, page 327.

6. How would you obtain series solutions of the Frobenius type for a differential equation such as $2(\sin x)y'' + y' - xy = 0$?

7. (a) Generalize the ideas of the method of Frobenius to third-order linear differential equations. (b) Illustrate (a) by showing how to solve $xy''' + y = 0$.

3 Series Solutions of Some Important Differential Equations

On page 302, (and C Exercise 3, page 317) we pointed out that even if we had never heard of sin x or cos x we could have been led to them and their properties by means of the differential equation

$$\frac{d^2y}{dx^2} + y = 0$$

which arises, for example, in a vibrating spring problem. This leads to the idea that by "making up" differential equations and then solving them we might conceivably come up with some *special functions* of interest which are not related to functions already familiar.* While this can be done, of course the procedure takes on even greater significance when the differential equation happens to arise in an important situation, for example an applied problem. This explains why many discoveries of mathematical importance, especially in the 18th and 19th centuries, were made by scientists and engineers.

In the process of formulating various applied problems, several important differential equations have been found leading to special functions named after their discoverers. One such equation is called *Bessel's differential equation*, after the German astronomer *Friedrich Wilhelm Bessel*, who discovered it in formulating a problem in planetary motion (see C Exercise 10, page 342). Another is called *Legendre's differential equation*, after the French mathematician and astronomer *Adrien Marie Legendre*. In this section we shall obtain solutions to these differential equations and examine some of their interesting properties. Important applications of the results obtained in this chapter will be given in later chapters, especially in Chapter Fourteen.

3.1 BESSEL'S DIFFERENTIAL EQUATION

The equation $\qquad x^2y'' + xy' + (x^2 - n^2)y = 0 \qquad$ (1)

where n may have any value but is usually taken to be an integer, is known as *Bessel's equation of order n.*† This equation has a regular singular point at $x = 0$ and no other singular point. Thus by Theorem 1, page 320, there exists a Frobenius-type solution convergent for all x given by

$$y = \sum a_j x^{j+c}, \qquad a_j = 0 \quad \text{for} \quad j < 0 \qquad (2)$$

If we substitute (2) into the given equation, we obtain

$$x^2 \sum (j + c)(j + c - 1)a_j x^{j+c-2} + x \sum (j + c)a_j x^{j+c-1}$$
$$+ x^2 \sum a_j x^{j+c} - n^2 \sum a_j x^{j+c} = 0$$

---

* Various books are available which deal with special functions such as [25], [28] or [29].

† Unfortunately, the same word is sometimes used in mathematics with different meanings. In this case, *order* refers to the parameter n but no confusion with the terminology of page 4 should arise.

or

$$\sum (j + c)(j + c - 1)a_j x^{j+c} + \sum (j + c)a_j x^{j+c} + \sum a_j x^{j+c+2} - \sum n^2 a_j x^{j+c} = 0$$

This can be written

$$\sum [(j + c)(j + c - 1)a_j + (j + c)a_j + a_{j-2} - n^2 a_j]x^{j+c} = 0$$

or

$$\sum [\{(j + c)^2 - n^2\}a_j + a_{j-2}]x^{j+c} = 0$$

Thus we have

$$\{(j + c)^2 - n^2\}a_j + a_{j-2} = 0 \tag{3}$$

Putting $j = 0$ in (3) and noting that $a_{-2} = 0$ while $a_0 \neq 0$, this becomes

$$(c^2 - n^2)a_0 = 0 \quad \text{or} \quad c^2 - n^2 = 0$$

which yields the required indicial roots $c = \pm n$. Let us consider the two cases, $c = n, n \geq 0$ and $c = -n, n > 0$.

Case 1, $c = n, n \geq 0$. In this case (3) becomes

$$j(2n + j)a_j + a_{j-2} = 0 \quad \text{i.e.,} \quad a_j = \frac{-a_{j-2}}{j(2n + j)} \tag{4}$$

Putting $j = 1$, (4) shows that $a_1 = 0$ since $a_{-1} = 0$. Furthermore, (4) also shows that $a_3 = 0, a_5 = 0, a_7 = 0, \ldots$, or all a's with odd subscripts are zero.
Putting $j = 2, 4, 6, \ldots$ in succession yields

$$a_2 = \frac{-a_0}{2(2n + 2)}, \qquad a_4 = \frac{-a_2}{4(2n + 4)} = \frac{a_0}{2 \cdot 4(2n + 2)(2n + 4)},$$

$$a_6 = \frac{-a_4}{6(2n + 6)} = \frac{-a_0}{2 \cdot 4 \cdot 6(2n + 2)(2n + 4)(2n + 6)}, \cdots$$

where the rule of formation is apparent. The required series solution is thus given by

$$y = \sum a_j x^{j+n}$$

$$= a_0 x^n \left[1 - \frac{x^2}{2(2n + 2)} + \frac{x^4}{2 \cdot 4(2n + 2)(2n + 4)} \right.$$

$$\left. - \frac{x^6}{2 \cdot 4 \cdot 6(2n + 2)(2n + 4)(2n + 6)} + \cdots \right]$$

which can also be written in either of the forms

$$y = a_0 x^n \left[1 - \frac{x^2}{2^2(n + 1)} + \frac{x^4}{2^4 1 \cdot 2(n + 1)(n + 2)} - \frac{x^6}{2^6 1 \cdot 2 \cdot 3(n + 1)(n + 2)(n + 3)} + \cdots \right]$$

$$\text{or } y = a_0 x^n \left[1 - \frac{(x/2)^2}{1!(n + 1)} + \frac{(x/2)^4}{2!(n + 1)(n + 2)} - \frac{(x/2)^6}{3!(n + 1)(n + 2)(n + 3)} + \cdots \right] \tag{5}$$

On looking at the terms in the last series we note that the denominators contain factorials, i.e., $1!, 2!, 3!, \ldots$. Also present in these denominators are $n + 1$, $(n + 1)(n + 2)$, $(n + 1)(n + 2)(n + 3), \ldots$, which would become factorials, i.e., $(n + 1)!, (n + 2)!, (n + 3)!, \ldots$ if we multiplied each of them by $n!$. Another thing that we notice is that the terms of the series all involve $x/2$, while the multiplying factor in front involves only x. A particular solution supplying these factorials and introducing $x/2$ instead of x in the factor is obtained by choosing

$$a_0 = \frac{1}{2^n n!}$$

in which case (5) becomes

$$(x/2)^n \left[\frac{1}{n!} - \frac{(x/2)^2}{1!(n + 1)!} + \frac{(x/2)^4}{2!(n + 2)!} - \frac{(x/2)^6}{3!(n + 3)!} + \cdots \right] \tag{6}$$

This is a solution for $n = 0, 1, 2, \ldots$, where we define $0! = 1$ so as to make (6) agree with (5) for $n = 0$.

In order to allow for the possibility that n is any positive number, we can use the generalization of the factorial called the *gamma function*, which was considered on page 261. This function is defined by

$$\Gamma(n) = \int_0^\infty u^{n-1} e^{-u} \, du \qquad n > 0 \tag{7}$$

As on page 262, we have the *recurrence formula*

$$\Gamma(n + 1) = n\Gamma(n), \qquad \Gamma(1) = 1 \tag{8}$$

Thus if n is a positive integer, say $1, 2, 3, \ldots$, then

$$\Gamma(2) = 1\Gamma(1) = 1!, \qquad \Gamma(3) = 2\Gamma(2) = 2 \cdot 1! = 2!$$

and in general $\qquad\qquad \Gamma(n + 1) = n!, \qquad n = 1, 2, 3, \ldots \tag{9}$

We have the special value (see page 262) $\Gamma(\tfrac{1}{2}) = \sqrt{\pi}$ \hfill (10)

so that on using (8)

$$\Gamma(\tfrac{3}{2}) = \tfrac{1}{2}\Gamma(\tfrac{1}{2}) = \tfrac{1}{2}\sqrt{\pi}, \qquad \Gamma(\tfrac{5}{2}) = \tfrac{3}{2}\Gamma(\tfrac{3}{2}) = \tfrac{3}{2} \cdot \tfrac{1}{2}\sqrt{\pi}, \ldots$$

To determine the gamma function for any positive number, it suffices because of (8) to know its value for numbers between 0 and 1. These are available in tables. With this use of the gamma function to generalize factorials we can write (6) as

$$(x/2)^n \left[\frac{1}{\Gamma(n + 1)} - \frac{(x/2)^2}{1!\Gamma(n + 2)} + \frac{(x/2)^4}{2!\Gamma(n + 3)} - \frac{(x/2)^6}{3!\Gamma(n + 4)} + \cdots \right] \tag{11}$$

which is a solution of Bessel's equation for all $n \geq 0$. Now just as we give names to people so we can talk about them, so we give names to important functions. We call (11) by the name $J_n(x)$. Thus

$$J_n(x) = \frac{(x/2)^n}{\Gamma(n + 1)} - \frac{(x/2)^{n+2}}{1!\Gamma(n + 2)} + \frac{(x/2)^{n+4}}{2!\Gamma(n + 3)} - \frac{(x/2)^{n+6}}{3!\Gamma(n + 4)} + \cdots \tag{12}$$

Case 2, $c = -n$, $n > 0$. To treat this case, we could go back to the result (3) and rework all the coefficients. However, this is not necessary since all we have to do is replace n by $-n$ in (11) wherever it occurs. This leads us to

$$J_{-n}(x) = \frac{(x/2)^{-n}}{\Gamma(-n+1)} - \frac{(x/2)^{-n+2}}{1!\Gamma(-n+2)} + \frac{(x/2)^{-n+4}}{2!\Gamma(-n+3)} - \frac{(x/2)^{-n+6}}{3!\Gamma(-n+4)} + \cdots \quad (13)$$

A question immediately arises as to how we interpret the gamma function for a negative number. Thus, for example, if $n = \frac{5}{2}$, the first two denominators in (13) are $\Gamma(-\frac{3}{2})$ and $\Gamma(-\frac{1}{2})$. The definition (7) cannot be used since it is applicable only for $n > 0$ where the integral converges. We can however agree to extend the range of definition to $n < 0$ by use of (8), i.e.,

$$\Gamma(n+1) = n\Gamma(n) \qquad \text{for all } n \quad (14)$$

Thus, for example, on putting $n = -\frac{1}{2}$ and $n = -\frac{3}{2}$ in (14) we find

$$\Gamma(-\tfrac{1}{2}) = \frac{\Gamma(\tfrac{1}{2})}{-\tfrac{1}{2}} = -2\sqrt{\pi}, \qquad \Gamma(-\tfrac{3}{2}) = \frac{\Gamma(-\tfrac{1}{2})}{-\tfrac{3}{2}} = \frac{-2\sqrt{\pi}}{-\tfrac{3}{2}} = \frac{4\sqrt{\pi}}{3}$$

If we put $n = 0$ in (14), we have formally $1 = 0\Gamma(0)$ which leads us to define $1/\Gamma(0) = 0$ or $\Gamma(0)$ as infinite. Similarly, if we put $n = -1$ in (14), we have $\Gamma(0) = -1\Gamma(-1)$ or $1/\Gamma(-1) = 0$. The process can be continued and we find

$$\frac{1}{\Gamma(0)} = 0, \qquad \frac{1}{\Gamma(-1)} = 0, \qquad \frac{1}{\Gamma(-2)} = 0, \qquad \frac{1}{\Gamma(-3)} = 0, \dots \quad (15)$$

Using these interpretations, (13) becomes meaningful for all $n > 0$ and represents a solution of (1).

Now if n is positive but is not an integer, the solution (13) is not bounded at $x = 0$ while (11) is bounded (and in fact is zero) for $x = 0$. This implies that the two solutions are linearly independent so that the general solution of (1) is

$$y = c_1 J_n(x) + c_2 J_{-n}(x), \qquad n \neq 0, 1, 2, 3, \dots \quad (16)$$

where we have also included $n \neq 0$ in (16), since for $n = 0$, $J_n(x)$ and $J_{-n}(x)$ both reduce to $J_0(x)$, and (16) clearly does not give the general solution.

If n is an integer, $J_n(x)$ and $J_{-n}(x)$ are linearly dependent. In fact we have

$$J_{-n}(x) = (-1)^n J_n(x), \qquad n = 0, 1, 2, 3, \dots \quad (17)$$

For example if $n = 3$, (13) becomes on using (15)

$$J_{-3}(x) = -\frac{(x/2)^3}{3!} + \frac{(x/2)^5}{4!1!} - \frac{(x/2)^7}{5!2!} + \cdots \quad (18)$$

Also from (12) we have if $n = 3$

$$J_3(x) = \frac{(x/2)^3}{3!} - \frac{(x/2)^5}{1!4!} + \frac{(x/2)^7}{2!5!} - \cdots \quad (19)$$

Comparing (18) and (19), we see that $J_{-3}(x) = -J_3(x)$ agreeing with (17) for $n = 3$. Similarly (17) can be established for all integers n.

To obtain a solution which is linearly independent of $J_n(x)$ in case n is an integer, we can use Theorems 3 or 8, pages 175 or 186. We find for the general solution

$$y = AJ_n(x) + BJ_n(x) \int \frac{dx}{x[J_n(x)]^2} \tag{20}$$

This is the general solution regardless of whether n is an integer or not and thus includes (16). A disadvantage of the second solution in (20) is that we do not have it in explicit form so that we cannot graph it, etc.

To obtain a second solution in explicit form, we can resort to the following interesting device. If n is not an integer, it follows from the fact that $J_n(x)$ and $J_{-n}(x)$ are both solutions of (1) that a solution is also given by

$$Y_n(x) = \frac{\cos n\pi J_n(x) - J_{-n}(x)}{\sin n\pi} \tag{21}$$

and that this solution is linearly independent of $J_n(x)$. If on the other hand n is an integer, (21) assumes an indeterminate form $0/0$ because of (17) since $\cos n\pi = (-1)^n$. We are thus led to consider as a possible solution for n equal to an integer

$$Y_n(x) = \lim_{p \to n} \frac{\cos p\pi J_p(x) - J_{-p}(x)}{\sin p\pi} \tag{22}$$

This limit can be found by using L'Hôpital's rule as in calculus and yields a solution, the details of which are left to C Exercise 6. We call this second solution, which is linearly independent of $J_n(x)$ whether n is or is not an integer, the *Bessel function of the second kind of order n*, reserving the name *Bessel function of the first kind of order n* for $J_n(x)$.* Using this we can thus write the general solution of (1) for all n as

$$y = AJ_n(x) + BY_n(x) \tag{23}$$

Let us now attempt to gain some familiarity with these Bessel functions. We can start out with the simplest case where $n = 0$. The Bessel function of the first kind in this case is given by

$$J_0(x) = 1 - \frac{x^2}{2^2} + \frac{x^4}{2^2 4^2} - \frac{x^6}{2^2 4^2 6^2} + \cdots \tag{24}$$

Now we can do just as much with (24) as we can with e^x. For example, after laborious calculations we can tabulate the results:

$$J_0(0) = 1, \ J_0(1) = 0.77, \ J_0(2) = -0.22, \ J_0(3) = -0.26, \ J_0(4) = -0.40$$

We may graph $J_0(x)$ for $x \geq 0$ and obtain that shown in Fig. 7.3. The graph for $x < 0$ is easily obtained since it is symmetric to the y axis.

It will be seen that the graph is oscillatory in character, resembling the graphs of the damped vibrations of Chapter Five. The graph also reveals that there are

* The function $Y_n(x)$ is also called the *Neumann function* after the mathematician who investigated its properties.

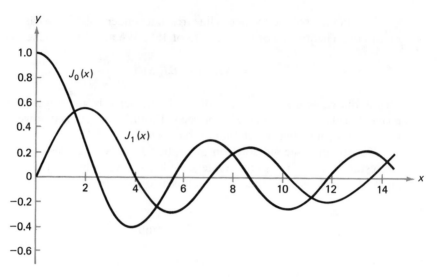

Figure 7.3

roots of the equation $J_0(x) = 0$, also called *zeros* of $J_0(x)$, obtained as points of intersection of the graph with the x axis. Investigation reveals that there are infinitely many roots which are all real and positive.

In a similar manner for $J_1(x)$ we have

$$J_1(x) = \frac{x}{2} - \frac{x^3}{2^2 4} + \frac{x^5}{2^2 4^2 6} - \cdots \tag{25}$$

the graph of which is also shown in Fig. 7.3. It should be noted that the roots of $J_1(x) = 0$ lie between those of $J_0(x) = 0$.

The graphs of $J_n(x)$ for other values of n are similar to those in Fig. 7.3. It is possible to show that the roots of $J_n(x) = 0$ are all real, and that between any two successive positive roots of $J_n(x) = 0$ there is one and only one root of $J_{n+1}(x) = 0$. This is illustrated in Table 7.1 in which we have listed for purposes of reference the first few positive zeros of $J_0(x), J_1(x), J_2(x), J_3(x)$.

Table 7.1 Positive Zeros of Some Bessel Functions

Zeros of	1	2	3	4	5
$J_0(x)$	2.4048	5.5201	8.6537	11.7915	14.9309
$J_1(x)$	3.8317	7.0156	10.1735	13.3237	16.4706
$J_2(x)$	5.1356	8.4172	11.6198	14.7960	17.9598
$J_3(x)$	6.3802	9.7610	13.0152	16.2235	19.4094

It is of interest that if we take the differences between successive zeros of $J_0(x)$ we obtain 3.1153, 3.1336, 3.1378, 3.1394, ..., suggesting the conjecture that these differences approach $\pi = 3.14 \ldots$. A similar observation on the differences of successive zeros of $J_1(x)$, $J_2(x)$, ... also leads to such a conjecture. The conjecture can actually be proved (see C Exercise 5), and since the successive zeros of $\sin x$ or $\cos x$ differ by π, the approximate description of the Bessel function $J_n(x)$ as a "damped sine wave" is further warranted.

The wave characteristics of Bessel functions seem very much like the shapes of water waves which could be generated for example by dropping a stone into the middle of a large puddle of water. The water waves thus generated would resemble the surface of revolution generated by revolving the curves of Fig. 7.3 about the y axis. Indeed it does turn out that in the theory of hydrodynamics such waves having the shapes of Bessel functions do arise.

In like manner the Bessel functions of the second kind $Y_0(x)$, $Y_1(x)$ are shown in Fig. 7.4. Note that these are not bounded as $x \to 0$ as we have already indicated above.

On putting $n = \frac{1}{2}$ and $n = -\frac{1}{2}$ in (11), we find

$$
\begin{aligned}
J_{1/2}(x) &= \frac{(x/2)^{1/2}}{\Gamma(\frac{3}{2})} - \frac{(x/2)^{5/2}}{1!\Gamma(\frac{5}{2})} + \frac{(x/2)^{9/2}}{2!\Gamma(\frac{7}{2})} - \cdots \\
&= \sqrt{\frac{2}{\pi x}} \left(x - \frac{x^3}{3!} + \frac{x^5}{5!} - \cdots \right) = \sqrt{\frac{2}{\pi x}} \sin x
\end{aligned}
\tag{26}
$$

$$
\begin{aligned}
J_{-1/2}(x) &= \frac{(x/2)^{-1/2}}{\Gamma(\frac{1}{2})} - \frac{(x/2)^{3/2}}{1!\Gamma(\frac{3}{2})} + \frac{(x/2)^{7/2}}{2!\Gamma(\frac{5}{2})} - \cdots \\
&= \sqrt{\frac{2}{\pi x}} \left(1 - \frac{x^2}{2!} + \frac{x^4}{4!} - \cdots \right) = \sqrt{\frac{2}{\pi x}} \cos x
\end{aligned}
\tag{27}
$$

which shows that $J_{1/2}(x)$, $J_{-1/2}(x)$ are in fact elementary functions obtained in terms of $\sin x$ and $\cos x$. It is interesting to note that in such case the differences between successive zeros are *exactly* equal to π.

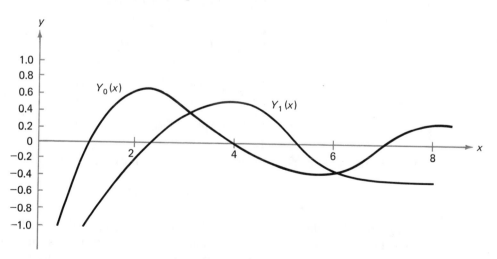

Figure 7.4

There are many identities which connect the various Bessel functions. A few of the important ones are as follows:

1. $\dfrac{d}{dx}[x^n J_n(x)] = x^n J_{n-1}(x), \qquad \dfrac{d}{dx}[x^{-n} J_n(x)] = -x^{-n} J_{n+1}(x)$ (28)

2. $J_{n+1}(x) = \dfrac{2n}{x} J_n(x) - J_{n-1}(x)$ (29)

This is a *recurrence formula*, which enables us to obtain $J_{n+1}(x)$ when we are given $J_n(x)$ and $J_{n-1}(x)$.

3. $e^{(x/2)(t - 1/t)} = \displaystyle\sum_{n=-\infty}^{\infty} J_n(x) t^n$ (30)

This is often called the *generating function* for Bessel functions. Other interesting and important properties and applications of Bessel functions are given in the exercises and in later chapters.

ILLUSTRATIVE EXAMPLE 1

Prove the recurrence formula (29) for the case where n is an integer.

Solution   To obtain various properties of $J_n(x)$ for integer n, the generating function (30) is often employed. In the present case if we differentiate both sides of (30) with respect to t, keeping x constant, we find, omitting the summation limits for brevity,

$$[e^{(x/2)(t - 1/t)}]\left[\frac{x}{2}\left(1 + \frac{1}{t^2}\right)\right] = \sum n J_n(x) t^{n-1}$$ (31)

Using (30) in the first factor on the left of (31), we obtain

$$\frac{x}{2}\left(1 + \frac{1}{t^2}\right)\sum J_n(x) t^n = \sum n J_n(x) t^{n-1}$$

or $$\sum \frac{x}{2} J_n(x) t^n + \sum \frac{x}{2} J_n(x) t^{n-2} = \sum n J_n(x) t^{n-1}$$

i.e., $$\sum \frac{x}{2} J_n(x) t^n + \sum \frac{x}{2} J_{n+2}(x) t^n = \sum (n + 1) J_{n+1}(x) t^n$$

so that $$\sum \frac{x}{2}[J_n(x) + J_{n+2}(x)] t^n = \sum (n + 1) J_{n+1}(x) t^n$$

Since corresponding coefficients of t^n on both sides must be equal, we have

$$\frac{x}{2}[J_n(x) + J_{n+2}(x)] = (n + 1) J_{n+1}(x)$$

which yields the result (29) on replacing n by $n - 1$. This technique is often useful in arriving at results valid for integer values of n. Once we have such results we may be able to show that they are also true for other values of n by using the series (12) directly. This is the case for (29), which in fact holds for all n (see A Exercise 10).

A EXERCISES

1. Show that $J_0(x) = 1 - \dfrac{(x/2)^2}{1!^2} + \dfrac{(x/2)^4}{2!^2} - \dfrac{(x/2)^6}{3!^2} + \cdots$.

2. Show that (a) $J_0'(x) = -J_1(x)$. (b) $\dfrac{d}{dx}[xJ_1(x)] = xJ_0(x)$.

3. Solve Bessel's equation $x^2y'' + xy' + (x^2 - \frac{1}{9})y = 0$ directly using the method of Frobenius.

4. Show that (a) $J_{-1}(x) = -J_1(x)$, $J_{-2}(x) = J_2(x)$ and, in general, $J_{-n}(x) = (-1)^n J_n(x)$. (b) $J_n(-x) = (-1)^n J_n(x)$, where n is an integer.

5. Write the general solution of (a) $x^2y'' + xy' + (x^2 - 9)y = 0$. (b) $x^2y'' + xy' + (x^2 - 8)y = 0$.

6. Show that (a) $J_{3/2}(x) = \sqrt{\dfrac{2}{\pi x}}\left(\dfrac{\sin x}{x} - \cos x\right)$. (b) $J_{-3/2}(x) = -\sqrt{\dfrac{2}{\pi x}}\left(\sin x + \dfrac{\cos x}{x}\right)$.

7. Show that if x is replaced by λx, where λ is a constant, then Bessel's equation becomes

$$x^2y'' + xy' + (\lambda^2x^2 - n^2)y = 0$$

8. Use Exercise 7 to solve (a) $x^2y'' + xy' + (3x^2 - 4)y = 0$. (b) $4x^2y'' + 4xy' + (2x^2 - 1)y = 0$.

9. Show that $Y_{1/2}(x) = -\sqrt{\dfrac{2}{\pi x}}\cos x$.

10. Prove that (29) holds for all n.

B EXERCISES

1. Find the solution of $x^2y'' + xy' + (4x^2 - 1)y = 0$, which is bounded at $x = 0$ and satisfies the condition $y(2) = 5$.

2. Derive the results (a) $\dfrac{d}{dx}[x^n J_n(x)] = x^n J_{n-1}(x)$. (b) $\dfrac{d}{dx}[x^{-n}J_n(x)] = -x^{-n}J_{n+1}(x)$.

3. Show that (a) $J_n'(x) = \frac{1}{2}[J_{n-1}(x) - J_{n+1}(x)]$. (b) $J_n''(x) = \frac{1}{4}[J_{n-2}(x) - 2J_n(x) + J_{n+2}(x)]$.

4. Show that (a) $\displaystyle\int x^n J_{n-1}(x)dx = x^n J_n(x) + c$. (b) $\displaystyle\int \dfrac{J_{n+1}(x)}{x^n}dx = -\dfrac{J_n(x)}{x^n} + c$.

5. Evaluate (a) $\displaystyle\int xJ_0(x)dx$. (b) $\displaystyle\int x^3 J_2(x)dx$. (c) $\displaystyle\int x^{-4}J_5(x)dx$. (d) $\displaystyle\int x^{1/2}J_{1/2}(x)dx$.

6. Evaluate $\displaystyle\int x^3 J_0(x)dx$. [Hint: Use Exercise 4(a) and integration by parts.]

7. Show that if m and n are integers then $\displaystyle\int x^m J_n(x)dx$ can be evaluated in closed form if $m + n$ is odd and in terms of $\displaystyle\int J_0(x)dx$ (which cannot be integrated in closed form) if $m + n$ is even. Illustrate by working (a) $\displaystyle\int J_3(x)dx$. (b) $\displaystyle\int \dfrac{J_2(x)}{x^2}dx$.

8. Show that

(a) $\dfrac{d}{dx}[J_n^2(x)] = \dfrac{x}{2n}[J_{n-1}^2(x) - J_{n+1}^2(x)]$. (b) $\dfrac{d}{dx}[xJ_n(x)J_{n+1}(x)] = x[J_n^2(x) - J_{n+1}^2(x)]$.

9. Demonstrate graphically that the differences of successive zeros of $J_{3/2}(x)$ approach π. (Hint: Use A Exercise 6(a), and the fact that the intersections of the graphs of $y = x$ and $y = \tan x$ give the zeros.)

10. Prove that $\int x^k J_0(x)dx = x^k J_1(x) + (k-1)x^{k-1}J_0(x) - (k-1)^2 \int x^{k-2}J_0(x)dx$ and thus find $\int x^5 J_0(x)dx$.

C EXERCISES

1. Derive the generating function (30), page 340. [*Hint*: Write $e^{(x/2)[t-1/t]} = e^{(xt/2)} \cdot e^{-x/2t}$ and then use the series expansion for e^u.]

2. Prove that between any two consecutive zeros of $J_0(x)$ there is one and only one zero of $J_1(x)$. [*Hint*: Use A Exercise 2(a) and *Rolle's Theorem* of elementary calculus, which states that between any two consecutive real zeros of a differentiable function $f(x)$ there is at least one zero of the derivative $f'(x)$.]

3. Generalize Exercise 2 to the zeros of $J_n(x)$ and $J_{n+1}(x)$. [*Hint*: Use B Exercise 2(a) and Rolle's Theorem.]

4. Show that by the change of dependent variable from y to v given by $y = x^{-1/2}v$ Bessel's differential equation becomes

$$v'' + \left(1 - \frac{4n^2 - 1}{4x^2}\right)v = 0$$

Use the result to solve Bessel's equation for the case $n = \frac{1}{2}$.

5. Show that for large values of x the solutions of Bessel's equation have the form

$$\frac{c_1 \sin x + c_2 \cos x}{\sqrt{x}}$$

Explain how this can be used to demonstrate that differences of successive zeros of $J_n(x)$ approach π. [*Hint*: Use Exercise 4.]

6. Find the limit expressed in (22), page 337, for the case $n = 0$. Compare the result obtained in C Exercise 1, Illustrative Example 6 (Case A), page 330, where the same differential equation is solved in another way.

7. A differential equation arising in various problems on electrical engineering is $xy'' + y' - ixy = 0$, where $i = \sqrt{-1}$ is the imaginary unit. (a) Show that a solution bounded at $x = 0$ is given by $J_0(i^{3/2}x)$. (b) The real and imaginary parts of $J_0(i^{3/2}x)$ are known by $Ber(x)$ and $Bei(x)$. Show that

$$Ber(x) = 1 + \sum_{j=1}^{\infty}(-1)^j \frac{x^{4j}}{2^2 4^2 \cdots (4j)^2}, \qquad Bei(x) = \sum_{j=1}^{\infty}(-1)^{j-1}\frac{x^{4j-2}}{2^2 4^2 \cdots (4j-2)^2}$$

8. Show that the general solution of $y'' + xy = 0$ often called *Airy's equation* is

$$y = Ax^{1/3}J_{1/3}(\tfrac{2}{3}x^{3/2}) + Bx^{1/3}J_{-1/3}(\tfrac{2}{3}x^{3/2})$$

9. Show that the general solution of $xy'' + y = 0$ is $y = A\sqrt{xJ_1(2\sqrt{x})} + B\sqrt{xY_1(2\sqrt{x})}$.

Compare with the solution found in C Exercise 1, Illustrative Example 6 (Case B), page 331.

10. In a problem on planetary orbits Bessel arrived at the function of x defined by the integral

$$y = \frac{1}{2\pi}\int_0^{2\pi} \cos(n\theta - x\sin\theta)d\theta$$

Show that for $n = 0$ we have $y = J_0(x)$. (*Hint:* Use the Maclaurin series for cos u where $u = x \sin \theta$ and then integrate term by term.)

11. Without using the method indicated in the hint of Exercise 10, show that for $n = 0$ the function defined by the integral is a solution of Bessel's differential equation of order zero, i.e.,

$$xy'' + y' + xy = 0$$

12. The integral given in Exercise 10 in the case where n is a non-negative integer is equal to $J_n(x)$. Can you prove this?

3.2 LEGENDRE'S DIFFERENTIAL EQUATION

The equation $\qquad (1 - x^2)y'' - 2xy' + n(n + 1)y = 0 \qquad (32)$

is known as *Legendre's equation of order n.** It is seen that $x = 0$ is an ordinary point of the equation, so that we can obtain solutions of the form

$$y = \sum a_j x^j \qquad (33)$$

We could also use a Frobenius-type solution but would come out with $c = 0$, which is equivalent to (33). Since $x = \pm 1$ are singular points of (32), the series (33) which is obtained should at least converge in the interval $-1 < x < 1$.

Substituting (33) into (32) we have

$$\sum j(j - 1)a_j x^{j-2} - \sum j(j - 1)a_j x^j - \sum 2ja_j x^j + \sum n(n + 1)a_j x^j = 0$$

i.e., $\qquad \sum [(j + 2)(j + 1)a_{j+2} - j(j - 1)a_j - 2ja_j + n(n + 1)a_j]x^j = 0$

or $\qquad \sum \{(j + 2)(j + 1)a_{j+2} + [n(n + 1) - j(j + 1)]a_j\}x^j = 0$

so that $\qquad (j + 2)(j + 1)a_{j+2} + [n(n + 1) - j(j + 1)]a_j = 0 \qquad (34)$

Putting $j = -2$ in (34) shows that a_0 is arbitrary. Putting $j = -1$ in (34) shows that a_1 is arbitrary. Thus we can expect to get two arbitrary constants and therefore the general solution.

From (34), $\qquad a_{j+2} = -\dfrac{[n(n + 1) - j(j + 1)]}{(j + 2)(j + 1)} a_j$

Putting $j = 0, 1, 2, 3, \ldots$ in succession we find

$$a_2 = -\frac{n(n + 1)}{2!} a_0, \qquad a_3 = -\frac{[n(n + 1) - 1 \cdot 2]}{3!} a_1,$$

$$a_4 = -\frac{[n(n + 1) - 2 \cdot 3]}{4 \cdot 3} a_2 = \frac{n(n + 1)[n(n + 1) - 2 \cdot 3]}{4!} a_0,$$

$$a_5 = -\frac{[n(n + 1) - 3 \cdot 4]}{5 \cdot 4} a_3 = \frac{[n(n + 1) - 1 \cdot 2][n(n + 1) - 3 \cdot 4]}{5!} a_1, \qquad \text{etc.}$$

---

* See the second footnote on page 333.

Then (33) becomes

$$y = a_0 \left\{ 1 - \frac{n(n+1)}{2!}x^2 + \frac{n(n+1)[n(n+1)-2\cdot 3]}{4!}x^4 - \cdots \right\}$$

$$+ a_1 \left\{ x - \frac{[n(n+1)-1\cdot 2]}{3!}x^3 + \frac{[n(n+1)-1\cdot 2][n(n+1)-3\cdot 4]}{5!}x^5 - \cdots \right\} \quad (35)$$

If n is not an integer, both of these series converge for $-1 < x < 1$, but they can be shown to diverge for $x = \pm 1$. If n is a positive integer or zero, one of the series terminates, i.e., is a polynomial, while the other series converges for $-1 < x < 1$ but diverges for $x = \pm 1$.

Let us find only the polynomial solutions. For $n = 0, 1, 2, 3, \ldots$, we obtain

$$1, \quad x, \quad 1 - 3x^2, \quad x - \tfrac{5}{3}x^3, \ldots$$

which are polynomials of degree 0, 1, 2, 3, ... It is convenient to multiply each of these by a constant so chosen that the resulting polynomial has the value 1 when $x = 1$. The resulting polynomials are called *Legendre polynomials* and are denoted by $P_n(x)$. The first few are given by

$$P_0(x) = 1, \quad P_1(x) = x, \quad P_2(x) = \tfrac{1}{2}(3x^2 - 1), \quad P_3(x) = \tfrac{1}{2}(5x^3 - 3x),$$

$$P_4(x) = \tfrac{1}{8}(35x^4 - 30x^2 + 3), \quad P_5(x) = \tfrac{1}{8}(63x^5 - 70x^3 + 15x)$$

For any Legendre polynomial we have

$$P_n(-x) = (-1)^n P_n(x), \quad P_n(1) = 1, \quad P_n(-1) = (-1)^n$$

It should be noted that the Legendre polynomials are the only solutions of Legendre's equation which are bounded in the interval $-1 \le x \le 1$, since the series giving all other solutions diverge for $x = \pm 1$.

Since we know that $P_n(x)$ is a solution of Legendre's equation, we can use Theorems 3 or 8, pages 175 or 186, to find the general solution. We find

$$y = AP_n(x) + BP_n(x) \int \frac{dx}{(x^2 - 1)[P_n(x)]^2} \quad (36)$$

This second solution is related to the non-terminating series in (35). If we denote the second solution by $Q_n(x)$, the general solution can be written

$$y = AP_n(x) + BQ_n(x) \quad (37)$$

These functions are often known collectively as *Legendre functions*.

Some important results involving Legendre polynomials are as follows:

1. $$P_{n+1}(x) = \frac{2n+1}{n+1}xP_n(x) - \frac{n}{n+1}P_{n-1}(x) \quad (38)$$

This is a *recurrence formula* for the Legendre polynomials.

2. $$P_n(x) = \frac{1}{2^n n!}\frac{d^n}{dx^n}(x^2 - 1)^n \quad (39)$$

This is known as *Rodrigues' formula* for Legendre polynomials.

3. $$\frac{1}{\sqrt{1 - 2tx + t^2}} = \sum_{n=0}^{\infty} P_n(x)t^n \qquad (40)$$

This is called the *generating function* for Legendre polynomials.

If we make the definition $P_n(x) = 0$ for $n = -1, -2, \ldots$, (40) can be written in the convenient form

$$\frac{1}{\sqrt{1 - 2tx + t^2}} = \sum P_n(x)t^n \qquad (41)$$

where the summation is taken from $n = -\infty$ to ∞.

Other interesting and important properties and applications of Legendre polynomials are given in the exercises and in later chapters.

ILLUSTRATIVE EXAMPLE 2

Prove the recurrence formula (38) for the Legendre polynomials.

Solution   Differentiating both sides of the generating function (41) with respect to t, keeping x constant, we have

$$\frac{x - t}{(1 - 2tx + t^2)^{3/2}} = \sum nP_n(x)t^{n-1}$$

Multiplying both sides by $(1 - 2tx + t^2)$ and using (41) gives

$$(x - t) \sum P_n(x)t^n = (1 - 2tx + t^2) \sum nP_n(x)t^{n-1}$$

or $\quad \sum xP_n(x)t^n - \sum P_n(x)t^{n+1} = \sum nP_n(x)t^{n-1} - \sum 2nxP_n(x)t^n + \sum nP_n(x)t^{n+1}$

i.e., $\sum [xP_n(x) - P_{n-1}(x)]t^n = \sum [(n + 1)P_{n+1}(x) - 2nxP_n(x) + (n - 1)P_{n-1}(x)]t^n$

Equating corresponding coefficients of t^n gives

$$xP_n(x) - P_{n-1}(x) = (n + 1)P_{n+1}(x) - 2nxP_n(x) + (n - 1)P_{n-1}(x)$$

which leads to the required result (38).

3.3 OTHER SPECIAL FUNCTIONS

Series solutions of differential equations provide a significant source of special functions, some of which have important applications and are associated with particular mathematicians who studied them. The following are three examples.

(a) Hermite's Differential Equation: $y'' - 2xy' + 2ny = 0$.

This equation has solutions called *Hermite polynomials* in case $n = 0, 1, 2, \ldots$. These polynomials denoted by $H_n(x)$ have many important properties analogous to those of Bessel functions and Legendre polynomials. This analogy is demonstrated by the following results, which the student might compare with the results on page 340 and (38)–(40) above.

1. $H_{n+1}(x) = 2xH_n(x) - 2nH_{n-1}(x)$ $\qquad\qquad$ (recurrence formula)

2. $H_n(x) = (-1)^n e^{x^2} \dfrac{d^n}{dx^n} (e^{-x^2})$ (Rodrigues' formula)

3. $e^{2tx-t^2} = \displaystyle\sum_{n=0}^{\infty} \dfrac{H_n(x)t^n}{n!}$ (generating function)

The first few Hermite polynomials are

$$H_0(x) = 1, \qquad H_1(x) = 2x, \qquad H_2(x) = 4x^2 - 2, \qquad H_3(x) = 8x^3 - 12x$$

(b) Laguerre's Differential Equation: $xy'' + (1 - x)y' + ny = 0$.

This equation has solutions called *Laguerre polynomials* denoted by $L_n(x)$ where $n = 0, 1, 2, \ldots$. These have the properties

1. $L_{n+1}(x) = (2n + 1 - x)L_n(x) - n^2 L_{n-1}(x)$ (recurrence formula)

2. $L_n(x) = e^x \dfrac{d^n}{dx^n} (x^n e^{-x})$ (Rodrigues' formula)

3. $\dfrac{e^{-xt/(1-t)}}{1 - t} = \displaystyle\sum_{n=0}^{\infty} \dfrac{L_n(x)t^n}{n!}$ (generating function)

The first few Laguerre polynomials are

$$L_0(x) = 1, \qquad L_1(x) = -x + 1, \qquad L_2(x) = x^2 - 4x + 2,$$

$$L_3(x) = -x^3 + 9x^2 - 18x + 6$$

(c) Gauss's Differential Equation: $x(1-x)y'' + [\gamma - (\alpha + \beta + 1)x]y' - \alpha\beta y = 0$.

This equation in which α, β, γ are given constants is also called the *hypergeometric differential equation*, and its solutions are called *hypergeometric functions* often denoted by $F(\alpha, \beta, \gamma; x)$. A series solution is given by

$$y = 1 + \frac{\alpha \cdot \beta}{1 \cdot \gamma}x + \frac{\alpha(\alpha + 1)\beta(\beta + 1)}{1 \cdot 2 \cdot \gamma(\gamma + 1)}x^2 + \cdots \tag{42}$$

A EXERCISES

1. Verify that the first few Legendre polynomials given on page 344 are solutions of their corresponding differential equations.

2. Find a solution of $(1 - x^2)y'' - 2xy' + 6y = 0$, which is bounded for $-1 \leqq x \leqq 1$ and satisfies the condition $y(\tfrac{1}{2}) = 10$.

3. Find the first few Legendre polynomials by using (a) the recurrence formula (38), page 344, with $P_0(x) = 1, P_1(x) = x$. (b) Rodrigue's formula (39), page 344.

4. Find the second solution (i.e., the non-polynomial solution) of Legendre's equation for the cases (a) $n = 0$, (b) $n = 1$, and thus write the general solution.

5. Verify that the second solutions found in Exercise 4 are unbounded for $x = \pm 1$.

6. Derive the result (36), page 344.

1. Find a solution of $(1 - x^2)y'' - 2xy' + 2y = 0$ in the interval $-\frac{1}{2} \le x \le \frac{1}{2}$ such that $y(0) = 3$, $y'(0) = 4$.

2. Show that $(2n + 1)P_n(x) = P'_{n+1}(x) - P'_{n-1}(x)$ so that $\int P_n(x)dx = \dfrac{P_{n+1}(x) - P_{n-1}(x)}{2n + 1} + c.$

3. Obtain the first few Legendre polynomials by using the generating function (40), page 345.

4. Show that (a) $P'_{n+1}(x) - P'_n(x) = (n + 1)P_n(x)$. (b) $xP'_n(x) - P'_{n-1}(x) = nP_n(x)$.

5. Show that $\dfrac{1}{2}\csc\dfrac{\theta}{2} = \displaystyle\sum_{k=0}^{\infty} P_k(\cos\theta)$, $0 < \theta < \pi$.

6. Derive the generating function (40), page 345.

7. Derive Rodrigues' formula (39), page 344.

8. Show that (a) $P_{2n+1}(0) = 0$. (b) $P_{2n}(0) = (-1)^n \dfrac{1 \cdot 3 \cdots (2n-1)}{2 \cdot 4 \cdots 2n}.$

1. Use Rodrigues' formula for the Hermite polynomials on page 346 to obtain $H_0(x)$, $H_1(x)$, $H_2(x)$, $H_3(x)$, and verify that they satisfy their corresponding differential equations.

2. Find the first few Hermite polynomials by using the generating function on page 346.

3. Derive the recurrence formula on page 345 for the Hermite polynomials, and show how the polynomials can be obtained by use of it.

4. Obtain the first few Laguerre polynomials and verify that they satisfy their corresponding differential equations.

5. Write the general solutions of (a) Hermite's equation; (b) Laguerre's equation.

6. Obtain a solution of $x(1 - x)y'' + (1 - 3x)y' - y = 0$ by recognizing it as a special case of the hypergeometric differential equation. Can you find the general solution?

7. Show that (42) is a solution of Gauss's differential equation.

eight

◆ orthogonal functions and Sturm-Liouville problems

1 Orthogonal Functions

1.1 FUNCTIONS AS VECTORS

The student will recall from analytic geometry and calculus that a vector in three-dimensional space can be denoted by

$$\vec{A} = A_1\vec{i} + A_2\vec{j} + A_3\vec{k} \tag{1}$$

where $\vec{i}, \vec{j}, \vec{k}$ are unit vectors in the positive directions of the x, y, z axes of a rectangular coordinate system (see Fig. 8.1). Since vector (1) is known when the coefficients A_1, A_2, A_3 are known and conversely, we could just as well represent the vector $\vec{A}$ by the *ordered triple* (A_1, A_2, A_3). This is simply a set of three real numbers which we shall call *components* of the vector* in the x, y, and z directions respectively. Similarly if we have a vector in the xy plane we can represent it by $\vec{A} = A_1\vec{i} + A_2\vec{j}$ or simply the *ordered pair* (A_1, A_2).

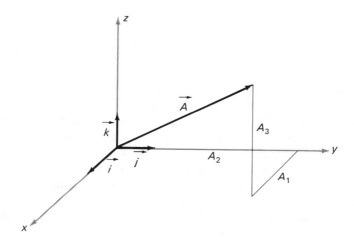

Figure 8.1

Now mathematicians love to generalize and from such generalizations they may often arrive at interesting and important results. In the case of a vector for example a mathematician might be easily led to ask the following:

Question. Since a three-dimensional vector can be represented by an ordered triple of three real numbers, why not represent an n-dimensional vector by an ordered n-tuple of real numbers given by $(A_1, A_2, \ldots, A_n)$? The question of what an n-dimensional vector could possibly mean *physically* for $n > 3$ is, strictly speaking, of no concern from a *mathematical* point of view. There are however instances where

---

* In such case the vectors $A_1\vec{i}$, $A_2\vec{j}$, $A_3\vec{k}$ are called the *component vectors* in the x, y, z directions. Because (A_1, A_2, A_3) represents the terminal point of the vector $\vec{A}$, some authors prefer to call A_1, A_2, A_3 the *coordinates* of the vector.

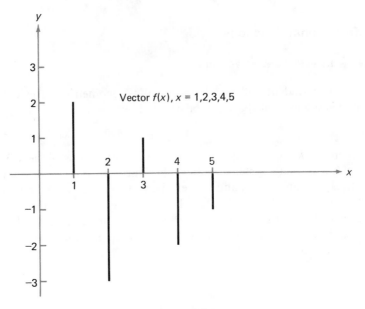

Figure 8.2

physical interpretations can be given to vectors in more than three dimensions, such as for example in *Einstein's theory of relativity*, according to which we live in a four-dimensional universe made up of the usual three space coordinates x, y, z and the time t. A vector in such a space could then be denoted by (x, y, z, t).

Although there is no way of representing vectors physically as in Fig. 8.1 for $n > 3$, there is a way of representing them graphically. To see this, suppose for example that we have a five-dimensional vector $\vec{A}$ whose components are given by $2, -3, 1, -2, -1$. We can represent these components by the ordinates y in Fig. 8.2 corresponding to $x = 1, 2, 3, 4, 5$, respectively. We can call the value of y corresponding to $x = 1$ the *first component* or *component in the "x_1 direction,"* the value of y corresponding to $x = 2$ the *second component* or *component in the "x_2 direction,"* etc. Thus the fourth component or component in the "x_4 direction" is -2. We see that if y is the component corresponding to x then y is a function of x so that, corresponding to Fig. 8.2, we can write $y = f(x)$, where $x = 1, 2, 3, 4, 5$.

It does not take much imagination now to see that we can think of a vector as a function and, conversely, a function as a vector.* In Fig. 8.3, for example, which shows the graph of $y = f(x)$ for $a \leq x \leq b$, we can think of $f(c)$ where c is some value between a and b as representing the component of the vector $f(x)$ in the "x_c direction."

If we have a vector such as shown in Fig. 8.2 in which there is a finite number of components, we speak of it as a *finite-dimensional vector*. By extending Fig. 8.2 to include components for $x = 6, 7, \ldots$, we would have a vector with infinitely many components; but because these can be counted (i.e., $1, 2, \ldots$), we would speak of it

---

* We shall assume unless otherwise stated that functions considered are real.

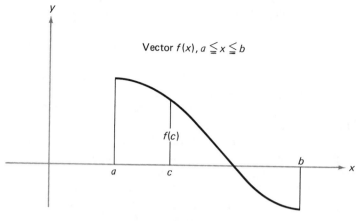

Vector $f(x)$, $a \leqq x \leqq b$

$f(c)$

a c b

Figure 8.3

as a *countably infinite dimensional vector*. The vector of Fig. 8.3, on the other hand, which has infinitely many components which *cannot* be counted, is spoken of as a *non-countably infinite dimensional vector.**

1.2 ORTHOGONALITY

Since the student already knows many results involving vectors such as (1) in three dimensions, it is natural to seek generalizations of these results to higher dimensions. Although many such generalizations are possible, some of which are given in the exercises, we shall concentrate on only two.

The first generalization which we shall seek is that of the perpendicularity, or as we often say *orthogonality*, of vectors. In the case of three dimensions the student will recall that if we have two vectors

$$\vec{A} = A_1 \vec{i} + A_2 \vec{j} + A_3 \vec{k}, \qquad \vec{B} = B_1 \vec{i} + B_2 \vec{j} + B_3 \vec{k}$$

then their *scalar* or *dot product*, denoted by $(\vec{A}, \vec{B})$ or $\vec{A} \cdot \vec{B}$ is given by

$$\vec{A} \cdot \vec{B} = (\vec{A}, \vec{B}) = A_1 B_1 + A_2 B_2 + A_3 B_3 \tag{2}$$

The student will also recall that the vectors $\vec{A}$ and $\vec{B}$ are *orthogonal* if their scalar or dot product is zero. Conversely, if this product is zero, the vectors are orthogonal. We can thus say that $\vec{A}$ and $\vec{B}$ are orthogonal if and only if their scalar product

$$A_1 B_1 + A_2 B_2 + A_3 B_3 = 0 \tag{3}$$

Now it is easy to generalize (2) to the case where $\vec{A}$ and $\vec{B}$ each have n components, since it is natural to make the following

Definition 1. The scalar product of $\vec{A}$ and $\vec{B} = \vec{A} \cdot \vec{B} = (\vec{A}, \vec{B}) = \sum_{j=1}^{n} A_j B_j$ (4)

* The fact that the components in this case cannot be counted is proved in advanced calculus. See [26], for example.

Corresponding to (3) we thus have the statement

$$\vec{A} \text{ and } \vec{B} \text{ are orthogonal if and only if } \sum_{j=1}^{n} A_j B_j = 0 \tag{5}$$

Although (4) and (5) hold for finite-dimensional vectors $\vec{A}$ and $\vec{B}$ of the same dimension n, it is easy to extend them to countably infinite dimensional vectors by simply taking n as infinite. In such case, however, care must be taken to examine the convergence of the resulting infinite series.

A little more thought is required to arrive at a generalization to non-countably infinite dimensional vectors. If we are given two vectors, say $f(x)$ and $g(x)$ where $a \leq x \leq b$, we might be led by analogy with (4) to define the scalar product of f and g as $\sum f(x)g(x)$ where the sum is taken over all values of x in the interval $a \leq x \leq b$ where $b > a$. However, such a definition would be unsatisfactory because even in the simple case where $f(x) > 0$, $g(x) > 0$, the sum would not exist. A way out of the difficulty is provided by our experience with integral calculus where integrals evolve from limits of sums. This idea suggests that we adopt the following definition for the scalar product of f and g:

Definition 2. The scalar product of f and $g = (f, g) = \int_a^b f(x)g(x)dx$ (6)

Note that in (6) we have avoided the dot notation $f \cdot g$ for scalar product suggested by $\vec{A} \cdot \vec{B}$ since this could be confused with ordinary multiplication of f and g. Corresponding to (5) we would then have the following statement

The functions f and g are orthogonal in $a \leq x \leq b$ if and only if

$$(f, g) = \int_a^b f(x)g(x)dx = 0 \tag{7}$$

As an illustration of these ideas, let us consider the following

ILLUSTRATIVE EXAMPLE 1

Let $f(x) = 2x$, $g(x) = 3 + cx$ for $0 \leq x \leq 1$. Find the value of the constant c so that the two functions are orthogonal in the interval.

Solution The scalar product of f and g is given by

$$(f, g) = \int_0^1 (2x)(3 + cx)dx = 3 + \tfrac{2}{3}c$$

Then the functions are orthogonal in the interval $0 \leq x \leq 1$ when the scalar product is zero, i.e., $3 + \tfrac{2}{3}c = 0$ or $c = -\tfrac{9}{2}$. The functions in this case are $f(x) = 2x$, $g(x) = 3 - \tfrac{9}{2}x$.

1.3 LENGTH OR NORM OF A VECTOR. ORTHONORMALITY

The second generalization which we shall seek is that of the *length* of a vector which in spaces of dimension higher than three is often referred to as the *norm*. The length of the vector (1) can be written as

$$l_A = \sqrt{\vec{A} \cdot \vec{A}} = \sqrt{A_1^2 + A_2^2 + A_3^2} \tag{8}$$

a fact which can be surmised directly from Fig. 8.1 by using the Pythagorean theorem. By a natural extension of this we have the following

Definition 3. The length or norm of an n-dimensional vector is given by

$$l_A = \sqrt{A_1^2 + A_2^2 + \cdots + A_n^2} \tag{9}$$

If n is countably infinite this corresponds to the square root of an infinite series.
Similarly, using (6) we can arrive at the following definition for the norm N_f of $f(x)$ where $a \leq x \leq b$.

Definition 4. The norm of $f = N_f = \sqrt{(f, f)} = \sqrt{\int_a^b |f(x)|^2 \, dx} \tag{10}$

In three dimensions a unit vector is a vector whose length is equal to one. If we are given a vector $\vec{A}$, then we can find a unit vector $\vec{u}$ in the same direction as $\vec{A}$ by dividing $\vec{A}$ by the length of $\vec{A}$, i.e., l_A. The process of finding this unit vector is often called *normalization, and $\vec{u}$ is called a *normalized vector.*

Example 1. If $\vec{A} = 3\vec{i} - 2\vec{j} + 5\vec{k}$, then $l_A = \sqrt{(3)^2 + (-2)^2 + (5)^2} = \sqrt{38}$

so that
$$\vec{u} = \frac{\vec{A}}{l_A} = \frac{3}{\sqrt{38}} \vec{i} - \frac{2}{\sqrt{38}} \vec{j} + \frac{5}{\sqrt{38}} \vec{k}$$

In the same way, unit vectors or normalized vectors in higher dimensions can be obtained by dividing the vectors by their norms.

ILLUSTRATIVE EXAMPLE 2

(a) Find the norms of the functions (vectors) f and g in Illustrative Example 1, page 352, and (b) obtain the corresponding normalized functions.

Solution (a) The squares of the norms of f and g are given, respectively, by

$$N_f^2 = (f, f) = \int_0^1 (2x)^2 \, dx = \tfrac{4}{3}, \qquad N_g^2 = (g, g) = \int_0^1 (3 - \tfrac{9}{2}x)^2 \, dx = \tfrac{9}{4}$$

Thus
$$N_f = \sqrt{\tfrac{4}{3}}, \qquad N_g = \tfrac{3}{2}$$

(b) The normalized functions corresponding to f and g are given by

$$\phi_1(x) = \frac{f(x)}{N_f} = \frac{2x}{\sqrt{\tfrac{4}{3}}} = \sqrt{3}x, \qquad \phi_2(x) = \frac{g(x)}{N_g} = \frac{3 - 9x/2}{\tfrac{3}{2}} = 2 - 3x$$

In case two vectors or functions are orthogonal and in addition are normalized, we often say that they are *orthonormal* vectors or functions.

Example 2. The functions $\phi_1(x)$ and $\phi_2(x)$ of Illustrative Example 2 are normalized and also orthogonal in the interval so that they are orthonormal.

In case we have a set of functions, say $\phi_1(x), \phi_2(x), \ldots,$ in some interval $a \leq x \leq b$ which are all normalized and such that any pair, say $\phi_j(x)$ and $\phi_k(x)$, are orthogonal, we can say that the set is a *mutually orthogonal* and *normalized set*, or briefly *an orthonormal set*. In such case we have

$$\int_a^b \phi_j(x)\phi_k(x)dx = \begin{cases} 0, & j \neq k \\ 1, & j = k \end{cases} \tag{11}$$

Remark. The non-countably infinite dimensional space which consists of all functions (vectors) $f(x)$, $a \leq x \leq b$, which have finite norm (length) is often called *Hilbert space* after the mathematician who investigated the properties of such functions.

A EXERCISES

1. Graph the following vectors (functions) and state whether they are finite, countably infinite, or non-countably infinite dimensional vectors.

(a) $(2, -1, 0, 5)$.

(b) $f(x) = \sin x, 0 \leq x \leq \pi$. (c) $f(x) = \begin{cases} 3, & x = 2 \\ -2, & x = 5 \\ 4, & x = 7 \end{cases}$

(d) $f(x) = \begin{cases} 1, & x \leq 0 \\ -1, & x > 0 \end{cases}$ (e) $f(x) = \dfrac{1}{n}$ for $x = n$, where $n = 1, 2, 3, \ldots$.

2. Determine the component of each of the following vectors in the indicated "directions."
   (a) $(-5, 3, 2, -6)$; x_2 directions. (b) $f(x) = 6 \cos \pi x, 0 < x < 1$; $x_{1/3}, x_{1/2},$ and $x_{2/3}$ directions.

(c) $f(x) = \begin{cases} 4, & 0 \leq x \leq 2 \\ -1, & 2 < x \leq 4 \end{cases}$; $x_{\sqrt{2}}$ and x_π directions.

3. Find the scalar product of each of the following: (a) $\vec{A} = (2, -1, 4, 3)$ and $\vec{B} = (3, 2, -1, 0)$.
   (b) $f(x) = 2x - 1, g(x) = x + 3$, where $0 \leq x \leq 1$. (c) $f(x) = 2e^{-3x}, g(x) = 6e^{-2x}, x \geq 0$.

4. Find the length or norm of each of the following:
   (a) $\vec{A} = (2, -1, 4, 3)$.   (b) $f(x) = 2x - 1, 0 \leq x \leq 1$.
   (c) $f(x) = 4e^{-x}, x \geq 0$.   (d) $f(x) = \begin{cases} 3, & 0 \leq x \leq 2 \\ -4, & 2 < x \leq 3 \end{cases}$.

5. Find the constant c so that each of the following sets are orthogonal.
   (a) $\vec{A} = (5, 1, 2, 3c, -2), \vec{B} = (-c, -2, 4, 1, -1)$.
   (b) $f(x) = cx^2 - 1, g(x) = x + 2, 0 \leq x \leq 3$.
   (c) $f(x) = 2cx, g(x) = x + c, 1 \leq x \leq 2$.

6. Find normalized or unit vectors (functions) corresponding to each of the following:
   (a) $\vec{A} = (-4, 3, 5, -2)$.   (b) $f(x) = 2x - 3, 0 \leq x \leq 2$.
   (c) $f(x) = 20e^{-5x}, x \geq 0$.   (d) $f(x) = \begin{cases} x, & 0 \leq x < 3 \\ 4, & 3 \leq x \leq 4 \end{cases}$

7. (a) Determine the constants c_1, c_2, c_3 so that $f_1(x) = c_1 x + 2, f_2(x) = c_2 x^2 + c_3 x + 1,$ and $f_3(x) = x - 1$ are mutually orthogonal in $0 \leq x \leq 1$. (b) Obtain a corresponding orthonormal set.

1. Given the vectors $\vec{A} = (a_1, a_2, a_3, a_4)$, $\vec{B} = (b_1, b_2, b_3, b_4)$. By using the analogy with three-dimensional vectors, explain how you would define (a) $\vec{A} + \vec{B}$. (b) $\vec{A} - \vec{B}$. (c) $m\vec{A}$ and $n\vec{A}$, where m and n are any real numbers (or *scalars*). (d) $m\vec{A} + n\vec{B}$ Explain how you might generalize these ideas.

2. Show that the functions $f_1(x) = a_1 + b_1 x$, $f_2(x) = a_2 + b_2 x$, $f_3(x) = a_3 + b_3 x$, where the a's and b's are constants, cannot be mutually orthogonal in any given interval unless at least one of the functions is identically zero.

3. (a) Show that the functions $f_1(x) = \sin x$, $f_2(x) = \sin 2x$, $f_3(x) = \sin 3x$ are mutually orthogonal in the interval $0 \leq x \leq \pi$. (b) Obtain a corresponding orthonormal set.

4. Generalize Exercise 3 by showing that the functions $f_n(x) = \sin nx$, $n = 1, 2, 3, \ldots$, are mutually orthogonal in the interval $0 \leq x \leq \pi$, and obtain a corresponding orthonormal set.

5. Given the vectors $\vec{A} = (1, \frac{1}{2}, \frac{1}{3}, \ldots)$, $\vec{B} = (1, 1/\sqrt{2}, 1/\sqrt{3}, \ldots)$. Show that (a) the norm of $\vec{A}$ is finite, (b) the norm of $\vec{B}$ is infinite, (c) the scalar product is finite.

6. Let $f(x) = 1/\sqrt{x}$, $g(x) = 1/\sqrt[4]{x}$ where $0 \leq x \leq 1$. Show that (a) the norm of f does not exist, (b) the norm of g exists, (c) the scalar product exists. Compare with Exercise 5.

C EXERCISES

1. In elementary calculus the dot or scalar product $\vec{A} \cdot \vec{B}$ of two vectors $\vec{A} = (a_1, a_2, a_3)$ and $\vec{B} = (b_1, b_2, b_3)$ is often defined as the product of the length of $\vec{A}$, the length of $\vec{B}$, and the cosine of the angle θ between them. (a) Show that according to this we must have

$$\cos \theta = \frac{a_1 b_1 + a_2 b_2 + a_3 b_3}{\sqrt{a_1^2 + a_2^2 + a_3^2} \sqrt{b_1^2 + b_2^2 + b_3^2}}$$

and that in particular $\vec{A}$ and $\vec{B}$ are orthogonal if and only if $\cos \theta = 0$. (b) Show that the result in (a) implies the inequality $|a_1 b_1 + a_2 b_2 + a_3 b_3|^2 \leq (a_1^2 + a_2^2 + a_3^2)(b_1^2 + b_2^2 + b_3^2)$ often called *Schwarz's inequality*. Illustrate these results by choosing particular vectors.

2. Can you generalize the ideas of Exercise 1 to the case of vectors in four dimensions such as $\vec{A} = (a_1, a_2, a_3, a_4)$, $\vec{B} = (b_1, b_2, b_3, b_4)$ or higher? What difficulties would be involved in such generalization? Do you believe that Schwarz's inequality would hold in such case? Do your conclusions contradict B Exercise 5? Explain.

3. If you wished to extend the ideas of Exercises 1 and 2 to functions $f(x)$, $g(x)$, where $a \leq x \leq b$, explain why it would be appropriate to take

$$\cos \theta = \frac{(f, g)}{\sqrt{(f, f)} \sqrt{(g, g)}}$$

Show that Schwarz's inequality can in such case be written

$$\left| \int_a^b f(x) g(x) dx \right|^2 \leq \int_a^b |f(x)|^2 \, dx \int_a^b |g(x)|^2 \, dx$$

Would this contradict B Exercise 6?

4. Justify each of the following steps providing a proof of *Schwarz's inequality* of Exercise 3.

(a) If α is any real parameter (scalar), then $(f + \alpha g, f + \alpha g) = (f, f) + 2\alpha(f, g) + \alpha^2(g, g) \geq 0$.

(b) The result in (a) can be written $\left(\alpha + \dfrac{B}{A}\right)^2 + \dfrac{AC - B^2}{A^2} \geq 0$, for all α

where $A = (g, g) > 0$, $B = (f, g)$, $C = (f, f) > 0$

(c) From the result in (b) we must have $AC - B^2 \geq 0$, which is Schwarz's inequality. State appropriate conditions on the functions f and g.

5. Use the method of proof outlined in Exercise 4 to prove the Schwarz inequality

$$|a_1 b_1 + a_2 b_2 + \cdots + a_n b_n|^2 \leq (a_1^2 + a_2^2 + \cdots + a_n^2)(b_1^2 + b_2^2 + \cdots + b_n^2)$$

Discuss the case where n is infinite.

6. Prove that if the norms of two functions f and g exist in an interval then their scalar product also exists in the interval.

7. (a) If $\vec{A} = (a_1, a_2, a_3)$, $\vec{B} = (b_1, b_2, b_3)$, $\vec{C} = (c_1, c_2, c_3)$ are vectors which form the sides of a triangle, show that $\sqrt{a_1^2 + a_2^2 + a_3^2} + \sqrt{b_1^2 + b_2^2 + b_3^2} \geq \sqrt{c_1^2 + c_2^2 + c_3^2}$.

(b) What generalizations of the result in (a) would you expect to hold for higher-dimensional spaces? Can you prove them?

8. Three vectors $\vec{A}_1$, $\vec{A}_2$, $\vec{A}_3$ in three-dimensional space are said to be *linearly dependent* if there exist constants c_1, c_2, c_3 not all zero such that $c_1 \vec{A}_1 + c_2 \vec{A}_2 + c_3 \vec{A}_3 = 0$ identically, or in other words if one of the vectors can be expressed as a linear combination of the other two. Otherwise, the vectors are *linearly independent*. (a) Show that the vectors $\vec{A}_1 = (2, -1, 4)$, $\vec{A}_2 = (1, 2, -3)$, $\vec{A}_3 = (-1, 3, -7)$ are linearly dependent and give a geometric interpretation. (b) Show that the vectors $\vec{A}_1 = (4, -2, 1)$, $\vec{A}_2 = (2, 0, 3)$, $\vec{A}_3 = (-2, 3, -8)$ are linearly independent and interpret geometrically. (c) Can you find a condition involving third-order determinants which will indicate linear dependence or independence? Illustrate by using (a) and (b).

9. Generalize the ideas in Exercise 8 to higher-dimensional spaces. In particular, discuss these ideas in relation to the concepts of linear independence and Wronskians given on pages 181–189. Can you give a geometric interpretation?

2 Sturm–Liouville Problems

2.1 MOTIVATION FOR STURM–LIOUVILLE PROBLEMS. EIGENVALUES AND EIGENFUNCTIONS

Since $y = f(x)$ where $a \leq x \leq b$ is a vector in a non-countably infinite dimensional space, we can try to let our imaginations run wild and think of the various possible things that can be done to ordinary three-dimensional vectors and then attempt to generalize. One thing which can be done to an ordinary vector would be to *transform* it or *operate* on it. For example, we could rotate it, thereby changing its direction, or we could stretch it or contract it, thereby changing its magnitude, or we could combine both, i.e., rotation and stretching (or contracting).

We have already encountered operators, actually *differential operators*, earlier in this book. However, with this new interpretation of a function as a vector it is natural to ask what we are doing when, for example, we take the derivative of $\sin 2x$ to get $2 \cos 2x$, i.e., $D \sin 2x = 2 \cos 2x$. We can think of $\sin 2x$ as a vector in a non-countably infinite dimensional space and D as some transformation being applied to this vector $\sin 2x$ producing another vector $2 \cos 2x$ in this space. If we apply the operation twice to $\sin 2x$, symbolized by $D^2 \sin 2x$, a rather remarkable thing happens, since we get $-4 \sin 2x$, which we can think of as a vector in this space which has a "direction" opposite to that of $\sin 2x$, but a "magnitude" four times as large. Thus when D^2 acts on $\sin 2x$ it produces a rotation and stretching.*

Suppose now we consider the differential equation

$$\phi(D)y = F(x) \tag{1}$$

as we have often done. The vector interpretation is as follows: Given the vector $F(x)$ in a non-countably infinite dimensional space, find all vectors y in the space which are transformed into $F(x)$ by $\phi(D)$. We can of course find these vectors by solving the differential equation, which in many cases we can already do. Thus, although the interpretation is new, the method is old.

We have already given a vector interpretation to the transformation $D^2 \sin 2x = -4 \sin 2x$. This is actually a rather unusual situation since, for example, if we apply D once, twice, or any number of times to some other function (vector) such as $\tan x$, we would never get back to a multiple of $\tan x$. This leads us to ask the following

Question. Given an operator, say $\phi(D)$, can we determine those functions (vectors) y which are transformed by $\phi(D)$ into constant multiples of themselves?

We can express this in the form

$$\phi(D)y = -\lambda y \tag{2}$$

where we have used $-\lambda y$ to denote the constant multiple of y. The minus sign has been used simply because we plan to transpose $-\lambda y$ to the left side of the equation, and when we do we would like the term to have an associated positive sign.

For the sake of discussion, let us assume that $\phi(D)$ is a second-order differential operator given by $a_0 D^2 + a_1 D + a_2$, where a_0, a_1, a_2 may be functions of x. In such case (2) can be written

$$(a_0 D^2 + a_1 D + a_2)y = -\lambda y \quad \text{or} \quad a_0 \frac{d^2 y}{dx^2} + a_1 \frac{dy}{dx} + (a_2 + \lambda)y = 0$$

Dividing by a_0, which is assumed to be different from zero, we obtain

$$\frac{d^2 y}{dx^2} + \frac{a_1}{a_0} \frac{dy}{dx} + \left(\frac{a_2}{a_0} + \frac{\lambda}{a_0}\right) y = 0$$

---

* We shall use the word *stretching* to include the special cases where the length or norm is decreased (contracting) or unchanged (invariant) unless otherwise specified.

If we now multiply this equation by $e^{\int (a_1/a_0)dx}$, the sum of the first two terms on the left can be written as an exact derivative, i.e.,

$$\frac{d}{dx}\left[e^{\int (a_1/a_0)dx}\frac{dy}{dx}\right] + \left(\frac{a_2}{a_0} + \frac{\lambda}{a_0}\right) e^{\int (a_1/a_0)dx}y = 0 \tag{3}$$

where the analogy with the integrating factor of page 53 is noted.

We can write (3) in the form

$$\frac{d}{dx}\left[P(x)\frac{dy}{dx}\right] + [Q(x) + \lambda R(x)]y = 0 \tag{4}$$

where $\quad P(x) = e^{\int (a_1/a_0)dx}, \qquad Q(x) = \dfrac{a_2}{a_0} e^{\int (a_1/a_0)dx}, \qquad R(x) = \dfrac{e^{\int (a_1/a_0)dx}}{a_0} \tag{5}$

We call the differential equation (4) a *Sturm–Liouville differential equation* after the mathematicians who investigated its properties.

ILLUSTRATIVE EXAMPLE 1

Express in Sturm–Liouville form the differential equation

$$x\frac{d^2y}{dx^2} - 3\frac{dy}{dx} + (x^2 + \lambda)y = 0$$

Solution   *Method 1.*   Divide the given equation by x to obtain

$$\frac{d^2y}{dx^2} - \frac{3}{x}\frac{dy}{dx} + \left(x + \frac{\lambda}{x}\right)y = 0$$

To make the sum of the first two terms an exact derivative, multiply by the integrating factor $e^{\int (-3/x)dx} = e^{-3\ln x} = e^{\ln (x^{-3})} = x^{-3}$, assuming $x > 0$. Then the equation can be written in the required Sturm–Liouville form:

$$\frac{d}{dx}\left[x^{-3}\frac{dy}{dx}\right] + [x^{-2} + \lambda x^{-4}]y = 0 \tag{6}$$

which is the same as (4) with $P(x) = x^{-3}, \qquad Q(x) = x^{-2}, \qquad R(x) = x^{-4}$

Method 2.   Put $a_0 = x, a_1 = -3, a_2 = x^2$ in (5). Then

$$P(x) = e^{\int (-3/x)dx} = x^{-3}, \qquad Q(x) = \frac{x^2}{x} e^{\int (-3/x)dx} = x^{-2}, \qquad R(x) = \frac{1}{x} e^{\int (-3/x)dx} = x^{-4}$$

which when used in (4) yields (6).

One solution of the Sturm–Liouville equation (4) is of course given by $y = 0$, which constitutes the *null or zero vector* of non-countably infinite dimensional space. Clearly, this is a *trivial solution* which does not interest us. We are interested in *non-trivial solutions*, or as we sometimes say *proper* ones, for which $y \neq 0$. Now since much of the research on equation (4) was written in German, and since the German word for proper is *eigen*, it has become the fashion to refer to the non-trivial or proper functions which satisfy (4) as *eigenfunctions* or, thinking of functions as vectors, *eigenvectors* combining the German and English words. The corresponding constants λ are then called *eigenvalues.**

As an example, suppose that we consider the special Sturm–Liouville differential equation

$$D^2 y = -\lambda y \quad \text{or} \quad \frac{d^2 y}{dx^2} + \lambda y = 0 \tag{7}$$

If $y = \sin 2x$, (7) becomes $-4 \sin 2x = -\lambda \sin 2x$, so that $\lambda = 4$. Thus $\sin 2x$ is an eigenfunction and $\lambda = 4$ a corresponding eigenvalue. Similarly, if $y = \sin 3x$, then (7) becomes $-9 \sin 3x = -\lambda \sin 3x$, so that $\lambda = 9$. Thus $y = \sin 3x$ is an eigenfunction and $\lambda = 9$ a corresponding eigenvalue.

Let us suppose now that we have two different eigenfunctions y_j and y_k (i.e., $j \neq k$) which satisfy (4) with corresponding different eigenvalues λ_j and λ_k, respectively. Then we would have

$$\frac{d}{dx}[P(x)y_j'] + [Q(x) + \lambda_j R(x)]y_j = 0, \quad \frac{d}{dx}[P(x)y_k'] + [Q(x) + \lambda_k R(x)]y_k = 0$$

where we have used y_j', y_k' in place of dy_j/dx, dy_k/dx. If we multiply the first equation by y_k, the second by y_j, and then subtract, we find

$$y_k \frac{d}{dx}[P(x)y_j'] - y_j \frac{d}{dx}[P(x)y_k'] + (\lambda_j - \lambda_k)R(x)y_j y_k = 0$$

in which we notice that $Q(x)$ has been eliminated. We can write this as

$$(\lambda_j - \lambda_k)R(x)y_j y_k = y_j \frac{d}{dx}[P(x)y_k'] - y_k \frac{d}{dx}[P(x)y_j'] \tag{8}$$

The right side of (8) can be expressed as an exact derivative so that

$$(\lambda_j - \lambda_k)R(x)y_j y_k = \frac{d}{dx}[P(x)(y_j y_k' - y_k y_j')] \tag{9}$$

as can be verified by direct differentiation.

If we integrate both sides of (9) from $x = a$ to $x = b$, we obtain

$$(\lambda_j - \lambda_k) \int_a^b R(x)y_j y_k \, dx = P(b)[y_j(b)y_k'(b) - y_k(b)y_j'(b)]$$
$$- P(a)[y_j(a)y_k'(a) - y_k(a)y_j'(a)] \tag{10}$$

Now it would certainly be nice if the right-hand side of (10) were equal to zero, because in such case since $\lambda_j \neq \lambda_k$ we could divide by $\lambda_j - \lambda_k$ to obtain

$$\int_a^b R(x)y_j y_k \, dx = 0, \quad j \neq k \tag{11}$$

or if $R(x) \geq 0$,

$$\int_a^b [\sqrt{R(x)}y_j][\sqrt{R(x)}y_k] dx = 0, \quad j \neq k$$

* Eigenfunctions and eigenvalues are also called *characteristic functions* and *characteristic values* respectively.

This would show that the eigenfunctions $\sqrt{R(x)}\,y_j$ and $\sqrt{R(x)}\,y_k$ are *orthogonal* in the interval $a \leq x \leq b$.

Remark. Using the interpretation of a function as a vector where the values of the function at x represent components of the vector in the x'th direction (see page 350), we can interpret $R(x)$ as a *weighting factor* or *weight factor* which weights different directions. Because of this, when the property (11) holds, we shall often say that the functions y_j and y_k are *orthogonal* in the interval $a \leq x \leq b$ *with respect to the weight factor* (or *weight function*) $R(x)$. The idea can also be extended to mutual orthogonality and orthonormality with respect to weight factors. We can summarize the remarks in the following

Definition. Let $y_1, y_2, y_3, \ldots$ be a set of functions such that

$$\int_a^b R(x)y_j y_k \, dx = 0, \qquad j \neq k \tag{12}$$

for any pair of functions y_j, y_k. Then the functions are said to be *mutually orthogonal* in the interval with respect to the weight factor $R(x) \geq 0$. In case we also have

$$\int_a^b R(x)y_j^2 \, dx = 1 \tag{13}$$

for all the functions in the set, we say that the functions are *orthonormal in the interval with respect to the weight factor*. In the case where (13) holds, we also say that the functions y_j are *normalized with respect to the weight factor*.

Let us now examine the various cases under which the right-hand side of (10) will be zero. There are four important cases which can arise.

Case 1. $y_j(a)y_k'(a) - y_k(a)y_j'(a) = 0, \qquad y_j(b)y_k'(b) - y_k(b)y_j'(b) = 0 \tag{14}$

The first of these is equivalent to

$$a_1 y_j(a) + a_2 y_j'(a) = 0, \qquad a_1 y_k(a) + a_2 y_k'(a) = 0 \tag{15}$$

where a_1, a_2 are given constants. To see this we have only to solve for a_1 or a_2 in one of these equations and substitute it in the other. Similarly, the second is equivalent

to $\qquad\qquad b_1 y_j(b) + b_2 y_j'(b) = 0, \qquad b_1 y_k(b) + b_2 y_k'(b) = 0 \tag{16}$

where b_1, b_2 are given constants. We can also express (15) and (16) in the form

$$a_1 y(a) + a_2 y'(a) = 0, \qquad b_1 y(b) + b_2 y'(b) = 0, \qquad \text{for } y = y_j \text{ and } y = y_k \tag{17}$$

We shall refer to this case as the *ordinary case*.

Case 2. $P(a) = 0, \qquad y_j(b)y_k'(b) - y_k(b)y_j'(b) = 0 \tag{18}$

The last condition as in Case 1 is equivalent to

$$b_1 y(b) + b_2 y'(b) = 0, \qquad \text{for } y = y_j \text{ and } y = y_k \tag{19}$$

Since $P(a) = 0$ is equivalent to the situation where the Sturm–Liouville differential equation has a singular point at $x = a$, we shall refer to this case as the *one singular point case.* Note that we can interchange the a's and b's in (18) and (19). This would amount to having the singular point at $x = b$ instead of $x = a$.

Case 3. $P(a) = 0$, $P(b) = 0$ $\qquad\qquad\qquad\qquad\qquad\qquad\qquad$ (20)

This case would occur if, for example, $P(x) = 1 - x^2$, $a = -1$, $b = 1$. Since (20) is equivalent to the situation where the Sturm–Liouville differential equation has two singular points, i.e., at $x = a$ and $x = b$, we shall refer to this case as the *two singular points case.*

Case 4. $P(a) = P(b) \neq 0$, $y(a) = y(b)$, $y'(a) = y'(b)$, for $y = y_j$ and $y = y_k$ $\quad$ (21)

We shall refer to this case as the *periodic case,* since the values of $P(x)$, $y(x)$, and $y'(x)$ are the same at $x = a$ and $x = b$. It often also happens that $Q(a) = Q(b)$ and $R(a) = R(b)$.

The above four cases each suggest a boundary value problem involving the Sturm–Liouville differential equation

$$\frac{d}{dx}\left[P(x)\frac{dy}{dx}\right] + [Q(x) + \lambda R(x)]y = 0 \qquad\qquad (22)$$

and one of the following four associated boundary conditions corresponding to these cases.

A. Ordinary case, $P(a) \neq 0$, $P(b) \neq 0$.

$$a_1 y(a) + a_2 y'(a) = 0, \qquad b_1 y(b) + b_2 y'(b) = 0 \qquad\qquad (23)$$

B. One singular point case, $P(a) = 0$.

$$y \text{ and } y' \text{ bounded at } x = a, \quad b_1 y(b) + b_2 y'(b) = 0 \qquad\qquad (24)$$

An analogous boundary-value problem results if we interchange a and b.

C. Two singular points case, $P(a) = P(b) = 0$.

$$y \text{ and } y' \text{ bounded at } x = a \text{ and } x = b \qquad\qquad\qquad\qquad (25)$$

D. Periodic case, $P(a) = P(b) \neq 0$.

$$y(a) = y(b), \qquad y'(a) = y'(b) \qquad\qquad\qquad\qquad\qquad (26)$$

Each of these boundary value problems is called a *Sturm–Liouville boundary value problem* or briefly a *Sturm–Liouville problem.* We shall consider various examples of such problems in this chapter.*

---

* The first condition in (24) is in many practical cases more than is required, e.g., we may only need y bounded at $x = a$. More precisely, we should have

$$\lim_{x \to a+} P(x)[y_j(x)y_k'(x) - y_k(x)y_j'(x)] = 0$$

where $x \to a+$ means that x approaches a from the right [see (10), page 359]. The same remark applies to the conditions in (25).

As an example of a Sturm–Liouville problem with boundary conditions given in A above, we consider the following Illustrative Example. Illustrations of the Sturm–Liouville problems B, C, and D will be given in later sections.

ILLUSTRATIVE EXAMPLE 2

Given the boundary value problem

$$\frac{d^2y}{dx^2} + \lambda y = 0, \qquad y(0) = 0, \ y(1) = 0 \tag{27}$$

(a) Show that it is a Sturm–Liouville problem. (b) Find the eigenvalues and eigenfunctions. (c) Obtain a set of functions which are mutually orthogonal in the interval $0 \leq x \leq 1$. (d) Obtain a corresponding set of functions orthonormal in the interval $0 \leq x \leq 1$.

Solution (a) The differential equation and boundary conditions (27) are special cases of (22) and (23) with

$$P(x) = 1, \qquad Q(x) = 0, \qquad R(x) = 1, \qquad a = 0, \qquad b = 1,$$

$$a_1 = 1, \qquad a_2 = 0, \qquad b_1 = 1, \qquad b_2 = 0$$

so that (27) is a Sturm–Liouville problem. (b) The differential equation in (27) has the general solution

$$y = A \cos \sqrt{\lambda} x + B \sin \sqrt{\lambda} x \tag{28}$$

where A and B are arbitrary constants. To satisfy the first condition in (27), i.e., $y(0) = 0$, we see from (28) that $A = 0$. Thus (28) becomes

$$y = B \sin \sqrt{\lambda} x \tag{29}$$

To satisfy the second condition in (27), i.e., $y(1) = 0$, we see from (29) that we must have $B \sin \sqrt{\lambda} = 0$. There are two possibilities, either $B = 0$ or $\sin \sqrt{\lambda} = 0$. If $B = 0$, then the solution (29) reduces to the trivial one $y = 0$ which we do not want. We thus require $\sin \sqrt{\lambda} = 0$ from which we have

$$\sqrt{\lambda} = \pm n\pi \quad \text{or} \quad \lambda = \lambda_n = n^2\pi^2, \qquad n = 0, 1, 2, \ldots$$

The corresponding solutions (29) are then given by

$$y = y_n = B_n \sin n\pi x, \qquad n = 0, 1, 2, \ldots$$

where we have used different values of B given by $B_0, B_1, B_2, \ldots$, since multiples of solutions are also solutions. Since $n = 0$ yields the trivial solution $y = 0$, we exclude it. Thus the required eigenvalues and corresponding eigenfunctions are given by

$$\lambda = \lambda_n = n^2\pi^2, \qquad y = y_n = B_n \sin n\pi x, \qquad n = 1, 2, 3, \ldots \tag{30}$$

(c) From the theory given on page 360, the eigenfunctions must be mutually orthogonal in the interval $0 \leq x \leq 1$. However, if we like we can verify this directly, for we have from (30) $y_j = B_j \sin j\pi x$, $y_k = B_k \sin k\pi x$ for any positive integers j and k. Thus if $j \neq k$

$$\int_0^1 y_j y_k \, dx = \int_0^1 (B_j \sin j\pi x)(B_k \sin k\pi x) dx = B_j B_k \int_0^1 \sin j\pi x \sin k\pi x \, dx$$

$$= B_j B_k \int_0^1 \sin j\pi x \sin k\pi x \, dx$$

$$= \tfrac{1}{2} B_j B_k \int_0^1 [\cos (j - k)\pi x - \cos (j + k)\pi x] dx$$

$$= \tfrac{1}{2} B_j B_k \left[\frac{\sin (j - k)\pi x}{(j - k)\pi} - \frac{\sin (j + k)\pi x}{(j + k)\pi} \right] \Bigg|_0^1 = 0$$

where we have used the identity $\sin \theta_1 \sin \theta_2 = \tfrac{1}{2}[\cos(\theta_1 - \theta_2) - \cos(\theta_1 + \theta_2)]$.

(d) To obtain an orthonormal set we must determine the constants B_n in (30) so that the scalar product of y_n with itself is 1. This leads to

$$\int_0^1 (B_n \sin n\pi x)^2 \, dx = 1 \quad \text{or} \quad B_n^2 \int_0^1 \sin^2 n\pi x \, dx = 1 \tag{31}$$

Since
$$\int_0^1 \sin^2 n\pi x \, dx = \tfrac{1}{2} \int_0^1 (1 - \cos 2n\pi x) dx = \tfrac{1}{2}$$

(31) yields $B_n^2 = 2$ or $B_n = \sqrt{2}$, taking positive values for B_n. This leads to the orthonormal set of functions $\phi_n = \sqrt{2} \sin n\pi x$, $n = 1, 2, 3, \ldots$.

The results obtained in the preceding example are typical for Sturm–Liouville problems under very mild restrictions on the functions $P(x)$, $Q(x)$, $R(x)$, which are usually satisfied in practice. We can summarize the main results as follows:

1. There will be an infinite set of eigenvalues which are all real and non-negative. They can be denoted in increasing order by $\lambda_1, \lambda_2, \lambda_3, \ldots$, where $\lambda_n \to \infty$ as $n \to \infty$.
2. There will be an infinite set of eigenfunctions corresponding to the eigenvalues in statement 1. They can be denoted by $y_1, y_2, y_3, \ldots$.
3. The functions $\sqrt{R(x)}y_1, \sqrt{R(x)}y_2, \sqrt{R(x)}y_3, \ldots$ are mutually orthogonal in the interval $a \leq x \leq b$. By normalizing these functions they can be converted into the functions $\phi_1(x), \phi_2(x), \ldots$ which comprise an orthonormal set in the interval. Alternatively, we can say that the functions $y_1, y_2, y_3, \ldots$ are mutually orthogonal (and can be made orthonormal) in the interval $a \leq x \leq b$ with respect to the weight factor $R(x)$.

2.2 AN APPLICATION TO THE BUCKLING OF BEAMS

Suppose that we have a beam or column OR of length L (see Fig. 8.4) which is pinned at points O and R. This beam is initially straight so that its axis (or elastic curve) coincides with the x axis taken in the downward direction as shown in the figure. Due to an axial load of constant magnitude P suppose that the beam undergoes a deflection as shown exaggerated in the figure. Let us try to find the magnitude of this deflection.

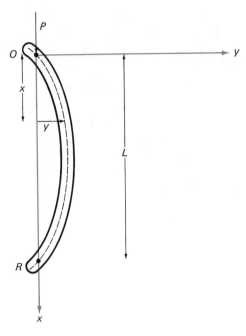

Figure 8.4

Mathematical Formulation. According to the theory given on page 139, the deflection y of the beam is given by

$$EI \frac{d^2y}{dx^2} = M(x) \qquad (32)$$

where E and I are constants and $M(x)$ is the bending moment at a section of the beam at distance x from the end O. This bending moment is seen from the figure to have a magnitude equal to the force P multiplied by the distance y of the section to the line of action of the force. However, since the rate of change of the slope (i.e., d^2y/dx^2) is seen to be negative, we must have $M(x) = -Py$ so that (32) becomes

$$EI \frac{d^2y}{dx^2} = -Py \quad \text{or} \quad \frac{d^2y}{dx^2} + \frac{P}{EI} y = 0 \qquad (33)$$

Since the ends of the beam have zero deflection at $x = 0$ and $x = L$, we must also have

$$y(0) = 0, \qquad y(L) = 0 \qquad (34)$$

Solution The differential equation (33) and boundary conditions (34) constitute a Sturm–Liouville problem. If we solve (33), we get

$$y = A \cos \sqrt{P/EI}x + B \sin \sqrt{P/EI}x \qquad (35)$$

Using the first condition in (34) yields $A = 0$, while the second condition yields $B \sin \sqrt{P/EI}L = 0$. However, since $B \neq 0$ (otherwise there is no deflection), we must have $\sin \sqrt{P/EIL} = 0$, from which it follows that

$$\sqrt{P/EIL} = n\pi \quad \text{or} \quad P = \frac{n^2\pi^2 EI}{L^2}, \qquad n = 1, 2, 3, \ldots \qquad (36)$$

These constitute the eigenvalues of the Sturm–Liouville problem. The corresponding eigenfunctions, obtained from (35) with $A = 0$, are given by

$$y = B_n \sin \frac{n\pi x}{L}, \qquad n = 1, 2, 3, \ldots \qquad (37)$$

where the coefficients which can depend on n are denoted by B_n.

Interpretation. Physically, the values of P (eigenvalues) given by (36) represent *critical loads* $P_1, P_2, P_3, \ldots$ for which the corresponding deflections (eigenfunctions) are given by (37). For a given load $P < P_1$, which is the smallest critical load, the deflection is zero, so the beam can support such a load. For $P = P_1$ the beam will buckle as indicated in Fig. 8.5 (or in the opposite direction), i.e., the beam will fail to support the load P_1. To prevent such buckling it is necessary to provide a constraint at the midpoint $x = L/2$ of the beam so that no deflection occurs. If this is done the beam will not buckle as shown in Fig. 8.5 until the critical load P_2 is reached. The load P_2 causes the beam to buckle in the manner shown in Fig. 8.6 so that the beam fails to support the load P_2. To prevent this buckling, however, we can provide two additional constraints at $x = L/3$ and $x = 2L/3$. If this is done the beam will not buckle until the critical load P_3 is reached, and this in turn causes buckling as shown in Fig. 8.7. Continuing the process of further constraints allows larger critical loads.

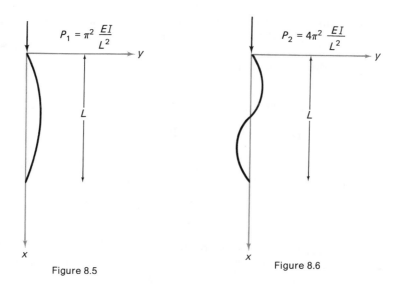

Figure 8.5

Figure 8.6

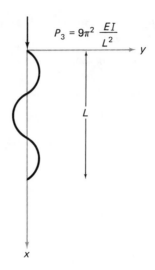

$$P_3 = 9\pi^2 \, \frac{EI}{L^2}$$

Figure 8.7

3 Orthogonality of Bessel and Legendre Functions

At the end of Chapter Seven we discussed the solutions, in the form of series, of Bessel's and Legendre's differential equations. Since the equations involved are linear and of second order, it is reasonable to ask whether we can use the Sturm–Liouville theory developed in this chapter to obtain further properties of Bessel and Legendre functions. In particular we would like to know whether boundary value problems involving these equations have eigenvalues and eigenfunctions, whether the eigenfunctions are mutually orthogonal, etc.

It turns out that we can apply Sturm–Liouville theory to these special equations, and that Bessel and Legendre functions provide further examples of orthogonal functions of great importance in later applications. We first consider the orthogonality of Bessel functions.

3.1 ORTHOGONALITY OF BESSEL FUNCTIONS

In Chapter Seven we found that Bessel's differential equation

$$x^2 y'' + xy' + (x^2 - n^2)y = 0 \tag{1}$$

has a solution for $n \geq 0$ given by the Bessel function

$$J_n(x) = \frac{x^n}{2^n \Gamma(n + 1)} \left\{ 1 - \frac{x^2}{2(2n + 2)} + \frac{x^4}{2 \cdot 4(2n + 2)(2n + 4)} - \cdots \right\} \tag{2}$$

For $n \neq 0, 1, 2, 3, \ldots$, a solution which is linearly independent of (2) is given by $J_{-n}(x)$, but for $n = 0, 1, 2, 3, \ldots$ (and in fact for all n), a solution which is linearly independent of (2) is given by $Y_n(x)$ (see page 337). Both of the functions $J_{-n}(x)$ and

$Y_n(x)$ become infinite as $x \to 0$, so that (2) is the only solution of (1) which is bounded at $x = 0$.

To investigate the Sturm–Liouville theory and the possible orthogonality of Bessel functions, the idea might occur of introducing the parameter λ into the last term on the left in (1), for example, by replacing x by $\sqrt{\lambda}x$. We can do this by making the change of independent variable $x = \sqrt{\lambda}u$ in (1) and then formally replacing u by x. We leave the details of this to the student (see A Exercise 8) and merely note that the result is

$$x^2 \frac{d^2y}{dx^2} + x \frac{dy}{dx} + (\lambda x^2 - n^2)y = 0 \tag{3}$$

By using either of the methods in Illustrative Example 1, page 358, equation (3) can be written in Sturm–Liouville form as

$$\frac{d}{dx}\left[x \frac{dy}{dx} \right] + \left(\lambda x - \frac{n^2}{x} \right) y = 0 \tag{4}$$

which corresponds to (22), page 361, with

$$P(x) = x, \qquad Q(x) = -n^2/x, \qquad R(x) = x \tag{5}$$

The question now arises as to what boundary conditions to impose. If we want a boundary condition at $x = 0$, we will have to exclude solutions such as $J_{-n}(x)$ and $Y_n(x)$, which become infinite at $x = 0$. Since $P(x) = x$, we have $P(0) = 0$, but there are no other values of x for which $P(x) = 0$ i.e., there is only one singular point. This suggests that we use the boundary conditions associated with the Sturm–Liouville problem type B on page 361. In such case we have

$$\frac{d}{dx}\left[x \frac{dy}{dx} \right] + \left(\lambda x - \frac{n^2}{x} \right) y = 0,$$

$$y \text{ and } y' \text{ are bounded at } x = 0, \qquad b_1 y(b) + b_2 y'(b) = 0 \tag{6}$$

For simplicity let us choose $b = 1$. The theory of Section 2 indicates that we should expect to find non-trivial solutions of this problem only for certain values of λ, i.e., eigenvalues, and that the corresponding eigenfunctions should be orthogonal in the interval $0 \leq x \leq 1$ with respect to the weight function $R(x) = x$. Let us attempt to confirm this for the special case where $b_1 = 1$, $b_2 = 0$, $b = 1$ in (6). In such case the last condition in (6) becomes

$$y(1) = 0 \tag{7}$$

Now the general solution of (3) is obtained from (16) or (23), pages 336–337, on replacing x by $\sqrt{\lambda}x$. From the first condition in (6) we require $c_2 = 0$ or $B = 0$ so as to exclude $J_{-n}(x)$ or $Y_n(x)$, which are unbounded for $x = 0$. The required solution is thus

$$y = AJ_n(\sqrt{\lambda}x) \tag{8}$$

Now to satisfy the condition (7) we have from (8)

$$J_n(\sqrt{\lambda}) = 0 \tag{9}$$

since $A \neq 0$ (otherwise we would have the trivial solution $y = 0$). Now as we have already mentioned on page 338, there are infinitely many positive roots of $J_n(x) = 0$ given by $x = r_1, r_2, r_3, \ldots$. Thus (9) has the solutions

$$\sqrt{\lambda} = r_1, r_2, r_3, \ldots \quad \text{or} \quad \lambda = r_1^2, r_2^2, r_3^2, \ldots \tag{10}$$

which are the eigenvalues. The corresponding eigenfunctions are given by

$$J_n(r_1 x), J_n(r_2 x), J_n(r_3 x), \ldots \tag{11}$$

Using (10), page 359, it follows that $\displaystyle\int_0^1 x J_n(r_j x) J_n(r_k x) dx = 0, \quad j \neq k \tag{12}$

which shows that the functions $\quad \sqrt{x} J_n(r_1 x), \sqrt{x} J_n(r_2 x), \sqrt{x} J_n(r_3 x), \ldots \tag{13}$

are mutually orthogonal in the interval $0 \leq x \leq 1$, or equivalently that $J_n(r_1 x)$, $J_n(r_2 x), J_n(r_3 x), \ldots$ are mutually orthogonal in the interval $0 \leq x \leq 1$ with respect to the weight function x.

We can also convert the set (13) into an orthonormal set by normalizing each function. To do this we consider $\phi_k(x) = c_k \sqrt{x} J_n(r_k x)$ and seek to determine the constant c_k so that

$$\int_0^1 [\phi_k(x)]^2 \, dx = \int_0^1 [c_k \sqrt{x} J_n(r_k x)]^2 \, dx = 1$$

This yields $\qquad\qquad c_k = \left[\int_0^1 x J_n^2(r_k x) dx \right]^{-1/2} \tag{14}$

assuming $c_k > 0$. The integral in (14) is not very easy to evaluate directly. However, we can use the following device.

1. Evaluate the integral $\qquad \displaystyle\int_0^1 x J_n(\alpha x) J_n(\beta x) dx \tag{15}$

This can be accomplished by using (10) on page 359.

2. Assuming that α is a variable and β a constant, let α approach β in (15). This will yield the integral

$$\int_0^1 x J_n^2(\beta x) dx \tag{16}$$

3. Use the value $\beta = r_k$ in (16).

Let us see what we get by this procedure. From (10), page 359, with $P(x) = x$, $R(x) = x$, $a = 0$, $b = 1$, $y = J_n(\sqrt{\lambda} x)$, so that $y' = \sqrt{\lambda} J_n'(\sqrt{\lambda} x)$, we obtain

$$(\lambda_j - \lambda_k) \int_0^1 x J_n(\sqrt{\lambda_j} x) J_n(\sqrt{\lambda_k} x) dx = J_n(\sqrt{\lambda_j}) [\sqrt{\lambda_k} J_n'(\sqrt{\lambda_k})] - J_n(\sqrt{\lambda_k}) [\sqrt{\lambda_j} J_n'(\sqrt{\lambda_j})]$$

Then writing $\lambda_j = \alpha^2$, $\lambda_k = \beta^2$, we find

$$\int_0^1 x J_n(\alpha x) J_n(\beta x) dx = \frac{\beta J_n(\alpha) J_n'(\beta) - \alpha J_n(\beta) J_n'(\alpha)}{\alpha^2 - \beta^2} \tag{17}$$

Now if we take the limit as $\alpha \to \beta$, the right-hand side assumes the indeterminate form $0/0$, which suggests that we use L'Hôpital's rule of elementary calculus. This rule states that the desired limit, if it exists, is equal to the limit as $\alpha \to \beta$ of the fraction obtained by differentiating the numerator with respect to α and the denominator with respect to α. If we perform these differentiations, (17) becomes

$$\int_0^1 xJ_n^2(\beta x)dx = \lim_{\alpha \to \beta} \frac{\beta J_n'(\alpha)J_n'(\beta) - \alpha J_n(\beta)J_n''(\alpha) - J_n(\beta)J_n'(\alpha)}{2\alpha}$$

or

$$\int_0^1 xJ_n^2(\beta x)dx = \frac{\beta J_n'^2(\beta) - \beta J_n(\beta)J_n''(\beta) - J_n(\beta)J_n'(\beta)}{2\beta} \tag{18}$$

Now since $J_n(x)$ satisfies Bessel's differential equation, we have

$$x^2 J_n''(x) + x J_n'(x) + (x^2 - n^2)J_n(x) = 0 \tag{19}$$

so that for $x = \beta$,

$$\beta^2 J_n''(\beta) + \beta J_n'(\beta) + (\beta^2 - n^2)J_n(\beta) = 0 \tag{20}$$

Using the value of $J_n''(\beta)$ obtained from (20) in (18), we find on simplifying

$$\int_0^1 xJ_n^2(\beta x)dx = \frac{\beta^2 J_n'^2(\beta) + (\beta^2 - n^2)J_n^2(\beta)}{2\beta^2} \tag{21}$$

This result holds for all values of β. In the special case where $\beta = r_k$ is a root of $J_n(x) = 0$ so that $J_n(r_k) = 0$, (21) assumes the simple form

$$\int_0^1 xJ_n^2(r_k x)dx = \tfrac{1}{2}J_n'^2(r_k) \tag{22}$$

From this it follows that the set of functions

$$\phi_k(x) = \frac{\sqrt{2x}J_n(r_k x)}{J_n'(r_k)}, \qquad k = 1, 2, 3, \ldots \tag{23}$$

is orthonormal in the interval $0 \leq x \leq 1$.

In the case of the equation $\qquad J_n'(x) = 0 \tag{24}$

there are also infinitely many positive roots which we also denote by r_k, although these roots are of course different from those of $J_n(x) = 0$. In such case $J_n'(r_k) = 0$, and so we have from (21) with $\beta = r_k$

$$\int_0^1 xJ_n^2(r_k x)dx = \frac{(r_k^2 - n^2)J_n^2(r_k)}{2r_k^2} \tag{25}$$

This shows that $\qquad \phi_k(x) = \frac{\sqrt{2x}r_k J_n(r_k x)}{\sqrt{r_k^2 - n^2}J_n(r_k)} \tag{26}$

is an orthonormal set in the interval $0 \leq x \leq 1$.

Let us now consider the case $a = 0, b = 1, b_1 = \mu, b_2 = 1$, where μ is a positive constant. Then the last condition in (6) becomes

$$\mu y(1) + y'(1) = 0 \tag{27}$$

The solution bounded at $x = 0$ must as on page 367 be given by

$$y = AJ_n(\sqrt{\lambda}x) \tag{28}$$

To satisfy (27) we thus must have

$$\mu J_n(\sqrt{\lambda}) + \sqrt{\lambda} J_n'(\sqrt{\lambda}) = 0 \tag{29}$$

Now just as it is true that $J_n(x) = 0$ and $J_n'(x) = 0$ have infinitely many positive roots, so it is also true that

$$\mu J_n(x) + x J_n'(x) = 0 \tag{30}$$

for any positive number μ has infinitely many positive roots. We shall denote these roots also by r_k, but it must of course be realized that the roots of (30) are different from the roots of $J_n(x) = 0$ or $J_n'(x) = 0$. It follows that the roots of (29) are given by

$$\sqrt{\lambda} = r_1, r_2, r_3, \ldots \quad \text{or} \quad \lambda = r_1^2, r_2^2, r_3^2, \ldots \tag{31}$$

From (13), page 368, it thus follows that the functions

$$\sqrt{x} J_n(r_1 x), \quad \sqrt{x} J_n(r_2 x), \quad \sqrt{x} J_n(r_3 x), \ldots \tag{32}$$

where r_k are the roots of (30) are orthogonal in the interval $0 \leq x \leq 1$. To normalize these we can proceed as on page 369. Putting $\beta = r_k$ in (21), we have

$$\int_0^1 x J_n^2(r_k x) dx = \frac{r_k^2 J_n'^2(r_k) + (r_k^2 - n^2) J_n^2(r_k)}{2 r_k^2} \tag{33}$$

Since r_k are roots of (30),

$$\mu J_n(r_k) + r_k J_n'(r_k) = 0 \tag{34}$$

Using this in (33),

$$\int_0^1 x J_n^2(r_k x) dx = \frac{(\mu^2 + r_k^2 - n^2) J_n^2(r_k)}{2 r_k^2} \tag{35}$$

This can be used to normalize the functions (32) and leads to the set of orthonormal functions in the interval $0 \leq x \leq 1$ given by

$$\phi_k(x) = \frac{\sqrt{2x} r_k J_n(r_k x)}{\sqrt{\mu^2 + r_k^2 - n^2} J_n(r_k)} \tag{36}$$

For purposes of reference, we summarize in Table 8.1 the important results arrived at above. They will be very useful to us in our later work.

Table 8.1

Equation having roots r_k	Value of $\int_0^1 x J_n^2(r_k x) dx$
1. $J_n(x) = 0$	$\frac{1}{2} J_n'^2(r_k)$
2. $J_n'(x) = 0$	$\frac{(r_k^2 - n^2) J_n^2(r_k)}{2 r_k^2}$
3. $\mu J_n(x) + x J_n'(x) = 0$	$\frac{(\mu^2 + r_k^2 - n^2) J_n^2(r_k)}{2 r_k^2}$

Remarks similar to those made for Bessel functions can also be made for Legendre functions. In Chapter Seven we solved Legendre's differential equation

$$(1 - x^2)\frac{d^2y}{dx^2} - 2x\frac{dy}{dx} + n(n + 1)y = 0 \tag{37}$$

and found that for $n = 0, 1, 2, \ldots$ the equation has polynomial solutions called *Legendre polynomials*, while for $n \neq 0, 1, 2, \ldots$ it has power series solutions about $x = 0$ which converge for $-1 < x < 1$, but diverge for $x = \pm1$. We can write (37) in Sturm–Liouville form as

$$\frac{d}{dx}\left[(1 - x^2)\frac{dy}{dx}\right] + n(n + 1)y = 0 \tag{38}$$

which is a special case of (22), page 361, with

$$P(x) = 1 - x^2, \qquad Q(x) = 0, \qquad R(x) = 1, \qquad \lambda = n(n + 1) \tag{39}$$

Since $P(x) = 1 - x^2$, we have $P(x) = 0$ where $x = \pm1$, i.e., two singular points, and this suggests that we choose $a = -1, b = 1$. Also the fact that $P(a) = 0, P(b) = 0$ for $a = -1$, $b = 1$ suggests the Sturm–Liouville problem of type C, page 361, involving the boundary conditions

$$y \text{ and } y' \text{ are bounded at } x = \pm1 \tag{40}$$

A solution of (38) which satisfies (40) is given by the Legendre polynomials

$$y = AP_n(x) \tag{41}$$

The eigenvalues are already indicated by

$$\lambda = n(n + 1), \qquad n = 0, 1, 2, 3, \ldots \tag{42}$$

and the corresponding eigenfunctions are thus given by

$$P_0(x), P_1(x), P_2(x), P_3(x), \ldots \tag{43}$$

Using (10) on page 359 with $P(a) = P(b) = 0$, it follows that

$$\int_{-1}^{1} P_j(x)P_k(x)dx = 0, \qquad j \neq k \tag{44}$$

which shows that $P_0(x), P_1(x), P_2(x), \ldots$ are mutually orthogonal in the interval $-1 \leq x \leq 1$, the weight function being $R(x) = 1$. The functions (43) can be normalized. To do this we write $\phi_k(x) = A_k P_k(x)$ and seek to find the constant A_k so that

$$\int_{-1}^{1} [\phi_k(x)]^2 \, dx = \int_{-1}^{1} [A_k P_k(x)]^2 \, dx = 1$$

This yields

$$A_k = \left[\int_{-1}^{1} P_k^2(x)dx\right]^{-1/2} \tag{45}$$

on choosing $A_k > 0$. Unfortunately, we cannot use the same device given on page 368 for Bessel functions since j and k are integers. One way out is to evaluate the integral in (45) for different values of k since we know, for example, that $P_0(x) = 1$, $P_1(x) = x$, $P_2(x) = \frac{1}{2}(3x^2 - 1)$, $P_3(x) = \frac{1}{2}(5x^3 - 3x)$ (see page 344). If we do we find

$$\int_{-1}^{1} P_0^2(x)dx = 2, \qquad \int_{-1}^{1} P_1^2(x)dx = \frac{2}{3}, \int_{-1}^{1} P_2^2(x)dx = \frac{2}{5}, \qquad \int_{-1}^{1} P_3^2(x)dx = \frac{2}{7}, \ldots$$

The pattern strongly suggests the result

$$\int_{-1}^{1} P_k^2(x)dx = \frac{2}{2k + 1}, \qquad k = 0, 1, 2, 3, \ldots \qquad (46)$$

which is confirmed by taking larger values of k. However, merely taking a finite set of values for k and finding the result to be true does not constitute a mathematical proof (although such evidence is often used in science and engineering as a "proof").*

A proof can be supplied by using the generating function for Legendre polynomials (see page 345) given by

$$\sum_{k=0}^{\infty} P_k(x)t^k = \frac{1}{\sqrt{1 - 2tx + x^2}} \qquad (47)$$

If we square both sides of (47) we obtain

$$\sum_{k=0}^{\infty} P_k^2(x)t^{2k} + \sum_{j \neq k} P_j(x)P_k(x)t^{j+k} = \frac{1}{1 - 2tx + x^2} \qquad (48)$$

where the second sum on the left is the sum of all the cross product terms which appear in the squaring process. Integration of (48) from $x = -1$ to $x = 1$ yields

$$\sum_{k=0}^{\infty} \left[\int_{-1}^{1} P_k^2(x)dx \right] t^{2k} + \sum_{j \neq k} \left[\int_{-1}^{1} P_j(x)P_k(x)dx \right] t^{j+k} = \int_{-1}^{1} \frac{dx}{1 - 2tx + x^2} \qquad (49)$$

Since the integral in the second sum on the left is zero because of the orthogonality of the Legendre polynomials, we can write (49) as

$$\sum_{k=0}^{\infty} I_k t^{2k} = -\frac{1}{2t} \ln(1 - 2tx + t^2) \Big|_{x=-1}^{1} = \frac{1}{t}[\ln(1 - t) - \ln(1 + t)] \qquad (50)$$

where I_k denotes the value of the integral (46) which we are seeking. Multiplying both sides of (50) by t gives

$$\sum_{k=0}^{\infty} I_k t^{2k+1} = \ln(1 - t) - \ln(1 + t) \qquad (51)$$

If we now differentiate both sides of (51) with respect to t, we arrive at

$$\sum_{k=0}^{\infty} (2k + 1)I_k t^{2k} = \frac{2}{1 - t^2} = 2(1 + t^2 + t^4 + \cdots) = \sum_{k=0}^{\infty} 2t^{2k} \qquad (52)$$

* An interesting example of this is the "formula for generating prime numbers" given by $p = k^2 - k + 41$. By putting $k = 1, 2, 3, \ldots$, we obtain only prime numbers (i.e., numbers p having only ± 1 and $\pm p$ as factors). The formula however breaks down on reaching $k = 41$.

Thus on equating coefficients of t^{2k} in the two sums we find

$$(2k + 1)I_k = 2 \quad \text{or} \quad I_k = \int_{-1}^{1} P_k^2(x)dx = \frac{2}{2k + 1} \tag{53}$$

3.3 MISCELLANEOUS ORTHOGONAL FUNCTIONS

There are many other examples of orthogonal functions which could be studied. This is to be expected because there are many special cases of the Sturm–Liouville differential equation. Of course not all of these are important in applications, but the relatively few which are important warrant special investigation. Two important sets of orthogonal functions are the Hermite and Laguerre polynomials $H_n(x)$ and $L_n(x)$, respectively, already discussed on pages 345–346. Using the differential equations for these functions, we can prove the following results:

1. **Hermite polynomials** $\quad \displaystyle\int_{-\infty}^{\infty} e^{-x^2}H_j(x)H_k(x)dx = \begin{cases} 0, & j \neq k \\ 2^k k! \sqrt{\pi}, & j = k \end{cases}$

This shows that the Hermite polynomials $H_n(x)$ are orthogonal with respect to the weight function e^{-x^2} in the interval $-\infty < x < \infty$. An orthonormal set of functions can be obtained as well.

2. **Laguerre polynomials** $\quad \displaystyle\int_{0}^{\infty} e^{-x}L_j(x)L_k(x)dx = \begin{cases} 0, & j \neq k \\ (k!)^2, & j = k \end{cases}$

This shows that the Laguerre polynomials $L_n(x)$ are orthogonal with respect to the weight function e^{-x} in the interval $0 \leq x < \infty$. A corresponding orthonormal set of functions can be found.

ILLUSTRATIVE EXAMPLE

Prove that the Hermite polynomials are orthogonal with respect to the weight function e^{-x^2} in $-\infty < x < \infty$.

Solution   We know that $H_j(x)$ and $H_k(x)$ satisfy the respective equations

$$H_j'' - 2xH_j' + 2jH_j = 0, \qquad H_k'' - 2xH_k' + 2kH_k = 0$$

If we multiply the first equation by H_k, the second by H_j, and subtract, we find

$$H_jH_k'' - H_kH_j'' - 2x(H_jH_k' - H_kH_j') + (2k - 2j)H_jH_k = 0$$

or $\qquad \dfrac{d}{dx}(H_jH_k' - H_kH_j') - 2x(H_jH_k' - H_kH_j') + (2k - 2j)H_jH_k = 0$

Multiplying this by the "integrating factor" $e^{\int(-2x)dx} = e^{-x^2}$ we can write this as

$$\frac{d}{dx}[e^{-x^2}(H_jH_k' - H_kH_j')] = (2j - 2k)e^{-x^2}H_jH_k$$

Integrating from $-\infty$ to ∞ we thus have

$$(2j - 2k)\int_{-\infty}^{\infty} e^{-x^2}H_jH_k \, dx = e^{-x^2}(H_jH_k' - H_kH_j')\Big|_{-\infty}^{\infty} = 0$$

from which we obtain the required result on dividing by $2j - 2k \neq 0$.

We could also have obtained the result by first writing the differential equation in Sturm–Liouville form and using the technique on page 371.

1. Give a geometric intrepretation involving vectors to each of the following operations.
   (a) $(D^2 + 4)(c_1 \sin 2x + c_2 \cos 2x)$; c_1, c_2 are constants. (b) $D^2 \cos 4x$. (c) $D^k e^{mx} = m^k e^{mx}$, where m is a constant and $k = 1, 2, 3, \ldots$.

2. Interpret geometrically the results $D^{2k}u = -u$, $D^{4k}u = u$, where $u = c_1 \sin x + c_2 \cos x$, c_1, c_2 are constants and $k = 1, 2, 3, \ldots$.

3. (a) Find all vectors y in a non-countably infinite dimensional space which are transformed into the vector $\sin x$ by the transformation $D^2 + 1$. (b) Which of these vectors are such that $y(0) = 0$, $y'(\pi) = 0$. Give a geometric interpretation.

4. Given that $Dy = -\lambda y$, show that there are infinitely many eigenfunctions given by $y = ce^{-mx}$ with infinitely many corresponding eigenvalues $\lambda = m$. Interpret geometrically.

5. Write each of the following equations in Sturm–Liouville form.
   (a) $xy'' + 2y' + (\lambda + x)y = 0$.   (b) $y'' - y' + (\lambda + e^{-x})y = 0$.
   (c) $y'' + (3 + \lambda \cos x)y = 0$.   (d) $y'' - (\tan x)y' + (1 + \lambda \tan x)y = 0$.

6. Show that each of the following are Sturm–Liouville boundary value problems. In each case find (i) the eigenvalues and eigenfunctions, (ii) a set of functions which are mutually orthogonal in the interval (check the orthogonality by direct integration), and (iii) an orthonormal set of functions in the interval. (a) $y'' + \lambda y = 0$; $y(0) = 0$, $y(4) = 0$. (b) $y'' + \lambda y = 0$; $y'(0) = 0$, $y(\pi) = 0$. (c) $y'' + \lambda y = 0$; $y'(0) = 0$, $y'(2) = 0$.

7. Discuss the buckling of the beam on page 365 for the cases (a) $n = 4$; (b) $n = 5$.

8. Show that if x is replaced by $\sqrt{\lambda}\, x$ in equation (1) page 366 then equation (3) is obtained.

B EXERCISES

1. Given the boundary value problem $y'' + \lambda y = 0$; $y(-1) = y(1)$, $y'(-1) = y'(1)$. (a) Show that it is a Sturm–Liouville problem and classify it according to type as indicated on page 361. (b) Find the eigenvalues and eigenfunctions. (c) Verify that the eigenfunctions are mutually orthogonal in the interval $-1 \leq x \leq 1$. (d) Find a corresponding orthonormal set of functions.

2. Discuss the boundary value problem $y'' + \lambda y = 0$; $y(0) + y(1) = 0$, $y'(0) = 0$, answering in particular the following questions: (a) Is it a Sturm–Liouville problem? (b) Does it have eigenvalues and eigenfunctions? (c) If eigenfunctions do exist, are they mutually orthogonal? Explain.

3. Work Exercise 2 for $y'' + \lambda y = 0$; $y(-1) = -y(1)$, $y'(-1) = -y'(1)$. Compare with Exercise 1.

4. Work the problem of the buckling beam on page 364 if the boundary conditions (34) are replaced by (a) $y(0) = 0$, $y'(L) = 0$; (b) $y'(0) = 0$, $y'(L) = 0$. Discuss the physical significance if any.

5. Determine the normalized eigenfunctions and equation defining the eigenvalues for the boundary value problem $xy'' + y' + \lambda xy = 0$ with conditions y bounded at $x = 0$ and (a) $y(1) = 0$; (b) $y'(1) = 0$; (c) $y(1) + 2y'(1) = 0$. Is y' bounded at $x = 0$? Explain.

6. Find the eigenvalues and eigenfunctions for the boundary value problems:
(a) $x^2y'' + xy' + (\lambda x^2 - 4)y = 0$; where y is bounded at $x = 0$ and $y'(1) = 0$.
(b) $x^2y'' + xy' + (\lambda x^2 - 25)y = 0$; where y is bounded at $x = 0$ and $2y(1) + 3y'(1) = 0$.

7. Find the eigenvalues and normalized eigenfunctions for the boundary value problem consisting of the differential equation $x^2y'' + xy' + (\lambda x^2 - \frac{1}{4})y = 0$ with the condition that y is bounded at $x = 0$ and (a) $y(1) = 0$; (b) $y'(1) = 0$; (c) $y(1) + 5y'(1) = 0$.

8. Verify the result (18) on page 369 by using L'Hôpital's rule.

9. Show that (a) $\int_{-1}^{1} xP_n(x)P_{n-1}(x)dx = \dfrac{2n}{4n^2 - 1}$. (b) $\int_{-1}^{1} (1 - x^2)[P'_n(x)]^2\, dx = \dfrac{2n(n + 1)}{2n + 1}$.

10. Prove that the Laguerre polynomials are orthogonal with respect to the weight function e^{-x} in the interval $0 \le x < \infty$.

C EXERCISES

1. Given the boundary value problem $y'' + \lambda y = 0$, $y(0) = 0$, $y'(1) + \mu y'(1) = 0$, where μ is a constant. (a) Show that there are infinitely many positive eigenvalues $\lambda_1, \lambda_2, \lambda_3, \ldots$, where $\lambda_n - \frac{1}{4}(2n - 1)^2\pi^2 \to 0$ as $n \to \infty$. (b) Obtain the corresponding normalized eigenfunctions and verify that they are mutually orthogonal.

2. Solve the boundary value problem $x^2y'' + 5xy' + (\lambda + y)y = 0$, $y(1) = 0$, $y(e) = 0$, giving eigenvalues and eigenfunctions. What is the weight factor with respect to which the solutions are orthogonal?

3. Let $T_n(x)$, $n = 0, 1, 2, \ldots$ denote solutions of the differential equation

$$(1 - x^2)y'' - xy' + n^2y = 0$$

(a) Show that solutions for the first few values of n are given by

$$T_0(x) = 1, \qquad T_1(x) = x, \qquad T_2(x) = 2x^2 - 1, \qquad T_3(x) = 4x^3 - 3x$$

(b) Show that the solutions are given by

$$T_n = \cos(n\cos^{-1}x), \qquad n = 1, 2, 3, \ldots; \qquad T_0(x) = 1$$

and that these are polynomials of degree n. [*Hint:* Use $(\cos\theta + i\sin\theta)^n = \cos n\theta + i\sin n\theta$ and take $x = \cos\theta$.] (c) Show that the set of functions is orthogonal with respect to the weight function $(1 - x^2)^{-1/2}$ in the interval $-1 < x < 1$. The polynomials $T_n(x)$ are called *Chebyshev polynomials*.

4 Orthogonal Series

4.1 INTRODUCTION

We now have various examples of sets of mutually orthogonal functions which can be made into orthonormal sets of functions by appropriate normalization. We can denote an orthonormal set of functions by

$$\phi_1(x), \phi_2(x), \phi_3(x), \ldots \tag{1}$$

where

$$\int_a^b \phi_j(x)\phi_k(x)dx = \begin{cases} 0, & \text{if } j \neq k \\ 1, & \text{if } j = k \end{cases} \qquad (2)$$

Such functions are very much reminiscent of the $\vec{i}, \vec{j}, \vec{k}$ vectors of ordinary three-dimensional space which also constitute an orthonormal set.

Since any three-dimensional vector $\vec{A}$ can be expanded in terms of the $\vec{i}, \vec{j}, \vec{k}$ vectors in the form

$$\vec{A} = A_1\vec{i} + A_2\vec{j} + A_3\vec{k} \qquad (3)$$

for suitable constants A_1, A_2, A_3 (which represent the components of $\vec{A}$ in the $\vec{i}, \vec{j}$, and $\vec{k}$ directions, respectively), it is natural to ask whether we can expand a function $f(x)$ into a series of orthonormal functions, i.e.,

$$f(x) = c_1\phi_1(x) + c_2\phi_2(x) + \cdots = \sum_{j=1}^{\infty} c_j\phi_j(x) \qquad (4)$$

for suitable constants $c_1, c_2, \ldots$ (which by analogy represent components of $f(x)$ in the "directions" of $\phi_1(x), \phi_2(x), \ldots$). We assume here of course that the functions are all defined in the same interval $a \leq x \leq b$, which amounts to saying that they are all vectors in the same space.

Formally, it is easy to obtain the constants $c_1, c_2, \ldots$. For let us multiply equation (4) by $\phi_k(x)$ and integrate term by term from $x = a$ to $x = b$ to obtain

$$\int_a^b f(x)\phi_k(x)dx = \sum_{j=1}^{\infty} c_j \int_a^b \phi_j(x)\phi_k(x)dx \qquad (5)$$

Now if we make use of (2), the series on the right of (5) reduces to only one term, the one for which $j = k$, since all other terms are zero. The value of this remaining term is c_k. Thus we have

$$c_k = \int_a^b f(x)\phi_k(x)dx \qquad (6)$$

Note that c_k is the scalar product of $f(x)$ and $\phi_k(x)$. This is analogous to A_1, A_2, A_3 being equal to the scalar products of $\vec{A}$ with $\vec{i}, \vec{j}, \vec{k}$, respectively, i.e., $A_1 = \vec{A} \cdot \vec{i}$, $A_2 = \vec{A} \cdot \vec{j}, A_3 = \vec{A} \cdot \vec{k}$ so that

$$\vec{A} = A_1\vec{i} + A_2\vec{j} + A_3\vec{k} = (\vec{A} \cdot \vec{i})\vec{i} + (\vec{A} \cdot \vec{j})\vec{j} + (\vec{A} \cdot \vec{k})\vec{k} \qquad (7)$$

There are several questions which should be asked.

Question 1. How do we know whether the assumed series on the right of (4) contains *all* of the functions needed for the expansion. To put the question another way, what would happen if we left out one or more of the functions on the right? This involves what is called the *completeness* of the orthonormal set of functions, and, as we might expect, unless the set is complete we would certainly not expect (4) to hold.

Question 2. Even if we do have a complete orthonormal set, how do we know that the series on the right of (4) with coefficients c_k given by (6) converges and even if it converges how do we know that it converges to $f(x)$?

In order to appreciate Question 1, suppose that we were only given the unit vectors $\vec{i}$ and $\vec{j}$ and were asked to expand an arbitrary three-dimensional vector $\vec{A}$ in terms of them, i.e., to find A_1 and A_2 so that

$$\vec{A} = A_1\vec{i} + A_2\vec{j} \tag{8}$$

Clearly, we cannot do this since the set $\vec{i}, \vec{j}$ is not a complete orthonormal set and to complete it we would need the unit vector $\vec{k}$. It is interesting to note that in the case (3) where we do have completeness we also have

$$A^2 = \vec{A} \cdot \vec{A} = (\vec{A} \cdot \vec{i})^2 + (\vec{A} \cdot \vec{j})^2 + (\vec{A} \cdot \vec{k})^2 = A_1^2 + A_2^2 + A_3^2 \tag{9}$$

which is the Pythagorean theorem in three dimensions.

From this observation it would seem reasonable by analogy to make the definition that the orthonormal set of functions $\phi_1(x)$, $\phi_2(x)$, ... is *complete* in the interval $a \leqq x \leqq b$ if the equality

$$\int_a^b [f(x)]^2 \, dx = \sum_{k=1}^{\infty} c_k^2 \tag{10}$$

where

$$c_k = \int_a^b f(x)\phi_k(x)dx \tag{11}$$

holds for every function $f(x)$ for which the left side of (10) exists. The equality (10), which amounts to the Pythagorean theorem in countably infinite dimensional space, is often called *Parseval's equality* or *Parseval's identity*.

The result (11) is easily obtained formally if we have

$$f(x) = \sum_{k=1}^{\infty} c_k\phi_k(x) \tag{12}$$

because if we multiply both sides of (12) by $f(x)$ and integrate term by term from $x = a$ to $x = b$ we find

$$\int_a^b [f(x)]^2 \, dx = \sum_{k=1}^{\infty} c_k \int_a^b f(x)\phi_k(x)dx \tag{13}$$

which reduces to (10) on using (11). Unfortunately, the reverse does not follow, i.e., we cannot obtain (12) from (10). In fact, it could very well happen that (10) holds while (12) does not hold.

One disadvantage in using (10) as a definition for completeness is that there is no obvious relationship between $f(x)$ and the orthonormal set providing the basis for expansion as given in (12). This situation can, however, be remedied because we can show that (10) is equivalent to the statement

$$\lim_{n \to \infty} \int_a^b \left[f(x) - \sum_{j=1}^{n} c_j\phi_j(x) \right]^2 \, dx = 0 \tag{14}$$

where it is noted that the integrand contains the difference between $f(x)$ and the sum of the first n terms of the series on the right of (12). The equivalence is not difficult to establish. To do so we must first square

$$f(x) - \sum_{j=1}^{n} c_j \phi_j(x) \tag{15}$$

This is done by squaring each term in (15), finding the cross product terms, and then adding. The result is

$$[f(x)]^2 + \sum_{j=1}^{n} c_j^2 [\phi_j(x)]^2 - 2 \sum_{j=1}^{n} c_j f(x) \phi_j(x) + \sum_{j \neq k} c_j c_k \phi_j(x) \phi_k(x) \tag{16}$$

If we now integrate (16) from $x = a$ to b we can write the result (14) as

$$\lim_{n \to \infty} \left\{ \int_a^b [f(x)]^2 \, dx + \sum_{j=1}^{n} c_j^2 \int_a^b [\phi_j(x)]^2 \, dx - 2 \sum_{j=1}^{n} c_j \int_a^b f(x) \phi_j(x) dx \right.$$

$$\left. + \sum_{j \neq k} c_j c_k \int_a^b \phi_j(x) \phi_k(x) dx \right\} = 0$$

i.e.,

$$\lim_{n \to \infty} \left\{ \int_a^b [f(x)]^2 \, dx + \sum_{k=1}^{n} c_k^2 - 2 \sum_{k=1}^{n} c_k^2 + 0 \right\} = \lim_{n \to \infty} \left\{ \int_a^b [f(x)]^2 \, dx - \sum_{k=1}^{n} c_k^2 \right\} = 0$$

which is equivalent to (10).

Now suppose that we consider

$$E_n = \frac{1}{b-a} \int_a^b \left[f(x) - \sum_{j=1}^{n} c_j \phi_j(x) \right]^2 dx \tag{17}$$

We can think of the finite sum

$$\sum_{j=1}^{n} c_j \phi_j(x) \tag{18}$$

as being an approximation to $f(x)$ and (15) as being the error made. Then (17) can be interpreted as the *mean square error* in the approximation and the square root of this as the *root mean square error* or *rms error* with which the scientist and engineer may be familiar from electric circuit analysis, physics, etc., in connection with root mean square currents, speeds, etc.

These remarks lead us to make the following definition of completeness.

Definition. The orthonormal set of functions $\phi_1(x)$, $\phi_2(x)$, ... is said to constitute a *complete orthonormal set* in the interval $a \leqq x \leqq b$ if for all functions $f(x)$ such that the left side of (10) converges we have

$$\lim_{n \to \infty} E_n = \lim_{n \to \infty} \frac{1}{b-a} \int_a^b \left[f(x) - \sum_{j=1}^{n} c_j \phi_j(x) \right]^2 dx = 0 \tag{19}$$

where the coefficients c_j are given by (11), i.e., the root mean square error in the approximation (18) approaches zero as $n \to \infty$.

The result (19) is often written in the rather suggestive form

$$\underset{n \to \infty}{\text{l.i.m.}} \sum_{j=1}^{n} c_j \phi_j(x) = f(x) \tag{20}$$

which is read $f(x)$ is the *limit in mean* of (18) as $n \to \infty$, or that (18) *converges in the mean* to $f(x)$ as $n \to \infty$. It should be noted that such *mean convergence*, as it is called, is different from *ordinary convergence*, i.e.,

$$\lim_{n \to \infty} \sum_{j=1}^{n} c_j \phi_j(x) = f(x) \tag{21}$$

which corresponds to (12).

Although questions of ordinary and mean convergence are beyond the scope of this course, it is of interest that the student see some of the ideas and perhaps be motivated to pursue more advanced courses in which such topics are treated. Some of the concepts will be considered in the exercises for the interested student.

Fortunately, from the applied point of view it turns out that in most cases which arise in practice we have the ordinary convergence (21), except at some isolated values of x and also the mean convergence (20). Illustrations will be given later.

Remark. In the above discussion we assumed the expansion of a function in terms of orthonormal functions $\phi_1(x)$, $\phi_2(x)$, ... This is desirable especially from a theoretical point of view, since results obtained such as (10) and (11) are greatly simplified. From a practical point of view, however, this requires first normalization of the orthogonal functions and then expansion of the function into the resulting orthonormal series. Instead of doing this, it is often simpler to assume the series expansion in terms of the orthogonal functions and determine the coefficients directly. Thus, for example, suppose that we are given the set of orthogonal functions $u_1(x)$, $u_2(x)$, ... (which hopefully, on proper normalization, will correspond to a complete orthonormal set), and are required to expand a given function $f(x)$ in the series

$$f(x) = a_1 u_1(x) + a_2 u_2(x) + \cdots = \sum_{j=1}^{\infty} a_j u_j(x) \tag{22}$$

On multiplying both sides by $u_k(x)$ and integrating from a to b, we find

$$\int_a^b f(x) u_k(x) dx = \sum_{j=1}^{\infty} a_j \int_a^b u_j(x) u_k(x) dx \tag{23}$$

Then on using

$$\int_a^b u_j(x) u_k(x) dx = 0, \qquad j \neq k \tag{24}$$

(23) becomes

$$\int_a^b f(x) u_k(x) dx = a_k \int_a^b [u_k(x)]^2 dx \tag{25}$$

which yields

$$a_k = \frac{\int_a^b f(x) u_k(x) dx}{\int_a^b [u_k(x)]^2 \, dx} \tag{26}$$

The resulting series (22) will of course be the same as if we had started out with normalized functions.

4.2 FOURIER SERIES

On pages 360–361 we mentioned four types of Sturm–Liouville problems A, B, C, and D. Thus far we have given examples of all of these except for D, the *periodic case*. We now examine this case. We choose the simplest Sturm–Liouville differential equation

$$\frac{d^2y}{dx^2} + \lambda y = 0 \tag{27}$$

and shall take the periodic boundary conditions as

$$y(-L) = y(L), \qquad y'(-L) = y'(L) \tag{28}$$

where we have used $a = -L$ and $b = L$, where L is a given constant. With this choice it turns out that the results are greatly simplified.

Now the general solution of the differential equation (27) is

$$y = A\cos\sqrt{\lambda}\,x + B\sin\sqrt{\lambda}\,x \tag{29}$$

The boundary conditions (28) yield

$$A\cos\sqrt{\lambda}L - B\sin\sqrt{\lambda}L = A\cos\sqrt{\lambda}L + B\sin\sqrt{\lambda}L \tag{30}$$

$$A\sqrt{\lambda}\sin\sqrt{\lambda}L + B\sqrt{\lambda}\cos\sqrt{\lambda}L = -A\sqrt{\lambda}\sin\sqrt{\lambda}L + B\sqrt{\lambda}\cos\sqrt{\lambda}L \tag{31}$$

These equations lead respectively to

$$2B\sin\sqrt{\lambda}L = 0, \qquad 2A\sqrt{\lambda}\sin\sqrt{\lambda}L = 0 \tag{32}$$

There are two possibilities which can occur:

(a) $\sin\sqrt{\lambda}L = 0$, from which we obtain

$$\lambda = \frac{k^2\pi^2}{L^2}, \, k = 0, 1, 2, 3, \dots \tag{33}$$

which are the eigenvalues. The corresponding eigenfunctions obtained from (29) are then given by

$$y_k = a_k\cos\frac{k\pi x}{L} + b_k\sin\frac{k\pi x}{L}, \qquad k = 0, 1, 2, 3, \dots \tag{34}$$

where a_k, b_k are constants.

(b) $\sin\sqrt{\lambda}L \neq 0$, from which $B = 0$, $A\sqrt{\lambda} = 0$. This means that either $B = 0$, $A = 0$, which gives the trivial solution $y = 0$, or $B = 0$, $\lambda = 0$, which leads to a rather uninteresting case of one eigenvalue and one eigenfunction so that orthogonality becomes meaningless.

Obviously, the possibility (a) is the only one of significance. The eigenvalues are $\sqrt{\lambda} = 0, \pi/L, 2\pi/L, \dots$, and the corresponding eigenfunctions are

$$a_0, \quad a_1\cos\frac{\pi x}{L} + b_1\sin\frac{\pi x}{L}, \quad a_2\cos\frac{2\pi x}{L} + b_2\sin\frac{2\pi x}{L}, \dots \tag{35}$$

These are of course, from the manner in which they were derived, mutually orthogonal, i.e.,

$$\int_{-L}^{L} \left(a_j \cos\frac{j\pi x}{L} + b_j \sin\frac{j\pi x}{L} \right) \left(a_k \cos\frac{k\pi x}{L} + b_k \sin\frac{k\pi x}{L} \right) dx = 0, \qquad j \neq k \quad (36)$$

In particular, since (36) is true for any values of the a's and b's, we have

$$\int_{-L}^{L} \cos\frac{j\pi x}{L} \cos\frac{k\pi x}{L} \, dx = 0, \qquad \int_{-L}^{L} \sin\frac{j\pi x}{L} \sin\frac{k\pi x}{L} \, dx = 0,$$

$$\int_{-L}^{L} \sin\frac{j\pi x}{L} \cos\frac{k\pi x}{L} \, dx = 0, \qquad j \neq k$$

(37)

which can be verified by direct integration.

If $f(x)$ is a function defined in the interval $-L \leq x \leq L$, we are naturally led to the expansion of $f(x)$ in the series

$$f(x) = a_0 + \sum_{j=1}^{\infty} \left(a_j \cos\frac{j\pi x}{L} + b_j \sin\frac{j\pi x}{L} \right) \tag{38}$$

To obtain the a's, we first multiply both sides of (38) by $\cos (k\pi x/L)$ and integrate from $-L$ to L. Then

$$\int_{-L}^{L} f(x) \cos\frac{k\pi x}{L} \, dx = a_0 \int_{-L}^{L} \cos\frac{k\pi x}{L} \, dx + \sum_{j=1}^{\infty} \left(a_j \int_{-L}^{L} \cos\frac{j\pi x}{L} \cos\frac{k\pi x}{L} \, dx \right.$$

$$\left. + b_j \int_{-L}^{L} \sin\frac{j\pi x}{L} \cos\frac{k\pi x}{L} \, dx \right)$$

Because of (37), all the terms on the right are zero except one, that for which $j = k$. Thus we have

$$\int_{-L}^{L} f(x) \cos\frac{k\pi x}{L} \, dx = a_k \int_{-L}^{L} \cos^2\frac{k\pi x}{L} \, dx \quad \text{or} \quad a_k = \frac{\displaystyle\int_{-L}^{L} f(x) \cos\frac{k\pi x}{L} \, dx}{\displaystyle\int_{-L}^{L} \cos^2\frac{k\pi x}{L} \, dx} \tag{39}$$

The denominator in (39) is easily evaluated. We have

$$k = 0: \qquad \int_{-L}^{L} \cos^2\frac{k\pi x}{L} \, dx = \int_{-L}^{L} 1 \, dx = 2L$$

$$k = 1, 2, 3, \ldots : \qquad \int_{-L}^{L} \cos^2\frac{k\pi x}{L} \, dx = \int_{-L}^{L} \frac{1}{2}\left(1 + \cos\frac{2k\pi x}{L} \right) dx = L$$

Thus (39) becomes

$$a_0 = \frac{1}{2L} \int_{-L}^{L} f(x) \, dx, \qquad a_k = \frac{1}{L} \int_{-L}^{L} f(x) \cos\frac{k\pi x}{L} \, dx, \qquad k = 1, 2, 3, \ldots \tag{40}$$

Similarly, to obtain the b's, we multiply both sides of (38) by $\sin(k\pi x/L)$ and integrate from $-L$ to L. Thus

$$\int_{-L}^{L} f(x) \sin \frac{k\pi x}{L} dx = a_0 \int_{-L}^{L} \sin \frac{k\pi x}{L} dx + \sum_{j=1}^{\infty} \left(a_j \int_{-L}^{L} \cos \frac{j\pi x}{L} \sin \frac{k\pi x}{L} dx \right.$$

$$\left. + b_j \int_{-L}^{L} \sin \frac{j\pi x}{L} \sin \frac{k\pi x}{L} dx \right)$$

Because of (37) all the terms on the right but one give zero, so that

$$\int_{-L}^{L} f(x) \sin \frac{k\pi x}{L} dx = b_k \int_{-L}^{L} \sin^2 \frac{k\pi x}{L} dx \quad \text{or} \quad b_k = \frac{\int_{-L}^{L} f(x) \sin \frac{k\pi x}{L} dx}{\int_{-L}^{L} \sin^2 \frac{k\pi x}{L}} \tag{41}$$

The denominator in (41) is easily evaluated. We find

$$\int_{-L}^{L} \sin^2 \frac{k\pi x}{L} dx = \int_{-L}^{L} \frac{1}{2}\left(1 - \cos \frac{2k\pi x}{L} \right) dx = L, \quad k = 1, 2, 3, \ldots$$

so that (41) becomes

$$b_k = \frac{1}{L} \int_{-L}^{L} f(x) \sin \frac{k\pi x}{L} dx, \quad k = 1, 2, 3, \ldots \tag{42}$$

The series expansion (38) with coefficients given by (40) and (42) is called a *Fourier series* after the man who first discovered it in his researches on heat flow involving partial differential equations in the early part of the 19th century. We shall see how such series arise from applications to partial differential equations in Chapter Thirteen and shall have occasion to work many applied problems making use of them.

It is observed that there are two separate formulas for the a's given in (40), one for a_0 and one for a_k, $k = 1, 2, \ldots$. It is possible and desirable to replace these by one formula. To see how this can be done, let us put $k = 0$ in the second formula of (40). Then we find

$$a_0 = \frac{1}{L} \int_{-L}^{L} f(x) dx \tag{43}$$

This same result is achieved if we replace a_0 in the first result of (40) by $a_0/2$. If we do this, we must also of course replace the a_0 in (38) by $a_0/2$. We are thus led to the following summary of the basic procedure for the expansion of a function in a Fourier series.

Summary of Procedure for Finding Fourier Series

1. Given the function $f(x)$ defined in the interval $-L \leq x \leq L$, evaluate the coefficients, often called *Fourier coefficients*, given by

$$a_k = \frac{1}{L} \int_{-L}^{L} f(x) \cos \frac{k\pi x}{L} dx, \quad b_k = \frac{1}{L} \int_{-L}^{L} f(x) \sin \frac{k\pi x}{L} dx \tag{44}$$

2. Using these coefficients and assuming that the series converges to $f(x)$, the required Fourier series is given by

$$f(x) = \frac{a_0}{2} + \sum_{k=1}^{\infty} \left(a_k \cos \frac{k\pi x}{L} + b_k \sin \frac{k\pi x}{L} \right) \tag{45}$$

As we have already mentioned, there is a question as to whether the series on the right of (45) actually converges to $f(x)$ so that the equality is justified. For this purpose we shall have to examine the question of convergence of Fourier series. Before doing so, however, let us work an example illustrating how a Fourier series for a given function $f(x)$ can be found.

ILLUSTRATIVE EXAMPLE 1

Find the Fourier series corresponding to the function $f(x) = 1 - x$ in the interval $-1 \leq x \leq 1$.

Solution Since the interval is $-1 \leq x \leq 1$, we have $L = 1$. Then the Fourier coefficients a_k, b_k are given by

$$a_k = \frac{1}{1} \int_{-1}^{1} (1 - x) \cos \frac{k\pi x}{1} \, dx, \qquad b_k = \frac{1}{1} \int_{-1}^{1} (1 - x) \sin \frac{k\pi x}{1} \, dx$$

From these we find

$$a_0 = \int_{-1}^{1} (1 - x) dx = x - \frac{x^2}{2} \Big|_{-1}^{1} = 2$$

$$a_k = \int_{-1}^{1} (1 - x) \cos k\pi x \, dx = (1 - x) \left(\frac{\sin k\pi x}{k\pi} \right) - (-1) \left(-\frac{\cos k\pi x}{k^2 \pi^2} \right) \Big|_{-1}^{1} = 0$$

$$b_k = \int_{-1}^{1} (1 - x) \sin k\pi x \, dx = (1 - x) \left(-\frac{\cos k\pi x}{k\pi} \right) - (-1) \left(-\frac{\sin k\pi x}{k^2 \pi^2} \right) \Big|_{-1}^{1}$$

$$= \frac{2 \cos k\pi}{k\pi}$$

Thus $a_0 = 2$, $a_1 = a_2 = \cdots = 0$, $b_1 = -\dfrac{2}{\pi}$, $b_2 = \dfrac{2}{2\pi}$, $b_3 = -\dfrac{3}{2\pi}, \ldots$ and the Fourier series can be written formally as

$$1 - x = 1 - \frac{2}{\pi} \sin \pi x + \frac{2}{2\pi} \sin 2\pi x - \frac{2}{3\pi} \sin 3\pi x + \cdots \tag{46}$$

assuming that there is convergence of the series on the right to $f(x) = 1 - x$ for $-1 \leq x \leq 1$.

When scientists, engineers, or mathematicians arrive at possible results by somewhat devious means, they have recourse to either proving them rigorously or at least checking them to see if they are reasonable. The last is usually easier, and in many instances a proof is sought only after enough evidence has been obtained to conjecture that the results have validity. With this in mind let us seek to check the result (46) obtained in the example above. Since we have obtained this result assuming $-1 \leq x \leq 1$, let us see if the result holds for various values in this interval. If we try $x = 1$ and -1, i.e., the endpoints of the interval, we obtain formally $0 = 1$ and $2 = 1$, respectively, evidence which may cause some to give up immediately and claim that the result must be wrong. Others, having more faith in the analysis, may be inclined to rationalize that perhaps the result only holds for $-1 < x < 1$, in which case it is natural for us to get a wrong result for $x = \pm 1$. With this in mind, we can try a point in the middle of the interval, i.e., $x = 0$. Using this in (46) we are led to $1 = 1$, which may be a simple coincidence. Continuing further, we let $x = \frac{1}{2}$, in which case (46) becomes

$$\frac{1}{2} = 1 - \frac{2}{\pi} + \frac{2}{3\pi} - \frac{2}{5\pi} + \frac{2}{7\pi} - \cdots = 1 - \frac{2}{\pi}\left(1 - \frac{1}{3} + \frac{1}{5} - \frac{1}{7} + \cdots\right)$$

or

$$1 - \frac{1}{3} + \frac{1}{5} - \frac{1}{7} + \cdots = \frac{\pi}{4}$$

This, if true, would seem to be more than a coincidence. Using a calculator, we see that it seems to have validity. Actually, it is not difficult to prove the result which is based on the easily obtained Taylor series expansion (using methods employed in elementary calculus)

$$\tan^{-1} t = t - \frac{t^3}{3} + \frac{t^5}{5} - \frac{t^7}{7} + \cdots$$

and simply taking $t = 1$ on both sides.

There is still a question as to why the result failed at $x = \pm 1$. To seek a possible explanation, we note that the terms on the right in (46) are all periodic functions, and in fact the period of all of them is 2. Thus, if (46) is to be valid, the left side must also have period 2. We can see this better if we graph $f(x) = 1 - x$ for $-1 \leq x \leq 1$ and repeat this graph to the right and left, as shown in Fig. 8.8.

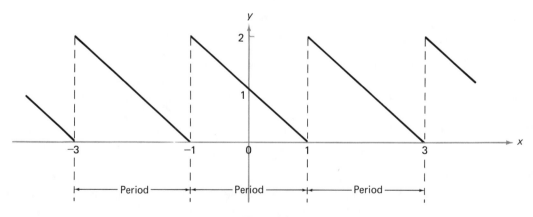

Figure 8.8

The graph serves to clear up our dilemma because we see from it that there are points of discontinuity at $x = \pm 1, \pm 2, \pm 3, \ldots$. Since the value of the series is 0 or 2 near a point of discontinuity, depending on which side of the discontinuity we consider, the series in its attempt to be "unbiased" gives the average value $\frac{1}{2}(0 + 2) = 1$ at the point of discontinuity.

As we have already remarked, the question of the convergence of Fourier series involves some rather complicated mathematical analysis which we shall omit. It turns out, however, that there are some very simple conditions under which we can tell when a Fourier series converges and also to what it converges. Fourier himself did not develop these conditions, and it was left to the mathematician *Dirichlet* to perform this task.

Before stating the basic theorem of Dirichlet, let us make some further observations. First, it is quite clear that if the series (45) is to hold then, since each of the functions on the right has the period $2L$, the sum should also have the period $2L$, and thus $f(x)$ should have the period $2L$. We can indicate this by writing

$$f(x + 2L) = f(x)$$

Second, it is clear that the function $f(x)$ must be such that the integrals in (44) exist. One simple condition for this is that $f(x)$ be continuous in the interval $-L \leq x \leq L$. This is a little more restrictive than we need, however, since it is possible for a function to have discontinuities and still be integrable. One possibility is to use piecewise continuous functions, which we have already encountered in the chapter on Laplace transforms. Let us review this in the following

Definition. A function is said to be *piecewise continuous* in an interval if the function has only a finite number of discontinuities in the interval and if both the right and left hand limits at each discontinuity exist.

An example of a function which has period $2L$ and which is piecewise continuous is shown in Fig. 8.9.

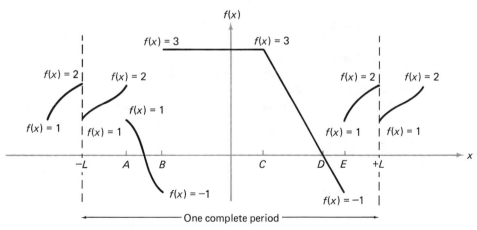

Figure 8.9

We are now ready to state conditions known as *Dirichlet conditions* under which a Fourier series is guaranteed to converge and to what it converges in the following

Theorem 1 (Dirichlet's Theorem). (a) Let $f(x)$ be a function of x defined in the interval $-L < x < L$ and defined by periodic extension by the relation $f(x + 2L) = f(x)$, i.e., $f(x)$ has period $2L$. (b) Let $f(x)$ be piecewise continuous in the interval $-L < x < L$. (c) Let the derivative $f'(x)$ be piecewise continuous in the interval $-L < x < L$. Then the series on the right of (45) with coefficients (44) converges to $f(x)$ if x is a point of continuity and converges to

$$\tfrac{1}{2}[f(x + 0) + f(x - 0)] \tag{47}$$

if x is a point of discontinuity where (47) represents the arithmetic mean of the limits of $f(x)$ from both sides of the point of discontinuity.* The conditions in (a), (b), and (c) are called *Dirichlet conditions*.

This theorem is of great value since it guarantees that a function such as that shown in Fig. 8.9 possesses a convergent Fourier series. If the function sketched in the graph were used to arrive at a Fourier series, the resulting series would converge to the following values:

At A the series would converge to $\dfrac{2 + 1}{2} = 1.5$.

At B the series would converge to $\dfrac{3 + (-1)}{2} = 1$.

At C the series would converge to 3, since x is a point of continuity.

At D the series would converge to 0.

At E the series would converge to $\dfrac{1 + (-1)}{2} = 0$.

At $-L$ and $+L$ the series would converge to $\dfrac{2 + 1}{2} = 1.5$.

The theorem serves to explain the results achieved in Illustrative Example 1, page 383.

Before going on to more examples and the use of the above theorem, there are several things which should be pointed out.

1. Sometimes the function $f(x)$ is specified in an interval of length $2L$ other than $-L < x < L$, such as for example $c < x < c + 2L$, where c can be any real number. In such case the Fourier series is the same as given in (45), but the Fourier coefficients (44) are replaced by

$$a_k = \frac{1}{L} \int_c^{c+2L} f(x) \cos \frac{k\pi x}{L} \, dx, \qquad b_k = \frac{1}{L} \int_c^{c+2L} f(x) \sin \frac{k\pi x}{L} \, dx \tag{48}$$

In the special space where $c = -L$, (48) reduces to (44).

* See the footnote on page 263.

2. Two important special cases of Fourier series occur in which there are (a) only cosine terms present, or (b) only sine terms present. We refer to these as *Fourier cosine series* and *Fourier sine series*, respectively. It is often possible to predict from the graph of the function whether we will get a Fourier cosine series, a Fourier sine series, or a regular series involving both cosines and sines. To see how, let us suppose first that we do have a cosine series, i.e.,

$$f(x) = \frac{a_0}{2} + a_1 \cos \frac{\pi x}{L} + a_2 \cos \frac{2\pi x}{L} + \cdots \tag{49}$$

where all the b's are zero. If we replace x by $-x$ in (49) and realize that $\cos(-\theta) = \cos \theta$, (49) becomes

$$f(-x) = \frac{a_0}{2} + a_1 \cos \frac{\pi x}{L} + a_2 \cos \frac{2\pi x}{L} + \cdots \tag{50}$$

i.e., $f(-x) = f(x)$. A function which has this property is often called an *even function*, the terminology probably originating from the fact that if n is an even integer then $(-x)^n = x^n$. Examples of graphs of even functions are shown in Fig. 8.10. In each of these the value of the function at $-x$, i.e., $f(-x)$, is equal to the value of the function at x, i.e., $f(x)$. The graph is thus symmetric about the y axis so that the y axis acts like a mirror in which the part of the graph to the right is the object and the part to the left is the image.

If only sine terms are present, i.e., we have a Fourier sine series, then

$$f(x) = b_1 \sin \frac{\pi x}{L} + b_2 \sin \frac{2\pi x}{L} + \cdots \tag{51}$$

where all the a's are zero. If we replace x by $-x$ in (51), and realize that $\sin(-\theta) = -\sin \theta$, (51) becomes

$$f(-x) = -\left(b_1 \sin \frac{\pi x}{L} + b_2 \sin \frac{2\pi x}{L} + \cdots \right) = -f(x) \tag{52}$$

i.e., $f(x) = -f(x)$. A function which has this property is called an *odd function*, probably from the fact that, for n equal to an odd integer, $(-x)^n = -x^n$. Examples of graphs of odd functions are shown in Fig. 8.11. In each of these the value of the function at $-x$, i.e., $f(-x)$, is equal to the negative of the value of the function at x, i.e., $-f(x)$. If we assume that the part of the graph to the right of the y axis is the object, then the required image is obtained by reflections in two mirrors, one being the y axis and the other the negative x axis.

Conversely, we can show that even functions have corresponding Fourier cosine series, while odd functions have Fourier sine series. If therefore we know in advance that we have an even, or odd, function, we can cut the labor of finding the Fourier coefficients in half, since we only have to find the a's or b's, respectively.

If $f(x)$ is even, then so also is $f(x) \cos k\pi x/L$, so that (44) is replaced by

$$a_k = \frac{2}{L} \int_0^L f(x) \cos \frac{k\pi x}{L} \, dx, \qquad b_k = 0 \tag{53}$$

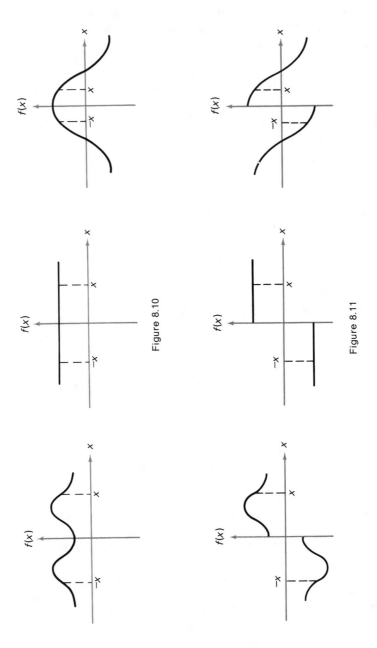

Figure 8.10

Figure 8.11

Similarly, in case $f(x)$ is odd, then $f(x) \sin k\pi x/L$ is even, so that the result (44) is replaced by

$$a_k = 0, \qquad b_k = \frac{2}{L} \int_0^L f(x) \sin \frac{k\pi x}{L} \, dx \qquad (54)$$

Sometimes we are given the function $f(x)$ in an interval $0 < x < L$ of length L, i.e., half of the usual interval, and are asked to find the Fourier cosine series for the function. In such case we simply use (53). The result is of course the same as if we were to extend the definition of the function to the other half of the interval, $-L < x < 0$ so that the resulting function in the complete interval $-L < x < L$ is an even function.

Similarly, if we are given the function $f(x)$ in $0 < x < L$ and are asked to find the Fourier sine series for the function, we use (54). The result is the same as if we were to extend the definition of the function to the other half of the interval $-L < x < 0$ so that the resulting function in the complete interval $-L < x < L$ is an odd function. Such cases are often referred to for obvious reasons as *half-range Fourier cosine series* and *half-range Fourier sine series*, respectively.

Let us illustrate the above remarks by working a few more examples.

ILLUSTRATIVE EXAMPLE 2

Find a Fourier series for the function $f(x) = \begin{cases} -1, & -2 < x < 0 \\ 1, & 0 < x < 2 \end{cases}$

and having period 4.

Solution The graph of this function appears in Fig. 8.12. It is seen that the function is odd. Hence we expect a Fourier sine series. Since the period is 4, we have $2L = 4$, $L = 2$. Hence from (54),

$$a_k = 0, \; b_k = \frac{2}{2} \int_0^2 f(x) \sin \frac{k\pi x}{2} \, dx = \int_0^2 (1) \sin \frac{k\pi x}{2} \, dx$$

Integration yields
$$b_k = -\frac{2}{k\pi} \cos \frac{k\pi x}{2} \Big|_0^2 = \frac{2}{k\pi} (1 - \cos k\pi)$$

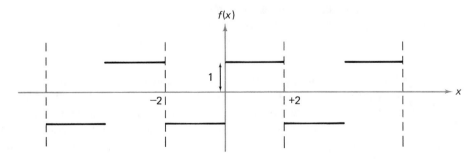

$f(x)$

-2

$+2$

1

x

Figure 8.12

from which we obtain $\qquad b_1 = \dfrac{4}{\pi}, \qquad b_2 = 0, \qquad b_3 = \dfrac{4}{3\pi}, \dots$

Thus, the required Fourier series is

$$\frac{4}{\pi} \sin \frac{\pi x}{2} + \frac{4}{3\pi} \sin \frac{3\pi x}{2} + \frac{4}{5\pi} \sin \frac{5\pi x}{2} + \cdots = \sum_{k=1}^{\infty} \frac{2}{k\pi} (1 - \cos k\pi) \sin \frac{k\pi x}{2}$$

From Dirichlet's theorem we see that at points of discontinuity such as $x = 0$, $2, 4, \dots$ the series converges to zero. At a place such as $x = 1$, which is a point of continuity, the series should converge to the value of $f(x)$ for $x = 1$ which is 1. It follows that

$$\frac{4}{\pi}\left(1 - \frac{1}{3} + \frac{1}{5} - \frac{1}{7} + \cdots\right) = 1$$

which is the same result obtained on page 384.

We can if we like define a new function $F(x)$ of period 4 such that

$$F(x) = \begin{cases} -1, & -2 < x < 0 \\ 1, & 0 < x < 2 \\ 0, & x = \pm 2 \end{cases}$$

In such case we would have for all x

$$F(x) = \sum_{k=1}^{\infty} \frac{2}{k\pi} (1 - \cos k\pi) \sin \frac{k\pi x}{2}.$$

ILLUSTRATIVE EXAMPLE 3

Expand in a Fourier series $\quad f(x) = \begin{cases} 0.1x, & 0 \le x \le 10 \\ 0.1(20 - x), & 10 \le x \le 20 \end{cases}$.

Solution Since no mention has been made of the period for this function, there are at first sight three possible cases which could arise.

(a) The period could be taken as 20 (and the series could have both sines and cosines present).
(b) We could expand the function into a Fourier sine series, in which case the period would be taken as 40 (i.e., the interval $0 \le x \le 20$ would be half of the range).
(c) We could expand the function into a Fourier cosine series, in which case the period would be taken as 40 (i.e., the interval $0 \le x \le 20$ would be half of the range).

All of these series would serve the purpose of representing the function in the interval $0 \le x \le 20$. The fact that we can get various possibilities shows the need for specifying the period or at least mentioning whether a sine series or cosine series is wanted.

Let us assume that it is the case (a) which is wanted, i.e., the period of the function is 20. In such case the graph of $f(x)$ appears as in Fig. 8.13. Since this graph is an even function, the Fourier series will have only cosines in it, i.e., it will be a Fourier cosine series. This would turn out to be the same as the case (c) above, so that there

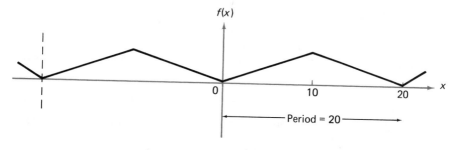

Figure 8.13

are really only two possibilities instead of three. Since the period is 20, we have $2L = 20$ or $L = 10$. Thus

$$a_k = \frac{2}{10} \int_0^{10} (0.1x) \cos \frac{k\pi x}{10} \, dx, \qquad b_k = 0$$

For $k = 0$ we have

$$a_0 = \frac{2}{10} \int_0^{10} 0.1x \, dx = 1$$

For $k = 1, 2, 3, \ldots$ we have, on integrating by parts,

$$a_k = \frac{2}{k^2 \pi^2} (\cos k\pi - 1)$$

Then the required Fourier series is given by

$$\frac{1}{2} + \sum_{k=1}^{\infty} \frac{2}{k^2 \pi^2} (\cos k\pi - 1) \cos \frac{k\pi x}{10} \tag{55}$$

or

$$\frac{1}{2} - \frac{4}{\pi^2} \left(\cos \frac{\pi x}{10} + \frac{1}{3^2} \cos \frac{3\pi x}{10} + \frac{1}{5^2} \cos \frac{5\pi x}{10} + \cdots \right) \tag{56}$$

Since $f(x)$ has no points of discontinuity, we see by Dirichlet's theorem that this series converges to $f(x)$ for all values of x. We can thus legitimately write the equality

$$f(x) = \frac{1}{2} + \sum_{k=1}^{\infty} \frac{2}{k^2 \pi^2} (\cos k\pi - 1) \cos \frac{k\pi x}{10}$$

In particular, for $x = 0$ the series converges to zero. Thus putting $x = 0$ we have

$$0 = \frac{1}{2} - \frac{4}{\pi^2} \left(\frac{1}{1^2} + \frac{1}{3^2} + \frac{1}{5^2} + \cdots \right) \quad \text{or} \quad \frac{1}{1^2} + \frac{1}{3^2} + \frac{1}{5^2} + \cdots = \frac{\pi^2}{8} \tag{57}$$

an interesting series which is difficult to sum by other methods.

Remark 2. It should be noted that if for some reason we did not notice that $f(x)$ is an even function we could have gotten the same result (55) by using

$$a_k = \frac{1}{10} \int_0^{10} 0.1x \cos \frac{k\pi x}{10} \, dx + \frac{1}{10} \int_{10}^{20} 0.1(20 - x) \cos \frac{k\pi x}{10} \, dx$$

$$b_k = \frac{1}{10} \int_0^{10} 0.1x \sin \frac{k\pi x}{10} dx + \frac{1}{10} \int_{10}^{20} 0.1(20 - x) \sin \frac{k\pi x}{10} dx$$

The remaining case (b) mentioned above is treated in the following example.

ILLUSTRATIVE EXAMPLE 4

Expand the function of Illustrative Example 3 in a Fourier sine series.

Solution In order to do this we must use the function defined in Illustrative Example 3 for the interval $0 \leq x \leq 20$, and then extend the definition to the left of $x = 0$ so that we will have an odd function as indicated in Fig. 8.14. In such case the period is 40 so that $2L = 40$, $L = 20$. We thus have $a_k = 0$ and

$$b_k = \frac{2}{20} \int_0^{20} f(x) \sin \frac{k\pi x}{20} dx = \frac{1}{10} \int_0^{10} 0.1x \sin \frac{k\pi x}{20} dx + \frac{1}{10} \int_{10}^{20} 0.1(20 - x) \sin \frac{k\pi x}{20} dx$$

Integration by parts yields
$$b_k = \frac{8 \sin k\pi/2}{k^2 \pi^2}$$

so that the required Fourier series is

$$\sum_{k=1}^{\infty} \frac{8 \sin k\pi/2}{k^2 \pi^2} \sin \frac{k\pi x}{20} \tag{58}$$

or
$$\frac{8}{\pi^2} \left(\sin \frac{\pi x}{20} - \frac{1}{3^2} \sin \frac{3\pi x}{20} + \frac{1}{5^2} \sin \frac{5\pi x}{20} - \cdots \right) \tag{59}$$

This series converges for all x to $f(x)$ by Dirichlet's theorem, since $f(x)$ is continuous for all x. In particular, for $x = 10$ the series converges to 1. Thus we have

$$1 = \frac{8}{\pi^2} \left(1 + \frac{1}{3^2} + \frac{1}{5^2} + \cdots \right) \tag{60}$$

which yields the same result as (57). Note that the series (55) and (58), although they *look different*, actually represent the same function for $0 \leq x \leq 20$.

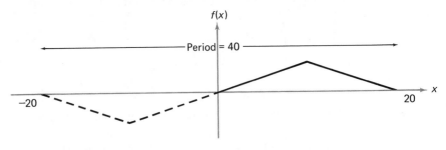

Figure 8.14

Find a Fourier series of period 8 for the function $f(x) = \begin{cases} x, & 0 < x \le 4 \\ 4, & 4 \le x < 8 \end{cases}$.

Solution   The graph of the function is shown in Fig. 8.15, and it is seen to be neither even nor odd. Thus the Fourier series should have both sines and cosines present. Since the period is 8, we have $2L = 8, L = 4$. Then using (48), page 386, with $c = 0$, we have

$$a_k = \frac{1}{4} \int_0^8 f(x) \cos \frac{k\pi x}{4} \, dx = \frac{1}{4} \int_0^4 x \cos \frac{k\pi x}{4} \, dx + \frac{1}{4} \int_4^8 4 \cos \frac{k\pi x}{4} \, dx$$

$$b_k = \frac{1}{4} \int_0^8 f(x) \sin \frac{k\pi x}{4} \, dx = \frac{1}{4} \int_0^4 x \sin \frac{k\pi x}{4} \, dx + \frac{1}{4} \int_4^8 4 \sin \frac{k\pi x}{4} \, dx$$

For $k = 0$ we have $a_0 = \dfrac{1}{4} \int_0^4 x \, dx + \dfrac{1}{4} \int_4^8 4 \, dx = 6$

For $k > 0$ we have on integrating by parts

$$a_k = \frac{4}{k^2\pi^2} (\cos k\pi - 1), \qquad b_k = \frac{8}{k\pi} (\cos k\pi - 1)$$

Then the required Fourier series is

$$3 + \sum_{k=1}^{\infty} \left[\frac{4}{k^2\pi^2} (\cos k\pi - 1) \cos \frac{k\pi x}{4} + \frac{8}{k\pi} (\cos k\pi - 1) \sin \frac{k\pi x}{4} \right] \qquad (61)$$

Remark 3. For the points of discontinuity $x = 0, \pm 8, \pm 16, \ldots$ the series converges to $\frac{1}{2}(4 + 0) = 2$ by Dirichlet's theorem. For any other value of x the series converges to $f(x)$. If we like we can define a new function

$$F(x) = \begin{cases} x, & 0 < x \le 4 \\ 4, & 4 \le x < 8 \\ 2, & x = 0, 8 \end{cases}$$

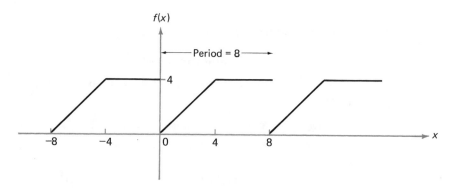

Figure 8.15

We can then say that the series (61) is the Fourier series for $F(x)$ and that it converges to $F(x)$ for all x in the interval $0 \leq x \leq 8$.

The general discussion regarding orthogonal series on pages 375–379 applies of course to Fourier series in particular. One of the questions raised in that discussion involved the completeness of an orthogonal (or orthonormal) set of functions. For the case of Fourier series we can prove the following

Theorem 2. The set of functions $1, \cos \dfrac{k\pi x}{L}, \sin \dfrac{k\pi x}{L}$, where $k = 1, 2, \ldots$ is a complete orthogonal set in any interval of length $2L$, and if $f(x)$ satisfies the Dirichlet conditions of the theorem on page 386, then *Parseval's identity* given by

$$\frac{1}{L} \int_c^{c+2L} [f(x)]^2 \, dx = \frac{a_0^2}{2} + \sum_{k=1}^{\infty} (a_k^2 + b_k^2) \tag{62}$$

where c is any real number, is valid.

Formally, we can deduce (62) by first writing the Fourier series for $f(x)$, i.e.,

$$f(x) = \frac{a_0}{2} + \sum_{k=1}^{\infty} \left(a_k \cos \frac{k\pi x}{L} + b_k \sin \frac{k\pi x}{L} \right) \tag{63}$$

Let us now multiply both sides of (63) by $f(x)$ and integrate from c to $c + 2L$ to obtain

$$\int_c^{c+2L} [f(x)]^2 \, dx = \frac{a_0}{2} \int_c^{c+2L} f(x) dx + \sum_{k=1}^{\infty} a_k \int_c^{c+2L} f(x) \cos \frac{k\pi x}{L} \, dx$$
$$+ \sum_{k=1}^{\infty} b_k \int_c^{c+2L} f(x) \sin \frac{k\pi x}{L} \, dx \tag{64}$$

From (48), page 386, we have

$$\int_c^{c+2L} f(x) dx = \frac{La_0}{2}, \quad \int_c^{c+2L} f(x) \cos \frac{k\pi x}{L} \, dx = La_k, \quad \int_c^{c+2L} f(x) \sin \frac{k\pi x}{L} \, dx = Lb_k$$

Then (64) becomes

$$\int_c^{c+2L} [f(x)]^2 \, dx = \frac{La_0^2}{2} + \sum_{k=1}^{\infty} (La_k^2 + Lb_k^2)$$

which yields (62) on dividing by L.

ILLUSTRATIVE EXAMPLE 6

Write Parseval's identity corresponding to the Fourier series of Illustrative Example 4, page 392.

Solution   We have $c = 0, L = 10$, and $a_k = 0, b_k = \dfrac{8 \sin k\pi/2}{k^2 \pi^2}$

Then (62) becomes

$$\frac{1}{10} \int_0^{20} [f(x)]^2 \, dx = \sum_{k=1}^{\infty} \frac{64 \sin^2 k\pi/2}{k^4 \pi^4} = \frac{64}{\pi^4} \left(\frac{1}{1^4} + \frac{1}{3^4} + \frac{1}{5^4} + \frac{1}{7^4} + \cdots \right) \tag{65}$$

But since $\int_0^{20} [f(x)]^2 \, dx = \int_0^{10} (0.1x)^2 \, dx + \int_{10}^{20} [0.1(20 - x)]^2 \, dx = \dfrac{20}{3}$

(65) becomes $\qquad \dfrac{2}{3} = \dfrac{64}{\pi^4} \left(\dfrac{1}{1^4} + \dfrac{1}{3^4} + \dfrac{1}{5^4} + \dfrac{1}{7^4} + \cdots \right)$

which is the required Parseval's identity. This yields the interesting result

$$\frac{1}{1^4} + \frac{1}{3^4} + \frac{1}{5^4} + \frac{1}{7^4} + \cdots = \frac{\pi^4}{96} \tag{66}$$

A EXERCISES

1. Obtain Fourier series expansions corresponding to each of the following functions defined in the indicated interval and having indicated period outside that interval. For each case (i) graph the function, (ii) state whether it is even or odd, (iii) determine the points of discontinuity if any, and (iv) give the values to which the series converges.

(a) $f(x) = \begin{cases} 2, & -\pi < x < 0 \\ -2, & 0 < x < \pi \end{cases}; \ 2\pi$
(b) $f(x) = 1, \quad -2 < x < 2; \ 4$

(c) $f(x) = \begin{cases} 0, & -3 < x < 0 \\ 2, & 0 < x < 3 \end{cases}; \ 6$
(d) $f(x) = \begin{cases} 0, & -1 < x < 0 \\ x, & 0 < x < 1 \end{cases}; \ 2$

(e) $f(x) = \begin{cases} -x, & -\pi < x < 0 \\ x, & 0 < x < \pi \end{cases}; \ 2\pi$
(f) $f(x) = \begin{cases} 3, & 0 < x < 4 \\ -3, & 4 < x < 8 \end{cases}; \ 8$

(g) $f(x) = 4x - x^2, \quad 0 < x < 4; \ 4$
(h) $f(x) = \begin{cases} 1, & 0 < x < 2\pi/3 \\ 0, & 2\pi/3 < x < 4\pi/3; \ 2\pi \\ -1, & 4\pi/3 < x < 2\pi \end{cases}$

2. Expand each function defined in the indicated interval into Fourier sine or Fourier cosine series as indicated. Give the values to which each series converges.

(a) $f(x) = 5, \quad 0 < x < 3 \qquad$ sine
(b) $f(x) = \begin{cases} 1, & 0 < x < \pi/2 \\ 0, & \pi/2 < x < \pi \end{cases}$ cosine

(c) $f(x) = \begin{cases} x, & 0 < x < 5 \\ 0, & 5 < x < 10 \end{cases}$ sine
(d) $f(x) = 2, \quad 0 < x < 1 \qquad$ cosine

(e) $f(x) = 4x - x^2, \quad 0 < x < 4$ sine
(f) $f(x) = \begin{cases} -1, & 0 < x < 2 \\ 1, & 2 < x < 4 \end{cases}$ cosine

3. Write Parseval's identity corresponding to the series obtained in (a) Exercise 1(a); (b) Exercise 1(d); (c) Exercise 1(g).

4. Obtain the results (37), page 381, by direct integration.

B EXERCISES

1. Show that $x^2 = \dfrac{\pi^2}{3} - 4 \left(\cos x - \dfrac{\cos 2x}{2^2} + \dfrac{\cos 3x}{3^2} - \cdots \right), \qquad -\pi < x < \pi.$

Show by Dirichlet's theorem that the equality holds at $x = \pm \pi$ and, hence, that

$$\frac{1}{1^2} + \frac{1}{2^2} + \frac{1}{3^2} + \cdots = \frac{\pi^2}{6}$$

2. Show that if $a \neq 0, \pm 1, \pm 2, \ldots$,

$$\frac{\pi \cos ax}{2a \sin a\pi} = \frac{1}{2a^2} + \frac{\cos x}{1^2 - a^2} - \frac{\cos 2x}{2^2 - a^2} + \frac{\cos 3x}{3^2 - a^2} - \cdots, \qquad -\pi < x < \pi$$

Show that the equality holds for $x = \pm \pi$ and hence prove that

$$\frac{1}{1^2 - a^2} + \frac{1}{2^2 - a^2} + \frac{1}{3^2 - a^2} + \cdots = \frac{1}{2a^2} - \frac{\pi}{2a} \cot \pi a$$

3. Expand
$$f(x) = \begin{cases} -x^2, & -L < x < 0 \\ x^2, & 0 < x < L \end{cases}$$

in a Fourier series of period $2L$. What happens at $x = \pm L$?

4. Show that $e^x = \dfrac{e^{2\pi} - 1}{\pi}\left(\dfrac{1}{2} + \displaystyle\sum_{n=1}^{\infty} \dfrac{\cos nx - n\sin nx}{n^2 + 1}\right)$, $\qquad 0 < x < 2\pi$.

5. (a) By replacing x by $2\pi - x$ in Exercise 4, show that $\dfrac{\pi}{2} \cdot \dfrac{\cosh(\pi - x)}{\sinh \pi} = \dfrac{1}{2} + \displaystyle\sum_{n=1}^{\infty} \dfrac{\cos nx}{n^2 + 1}$.

(b) Prove that the equality holds for $x = 0$, and hence evaluate $\displaystyle\sum_{n=1}^{\infty} \dfrac{1}{n^2 + 1}$.

6. Show by using Parseval's identity on Exercise 1 that $\dfrac{1}{1^4} + \dfrac{1}{2^4} + \dfrac{1}{3^4} + \cdots = \dfrac{\pi^4}{90}$.

7. Show that any orthogonal set must be linearly independent.

C EXERCISES

1. An inductor L and a capacitor C are in series with an emf of period $2T$ given by

$$E(t) = \begin{cases} E_0, & 0 < t < T \\ -E_0, & T < t < 2T \end{cases}$$

If the charge on the capacitor and the current are zero at $t = 0$ and if L, C, E_0, T are constants, show that the charge is given by

$$Q = \frac{4E_0}{\pi L\omega} \sum_{n=1}^{\infty} \frac{\omega \sin \omega_n t - \omega_n \sin \omega t}{(2n - 1)(\omega^2 - \omega_n^2)}$$

where $\omega_n = (2n - 1)\pi/T$, $\omega = 1/\sqrt{LC}$. Determine the current at any time. What happens when $\omega_n = \omega$ for some n?

2. Find the sum of the series $\displaystyle\sum_{n=1}^{\infty} \dfrac{1}{(n^2 + 1)^2}$. (*Hint:* Use B Exercise 5.)

3. Let $\phi_k(x)$, $k = 1, 2, \ldots$ be an orthonormal set in the interval $a \leq x \leq b$ and let α_k, $k = 1, 2, \ldots$ be any constants. Show that if a given function $f(x)$ is approximated in the interval by

$$\sum_{k=1}^{n} \alpha_k \phi_k(x)$$

then the mean square error
$$E_n = \frac{1}{b - a} \int_a^b \left[f(x) - \sum_{k=1}^{n} \alpha_k \phi_k(x) \right]^2 dx$$

is a minimum when α_k are given by

$$\alpha_k = c_k = \int_a^b f(x)\phi_k(x)dx$$

i.e., the constants determined from the series expansion of $f(x)$ in terms of $\phi_k(x)$. What assumptions should you make about the functions? (*Hint*: Set each partial derivative of E_n with respect to α_k equal to zero.)

4. (a) Use the fact that $E_n \geq 0$ in Exercise 3 to show that $\displaystyle\sum_{k=1}^n c_k^2 \leq \int_a^b [f(x)]^2\, dx$.

(b) By taking the limit as $n \to \infty$ in (a), deduce that $\displaystyle\sum_{k=1}^\infty c_k^2 \leq \int_a^b [f(x)]^2\, dx$.

The result is often called *Bessel's inequality*, whereas the special case of equality is *Parseval's equality*, the first being true even if the orthonormal set is not necessarily complete.

(c) Using (b) show that the infinite series converges, and thus

$$\lim_{k\to\infty} c_k = \lim_{k\to\infty} \int_a^b f(x)\phi_k(x)dx = 0$$

5. How would you prove that $\displaystyle\lim_{n\to\infty}\int_{-\pi}^\pi \frac{\sin nx}{1 + x^2}\, dx = 0$?

6. Show that there cannot be any function $f(x)$ in the interval $-\pi < x < \pi$ such that the integral $\displaystyle\int_{-\pi}^\pi [f(x)]^2\, dx$ is finite and whose Fourier series is $\displaystyle\sum_{n=1}^\infty \frac{\sin nx}{\sqrt{n}}$.

4.3 BESSEL SERIES

From page 368, we know that the Bessel functions

$$J_n(r_1x),\ J_n(r_2x),\ J_n(r_3x),\ \ldots \tag{67}$$

are mutually orthogonal in the interval $0 \leq x \leq 1$ with respect to the weight function x, where r_k, $k = 1, 2, 3, \ldots$ are the successive positive roots of either of the equations

$$J_n(x) = 0, \qquad J_n'(x) = 0, \qquad \mu J_n(x) + xJ_n'(x) = 0 \tag{68}$$

where we shall assume that $n \geq 0$ and $\mu > 0$. We would therefore expect to be able to expand a function $f(x)$ into a series of these Bessel functions, or as we shall say a *Bessel series*, having the form

$$f(x) = a_1 J_n(r_1x) + a_2 J_n(r_2x) + \cdots = \sum_{j=1}^\infty a_j J_n(r_jx) \tag{69}$$

The method for obtaining the coefficients $a_1, a_2, \ldots$ consists of multiplying (69) by $xJ_n(r_kx)$ and then integrating from $x = 0$ to $x = 1$ to obtain

$$\int_0^1 xf(x)J_n(r_kx)dx = \sum_{j=1}^\infty a_j \int_0^1 xJ_n(r_jx)J_n(r_kx)dx$$

By the orthogonality the integral on the right is zero except for $j = k$, so that the series on the right reduces to one term and we find

$$\int_0^1 xf(x)J_n(r_kx)dx = a_k \int_0^1 xJ_n^2(r_kx)dx \ \text{ or } \ a_k = \frac{\displaystyle\int_0^1 xf(x)J_n(r_kx)dx}{\displaystyle\int_0^1 xJ_n^2(r_kx)dx} \tag{70}$$

The value of the denominator in (70) can be found by using Table 8.1. The results can be justified by the following theorem, which is analogous to Dirichlet's theorem for Fourier series on page 386.

Theorem 3. Let $f(x)$ and $f'(x)$ be piecewise continuous in the interval $0 \le x \le 1$. Then the series on the right of (69) with coefficients (70) converges to $f(x)$ if x is point of continuity and converges to $\frac{1}{2}[f(x + 0) + f(x - 0)]$ if x is a point of discontinuity. At the endpoints $x = 0, 1$ the series may or may not converge to $f(x)$, and this requires special investigation.

ILLUSTRATIVE EXAMPLE 7

Expand $f(x) = 1, 0 \le x \le 1$, in terms of the functions $J_0(r_k x)$, $k = 1, 2, 3, \ldots$, where r_k are the roots of $J_0(x) = 0$. Investigate the convergence at the endpoints.

Solution Using $f(x) = 1$, $n = 0$, the coefficients (70) become

$$a_k = \frac{\int_0^1 x J_0(r_k x)dx}{\int_0^1 x J_0^2(r_k x)dx} = \frac{2}{J_0'^2(r_k)} \int_0^1 x J_0(r_k x)dx = \frac{2}{J_1^2(r_k)} \int_0^1 x J_0(r_k x)dx \tag{71}$$

where we have used entry 1 in Table 8.1 and also the result $J_0'(x) = -J_1(x)$ given in A Exercise 2(a), page 341. To evaluate the integral at the right of (71), let $r_k x = u$. Then using A Exercise 2(b), page 341,

$$\int_0^1 x J_0(r_k x)dx = \frac{1}{r_k^2} \int_0^{r_k} u J_0(u)du = \frac{1}{r_k^2} u J_1(u) \Big|_0^{r_k} = \frac{J_1(r_k)}{r_k}$$

so that (71) yields

$$a_k = \frac{2}{r_k J_1(r_k)} \tag{72}$$

Thus (69) reduces to the interesting result

$$1 = \frac{2J_0(r_1 x)}{r_1 J_1(r_1)} + \frac{2J_0(r_2 x)}{r_2 J_1(r_2)} + \cdots = 2 \sum_{k=1}^{\infty} \frac{J_0(r_k x)}{r_k J_1(r_k)} \tag{73}$$

By Theorem 3, this is valid for all x such that $0 < x < 1$, since there are no points of discontinuity in the interval. For $x = 1$ the series converges to 0, while for $x = 0$ it can be shown to converge to 1.

Remark 4. Although we have used the basic interval for Bessel series as $0 < x < 1$ with possible convergence at the endpoints in accordance with Theorem 3 given above, the actual interval of convergence may be larger. Thus, for instance, in Illustrative Example 7 the series (73) converges for $0 \le x < 1$. However, since $J_0(x)$ is an even function it is easily seen that (73) converges for $-1 < x < 1$.

ILLUSTRATIVE EXAMPLE 8

Expand $f(x) = 1 - x^2, 0 \le x \le 1$ in a series of functions $J_2(r_k x)$, where r_k are the roots of $3J_2(x) + x J_2'(x) = 0$.

Solution In this case $n = 2$ and the numerator of (70) becomes

$$\int_0^1 x(1 - x^2)J_2(r_kx)dx = \int_0^1 xJ_2(r_kx)dx - \int_0^1 x^3J_2(r_kx)dx$$

$$= \frac{1}{r_k^2}\int_0^{r_k} uJ_2(u)du - \frac{1}{r_k^4}\int_0^{r_k} u^3J_2(u)du$$

on letting $r_kx = u$. Now if we use the results

$$\int_0^{r_k} uJ_2(u)du = -2J_0(u) - uJ_1(u)\Big|_0^{r_k} = 2 - 2J_0(r_k) - r_kJ_1(r_k) \tag{74}$$

$$\int_0^{r_k} u^3J_2(u)du = u^3J_3(u)\Big|_0^{r_k} = r_k^3J_3(r_k) \tag{75}$$

we have $\int_0^1 x(1 - x^2)J_2(r_kx)dx = \frac{1}{r_k^2}[2 - 2J_0(r_k) - r_kJ_1(r_k)] - \frac{J_3(r_k)}{r_k}$ $\quad$ (76)

The denominator of (70) can be obtained from entry 3 of Table 8.1, with $n = 2$, $\mu = 3$. We find

$$\int_0^1 xJ_2^2(r_kx)dx = \frac{(r_k^2 + 5)J_2^2(r_k)}{2r_k^2} \tag{77}$$

From (76) and (77) we can find a_k, and using this in (69) we obtain the required series expansion

$$f(x) = \sum_{k=1}^{\infty} \frac{4 - 4J_0(r_k) - 2r_kJ_1(r_k) - 2r_kJ_3(r_k)}{(r_k^2 + 5)J_2^2(r_k)} J_2(r_kx)$$

for $0 < x \leq 1$. For $x = 0$ the series converges to zero.

Parseval's identity can also be written for Bessel series since it can be shown that the functions (67), page 397, constitute a complete set. We can obtain this identity formally by squaring both sides of (69), multiplying by x, and then integrating with respect to x from 0 to 1 using the orthogonality of the Bessel functions. The result is given in the following

Theorem 4. Let $f(x), 0 \leq x \leq 1$, satisfy the conditions of Theorem 3, page 398. Then we have *Parseval's identity*

$$\int_0^1 x[f(x)]^2 \, dx = \sum_{k=1}^{\infty} d_ka_k^2 \tag{78}$$

where the coefficients a_k are obtained from (70) and d_k is the denominator of (70) given by

$$d_k = \int_0^1 xJ_n^2(r_kx)dx \tag{79}$$

In any particular problem it is usually best to obtain Parseval's identity corresponding to a given series expansion directly by squaring, multiplying by x, and then integrating from $x = 0$ to 1.

Write Parseval's identity corresponding to the series of Illustrative Example 7.

Solution   Square both sides of (73), page 398, multiply by x, and integrate from $x = 0$ to 1. Then we obtain

$$\int_0^1 x(1)^2 \, dx = 4 \sum_{k=1}^{\infty} \frac{1}{r_k^2 J_1^2(r_k)} \int_0^1 x J_0^2(r_k x) dx \tag{80}$$

where we have omitted the cross product terms since they are equal to zero because of the orthogonality of the Bessel functions. Since

$$\int_0^1 x J_0^2(r_k x) dx = \tfrac{1}{2} J_0'^2(r_k) = \tfrac{1}{2} J_1^2(r_k)$$

(80) yields the interesting result $\quad \dfrac{1}{2} = 2 \sum_{k=1}^{\infty} \dfrac{1}{r_k^2} \quad$ or $\quad \dfrac{1}{r_1^2} + \dfrac{1}{r_2^2} + \dfrac{1}{r_3^2} + \cdots = \dfrac{1}{4} \quad$ (81)

where $r_1, r_2, r_3, \ldots$ are the successive positive roots of $J_0(x) = 0$.

A EXERCISES

1. (a) Expand $f(x) = x, 0 < x < 1$, in a series of Bessel functions $J_1(r_k x)$, where $r_k, k = 1, 2, \ldots,$ are the positive roots of $J_1(x) = 0$ to obtain

$$x = 2 \sum_{k=1}^{\infty} \frac{J_1(r_k x)}{r_k J_2(r_k)}$$

   (b) Show that the series in (a) converges to $f(x) = x$ if $-1 < x < 1$ and 0 if $x = \pm 1$.

2. Let $r_k, k = 1, 2, \ldots,$ be the positive roots of $J_0(x) = 0$. Show that for $0 \leqq x \leqq 1$ we have

   (a) $\dfrac{x}{4} = \sum_{k=1}^{\infty} \dfrac{J_1(r_k x)}{r_k^2 J_2(r_k)}.$ (b) $\dfrac{x^2}{16} = \sum_{k=1}^{\infty} \dfrac{J_2(r_k x)}{r_k^3 J_1(r_k)}.$

3. Expand $f(x) = x^2, 0 < x < 1$, in a series of Bessel functions $J_2(r_k x)$, where $r_k, k = 1, 2, \ldots,$ are the positive roots of $J_2(x) = 0$, and show that the resulting series also converges to $f(x)$, where $-1 < x < 1$.

4. Show that for $-1 \leqq x \leqq 1$ $\dfrac{x}{2} = \sum_{k=1}^{\infty} \dfrac{r_k J_2(r_k) J_1(r_k x)}{(r_k^2 - 1) J_1^2(r_k)}$

   where $r_k, k = 1, 2, \ldots,$ are the positive roots of $J_1'(x) = 0$.

5. Expand $f(x) = 1, 0 < x < 1$, in a series of Bessel functions $J_0(r_k x)$, where $r_k, k = 1, 2, \ldots,$ are the positive roots of $x J_0'(x) + J_0(x) = 0$.

6. By using Parseval's identity in Exercise 1, show that $\sum_{k=1}^{\infty} \dfrac{J_1'^2(r_k)}{r_k^2 J_2^2(r_k)} = \dfrac{1}{8}$

7. Write Parseval's identity corresponding to the series obtained in (a) Exercise 2(a); (b) Exercise 3.

8. (a) If $r_k, k = 1, 2, \ldots,$ are the positive roots of $J_3(x) = 0$, show that for $0 \leqq x < 1$

$$x^3 = 2 \sum_{k=1}^{\infty} \frac{J_3(r_k x)}{r_k J_4(r_k)}$$

   (b) Write Parseval's identity corresponding to the result in (a).

9. (a) If r_k, $k = 1, 2, \ldots$, are the positive roots of $J_0(x) = 0$, show that for $0 < x \le 1$

$$x^2 = 2 \sum_{k=1}^{\infty} \frac{(r_k^2 - 4)J_0(r_k x)}{r_k^3 J_1(r_k)}$$

(b) Write Parseval's identity corresponding to the result in (a).

B EXERCISES

1. (a) If r_k, $k = 1, 2, \ldots$, are the positive roots of $J_0(x) = 0$, show that for $-1 \le x \le 1$

$$\frac{1 - x^2}{8} = \sum_{k=1}^{\infty} \frac{J_0(r_k x)}{r_k^3 J_1(r_k)}$$

(b) Show that $\displaystyle\sum_{k=1}^{\infty} \frac{1}{r_k^3 J_1(r_k)} = \frac{1}{8}$.

(c) By differentiating the result in (a), obtain the result in A Exercise 1.

2. (a) Show that for $-1 \le x \le 1$
$$\frac{x^3 - x}{16} = \sum_{k=1}^{\infty} \frac{J_1(r_k x)}{r_k^3 J_0(r_k)}$$

where r_k, $k = 1, 2$, are the positive roots of $J_1(x) = 0$.

(b) Deduce that $\displaystyle\sum_{k=1}^{\infty} \frac{J_1(r_k/2)}{r_k^3 J_0(r_k)} = -\frac{3}{128}$.

3. Expand $f(x) = \begin{cases} 1, & 0 < x < \frac{1}{2} \\ 0, & \frac{1}{2} < x < 1 \end{cases}$ in a series of Bessel functions $J_0(r_k x)$, where r_k, $k = 1, 2, \ldots$, are the positive roots of $J_0(x) = 0$, and use the result to show that

$$\sum_{k=1}^{\infty} \frac{J_0(r_k/2)J_1(r_k/2)}{r_k J_1^2(r_k)} = \frac{1}{2}$$

4. (a) Show that for $0 < x \le 1$
$$\ln x = -2 \sum_{k=1}^{\infty} \frac{J_0(r_k x)}{r_k^2 J_1^2(r_k)}$$

where r_k, $k = 1, 2, \ldots$, are the positive roots of $J_0(x) = 0$. **(b)** Use Parseval's identity to show that

$$\sum_{k=1}^{\infty} \frac{1}{r_k^4 J_1^2(r_k)} = \frac{1}{8}$$

5. Expand $f(x) = x^2$, $0 \le x \le 3$, in a series of Bessel functions $J_0(r_k x)$, where r_k, $k = 1, 2, \ldots$, are the positive roots of $J_0(3x) = 0$.

C EXERCISES

1. Let r_k, $k = 1, 2, \ldots$, be the positive roots of $J_n(x) = 0$, $n \ge 0$. **(a)** Show that for $0 \le x \le 1$

$$\frac{x^{n+1}}{4(n+1)} = \sum_{k=1}^{\infty} \frac{J_{n+1}(r_k x)}{r_k^2 J_{n+1}(r_k)}$$

(b) Deduce that $\displaystyle\sum_{k=1}^{\infty} \frac{1}{r_k^2} = \frac{1}{4(n+1)}$ and thus arrive at a generalization of (81), page 400.

2. Show that for $0 \leq x < 1$

$$\frac{J_0(ax)}{2J_0(a)} = \sum_{k=1}^{\infty} \frac{r_k J_0(r_k x)}{(r_k^2 - a^2) J_1(r_k)}$$

where r_k, $k = 1, 2, \ldots$, are the positive roots of $J_0(x) = 0$.

3. Use Parseval's identity on the result of Exercise 2 to show that

$$\sum_{k=1}^{\infty} \frac{r_k^2}{(r_k^2 - a^2)^2} = \frac{J_0^2(a) + J_1^2(a)}{4}$$

4. Deduce from Exercise 3 that if r_k, $k = 1, 2, \ldots$ are the positive roots of $J_0(x) = 0$, then

(a) $\displaystyle\sum_{k=1}^{\infty} \frac{1}{r_k^2} = \frac{1}{4}$.   (b) $\displaystyle\sum_{k=1}^{\infty} \frac{1}{r_k^4} = \frac{1}{32}$.   (c) $\displaystyle\sum_{k=1}^{\infty} \frac{1}{r_k^6} = \frac{1}{192}$.

[*Hint:* For (a), let $a = 0$; for (b) and (c), differentiate the result of Exercise 3 with respect to a once and twice, respectively, and then let $a = 0$.]

5. Use Exercise 4(c) to show that the least positive zero of $J_0(x)$ is approximately $\sqrt[6]{192}$ and compare with the value given on page 338.

4.4 LEGENDRE SERIES

Since the Legendre polynomials $P_0(x)$, $P_1(x)$, $P_2(x)$, ... are mutually orthogonal in the interval $-1 \leq x \leq 1$, and can be shown to constitute a complete set we should be able to expand a function $f(x)$ in a series of these polynomials given by

$$f(x) = a_0 P_0(x) + a_1 P_1(x) + a_2 P_2(x) + \cdots = \sum_{j=0}^{\infty} a_j P_j(x) \tag{82}$$

and called a *Legendre series*. The coefficients a_j should then be determined by a now familiar method, i.e., multiplying the series by $P_k(x)$ and integrating term by term from $x = -1$ to 1. We then find

$$\int_{-1}^{1} f(x) P_k(x) dx = \sum_{j=0}^{\infty} a_j \int_{-1}^{1} P_j(x) P_k(x) dx$$

Since the integral on the right is zero for $j \neq k$, the series reduces to the single term where $j = k$ so that

$$\int_{-1}^{1} f(x) P_k(x) dx = a_k \int_{-1}^{1} [P_k(x)]^2 dx \quad \text{or} \quad a_k = \frac{\int_{-1}^{1} f(x) P_k(x) dx}{\int_{-1}^{1} P_k^2(x) dx} \tag{83}$$

The denominator in (83) has already been evaluated on page 372 so that we have

$$a_k = \frac{2k+1}{2} \int_{-1}^{1} f(x) P_k(x) dx \tag{84}$$

The above results can be justified by the following theorem, which is analogous to Dirichlet's theorem for Fourier series on page 386.

Theorem 5.   Let $f(x)$ and $f'(x)$ be piecewise continuous in the interval $-1 \leq x \leq 1$. Then the series on the right of (82) with coefficients (84) converges to $f(x)$ if x is a point of continuity, and converges to

$$\tfrac{1}{2}[f(x + 0) + f(x - 0)]$$

if x is a point of discontinuity. At the endpoints $x = \pm 1$ the series may or may not converge to $f(x)$ and requires special investigation.

ILLUSTRATIVE EXAMPLE 10

Expand $f(x) = \begin{cases} 1, & 0 < x \le 1 \\ -1, & -1 \le x < 0 \end{cases}$ in a series of Legendre polynomials and examine the convergence.

Solution From (84) we find that

$$
\begin{aligned}
a_k &= \frac{2k + 1}{2}\left[\int_{-1}^{0}(-1)P_k(x)dx + \int_{0}^{1}(1)P_k(x)dx\right] \\
&= \frac{2k + 1}{2}\left[\int_{0}^{1} P_k(x)dx - \int_{-1}^{0} P_k(x)dx\right]
\end{aligned}
\tag{85}
$$

If we like we could now put $k = 0, 1, 2, \ldots$, and use the first few polynomials on page 344 to find $a_0, a_1, a_2, \ldots$. We can, however, save some of the labor involved if we notice that $P_k(x)$ is an even function for even k, i.e., $k = 0, 2, 4, \ldots$, and an odd function for odd k, i.e., $k = 1, 3, 5, \ldots$, so that

$$\int_{0}^{1} P_k(x)dx = \int_{-1}^{0} P_k(x)dx \qquad \text{for even } k,$$

$$\int_{0}^{1} P_k(x)dx = -\int_{-1}^{0} P_k(x)dx \qquad \text{for odd } k$$

It thus follows from (85) that all a_k where k is even are equal to 0, and so we need only be concerned with a_k for odd k. In such case (85) gives

$$a_k = (2k + 1)\int_{0}^{1} P_k(x)dx, \qquad k = 1, 3, 5, \ldots \tag{86}$$

Putting $k = 1, 3, 5, \ldots$ in (86) yields

$$a_1 = \tfrac{3}{2}, \qquad a_3 = -\tfrac{7}{8}, \qquad a_5 = \tfrac{11}{16}, \qquad a_7 = -\tfrac{75}{128}, \ldots$$

Thus the required expansion is

$$f(x) = \tfrac{3}{2}P_1(x) - \tfrac{7}{8}P_3(x) + \tfrac{11}{16}P_5(x) - \tfrac{75}{128}P_7(x) + \cdots \tag{87}$$

By Theorem 5, the series converges to 1 for $0 < x < 1$, and -1 for $-1 < x < 0$, and converges to $\tfrac{1}{2}[(1) + (-1)] = 0$ for $x = 0$.

Remark 5. The coefficients a_k in (86) can be determined exactly by using the generating function for Legendre polynomials given by

$$\sum_{k=0}^{\infty} P_k(x)t^k = \frac{1}{\sqrt{1 - 2tx + t^2}} \tag{88}$$

Integrating (88) from $x = 0$ to 1 we find

$$\sum_{k=0}^{\infty} \left[\int_0^1 P_k(x)dx \right] t^k = \int_0^1 \frac{dx}{\sqrt{1 - 2tx + t^2}} \tag{89}$$

so that on evaluating the integral on the right

$$\sum_{k=0}^{\infty} \left[\int_0^1 P_k(x)dx \right] t^k = -\frac{1}{t}(1 - 2tx + t^2)^{1/2} \Big|_{x=0}^{1} = 1 + \frac{(1 + t^2)^{1/2} - 1}{t} \tag{90}$$

Now by using the binomial theorem (or equivalently Maclaurin series expansion), we have

$$(1 + u)^{1/2} = 1 + \tfrac{1}{2}u + \frac{(\tfrac{1}{2})(\tfrac{1}{2} - 1)}{2!}u^2 + \frac{\tfrac{1}{2}(\tfrac{1}{2} - 1)(\tfrac{1}{2} - 2)}{3!}u^3 + \cdots$$

$$= 1 + \tfrac{1}{2}u - \frac{1}{2 \cdot 4}u^2 + \frac{1 \cdot 3}{2 \cdot 4 \cdot 6}u^3 - \frac{1 \cdot 3 \cdot 5}{2 \cdot 4 \cdot 6 \cdot 8}u^4 + \cdots \tag{91}$$

If we now let $u = t^2$, we see that the left side of (90) is equal to

$$1 + \tfrac{1}{2}t - \frac{1}{2 \cdot 4}t^3 + \frac{1 \cdot 3}{2 \cdot 4 \cdot 6}t^5 - \frac{1 \cdot 3 \cdot 5}{2 \cdot 4 \cdot 6 \cdot 8}t^7 + \cdots \tag{92}$$

Then equating like powers of t on both sides leads to the result

$$\int_0^1 P_k(x)dx = \begin{cases} (-1)^{(k-1)/2} \dfrac{1 \cdot 3 \cdots (k - 2)}{2 \cdot 4 \cdots (k + 1)} & \text{if } k = 3, 5, 7, \ldots \\ 1 \quad \text{if } k = 0, \quad \tfrac{1}{2} \text{ if } k = 1, \quad 0 \text{ if } k = 2, 4, 6, \ldots \end{cases} \tag{93}$$

Using this we can write (87) as

$$f(x) = \tfrac{3}{2}P_1(x) + \sum_{k=3,5,\ldots} (-1)^{(k-1)/2}(2k + 1)\frac{1 \cdot 3 \cdots (k - 2)}{2 \cdot 4 \cdots (k + 1)} P_k(x) \tag{94}$$

where the sum on the right is taken as indicated over the odd positive integers 3, 5, ...

In case we wish to expand a polynomial in a series of Legendre polynomials the result is a finite rather than infinite series. While the above method can be used in such case, it is also possible to proceed as in the following

ILLUSTRATIVE EXAMPLE 11

Expand $f(x) = 2x^3 - 5x^2 + 3x - 10$, $-1 \leq x \leq 1$, in a Legendre series.

Solution We already know that

$$P_0(x) = 1, \quad P_1(x) = x, \quad P_2(x) = \tfrac{3}{2}x^2 - \tfrac{1}{2}, \quad P_3(x) = \tfrac{5}{2}x^3 - \tfrac{3}{2}x$$

From these we have

$$1 = P_0(x), \quad x = P_1(x), \quad x^2 = \tfrac{1}{3}P_0(x) + \tfrac{2}{3}P_2(x), \quad x^3 = \tfrac{3}{5}P_1(x) + \tfrac{2}{5}P_3(x)$$

Thus

$$2x^3 - 5x^2 + 3x - 10 = 2[\tfrac{3}{5}P_1(x) + \tfrac{2}{5}P_3(x)] - 5[\tfrac{1}{3}P_0(x) + \tfrac{2}{3}P_2(x)] + 3P_1(x) - 10P_0(x)$$

Simplification of the right side leads to the required Legendre series:

$$f(x) = -\tfrac{35}{3}P_0(x) + \tfrac{21}{5}P_1(x) - \tfrac{10}{3}P_2(x) + \tfrac{4}{5}P_3(x) \tag{95}$$

We can write Parseval's identity for Legendre series since the set of Legendre polynomials is a complete orthogonal set. The theorem involved is as follows.

Theorem 6. Let $f(x)$, $-1 \leq x \leq 1$, satisfy the conditions of Theorem 5, page 402. Then we have *Parseval's identity*

$$\int_{-1}^{1} [f(x)]^2\, dx = \sum_{k=0}^{\infty} \left(\frac{2}{2k+1}\right) a_k^2 \tag{96}$$

where the coefficients a_k are given by (84).

ILLUSTRATIVE EXAMPLE 12

Write Parseval's identity corresponding to the series of Illustrative Example 10.

Solution The required identity can be obtained formally by squaring the series (94) and integrating term by term from $x = -1$ to $x = 1$, using the fact that the integrals of cross product terms are zero because of the orthogonality of the polynomials. We thus obtain

$$\int_{-1}^{1} [f(x)]^2\, dx = \frac{9}{4}\int_{-1}^{1} P_1^2(x)dx + \sum_{k=3,5,\ldots} (2k+1)^2 \left[\frac{1\cdot 3\cdots(k-2)}{2\cdot 4\cdots(k+1)}\right]^2 \int_{-1}^{1} P_k^2(x)dx \tag{97}$$

or

$$\sum_{k=3,5,\ldots} (2k+1)\left[\frac{1\cdot 3\cdots(k-2)}{2\cdot 4\cdots(k+1)}\right]^2 = \frac{1}{4} \tag{98}$$

4.5 MISCELLANEOUS ORTHOGONAL SERIES

Orthogonal series other than those of Fourier, Bessel, and Legendre also arise in applications. For example, we can have *Hermite series* and *Laguerre series* corresponding to the orthogonal polynomials of Hermite and Laguerre on pages 345–346. The ideas of expansion of functions into such series are basically the same as given above. We can also write Parseval's identity for such series. As an illustration, let us consider the following

ILLUSTRATIVE EXAMPLE 13

Expand $f(x) = x(1 - x)$ in a series of Hermite polynomials.

Solution *Method 1.* Write $f(x) = a_0 H_0(x) + a_1 H_1(x) + \cdots = \sum_{j=0}^{\infty} a_j H_j(x)$.

Multiplying both sides by $e^{-x^2}H_k(x)$ and integrating from $-\infty$ to ∞, we have

$$\int_{-\infty}^{\infty} e^{-x^2}f(x)H_k(x)dx = \sum_{j=0}^{\infty} a_j \int_{-\infty}^{\infty} e^{-x^2}H_j(x)H_k(x)dx$$

Because of the results on page 373, the right side reduces to only one term, the one for which $j = k$. Thus we have

$$\int_{-\infty}^{\infty} e^{-x^2}f(x)H_k(x)dx = a_k \int_{-\infty}^{\infty} e^{-x^2}[H_k(x)]^2\, dx = 2^k k!\sqrt{\pi}a_k$$

or

$$a_k = \frac{1}{2^k k!\sqrt{\pi}} \int_{-\infty}^{\infty} e^{-x^2}f(x)H_k(x)dx$$

Putting $k = 0, 1, 2, \ldots$ (noting that $0! = 1$), we obtain

$$a_0 = \frac{1}{\sqrt{\pi}} \int_{-\infty}^{\infty} e^{-x^2}x(1 - x)dx, \qquad a_1 = \frac{1}{\sqrt{\pi}} \int_{-\infty}^{\infty} e^{-x^2}x^2(1 - x)dx$$

$$a_2 = \frac{1}{8\sqrt{\pi}} \int_{-\infty}^{\infty} e^{-x^2}x(1 - x)(4x^2 - 2)dx, \ldots$$

This leads to $a_0 = -\frac{1}{2}, a_1 = \frac{1}{2}, a_2 = -\frac{1}{4}$ and $a_k = 0$ for $k > 2$. Thus

$$f(x) = -\tfrac{1}{2}H_0(x) + \tfrac{1}{2}H_1(x) - \tfrac{1}{4}H_2(x)$$

Method 2. Since $H_0(x) = 1$, $\qquad H_1(x) = 2x$, $\qquad H_2(x) = 4x^2 - 2$

it follows that $\qquad 1 = H_0(x)$, $\qquad x = \tfrac{1}{2}H_1(x)$, $\qquad x^2 = \tfrac{1}{2}H_0(x) + \tfrac{1}{4}H_2(x)$

Thus $\qquad x(1 - x) = x - x^2 = \tfrac{1}{2}H_1(x) - [\tfrac{1}{2}H_0(x) + \tfrac{1}{4}H_2(x)]$

or $\qquad f(x) = -\tfrac{1}{2}H_0(x) + \tfrac{1}{2}H_1(x) - \tfrac{1}{4}H_2(x)$

Clearly, Method 2 is superior in the case where a polynomial is to be expanded in a Hermite series. For other functions, however, Method 1 must be used.

A EXERCISES

1. Expand each of the following functions in a Legendre series and examine the convergence.
   (a) $f(x) = 4x^2, -1 \leq x \leq 1$.
   (b) $f(x) = 2x^3 + 5x^2 - x + 4, -1 \leq x \leq 1$.

   (c) $f(x) = \begin{cases} 4, & 0 < x \leq 1 \\ 0, & -1 \leq x < 0 \end{cases}$.
   (d) $f(x) = \begin{cases} 2, & -1 \leq x < 0 \\ 3, & 0 < x \leq 1 \end{cases}$.

2. Expand $f(x) = \begin{cases} x, & 0 < x \leq 1 \\ 0, & -1 \leq x < 0 \end{cases}$ in a Legendre series and examine the convergence.

3. Show that the expansion of $f(x) = \begin{cases} x, & 0 < x \leq 1 \\ -x, & -1 \leq x < 0 \end{cases}$ in a Legendre series is given by

$$f(x) = \frac{1}{2} + \sum_{k=1}^{\infty} \frac{(-1)^{k-1}(4k + 1)}{2^{k+1}(k + 1)!} P_{2k}(x)$$

and examine the convergence.

4. Write Parseval's identity corresponding to (a) Exercise 1(a), and (b) Exercise 1(b).

5. Write Parseval's identity corresponding to Exercise 3.

1. Expand each of the following into a Hermite series. (a) $f(x) = x^2 + 1$. (b) $f(x) = 5x^3$.

2. Expand each of the following into a Laguerre series. (a) $f(x) = 3x^2 - 2x$. (b) $f(x) = 2x^3 + x^2 + 1$.

3. Write Parseval's identity corresponding to (a) Exercise 1(a), and (b) Exercise 2(a).

4. Show that $P'_n(x) = (2n - 1)P_{n-1}(x) + (2n - 5)P_{n-3}(x) + (2n - 9)P_{n-5}(x) + \cdots$.

5. Show that $xP'_n(x) = nP_n(x) + (2n - 3)P_{n-2}(x) + (2n - 7)P_{n-4}(x) + \cdots$.

6. If m and n are integers which are both even or both odd and $m \leq n$, show that

$$\int_{-1}^{1} P'_m(x)P'_n(x)dx = m(m + 1)$$

7. Show that the expansion of $f(x) = \begin{cases} x^2, & 0 \leq x \leq 1 \\ -x^2, & -1 \leq x \leq 0 \end{cases}$ into a Legendre series is given by

$$f(x) = \tfrac{3}{4}P_1(x) + \sum_{k=1}^{\infty} (-1)^{k-1} \frac{1 \cdot 3 \cdot 5 \cdots (2k - 3)}{2 \cdot 4 \cdot 6 \cdots 2k} \frac{4k + 3}{k + 2} P_{2k+1}(x)$$

and examine the convergence.

1. If $m > -1$ and $n = 2, 3, 4, \ldots$, show that

$$\int_0^1 x^m P_n(x)dx = \frac{m(m - 1) \cdots (m - n + 2)}{(m + n + 1)(m + n - 1) \cdots (m - n + 3)}$$

2. If $k = 0, 1, 2, \ldots$, show that

$$x^k = \frac{k!}{1 \cdot 3 \cdot 5 \cdots (2k + 1)} \left[(2k + 1)P_k(x) + \frac{2k + 1}{2 \cdot 1!}(2k - 3)P_{k-2}(x) \right.$$

$$\left. + \frac{(2k + 1)(2k - 1)}{2^2 \cdot 2!}(2k - 7)P_{k-4}(x) + \cdots \right]$$

3. Show that
$$nP_n(\cos \theta) = \sum_{k=1}^{n} \cos k\theta P_{n-k}(\cos \theta)$$

4. Use the results of C Exercise 3, page 375, to show how to expand a function into a *Chebyshev series*, and write the corresponding Parseval identity. Illustrate by using a particular function.

5 Some Special Topics

5.1 SELF-ADJOINT DIFFERENTIAL EQUATIONS

Consider the differential equation

$$a_0 y'' + a_1 y' + a_2 y = F(x) \tag{1}$$

where a_0, a_1, a_2 are functions of x. If we can write the left side of (1) as an exact derivative, i.e.,

$$a_0 y'' + a_1 y' + a_2 y = D(b_0 y' + b_1 y) \qquad (2)$$

for suitable functions of x denoted by b_0, b_1, then (1) will be immediately integrable. Let us see if we can find a condition under which this can be done. Expanding the right side of (2) we have

$$a_0 y'' + a_1 y' + a_2 y = b_0 y'' + (b_0' + b_1)y' + b_1' y$$

Since this must be an identity, it follows that

$$a_0 = b_0, \qquad a_1 = b_0' + b_1, \qquad a_2 = b_1' \qquad (3)$$

Differentiating the second equation yields $a_1' = b_0'' + b_1'$, from which we obtain, on substituting $b_0 = a_0$, $b_1' = a_2$,

$$a_0'' - a_1' + a_2 = 0 \qquad (4)$$

Conversely, if we are given (4), then we can write

$$a_0 y'' + a_1 y' + a_2 y = D[a_0 D + (a_1 - a_0')]y \qquad (5)$$

By analogy with first-order linear differential equations, we call the differential equation (1), where the left side can be written as an exact derivative, an *exact differential equation*. Because of the above remarks we have the following

Theorem 1. The differential equation

$$a_0 y'' + a_1 y' + a_2 y = F(x) \qquad (6)$$

is exact if and only if $\qquad\qquad a_0'' - a_1' + a_2 = 0 \qquad (7)$

where the indicated derivatives of a_0 and a_1 are assumed to exist.

ILLUSTRATIVE EXAMPLE 1

Solve the differential equation $xy'' + (x + 2)y' + y = 4x$.

Solution Comparing this equation with (1), we have $a_0 = x$, $a_1 = x + 2$, $a_2 = 1$, so that

$$a_0'' - a_1' + a_2 = 0$$

and the equation is exact. Thus it can be written $D[xD + (x + 1)]y = 4x$. Integrating we find

$$x\frac{dy}{dx} + (x + 1)y = 2x^2 + c_1 \quad \text{or} \quad \frac{dy}{dx} + \left(\frac{x + 1}{x}\right)y = 2x + \frac{c_1}{x}$$

This last equation has integrating factor $e^{\int (x+1)/x \, dx} = xe^x$ so that it can be written

$$\frac{d}{dx}(xe^x y) = 2x^2 e^x + c_1 e^x$$

Integrating this gives

$$xe^x y = 2(x^2 - 2x + 2)e^x + c_1 e^x + c_2 \quad \text{or} \quad y = \frac{2(x^2 - 2x + 2)}{x} + \frac{c_1}{x} + \frac{c_2 e^{-x}}{x}$$

In case (1) is not exact, it may be possible to multiply the equation by some suitable function of x denoted by μ so that the resulting equation

$$(a_0 \mu)y'' + (a_1 \mu)y' + (a_2 \mu)y = \mu F(x) \tag{8}$$

is exact. By analogy with first-order equations, we call μ an *integrating* factor. From Theorem 1 we see that (8) is exact if and only if

$$(a_0 \mu)'' - (a_1 \mu)' + a_2 \mu = 0 \tag{9}$$

We call this differential equation for the unknown function μ the *adjoint* of (1). We can summarize this in the following

Definition 1.   The *adjoint* of the differential equation

$$a_0 y'' + a_1 y' + a_2 y = F(x) \tag{10}$$

is

$$(a_0 \mu)'' - (a_1 \mu)' + a_2 \mu = 0 \tag{11}$$

A consequence of this is that if we can find a non-zero solution of the adjoint (11) then we can solve equation (10).

ILLUSTRATIVE EXAMPLE 2

Given

$$y'' + y = \csc x \tag{12}$$

(a) Write its adjoint. (b) Use (a) to find an integrating factor and thus solve (12).

Solution   (a) Comparing (12) with (1) we see that $a_0 = 1$, $a_1 = 0$, $a_2 = 1$. Then from (11) the adjoint is

$$\mu'' + \mu = 0 \tag{13}$$

(b) A particular solution of (13) is $\sin x$, which we can use as an integrating factor of (12). Multiplying (12) by $\sin x$ we find

$$(\sin x)y'' + (\sin x)y = 1$$

which is easily seen to be exact. We can write it as

$$D[(\sin x)y' - (\cos x)y] = 1$$

Integration yields $(\sin x)y' - (\cos x)y = x + c_1$

Dividing by $\sin x$, we see that $e^{-\int \cot x \, dx} = e^{-\ln \sin x} = \csc x$ is an integrating factor. Thus we have

$$\frac{d}{dx}[y \csc x] = x \csc^2 x + c_1 \csc^2 x$$

Integrating, $y \csc x = -x \cot x + \ln \sin x - c_1 \cot x + c_2$

or $y = -x \cos x + \sin x \ln \sin x - c_1 \cos x + c_2 \sin x$

In expanded form the differential equation (11) can be written as

$$a_0\mu'' + (2a_0' - a_1)\mu' + (a_0'' - a_1' + a_2)\mu = 0 \tag{14}$$

or
$$[a_0D^2 + (2a_0' - a_1)D + a_0'' - a_1' + a_2]\mu = 0 \tag{15}$$

This leads us to the following

Definition 2. The *adjoint* of the operator $a_0D^2 + a_1D + a_2$ is the operator

$$a_0D^2 + (2a_0' - a_1)D + a_0'' - a_1' + a_2 \tag{16}$$

In case these two operators are the same, they are called *self-adjoint* operators.

Clearly, the operators in this definition are self-adjoint if and only if

$$a_0D^2 + a_1D + a_2 = a_0D^2 + (2a_0' - a_1)D + a_0'' - a_1' + a_2$$

or
$$a_1 = 2a_0' - a_1, \qquad a_2 = a_0'' - a_1' + a_2$$

i.e.,
$$a_1 = a_0'$$

In such case the operators can be written

$$a_0D^2 + a_0'D + a_2 = D(a_0D) + a_2 \tag{17}$$

We thus have the following

Theorem 2. An operator $a_0D^2 + a_1D + a_2$ is a *self-adjoint operator* if and only if $a_1 = a_0'$, i.e., the operator can be written in the form

$$D(a_0D) + a_2 \quad \text{or} \quad \frac{d}{dx}\left(a_0 \frac{d}{dx}\right) + a_2 \tag{18}$$

A second-order linear differential equation in which the operator is self-adjoint is often for obvious reasons called a *self-adjoint differential equation*. It is seen that the Sturm–Liouville equation

$$\frac{d}{dx}\left[P(x)\frac{dy}{dx}\right] + Q(x)y = -\lambda y \tag{19}$$

is a self-adjoint differential equation, Thus the whole theory already developed concerning eigenvalues and eigenfunctions, orthogonality and orthogonal series, etc., is related to the theory of self-adjoint operators.

5.2 THE GRAM–SCHMIDT ORTHONORMALIZATION METHOD

Suppose that we are given a set of linearly independent functions of x defined in interval $a \leq x \leq b$ and denoted briefly by

$$\psi_1, \psi_2, \psi_3, \ldots \tag{20}$$

Then it is possible to generate from them a new set of functions orthonormal in the interval. The method of doing this, which is quite simple, is often called the *Gram–Schmidt orthonormalization method*. The first step in the method is to construct from

(20) the new set of functions given by

$$\phi_1 = c_{11}\psi_1, \qquad \phi_2 = c_{21}\psi_1 + c_{22}\psi_2, \qquad \phi_3 = \phi_{31}\psi_1 + c_{32}\psi_2 + c_{33}\psi_3, \ldots \quad (21)$$

where the c's are constants. Note that to get the nth function in (21) we multiply each of the first n functions in (20) by constants and add the results. We must now determine the constants in (21) so that the functions are mutually orthogonal and normalized in the interval given. We can do this systematically by writing down the condition that the kth function ϕ_k in (21) be orthogonal to each of the preceding $k - 1$ functions $\phi_1, \phi_2, \ldots, \phi_{k-1}$, which leads to relationships among the constants. Finally, we normalize each function. Let us illustrate the procedure in the following

ILLUSTRATIVE EXAMPLE 3

Use the Gram–Schmidt method on the set of functions $1, x, x^2, x^3, \ldots$ to generate a set of functions orthonormal in the interval $-1 \leq x \leq 1$.

Solution As in (21) we construct from the given set of functions the set

$$\phi_1 = c_{11}, \qquad \phi_2 = c_{21} + c_{22}x, \qquad \phi_3 = c_{31} + c_{32}x + c_{33}x^2, \ldots \quad (22)$$

In order that ϕ_2 be orthogonal to ϕ_1 we must have

$$\int_{-1}^{1} \phi_1\phi_2 \, dx = \int_{-1}^{1} c_{11}(c_{21} + c_{22}x)dx = 2c_{11}c_{21} = 0$$

Since $c_{11} \neq 0$, otherwise we would have $\phi_1 = 0$, we must have $c_{21} = 0$ so that

$$\phi_2 = c_{22}x$$

In order that ϕ_3 be orthogonal to ϕ_1 and ϕ_2, we must have

$$\int_{-1}^{1} \phi_1\phi_3 \, dx = \int_{-1}^{1} c_{11}(c_{31} + c_{32}x + c_{33}x^2)dx = 2c_{11}(c_{31} + \tfrac{1}{3}c_{33}) = 0$$

$$\int_{-1}^{1} \phi_2\phi_3 \, dx = \int_{-1}^{1} c_{22}x(c_{31} + c_{32}x + c_{33}x^2)dx = \tfrac{2}{3}c_{22}c_{32} = 0$$

Since $c_{22} \neq 0$, otherwise we would have $\phi_2 = 0$, we see from these that $c_{31} = -\tfrac{1}{3}c_{33}$ and $c_{32} = 0$ so that

$$\phi_3 = c_{33}(x^2 - \tfrac{1}{3})$$

Continuing in this manner we can find

$$\phi_4 = c_{44}(x^3 - \tfrac{3}{5}x), \qquad \phi_5 = c_{55}(x^4 - \tfrac{6}{7}x^2 + \tfrac{3}{35}), \ldots$$

We now normalize the functions by choosing the constants so that

$$\int_{-1}^{1} \phi_k^2 \, dx = 1, \qquad k = 1, 2, 3, \ldots$$

If this is done, we find

$$\phi_1 = \pm\sqrt{\frac{1}{2}}, \qquad \phi_2 = \pm\sqrt{\frac{3}{2}}\, x, \qquad \phi_3 = \pm\sqrt{\frac{5}{2}}\left(\frac{3x^2 - 1}{2}\right),$$

$$\phi_4 = \pm\sqrt{\frac{7}{2}}\left(\frac{5x^3 - 3x}{2}\right), \qquad \phi_5 = \pm\sqrt{\frac{9}{2}}\left(\frac{35x^4 - 30x^2 + 3}{8}\right), \ldots$$

The polynomials which appear here, apart from the multiplicative constants involving the square root, are the *Legendre polynomials* already obtained in a different manner on page 343.

A EXERCISES

1. Show that each of the following differential equations is exact and thus obtain its solution.
(a) $x^2 y'' + 4xy' + 2y = 12x$. (b) $xy'' + (2x + 1)y' + 2y = 3x - 5$. (c) $(\sin x)y'' + 3(\cos x)y' - 2(\sin x)y = 5\cos x$.

2. Show that each of the following differential equations is not exact but can be made exact on multiplying by the integrating factor indicated. Thus solve each equation.
(a) $xy'' + (x + 2)y' + 2y = 6x$; x.   (b) $y'' + y = \cot x$; $\sin x$.

3. (a) Find the adjoint of the equation $xy'' + 2y' + xy = 0$. (b) By determining a solution of this adjoint equation, solve the original equation.

4. Show that the equation of Exercise 1(a) is also exact if it is multiplied by x, and obtain its solution by this method.

5. Show that if Bessel's differential equation is written as $xy'' + y' + xy = 0$ then it is self-adjoint, but if it is written as $y'' + (1/x)y' + y = 0$, it is not self-adjoint. Explain.

6. Given the set of functions $1, x^2, x^4, \ldots$. Obtain from this a set of functions orthonormal in the interval $0 \leq x \leq 1$. Are the functions related to the Legendre polynomials?

7. Work Exercise 6 for the set of functions $x, x^3, x^5, \ldots$.

B EXERCISES

1. Prove that the adjoint of the adjoint of a second-order operator $\phi(D)$ is $\phi(D)$.

2. (a) Find the adjoint of $y'' - x^2 y' + xy = 0$. (b) By noting that $y = x$ is a solution to the equation in (a), solve the adjoint equation.

3. Generalize Exercise 2 by considering the equation $y'' - xg(x)y' + g(x)y = 0$.

4. Solve $y'' + (x \sin x)y' + (x \cos x + 2 \sin x)y = 0$.

5. Find an integrating factor of the form x^p for the equation $x^5 y'' + (2x^4 - 8x)y' + (8 - 2x^3)y = 0$ and thus obtain its solution.

6. Given the set of functions $1, x, x^2, x^3, \ldots$. (a) Obtain from this a set of functions mutually orthogonal in $0 \leq x < \infty$ with respect to the weight function e^{-x} and give the corresponding orthonormal set. (b) Can you recognize the set of functions?

7. Work Exercise 6 in case the interval is $-\infty < x < \infty$ and the weight function is e^{-x^2}.

C EXERCISES

1. Can a linear second-order differential equation have more than two linearly independent integrating factors? Justify your conclusion.

2. Given the third-order linear differential equation $a_0 y''' + a_1 y'' + a_2 y' + a_3 y = F(x)$ where $a_0, a_1, a_2, F(x)$ are functions of x. As in the second-order case this equation is called *exact* if the left side is an exact derivative, i.e., can be written as $D(b_0 y'' + b_1 y' + b_2 y)$ for suitable

functions b_0, b_1, b_2. Prove that this third-order equation is exact if and only if

$$a_0''' - a_1'' + a_2' - a_3 \equiv 0$$

3. Use Exercise 2 to solve $xy''' + (x^2 + 2)y'' + 4xy' + 2y = 24x$.

4. If the third-order equation of Exercise 2 is not exact, it may be possible to make it exact by multiplying by an integrating factor μ. Show that μ must be a solution of the equation

$$(a_0\mu)''' - (a_1\mu)'' + (a_2\mu)' - a_3\mu = 0$$

which is called the *adjoint* of the given equation. Use this to solve

$$(x^3 - x)y''' + (7x^2 - 3)y'' + 10xy' + 2y = 0$$

5. Generalize the concept of exactness, integrating factor, and adjoint to any nth-order linear differential equation.

6. Answer Exercise 1 for nth-order linear differential equations.

7. Investigate the idea of self-adjoint linear operators and differential equations for the cases where the order is 3, 4, or higher.

8. Explain how you could generate the *Chebyshev polynomials* described in C Exercise 3, page 375, by using the Gram–Schmidt orthonormalization method. Obtain the first few polynomials using the procedure.

nine

the numerical solution of differential equations

In many fields of scientific investigation, the end product is a number or a table of values. Since differential equations play an important part in scientific investigations, it would naturally seem desirable to learn how differential equations can be solved numerically. This is of even greater value when we realize that remarkable computing machines are now available which help extraordinarily in the laborious tasks of numerical work. Some of these machines compute tables of values and may even graph results in a very small percentage of the time that it would take an ordinary computer. The fact that machines lessen the labor, however, does not mean that the operator need know less about numerical methods. On the contrary he should know much about the various methods since he must know the most efficient way of "feeding" the mathematics into the machine.

A study of techniques of numerical analysis is an extensive field in itself. In a book such as this we can give but a brief introduction to this important subject.

1 Numerical Solution of $y' = f(x, y)$

In this section we restrict ourselves to a study of solving numerically the first-order differential equation* $y' = f(x, y)$. We ask the

Question. Given that a solution of $y' = f(x, y)$ is such that y is to equal c where $x = a$, how can we determine the value of y when $x = b$?

By integration of the differential equation with respect to x we have[†]

$$y = c + \int_a^x f(x, y)dx \tag{1}$$

and it is clear that $y = c$ when $x = a$ so that (1) satisfies the required condition. The value of y when $x = b$ must be given by

$$y = c + \int_a^b f(x, y)dx \tag{2}$$

Unfortunately, since y occurs under the integral sign of (2), we cannot proceed further without some sort of approximation. Each type of approximation used in (2) determines one method of numerical analysis. We first examine one of these methods, which we call the *constant slope* or *Euler method*.

---

* We suppose that $f(x, y)$ satisfies the conditions of the fundamental existence and uniqueness theorem of page 24. If the solution does not exist or is not unique, there is no point in attempting a numerical solution.

[†] In the integral (1) we are using the symbol x as a dummy symbol in the integration as well as the independent variable. We could of course have used a different symbol, for example u, to denote the dummy variable and written

$$y = c + \int_a^x f(u, y)du$$

However, no confusion should arise. Compare the footnote on page 315.

Let us assume that the interval from $x = a$ to $x = b$ is subdivided into n equal parts, each of length h, so that

$$h = \frac{b - a}{n} \quad \text{or} \quad b = a + nh \tag{3}$$

We call h the *step size* and n the *number of steps*. Then (2) becomes

$$y = c + \int_a^{a+nh} f(x, y)dx \tag{4}$$

If we use only one step, i.e., $n = 1$, this becomes

$$y = c + \int_a^{a+h} f(x, y)dx \tag{5}$$

The simplest approximation to take in (5) is to assume that the slope $f(x, y)$ is constant over the interval $a \leq x \leq a + h$ and equal to the slope at the point where $x = a$, $y = c$, i.e., $f(a, c)$. In this case (5) becomes

$$y = c + \int_a^{a+h} f(a, c)dx = c + hf(a, c) \tag{6}$$

This is called the *constant slope method* or, since it was first used by Euler, the *Euler method*. Clearly, (6) will give a good approximation to the value of y at $x = a + h$ only if h is small. The degree of smallness evidently depends on the degree of accuracy desired. Hence the word "small" must necessarily be vague until further information is available.

The graphical interpretation of (6) is seen from Fig. 9.1. The true solution is represented by the dashed curve AE. Since distance $AD = h$ it is easy to see that the value of y corresponding to (6) is represented by the ordinate QB. The error made is given by BE. This gets smaller as h gets smaller. If h is large, the error made is large. If the length of the interval from a to b is large, it would seem natural to take smaller

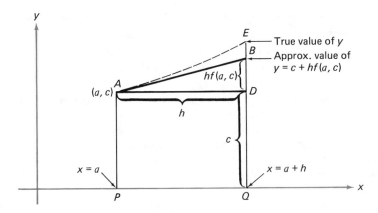

Figure 9.1

values of h corresponding to an increase in the number of steps, hoping in this manner to decrease the error involved. With this idea in mind we are led to write (4) as

$$y = c + \int_a^{a+h} f(x, y)dx + \int_{a+h}^{a+2h} f(x, y)dx + \cdots + \int_{a+(n-1)h}^{a+nh} f(x, y)dx \quad (7)$$

Using the approximation described on page 416 for each of the integrals in (7), we see that an approximation to (7) is given by

$$y = c + hf(a, c) + hf(a + h, c_1) + hf(a + 2h, c_2) + \cdots + hf(a + (n - 1)h, c_{n-1}) \quad (8)$$

where c_j is the value of y when $x = a + jh, j = 1, 2, \ldots, n - 1$.

The geometric interpretation of (8) is given in Fig. 9.2. By an application of (6), we see that $A_1B_1 = hf(a, c)$, the ordinate of point B_1 being given by

$$c_1 = c + hf(a, c)$$

A new slope is now computed corresponding to point B_1, whose coordinates are $(a + h, c_1)$; the value of this slope is given by $f(a + h, c_1)$. Using this, we arrive at point B_2, the distance A_2B_2 given by $hf(a + h, c_1)$. Since the ordinate of B_2 is the ordinate of B_1 plus the distance A_2B_2, the ordinate of B_2 is

$$c_2 = c + hf(a, c) + hf(a + h, c_1)$$

Similarly, the ordinate of point B_j is

$$c_j = c + hf(a, c) + hf(a + h, c_1) + \cdots + hf(a + (j - 1)h, c_{j-1})$$

and in particular

$$c_n = c + hf(a, c) + hf(a + h, c_1) + \cdots + hf(a + (n - 1)h, c_{n-1})$$

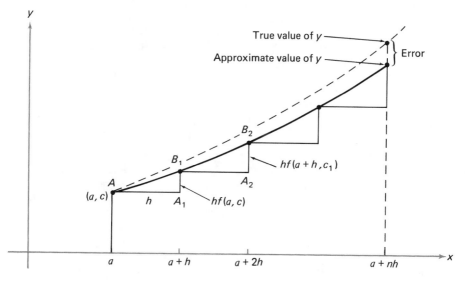

Figure 9.2

is the ordinate of the point reached after n steps, which is the value of y given in (8). If we use the notation $a_j = a + jh, j = 1, 2, \ldots$, this can be written simply as

$$c_n = c + hf(a, c) + hf(a_1, c_1) + \cdots + hf(a_{n-1}, c_{n-1})$$

In spite of the simplicity of this method results obtained can be good, the accuracy becoming better in general as n is chosen larger and larger. For larger n, however, although accuracy may be greater, the computation becomes more laborious and a compromise must thus be reached. The method is well adapted to computers and is not difficult to set up. Let us illustrate the method by the following

ILLUSTRATIVE EXAMPLE 1

Given $y' = x + y$, find the value of y corresponding to $x = 1$ if $y = 1$ where $x = 0$.

Solution   Here $a = 0$, $b = 1$. We would like to choose n so that $h = (b - a)/n$ is small. It is convenient to choose $n = 10$ so that $h = 0.1$. The computation can then be arranged as in Table 9.1.

The initial condition $x = 0$, $y = 1$ determines a slope $y' = x + y$ (first line of table). Since the increment in x is 0.1, the new value of y, which we denote by y_{new}, is obtained from the old value of y denoted by y_{old} as

$$y_{new} = y_{old} + 0.1(\text{slope}) = 1.00 + 0.1(1.00) = 1.10$$

Table 9.1

x	y	$y' = x + y$	$y_{old} + 0.1 \text{ (slope)} = y_{new}$
0.00	1.00	1.00	$1.00 + 0.1(1.00) = 1.10$
0.10	1.10	1.20	$1.10 + 0.1(1.20) = 1.22$
0.20	1.22	1.42	$1.22 + 0.1(1.42) = 1.36$
0.30	1.36	1.66	$1.36 + 0.1(1.66) = 1.53$
0.40	1.53	1.93	$1.53 + 0.1(1.93) = 1.72$
0.50	1.72	2.22	$1.72 + 0.1(2.22) = 1.94$
0.60	1.94	2.54	$1.94 + 0.1(2.54) = 2.19$
0.70	2.19	2.89	$2.19 + 0.1(2.89) = 2.48$
0.80	2.48	3.28	$2.48 + 0.1(3.28) = 2.81$
0.90	2.81	3.71	$2.81 + 0.1(3.71) = 3.18$
1.00	3.18		

This value of y is then transferred to the second line of the table and the process is repeated. In Table 9.1 we have kept three significant figures; the value obtained for y corresponding to $x = 1$ is 3.18. By solving exactly, it may be verified that the true value of y where $x = 1$ is 3.44; the error is thus about 8 per cent. If we had used $n = 20$, the accuracy would have been considerably increased but there would be twice as much computation involved.

1.2 THE AVERAGE SLOPE OR MODIFIED EULER METHOD

In the preceding method the slope $f(x, y)$ over the interval $a \leq x \leq a + h$ was replaced by $f(a, c)$ so that the value of y at $x = a + h = a_1$ turned out to be

$$c_1 = c + hf(a, c) \tag{9}$$

A better approximation is obtained if we replace $f(x, y)$ by the average of the slopes at the endpoints corresponding to $x = a$ and $x = a_1 = a + h$, which are given, respectively, by $f(a, c)$ and $f(a_1, c_1)$. Thus

$$\text{average slope} = \frac{f(a, c) + f(a_1, c_1)}{2} \tag{10}$$

where $a_1 = a + h$ and c_1 is given by (9). Using (10) as the approximate value of $f(x, y)$, the value of y at $x = a + h = a_1$ is given by

$$y = c + \int_a^{a+h} \left[\frac{f(a, c) + f(a_1, c_1)}{2} \right] dx = c + h \left[\frac{f(a, c) + f(a_1, c_1)}{2} \right] \tag{11}$$

This process of using average slopes can be continued for the successive intervals $a + h \leq x \leq a + 2h$, $a + 2h \leq x \leq a + 3h$, etc., until finally the value of y for $x = a + nh = b$ is reached. For example, in the interval $a + h \leq x \leq a + 2h$, which we write as $a_1 \leq x \leq a_2$, (10) is replaced by

$$\text{average slope} = \frac{f(a_1, c_1) + f(a_2, c_2)}{2} \tag{12}$$

where
$$c_2 = c_1 + hf(a_1, c_1) \tag{13}$$

and the value of y at $x = a + 2h = a_2$ is given by

$$c_1 + h \left[\frac{f(a_1, c_1) + f(a_2, c_2)}{2} \right] \tag{14}$$

Similar results can be written for later intervals.

For obvious reasons this method is called the *average slope method*, but it is also referred to as the *modified Euler method*.

ILLUSTRATIVE EXAMPLE 2

Work Illustrative Example 1, page 418, by using the average slope method.

Solution In this case let us also choose $n = 10$ so that $h = 0.1$, The computations can be arranged as in Table 9.2 in which the first four column headings are the same

Table 9.2

x	y	Left-hand slope	y_{new}	Right-hand slope $f(x_{new}, y_{new})$	Average slope y'_{av}	$y_{old} + 0.1(y'_{av}) = y_{corr}$
0.00	1.00	1.00	1.10	1.20	1.10	$1.00 + 0.1(1.10) = 1.11$
0.10	1.11	1.21	1.23	1.43	1.32	$1.11 + 0.1(1.32) = 1.24$
0.20	1.24	1.44	1.38	1.68	1.56	$1.24 + 0.1(1.56) = 1.40$
0.30	1.40	1.70	1.57	1.97	1.84	$1.40 + 0.1(1.84) = 1.58$
0.40	1.58	1.98	1.78	2.28	2.13	$1.58 + 0.1(2.13) = 1.79$
0.50	1.79	2.29	2.02	2.62	2.46	$1.79 + 0.1(2.46) = 2.04$
0.60	2.04	2.64	2.30	3.00	2.82	$2.05 + 0.1(2.82) = 2.33$
0.70	2.33	3.03	2.63	3.43	3.23	$2.33 + 0.1(3.23) = 2.65$
0.80	2.65	3.45	2.99	3.89	3.67	$2.65 + 0.1(3.67) = 3.02$
0.90	3.02	3.92	3.41	4.41	4.16	$3.02 + 0.1(4.16) = 3.44$
1.00	3.44					

as that of the Table 9.1. However, since we require two slopes, one at the left-hand endpoint of an interval and the other at the right-hand endpoint, in order to get the average slope, the heading of the third column of Table 9.1 has been relabeled *left-hand slope*. The remaining three columns in the Table 9.2 are used to find respectively, the right-hand slope, the average of the left- and right-hand slopes denoted by y'_{av} and the corrected value of y, denoted by y_{corr}.

To see how the computation proceeds, let us obtain the entries in the first row of Table 9.2. The first four entries are obtained exactly as in Table 9.1. The right-hand slope (fifth column entry) is obtained by calculating $y' = f(x, y) = x + y$, taking x as the right-hand endpoint $a + h = a_1$, which we call x_{new}, and y as the corresponding approximate value of y, which we have called y_{new} (already obtained in the fourth column entry). This right-hand slope is given by $f(x_{new}, y_{new}) = x_{new} + y_{new} = 0.10 + 1.10 = 1.20$.

Having found the right- and left-hand slopes, we can determine the average slope (sixth column entry) as $\frac{1}{2}(1.00 + 1.20) = 1.10$. Finally the corrected value of y denoted by y_{corr} (seventh column entry) is the new value of y (fourth column) plus 0.1 times the average slope, i.e., $y_{corr} = 1.00 + 0.1(1.10) = 1.11$. This corrected value of y is transferred to the second row of the table as indicated.

The same process used to obtain the entries in the first row can now be used to obtain the entries in the second row and all succeeding rows until the value of y corresponding to $x = 1$ is obtained. As Table 9.2 shows, this value is 3.44, which remarkably happens to *agree exactly* with the true value. This would seem to indicate that all the remaining pairs of values (x, y) in Table 9.2 are also correct, and we can use these to obtain a graph of the solution in the interval $0 \leq x \leq 1$. If we like we could continue the calculation for values of x beyond 1, but there is of course no assurance that the same accuracy will continue.

1.3 COMPUTER DIAGRAMS

It is of interest to present schematic diagrams depicting the successive steps in calculations for the constant slope and average slope methods. Such diagrams are called *computer diagrams* or *flow charts*. Figure 9.3 gives the computer diagram for the constant slope method. The initial pair of values (x, y) provides an *input* which is *fed into* the first box, which calculates the slope at (x, y). This slope is fed into another box which calculates the new value of y. The final *output* consists of the new pair of values (x_{new}, y_{new}). This provides a new input or *feedback*, and the same process is repeated over and over again until the final required results are achieved.

In a similar manner we can construct a computer diagram depicting the average slope method as in Fig. 9.4. The first row in this figure leading to (x_{new}, y_{new}) is of course identical with that of Fig. 9.3. The modification consists of three additional boxes, the first providing computation of the right-hand slope from (x_{new}, y_{new}), the second obtaining the average slope from the left- and right-hand slopes as indicated, and the third providing the corrected value of y. The pair of values (x_{new}, y_{corr}) is the

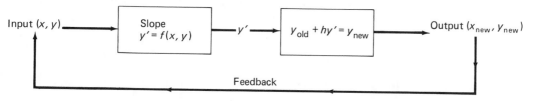

Figure 9.3 Computer diagram for constant slope method

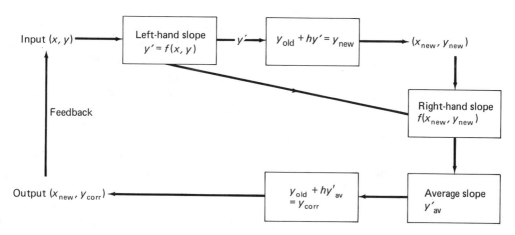

Figure 9.4 Computer diagram for average slope method

output, which serves as a new input, and the process is repeated over and over again as before.

1.4 ERROR ANALYSIS

Whenever a formula is developed for obtaining approximate results, it is natural to seek some estimate of the error which may be involved in using it. The process of estimating such errors is often called *error analysis*. In dealing with these errors, we shall not be concerned with *random errors*, such as *rounding errors*, e.g., rounding 2.84316 to 2.8432, which may for example be due to limitations in the storage capacity of a computer. Instead we shall be interested only in the error produced by using the particular formula. If one formula produces more accurate results than another, we would certainly expect this increased accuracy to show up in an error analysis. We shall illustrate this by an error analysis of the constant slope and average slope methods.

(a) *Error analysis for the constant slope method.* From page 416 we see that an approximate solution of the initial value problem

$$y' = f(x, y), \qquad y(a) = c \tag{15}$$

for $y(a + h)$ is given by

$$y(a + h) = c + hf(a, c) \tag{16}$$

However, by use of Taylor series, assuming appropriate differentiability conditions on $f(x, y)$, the true value of $y(a + h)$ is

$$y(a) + hy'(a) + \frac{h^2}{2!} y''(a) + \cdots = c + hf(a, c) + \frac{h^2}{2!} y''(a) + \cdots \tag{17}$$

Comparing (16) and (17), we see that the error made in using (16) instead of (17) is

$$E = \frac{h^2}{2!} y''(a) + \frac{h^3}{3!} y'''(a) + \cdots \tag{18}$$

where the values $y''(a), y'''(a), \ldots$ are found by differentiating successively the differential equation in (15). This error occurs due to the fact that the series beyond the first two terms in (17) has been cut off. For this reason the error is often called a *truncation error* (from the Latin *truncare*, meaning *to cut off*).

For values of h which are small enough, we would expect the error (18) to be very nearly equal to the first term, i.e., $\frac{1}{2}h^2 y''(a)$ with the rest of the terms negligible. However, the student may rightfully be disturbed about neglecting an infinite number of terms without proper justification. Fortunately, there is another way out of the predicament by making use of *Taylor's theorem with a remainder*, which the student may have studied in calculus. This theorem states that if y is at least twice differentiable [or $f(x, y)$ is at least once differentiable], then

$$y(a + h) = y(a) + hy'(a) + \frac{h^2}{2!} y''(r), \qquad \text{where } r \text{ is between } a \text{ and } a + h \tag{19}$$

Here the last term on the right represents the remainder of the series after the first two terms, and (19) provides a decided improvement over (17), which requires the existence of all derivatives. Using (19), the error can be written

$$E = \frac{h^2}{2!} y''(r), \qquad \text{where } r \text{ is between } a \text{ and } a + h \tag{20}$$

Since this error is proportional to h^2, or as we sometimes say is *of the order* h^2, abbreviated $O(h^2)$, we see that by reducing the size of h we can considerably reduce the size of the error. Since $y''(r)$ can be positive or negative the error (20) can also be positive or negative. If we denote by M a positive constant such that $|y''(r)| < M$ for r between a and $a + h$, then

$$|E| < \frac{Mh^2}{2} \tag{21}$$

the right-hand side representing an upper bound for the error.

Since the error is given by (20), if we have $y(a)$ and seek $y(a + h)$, the question quite naturally arises as to what would be the *cumulative error* if we proceed by n steps to the calculation of $y(b)$, where $b = a + nh$. Further analysis, which is rather tedious and which we shall not pursue here, shows that the maximum error is n times that given in (21), with perhaps a different value of M, which we shall call K; i.e., for n steps

$$|E_n| < \frac{Knh^2}{2} \tag{22}$$

or since $n = (b - a)/h$,

$$|E_n| < \frac{K(b - a)h}{2} \tag{23}$$

which shows that the cumulative error is of the order h.

(b) *Error analysis for the average slope method.* From page 419 we see that an approximate solution of the initial value problem (15) for $y(a + h)$ is

$$y(a + h) = c + \frac{h}{2}[f(a, c) + f(a + h, c + hf(a, c))] \tag{24}$$

which is to be compared with the true value (17) or (19). The error made in this case is

$$E = y(a + h) - \left\{ c + \frac{h}{2}[f(a, c) + f(a + h, c + hf(a, c))] \right\} \tag{25}$$

In order to see how the right side of (25) depends on h, we again use Taylor series. We shall use the infinite series rather than the one with a remainder for simplicity in notation. As before we have

$$y(a + h) = y(a) + hy'(a) + \frac{h^2}{2!} y''(a) + \frac{h^3}{3!} y'''(a) + \cdots \tag{26}$$

To find a series for the last term on the right of (25), we use the Taylor series for the two-variable case, which is analogous to the one-variable case. This is given by

$$f(a + h, c + k) = f(a, c) + hf_x(a, c) + kf_y(a, c)$$

$$+ \frac{1}{2!}[h^2 f_{xx}(a, c) + 2hk f_{xy}(a, c) + k^2 f_{yy}(a, c)] + \cdots \quad (27)$$

where $f_x(a, c)$ denotes the partial derivative of $f(x, y)$ with respect to x evaluated at $x = a, y = c$, $f_{xx}(a, c)$ denotes the second partial derivative of $f(x, y)$ with respect to x evaluated at $x = a, y = c$, $f_{xy}(a, c)$ denotes the partial derivative of $f(x, y)$ with respect to both x and y evaluated at $x = a, y = c$, etc.

Taking $k = hf(a, c)$ in (27) we find

$$f(a + h, c + hf(a, c)) = f(a, c) + hf_x(a, c) + hf(a, c)f_y(a, c)$$

$$+ \frac{1}{2!}[h^2 f_{xx}(a, c) + 2h^2 f(a, c)f_{xy}(a, c) + h^2\{f(a, c)\}^2 f_{yy}(a, c)] + \cdots \quad (28)$$

Substituting (26) and (28) into (25) yields

$$E = [y(a) - c] + h[y'(a) - f(a, c)] + \frac{h^2}{2}[y''(a) - f_x(a, c) - f_y(a, c)f(a, c)]$$

$$+ \text{ terms involving } h^3 \text{ and higher} \quad (29)$$

Since $y(a) = c$ from the initial condition in (15), while $y'(a) = f(a, c)$ from the differential equation in (15), the first two terms in (29) are zero. By taking the derivative of both sides of the differential equation in (15), we have on using the chain rule of elementary calculus

$$y'' = \frac{\partial f}{\partial x} + \frac{\partial f}{\partial y}\frac{dy}{dx} = f_x + f_y y' \quad (30)$$

From this we have on evaluating the derivatives at $x = a, y = c$

$$y''(a) = f_x(a, c) + f_y(a, c)f(a, c) \quad (31)$$

so that the third term in (29) is also zero. The result shows that E involves only terms in h^3 or higher. By using Taylor series with a remainder as in the constant slope case, it follows that the error is of the order h^3, i.e., $E = O(h^3)$. Since $E = O(h^2)$ for the constant slope method, while $E = O(h^3)$ for the average slope method, we can readily see why the second method is much more accurate.

As an illustration of how the above ideas of error analysis can be used in estimation of errors, let us consider the following

ILLUSTRATIVE EXAMPLE 3

Given the initial value problem $y' = x + y, y(0) = 1$. Estimate the error made in calculating $y(0.1)$ using the constant slope method with $h = 0.1$.

Solution Since $y' = f(x, y) = x + y$, we have $y'' = 1 + y' = 1 + x + y$ so that

$$y''(r) = 1 + r + y, \quad \text{where } 0 < r < 0.1$$

It is reasonable to suppose that $y < 2$. Thus we have

$$|y''(r)| < 1 + 0.1 + 2 = 3.1$$

and from (20) or (21) we see that

$$|E| < \frac{(0.1)^2(3.1)}{2}, \quad \text{i.e.,} \quad |E| < 0.016 \text{ approx.}$$

Since the constant slope method gives 0.10 (see Table 9.1) while the true value is 0.11 the actual error is 0.01, which is in agreement with the above estimate.

1.5 SOME PRACTICAL GUIDELINES FOR NUMERICAL SOLUTION

There are many methods available for the numerical solution of differential equations as the student may learn by reading the literature on the subject.* For most methods an error analysis is difficult, as might be surmised by the derivation on page 424 for the relatively simple case of the average slope method. Also, even when an error analysis is available it does not provide a simple error term, such as given in (20), page 423, but only its order as a function of step size, as for example $O(h^4)$, $O(h^5)$, etc. A further complication is that accuracy achieved is limited by the particular characteristics of the differential equation considered. Thus, for instance, if a solution is desired near a singularity, even the "best method" might provide poor accuracy. Because of these difficulties there is no simple panacea for numerical solution (which may serve to indicate why so many methods are available). In spite of this there are some practical guidelines which a scientist can follow.

1. Choose a particular method involving a step size h which looks reasonable in view of the type of differential equation involved. For example, if the differential equation involves a singularity such as $x = 1$ in the initial value problem

$$\frac{dy}{dx} = \frac{y}{1 - x}, \quad y(0) = 1$$

where we seek $y(0.9)$ for instance, smaller step sizes are needed near $x = 0.9$ than near $x = 0$. In such case we may wish to divide the interval from $x = 0$ to $x = 0.9$ into intervals of unequal length or step size.

2. To check the accuracy of the numerical value found in 1, repeat the method using a smaller step size, for example half the step size used before. If this procedure results in only a minor change in the significant figures we can be sure of those which do not change. Thus, if the repeated value is for example 5.2435, rather than 5.2408, we can at least be sure of the accuracy to three significant figures given by 5.24. If there is a major discrepancy, we may have to reduce the step size further.

* See for example [12].

3. Reduction of step size, resulting in an increased number of steps, may lead to *cumulative rounding errors* which affect the accuracy. Because of this, calculations must be made with sufficiently many significant figures. Unfortunately, no rules can be given since this again depends on the kinds of calculations involved. Thus, for example, even though we may have $x = 7.41563$ and $y = 7.41492$ each accurate to six significant figures, most of these are lost in taking the difference $x - y = 0.00071$. However, in most cases which arise in practice such losses in significant figures do not occur. In these "routine" cases a reasonable rule to follow is to use at least two more figures than are required in the answer.

A EXERCISES

Use (a) the constant slope method and (b) the average slope method to determine the indicated value of y for each of the following initial value problems by taking the indicated number of subdivisions n. If possible compare with the exact value.

1. $y' = 2x + y$; $y(0) = 0$. Find $y(0.5)$; use $n = 5$.

2. $y' = x^2 - y$; $y(1) = 0$. Find $y(1.6)$; use $n = 6$.

3. $y' = (y + 1)/x$; $y(2) = 3$. Find $y(2.8)$; use $n = 4$ and $n = 8$.

4. $y' = (x - y)/(x + y)$; $y(3) = 2$. Find $y(1)$; use $n = 5$ and $n = 10$.

5. $y' = x^2 + y^2$; $y(1) = 2$. Find $y(0.5)$; use $n = 5$ and $n = 10$.

6. $y' = \sqrt{x + y}$; $y(5) = 4$. Find $y(4)$; use $n = 5$ and $n = 10$.

7. $y' = \sin y$; $y(0) = 1$. Find $y(1)$; use $n = 5$ and $n = 10$.

B EXERCISES

1. Given $y' = y$; $y(0) = 1$, find $y(1)$ numerically. How can you use your results to compute e?

2. If $y' = \tan^{-1} x$; $y(0) = 1$, find $y(1)$.

3. (a) Use the initial value problem $y' = -2y$, $y(0) = 1$ to calculate $y(0.1)$ by the constant slope method with $h = 0.1$. (b) Estimate the error of the calculation in (a) by using (20), page 423, and compare with the true value.

4. Given the initial value problem $y' = \dfrac{1}{1 + x^2}$, $y(0) = 0$ (a) Calculate $y(0.2)$ by the constant slope method using an appropriate value of h. (b) Estimate the error in the calculation and compare with the true value.

5. In Exercise 3 calculate $y(0.2)$ using the constant slope method with $h = 0.1$. How would you estimate the error made?

6. (a) The differential equation $y' = (x + y)^{-2}$ is to be solved numerically for $y(2)$, given $y(1) = 1$. What are the numerical values obtained by choosing $n = 2, 5, 10$? How can one be sure of a solution accurate to two decimal places? (b) Solve the differential equation of (a) exactly by using the transformation $x + y = v$. Thus find $y(2)$ exactly and compare with (a).

7. Given $y' = e^{-xy}$; $y(0) = 1$, find $y(1)$ by using a calculator or handbook which tabulates the various values of e^u. Obtain accuracy to at least two decimal places.

8. Given $y' = y(x + y)$; $y(0) = 0.5$, find $y(0.5)$, using $n = 5$. Compare with the exact solution. What accuracy is obtained by using $n = 20$?

9. Given the initial value problem $\dfrac{dy}{dx} = \dfrac{y}{1 - x}$, $y(0) = 1$ find $y(0.9)$ and compare with the true value (see page 425).

10. Given $y' = 1/(1 + x^2)$; $y(0) = 0$, find $y(1)$ numerically. How can you use your results to compute π?

C EXERCISES

1. The method of this section was limited to the numerical solution of a first-order differential equation. The second-order differential equation $y'' = f(x, y, y')$ subject to the conditions $y = c_1$, $y' = c_2$ where $x = a$ may be written as two simultaneous first-order equations

$$y' = v, \qquad v' = f(x, y, v)$$

subject to the conditions $y = c_1$, $v = c_2$ where $x = a$. Can you devise a procedure for finding y and y' when $x = a + h$? If so, use the method on the equation $y'' = x + y$ subject to the conditions $y(0) = y'(0) = 0$ to obtain $y(1)$. Compare with the exact solution.

2. Use the method discovered in Exercise 1 to find $y(0.8)$ and $y'(0.8)$ for the differential equation $y'' = \sqrt{x + y}$ subject to the conditions $y(0) = 1$, $y'(0) = 0$.

3. Suppose that in the integral of equation (5), page 416, we approximate $f(x, y)$ by its value at $x = a + \frac{1}{2}h$ and $y = c + \frac{1}{2}hf(a, c)$ which are the average values of x and y as obtained by the constant slope method. (a) Show that

$$y = c + \frac{h}{2} f(a + \tfrac{1}{2}h, c + \tfrac{1}{2}hf(a, c))$$

(b) By use of error analysis compare the accuracy of the result in (a) with that obtained by the constant slope and average slope methods. (c) Work some of the A Exercises on page 426 by this method and compare with results of other methods. (d) Construct a computer diagram for the method which is sometimes called *Runge's method*.

2 The Runge–Kutta Method

As we have already seen (page 416), if we are given the differential equation

$$\frac{dy}{dx} = f(x, y), \qquad \text{where } y(a) = c$$

then on taking $n = 1$ so that $b = a + h$, we find

$$y(a + h) = c + \int_a^{a+h} f(x, y)dx \tag{1}$$

Also from the Taylor series expansion we have

$$y(a + h) = y(a) + hy'(a) + \frac{h^2}{2!} y''(a) + \frac{h^2}{3!} y'''(a) + \cdots \tag{2}$$

By expressing the various derivatives indicated in (2) in terms of $f(x, y)$, Runge and Kutta were able to arrive at various formulas for approximating the series in (2). One such formula, which is found to agree with (2) up to and including the term involving h^4 is given by

$$y(a + h) = c + \tfrac{1}{6}(m_1 + 2m_2 + 2m_3 + m_4)$$

where
$$m_1 = hf(a, c), \qquad m_2 = hf(a + \tfrac{1}{2}h, c + \tfrac{1}{2}m_1)$$
$$m_3 = hf(a + \tfrac{1}{2}h, c + \tfrac{1}{2}m_2), \qquad m_4 = hf(a + h, c + m_3)$$

The verification of this formula is tedious but not difficult (see C Exercise 2). Another formula is given in C Exercise 1. To see an application of this *Runge–Kutta method*, as it is often called, let us consider the following

ILLUSTRATIVE EXAMPLE 1

Given $y' = x + y$, $y(0) = 1$, find $y(1)$.

Solution We have in this case $f(x, y) = x + y$, $a = 0$, $c = 1$. If we choose $h = 1$, we find

$$m_1 = hf(a, c) = f(0, 1) = 1, \qquad m_2 = hf(a + \tfrac{1}{2}h, c + \tfrac{1}{2}m_1) = f(\tfrac{1}{2}, \tfrac{3}{2}) = 2$$
$$m_3 = hf(a + \tfrac{1}{2}h, c + \tfrac{1}{2}m_2) = f(\tfrac{1}{2}, 2) = \tfrac{5}{2}, \qquad m_4 = hf(a + h, c + m_3) = f(1, \tfrac{7}{2}) = \tfrac{9}{2}$$

and so
$$y(1) = 1 + \tfrac{1}{6}(1 + 4 + 5 + \tfrac{9}{2}) = 3.42$$

The fact that this is in such good agreement with the true value 3.44, even though the relatively large step size $h = 1$ has been used, serves to indicate the distinct superiority of the Runge–Kutta method over the average slope method (and certainly the constant slope method), which required more calculation, as seen in Illustrative Examples 1 and 2 on pages 418–421. Increased accuracy in the above example is obtained by using smaller values of h and applying the method more than once. Thus, for instance, if we choose $h = 0.5$ and apply the method twice, we arrive at the true value 3.44.

A EXERCISES

For each of the following differential equations and conditions, determine the indicated value of y by using the Runge–Kutta method. If possible compare with values obtained by solving the equation exactly.

1. $y' = 2x + y$; $y(0) = 1$. Find $y(0.5)$. 　 2. $y' = x^2 - y$; $y(1) = 0$. Find $y(1.6)$.

3. $y' = (y + 1)/x$; $y(2) = 3$. Find $y(2.8)$. 　 4. $y' = x - y$; $y(1) = 2$. Find $y(0.5)$.

5. Work A Exercises 1–7 on page 426 by using the Runge–Kutta method and compare the accuracy achieved.

B EXERCISES

1. Work A Exercise 1 by using two applications of the Runge–Kutta method. Discuss the results.

2. Use (a) two, (b) three, and (c) four applications of the Runge–Kutta method to solve A Exercise 2. What are the advantages of using successive applications of the method? How could you determine the number of applications needed?

3. Use two applications of the Runge–Kutta method to solve B Exercise 1 on page 426. Compare the accuracy of this method with that of the constant slope method.

4. (a) Using $h = 0.2$, find $y(1.2)$ by the Runge–Kutta method for the equation $y' = x + y$ subject to $y(1) = 0$. (b) Using $h = 0.1$, find $y(1.1)$ by the Runge–Kutta method for the equation in (a). Using this result, then compute $y(1.2)$. Compare with the accuracy of (a) and with the exact answer.

5. Use the Runge–Kutta method to solve the initial value problem $y' = f(x)$, $y(a) = c$ by showing that

$$y(a + h) = c + \frac{h}{6}\left[f(a) + 4f\left(a + \frac{h}{2} \right) + f(a + h) \right]$$

This is *Simpson's rule* used in calculus for evaluating definite integrals.

6. Use Exercise 5 to solve $y' = 1/(1 + x^2)$, $y(0) = 0$ for $y(1)$ and compare with the true value.

C EXERCISES

1. Another method due to Runge consists in computing the following

$$m_1 = hf(a, c), \qquad m_2 = hf(a + h, b + m_1)$$
$$m_3 = hf(a + h, b + m_2), \qquad m_4 = hf(a + \tfrac{1}{2}h, b + \tfrac{1}{2}m_1)$$

and then using $\qquad y(a + h) = c + \tfrac{1}{6}(m_1 + 4m_4 + m_3)$

By solving various differential equations using this method, for example the ones in the A and B Exercises above, decide whether this method is better or worse than the method given in the text. Can you give a possible explanation for your conjecture?

2. Derive the Runge–Kutta formula of the text. [*Hint:* Use Taylor's theorem for functions of two variables and the trapezoidal rule for approximating integrals. Then compare the terms of equations (1) and (2) of page 427 up to and including the terms involving h^4.]

3. Explain how the formula of Exercise 1 may be obtained.

II

systems of ordinary differential equations

ten

systems of differential equations and their applications

1 Systems of Differential Equations

1.1 MOTIVATION FOR SYSTEMS OF DIFFERENTIAL EQUATIONS

Up to now we have studied differential equations involving one independent and one dependent variable. Often in applications we encounter equations containing one independent but two or more dependent variables. As an example, let us consider the following.

PROBLEM FOR DISCUSSION

Suppose that a particle of mass m moves in the xy plane (see Fig. 10.1) due to some force acting on it. Let us assume that the direction of the force is in the xy plane and that its magnitude depends only on the instantaneous position (x, y) of the particle and on the time t. Assuming that the particle is initially at rest at some point, say the origin, it is natural for us to ask where the particle will be at any later time.

Mathematical Formulation. Let us resolve the force into two components F_1 and F_2 in the positive x and y directions as indicated in the figure. Then by Newton's law we have

$$m\frac{d^2x}{dt^2} = F_1(x, y, t), \qquad m\frac{d^2y}{dt^2} = F_2(x, y, t) \qquad (1)$$

since d^2x/dt^2 and d^2y/dt^2 are the components of the acceleration in the positive x and y directions, respectively. In (1) we have specifically indicated the dependence of the components on the instantaneous position (x, y) of the particle and the time t. Since the particle is assumed to be at rest initially at the origin, we have as initial conditions

$$x = 0, \quad y = 0, \quad \frac{dx}{dt} = 0, \quad \frac{dy}{dt} = 0 \quad \text{at } t = 0$$

or

$$x(0) = y(0) = x'(0) = y'(0) = 0 \qquad (2)$$

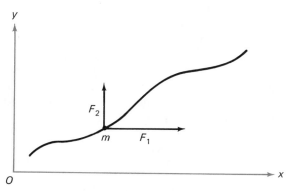

Figure 10.1

As a specific example, suppose that the particle is of unit mass, i.e., $m = 1$, and that the force is such that the components are given by

$$F_1 = y + 4e^{-2t}, \qquad F_2 = x - e^{-2t} \qquad (3)$$

Then from (1), $\qquad \dfrac{d^2x}{dt^2} = y + 4e^{-2t}, \qquad \dfrac{d^2y}{dt^2} = x - e^{-2t} \qquad (4)$

while the conditions on x and y are given as before by (2).

The equations (1) or (4) involve one independent variable t and two dependent variables x and y which depend on t. Since they involve ordinary derivatives with respect to t, it is natural to call them *ordinary differential equations*.

To determine where the particle will be at any time $t > 0$, we must try to find a pair of functions x and y depending on t which when substituted into (4) will satisfy each equation (i.e., reduce each equation to an identity), and which will also satisfy the conditions (2). It is natural to call such a pair of functions x and y a *solution* of the *initial value problem* described by the equations (4) and conditions (2).

Now it is certainly easy to find pairs of functions which satisfy the first equation in (4). For instance, if we let $x = t^2$ in this first equation, we find

$$y = 2 - 4e^{-2t}$$

Thus the pair $x = t^2$, $y = 2 - 4e^{-2t}$ is a solution of this first equation. However, it is found by substitution in the second equation of (4) that this pair is not a solution of that equation. Our problem of course is to find a pair of functions which will satisfy *simultaneously* the equations (4). For this reason we sometimes call the combination of equations (1) or (4) *simultaneous differential equations*. We also refer to them as a *system of differential equations*.

The problem of determining a solution to simultaneous differential equations may recall to mind a similar situation encountered in algebra. Given the linear equation $x + 2y = 6$, for example, there are infinitely many pairs of solutions. Thus $x = 6$, $y = 0$; $x = 3$, $y = \frac{3}{2}$, etc., are solutions. If, however, we have another equation which must be satisfied simultaneously, say $2x + y = 3$, only the pair $x = 0$, $y = 3$ is a solution.

Now we know from algebra that there are various ways in which solutions to simultaneous equations can be found. One of the simplest is that of eliminating one of the unknowns. Thus, given the pair $x + 2y = 6$, $2x + y = 3$, we obtain from the second equation $y = 3 - 2x$. Substituting this into the first equation yields $x + 2(3 - 2x) = 6$, i.e., $x = 0$, so that $y = 3$. This *method of elimination* which can be applied in cases of three equations in three unknowns, four equations in four unknowns, etc., turns out also to be a basic procedure in solving simultaneous differential equations, as we shall now illustrate.

1.2 METHOD OF ELIMINATION FOR SOLVING SYSTEMS OF DIFFERENTIAL EQUATIONS

To show how the method of elimination can be used in differential equations, let us return to the system of equations (4). We can, because of the particular form of the equations, solve the first equation for y in terms of x and t to obtain

$$y = \frac{d^2x}{dt^2} - 4e^{-2t} \tag{5}$$

If we now substitute this into the second equation of (4) and simplify, we find

$$\frac{d^4x}{dt^4} - x = 15e^{-2t} \tag{6}$$

Since this is a linear differential equation for the single unknown x, we can use methods of Chapter Four to obtain

$$x = c_1 \cos t + c_2 \sin t + c_3 e^t + c_4 e^{-t} + e^{-2t} \tag{7}$$

Substitution of this in (5) yields

$$y = -c_1 \cos t - c_2 \sin t + c_3 e^t + c_4 e^{-t} \tag{8}$$

The pair (7) and (8) is thus the required solution of the system (4). We can now use the conditions (2) to determine the constants c_1, c_2, c_3, c_4. We shall need dx/dt and dy/dt. From (7) and (8) we have

$$\frac{dx}{dt} = -c_1 \sin t + c_2 \cos t + c_3 e^t - c_4 e^{-t} - 2e^{-t} \tag{9}$$

$$\frac{dy}{dt} = c_1 \sin t - c_2 \cos t + c_3 e^t - c_4 e^{-t} \tag{10}$$

Putting $t = 0$ in (7), (8), (9), and (10), we see from (2) that the corresponding values of x, y, dx/dt and dy/dt are all zero which leads to the simultaneous equations

$$\left. \begin{array}{ll} c_1 + c_3 + c_4 = -1, & -c_1 + c_3 + c_4 = 0, \\ c_2 + c_3 - c_4 = 2, & -c_2 + c_3 - c_4 = 0 \end{array} \right\} \tag{11}$$

From these we find $c_1 = -\frac{1}{2}$, $c_2 = 1$, $c_3 = \frac{1}{4}$, $c_4 = -\frac{3}{4}$ $\tag{12}$

so that

$$x = -\tfrac{1}{2} \cos t + \sin t + \tfrac{1}{4}e^t - \tfrac{3}{4}e^{-t} + e^{-2t} \tag{13}$$

$$y = \tfrac{1}{2} \cos t - \sin t + \tfrac{1}{4}e^t - \tfrac{3}{4}e^{-t} \tag{14}$$

Equations (13) and (14) represent the parametric equations of the path taken by the particle as it moves in the xy plane and from these equations we can determine the position (x, y) of the particle at any time as was required.

Although the method of elimination described above is easy in principle, it is unfortunately not always a simple matter to apply. To show this, let us suppose that in the problem of the particle considered on page 433, the components of force acting on the particle depend on the components of the velocity, i.e., dx/dt, dy/dt, as well as the position (x, y) and the time t. In such case equations (1) are replaced by

$$m\frac{d^2x}{dt^2} = F_1\left(x, y, \frac{dx}{dt}, \frac{dy}{dt}, t\right), \quad m\frac{d^2y}{dt^2} = F_2\left(x, y, \frac{dx}{dt}, \frac{dy}{dt}, t\right) \tag{15}$$

If in particular $m = 1$ and the force components are given by

$$F_1 = y + 3x - \frac{dy}{dt}, \qquad F_2 = -4x - 2y - \frac{dx}{dt} \tag{16}$$

equations (15) become

$$\frac{d^2x}{dt^2} = y + 3x - \frac{dy}{dt}, \qquad \frac{d^2y}{dt^2} = -4x - 2y - \frac{dx}{dt} \tag{17}$$

To appreciate the difficulties which can occur in applying the method of elimination, let us attempt to find the solution to the system (17) as in the following

ILLUSTRATIVE EXAMPLE 1

Solve the system (a) $\dfrac{d^2x}{dt^2} + \dfrac{dy}{dt} - 3x = y,$ (b) $\dfrac{d^2y}{dt^2} + \dfrac{dx}{dt} + 4x + 2y = 0.$

Solution The given equations are rearrangements of equations (17). Let us try to eliminate y from the two equations. Differentiating (a) we find

$$\frac{d^3x}{dt^3} + \frac{d^2y}{dt^2} - 3\frac{dx}{dt} = \frac{dy}{dt} \tag{18}$$

Subtracting equation (b), we eliminate d^2y/dt^2, so that

$$\frac{d^3x}{dt^3} - 4\frac{dx}{dt} - 4x - 2y = \frac{dy}{dt} \tag{19}$$

We must still eliminate y and dy/dt. If we substitute dy/dt from (19) into (a), we find

$$\frac{d^3x}{dt^3} + \frac{d^2x}{dt^2} - 4\frac{dx}{dt} - 7x - 3y = 0 \tag{20}$$

so that only y needs to be eliminated now. To do this let us solve for y in (20) and substitute the result into (a) or (b). This gives

$$\frac{d^4x}{dt^4} - 2\frac{d^2x}{dt^2} - 3\frac{dx}{dt} - 2x = 0 \tag{21}$$

so that we have finally arrived at an equation for x. From this point we solve for x, then go back and find y, and finally satisfy the given initial conditions.

We shall not carry through the details of finding the solution since the main purpose of the example was to show how tedious the method of elimination can be. The example given certainly should provide us with the motivation to seek a shorter approach. We shall now show one such approach which makes use of operators.

1.3 THE USE OF OPERATORS IN ELIMINATING UNKNOWNS

Using the notation of the symbolic operator $D \equiv d/dt$, equations (a) and (b) of Illustrative Example 1 can be written respectively as

$$(D^2 - 3)x + (D - 1)y = 0 \tag{22}$$

$$(D + 4)x + (D^2 + 2)y = 0 \tag{23}$$

The temptation to eliminate x or y by treating D as nothing more than an algebraic symbol is great. If we should yield to this urge realizing that in so doing there still remains the task of justification, we would proceed as follows: "multiply" (22) by $(D^2 + 2)$ and (23) by $(D - 1)$ to obtain

$$(D^2 + 2)(D^2 - 3)x + (D^2 + 2)(D - 1)y = 0$$

$$(D - 1)(D + 4)x + (D - 1)(D^2 + 2)y = 0$$

Subtracting, assuming $(D^2 + 2)(D - 1)y \equiv (D - 1)(D^2 + 2)y$, we have

$$(D^2 + 2)(D^2 - 3)x - (D - 1)(D + 4)x = 0$$

i.e.,

$$(D^4 - 2D^2 - 3D - 2)x = 0$$

which is the same equation as (21), which we obtained with more labor.

To justify this procedure, all we must realize is that "multiplication" by $(D^2 + 2)$, for example, merely means that a certain operation is to be performed. Thus $(D^2 + 2)Y$ says (1) take the second derivative of Y, (2) multiply Y by 2 and (3) add the results of (1) and (2). In a similar manner we can show that

$$(D^2 + 2)(D - 1)y \equiv (D - 1)(D^2 + 2)y$$

since $(D^2 + 2)(D - 1)y \equiv \left(\dfrac{d^2}{dt^2} + 2\right)\left(\dfrac{dy}{dt} - y\right) \equiv \dfrac{d^2}{dt^2}\left(\dfrac{dy}{dt} - y\right) + 2\left(\dfrac{dy}{dt} - y\right)$

$$\equiv \dfrac{d^3y}{dt^3} - \dfrac{d^2y}{dt^2} + 2\dfrac{dy}{dt} - 2y$$

$(D - 1)(D^2 + 2)y \equiv \left(\dfrac{d}{dt} - 1\right)\left(\dfrac{d^2y}{dt^2} + 2y\right) \equiv \dfrac{d}{dt}\left(\dfrac{d^2y}{dt^2} + 2y\right) - \left(\dfrac{d^2y}{dt^2} + 2y\right)$

$$\equiv \dfrac{d^3y}{dt^3} - \dfrac{d^2y}{dt^2} + 2\dfrac{dy}{dt} - 2y$$

Thus, the use of the operator notation is essentially a short cut.

To show its use in another example, which we carry out in detail, let us consider

ILLUSTRATIVE EXAMPLE 2

Solve the system (a) $\dfrac{dx}{dt} - 3x - 6y = t^2$, (b) $\dfrac{dy}{dt} + \dfrac{dx}{dt} - 3y = e^t$.

Solution Write the equations in operator form as

(c) $(D - 3)x - 6y = t^2$, (d) $Dx + (D - 3)y = e^t$

"Multiply" (c) by D, (d) by $(D - 3)$, to find

$$D(D - 3)x - 6Dy = Dt^2 = \dfrac{d}{dt}(t^2) = 2t$$

$$(D - 3)Dx + (D - 3)^2y = (D - 3)e^t = De^t - 3e^t = -2e^t$$

Hence, by subtraction, we have after simplifying, $(D^2 + 9)y = -2e^t - 2t$ whose general solution is

$$y = c_1 \cos 3t + c_2 \sin 3t - \tfrac{1}{5}e^t - \frac{2t}{9} \qquad (24)$$

Two alternatives are possible: we may substitute y into either of the given equations or we may eliminate y between the two given equations to produce an equation involving x alone. We choose the latter. "Multiplying" equation (c) by $(D - 3)$ and equation (d) by 6 and subtracting, we find

$$(D^2 + 9)x = 6e^t - 3t^2 + 2t$$

or $$x = c_3 \cos 3t + c_4 \sin 3t + \tfrac{3}{5}e^t - \tfrac{1}{3}t^2 + \tfrac{2}{9}t + \tfrac{2}{27} \qquad (25)$$

We must now check to see whether we have too many arbitrary constants (usually we do). Substituting x and y from (24) and (25) into either of the given equations, say (c), we find

$$(\sin 3t)(-3c_3 - 3c_4 - 6c_2) + (\cos 3t)(3c_4 - 3c_3 - 6c_1) + t^2 = t^2$$

This must be an identity, and so we must have

$$-3c_3 - 3c_4 - 6c_2 = 0, \qquad 3c_4 - 3c_3 - 6c_1 = 0 \qquad (26)$$

Equations (26) yield $\quad c_1 = \tfrac{1}{2}c_4 - \tfrac{1}{2}c_3, \qquad c_2 = -\tfrac{1}{2}c_3 - \tfrac{1}{2}c_4$

Using these in (24) we have as the required simultaneous solution,

$$\left.\begin{array}{l} x = c_3 \cos 3t + c_4 \sin 3t + \tfrac{3}{5}e^t - \tfrac{1}{3}t^2 + \tfrac{2}{9}t + \tfrac{2}{27} \\[2mm] y = (\tfrac{1}{2}c_4 - \tfrac{1}{2}c_3) \cos 3t + (-\tfrac{1}{2}c_3 - \tfrac{1}{2}c_4) \sin 3t - \tfrac{1}{5}e^t - \dfrac{2t}{9} \end{array}\right\} \qquad (27)$$

These will also check in the other given equation as is easily verified.

The student should not find it difficult to realize that the operator method is applicable for a system of differential equations which can be written in the form

$$\left.\begin{array}{l} \phi_1(D)x + \psi_1(D)y = F_1(t) \\ \phi_2(D)x + \psi_2(D)y = F_2(t) \end{array}\right\} \qquad (28)$$

where $\phi_1(D), \psi_1(D), \phi_2(D), \psi_2(D)$ are linear operators in $D \equiv d/dt$.* Because the system (28) is a logical generalization of the linear differential equation

$$\phi(D)y = F(t) \qquad (29)$$

considered in previous chapters, it is natural to call (28) a *linear system of differential equations*, while a system of two equations in x and y which cannot be written in

* We shall assume unless otherwise stated that the operators are polynomials in D with constant coefficients. Otherwise the operators do not commute in general, i.e., $\phi_1(D)\phi_2(D) \neq \phi_2(D)\phi_1(D)$, for example, and so the method of elimination using operators given on page 437 will not apply in general (see C Exercise 1, page 172). However, it may still be possible to employ other methods of elimination.

the form (28) is a *non-linear system.* Of course, generalizations of these ideas to more than two dependent variables are easily made.

Example 1. The system of differential equations $\dfrac{dx}{dt} + 3\dfrac{dy}{dt} + y = e^t$, $\dfrac{dy}{dt} = x + y$

is a linear system since it can be written as

$$\left. \begin{array}{l} Dx + (3D + 1)y = e^t \\ -x + (D - 1)y = 0 \end{array} \right\} \tag{30}$$

which has the form (28).

Example 2. The system of equations

$$3\left(\frac{dx}{dt}\right)^2 + 4\frac{dy}{dt} = t^2 + 1, \qquad \frac{dx}{dt} - 2\frac{dy}{dt} = e^{-3t} \tag{31}$$

is a *non-linear system* since not both of these have the form (28).

Remark 1. Given a linear system of differential equations, a question arises as to the total number of arbitrary constants which should show up in the simultaneous solution. This is of importance since the number should match up with the number of conditions available in the problem from which the system is formulated. We can show that the *total* number of arbitrary constants which are to be present in the simultaneous solution of the system (28) is the same as the degree of the polynomial operator obtained from the determinant of the coefficients of x and y in (28), i.e.,

$$\begin{vmatrix} \phi_1(D) & \psi_1(D) \\ \phi_2(D) & \psi_2(D) \end{vmatrix} = \phi_1(D)\psi_2(D) - \psi_1(D)\phi_2(D) \tag{32}$$

This can be generalized to the case of n linear equations in n unknowns, in which case (32) is repaced by an nth-order determinant.

Example 3. In Illustrative Example 2, the determinant is

$$\begin{vmatrix} D - 3 & -6 \\ D & D - 3 \end{vmatrix} = (D - 3)(D - 3) - (-6)(D) = D^2 + 9$$

which is of second degree, and so the total number of arbitrary constants should be two, as we have, in fact, found.

Remark 2. There are various other methods available for solving linear systems of differential equations besides the method of elimination. While some of these may have advantages in particular situations and may perhaps cut down the labor involved, the student will find as a general policy that the above method of elimination is adequate for most purposes arising in practice. However, in the interest of presenting these other possible methods, we indicate them as follows.

(a) *Short-cut operator methods.* These methods are identical with those used on page 204 for single linear differential equations. We shall give an example of the procedure involved on page 440.

(b) *Laplace transform methods.* These methods are the same as those used in Chapter Six. We shall give some examples at the end of this chapter (page 489). As before, an important advantage of the method of Laplace transforms is that initial conditions are automatically taken into account, and as a result the labor involved in the method of elimination may be reduced.

(c) *Method of complementary and particular solutions.* The standard method used in Chapter Four for linear differential equations of first finding the complementary solution, then finding a particular solution, and finally adding these to obtain the general solution can be generalized to systems of linear differential equations. Examples of the procedure involved are given on page 491.

(d) *Methods using matrices.* These methods are especially useful when we have systems of three or more linear differential equation where the method of elimination may become tedious. For those unfamiliar with matrices, it is of course necessary to "go off on a tangent" and build up the theory before we can see its application. In essence, matrix methods utilize ideas analogous to those given in method (c) above, where we have not used matrices. For the benefit of those who have the time and inclination to pursue such study, we have included a short chapter (Chapter Eleven, which summarizes the basic concepts needed. It should be emphasized, however, that this chapter is optional and may be omitted, although it should also be noted that the ideas involving matrices are both interesting and important from a theoretical as well as practical viewpoint. Of especial interest and importance from the theoretical viewpoint is the connection between the theory using matrices and the concepts of orthogonal functions and Sturm–Liouville problems of Chapter Eight. However, as we have already remarked, both chapters are optional and may be omitted in a short course.

1.4 SHORT-CUT OPERATOR METHODS

Methods analogous to those given on page 204 can also be used in solving linear systems. In such cases use of determinants serves to minimize the labor involved. To illustrate the procedure, let us consider the following

ILLUSTRATIVE EXAMPLE 3

Work Illustrative Example 2, page 437, using short-cut operator methods.

Solution   Writing the system in operator form, we have

$$(D - 3)x - 6y = t^2$$

$$Dx + (D - 3)y = e^t$$

Then using determinants we have*

$$x = \frac{\begin{vmatrix} t^2 & -6 \\ e^t & D - 3 \end{vmatrix}}{\begin{vmatrix} D - 3 & -6 \\ D & D - 3 \end{vmatrix}}, \qquad y = \frac{\begin{vmatrix} D - 3 & t^2 \\ D & e^t \end{vmatrix}}{\begin{vmatrix} D - 3 & -6 \\ D & D - 3 \end{vmatrix}}$$

* We assume here that Cramer's rules (see Appendix) are applicable to operators.

On expanding these determinants, making sure that in the expansions of the numerator the operator always precedes the function it operates upon, we find

$$x = \frac{(D-3)(t^2) - (-6)(e^t)}{(D-3)(D-3) - (-6)(D)} = \frac{2t - 3t^2 + 6e^t}{D^2 + 9} = \frac{1}{D^2 + 9}(2t - 3t^2) + \frac{1}{D^2 + 9}(6e^t)$$

$$= \frac{1}{9(1 + D^2/9)}(2t - 3t^2) + \frac{1}{(1)^2 + 9}(6e^t)$$

$$= \frac{1}{9}\left(1 - \frac{D^2}{9} + \cdots\right)(2t - 3t^2) + \tfrac{3}{5}e^t = \frac{2t}{9} - \frac{t^2}{3} + \frac{2}{27} + \tfrac{3}{5}e^t$$

$$y = \frac{(D-3)(e^t) - (D)(t^2)}{(D-3)(D-3) - (-6)(D)} = \frac{-2e^t - 2t}{D^2 + 9} = \frac{1}{D^2 + 9}(-2e^t) + \frac{1}{D^2 + 9}(-2t)$$

$$= \frac{1}{(1)^2 + 9}(-2e^t) + \frac{1}{9}\left(1 - \frac{D^2}{9} + \cdots\right)(-2t) = -\tfrac{1}{5}e^t - \frac{2t}{9}$$

As on page 208, the above yield only particular solutions. We must still add the complementary solutions, i.e., the general solutions of $(D^2 + 9)x = 0$ and $(D^2 + 9)y = 0$. Doing this we have

$$x = c_3 \cos 3t + c_4 \sin 3t + \frac{2t}{9} - \frac{t^2}{3} + \frac{2}{27} + \tfrac{3}{5}e^t, \quad y = c_1 \cos 3t + c_2 \sin 3t - \tfrac{1}{5}e^t - \frac{2t}{9}$$

as in (24) and (25), page 438. We must then find the relationships among the constants as before, thus arriving at (27).

2 Solutions of Non-Linear Systems of Ordinary Differential Equations

We have seen in preceding chapters that non-linear differential equations cannot be solved exactly in general, and when they can special techniques are usually required. In view of this it is not surprising that non-linear systems of differential equations also require special techniques if exact solution is possible. There are several methods of attack which can be employed, such as the following.

(a) *Method of elimination.* Eliminate all but one of the dependent variables from the system, as in the case of linear systems. This leads to a single non-linear equation with one dependent and one independent variable. Hopefully, we may be able to solve this equation and thus be led to the solution of the system. A difficulty of this procedure is that, unlike linear systems, the elimination is not always possible even when an exact solution exists.

(b) *Transformation of variables.* If the method of elimination fails, it may be possible to transform variables so as to produce a simpler system. The types of transformations which can be tried are often suggested by the particular form of the system, but ingenuity may also play an important role in making a strategic choice. For an illustration of this see page 452.

(c) *Method of linearization.* If the non-linear system cannot be solved exactly, it may be possible to replace it by a linear system which can serve as a reasonably good approximation. If the system arises from the mathematical formulation of a problem in science or engineering, this usually involves making assumptions about the mathematical model, which can serve as first approximation to the real situation. Thus, for example, if a projectile is launched from the earth's surface, we can as a first approximation assume that the earth is flat. However, this will only be a reasonably good approximation if the projectile does not travel too far. Otherwise, we may have to take into account the fact that the earth is spherical in shape (or more exactly is an oblate spheroid, i.e., slightly flattened at the poles), and that gravitational acceleration is not constant. For more precision we may have to take into account the motion of the earth and even the effects of the sun, moon, and other planets. An illustration of this technique is given in Section 7.

As an illustration of the method of elimination for a non-linear system, let us consider the following

ILLUSTRATIVE EXAMPLE

Solve
$$\frac{dx}{dt} = x^2 + 1, \qquad \frac{dy}{dt} = xy \tag{1}$$

Solution Method 1. We start with the first equation in (1), since it does not contain the dependent variable y. Writing it as

$$\frac{dx}{x^2 + 1} = dt$$

we have on integrating, $\tan^{-1} x = t + c_1$ or $x = \tan(t + c_1)$. Then the second equation in (1) becomes

$$\frac{dy}{dt} = y \tan(t + c_1) \quad \text{or} \quad \frac{dy}{y} = \tan(t + c_1)dt$$

and integration yields $\quad \ln y = \ln \sec(t + c_1) + \ln c_2 \quad$ or $\quad y = c_2 \sec(t + c_1)$
Thus the required solution is given by
$$x = \tan(t + c_1), \qquad y = c_2 \sec(t + c_1) \tag{2}$$

Method 2. Eliminate t from the given equations (1) by dividing their corresponding sides to obtain

$$\frac{dy/dt}{dx/dt} = \frac{xy}{x^2 + 1} \quad \text{or} \quad \frac{dy}{dx} = \frac{xy}{x^2 + 1}$$

which can be written
$$\frac{dy}{y} = \frac{x \, dx}{x^2 + 1}$$

Integration gives $\quad \ln y = \frac{1}{2} \ln(x^2 + 1) + \ln c \quad$ or $\quad y = c\sqrt{x^2 + 1} \tag{3}$

From the first equation of (1) we obtain, as in Method 1,

$$x = \tan(t + c_1) \tag{4}$$

Using (4) in (3) yields $\qquad y = c \sec(t + c_1)$

so that the same solution arrived at in Method 1 is obtained.

Remark. Systems of equations are sometimes written in other ways. For example, the system (1) of the above example could equivalently be written as

$$\frac{dx}{x^2 + 1} = \frac{dy}{xy} = dt$$

3 Differential Equations Expressed as First-Order Systems

We have seen from the above examples that the method of elimination applied to a system of differential equations results in a single differential equation. A question which arises naturally is whether the reverse is also true, i.e., can any differential equation be expressed as a system of differential equations? Not only can we show that the answer to this is "yes," but we can demonstrate even more as indicated in the following

Theorem. Any ordinary differential equation or system of ordinary differential equations can be expressed as a system of *first-order* differential equations.

This theorem, which also has some important theoretical value, is easy to prove. To do so, we note that any *n*th-order ordinary differential equation, linear or nonlinear, can be written in the form

$$y^{(n)} = g(t, y, y', \ldots, y^{(n-1)}) \tag{1}$$

where derivatives are with respect to t, provided that we can solve for the highest-order derivative in terms of the remaining ones. If we now make the following definitions

$$y' = u_1, \ y'' = u_1' = u_2, \ y''' = u_2' = u_3, \ldots, \ y^{(n-1)} = u_{n-2}' = u_{n-1} \tag{2}$$

equation (1) can be written

$$\frac{dy}{dt} = u_1, \quad \frac{du_1}{dt} = u_2, \ldots \quad \frac{du_{n-2}}{dt} = u_{n-1}, \quad \frac{du_{n-1}}{dt} = g(t, y, u_1, u_2, \ldots, u_{n-1}) \tag{3}$$

which is a system of n first-order equations in the n dependent variables $y, u_1, u_2, \ldots, u_{n-1}$. This proves the theorem for an ordinary differential equation. A proof for systems of ordinary differential equations is similar and is left as an exercise (C Exercise 4, page 446).

Let us illustrate the theorem by some examples.

ILLUSTRATIVE EXAMPLE 1

Express the differential equation

$$\frac{d^4x}{dt^4} - 2\frac{d^3x}{dt^3} + 5\frac{d^2x}{dt^2} + 3\frac{dx}{dt} - 8x = 6\sin 4t$$

as a system of first-order differential equations.

Solution   Let
$$\frac{dx}{dt} = u_1, \qquad \frac{d^2x}{dt^2} = \frac{du_1}{dt} = u_2, \qquad \frac{d^3x}{dt^3} = \frac{du_2}{dt} = u_3$$

Then the given differential equation can be written

$$\frac{du_3}{dt} - 2u_3 + 5u_2 + 3u_1 - 8x = 6 \sin 4t$$

Thus the given equation is equivalent to the first-order system

$$\frac{dx}{dt} = u_1, \qquad \frac{du_1}{dt} = u_2, \qquad \frac{du_2}{dt} = u_3, \qquad \frac{du_3}{dt} = 2u_3 - 5u_2 - 3u_1 + 8x + 6 \sin 4t$$

with dependent variables x, u_1, u_2, u_3.

ILLUSTRATIVE EXAMPLE 2

Express the system of differential equations in Illustrative Example 1, page 436, as a system of first-order differential equations.

Solution   Let
$$\frac{dx}{dt} = u_1, \qquad \frac{d^2x}{dt^2} = \frac{du_1}{dt}, \qquad \frac{dy}{dt} = u_2, \qquad \frac{d^2y}{dt^2} = \frac{du_2}{dt}$$

Then the given system of equations can be written

$$\frac{du_1}{dt} + u_2 - 3x = y, \qquad \frac{du_2}{dt} + u_1 + 4x + 2y = 0$$

which is equivalent to the first-order system

$$\frac{dx}{dt} = u_1, \qquad \frac{dy}{dt} = u_2, \qquad \frac{du_1}{dt} = 3x + y - u_2, \qquad \frac{du_2}{dt} = -4x - 2y - u_1$$

A EXERCISES

1. Find the solution of each system subject to any given conditions.

(a) $\dfrac{dy}{dt} = x, \dfrac{dx}{dt} = -y; y(0) = 0, x(0) = 1.$ (b) $\dfrac{du}{dx} = 2v - 1, \dfrac{dv}{dx} = 1 + 2u.$

(c) $\dfrac{dx}{dt} = x + y, \dfrac{dy}{dt} = x - y.$ (d) $\dfrac{d^2x}{dt^2} = -x, \dfrac{d^2y}{dt^2} = y.$

(e) $\dfrac{d^2y}{dt^2} = x - 2, \dfrac{d^2x}{dt^2} = y + 2.$

(f) $\dfrac{dy}{dt} + 6y = \dfrac{dx}{dt}, 3x - \dfrac{dx}{dt} = 2\dfrac{dy}{dt}; x = 2, y = 3$ at $t = 0.$

(g) $\begin{cases} (D + 1)x + 2y = 1, \\ 2x + (D - 2)y = t. \end{cases}$ (h) $\begin{cases} (D + 2)x + (D - 1)y = -\sin t, \\ (D - 3)x + (D + 2)y = 4 \cos t. \end{cases}$

2. Give a physical interpretation to (a) Exercise 1(a), (b) Exercise 1(e).

3. Solve the system of differential equations

$$\frac{d^2x}{dt^2} + 2\frac{dy}{dt} + 8x = 32t, \qquad \frac{d^2y}{dt^2} + 3\frac{dx}{dt} - 2y = 60e^{-t}$$

subject to the conditions $x(0) = 6$, $x'(0) = 8$, $y(0) = -24$, $y'(0) = 0$, and give a physical interpretation.

4. State which of the following systems are linear and which are non-linear.

(a) $\dfrac{dx}{dt} - 2\dfrac{dy}{dt} = e^t, \ 3\dfrac{dx}{dt} + \dfrac{dy}{dt} = \sqrt{t}.$

(b) $\dfrac{dx}{dt} + 3\dfrac{dy}{dt} = xy, \ 3\dfrac{dx}{dt} - \dfrac{dy}{dt} = \sin t.$

(c) $\dfrac{d^2r}{dt^2} = r + \phi, \ \dfrac{d^2\phi}{dt^2} = 5r - 3\phi + t^2.$

(d) $x\dfrac{dy}{dt} + y\dfrac{dx}{dt} = t^2, \ 2\dfrac{d^2x}{dt^2} - \dfrac{dy}{dt} = 5t.$

5. Write each of the following as a system of first-order differential equations.

(a) $\dfrac{d^2x}{dt^2} + 3\dfrac{dx}{dt} - 4x = 5 \sin 2t.$

(b) $(y'')^2 - (\sin x)y' = y \cos x.$

(c) $x\dfrac{d^2y}{dt^2} - y = 4t, \ 2\dfrac{d^2x}{dt^2} + \left(\dfrac{dy}{dt}\right)^2 = x.$

(d) $x^2y'' - 3xy' + 4y = 5 \ln x.$

6. Solve $\begin{cases} x'' + y' + x = y + \sin t \\ y'' + x' - y = 2t^2 - x \end{cases}$ subject to $x(0) = 2$, $x'(0) = -1$, $y(0) = -\frac{9}{2}$, $y'(0) = -\frac{7}{2}$.

7. Express as a system of first-order differential equations the system in (a) Exercise 1(d), (b) Exercise 3, (c) Exercise 6.

B EXERCISES

1. Solve $\dfrac{dx}{dt} = x + y, \ \dfrac{dy}{dt} = x - y, \ \dfrac{dz}{dt} = 2y$ if $x(0) = 2, y(0) = 0, z(0) = 0.$

2. Solve (a) $\dfrac{dx}{yz} = \dfrac{dy}{xz} = \dfrac{dz}{xy}.$

(b) $\dfrac{dx}{xy} = \dfrac{dy}{y^2 + 1} = \dfrac{dz}{z}.$

Are these linear or non-linear? Explain.

3. Solve the system $\begin{cases} x^2y'' + xz' + z = x \\ xy' + z = \ln x \end{cases}$ by first making the transformation $x = e^t$. Compare with the Euler equation, page 211.

4. Solve the system $\dfrac{dx}{dt} = y, \quad \dfrac{dy}{dt} = z, \quad \dfrac{dz}{dt} = x.$

C EXERCISES

1. The probability $P_n(t)$ of exactly n nuclear particles (such as appear in cosmic rays) registering in a counter (such as a Geiger counter) in a time t is determined from the system of differential equations,

$$P_n'(t) = \lambda[P_{n-1}(t) - P_n(t)], \qquad n \neq 0; \ P_0'(t) = -\lambda P_0(t)$$

where λ is a positive constant. Using suitable conditions, solve for $P_n(t)$, $n = 0, 1, 2, \ldots.$

From your results, show that

$$\sum_{n=0}^{\infty} P_n(t) = 1$$

What is the probability interpretation of this?

2. Solve the system $\dfrac{dx}{y+z} = \dfrac{dy}{z+x} = \dfrac{dz}{x+y}$.

3. Prove that the system obtained according to the theorem on page 443 is linear or non-linear according as the original differential equation is linear or non-linear.

4. Prove that every system of differential equations can be written as a system of first-order differential equations.

4 Applications to Mechanics

4.1 THE FLIGHT OF A PROJECTILE

Suppose that a projectile is fired from a cannon which is inclined at an angle θ_0 with the horizontal and which imparts to the projectile a muzzle velocity having magnitude v_0. Assuming no air resistance and a flat stationary earth, let it be required to describe the subsequent flight.

Mathematical Formulation. Let the cannon be located at the origin O of an xy coordinate system (Fig. 10.2). The dashed curve shown indicates the path of the projectile; OV represents the muzzle velocity, a vector having magnitude v_0 and a direction in the xy plane making an angle θ_0 with the positive x axis. The components of the velocity in the x and y directions have magnitudes $v_0 \cos \theta_0$ and $v_0 \sin \theta_0$, respectively. Since there is no force of air resistance, the only force acting on the projectile of mass m is its weight mg. Let us take "up" and "right" as the positive directions. According to Newton's law we have:

net force in x direction = mass times acceleration in x direction

net force in y direction = mass times acceleration in y direction

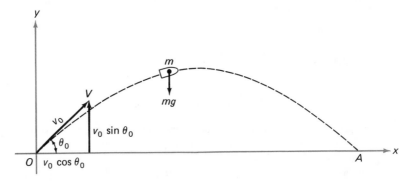

Figure 10.2

which we may write as $F_x = ma_x$, $F_y = ma_y$. Since the net force in the x direction is zero and $a_x = d^2x/dt^2$, we have

$$m\frac{d^2x}{dt^2} = 0 \quad \text{or} \quad \frac{d^2x}{dt^2} = 0 \tag{1}$$

Since the net force in the y direction is $-mg$ (because "down" is the negative direction) and since $a_y = d^2y/dt^2$, we have

$$m\frac{d^2y}{dt^2} = -mg \quad \text{or} \quad \frac{d^2y}{dt^2} = -g \tag{2}$$

Furthermore, from the conditions of the problem we have

$$x = 0, \quad y = 0, \quad \frac{dx}{dt} = v_0 \cos\theta_0, \quad \frac{dy}{dt} = v_0 \sin\theta_0, \quad \text{at} \quad t = 0 \tag{3}$$

Our complete mathematical formulation consists of the differential equations (1) and (2) subject to the conditions (3). From the differential equations it is seen that the motion does not depend on m, and hence on the size of the projectile, provided there is no air resistance.

Solution Upon integration of (1) we have $dx/dt = c_1$. Applying the condition that $dx/dt = v_0 \cos\theta_0$ at $t = 0$, we see that $c_1 = v_0 \cos\theta_0$, i.e., $dx/dt = (v_0 \cos\theta_0)$.* By another integration $x = (v_0 \cos\theta_0)t + c_2$, and since $x = 0$ at $t = 0$, $c_2 = 0$ and we have

$$x = (v_0 \cos\theta_0)t \tag{4}$$

In a similar manner, we have upon integration of (2) $\quad \dfrac{dy}{dt} = -gt + c_3$

and since $dy/dt = v_0 \sin\theta_0$ at $t = 0$, $c_3 = v_0 \sin\theta_0$, and we find

$$\frac{dy}{dt} = -gt + v_0 \sin\theta_0 \tag{5}$$

By another integration, using the fact that $y = 0$ at $t = 0$, we have

$$y = (v_0 \sin\theta_0)t - \tfrac{1}{2}gt^2 \tag{6}$$

The required solution is $\quad x = (v_0\cos\theta_0)t, \; y = (v_0 \sin\theta_0)t - \tfrac{1}{2}gt^2 \tag{7}$

These equations give the position (x, y) of the projectile at any time t after firing. From them we can discuss anything concerning the motion. For example, suppose we choose to ask the following questions:

1. What is the total time of flight from O to A (Fig. 10.2)?
2. What is the range (distance OA along x axis)?
3. What is the maximum height reached?
4. What type of curve does the projectile describe?

* Thus, the horizontal component of velocity remains constant.

Question 1 will be answered if we find the values of t which make $y = 0$. From (6) we see that this is so when

$$t[(v_0 \sin \theta_0) - \tfrac{1}{2}gt] = 0 \quad \text{or} \quad t = 0; \qquad t = \frac{2v_0 \sin \theta_0}{g}$$

The second value of t gives the time when the projectile is at A. Hence

$$\text{time of flight} = \frac{2v_0 \sin \theta_0}{g} \tag{8}$$

To answer Question 2 we calculate the value of x when $t = $ time of flight. From the first of equations (7) we therefore have

$$\text{range} = \frac{(v_0 \cos \theta_0)(2v_0 \sin \theta_0)}{g} = \frac{v_0^2 \sin 2\theta_0}{g} \tag{9}$$

From (9) it is clear that the range is greatest when $2\theta_0 = 90°$, i.e., $\theta_0 = 45°$ and the maximum range is v_0^2/g.

To answer Question 3 we must find when y is a maximum, i.e., when $dy/dt = 0$. This amounts to saying that at the highest point the velocity in the y direction is zero. From the equation (5) we have

$$\frac{dy}{dt} = v_0 \sin \theta_0 - gt = 0 \qquad \text{where} \qquad t = \frac{v_0 \sin \theta_0}{g}$$

(Note that this is half the total time of flight, which seems logical.) Placing this value of t in the equation (6) for y, we find

$$\text{maximum height} = \frac{v_0^2 \sin^2 \theta_0}{2g} \tag{10}$$

Question 4 is actually answered by (7), which represents the parametric equations of a parabola. The path of the projectile is thus a portion of a parabola. By eliminating the parameter t we find

$$y = x \tan \theta_0 - \frac{gx^2}{2v_0^2} \sec^2 \theta_0 \tag{11}$$

which is another form for the parabola.

A EXERCISES

1. A projectile is fired from a cannon which makes an angle of 60° with the horizontal. If the muzzle velocity is 160 ft/sec: (a) Write a system of differential equations and conditions for the motion. (b) Find the position of the projectile at any time. (c) Find the range, maximum height, and time of flight. (d) Determine the position and velocity of the projectile after it is in flight for 2 and 4 sec.

2. Determine what would have been the maximum range of the "Big Bertha" cannon of World War I which had a muzzle velocity of 1 mile per second, had air resistance been negligible. For this maximum range, what is the height reached and the total time of flight?

3. A stone is thrown horizontally from a cliff 256 ft high with a velocity of 50 ft/sec. (a) Find the time of flight. (b) At what distance from the base of the cliff will the stone land?

B EXERCISES

1. A projectile fired from a cannon located on a horizontal plane has a range of 2000 ft and achieves a maximum height of 1000 ft. Determine its muzzle velocity and the time of flight.

2. A cannon having muzzle velocity v_0 and making an angle θ with the horizontal is to be fired so as to hit an object located on the same level as the cannon and at a distance d from it. Show that this is possible for two values of θ given by $\theta = \pi/4 \pm \phi$, where

$$\phi = \frac{1}{2}\cos^{-1}\frac{gd}{v_0^2}, \qquad d < \frac{v_0^2}{g}$$

Discuss the cases $d \geq v_0^2/g$.

3. A plane is inclined at an angle α with the horizontal. A cannon on this plane makes an angle θ with the horizontal. Show that the range of the projectile on the inclined plane is

$$R = \frac{2v_0^2 \cos\theta \sin(\theta - \alpha)}{g\cos^2\alpha}$$

where v_0 is the muzzle velocity of the cannon. Hence show that the maximum range is obtained when $\theta = \alpha/2 + \pi/4$ and that its value is

$$R_{max} = \frac{v_0^2}{g(1 + \sin\alpha)}$$

Discuss the case $\alpha = 0°, 90°$.

C EXERCISES

1. A gun is located at a horizontal distance R from the base of a cliff of height H. The gun, which has muzzle velocity v_0, is to be inclined at an angle θ with the horizontal and aimed so as to hit a target at the top of the cliff. Show that this will be possible for two values of θ given by

$$\theta = \frac{\pi}{4} + \frac{1}{2}\tan^{-1}\frac{H}{R} \pm \frac{1}{2}\cos^{-1}\frac{H + gR^2/v_0^2}{\sqrt{R^2 + H^2}} \quad \text{if} \quad \frac{H + gR^2/v_0^2}{\sqrt{R^2 + H^2}} \leq 1$$

2. If in a given vertical plane projectiles are fired from a gun having muzzle velocity v_0 and located at O (Fig. 10.3), their paths will be parabolas as shown, one parabola corresponding to each angle of elevation (or depression) of the gun. (a) Determine the envelope of these parabolic trajectories, showing that this envelope is also a parabola. (b) If point O is at height H above a horizontal plane and the gun can be aimed in all possible directions, show that the volume over which the gun has "control" is

$$\frac{\pi v^2}{g}\left(H + \frac{v^2}{2g}\right)^2$$

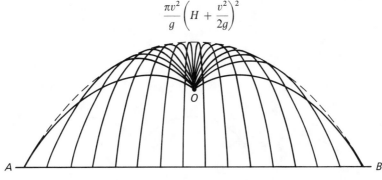

Figure 10.3

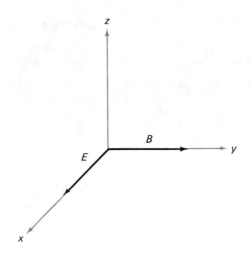

Figure 10.4

3. Discuss the motion of a projectile in which air resistance proportional to the instantaneous velocity is taken into account.

4. An electron of mass m and electric charge q moves in an electromagnetic field in which the electric and magnetic fields have constant magnitudes E and B, respectively, and directions as indicated in Fig. 10.4. The system of differential equations describing the motion is given by

$$m\frac{d^2x}{dt^2} = qB\frac{dy}{dt}, \qquad m\frac{d^2y}{dt^2} = -qB\frac{dx}{dt} + qE, \qquad m\frac{d^2z}{dt^2} = 0$$

At time $t = 0$ the electron is at rest at the origin. (a) Show that for $t > 0$ the electron moves in a path given by the *cycloid* located in the xy plane whose parametric equations are

$$x = \frac{Em}{qB^2}\left(1 - \cos\frac{qBt}{m}\right), \qquad y = \frac{Em}{qB^2}\left(\frac{qBt}{m} - \sin\frac{qBt}{m}\right), \qquad z = 0$$

(b) Show that this cycloidal path is described by the motion of a fixed point on a circle C of radius Em/qB^2 which rolls along the x axis at a speed E/B, as indicated in Fig. 10.5. The results were used by J. J. Thomson in 1897 to determine the ratio q/m of the charge on an electron to its mass.

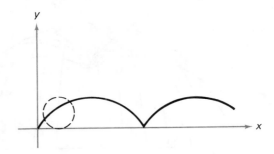

Figure 10.5

5. Describe the motion of the electron in Exercise 4 if it starts from the origin with an initial speed v_0 in the positive z direction.

6. If the electron of Exercise 5 has the components of its initial velocity in the x and y directions different from zero and zero in the z direction, show that the path is a *trochoid* described by the motion of a point on the spoke of a wheel which rolls along the x axis.

4.2 AN APPLICATION TO ASTRONOMY

According to Newton's famous universal law of gravitation, any two objects separated by a distance r and having masses M_1 and M_2, respectively, are attracted toward each other with a force having magnitude given by

$$F = \frac{GM_1 M_2}{r^2} \tag{12}$$

where G is a universal gravitation constant. It is interesting to make use of this law to describe the motion of planets in our solar system. We shall consider, in particular, the motion of the earth about the sun. In discussing this problem, we simplify our tasks tremendously by neglecting the effects of all other planets. Consequently the results are approximate but, nevertheless, they do represent to a high degree of accuracy, the true state of affairs as evidenced by experimental observations.

Mathematical Formulation. We assume the sun fixed at the origin of an xy coordinate system and that the earth is at point (x, y) at time t in its motion (Fig. 10.6). We take as positive directions of vector quantities the $+x$ and $+y$ directions. From the figure, the force F acting on the earth is seen to have x and y components of magnitudes $F \cos \phi$ and $F \sin \phi$, respectively. Letting m_s and m_e be the respective masses of sun and earth, we have, making use of (12),

$$m_e \frac{d^2 x}{dt^2} = -F \cos \phi = -\frac{Gm_e m_s}{r^2} \cos \phi \tag{13}$$

$$m_e \frac{d^2 y}{dt^2} = -F \sin \phi = -\frac{Gm_e m_s}{r^2} \sin \phi \tag{14}$$

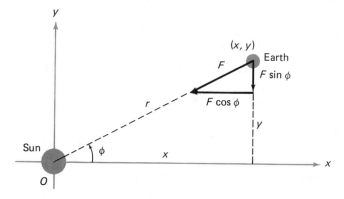

Figure 10.6

Since $\sin \phi = y/r$ and $\cos \phi = x/r$, equations (13) and (14) become

$$\frac{d^2x}{dt^2} = -\frac{kx}{r^3}, \qquad \frac{d^2y}{dt^2} = -\frac{ky}{r^3} \tag{15}$$

where $k = Gm_s$. Since $r = \sqrt{x^2 + y^2}$, equations (15) can be written

$$\frac{d^2x}{dt^2} = -\frac{kx}{(x^2 + y^2)^{3/2}}, \qquad \frac{d^2y}{dt^2} = -\frac{ky}{(x^2 + y^2)^{3/2}} \tag{16}$$

As initial conditions we assume that at $t = 0$, the earth is located on the x axis, a distance a from the sun, and is proceeding in the positive y direction with velocity v_0.

Thus

$$x = a, \qquad y = 0, \qquad \frac{dx}{dt} = 0, \qquad \frac{dy}{dt} = v_0 \quad \text{at} \quad t = 0 \tag{17}$$

If we can solve simultaneously equations (16) subject to conditions (17), we shall have the solution to our problem.

Solution   The system of differential equations (16) is a non-linear system, and a little experimentation with them soon reveals that it is difficult if not impossible to eliminate x or y. Upon noticing the presence of the combination $x^2 + y^2$, however, we may be led to consider a change of variables to polar coordinates. This is further evidenced by the realization that the position of the earth relative to the sun is perhaps better described by coordinates (r, ϕ) than by (x, y). Let us therefore transform equations (16) to polar coordinates.

Since the equations which provide the transformation from rectangular coordinates (x, y) to polar coordinates (r, ϕ) are given by

$$x = r \cos \phi, \qquad y = r \sin \phi$$

we have, on letting dots denote differentiations with respect to time t,

$$\dot{x} = \dot{r} \cos \phi - (r \sin \phi)\dot{\phi}$$
$$\ddot{x} = \ddot{r} \cos \phi - 2(\dot{r} \sin \phi)\dot{\phi} - (r \sin \phi)\ddot{\phi} - (r \cos \phi)\dot{\phi}^2$$
$$\dot{y} = \dot{r} \sin \phi + (r \cos \phi)\dot{\phi}$$
$$\ddot{y} = \ddot{r} \sin \phi + 2(\dot{r} \cos \phi)\dot{\phi} + (r \cos \phi)\ddot{\phi} - (r \sin \phi)\dot{\phi}^2$$

Thus,

$$\left. \begin{array}{l} \ddot{x} = (\ddot{r} - r\dot{\phi}^2) \cos \phi - (2\dot{r}\dot{\phi} + r\ddot{\phi}) \sin \phi \\ \ddot{y} = (\ddot{r} - r\dot{\phi}^2) \sin \phi + (2\dot{r}\dot{\phi} + r\ddot{\phi}) \cos \phi \end{array} \right\} \tag{18}$$

Equations (16) become, upon making use of equations (18),

$$(\ddot{r} - r\dot{\phi}^2) \cos \phi - (2\dot{r}\dot{\phi} + r\ddot{\phi}) \sin \phi = -\frac{k \cos \phi}{r^2} \tag{19}$$

$$(\ddot{r} - r\dot{\phi}^2) \sin \phi + (2\dot{r}\dot{\phi} + r\ddot{\phi}) \cos \phi = -\frac{k \sin \phi}{r^2} \tag{20}$$

Multiplying equation (19) by $\cos \phi$, equation (20) by $\sin \phi$, and adding,

$$\ddot{r} - r\dot{\phi}^2 = -\frac{k}{r^2} \tag{21}$$

Also multiplying (19) by sin ϕ, (20) by cos ϕ, and subtracting,

$$2\dot{r}\dot{\phi} + r\ddot{\phi} = 0 \tag{22}$$

The initial conditions (17) in rectangular form need to be replaced by corresponding conditions in polar form. Corresponding to the conditions (17), we have

$$r = a, \quad \phi = 0, \quad \dot{r} = 0, \quad \dot{\phi} = \frac{v_0}{a} \quad \text{at} \quad t = 0 \tag{23}$$

We must now solve (21) and (22) simultaneously subject to conditions (23). A simplification results if we notice that the left side of (22) is $\frac{1}{r} \cdot \frac{d}{dt}(r^2\dot{\phi})$ as is easily verified. Thus (22) can be replaced by

$$\frac{d}{dt}(r^2\dot{\phi}) = 0 \quad \text{or} \quad r^2\dot{\phi} = c_1 \tag{24}$$

This equation has an interesting interpretation. Suppose an object moves from P to Q along arc PQ (Fig. 10.7). Let A be the area bounded by lines OP, OQ, and arc PQ. Then from elementary calculus,

$$A = \frac{1}{2} \int_0^\phi r^2 \, d\phi \quad \text{i.e.,} \quad dA = \tfrac{1}{2}r^2 \, d\phi \quad \text{or} \quad \frac{dA}{dt} = \tfrac{1}{2}r^2 \frac{d\phi}{dt} = \tfrac{1}{2}r^2\dot{\phi}$$

The quantity dA/dt is called the *areal velocity*. Since $r^2\dot{\phi}$ is constant by (24), it follows that the areal velocity is constant. This is equivalent to saying that the object moves so that equal areas are described in equal times, or that the radius vector (line drawn from O to object) "sweeps" out equal areas in equal times. This *law of areas* is the first of Kepler's famous three laws. These laws were based on deductions from the voluminous observations of the astronomer Tycho Brahe who spent many years in the compilation of data. We now list

KEPLER'S LAWS

1. *Each of the planets moves in a plane curve about the sun, in such a way that the radius vector drawn from the sun to the planets describes equal areas in equal times.*

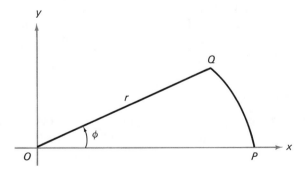

Figure 10.7

2. *The paths of the planets are ellipses, with the sun at one of the foci.*

3. *The squares of the periods (times for complete revolution about the sun) vary directly as the cubes of the major axes of the ellipses.*

It was essentially because of the work of Kepler in organizing the data of Tycho Brahe that Newton was able, with the aid of the calculus, to formulate his famous universal law of gravitation which is applied, not only to planets, but to all objects. In this section we have not used the historical approach but have instead started with Newton's universal law of gravitation, and deduced the first of Kepler's laws. In this section we will also deduce the second law. The third law is relegated to the B Exercises for the interested student. In addition, in the C Exercise 5, the manner in which Newton was able rigorously to deduce his universal law of gravitation from Kepler's laws is indicated.

We now return to the solution of (21) and (22) subject to (23). From (23) we see that $r = a$, $\phi = v_0/a$ at $t = 0$. Hence, from (24) $c_1 = av_0$. Thus, we may write $r^2 \dot{\phi} = av_0$. From this, $\dot{\phi} = av_0/r^2$ and (21) becomes

$$\ddot{r} = \frac{a^2 v_0^2}{r^3} - \frac{k}{r^2} \tag{25}$$

an equation in which ϕ does not appear. Equation (25) does not involve t explicitly. Hence, letting $\dot{r} = p$ the equation may be written

$$\frac{dp}{dt} = \frac{dp}{dr} \cdot \frac{dr}{dt} = p \frac{dp}{dr} = \frac{a^2 v_0^2}{r^3} - \frac{k}{r^2} \quad \text{or} \quad p \frac{dp}{dr} = \frac{a^2 v_0^2}{r^3} - \frac{k}{r^2}$$

Separating the variables and integrating this last equation yields

$$\frac{p^2}{2} = \frac{k}{r} - \frac{a^2 v_0^2}{2r^2} + c_2 \tag{26}$$

From (26) since $p = \dot{r} = 0$ where $r = a$, we have $\qquad c_2 = \frac{v_0^2}{2} - \frac{k}{a}$

Thus, $\dfrac{\dot{r}^2}{2} = \dfrac{k}{r} - \dfrac{a^2 v_0^2}{2r^2} + \dfrac{v_0^2}{2} - \dfrac{k}{a}$ or $\dfrac{dr}{dt} = \pm \sqrt{\left(v_0^2 - \dfrac{2k}{a} \right) + \dfrac{2k}{r} - \dfrac{a^2 v_0^2}{r^2}}$ $\qquad$ (27)

From this we may obtain r as a function of t (see C Exercise 8). Of more interest, perhaps, is a description of the path taken by the earth in its motion. To determine this, we would want an equation containing r and ϕ. From (27) and $\dot{\phi} = av_0/r^2$, we have the simultaneous equations

$$\frac{dr}{dt} = \pm \sqrt{\left(v_0^2 - \frac{2k}{a} \right) + \frac{2k}{r} - \frac{a^2 v_0^2}{r^2}}, \qquad \frac{d\phi}{dt} = \frac{av_0}{r^2} \tag{28}$$

The desired differential equation connecting r and ϕ but not containing t can be obtained by division of the equations in (28). We find

$$\frac{dr}{d\phi} = \pm r \sqrt{Ar^2 + 2Br - 1}, \qquad \text{where } A = \frac{1}{a^2} - \frac{2k}{a^3 v_0^2}, \ B = \frac{k}{a^2 v_0^2} \tag{29}$$

Thus

$$\int \frac{dr}{r\sqrt{Ar^2 + 2Br - 1}} = \pm \int d\phi = \pm \phi + c_3 \tag{30}$$

It is useful to make the substitution $r = 1/u$ in the integral on the left of (30), for we obtain

$$-\int \frac{du}{\sqrt{A + 2Bu - u^2}} = \pm \phi + c_3 \quad \text{or} \quad -\int \frac{du}{\sqrt{A + B^2 - (u - B)^2}} = \pm \phi + c_3$$

i.e.,

$$\cos^{-1} \frac{(u - B)}{\sqrt{A + B^2}} = \pm \phi + c_3$$

Then

$$u = B + \sqrt{A + B^2} \cos (\phi + c_4) = B[1 + \epsilon \cos (\phi + c_4)]$$

where

$$\epsilon = \frac{\sqrt{A + B^2}}{B} = \frac{a^2 v_0^2}{k} \left(\frac{1}{a^2} - \frac{2k}{a^2 v_0^2} + \frac{k^2}{a^4 v_0^2} \right)^{1/2} = \left| \frac{a v_0^2}{k} - 1 \right| \tag{31}$$

Since $u = 1/r$,

$$r = \frac{a^2 v_0^2/k}{1 + \epsilon \cos (\phi + c_4)} \tag{32}$$

There are two cases which can arise, $(a v_0^2/k) - 1 \geq 0$ and $(a v_0^2/k) - 1 < 0$.

Case 1, $(a v_0^2/k) - 1 \geq 0$. In this case

$$\epsilon = \frac{a v_0^2}{k} - 1 \geq 0 \tag{33}$$

and (32) can be written

$$r = \frac{a(1 + \epsilon)}{1 + \epsilon \cos (\phi + c_4)} \tag{34}$$

Since we must have $r = a$, $\phi = 0$ by condition (23), we find on using (34) that $\cos c_4 = 1$, $c_4 = 0$. Thus, (34) becomes

$$r = \frac{a(1 + \epsilon)}{1 + \epsilon \cos \phi} \tag{35}$$

From analytic geometry we know that (35) is the polar form of a conic section having eccentricity ϵ. It can represent four types of curves, as given in Table 10.1.

Table 10.1 Types of Orbits

Type 1	Ellipse	if	$0 < \epsilon < 1$	i.e.,	$k/a < v_0^2 < 2k/a$
Type 2	Circle	if	$\epsilon = 0$	i.e.,	$v_0^2 = k/a$
Type 3	Parabola	if	$\epsilon = 1$	i.e.,	$v_0^2 = 2k/a$
Type 4	Hyperbola	if	$\epsilon > 1$	i.e.,	$v_0^2 > 2k/a$

In our problem of the sun and the earth (or any planet in our solar system) we have $0 < \epsilon < 1$ so that the curve describing the orbit of the earth about the sun is of Type 1, i.e., an ellipse with the sun at one focus, as predicted by Kepler's second law. This ellipse is shown in Fig. 10.8. In this figure, a is the distance of the earth from the sun at its *nearest position P*, often called the *perihelion* (from the Greek

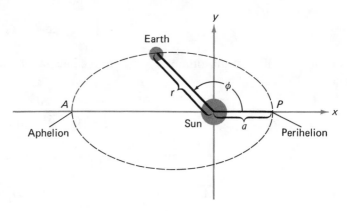

Figure 10.8

peri, meaning *near*, and *helios*, meaning *sun*). Since we have

$$\epsilon = \frac{av_0^2}{k} - 1, \qquad k = Gm_s \tag{36}$$

it follows that the speed at perihelion is given by

$$v_0^2 = \frac{Gm_s(1 + \epsilon)}{a} \tag{37}$$

The distance of the earth from the sun at its *furthest position A*, often called the *aphelion* (from the Greek *apo*, meaning *far from*) can be obtained from (35) by putting $\phi = \pi$. We find

$$\text{greatest distance from sun (aphelion)} = \frac{a(1 + \epsilon)}{1 - \epsilon} \tag{38}$$

Since $r^2\dot{\phi} = av_0$, the speed at aphelion denoted by v_A is given by $r\dot{\phi}$, where $r = a(1 + \epsilon)/(1 - \epsilon)$. This leads to

$$\text{speed at aphelion} = v_A = \frac{v_0(1 - \epsilon)}{1 + \epsilon} \tag{39}$$

This shows, as we might expect, that the speed at aphelion is less than that at perihelion.

Similar remarks can be made for all planets; i.e., they all have elliptical orbits with the sun at one focus, demonstrating Kepler's second law. Thus Fig. 10.8 applies to any other planet in place of the earth. Of course, the orbits all have different eccentricities, perihelions, aphelions, etc., but the above results are still valid.*

* In deriving the above results it was assumed of course that only two bodies, i.e., the sun and planet, are considered and that the effects of all other planets are negligible. While this is the case for most practical purposes, greater precision is achieved if this *two-body problem* is replaced by an *n-body problem*, where $n \geq 3$. Unfortunately, this problem cannot be solved exactly, but approximations are possible. Even with such approximations it was discovered in the late 19th century that certain discrepancies in the orbit of the planet Mercury could not be explained by Newtonian mechanics. However, *Einstein*, using his *theory of relativity*, was able to resolve them.

The following are further examples where the various types of curves given in Table 10.1 can arise.

Example 1. The Earth and the Moon. In this case the orbit of the moon is an ellipse with the earth at one focus. However, since the eccentricity is very nearly equal to zero, the orbit is very nearly circular (type 2).

Example 2. Recurring Comets. In this case a comet may have an orbit in the shape of an elongated ellipse with an eccentricity less than but close to 1. The time for a comet to reappear depends on the eccentricity of this elliptical path. One important example of a recurring comet is *Halley's comet*, which appears near the earth once every 76 years approximately. The last appearance was in 1910, and it is expected to appear next in 1986.

Example 3. Parabolic and Hyperbolic Orbits. We can also have objects with parabolic or hyperbolic orbits (types 3 and 4), which would appear only once, theoretically, never to return, as indicated in Fig. 10.9. Examples are meteors from outer space. Before the understanding of such events, the arrivals of these objects were probably classed as "miracles."

Other examples of parabolic and hyperbolic orbits occur in connection with experiments involving particles of atomic dimensions performed in *cloud chambers* at many of our universities. Such chambers are designed to reveal "fog tracks" of such high-speed atomic particles as they orbit in a parabolic or hyperbolic path around a more massive particle located at the focus. As might be expected, parabolic orbits are rare in practice because any slight disturbance can change the orbit from a parabola to an ellipse or hyperbola.

Case 2, $(av_0^2/k) - 1 < 0.$ In this case $\qquad 0 < \epsilon = 1 - \dfrac{av_0^2}{k} < 1 \qquad$ (40)

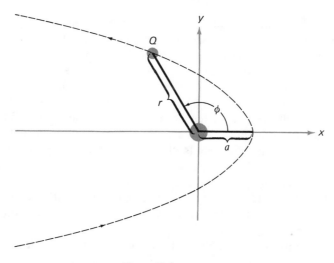

Figure 10.9

and (32) can be written
$$r = \frac{a(1 - \epsilon)}{1 + \epsilon \cos (\phi + c_4)} \tag{41}$$

Using $r = a$, $\phi = 0$ in (41), we find $\cos c_4 = -1$, $c_4 = \pi$, so that
$$r = \frac{a(1 - \epsilon)}{1 - \epsilon \cos \phi} \tag{42}$$

This is also an ellipse of eccentricity ϵ. However, as we see by putting $\phi = \pi$ in (42), the perihelion is at distance $a(1 - \epsilon)/(1 + \epsilon)$ from the focus while the aphelion is at distance a from the focus. Thus the ellipse is the reverse of that shown in Fig. 10.8. Since this case does not provide anything new, only Case 1 need be considered.

It should be noted that Kepler's first law, stated on page 453 for planetary motion in elliptical (or circular) orbits, is true as well for parabolic and hyperbolic orbits. The areal velocity given by $\frac{1}{2}r^2\dot{\phi}$ is thus a constant for any orbital curve of the type given in Table 10.1. This constant is often denoted by $\frac{1}{2}h$, so that we have

$$\text{areal velocity} = \tfrac{1}{2}r^2\dot{\phi} = \tfrac{1}{2}h = \tfrac{1}{2}av_0$$

where the last equality follows from page 454.

In terms of the areal velocity equation, (35) can be written as

$$r = \frac{h^2/k}{1 + \epsilon \cos \phi} \tag{43}$$

4.3 THE MOTION OF SATELLITES AND MISSILES

The above theory is also applicable with slight modification to the motion of *artificial or man-made satellites* launched by means of rockets from the earth's surface. There is in fact a theory which holds that the moon, which can of course be considered a natural satellite of the earth, was formed during early stages in the evolution of the earth when the earth in a molten state "threw off" a portion of itself. The same theory could also explain the origin of the earth and other planets being "thrown off" from the sun to become its satellites.

Artificial satellites have many applications, and as is often the case in science some of these work to the benefit of mankind and others may lead to its destruction. Of beneficial purpose is the use of satellites to predict weather patterns, provide scientific information regarding phenomena such as cosmic rays or solar radiation, and enable radio and television communication. Of potential destructive purpose is the use of satellites for spying and launching of nuclear weapons.

As an illustration of the use of the above theory in connection with artificial satellites, let us suppose that a satellite is launched by a rocket and placed into orbit around the earth. The details of such launching based on the principles of astronautics, although fascinating, will not be discussed here. The orbit of the satellite is, in general, an ellipse of eccentricity very nearly equal to zero so that its orbit is nearly circular. The motion is like that in Fig. 10.8, with the sun at the focus replaced by the earth and the earth replaced by the satellite, except that the ellipse is more like a circle.

The results already derived are easily adapted to yield corresponding results for satellite motion, Thus, for example, in satellite motion around the earth we must replace the mass of the sun m_s by the mass of the earth m_e. Similarly, if we have satellite motion around the moon, we must replace the mass of the sun by the mass of the moon.

As an illustration of the use of the theory already developed for planetary motion to work examples of satellite motion, let us consider the following

ILLUSTRATIVE EXAMPLE

A satellite launched from the earth's surface revolves in a nearly circular orbit at an altitude b above the earth's surface. Determine (a) its orbital speed, and (b) the time for one complete revolution around the earth.

Solution (a) We can adapt the result (37), page 456, to this problem by changing m_s to m_e, thus obtaining

$$v_0^2 = \frac{Gm_e(1 + \epsilon)}{a} \tag{44}$$

Since the orbit is nearly circular, we can take the eccentricity $\epsilon = 0$, and in this case the distance a from the earth's center will be very nearly constant and equal to

$$a = R_e + b \tag{45}$$

where R_e is the radius of the earth. To obtain the quantity Gm_e in (44), we can use the fact that on the earth's surface the force of attraction between earth and satellite is the weight mg of the satellite. Thus, using Newton's law of universal gravitation,

$$F = \frac{Gm_e m}{R_e^2} = mg \quad \text{or} \quad Gm_e = gR_e^2 \tag{46}$$

Using this in (44), $\quad v_0^2 = \dfrac{gR_e^2}{R_e + b} \quad \text{or} \quad v_0 = R_e\sqrt{\dfrac{g}{R_e + b}} \tag{47}$

Since b is in general very much smaller than the radius of the earth R_e, we can neglect b compared with R_e in (47) so that

$$v_0 = \sqrt{gR_e} \tag{48}$$

Taking $g = 9.8 \text{ m/sec}^2$ and $R_e = 6.6 \times 10^6$ m approximately, we find

$$v_0 = \sqrt{(9.8)(6.6 \times 10^6)} \text{ m/sec} = 8000 \text{ m/sec}$$

i.e., about 290,000 km/hr or 170,000 miles/hr.

(b) Since the circumference of a circle of radius $a = R_e + b$ is $2\pi(R_e + b)$, we see that the time for one complete revolution or *orbital period* is given by

$$T = \frac{2\pi(R_e + b)}{v_0} = \frac{2\pi(R_e + b)^{3/2}}{R_e\sqrt{g}} \tag{49}$$

if the result for v_0 given in (47) is used. However, if we assume as before that b is negligible compared with R_e, (49) becomes

$$T = 2\pi\sqrt{\frac{R_e}{g}} = 2\pi\sqrt{\frac{6.6 \times 10^6}{9.8}} \text{ sec} = 5200 \text{ sec}$$

or about $1\frac{1}{2}$ hr. This is actually the correct time for the orbital period of a satellite orbiting near the earth's surface as observed in our space program.

It should be emphasized that the speed and orbital period would vary considerably from the values given above if the altitude of the satellite above the earth's surface is not negligible compared to the earth's radius or if the satellite orbit is elliptic rather than almost circular. That this is so can be surmised from the fact that the moon, which can be considered an earth satellite, has an orbital period of about 28 days.

When a satellite is in elliptical orbit around the earth, the point at which it is *nearest* to the earth is often called the *perigee* (from the Greek *peri* for *near* and *geios* for *earth*). Similarly, the point at which it is farthest from the earth is called the *apogee* (from the Greek *apo*, meaning *far from*). These are to be compared with the words perihelion and aphelion. For convenience the words perigee and apogee are often used for satellite motion about any bodies besides the earth.

The basic ideas given above can be adapted for use in travel through space to the various planets of our solar system. From a more destructive point of view, they can also be used for launching a missile from one continent to another, i.e., an *intercontinental ballistic missile* (ICBM). The problem is similar to that in which we began our discussion on page 446, except that because the distances are much greater the curvature of the earth must be taken into account; i.e., we cannot use the simplification of a "flat earth" as on page 446. Instead of satellite motion in which the orbit encircles the earth, the path of a missile launched from some position P_1 on the earth would intersect the earth's surface at some other position P_2, as shown in Fig. 10.10.

A EXERCISES

1. Show that the areal velocity in rectangular coordinates is $\frac{1}{2}(x\dot{y} - y\dot{x})$.

2. (a) Use equations (16), page 452, to show that $x\ddot{y} - y\ddot{x} = 0$. (b) Demonstrate the fact that
$$x\ddot{y} - y\ddot{x} = \frac{d}{dt}(x\dot{y} - y\dot{x})$$
and thus deduce by use of Exercise 1 that the areal velocity is constant without making use of polar coordinates.

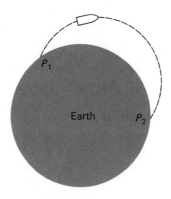

Figure 10.10

3. An object of mass m is attracted to a fixed point O with a force proportional to the instantaneous distance from O. Let O be the origin of a rectangular coordinate system and let (x, y)
represent the position of the object at any time. (a) Set up differential equations describing
the motion. (b) If the object starts on the x axis at distance a from O and is given an initial
velocity v_0 in the positive y direction, show that the path is an ellipse with center at O. Under
what conditions is the path a circle? (c) Show that the radius vector joining the mass and
and point O sweeps out equal areas in equal times, i.e., show that the areal velocity $\frac{1}{2}r^2\dot{\phi}$
is constant. What is the value of this constant?

4. If the force in Exercise 3 is one of repulsion instead of attraction, show that the path is a
portion of a hyperbola. Is the areal velocity constant in this case? Discuss the case $v_0 = 0$.

5. Find the maximum and minimum distances of the earth (or other planet) from the sun (focus)
by setting dr/dt in (27), page 454, equal to zero.

6. In the ellipse of Fig. 10.8, page 456, find (a) the length of the major axis; (b) the location of the
other focus.

7. If a planet moves around the sun in an elliptical orbit of eccentricity $\frac{1}{2}$, show that (a) the ratio
of the distance at aphelion to the distance at perihelion is $3:1$, while (b) the ratio of the
corresponding speeds is $1:3$.

8. An astronaut is in orbit around the earth in an elliptical path of eccentricity 0.1. At perigee
he is 1000 km above the earth's surface. Find (a) his orbital speed at perigee, (b) his altitude
at apogee, and (c) his orbital speed at apogee.

9. Work the Illustrative Example on page 459 if the satellite is in orbit around the moon.
Assuming that b is negligible compared with the moon's radius, what would be the (a) orbital
speed and (b) orbital period?

B EXERCISES

1. Show that the length of the minor axis of the ellipse of Fig. 10.8 is $2a\sqrt{\dfrac{1 + \epsilon}{1 - \epsilon}}$.

2. Show that the equation of the ellipse of Fig. 10.8 can be written as $\qquad r = \dfrac{\beta^2/\alpha}{1 + \epsilon\cos\phi}$

where α and β are the lengths of the semi-major and semi-minor axes, respectively. By comparison with equation (43) on page 458, show that the areal velocity is given by

$$h = \beta\sqrt{\frac{k}{\alpha}}$$

3. The time T for a planet to make one complete revolution in its elliptical path is the *period*
of the planet. Show that

$$T = \frac{\text{area of ellipse}}{\text{areal velocity}} = \frac{2\pi\alpha\beta}{h} = \frac{2\pi\alpha^{3/2}}{\sqrt{k}}$$

using the result of Exercise 2. Hence, prove Kepler's third law.

4. The period of revolution of the planet Jupiter around the sun is 11.9 earth years. Find the
length of the semi-major axis of its orbit and thus its approximate distance from the sun.

5. The eccentricity of the elliptical orbit of the planet Mercury is about 0.21, and its period of
rotation about the sun is about 88 earth days. (a) What are the lengths of the semi-major and
semi-minor axes of its orbit? (b) What are its distances from the sun at aphelion and perihelion?

6. Assuming that the radius of the earth is 3960 miles and that the orbital period of rotation of the moon around the earth is 27.3 days, find the distance from the center of the earth to the center of the moon.

7. The first United States two-man satellite to orbit the earth was Gemini 12, launched November 11, 1966. It had a perigee of 243 miles and an apogee of 310 miles. Calculate the orbital period and compare with the actually observed period of 89.9 min.

8. Explorer I, a satellite launched by the United States in 1958, achieved a maximum altitude above the earth of 2550 miles and a minimum altitude of 356 miles. Calculate the orbital period and compare with the actually observed period of 115 min.

9. An earth satellite launched by the United States in 1969 was designed to remain in orbit for more than 1 million years. The orbit is nearly circular and of radius 35,800 miles approximately. Calculate the orbital period, neglecting the moon's influence, and compare with the actual orbital period of very nearly 24 hr.

10. Suppose that at perigee a satellite suddenly undergoes an increase in its orbital velocity v_0 so that the orbit is enlarged and ultimately becomes parabolic. The least velocity v_p needed to do this is often called the *parabolic velocity* and amounts to an *escape velocity*. Show that $v_p = \sqrt{2}v_0$.

C EXERCISES

1. An object of mass m moves so that it is attracted to a fixed point O (taken as origin of an xy coordinate system) with a force $F(r)$, where r is the instantaneous distance of the mass from O. Such a force depending only on r is called a *central force*. (a) Show that the equations of motion are (dots denote derivatives with respect to t)

$$m\ddot{x} = -\frac{x}{r}F(r), \qquad m\ddot{y} = -\frac{y}{r}F(r)$$

(b) Prove that $x\ddot{y} - y\ddot{x} = 0$, and thus that $x\dot{y} - y\dot{x} = h$, where h is a constant. Using the result of A Exercise 2, show that the areal velocity is constant independent of the form of the central force.

2. Show that the differential equations of Exercise 1(a) may be replaced by

$$m(\ddot{r} - r\dot{\phi}^2) = -F(r), \qquad m(2\dot{r}\dot{\phi} + r\ddot{\phi}) = 0$$

Deduce that $r^2\dot{\phi} = h$ and thus show that the central force problem reduces to solving

$$\ddot{r} - \frac{h^2}{r^3} = -\frac{F(r)}{m}$$

3. Problems involving central forces are simplified by use of the transformation $u = 1/r$. Show that this transformation yields the result

$$\ddot{r} = -h^2u^2\frac{d^2u}{d\phi^2}$$

Thus transform the differential equation of Exercise 2 into $\quad \dfrac{d^2u}{d\phi^2} + u = \dfrac{G(u)}{mh^2u^2}$

where $F(1/u) \equiv G(u)$.

4. For a central force given by $F(r) = k/r^2$, show that the differential equation of Exercise 3 becomes

$$\frac{d^2u}{d\phi^2} + u = \frac{k}{mh^2} \quad \text{or} \quad u = A \cos \phi + B \sin \phi + \frac{k}{mh^2}$$

Compare with the result of the text.

5. Show that in order for an object to move along the ellipse $r = \dfrac{l}{1 + \epsilon \cos \phi}$ under the influence of a central force $F(r)$ located at the origin, we must have

$$F(r) \propto \frac{1}{r^2}$$

Historically, this is the manner in which Newton deduced from Kepler's law the "inverse square law" for planets and finally his universal law of gravitation for all objects.

6. Using the differential equation of Exercise 3, discuss the motion of an object moving in a central force field given by $F(r) = k/r^3$, assuming that it starts at $(a, 0)$, with initial velocity v_0 in the positive y direction. Show that if $k = mh^2 = ma^2v_0^2$, the mass spirals in toward the origin but never reaches it.

7. An object moves under the influence of a central force located at the origin of a coordinate system. If the path of the object is the circle $r = a \cos \phi$, determine the law of force.

8. Integrate equation (27) of the text and thus obtain the position of the earth (or other planet) with respect to the sun at any time.

9. Set up a system of differential equations describing the motion of three bodies of known masses, the only forces being their mutual gravitational attractions. This problem of determining the motion of three bodies, given their initial positions and velocities at some instant of time, is called the *three-body problem*; it has never been solved exactly except in certain very special cases (such as where the bodies are in the same straight line as on page 117). A similar *n-body problem* can be formulated where $n > 3$.

4.4 THE PROBLEM OF THE VIBRATING MASSES

A system consists of springs A, B, C, and objects D and E attached in a straight line on a horizontal frictionless table RS (Fig. 10.11), the ends of the springs A and C being fixed at O and P, respectively. The springs, of negligible mass, each have spring constant k, and the objects have equal mass M. The system is set into vibration by holding D in place, moving E to the right a distance $a > 0$, and then releasing both objects. The problem is to determine the positions of objects D and E at any time t thereafter.

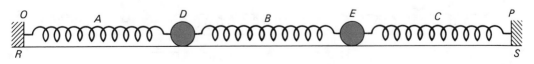

Figure 10.11

Mathematical Formulation. To determine the differential equations of motion, let us see what conditions prevail at time t. At this time D and E may be removed from their equilibrium positions somewhat, as shown in Fig. 10.12. Assume that at this time the objects are located at distances x_1 and x_2 from their respective equilibrium positions indicated by the dashed lines in the figure. We shall assume that directions to the right are positive. Let us consider the forces on D and E at time t:

1. Spring A is exerting a force on D to the left of magnitude kx_1.
2. Spring B is exerting a force on D to the right of magnitude $k(x_2 - x_1)$.
3. Spring C is exerting no direct force on D.

The net force to the right is $k(x_2 - x_1) - kx_1 = k(x_2 - 2x_1)$. Hence

$$M \frac{d^2x_1}{dt^2} = k(x_2 - 2x_1) \tag{50}$$

by Newton's law. Similarly,

1. Spring A is exerting no direct force on E.
2. Spring B is exerting a force on E to the left of magnitude $k(x_2 - x_1)$.
3. Spring C is exerting a force on E to the left of magnitude kx_2.

The net force to the right is $-k(x_2 - x_1) - kx_2 = k(x_1 - 2x_2)$. Hence

$$M \frac{d^2x_2}{dt^2} = k(x_1 - 2x_2) \tag{51}$$

by Newton's law. The initial conditions are given by

$$x_1 = 0, \quad x_2 = a, \quad \frac{dx_1}{dt} = 0, \quad \frac{dx_2}{dt} = 0, \quad \text{at } t = 0 \tag{52}$$

Our mathematical formulation consists of equations (50) and (51), which we must solve simultaneously subject to conditions (52).

Solution Letting $\omega^2 = k/M$, equations (50) and (51) can be written

$$(D^2 + 2\omega^2)x_1 - \omega^2 x_2 = 0, \qquad -\omega^2 x_1 + (D^2 + 2\omega^2)x_2 = 0 \tag{53}$$

where $D \equiv d/dt$. To eliminate x_2 from equations (53) let us operate on the first equation with $D^2 + 2\omega^2$ and multiply the second equation by ω^2 to obtain

$$(D^2 + 2\omega^2)^2 x_1 - \omega^2(D^2 + 2\omega^2)x_2 = 0, \qquad -\omega^4 x_1 + \omega^2(D^2 + 2\omega^2)x_2 = 0$$

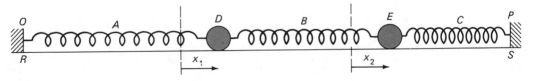

Figure 10.12

Addition of these two equations yields

$$[(D^2 + 2\omega^2)^2 - \omega^4]x_1 = 0 \qquad \text{or} \qquad (D^4 + 4\omega^2 D^2 + 3\omega^4)x_1 = 0 \qquad (54)$$

The auxiliary equation corresponding to this last differential equation is $m^4 + 4\omega^2 m^2 + 3\omega^4 = 0$, i.e., $(m^2 + \omega^2)(m^2 + 3\omega^2) = 0$ from which $m^2 = -\omega^2, -3\omega^2$ or $m = \pm i\omega, \pm i\sqrt{3}\omega$. Thus the solution of (54) is

$$x_1 = c_1 \cos \omega t + c_2 \sin \omega t + c_3 \cos \sqrt{3}\omega t + c_4 \sin \sqrt{3}\omega t \qquad (55)$$

In an exactly analogous manner we find by eliminating x_1 the equation

$$(D^4 + 4\omega^2 D^2 + 3\omega^4)x_2 = 0$$

which is the same as (54), and therefore has the solution

$$x_2 = c_5 \cos \omega t + c_6 \sin \omega t + c_7 \cos \sqrt{3}\omega t + c_8 \sin \sqrt{3}\omega t \qquad (56)$$

The determinant of the coefficients in equations (53) is

$$\begin{vmatrix} D^2 + 2\omega^2 & -\omega^2 \\ -\omega^2 & D^2 + 2\omega^2 \end{vmatrix} = D^4 + 4\omega^2 D^2 + 3\omega^4$$

a polynomial in D of fourth degree, so that there must be a total of four arbitrary constants. We have eight. If we substitute x_1 and x_2 from (55) and (56) in the original equations, we find the relations between the constants

$$c_5 = c_1, \qquad c_6 = c_2, \qquad c_7 = -c_3, \qquad c_8 = -c_4$$

From (55) and (56) we thus find

$$\left. \begin{aligned} x_1 &= c_1 \cos \omega t + c_2 \sin \omega t + c_3 \cos \sqrt{3}\omega t + c_4 \sin \sqrt{3}\omega t \\ x_2 &= c_1 \cos \omega t + c_2 \sin \omega t - c_3 \cos \sqrt{3}\omega t - c_4 \sin \sqrt{3}\omega t \end{aligned} \right\} \qquad (57)$$

Using the conditions (52) yields $\qquad c_1 = \dfrac{a}{2}, \qquad c_3 = -\dfrac{a}{2}, \qquad c_2 = c_4 = 0$

so that $\qquad x_1 = \dfrac{a}{2}(\cos \omega t - \cos \sqrt{3}\omega t), \qquad x_2 = \dfrac{a}{2}(\cos \omega t + \cos \sqrt{3}\omega t) \qquad (58)$

It is seen from the general motion (57) and in our special case (58) that there are two frequencies present, given by

$$f_1 = \frac{\omega}{2\pi} = \frac{1}{2\pi}\sqrt{\frac{k}{M}}, \qquad f_2 = \frac{\sqrt{3}\omega}{2\pi} = \frac{1}{2\pi}\sqrt{\frac{3k}{M}} \qquad (59)$$

These are referred to as the *normal or natural frequencies* of the system. In other special cases it is possible for the system to vibrate with only one of the frequencies given in (59). For example, if the frequency is f_1, we see from (57) that $c_3 = c_4 = 0$. This corresponds to $x_1 = x_2$, i.e., the case where the objects D and E are both moving in the same directions, as indicated in Fig. 10.13. On the other hand, if the frequency

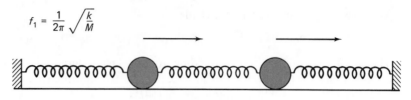

$$f_1 = \frac{1}{2\pi}\sqrt{\frac{k}{M}}$$

Figure 10.13

is f_2, we see from (57) that $c_1 = c_2 = 0$. This corresponds to $x_1 = -x_2$, i.e., the case where the objects D and E are both moving in opposite directions, as indicated in Fig. 10.14.

The special types of vibrations indicated in Figs. 10.13 and 10.14 in which the system vibrates with a single frequency are called *normal or natural modes of vibration* and are often simply referred to as the *normal modes*. More complex motions such as those indicated by (57) represent combinations of these normal modes because both frequencies are present.

The system of Figs. 10.11 and 10.12 is a generalization of the simple vibrating systems considered in Chapter Five for which there was a single *natural* or *normal* frequency. Further generalizations are possible. For example, we could have three objects connected by four springs, in which case there would be three normal frequencies and three normal modes. In general, complex structures such as buildings and bridges have many normal frequencies and modes of vibration.

In Chapter Five we found that, when a periodic external force having frequency equal (or almost equal in case of damping) to the natural frequency of a system was applied to a system, great disturbances could be set up in the system, a phenomenon which we called *resonance*. In more complex systems a larger number of normal frequencies may be present, and the chance for resonance taking place is increased. This makes even more plausible why it could be dangerous for a group to "keep in step" when crossing a bridge, since a bridge is a complicated structure with many natural or normal frequencies, and exciting one of them may prove disastrous.

Normal or natural frequencies are of great importance in science and technology. They appear in such seemingly unrelated fields as stress analysis (needed by civil and mechanical engineers in design of structures) and nuclear physics (where they are useful in explaining the theory of spectra as well as effects of atomic energy). They are also of great importance in electricity in connection with resonance, where they act in a constructive manner and are to be desired rather than avoided.

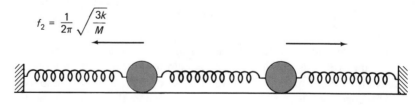

$$f_2 = \frac{1}{2\pi}\sqrt{\frac{3k}{M}}$$

Figure 10.14

1. Solve the problem of the text if the initial conditions are changed as follows: (a) D and E are both moved a distance $a > 0$ to the right and then released. (b) D is moved a distance $a > 0$ to the left, while E is moved a distance $a > 0$ to the right, and then both are released. Discuss the motion in each case.

2. In the problem of the text, suppose that D and E are initially at their equilibrium positions and are given initial velocities to the right of magnitudes v_1 and v_2, respectively. Describe the motion.

3. If the two end springs A and C of the text problem are replaced by ones with spring constant K, while spring B is left unchanged with spring constant k, show that

$$M \frac{d^2 x_1}{dt^2} = k(x_2 - x_1) - K x_1, \qquad M \frac{d^2 x_2}{dt^2} = k(x_1 - x_2) - K x_2$$

Prove that there are two frequencies f_1 and f_2 where $f_1 = \dfrac{1}{2\pi} \sqrt{\dfrac{K}{M}}$, $f_2 = \dfrac{1}{2\pi} \sqrt{\dfrac{K + 2k}{M}}$.

Describe the normal modes.

4. What physical significance if any can be given to the cases (a) $k = 0$ and (b) $K = 0$ in Exercise 3?

5. If the springs of the text problem are kept the same but the masses D and E are changed so as to have ratio $8:5$, show that the normal frequencies have ratio $\sqrt{3}:\sqrt{10}$.

1. A system consists of springs A, B and objects C, D attached in a straight line on a horizontal frictionless table RS (Fig. 10.15), one end of spring A being fixed at O. The springs of negligible mass each have spring constant k, and the objects have equal mass M. The system is set into vibration by holding D in place, moving C to the left a distance $a > 0$ and then releasing both objects.
(a) Show that the differential equations describing the motion are

$$M \frac{d^2 x_1}{dt^2} = k(x_2 - 2x_1), \qquad M \frac{d^2 x_2}{dt^2} = -k(x_2 - x_1)$$

where x_1 and x_2 are the displacements of C and D from their equilibrium positions at time t.

(b) Show that the system vibrates with frequencies $f_1 = \dfrac{(\sqrt{5} - 1)}{4\pi} \sqrt{\dfrac{k}{M}}$, $f_2 = \dfrac{(\sqrt{5} + 1)}{4\pi} \sqrt{\dfrac{k}{M}}$.

(c) Determine the positions of C and D at any time.
(d) Describe the normal modes of vibration.

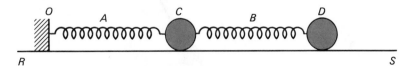

Figure 10.15

2. If the two springs of Exercise 1 have different spring constants k_1 and k_2, while the masses of C and D are kept equal, show that there are two normal frequencies f_1 and f_2, where

$$f_1 = \frac{1}{2\pi}\sqrt{\frac{k_1 + 2k_2 + \sqrt{k_1^2 + 4k_2^2}}{2M}}, \qquad f_2 = \frac{1}{2\pi}\sqrt{\frac{k_1 + 2k_2 - \sqrt{k_1^2 + 4k_2^2}}{2M}}$$

3. If the springs in Exercise 1 have the same spring constant, but the masses of C and D are changed so as to have ratio $3:2$, show that the normal frequencies have ratio $1:\sqrt{6}$.

4. A method often used in practice for determination of the normal frequencies is to substitute $x_1 = A_1 e^{i\omega t}$, $x_2 = A_2 e^{i\omega t}$, where A_1 A_2, ω are undetermined constants, into the differential equations. Then one sets up a condition for the quantities to be different from zero. Show that for the problem of the text this method leads to the condition that the determinant (called the *secular determinant*),

$$\begin{vmatrix} \dfrac{2k}{M} - \omega^2 & -\dfrac{k}{M} \\ -\dfrac{k}{M} & \dfrac{2k}{M} - \omega^2 \end{vmatrix} = 0$$

The rationale behind this method is given in Section 9, page 491.

5. Work the problem of the vibrating masses in the text if damping proportional to the instantaneous velocities is taken into account.

C EXERCISES

1. In the text problem, assume two additional forces to act, a damping force proportional to the instantaneous velocity of each of the masses, and an external force acting on each and given by $A \cos \alpha t$, $t \geq 0$ where A and α are constants. (a) Find the steady-state solution. (b) Show that there are two frequencies of the external force for which resonance occurs. In view of the fact that the ends of springs A and C are fixed, what is the effect of the resonance? (c) Discuss the solution if there is no damping. For what value of α will resonance occur in this case?

2. A mass M_1 is hung from a vertical spring (of constant k) which is supported at O (Fig. 10.16). From M_1 is hung a simple pendulum having a bob of mass M_2. Assume that M_1 can vibrate

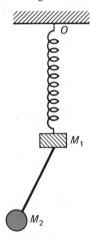

Figure 10.16

only vertically and that all motion takes place in the vertical plane. (a) Set up differential equations for the motion. (b) Find the positions of M_1 and M_2 at any time, assuming small vibrations and arbitrary initial conditions. (c) Determine normal frequencies and modes if any.

5 Applications to Electric Networks

We have already seen in Chapters Three and Five how Kirchhoff's law was used to solve problems involving single-loop electric circuits. In advanced engineering work, it is often essential to consider electric networks which involve more than one loop. An example of a two-loop network is shown in Fig. 10.17. In order to solve problems of electric networks involving two or more loops we need

KIRCHHOFF'S TWO LAWS

1. *The algebraic sum of the currents traveling toward any branch point (A or B in Fig. 10.17) is equal to zero.*
2. *The algebraic sum of the potential drops (or voltage drops) around any closed loop is equal to zero.*

To enable us to apply these laws consistently we adopt the following

CONVENTIONS

(a) If I is the current in one direction, $-I$ is the current in the opposite direction.
(b) In writing the algebraic sum of the potential drops around a closed loop, we consider a potential drop as *positive* if in describing the loop we travel in the *same* direction as the indicated current and negative if we travel in the opposite direction to the current.
(c) A potential rise (due to battery or generator for example) is considered to be the negative of a potential drop.

By using the above laws and conventions, we shall be able to formulate mathematically any problem in linear electric networks. To see the procedures involved,

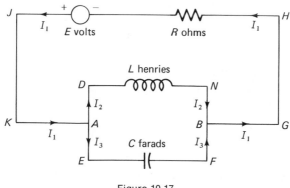

Figure 10.17

consider the network of Fig. 10.17, for example. There are three closed loops present, namely, *JHGBNDAKJ*, *JHGBFEAKJ*, and *NFEDN*. The second of Kirchhoff's laws applied to these three loops yields three differential equations. These equations, however, are not independent, since, as we will find, one differential equation can be obtained from the other two. We may thus say that there are only two independent loops. In practice, it is logical to choose those loops which are simplest. At present, for practice, we consider all three loops.

The first thing to do in formulating mathematically a problem in electric networks is to label currents in the various parts. The direction adopted for current flow is in general immaterial so long as we are consistent. In Fig. 10.17 we have adopted the directions shown for I_1, I_2, and I_3. It is clear that if I_2 flows in a direction from A to D, then it will continue from D to N and from N to B as indicated. The situation seems quite logical when we reason that current I_1 separates at branch A into currents I_2 and I_3. At branch B, I_2 and I_3 combine to give I_1 again. Since I_1 splits into I_2 and I_3, it is clear that $I_1 = I_2 + I_3$.

Another way of arriving at the same result is from Kirchhoff's first law. Since I_2 is flowing *away* from A, then by convention (a), $-I_2$ is flowing *toward* A. Similarly the current $-I_3$ is flowing toward A. The algebraic sum of the currents flowing toward A is thus given by $I_1 - I_2 - I_3$, which when equated to zero yields $I_1 = I_2 + I_3$ as before.

Let us now consider the application of Kirchhoff's second law.

For the loop JHGBNDAKJ:
   Voltage drop cross R is $-I_1R$ by convention (b).
   Voltage drop across L is $-L\,dI_2/dt$.
   Voltage drop across battery E is $+E$ by convention (c).

Hence,
$$-I_1R - L\frac{dI_2}{dt} + E = 0 \tag{1}$$

For the loop JHGBFEAKJ:
   Voltage drop across R is $-I_1R$.
   Voltage drop across C is $-Q_3/C$ (where Q_3 is the charge on capacitor C supplied by current I_3).
   Voltage drop across E is $+E$.

Hence,
$$-I_1R - \frac{Q_3}{C} + E = 0 \tag{2}$$

For the loop NFEDN:
   Voltage drop across L is $L\,dI_2/dt$ by convention (b).
   Voltage drop across C is $-Q_3/C$.

Hence,
$$L\frac{dI_2}{dt} - \frac{Q_3}{C} = 0 \tag{3}$$

It will be noted that, as mentioned before, equations (1), (2), and (3) are dependent. Thus, for instance, (3) is obtained by subtraction of (1) and (2). Realizing that only two differential equations are independent would have saved some labor. Also the loop *NFEDN* is slightly simpler than the others as evidenced by the simpler differential equation (3). Since we have a choice let us choose equations (2) and (3) together with the equation $I_1 = I_2 + I_3$. Since there are three equations and four unknowns, I_1, I_2, I_3, and Q_3, we need another equation. This is obtained by realizing that $I_3 = dQ_3/dt$. Using this value for I_3 and replacing I_1 by its equal $I_2 + I_3$, equations (2) and (3) can be written

$$-R\left(I_2 + \frac{dQ_3}{dt}\right) - \frac{Q_3}{C} + E = 0, \qquad L\frac{dI_2}{dt} - \frac{Q_3}{C} = 0$$

or in operator form with $D \equiv d/dt$,

$$RI_2 + \left(RD + \frac{1}{C}\right)Q_3 = E, \qquad LDI_2 - \frac{Q_3}{C} = 0 \qquad (4)$$

These equations may be solved by methods already taken up. Let us, for example, consider $E = E_0 \sin \omega t$ an alternating voltage having frequency $\omega/2\pi$. Operating on the first of equations (4) with LD, multiplying the second equation by R, and subtracting, we find

$$\left(LRD^2 + \frac{L}{C}D + \frac{R}{C}\right)Q_3 = LE_0 \omega \cos \omega t \qquad (5)$$

an equation which can be solved for Q_3. From here the method proceeds as usual.

A EXERCISES

1. In the electric network of Fig. 10.18, $E = 60$ volts. Determine the currents I_1 and I_2 as functions of time t, assuming that at $t = 0$, when the key K is closed, $I_1 = I_2 = 0$. Find the steady-state currents.

2. Work Exercise 1 if $E = 150 \sin 10t$. Find the steady-state currents.

3. At $t = 0$ the charge on the capacitor in the circuit of Fig. 10.19 is 1 coulomb, while the current in the inductor is zero. Determine the charge on the capacitor and the currents in the various branches at any time.

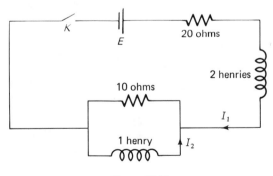

Figure 10.18

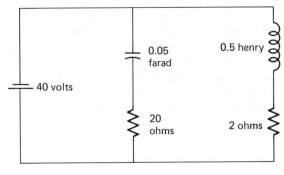

Figure 10.19

4. Work Exercise 3 if the emf is $100 \sin 2t$.

5. Complete the problem of the text if $R = 10$ ohms, $C = 10^{-3}$ farads, $L = 0.4$ henries, $E_0 = 100$ volts, $\omega = 50$. Assume that the charge on the capacitor and the currents in the various branches are initially zero.

B EXERCISES

1. At $t = 0$ the capacitors in the network of Fig. 10.20 are charged to potential V_0, and the keys K_1 and K_2 are closed. Show that the potential across the capacitors of higher and lower capacitances are given, respectively, by

$$\frac{V_0}{3}\left(2\cos\frac{t}{2\sqrt{LC}} + \cos\frac{t}{\sqrt{LC}}\right) \quad \text{and} \quad \frac{V_0}{3}\left(4\cos\frac{t}{2\sqrt{LC}} - \cos\frac{t}{\sqrt{LC}}\right)$$

2. Compare Exercise 1 with the problem of the vibrating masses on page 463, pointing out the various analogies. What mechanical system might correspond to the network of Fig. 10.20?

C EXERCISES

1. In the circuit of Fig. 10.21 the capacitor C has a charge Q_0 at time $t = 0$, while the currents through the inductors are zero at that time. Show that at any time $t > 0$ the charge on the

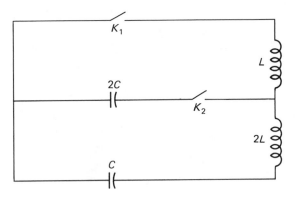

Figure 10.20

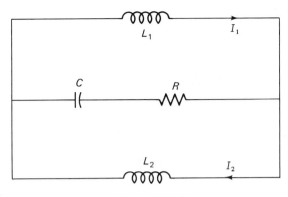

Figure 10.21

capacitor is given by

$$Q = Q_0 e^{-Rt/2L} \left(\cos \omega t + \frac{R}{2L\omega} \sin \omega t \right) \quad \text{where} \quad \frac{1}{L} = \frac{1}{L_1} + \frac{1}{L_2} \quad \text{and} \quad \omega = \sqrt{\frac{1}{LC} - \frac{R^2}{4L^2}}$$

Show also that the currents I_1 and I_2 through L_1 and L_2 are

$$I_1 = -\frac{Q_0}{\omega CL_1} e^{-Rt/2L} \sin \omega t, \qquad I_2 = \frac{Q_0}{\omega CL_2} e^{-Rt/2L} \sin \omega t$$

2. Discuss the previous problem if $1/LC \leqq R^2/4L^2$.

6 Applications to Biology

6.1 CONCENTRATION OF A DRUG IN A TWO-COMPARTMENT SYSTEM

We have already examined in Chapter Three some applications of differential equations to various problems in biology. We shall now consider biological problems which lead to systems of differential equations.

One interesting problem involves the determination of the concentration of some chemical, such as a drug, in a system consisting of two compartments separated by a membrane, as shown in Fig. 10.22. The drug can pass through this membrane

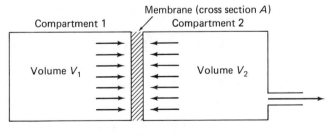

Figure 10.22

from compartment 1 to compartment 2, or conversely from compartment 2 to compartment 1. We shall also assume that the drug can escape to the external system through an opening in the second compartment, as indicated in the figure.

Mathematical Formulation. Let us suppose that the compartments have volumes V_1 and V_2, and that the cross-sectional area of the membrane is A. Denote by x_1 and x_2 the masses of the drug in compartments 1 and 2, respectively, at any time t. Using the word *rate* to mean *time rate* for brevity, we have

rate of change of drug mass in compartment 1

$$= \text{rate of flow of drug mass from compartment 2 into compartment 1}$$
$$- \text{rate of flow of drug mass from compartment 1 into 2} \tag{1}$$

We must now examine each of these terms. The first is easy because if the drug mass in compartment 1 is x_1 then

$$\text{rate of change of drug mass in compartment 1} = \frac{dx_1}{dt} \tag{2}$$

To obtain the second term, let us note that the rate of flow of the drug mass from compartment 2 to compartment 1 is proportional to the area A of the membrane and to the concentration of the drug in compartment 2, given by x_2/V_2. Thus

$$\text{rate of flow of drug mass from compartment 2 into 1} = \alpha_{21} A \frac{x_2}{V_2} \tag{3}$$

where we have used α_{21} to denote the constant of proportionality. The last term is obtained by similar reasoning; i.e., the rate of flow of the drug mass from compartment 1 into compartment 2 is proportional to the area A of the membrane and to the concentration of the drug in compartment 1 given by x_1/V_1. Thus

$$\text{rate of flow of drug mass from compartment 1 into 2} = \alpha_{12} A \frac{x_1}{V_1} \tag{4}$$

where we have used α_{12} to denote the constant of proportionality. The constants α_{12} and α_{21} need not be equal; i.e., the absorption rates of the drug on opposite sides of the membrane can be different. Using (2), (3), and (4), the result (1) becomes

$$\frac{dx_1}{dt} = \alpha_{21} A \frac{x_2}{V_2} - \alpha_{12} A \frac{x_1}{V_1} \tag{5}$$

A corresponding equation must now be obtained for compartment 2. We can write

rate of change of drug mass in compartment 2

$$= \text{rate of flow of drug mass from compartment 1 into 2}$$
$$- \text{rate of flow of drug mass from compartment 2 into 1}$$
$$- \text{rate of flow of drug mass from 2 into external system} \tag{6}$$

Since x_2 is the drug mass in compartment 2, we have for the first term in (6).

$$\text{rate of change of drug mass in compartment 2} = \frac{dx_2}{dt} \qquad (7)$$

The next two terms of (6) are exactly the same as given in (4) and (3), respectively. To obtain the final term in (6) we note that the rate of flow of the drug mass from compartment 2 is proportional to the concentration of the drug in compartment 2 given by x_2/V_2. Thus

$$\text{rate of flow of drug mass from 2 into external system} = \alpha \frac{x_2}{V_2} \qquad (8)$$

where we have used α to denote the constant of proportionality.

From these we can write (6) as

$$\frac{dx_2}{dt} = \alpha_{12} A \frac{x_1}{V_1} - \alpha_{21} A \frac{x_2}{V_2} - \alpha \frac{x_2}{V_2} \qquad (9)$$

Taking
$$\beta_{21} = \frac{\alpha_{21} A}{V_2}, \qquad \beta_{12} = \frac{\alpha_{12} A}{V_1}, \qquad \beta = \frac{\alpha}{V_2} \qquad (10)$$

equations (5) and (9) can be written

$$\frac{dx_1}{dt} = \beta_{21} x_2 - \beta_{12} x_1, \qquad \frac{dx_2}{dt} = \beta_{12} x_1 - \beta_{21} x_2 - \beta x_2 \qquad (11)$$

or $\qquad (D + \beta_{12})x_1 - \beta_{21} x_2 = 0, \qquad -\beta_{12} x_1 + (D + \beta_{21} + \beta)x_2 = 0 \qquad (12)$

The simultaneous differential equations (11) or (12) are to be solved subject to suitable conditions such as, for example,

$$x_1 = a, \qquad x_2 = b \qquad \text{at } t = 0 \qquad (13)$$

where a and b are constants.

Solution We can solve the system of equations by using any of the methods already given. Let us use the method of elimination. To eliminate x_2 operate on the first equation in (12) with $D + \beta_{21} + \beta$ and multiply the second equation in (12) by β_{21}. Addition of the resulting equations yields

$$[D^2 + (\beta_{12} + \beta_{21} + \beta)D + \beta\beta_{12}]x_1 = 0 \qquad (14)$$

The auxiliary equation corresponding to this is $m^2 + (\beta_{12} + \beta_{21} + \beta)m + \beta\beta_{12} = 0$

or $\qquad m = \dfrac{-(\beta_{12} + \beta_{21} + \beta) \pm \sqrt{(\beta_{12} + \beta_{21} + \beta)^2 - 4\beta\beta_{12}}}{2}$

If we let $\quad p = \frac{1}{2}(\beta_{12} + \beta_{21} + \beta), \qquad q = \frac{1}{2}\sqrt{(\beta_{12} + \beta_{21} + \beta)^2 - 4\beta\beta_{12}} \qquad (15)$

these roots are $m = -p \pm q$. Then we obtain the required solution

$$x_1 = e^{-pt}(c_1 e^{qt} + c_2 e^{-qt}), \quad x_2 = \frac{e^{-pt}}{\beta_{21}}[(\beta_{12} - p + q)c_1 e^{qt} + (\beta_{12} - p - q)c_2 e^{-qt}] \quad (16)$$

where the second equation for x_2 in (16) is found by substituting the result for x_1 into the first of equations (12). The constants c_1 and c_2 can be determined from conditions (13).

Interpretation. The value of q in (15) suggests that there are three cases which can arise corresponding to the cases where q is imaginary, $q > 0$, and $q = 0$. It turns out, however, that the first and last cases cannot arise. To see this, let us write

$$(\beta_{12} + \beta_{21} + \beta)^2 - 4\beta\beta_{12} = (\beta + \beta_{21} - \beta_{12})^2 + 4\beta_{12}\beta_{21}$$

Since this last quantity must always be positive if $\beta_{12} > 0$, $\beta_{21} > 0$, we must have $q > 0$. Using this and the fact that $p > q$ if $\beta > 0$, as seen from (15), it follows that the drug masses x_1 and x_2 approach zero steadily in a manner similar to that indicated by the graph of Fig. 5.6, page 231, corresponding to the *overdamped* case. We cannot have *damped oscillatory motion*, as in Fig. 5.5 on page 230.

6.2 THE PROBLEM OF EPIDEMICS WITH QUARANTINE

In the problem of epidemics worked on page 153, we did not take into consideration the possibility that students becoming infected would be removed from the community so as not to cause others to become infected. It is of interest to consider this more realistic model of disease in which students discovered to be infected are removed or, as we often say, *quarantined*. Several circumstances may arise which are mathematically equivalent to such quarantine, as follows:

(a) The disease may be such that those who develop it and recover cannot develop it again or transmit it to anyone else; i.e., they develop *immunity* and are not *carriers* of the disease.

(b) Those who develop the disease do not recover.

Mathematical Formulation. Let us introduce the following notation:

N_u = number of students who are uninfected at any time t; this does not include students who may have recovered as in situation (a) above and returned to the community

N_i = number of students who are infected at time t but have not yet been quarantined

N_q = number of students who are quarantined at time t [equivalent to the number of students corresponding to situations (a) and (b) above]

Clearly, we must have $\qquad N = N_u + N_i + N_q \qquad$ (17)

where N is the total number of students of the community, assumed constant.

As in the case of the mathematical model of page 154, we might expect that the time rate at which uninfected students become infected is proportional to the product of the numbers of infected and uninfected students, so that

$$\frac{dN_u}{dt} = -\kappa N_u N_i \qquad (18)$$

where κ is the constant of proportionality.

We might also expect that the rate at which infected students are quarantined is proportional to the number of infected students. This leads to

$$\frac{dN_q}{dt} = \lambda N_i \tag{19}$$

where λ is the constant of proportionality.

From (17), (18), and (19) it is easy to arrive at

$$\frac{dN_i}{dt} = \kappa N_u N_i - \lambda N_i \tag{20}$$

an equation which could have been surmised even without making use of (17), since it expresses the rather obvious fact that the rate of change in the number of infected students in the community at time t equals the rate at which they become infected minus the rate at which they are removed (quarantined) after discovery.

As possible initial conditions we can have

$$N_u = U_0, \qquad N_i = I_0, \qquad N_q = 0, \qquad \text{at } t = 0 \tag{21}$$

where U_0 and I_0 are constants. By using (17) we see that

$$N = U_0 + I_0 + 0 \quad \text{or} \quad U_0 + I_0 = N \tag{22}$$

Solution   Dividing equation (18) by (19) yields

$$\frac{dN_u}{dN_q} = -\frac{\kappa}{\lambda} N_u \tag{23}$$

whose solution on separating the variables is found to be

$$N_u = c_1 e^{-\kappa N_q/\lambda} \tag{24}$$

where c_1 is an arbitrary constant. Since $N_u = U_0$ when $N_q = 0$, as seen from (21), we have $c_1 = U_0$, so that

$$N_u = U_0 e^{-\kappa N_q/\lambda} \tag{25}$$

Now from (17) we see that

$$N_i = N - N_q - N_u = N - N_q - U_0 e^{-\kappa N_q/\lambda} \tag{26}$$

on making use of (25). Thus equation (19) becomes

$$\frac{dN_q}{dt} = \lambda(N - N_q - U_0 e^{-\kappa N_q/\lambda}) \quad \text{or} \quad \int \frac{dN_q}{N - N_q - U_0 e^{-\kappa N_q/\lambda}} = \int \lambda \, dt \tag{27}$$

on separating the variables. The integration on the left in (27) cannot be performed exactly. However, we can obtain a good approximation by using the first three terms of the series expansion for the exponential in (27) to obtain the approximation

$$N - N_q - U_0 e^{-\kappa N_q/\lambda} = N - N_q - U_0 + \frac{\kappa U_0 N_q}{\lambda} - \frac{\kappa^2 U_0 N_q^2}{2\lambda^2}$$

$$= I_0 - N_q + \frac{\kappa U_0 N_q}{\lambda} - \frac{\kappa^2 U_0 N_q^2}{2\lambda^2} = I_0 + \left(\frac{\kappa U_0}{\lambda} - 1\right) N_q - \frac{\kappa^2 U_0 N_q^2}{2\lambda^2}$$

where in the second equality we make use of the fact that $N - U_0 = I_0$ from (22). If we now change variables from N_q to w, where

$$\frac{\kappa\sqrt{U_0}}{\lambda\sqrt{2}} N_q = w \tag{28}$$

the last result can be written as

$$I_0 + \left(\frac{\kappa U_0}{\lambda} - 1\right)\frac{\lambda\sqrt{2}}{\kappa\sqrt{U_0}} w - w^2 = I_0 + 2\alpha w - w^2 = I_0 + \alpha^2 - (w - \alpha)^2$$

where for brevity we write

$$\alpha = \left(\frac{\kappa U_0}{\lambda} - 1\right)\frac{\lambda\sqrt{2}}{2\kappa\sqrt{U_0}} \tag{29}$$

Thus (27) can be written

$$\frac{\lambda\sqrt{2}}{\kappa\sqrt{U_0}} \int \frac{dw}{I_0 + \alpha^2 - (w - \alpha)^2} = \int \lambda \, dt \tag{30}$$

Making use of the elementary result

$$\int \frac{dx}{a^2 - x^2} = \frac{1}{2a} \ln\left(\frac{a + x}{a - x}\right) = \frac{1}{a} \tanh^{-1}\left(\frac{x}{a}\right)$$

(30) becomes

$$\frac{\sqrt{2}}{\kappa\sqrt{U_0}\sqrt{I_0 + \alpha^2}} \tanh^{-1}\left(\frac{w - \alpha}{\sqrt{I_0 + \alpha^2}}\right) = t + c_2 \tag{31}$$

where c_2 is a constant of integration. To find c_2 we use $N_q = 0$ or $w = 0$ at $t = 0$ in (31) to obtain

$$c_2 = \frac{\sqrt{2}}{\kappa\sqrt{U_0}\sqrt{I_0 + \alpha^2}} \tanh^{-1}\left(\frac{-\alpha}{\sqrt{I_0 + \alpha^2}}\right) = \frac{-\sqrt{2}}{\kappa\sqrt{U_0}\sqrt{I_0 + \alpha^2}} \tanh^{-1}\left(\frac{\alpha}{\sqrt{I_0 + \alpha^2}}\right)$$

Thus,

$$\tanh^{-1}\left(\frac{w - \alpha}{\sqrt{I_0 + \alpha^2}}\right) = \frac{\kappa\sqrt{U_0}\sqrt{I_0 + \alpha^2}\, t}{\sqrt{2}} - \tanh^{-1}\left(\frac{\alpha}{\sqrt{I_0 + \alpha^2}}\right)$$

from which

$$w = \alpha + \sqrt{I_0 + \alpha^2} \tanh(\beta t - \gamma) \tag{32}$$

where

$$\beta = \frac{\kappa\sqrt{U_0}\sqrt{I_0 + \alpha^2}}{\sqrt{2}}, \qquad \gamma = \tanh^{-1}\left(\frac{\alpha}{\sqrt{I_0 + \alpha^2}}\right) \tag{33}$$

The number of students removed by quarantine (or equivalent) given by N_q can be obtained from (32) on using (28). From this N_u can be obtained by use of (25) and N_i by use of (26). It should be emphasized of course that the results are approximate and valid only to the extent that the exponential term in (25) is closely approximated by the first three terms in its series expansion.

A EXERCISES

1. Verify the results (16) on page 475.

2. Suppose that in the two-compartment model we have $\beta = 0$. Suppose also that initially the amount of the drug present in compartment 1 is a but that the drug is not present in com-

partment 2. Show that (a) the amounts of the drug present in compartments 1 and 2 at time $t > 0$ are given respectively by

$$x_1 = \frac{a\beta_{21}}{\beta_{12} + \beta_{21}} + \frac{a\beta_{12}}{\beta_{12} + \beta_{21}} e^{-(\beta_{12}+\beta_{21})t}, \qquad x_2 = \frac{a\beta_{12}}{\beta_{12} + \beta_{21}} [1 - e^{-(\beta_{12}+\beta_{21})t}]$$

(b) Show that at all times $x_1 + x_2 = a$. (c) Show that the amounts of the drug present in the two compartments after a long time are given respectively by $a\beta_{21}/(\beta_{12} + \beta_{21})$ and $a\beta_{12}/(\beta_{12} + \beta_{21})$. (d) Interpret the results physically.

3. Suppose that in the two-compartment model of Exercise 2 we have (using the symbols on page 474) the volumes of the two compartments given by $V_1 = 25,000 \text{ cm}^3$, $V_2 = 40,000 \text{ cm}^3$, the area of membrane separating them 500 cm², and the proportionality constants $\alpha_{21} = 20$, $\alpha_{12} = 30$, $\alpha = 60$ (cgs units). Assuming that the amount of the drug present initially in compartment 1 to be 2.0 milligrams find (a) the amounts of the drug present in the compartments at any time $t > 0$, and (b) the amounts of the drug in the two compartments after a long time.

4. Work the problem of the two-compartment system if $\beta \neq 0$ for the cases (a) $\beta_{12} = 0$ (b) $\beta_{21} = 0$ and interpret physically.

5. Show that the disease model with quarantine given in the text reduces to the model given on page 154 if there is no quarantine.

B EXERCISES

1. Show that the number of students quarantined given by (32), page 478, increases as t increases and that

$$\lim_{t \to \infty} N_q = \frac{\lambda\sqrt{2}}{\kappa\sqrt{U_0}} (\alpha + \sqrt{I_0 + \alpha^2})$$

Does this agree with what you might expect for this limit? Explain.

2. Suppose that in the two-compartment model the drug is introduced into compartment 1 at a constant rate. (a) Set up a system of differential equations describing the process, and (b) solve the system subject to suitable conditions. Treat the cases $\beta = 0$ and $\beta \neq 0$ and interpret physically.

3. (a) Work Exercise 2 if the drug is introduced into compartment 1 at a rate which is a function of time t. (b) Use the result to obtain a solution to the case where the drug is introduced at a constant rate for a time interval T_1, stopped for the time interval T_2, and finally again introduced at the previous constant rate for the time interval T_1.

C EXERCISES

1. Set up a system of differential equations for the amounts of a drug in a three-compartment model, and thus obtain a generalization of the two-compartment model given in the text. Can you obtain the solution to this system subject to appropriate conditions?

2. Generalize the ideas formulated in Exercise 1 to an n-compartment model where $n > 3$.

3. How would you formulate the problem of epidemics given in the text if students who may be uninfected or at least not aware that they are infected "drop out" of the university at a rate proportional to the instantaneous number present?

Number

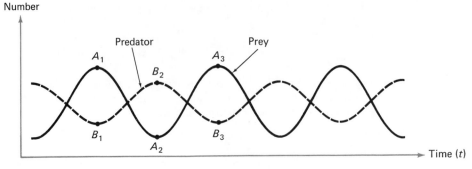

Figure 10.23

7 The Predator-Prey Problem: A Problem in Ecology

There are many instances in nature where one species of animal feeds on another species of animal, which in turn feeds on other things.

Example 1. Wolves in Alaska feed on caribou, which in turn feed on vegetation.

Example 2. Sharks in the oceans feed on small fish, which in turn feed on plants.

The first species is called the *predator* and the second species is called the *prey*.

Theoretically, the predator can destroy all the prey so that the latter become extinct. However, if this happens the predator will also become extinct since, as we assume, it depends on the prey for its existence.

What actually happens in nature is that a cycle develops where at some time the prey may be abundant and the predators few, as indicated by points A_1 and B_1, respectively, in Fig. 10.23. Because of the abundance of prey, the predator population grows and reduces the population of prey, leading to points A_2, B_2 of Fig. 10.23. This results in a reduction of predators and consequent increase of prey and the cycle continues. The graph of Fig. 10.23 should not of course be taken too literally, since we would not expect that the points of maximum prey coincide with points of minimum predators, but instead would expect some time lag or phase difference between them. Also, the actual curves might not be so symmetric as indicated in the figure, since in practice there are many other factors to contend with, such as climate.

An important problem of *ecology*, the science which studies the interrelationships of organisms and their environment, is to investigate the question of coexistence of the two species and to decide what mankind should do, if anything, to preserve this ecological balance of nature.

To answer this and other related questions, it is natural to seek a mathematical formulation of this predator-prey problem.*

* This and related problems were first investigated by *Lotka* and *Volterra* (See [18]).

7.1 MATHEMATICAL FORMULATION

Let us use the following notation:

x = number of the prey at any time t, y = number of the predator at any time t

Now if there were no predators, we would expect as a first approximation that the number of prey would increase at a rate proportional to their number at any time, so that

$$\frac{dx}{dt} = a_1 x \tag{1}$$

where $a_1 > 0$ is the constant of proportionality.

Similarly, if there were no prey, we would expect as a first approximation that the number of predators would decline at a rate proportional to their number so that

$$\frac{dy}{dt} = -b_1 y \tag{2}$$

where $b_1 > 0$ is the constant of proportionality.

Since (1) and (2) have respective solutions $x = c_1 e^{a_1 t}$ and $y = c_2 e^{-b_1 t}$, where c_1, c_2 are constants, we see that $x \to \infty$, $y \to 0$ as $t \to \infty$, which as expected does not agree with reality. To obtain a mathematical model of the actual situation, we must seek to modify equations (1) and (2) so as to take into account the interaction of the species on each other. To do this we should include in (1) an interaction term depending on x and y, say $F_1(x, y)$, which tends to diminish the rate at which x increases. Similarly, we should include in (2) an interaction term $F_2(x, y)$ which tends to increase the rate at which y decreases. The results are

$$\frac{dx}{dt} = a_1 x - F_1(x, y), \qquad \frac{dy}{dt} = -b_1 y + F_2(x, y) \tag{3}$$

A clue as to the forms of $F_1(x, y)$ and $F_2(x, y)$ is provided when we realize that they must be zero if either $x = 0$ or $y = 0$, i.e., if there are no predators or no prey. The simplest functions having this property are given by

$$F_1(x, y) = a_2 xy, \quad F_2(x, y) = b_2 xy \tag{4}$$

where a_2, b_2 are positive constants. Thus we are led to

$$\frac{dx}{dt} = a_1 x - a_2 xy, \qquad \frac{dy}{dt} = -b_1 y + b_2 xy \tag{5}$$

The assumptions (4) also seem reasonable from a physical point of view since we would expect the interaction terms to be proportional to the number of *meetings or encounters* of predator and prey, which is given by xy.

7.2 INVESTIGATION OF A SOLUTION

Although the non-linear system of equations (5), often called the *Lotka–Volterra equations*, look simple, no one has as yet been able to obtain an exact solution. We thus have recourse to use either series or numerical methods. It is possible, however, to gain some very important information from the equations even without solving them.

To begin with it is convenient to interpret equations (5) as giving the x and y components of the velocity of a particle moving in the xy plane. The curve, or as it is often called the *path*, *trajectory*, or *orbit*, in which the particle moves can be represented by the parametric equations

$$x = \phi_1(t, c_1, c_2), \qquad y = \phi_2(t, c_1, c_2) \tag{6}$$

where c_1, c_2 are constants which can be determined by prescribing a point through which the particle passes at some time, for example $t = 0$. The equations (6) represent solutions of the system (5) and also represent geometrically a family of curves or trajectories any one of which could be the path of the particle.

These ideas enable us to describe solutions of (5) by using concepts of mechanics and geometry and provide valuable insight into what is going on. The xy plane in which our "hypothetical particle" is moving is often called the *phase plane*, and analysis using such interpretation is appropriately called *phase plane analysis*. The curves of the family (6) are then often referred to as *phase curves*.*

Now just as a tourist may find a point at which he *stops* of more interest than a point through which he *travels*, so we may find points at which our particle stops moving of greater interest than points through which it moves. Such points where the particle stops moving, which we shall aptly call *rest points* or *equilibrium points*, occur where the velocity is zero, i.e., where $dx/dt = 0$ and $dy/dt = 0$. From equations (5) we see that

$$\frac{dx}{dt} = 0 \quad \text{for} \quad x = 0 \quad \text{or} \quad y = \frac{a_1}{a_2}, \qquad \frac{dy}{dt} = 0 \quad \text{for} \quad y = 0 \quad \text{or} \quad x = \frac{b_1}{b_2} \tag{7}$$

Thus there are two possible equilibrium points,

$$x = 0, \qquad y = 0 \quad \text{or} \quad (0, 0) \quad \text{and} \quad x = \frac{b_1}{b_2}, \qquad y = \frac{a_1}{a_2} \quad \text{or} \quad \left(\frac{b_1}{b_2}, \frac{a_1}{a_2}\right). \tag{8}$$

It is also interesting to note that these are also solutions of (5) which are independent of time and can be considered as special cases of (6) in which the curves degenerate into points. We shall now examine the two cases (8) and in particular seek to determine the behavior of solutions of (5) near the equilibrium points.

Case 1, Equilibrium Point $(0, 0)$. This is the case where there is neither predator nor prey. When the particle in our phase plane interpretation is sufficiently near this point, we can neglect the second terms $a_2 xy$ and $b_2 xy$ on the right of equations (5) in comparison to the first terms. Thus the equations become (1) and (2) on page 481 with which we began our discussion, i.e.,

$$\frac{dx}{dt} = a_1 x, \qquad \frac{dy}{dt} = -b_1 y \tag{9}$$

with solution
$$x = c_1 e^{a_1 t}, \qquad y = c_2 e^{-b_1 t} \tag{10}$$

_____

* The phase plane was used extensively by the mathematicians *Poincaré* and *Liapunov* in their researches on non-linear systems of differential equations.

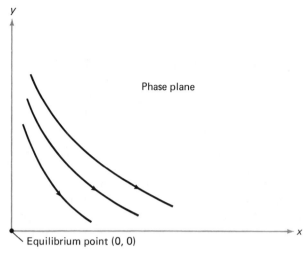

Phase plane

Equilibrium point (0, 0)

Figure 10.24

Since x and y are non-negative, we have $c_1 \geqq 0$, $c_2 \geqq 0$, and the family of curves described by (10) is as shown in Fig. 10.24, the particular case $c_1 = 0$, $c_2 = 0$ corresponding to the equilibrium point (degenerate curve) $x = 0$, $y = 0$. As t increases we see from (10) that x increases while y approaches zero, so that the particle moves in the direction shown in Fig. 10.24. It is apparent from these curves that if we should displace the particle slightly from the equilibrium point (0, 0) it tends to move *away* from the point, as shown in Fig. 10.24. For this reason we call the equilibrium point an *unstable equilibrium point* or a point of *instability*. If on the other hand a slight displacement from the equilibrium point resulted in a tendency for the particle to return or move toward the equilibrium point, as in the case of a mass on a stretched spring, we would call the point a *stable equilibrium point*. The situation can also be illustrated by referring to Figs. 10.25 and 10.26 below. In Fig. 10.25 a small hollow cylinder A rests on the uppermost point P of a larger hollow cylinder B. A slight displacement from point P results in the cylinder moving away and P is an unstable equilibrium point. In Fig. 10.26, in which A rests on the bottom of B at point P, any

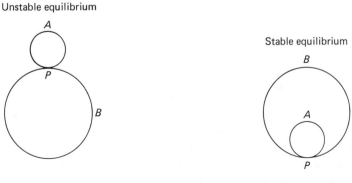

Unstable equilibrium

Stable equilibrium

Figure 10.25

Figure 10.26

slight displacement of cylinder A from point P results in it returning toward P so that P is a stable equilibrium point.

The interpretation is that if at some time there is a small number of predators and prey there will be a tendency for the number of prey to increase and the number of predators to decrease, as indicated in Fig. 10.24. In reality, however, x does not increase indefinitely since an increase in prey results in a subsequent increase in predator. This allows for the possibility that if the curves of Fig. 10.24 are extended they will tend to rise after some point.

Case 2, Equilibrium Point $(b_1/b_2, a_1/a_2)$. This is the case where predator and prey are in an equilibrium state in which their numbers do not change because $x = b_1/b_2$, $y = a_1/a_2$ is a solution of (5) independent of time t. It is of interest to ask the question, What happens if there is a slight departure from this equilibrium state, which can occur if for example hunters destroy either predator or prey (or both)?

To answer this question we make the transformation in (5) given by

$$x = \frac{b_1}{b_2} + u, \qquad y = \frac{a_1}{a_2} + v \tag{11}$$

so that
$$\frac{du}{dt} = -\frac{a_2 b_1}{b_2} v - a_2 uv, \qquad \frac{dv}{dt} = \frac{a_1 b_2}{a_2} u + b_2 uv \tag{12}$$

Now if the particle of our phase plane interpretation is close to the point $(b_1/b_2, a_1/a_2)$ of the xy plane, then u and v will be close to zero. In such case the second terms on the right of equations (12) can be neglected in comparison to the first terms so that we obtain the linearized system

$$\frac{du}{dt} = -\frac{a_2 b_1}{b_2} v, \qquad \frac{dv}{dt} = \frac{a_1 b_2}{a_2} u \tag{13}$$

Elimination of v yields
$$\frac{d^2 u}{dt^2} + a_1 b_1 u = 0 \tag{14}$$

with solution
$$u = c_1 \cos \sqrt{a_1 b_1} t + c_2 \sin \sqrt{a_1 b_1} t \tag{15}$$

so that
$$v = \frac{b_2}{a_2} \sqrt{\frac{a_1}{b_1}} (c_1 \sin \sqrt{a_1 b_1} t - c_2 \cos \sqrt{a_1 b_1} t) \tag{16}$$

Thus equations (11) become

$$x = \frac{b_1}{b_2} + c_1 \cos \sqrt{a_1 b_1} t + c_2 \sin \sqrt{a_1 b_1} t,$$

$$\left.\begin{array}{l} \\ y = \frac{a_1}{a_2} + \frac{b_2}{a_2} \sqrt{\frac{a_1}{b_1}} (c_1 \sin \sqrt{a_1 b_1} t - c_2 \cos \sqrt{a_1 b_1} t) \end{array}\right\} \tag{17}$$

The parametric equations (17) represent concentric ellipses having common center $(b_1/b_2, a_1/a_2)$ and direction counterclockwise, as shown in Fig. 10.27. This direction can be obtained from the solution (17) by noting how the point (x, y) moves as t increases or directly from the differential equations (13). To see how the direction is determined directly from the differential equations (13), let us refer to Fig. 10.28 in

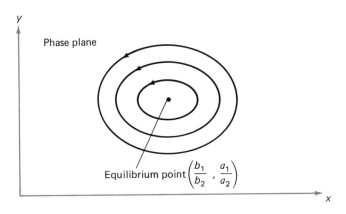

Figure 10.27

which we have shown a uv coordinate system with origin at the common center of the ellipses. From (13) we see that at a point in the first quadrant where $u > 0, v > 0$ the components of velocity are such that $du/dt < 0, dv/dt > 0$, so that the velocity has the counterclockwise direction indicated. This same direction can be confirmed by choosing points in other quadrants. Thus in the third quadrant where $u < 0$, $v < 0$, we have $du/dt > 0, dv/dt < 0$, showing again that the velocity is in the counterclockwise direction. In practice, it is often easier to determine the direction of the phase particle from the differential equation rather than from the solution.

The fact that the phase curves are concentric ellipses can also be deduced directly from equations (13) by dividing corresponding sides of the equations so as to eliminate t. We then obtain

$$\frac{du}{dv} = -\left(\frac{b_1 a_2^2}{a_1 b_2^2}\right)\frac{v}{u} \quad \text{or} \quad a_1 b_2^2 u \, du + b_1 a_2^2 v \, dv = 0$$

and by integration

$$\frac{u^2}{b_1 a_2^2} + \frac{v^2}{a_1 b_2^2} = c^2 \tag{18}$$

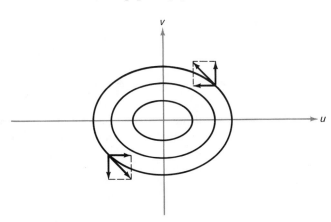

Figure 10.28

where we have chosen c^2 as the constant of integration. If we divide both sides of (18) by c^2, we see that it represents a family of concentric ellipses whose semimajor and minor axes have lengths $ca_2\sqrt{b_1}$ and $cb_2\sqrt{a_1}$.

The curves of Fig. 11.5 show that near the equilibrium point there is a periodicity in the numbers of prey and predator, and as we see from (17) the period is given by

$$T = \frac{2\pi}{\sqrt{a_1 b_1}} \tag{19}$$

In general, a closed curve in the phase plane which does not cross itself anywhere, often referred to as a *simple closed curve*, indicates a periodicity in the solution of the differential equations.

A question which naturally arises is whether we should consider the point $(b_1/b_2, a_1/a_2)$ a stable or an unstable equilibrium point. Since a slight displacement of our phase plane particle from the equilibrium point does not result in a movement either away or toward but only *around* the point, we might be inclined to call it *neither*, and maybe even make up a new word to describe the behavior. However, because a slight displacement does not cause the particle to get too far away, but instead to stay in the "neighborhood" of the point, we shall stretch our idea of stable equilibrium point to include as well this type of behavior.*

We can obtain some further interesting information from (17). To do so let $\theta = \sqrt{a_1 b_1} t$ and note that we can write

$$c_1 \cos \theta + c_2 \sin \theta = \sqrt{c_1^2 + c_2^2} \cos (\theta - \alpha) \tag{20}$$

where
$$\cos \alpha = \frac{c_1}{\sqrt{c_1^2 + c_2^2}}, \qquad \sin \alpha = \frac{c_2}{\sqrt{c_1^2 + c_2^2}} \tag{21}$$

Replacing θ by $\theta - (\pi/2)$ in (20), we then find

$$c_1 \sin \theta - c_2 \cos \theta = \sqrt{c_1^2 + c_2^2} \cos \left(\theta - \alpha - \frac{\pi}{2} \right) \tag{22}$$

Thus (17) can be written

$$\left. \begin{aligned} x &= \frac{b_1}{b_2} + \sqrt{c_1^2 + c_2^2} \cos \sqrt{a_1 b_1} \left(t - \frac{\alpha}{\sqrt{a_1 b_1}} \right) \\ y &= \frac{a_1}{a_2} + \frac{b_2}{a_2} \sqrt{\frac{a_1}{b_1}} \sqrt{c_1^2 + c_2^2} \cos \sqrt{a_1 b_1} \left(t - \frac{\alpha + \pi/2}{\sqrt{a_1 b_1}} \right) \end{aligned} \right\} \tag{23}$$

Equations (23) show that at time $t = \alpha/\sqrt{a_1 b_1}$, for example, the number of prey is a maximum. However, the number of predators does not reach a maximum until $t = (\alpha + \pi/2)/\sqrt{a_1 b_1}$, i.e., a time $\pi/2\sqrt{a_1 b_1}$ or one quarter of the period (19) later. The graphs of x and y versus t thus appear as in Fig. 10.29, which is what we surmised on page 480.

---

* It is possible to formulate a more precise definition of stable and unstable equilibrium points. See for example [3].

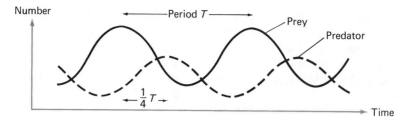

Figure 10.29

In the above analysis we examined the situation only near the equilibrium points. The question naturally arises as to the behavior of solutions elsewhere in the phase plane. Figures 10.24 and 10.28 suggest that the curves may appear as in Fig. 10.30. However, this would suggest that all the curves are in fact closed and that all solutions of (5) are periodic. By further analysis we can show that this is actually the case. The procedure involves elimination of t between equations (5) by dividing the two equations to obtain

$$\frac{dy}{dx} = \frac{(-b_1 + b_2 x)y}{(a_1 - a_2 y)x} \tag{24}$$

This equation, which is the differential equation of the family of curves or trajectories in the phase plane, has its variables separable and can be written

$$\left(\frac{b_1 - b_2 x}{x}\right) dx + \left(\frac{a_1 - a_2 y}{y}\right) dy = 0 \tag{25}$$

Integrating we find $\quad b_1 \ln x - b_2 x + a_1 \ln y - a_2 y = \text{constant} \tag{26}$

or $\qquad\qquad\qquad x^{b_1} y^{a_1} e^{-(b_2 x + a_2 y)} = A \tag{27}$

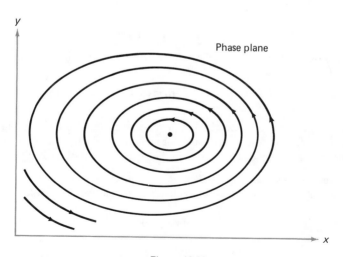

Figure 10.30

where A is an arbitrary constant. These can be shown to be closed curves, and as a consequence that solutions of (5) are periodic. However, one should not jump to the conclusion that the period for all such curves is given by (19). In actuality the period is a complicated function of the constants a_1, b_1, a_2, b_2 which reduces to (19) near the equilibrium point. The precise period for all cases has not been found.

7.3 SOME FURTHER APPLICATIONS

The ideas just presented are applicable in other connections. For example instead of having predator and prey we could have the situation where a species of insects depend for their survival on certain animals or plants by actually feeding on them. The former are often referred to as *parasites* and the latter as *hosts*. Whereas the prey is destroyed by the predator, the parasite finds it advantageous to keep the host alive so that it may continue to feed on it. As another example we can have more complicated situations arising when two or more species compete and may even destroy each other for the same prey.

The applications are not limited to ecology or the life sciences but can occur in other fields. In economics for instance, nations compete with each other for trade which in a sense is economic survival, producers compete with each other for consumers, and even consumers may compete with each other for products. In this connection it is interesting to note the similarity in appearance of the graph of Fig. 5.19, page 252, involving price and supply of a commodity with that of Fig. 10.29, page 487, involving predator and prey.

The concepts of the phase plane and stability described in the predator–prey problem can be extended to other non-linear systems

$$\frac{dx}{dt} = F(x, y), \qquad \frac{dy}{dt} = G(x, y)$$

However we shall not pursue this here but instead refer the interested students to references in the Bibliography.

A EXERCISES

1. Suppose that the system of differential equations describing a particular predator–prey model is given by

$$\frac{dx}{dt} = 10^{-2}x - 2 \cdot 10^{-5}xy, \qquad \frac{dy}{dt} = -4 \cdot 10^{-2}y + 10^{-5}xy$$

where y and x are the numbers of predator and prey, respectively, and t is the time in days. (a) Determine the numbers of predator and prey at equilibrium. (b) Find the period. (c) Write the equations of the phase curves in the neighborhood of the equilibrium point.

2. By making the change of independent variables $x = k_1 X$, $y = k_2 Y$, where k_1 and k_2 are constants show that the Lotka–Volterra equations can be transformed into the system

$$\frac{dX}{dt} = \alpha(XY - X), \qquad \frac{dY}{dt} = \beta(Y - XY)$$

and give the values of k_1, k_2, α, β in terms of a_1, a_2, b_1, b_2. What are the equilibrium values of X and Y? Discuss some advantages of this transformed system.

1. According to the Lotka–Volterra equations (5), if there are no predators the number of prey should increase indefinitely. Suppose however that we revise these equations so as to provide a mathematical model in which there is some theoretical maximum to the number of prey denoted by x_M. Show that a system suitable for such a model is given by

$$\frac{dx}{dt} = a_1 x \left(1 - \frac{x}{x_M} \right) - a_2 xy, \qquad \frac{dy}{dt} = -b_1 y + b_2 xy$$

2. Discuss how you might modify the Lotka–Volterra equations if prey or predator are introduced or removed from the environment at specified times. For what realistic situations might such a model serve?

3. Show how to arrive at the result (18), page 485, directly from (27), page 487.

4. (a) Show that a system of equations for describing the undamped motion of a mass m at the end of a vertical spring of constant k is given by

$$\frac{dx}{dt} = y, \qquad \frac{dy}{dt} = \frac{-kx}{m}$$

   (b) Show that in the phase plane the mass travels along an ellipse with equation

$$\tfrac{1}{2} kx^2 + \tfrac{1}{2} my^2 = E$$

   where E is a constant and explain the physical significance.

   (c) Explain why you would expect the motion to be periodic.

1. Carry through the analysis of the system of equations given in B Exercise 1 as far as you are able using the analogy of what was done in the text.

2. Discuss how you might proceed with the analysis of the equations obtained in B Exercise 2.

3. Use phase plane analysis to describe the motion of a simple pendulum if the angle made with the vertical is (a) small (b) not small.

4. Discuss the stability of the system

$$\frac{dx}{dt} = y, \qquad \frac{dy}{dt} = x^2 - 4x$$

   Is the motion periodic? Can you give a possible physical interpretation?

5. Given

$$\frac{dx}{dt} = \frac{x}{\sqrt{x^2 + y^2}} - x - y, \qquad \frac{dy}{dt} = \frac{y}{\sqrt{x^2 + y^2}} + x - y$$

   show that as t increases, the phase trajectories tend to approach the circle $x^2 + y^2 = 1$ as a limit. Such closed curves which trajectories tend to approach (or recede from) are called *limit cycles* and were investigated extensively by Poincaré and others. (*Hint:* Use polar coordinates.)

8 Solutions of Linear Systems by Laplace Transforms

The method of Laplace transforms already studied in Chapter Six can without any added difficulty be used to solve simultaneous linear differential equations, especially those with constant coefficients.

Solve $Y' + 6Y = X'$, $\qquad 3X - X' = 2Y'$, $\qquad X(0) = 2$, $\qquad Y(0) = 3$

where primes denote derivatives with respect to t.

Solution Taking the Laplace transform of each differential equation and using the given conditions we have if $\mathcal{L}\{X\} = x$ and $\mathcal{L}\{Y\} = y$

$$\{sy - Y(0)\} + 6y = \{sx - X(0)\}, \qquad 3x - \{sx - X(0)\} = 2\{sy - Y(0)\}$$

or $\qquad\qquad sx - (s + 6)y = 1, \qquad (3 - s)x - 2sy = -8$

Solving for x and y, we have

$$x = \frac{2s + 16}{(s - 2)(s + 3)} = \frac{4}{s - 2} - \frac{2}{s + 3}, \qquad y = \frac{3s - 1}{(s - 2)(s + 3)} = \frac{1}{s - 2} + \frac{2}{s + 3}$$

Thus $\qquad\qquad X = \mathcal{L}^{-1}\left\{\dfrac{4}{s - 2} - \dfrac{2}{s + 3}\right\} = 4e^{2t} - 2e^{-3t}$,

$$Y = \mathcal{L}^{-1}\left\{\frac{1}{s - 2} + \frac{2}{s + 3}\right\} = e^{2t} + 2e^{-3t}$$

As an illustration of Laplace transforms in an applied problem, consider

Work the problem of the vibrating masses on page 463 using Laplace transforms.

Solution The system of equations arrived at on page 464, together with initial conditions, are given by

$$x_1'' = \omega^2(x_2 - 2x_1), \qquad x_2'' = \omega^2(x_1 - 2x_2) \tag{1}$$

$$x_1(0) = 0, \qquad x_2(0) = a, \qquad x_1'(0) = 0, \qquad x_2'(0) = 0 \tag{2}$$

where the primes denote derivatives with respect to t and where $\omega^2 = k/M$. Taking the Laplace transform of each of the equations in (1), and using the notation $\mathcal{L}(x_1) = \bar{x}_1$, $\mathcal{L}(x_2) = \bar{x}_2$ we find

$$s^2\bar{x}_1 - sx_1(0) - x_1'(0) = \omega^2(\bar{x}_2 - 2\bar{x}_1), \qquad s^2\bar{x}_2 - sx_2(0) - x_2'(0) = \omega^2(\bar{x}_1 - 2\bar{x}_2)$$

or $\qquad\qquad (s^2 + 2\omega^2)\bar{x}_1 - \omega^2\bar{x}_2 = 0, \qquad -\omega^2\bar{x}_1 + (s^2 + 2\omega^2)\bar{x}_2 = sa$

which can be solved for $\bar{x}_1$ and $\bar{x}_2$ to yield

$$\bar{x}_1 = \frac{a\omega^2 s}{s^4 + 4\omega^2 s^2 + 3\omega^4} = \frac{a\omega^2 s}{(s^2 + \omega^2)(s^2 + 3\omega^2)} = \frac{a}{2}\left[\frac{1}{s^2 + \omega^2} - \frac{1}{s^2 + 3\omega^2}\right]$$

$$\bar{x}_2 = \frac{as(s^2 + 2\omega^2)}{s^4 + 4\omega^2 s^2 + 3\omega^4} = \frac{as(s^2 + 2\omega^2)}{(s^2 + \omega^2)(s^2 + 3\omega^2)} = \frac{a}{2}\left[\frac{1}{s^2 + \omega^2} + \frac{1}{s^2 + 3\omega^2}\right]$$

Then $\qquad x_1 = \dfrac{a}{2}(\cos \omega t - \cos \sqrt{3}\omega t), \qquad x_2 = \dfrac{a}{2}(\cos \omega t + \cos \sqrt{3}\omega t)$

in agreement with the results on page 465.

1. Solve each of the following systems of equations by Laplace transforms.
   (a) $X' = Y$, $Y' = -X$, $X(0) = 2$, $Y(0) = -1$.
   (b) $X' + X - 5Y = 0$, $Y' + 4X + 5Y = 0$, $X(0) = -1$, $Y(0) = 2$.
   (c) $X' + 3Y' + Y = e^t$, $Y' - X = Y$, $X(0) = 0$, $Y(0) = 1$.
   (d) $X' - 3X - 6Y = 27t^2$, $X' + Y' - 3Y = 5e^t$, $X(0) = 5$, $Y(0) = -1$.

2. Solve using Laplace transforms.

   (a) $\dfrac{d^2 X}{dt^2} = -2Y$, $\dfrac{dY}{dt} = Y - \dfrac{dX}{dt}$, $X(0) = 0$, $X'(0) = 10$, $Y(0) = 5$.

   (b) $Y'' = X - 2$, $X'' = Y + 2$, $X(0) = 0$, $X'(0) = 1$, $Y(0) = 2$, $Y'(0) = -3$.

3. Solve $X' + Y' = \cos t$, $X + Y'' = 2$, $X(\pi) = 2$, $Y(0) = 0$, $Y'(0) = \frac{1}{2}$ by Laplace transforms. [Hint: Let $X(0) = c$ and then solve for c.)

1. Work B Exercise 1, page 445, by use of Laplace transforms.

2. Solve $X' = X + Y + Z$, $Y' = 2X + 5Y + 3Z$, $Z' = 3X + 9Y + 5Z$ subject to the conditions $X(0) = -2$, $Y(0) = -1$, $Z(0) = 3$.

3. Work (a) A Exercise 1(g), page 444, (b) A Exercise 6, page 445, by using Laplace transforms.

4. Use Laplace transforms to work (a) A Exercise 1, page 471, (b) A Exercise 2, page 471, (c) A Exercise 3, page 471, (d) B Exercise 1, page 472.

5. Solve $\dfrac{d^2 x}{dt^2} = y + 4e^{-2t}$, $\dfrac{d^2 y}{dt^2} = x - e^{-2t}$ where $x(0) = y(0) = x'(0) = y'(0) = 0$.

1. Work C Exercise 1, page 445, by using Laplace transforms.

2. Use Laplace transforms to work (a) C Exercise 4, page 450, (b) C Exercise 5, page 451, (c) C Exercise 6, page 451.

3. Work C Exercise 1, page 472, by using Laplace transforms.

4. Solve the system of equations (11), page 475, subject to the condition (13) by using Laplace transforms.

9

Method of Complementary and Particular Solutions

When we solved a linear system of differential equations by the method of elimination (see page 434), we arrived at a single ordinary linear differential equation. Then we found the complementary and particular solutions of this single equation and from them its general solution. We then used this general solution to solve the given system of equations. The question naturally arises as to whether we can obtain complementary and particular solutions directly from the system *without* first using elimination. For the system of linear equations

$$\left.\begin{array}{l}\phi_1(D)x + \psi_1(D)y = F_1(t)\\ \phi_2(D)x + \psi_2(D)y = F_2(t)\end{array}\right\}$$ (1)

where $D \equiv d/dt$, this would amount to finding the general solution of the complementary system of equations

$$\left.\begin{array}{l}\phi_1(D)x + \psi_1(D)y = 0\\ \phi_2(D)x + \psi_2(D)y = 0\end{array}\right\}$$ (2)

[obtained from (1) by replacing the right-hand sides by zero], which we can call the complementary solution x_c, y_c of (1), and then finding a particular solution x_p, y_p of the given system (1). The required general solution of (1) should then be given by

$$x = x_c + x_p, y = y_c + y_p.$$

As in the case of a single ordinary linear differential equation, there are two theorems which are of fundamental importance.

Theorem 1. Let x_c, y_c be the general solution of the complementary system (2), and x_p, y_p any particular solution of the given system (1). Then the general solution of (1) is given by

$$x = x_c + x_p, \qquad y = y_c + y_p$$ (3)

Theorem 2 (Principle of Superposition). Let $x_1, y_1; x_2, y_2; \ldots$ be any number of solutions of the complementary system (2), and let $\alpha_1, \alpha_2, \ldots$ be arbitrary constants.

Then $\qquad x = \alpha_1 x_1 + \alpha_2 x_2 + \cdots, \qquad y = \alpha_1 y_1 + \alpha_2 y_2 + \cdots$ (4)

is also a solution of (2).

Proofs of these theorems are easy and very similar to those for single linear equations (see page 169).

Proof of Theorem 1. By the definitions of x_c, y_c and x_p, y_p, we have

(1) $\phi_1(D)x_c + \psi_1(D)y_c = 0,$ (3) $\phi_1(D)x_p + \psi_1(D)y_p = F_1(t)$

(2) $\phi_2(D)x_c + \psi_2(D)y_c = 0,$ (4) $\phi_2(D)x_p + \psi_2(D)y_p = F_2(t)$

Adding equations (1) and (3) and then equations (2) and (4), respectively, and making use of the linearity of the operators, we find

$$\phi_1(D)(x_c + x_p) + \psi_1(D)(x_c + x_p) = F_1(t),$$
$$\phi_2(D)(y_c + y_p) + \psi_2(D)(y_c + y_p) = F_2(t)$$ (5)

which shows that (3) is a solution of (1). This will be the general solution as we see from Remark 1, page 439, and the fact that the number of arbitrary constants in $x_c + x_p$, $y_c + y_p$ is the same as in x_c, y_c.

Proof of Theorem 2. By the definitions of $x_1, y_1; x_2, y_2; \ldots$, we have

$$\begin{aligned} \phi_1(D)x_1 + \psi_1(D)y_1 &= 0 \\ \phi_2(D)x_1 + \psi_2(D)y_1 &= 0 \end{aligned}\Big\}, \qquad \begin{aligned} \phi_1(D)x_2 + \psi_1(D)y_2 &= 0 \\ \phi_2(D)x_2 + \psi_2(D)y_2 &= 0 \end{aligned}\Big\}, \ldots \qquad (6)$$

Multiplying the first equation in each of these systems by $\alpha_1, \alpha_2, \ldots$, respectively, and adding, using the linearity of the operators, we have

$$\phi_1(D)(\alpha_1 x_1 + \alpha_2 x_2 + \cdots) + \psi_1(D)(\alpha_1 x_1 + \alpha_2 x_2 + \cdots) = 0 \qquad (7)$$

Doing the same thing to each of the second equations in (6), we have

$$\phi_2(D)(\alpha_1 x_1 + \alpha_2 x_2 + \cdots) + \psi_2(D)(\alpha_1 x_1 + \alpha_2 x_2 + \cdots) = 0 \qquad (8)$$

From (7) and (8) we see that (4) is a solution of (1), which proves the theorem.

Although we have stated the above results for the system (1) involving two equations in two unknowns x and y, they hold as well for systems involving n equations in n unknowns, where $n > 2$. To illustrate the procedures involved for finding solutions in the case of n unknowns, we can use the case of two unknowns since the procedures for both cases are essentially identical. We shall also concentrate on the cases where the operators involved are polynomials in $D \equiv d/dt$ with constant coefficients as in Chapter Four although the above theorems are valid as well for variable coefficients. Since finding the general solution of the complementary system (2) is basic to finding the general solution of (1), we first turn our attention to the question of finding the complementary solution.

9.1 HOW DO WE FIND THE COMPLEMENTARY SOLUTION?

Let us consider the following

PROBLEM FOR DISCUSSION

Solve the system of equations $\qquad \begin{aligned} (2D - 2)x + (D - 7)y &= 0 \\ (3D - 2)x + (2D - 8)y &= 0 \end{aligned}\Big\}. \qquad (9)$

For a single differential equation $\phi(D)y = 0, D \equiv d/dx$, we recall that we looked for solutions of the form $y = e^{mx}$ and then determined values of the constant m from the auxiliary equation. We do a similar thing for the system (9), assuming that there is a solution of the form

$$x = ae^{mt}, \qquad y = be^{mt} \qquad (10)$$

where m, a, b are constants undetermined at present. We use two constants a and b since it is highly unlikely that we would have the same coefficients in (10). If we substitute (10) into (9), we find after dividing by $e^{mt} \neq 0$,

$$\begin{aligned} (2m - 2)a + (m - 7)b &= 0 \\ (3m - 2)a + (2m - 8)b &= 0 \end{aligned}\Big\} \qquad (11)$$

Note that (11) is formally obtained from (9) on replacing D by m, x by a, and y by b, an observation which is often useful in saving time. Now equations (11) will have a non-trivial solution for a and b (i.e., one in which a and b are not both zero) if and only if the determinant of the coefficients of a and b in (11) is zero, i.e.,

$$\begin{vmatrix} 2m - 2 & m - 7 \\ 3m - 2 & 2m - 8 \end{vmatrix} = 0 \tag{12}$$

This gives $\qquad m^2 + 3m + 2 = 0 \quad$ or $\quad m = -1, -2$

Note that this corresponds to the auxiliary equation of Chapter Four and the roots of the auxiliary equation. For this case the roots happen to be real and distinct, but they could have been equal or imaginary. There are two cases to examine.

Case 1, $m = -1$. In this case equations (11) give $-4a - 8b = 0$, $-5a - 10b = 0$, both of which lead to $a = -2b$. If we let $b = c_1$, an arbitrary constant, then $a = -2c_1$. We are thus led to the solution

$$x = -2c_1 e^{-t}, \qquad y = c_1 e^{-t}$$

Case 2, $m = -2$. In this case equations (11) give $-6a - 9b = 0$, $-8a - 12b = 0$, both of which lead to $a = -\frac{3}{2}b$. To avoid fractions, let us write $b = 2c_2$, where c_2 is another arbitrary constant, so that $a = -3c_2$. We are thus led to

$$x = -3c_2 e^{-2t}, \qquad y = 2c_2 e^{-2t}$$

Because sums of solutions are also solutions, we are led to

$$x = -2c_1 e^{-t} - 3c_2 e^{-2t}, \qquad y = c_1 e^{-t} + 2c_2 e^{-2t} \tag{13}$$

which can be checked as the required solution by direct substitution in (9).
Some remarks should be made about the above procedure.

Remark 1. Unlike the method of elimination, the proper number of arbitrary constants in (13) is automatically obtained. Thus we do not have to bother to find relationships among constants, as required if the method of elimination is used.

Remark 2. As observed in Cases 1 and 2 above, a value of m leads to two equivalent equations involving a and b. We need obtain only one. Sometimes, however, it may be worthwhile to obtain both as a check on the work.
The technique just considered is clearly applicable whenever the roots are real and distinct. A question which we might very well ask now is, What happens if the roots are repeated. Let us examine this situation in the following

PROBLEM 2 FOR DISCUSSION

Solve the system of equations $\qquad \begin{aligned} Dx + (3D + 1)y &= 0 \\ (D + 2)x + (D + 3)y &= 0 \end{aligned} \Big\} . \tag{14}$

By the usual procedure of letting $x = ae^{mt}$, $y = be^{mt}$, we obtain

$$\begin{aligned} ma + (3m + 1)b &= 0 \\ (m + 2)a + (m + 3)b &= 0 \end{aligned} \Big\} \tag{15}$$

Thus, if we want non-trivial solutions for a and b, we must have

$$\begin{vmatrix} m & 3m + 1 \\ m + 2 & m + 3 \end{vmatrix} = 0, \quad \text{i.e.,} \quad m^2 + 2m + 1 = 0 \quad \text{and } m = -1, -1$$

Case 1, $m = -1$. Putting $m = -1$ in the first equation of (15), we find $-a - 2b = 0$ or $a = -2b$, which is also obtained from the second equation of (15). Thus taking $b = c_1$, $a = -2c_1$, we see that a solution is

$$x = -2c_1e^{-t}, \qquad y = c_1e^{-t} \tag{16}$$

Now obviously there is no point in considering $m = -1$ again as Case 2, so it would seem as if we are stuck at this point. However, on recalling the similar situation in Chapter Four, we may be led to multiply the results in (16) by t, using only a different constant c_2, and hope that

$$x = -2c_2te^{-t}, \qquad y = c_2te^{-t} \tag{17}$$

will also be a solution. Unfortunately, if we substitute the assumed solution (17) in the two equations of (14), we are led, respectively, to $c_2e^{-t} = 0$ and $-c_2e^{-t} = 0$, showing that (17) is not a solution. The fact that we got c_2e^{-t} and $-c_2e^{-t}$ in place of zero, however, suggests the possibility that we should add suitable constant multiples of e^{-t} to x and y in (17), i.e.,

$$x = -2c_2te^{-t} + K_1e^{-t}, \qquad y = c_2te^{-t} + K_2e^{-t} \tag{18}$$

and then seek to determine the constants K_1 and K_2 so that (18) is a solution of (14).

Then $\qquad c_2e^{-t} = K_1e^{-t} + 2K_2e^{-t}, \qquad -c_2e^{-t} = -K_1e^{-t} - 2K_2e^{-t} \tag{19}$

From each of the results in (19) we obtain

$$K_1 + 2K_2 = c_2 \tag{20}$$

which shows that it was necessary to substitute (18) into only one of the equations in (14). The result (20) also shows that we can use any values for K_1 and K_2 which are consistent with (20). For example, let us choose $K_2 = c_1$, $K_1 = c_2 - 2c_1$. For this case we obtain from (18)

$$x = -2c_1e^{-t} - 2c_2te^{-t} + c_2e^{-t}, \qquad y = c_1e^{-t} + c_2te^{-t} \tag{21}$$

which is the required general solution.

We have now treated the cases of real, distinct roots and repeated roots. By analogy with Chapter Four, we should now treat the final case where roots may be imaginary. To illustrate the procedure in this case, let us consider the following.

PROBLEM 3 FOR DISCUSSION

Solve the system of equations $\qquad \left. \begin{aligned} (4D + 1)x + (2D + 7)y &= 0 \\ (D - 1)x + (D + 1)y &= 0 \end{aligned} \right\}. \tag{22}$

Putting $x = ae^{mt}$, $y = be^{mt}$ in (22), we find

$$\left. \begin{aligned} (4m + 1)a + (2m + 7)b &= 0 \\ (m - 1)a + (m + 1)b &= 0 \end{aligned} \right\} \tag{23}$$

To obtain non-trivial solutions a and b, we must have

$$\begin{vmatrix} 4m + 1 & 2m + 7 \\ m - 1 & m + 1 \end{vmatrix} = 0, \quad \text{i.e.,} \quad m^2 + 4 = 0 \quad \text{and} \quad m = \pm 2i$$

Case 1, $m = 2i$. Putting $m = 2i$ in the second of equations (23) which appears to be simpler than the first, we get $(2i - 1)a + (2i + 1)b = 0$, so that

$$a = -\frac{2i + 1}{2i - 1}b = -\frac{(2i + 1)(-2i - 1)}{(2i - 1)(-2i - 1)}b = -\frac{3 - 4i}{5}b$$

This same result would of course be obtained if we used the first equation. Letting $b = 5c_1$, where c_1 is an arbitrary constant, we have $a = -(3 - 4i)c_1$. A solution corresponding to this is

$$x = -(3 - 4i)c_1 e^{2it}, \qquad y = 5c_1 e^{2it} \tag{24}$$

Case 2, $m = -2i$. We could now go through the same procedure as in Case 1. However, it is easier to note that the results would be the same as that of Case 1 with i replaced by $-i$, except that a different arbitrary constant, say c_2, would be used. This leads to the solution

$$x = -(3 + 4i)c_2 e^{-2it}, \qquad y = 5c_2 e^{-2it} \tag{25}$$

From these two cases we thus have as the required solution

$$x = -(3 - 4i)c_1 e^{2it} - (3 + 4i)c_2 e^{-2it}, \qquad y = 5c_1 e^{2it} + 5c_2 e^{-2it} \tag{26}$$

This is indeed the general solution. The only trouble is that it is not a *real* solution. We can, however, obtain a real solution from it. To do so we make use of Euler's identities (see page 178) to write (26) as

$$\left. \begin{aligned} x &= -(c_1 + c_2)(3 \cos 2t + 4 \sin 2t) - i(c_1 - c_2)(3 \sin 2t - 4 \cos 2t) \\ y &= 5(c_1 + c_2) \cos 2t + 5i(c_1 - c_2) \sin 2t \end{aligned} \right\} \tag{27}$$

Now since $c_1 + c_2$ and $i(c_1 - c_2)$ are also arbitrary constants, we can denote them, respectively, by A and B, so that (26) can be written in real form as

$$\left. \begin{aligned} x &= -A(3 \cos 2t + 4 \sin 2t) - B(3 \sin 2t - 4 \cos 2t) \\ y &= 5A \cos 2t + 5B \sin 2t \end{aligned} \right\} \tag{28}$$

which is the required general solution of (22).

Remark 3. There is a short-cut way to obtain (28) directly from (26). To see this we first observe that

$$\left. \begin{aligned} -(3 - 4i)e^{2it} &= -(3 - 4i)(\cos 2t + i \sin 2t) \\ &= -(3 \cos 2t + 4 \sin 2t) - i(3 \sin 2t - 4 \cos 2t) \\ 5e^{2it} &= 5(\cos 2t + i \sin 2t) = 5 \cos 2t + 5 i \sin 2t \end{aligned} \right\} \tag{29}$$

We now make use of the following

Theorem 3. If x_1, y_1 is a solution of (2), then

$$\text{Re}(x_1), \text{Re}(y_1) \quad \text{and} \quad \text{Im}(x_1), \text{Im}(y_1)$$

where Re and Im denote the real and imaginary parts, are two independent solutions of (2), and its general solution is thus given by

$$x = A \operatorname{Re}(x_1) + B \operatorname{Im}(x_1), \qquad y = A \operatorname{Re}(y_1) + B \operatorname{Im}(y_1) \tag{30}$$

where A, B are arbitrary constants.

From this theorem and (29) we see that the general solution of (22) is

$$x = -A(3 \cos 2t + 4 \sin 2t) - B(3 \sin 2t - 4 \cos 2t), \, y = 5A \cos 2t + 5B \cos 2t$$

agreeing with (28). A proof of Theorem 3 is not difficult and is left to A Exercise 3, page 499.

9.2 HOW DO WE FIND A PARTICULAR SOLUTION?

If the linear system of differential equations which is to be solved has the form (2), page 492, then the complementary solution which has already been considered is clearly the required general solution. However, for the system (1) in which the right-hand sides are not both zero, we must in addition find a particular solution. As in Chapter Four, there are three principal methods by which a particular solution may be found.

(a) Method of undetermined coefficients.
(b) Method of variation of parameters.
(c) Method of operators.

As we have seen, methods (a) and (c) work for only limited types of functions, such as polynomials, exponentials, etc.* However, the method of variation of parameters (b) works for all cases. We shall thus confine our attention to this method. To illustrate the procedure involved, let us consider the following

PROBLEM FOR DISCUSSION

Solve the system of equations $\qquad \left. \begin{array}{l} (2D - 2)x + (D - 7)y = t - 1 \\ (3D - 2)x + (2D - 8)y = e^{-t} \end{array} \right\}. \tag{31}$

We have already found the complementary solution of this system of equations on page 494 to be

$$x = -2c_1 e^{-t} - 3c_2 e^{-2t}, \qquad y = c_1 e^{-t} + 2c_2 e^{-2t} \tag{32}$$

The method of variation of parameters consists in replacing the arbitrary constants in this solution by arbitrary functions of the independent variable in the same way as in Chapter Four, and then finding these functions by substituting the assumed solution into (31). From (32) we obtain

$$x = -2\gamma_1 e^{-t} - 3\gamma_2 e^{-2t}, \qquad y = \gamma_1 e^{-t} + 2\gamma_2 e^{-2t} \tag{33}$$

where γ_1, γ_2 are functions of t which we want to determine. Using (33) in (31) and letting γ_1', γ_2' denote the derivatives of γ_1, γ_2 with respect to t, we find after simplifying

$$3\gamma_1' e^{-t} + 4\gamma_2' e^{-2t} = 1 - t, \qquad 4\gamma_1' e^{-t} + 5\gamma_2' e^{-2t} = -e^{-t} \tag{34}$$

---

* The operator method for determining particular solutions of systems of equations for such functions is essentially like that shown in Illustrative Example 3, page 440.

Note that all the terms involving γ_1 and γ_2 are absent, which is not surprising in view of the fact that for constant γ_1 and γ_2 (33) is the complementary solution of (31). This observation helps considerably in reducing the labor involved in obtaining the equations (34). If we solve for γ_1' and γ_2' from (34) we find

$$\gamma_1' = 5te^t - 5e^t - 4, \qquad \gamma_2' = 3e^t - 4te^{2t} + 4e^{2t}$$

Then by integrating, omitting the constant of integration,

$$\gamma_1 = 5te^t - 10e^t - 4t, \qquad \gamma_2 = 3e^t - 2te^{2t} + 3e^{2t} \tag{35}$$

Using (35) in (33) and adding the complementary solution (32), we find

$$x = -2c_1e^{-t} - 3c_2e^{-2t} + 8te^{-t} - 9e^{-t} - 4t + 11,$$

$$y = c_1e^{-t} + 2c_2e^{-2t} - 4te^{-t} + 6e^{-t} + t - 4 \tag{36}$$

which is the required general solution. It should be noted that if we had included the arbitrary constants of integration in (35) we would have obtained the complementary solution as well.

8.3 SUMMARY OF PROCEDURE

It is worthwhile summarizing the procedure to be used in solving linear systems of differential equations with constant coefficients by the complementary solution, particular solution method.

I. Write the Complementary Equation. This is done by replacing the right-hand side of each equation in the system by zero, assuming that it is not already zero.

II. Find the Complementary Solution. To do this, assume solutions of the form $x = ae^{mt}$, $y = be^{mt}$ obtaining a *determinant equation* for m. Then find the values of m which are roots of this equation and using them determine a and b in terms of arbitrary constants. Three situations can arise.

1. Roots are real and distinct. In this case the complementary solution can immediately be written.

2. Roots are repeated. If the roots are m_1, m_1 for example, we multiply the original solution $x = ae^{m_1 t}$, $y = be^{m_1 t}$ by t and add constant multiples of $e^{m_1 t}$ to obtain a new assumed solution

$$x = ate^{m_1 t} + K_1 e^{m_1 t}, \qquad y = bte^{m_1 t} + K_2 e^{m_1 t}$$

where K_1 and K_2 are to be determined by substitution in one of the complementary equations of the system.

3. Roots are imaginary. In this case we use the fact that the real and imaginary parts of solutions are also solutions.

III. Find a Particular Solution. This can be done using the method of *variation of parameters*, i.e., replacing the arbitrary constants by arbitrary functions of t, which must then be determined from the given system. The required general solution is then obtained by adding the complementary and particular solutions.

The above procedure is valid with minor modifications for systems in n linear equations in n unknowns where $n > 2$. However, for larger values of n, the calculations become rather laborious. A method using matrices is available, which we will describe in Chapter Eleven, which reduces the labor. However, one of the main reasons for using matrices is from a theoretical point of view, since the values of m turn out to be the *characteristic values*, or as we often say the *eigenvalues*, of a matrix. The theory is analogous to the *Sturm–Liouville theory* of Chapter Eight.

A EXERCISES

1. Solve each of the following systems and determine that solution satisfying any given conditions.

(a) $\left.\begin{array}{l}(D + 6)x + (3D + 2)y = 0 \\ (D + 5)x + (2D + 3)y = 0\end{array}\right\}$; $x(0) = 2$, $y(0) = 4$

(b) $\left.\begin{array}{l}(D - 1)x + (2D + 7)y = 0 \\ (2D + 1)x + (D + 5)y = 0\end{array}\right\}$

(c) $\left.\begin{array}{l}(D + 5)x + (3D - 11)y = 0 \\ (D + 3)x + (D - 7)y = 0\end{array}\right\}$

(d) $\left.\begin{array}{l}(D - 2)x + 4y = 0 \\ 3x + (2D + 1)y = 0\end{array}\right\}$

(e) $\left.\begin{array}{l}(D + 3)x + 2y = 0 \\ 3x + (D + 1)y = 0\end{array}\right\}$

(f) $\left.\begin{array}{l}(D + 4)x + (3D + 4)y = 0 \\ (D + 2)x + (2D + 2)y = 0\end{array}\right\}$; $x(0) = -1$, $y(0) = 6$

(g) $\left.\begin{array}{l}(D + 1)x + (2D + 3)y = 0 \\ (D - 2)x + 5Dy = 0\end{array}\right\}$

(h) $\left.\begin{array}{l}(D - 1)x - y = 0 \\ 5x + (D - 3)y = 0\end{array}\right\}$; $x(0) = 2$, $y(0) = 0$

(i) $\left.\begin{array}{l}2x - (D + 5)y = 0 \\ (D + 1)x + 2y = 0\end{array}\right\}$; $x(0) = 0$, $y(0) = -10$

(j) $\left.\begin{array}{l}(2D - 6)x + (3D - 2)y = 0 \\ (7D + 4)x + (7D + 20)y = 0\end{array}\right\}$

2. Solve each of the following systems using the complementary-particular solution method and determine that solution satisfying any given conditions.

(a) $\left.\begin{array}{l}(D + 1)x + 2y = 8 \\ 2x + (D - 2)y = 2e^{-t} - 8\end{array}\right\}$

(b) $\left.\begin{array}{l}Dx + 2y = 4e^{2t} \\ x + (D - 1)y = 2e^{2t}\end{array}\right\}$; $x(0) = 7$, $y(0) = 1$

(c) $\left.\begin{array}{l}(D - 1)x + (2D + 7)y = 3(t - 5) \\ (2D + 1)x + (D + 5)y = 9t - 7\end{array}\right\}$; $x(0) = 0$, $y(0) = -3$

(d) $\left.\begin{array}{l}(D + 3)x - (D + 1)y = 0 \\ (2D - 9)x + (D + 4)y = 15e^{-3t}\end{array}\right\}$

(e) $\left.\begin{array}{l}(D + 2)x + (D + 1)y = 10 \cos t \\ (D + y)x + (2D + 3)y = 6\end{array}\right\}$

(f) $\left.\begin{array}{l}3x - (D + 2)y = 8t \\ (D - 2)x + y = 16e^{-t}\end{array}\right\}$

(g) $\left.\begin{array}{l}(2D - 1)x - (D - 1)y = 4te^{-t} - 3e^{-t} \\ (D + 4)x - (2D + 4)y = 2te^{-t} - 6e^{-t}\end{array}\right\}$

(h) $\left.\begin{array}{l}(2D - 1)x + (7D + 3)y = 90 \sin 2t \\ (D - 5)x + (8D - 3)y = 0\end{array}\right\}$

3. Prove Theorem 3, page 496.

B EXERCISES

1. Prove (a) Theorem 1, (b) Theorem 2, and (c) Theorem 3 for systems of three or more linear differential equations.

2. Work (a) A Exercise 2(a), (b) A Exercise 2(d), (c) A Exercise 2(f) by using operators to get a particular solution.

3. Show how the method of undetermined coefficients can be used by working (a) A Exercise 2(a), (b) A Exercise 2(d), (c) A Exercise 2(f).

4. Work (a) A Exercise 2(b), (b) A Exercise 2(d), (c) A Exercise 2(f) by using operator methods or undetermined coefficients to find a particular solution.

5. Explain how the method of the text can be used to solve the systems:

(a) $\dfrac{d^2x}{dt^2} = y + 4e^{-2t}, \dfrac{d^2y}{dt^2} = x - e^{-2t}$ (see equations (4), page 434), (b) A Exercise 3, page 445.

C EXERCISES

1. Solve using the complementary-particular solution method.

$$(D - 5)x + Dy + 2z = 24e^{-t}, \qquad (D - 1)x - y = 0, \qquad (5D - 11)y + (2D - 2)z = 0$$

2. Work Exercise 1 by using the method of undetermined coefficients or the method of operators to find a particular solution.

3. Work (a) B Exercise 5(a); (b) B Exercise 5(b) by first expressing the systems as systems of first-order differential equations.

4. Solve A Exercise 6, page 445, using the complementary-particular solution method.

eleven

◆matrix eigenvalue methods for systems of linear differential equations

1 The Concept of a Matrix

1.1 INTRODUCTION

In Chapter Ten, page 491, we obtained solutions of linear systems of differential equations by the complementary and particular solution method. If the number of equations is large, the method can be rather laborious. However, this labor can be reduced by introducing the idea of *matrices*, which can be considered as generalizations of vectors.

Although such reduction of labor is from a practical viewpoint an initial motivation for the study of matrices, far greater benefits result from a theoretical viewpoint. This is often the case when we apply mathematics to a practical problem. The mathematics itself may turn out to be more interesting and rewarding than the original problem. Fortunately, the reverse is also true, and so perhaps the best thing is to enjoy the best of both worlds.

1.2 SOME SIMPLE IDEAS

Since not all students may have encountered matrices in earlier courses, we shall present the most salient features about them which are needed in connection with our discussion of linear differential equations. For students who have studied matrices, such a presentation can be considered as a review of the fundamentals.

Perhaps the best way to introduce the idea of matrices is to consider systems of linear equations such as studied in elementary algebra, for example,

$$\left.\begin{array}{rcl} 3x - 2y + 5z &=& 7 \\ 2x + y - z &=& -6 \\ 4x - 3y + 2z &=& -5 \end{array}\right\} \tag{1}$$

where x, y, z are unknowns which are to be determined. One method of solving such equations is the technique of *elimination*, the analogue of which for linear systems of differential equations was given on page 434. It turns out that the method is simplified considerably when the equations are expressed in matrix form. Let us describe this.

The solution of equations (1) clearly depends on the coefficients of x, y, z on the left sides of the equations, as well as the constants on the right. We can write these two sets of numbers, denoted by A and B, respectively, as

$$A = \begin{pmatrix} 3 & -2 & 5 \\ 2 & 1 & -1 \\ 4 & -3 & 2 \end{pmatrix}, \qquad B = \begin{pmatrix} 7 \\ -6 \\ -5 \end{pmatrix} \tag{2}$$

We call A and B *matrices*, each one separately being called a *matrix*. For obvious reasons A, which has as many *rows* of numbers as *columns*, is called a *square matrix*, while B, which has only a single column of numbers, is called a *column matrix*. The remaining numbers in (1), i.e., the unknowns x, y, z, can be written as a column matrix given by

$$X = \begin{pmatrix} x \\ y \\ z \end{pmatrix} \tag{3}$$

using the letter X to denote this unknown matrix. Each number in a matrix is often called an *element* or *entry* of the matrix.

More generally, we can think of a matrix as a rectangular array of numbers arranged in m rows and n columns, which we can indicate by

$$A = (a_{jk}) = \begin{pmatrix} a_{11} & a_{12} & \cdots & a_{1n} \\ a_{21} & a_{22} & \cdots & a_{2n} \\ & & \vdots & \\ a_{m1} & a_{m2} & \cdots & a_{mn} \end{pmatrix} \tag{4}$$

Here the number or element in the jth row and kth column is given by a_{jk}. We often call the matrix (4) an m *by* n or $m \times n$ matrix, and can refer to $m \times n$ as the *size* or *dimension* of the matrix. For example, A in (2) is a 3×3 matrix while B is a 3×1 matrix. We can of course have other matrices besides square and column matrices. For example we could have the matrices

$$R = (3 \quad -2 \quad 5), \qquad S = \begin{pmatrix} 2 & 1 & -1 \\ 4 & -3 & 2 \end{pmatrix} \tag{5}$$

where R is a 1×3 or *row matrix* and S is a 2×3 matrix. Note that an $m \times n$ matrix is not the same size as an $n \times m$ matrix unless $m = n$, i.e., unless the matrix is a square matrix.

In case A is a square matrix of size $n \times n$ we can associate with A the $n \times n$ or nth order determinant of A which we denote by $\det(A)$. We must of course distinguish between the square matrix and its determinant, the former being a set of numbers and the latter a single number.*

Example 1. If $\quad A = \begin{pmatrix} -3 & -4 \\ 5 & 2 \end{pmatrix}, \qquad B = \begin{pmatrix} 3 & -2 & 5 \\ 2 & 1 & -1 \\ 4 & -3 & 2 \end{pmatrix}$

then $\qquad\qquad \det(A) = 14, \qquad \det(B) = -37$

1.3 ROW AND COLUMN VECTORS

A vector in three-dimensional space can be characterized by its three components in the directions of the positive x, y, and z axes. We can thus think of a row or column matrix having three elements as representing a three-dimensional vector. For this reason, we often refer to such a row or column matrix as a *row* or *column vector*, respectively. By analogy an *n-dimensional vector* having n components

---

* For definitions and theorems on determinants see the Appendix.

can be represented by a row or column matrix with n elements and is called a row or column vector, respectively.

Example 2. Matrix B in (2) is a column vector representing a three-dimensional vector with components $7, -6, -5$. Matrix R in (5) is a row vector representing a three-dimensional vector with components $3, -2, 5$. The matrix $(-4 \quad 2 \quad -3 \quad 5)$ is a row vector and represents a four-dimensional vector.

In calculus various operations with vectors are performed. For example, to add two vectors (having the same dimensions), we simply add their components. In matrix terminology this amounts to

$$(a_1 \quad a_2 \quad a_3) + (b_1 \quad b_2 \quad b_3) = (a_1 + b_1 \quad a_2 + b_2 \quad a_3 + b_3) \tag{6}$$

with a similar result for column matrices. Similarly, to multiply a vector by a real number, often referred to as a *scalar*, we simply multiply each component by the number. In terms of matrices, this amounts to

$$c(a_1 \quad a_2 \quad a_3) = (ca_1 \quad ca_2 \quad ca_3) \tag{7}$$

with a similar result for column matrices. The rules (6) and (7) can easily be extended to higher dimensions.

Example 3.
$$(3 \quad -4 \quad 2) + (-2 \quad 4 \quad 3) = (1 \quad 0 \quad 5),$$
$$-2(2 \quad -1 \quad 4) = (-4 \quad 2 \quad -8)$$

Another important concept involving vectors is the concept of the *scalar product*, or as it is also called the *dot product*, of two vectors. The student will recall that this is defined as the sum of the products of corresponding components of the two vectors. From the viewpoint of matrices, it is convenient to express this scalar product for n-dimensional vectors in the form

$$(a_1 \quad a_2 \quad \cdots \quad a_n) \begin{pmatrix} b_1 \\ b_2 \\ \vdots \\ b_n \end{pmatrix} = a_1 b_1 + a_2 b_2 + \cdots + a_n b_n \tag{8}$$

where the first vector is written as a row vector and the second as a column vector.* Motivation for this definition is supplied when we look at the system of linear equations (1) on page 502. The left side $3x - 2y + 5z$, of the first equation is the scalar product of the first row of matrix A in (2) and the column matrix X in (3), i.e.,

$$(3 \quad -2 \quad 5) \begin{pmatrix} x \\ y \\ z \end{pmatrix} = 3x - 2y + 5z \tag{9}$$

---

* See also Chapter Eight, page 351.

Similar statements can be made for the left sides of the second and third equations in (1). Multiplication where a column vector precedes a row vector is undefined.

1.4 OPERATIONS WITH MATRICES

We can think of an $m \times n$ matrix with m rows and n columns as composed of m row vectors each of which is n-dimensional or n column vectors each of which is m-dimensional. Thus, just as we can have an algebra of row and column matrices, so we should have an algebra of matrices in general. Guided by the algebra for vectors, therefore, we present the following definitions.

(a) Equality. Two matrices, say A and B, having the same size are said to be *equal*, written $A = B$, if and only if their corresponding elements are equal. In symbols, if $A = (a_{jk})$, $B = (b_{jk})$, then $A = B$ if and only if $a_{jk} = b_{jk}$ for all j and k.

Example 4. $\begin{pmatrix} -2 & a & 1 \\ 3 & -2 & b \end{pmatrix} = \begin{pmatrix} x & 4 & y \\ z & w & 0 \end{pmatrix}$

if and only if $x = -2$, $a = 4$, $y = 1$, $z = 3$, $w = -2$, and $b = 0$.

(b) Multiplication by a real number (scalar). If A is a matrix and c a number (or scalar), then cA is a matrix whose elements are all c times the corresponding elements of A. In symbols, if $A = (a_{jk})$, then $cA = (ca_{jk})$.

Example 5. $-3 \begin{pmatrix} 4 & -1 \\ -2 & 0 \end{pmatrix} = \begin{pmatrix} (-3)(4) & (-3)(-1) \\ (-3)(-2) & (-3)(0) \end{pmatrix} = \begin{pmatrix} -12 & 3 \\ 6 & 0 \end{pmatrix}$

(c) Addition. If A and B are matrices of the same size, then their *sum*, denoted by $A + B$, is a matrix whose elements are the sums of corresponding elements of A and B. In symbols, if $A = (a_{jk})$, $B = (b_{jk})$, then $A + B = (a_{jk} + b_{jk})$.

Example 6. $\begin{pmatrix} -2 & 3 \\ 4 & -1 \end{pmatrix} + \begin{pmatrix} 1 & 0 \\ -2 & 5 \end{pmatrix} = \begin{pmatrix} -2+1 & 3+0 \\ 4-2 & -1+5 \end{pmatrix} = \begin{pmatrix} -1 & 3 \\ 2 & 4 \end{pmatrix}$

(d) Subtraction. If A and B are matrices of the same size, then their *difference*, denoted by $A - B$, is a matrix whose elements are the differences of corresponding elements of A and B. In symbols, if $A = (a_{jk})$, $B = (b_{jk})$, then $A - B = (a_{jk} - b_{jk})$. We can also define $A - B$ as $A + (-1)B$ and use (b) and (c).

Example 7. $\begin{pmatrix} -2 & 3 \\ 4 & -1 \end{pmatrix} - \begin{pmatrix} 1 & 0 \\ -2 & 5 \end{pmatrix} = \begin{pmatrix} -2-1 & 3-0 \\ 4-(-2) & -1-5 \end{pmatrix} = \begin{pmatrix} -3 & 3 \\ 6 & -6 \end{pmatrix}$

(e) Multiplication of matrices. Let A and B be matrices such that the number n of columns in A equals the number of rows in B. Then the product AB is a matrix whose element in the jth row and kth column equals the scalar product of the jth row vector in A and the kth column vector in B. In symbols, if $A = (a_{jk})$, $B = (b_{jk})$, and $C = AB = (c_{jk})$, then

$$c_{jk} = \sum_{r=1}^{n} a_{jr}b_{rk}$$

Example 8. $\begin{pmatrix} 3 & -4 & 2 \\ -1 & 0 & 1 \end{pmatrix} \begin{pmatrix} -2 & 1 \\ -3 & -1 \\ 1 & -2 \end{pmatrix}$

$= \begin{pmatrix} (3)(-2) + (-4)(-3) + (2)(1) & (3)(1) + (-4)(-1) + (2)(-2) \\ (-1)(-2) + (0)(-3) + (1)(1) & (-1)(1) + (0)(-1) + (1)(-2) \end{pmatrix} = \begin{pmatrix} 8 & 3 \\ 3 & -3 \end{pmatrix}$

Example 9. The product obtained by interchanging the matrices in Example 8 is not defined, because in the resulting product the number of columns of the first matrix would not be equal to the number of rows of the second matrix. This example shows that the commutative law for multiplication of matrices does not hold in general, i.e., $AB \neq BA$. This is true in general even when both AB and BA exist.

Remark. The definition of multiplication of matrices requires that if A is an $m \times n$ matrix and B a $p \times q$ matrix then we must have $n = p$ in order for the product AB to exist. In such case we say that A and B are *conformable*. The product AB is an $m \times q$ matrix. This can be indicated symbolically by the *cancellation rule*: $(m \times \acute{n})(\acute{n} \times q) \rightarrow m \times q$.

If a matrix is subtracted from itself, the result is a matrix whose elements are all equal to zero, called the *zero or null matrix* and denoted by the symbol O. The zero matrix can have any size or dimension.

Example 10. $O = \begin{pmatrix} 0 \\ 0 \\ 0 \end{pmatrix}$, $O = \begin{pmatrix} 0 & 0 \\ 0 & 0 \end{pmatrix}$, $O = \begin{pmatrix} 0 & 0 & 0 \\ 0 & 0 & 0 \end{pmatrix}$

The zero or null matrix acts in matrix algebra like the zero of ordinary algebra.

In order to have a matrix which acts like the one or unit of ordinary algebra, we introduce a *unit* or *identity matrix* defined by

$$I = \begin{pmatrix} 1 & 0 & \cdots & 0 \\ 0 & 1 & \cdots & 0 \\ & & \vdots & \\ 0 & 0 & \cdots & 1 \end{pmatrix} \tag{10}$$

which is a square matrix having ones in the diagonal from upper left to lower right, called the *main or principal diagonal*, and zeros elsewhere. If the products IA or AI exist, the result is A (see B Exercise 6).

Example 11.

$\begin{pmatrix} 1 & 0 \\ 0 & 1 \end{pmatrix} \begin{pmatrix} 4 & -2 \\ -1 & 3 \end{pmatrix} = \begin{pmatrix} (1)(4) + (0)(-1) & (1)(-2) + (0)(3) \\ (0)(4) + (1)(-1) & (0)(-2) + (1)(3) \end{pmatrix} = \begin{pmatrix} 4 & -2 \\ -1 & 3 \end{pmatrix}$

The following list gives some important results concerning matrices, which can be regarded as theorems. In all cases we assume that the operations indicated are defined and all letters denote matrices.

1. $A + B = B + A$ Commutative law for addition
2. $A + (B + C) = (A + B) + C$ Associative law for addition
3. $AB \neq BA$ in general   Commutative law for multiplication fails
4. $A(BC) = (AB)C$ Associative law for multiplication
5. $A(B + C) = AB + AC$ Distributive law
6. $A + O = A$
7. $AO = OA = O$
8. $AI = IA = A$

In many practical situations we have need for matrices whose elements are functions, such as for example

$$\begin{pmatrix} t^2 & e^{-t} & \sin 3t \\ 4 & 3t & 2t^3 \end{pmatrix}$$

in which the elements are functions of t. The rules for operating with such matrices are exactly the same as those already given. In addition, however, as might be expected in applications to differential equations, we need to define the derivative of a matrix. We make the following

Definition 1. If A is a matrix whose elements are differentiable functions of t, then the derivative of A with respect to t, denoted by dA/dt, is a matrix whose elements are the derivatives of the corresponding elements of A. In symbols, if $A = (a_{jk})$, then

$$\frac{dA}{dt} = \left(\frac{da_{jk}}{dt} \right) \tag{11}$$

Example 12. $\dfrac{d}{dt}\begin{pmatrix} x \\ y \\ z \end{pmatrix} = \begin{pmatrix} dx/dt \\ dy/dt \\ dz/dt \end{pmatrix}$, $\dfrac{d}{dt}\begin{pmatrix} t^2 & e^{-t} & \sin 3t \\ 4 & 3t & 2t^3 \end{pmatrix} = \begin{pmatrix} 2t & -e^{-t} & 3\cos 3t \\ 0 & 3 & 6t^2 \end{pmatrix}$

We can also define integrals for matrices.

Definition 2. If A is a matrix whose elements are integrable functions of t, then the integral of A with respect to t is a matrix whose elements are the integrals of the corresponding elements of A. In symbols, if $A = (a_{jk})$, then

$$\int A \, dt = \left(\int a_{jk} \, dt \right) \tag{12}$$

Example 13.   Let $A = (e^{-t} \quad -2\sin t \quad 3)$.

Then $\int A \, dt = (-e^{-t} + c_1 \quad 2\cos t + c_2 \quad 3t + c_3) = (-e^{-t} \quad 2\cos t \quad 3t) + C$

where $C = (c_1 \quad c_2 \quad c_3)$. C is sometimes called an *arbitrary constant matrix*.

Let us consider some examples illustrating various operations with matrices.

Express the system of linear equations (1), page 502, in matrix form.

Solution Let $A = \begin{pmatrix} 3 & -2 & 5 \\ 2 & 1 & -1 \\ 4 & -3 & 2 \end{pmatrix}$, $X = \begin{pmatrix} x \\ y \\ z \end{pmatrix}$, $B = \begin{pmatrix} 7 \\ -6 \\ -5 \end{pmatrix}$.

Then by using the definition of multiplication and equality of matrices, we can write the system in the form $AX = B$.

If $A = \begin{pmatrix} 2 & -1 \\ -6 & 1 \end{pmatrix}$ show that $A^2 - 3A - 4I = 0$.

Solution Taking $A^2 = AA$, we have

$$A^2 - 3A - 4I = \begin{pmatrix} 2 & -1 \\ -6 & 1 \end{pmatrix}\begin{pmatrix} 2 & -1 \\ -6 & 1 \end{pmatrix} - 3\begin{pmatrix} 2 & -1 \\ -6 & 1 \end{pmatrix} - 4\begin{pmatrix} 1 & 0 \\ 0 & 1 \end{pmatrix}$$

$$= \begin{pmatrix} 10 & -3 \\ -18 & 7 \end{pmatrix} + \begin{pmatrix} -6 & 3 \\ 18 & -3 \end{pmatrix} + \begin{pmatrix} -4 & 0 \\ 0 & -4 \end{pmatrix} = \begin{pmatrix} 0 & 0 \\ 0 & 0 \end{pmatrix} = O$$

Write in matrix form the system of differential equations

$$\frac{dx}{dt} + 2y - z = t^2 - 1, \quad \frac{dy}{dt} - x + y + 3z = 0, \quad \frac{dz}{dt} + x - 2y = 4e^{-2t} \quad (13)$$

Solution The given system (13) can be expressed according to the definition of equality of matrices as

$$\begin{pmatrix} Dx + 0x + 2y - z \\ Dy - x + y + 3z \\ Dz + x - 2y + 0z \end{pmatrix} = \begin{pmatrix} t^2 - 1 \\ 0 \\ 4e^{-2t} \end{pmatrix} \quad (14)$$

where for brevity we have used the operator $D \equiv d/dt$. Using the definition of addition of matrices, the left side of (14) can be written as a sum:

$$\begin{pmatrix} Dx \\ Dy \\ Dz \end{pmatrix} + \begin{pmatrix} 0x + 2y - z \\ -x + y + 3z \\ x - 2y + 0z \end{pmatrix} \quad \text{or} \quad D\begin{pmatrix} x \\ y \\ z \end{pmatrix} + \begin{pmatrix} 0 & 2 & -1 \\ -1 & 1 & 3 \\ 1 & -2 & 0 \end{pmatrix}\begin{pmatrix} x \\ y \\ z \end{pmatrix}$$

where we have used the definitions of derivatives and products of matrices. We can thus write the given system as

$$\frac{du}{dt} + Au = F \quad (15)$$

where $\qquad u = \begin{pmatrix} x \\ y \\ z \end{pmatrix}$, $A = \begin{pmatrix} 0 & 2 & -1 \\ -1 & 1 & 3 \\ 1 & -2 & 0 \end{pmatrix}$, $F = \begin{pmatrix} t^2 - 1 \\ 0 \\ 4e^{-2t} \end{pmatrix}$

The following is an important theorem which will be used often in this chapter.

Theorem. Suppose that we are given the system of linear equations

$$AX = O \tag{16}$$

where A is an $n \times n$ matrix whose elements are given constants, and X is an n-dimensional vector with unknown elements $x_1, x_2, \ldots, x_n$. Then the system has non-trivial solutions, i.e., $X \neq O$, if and only if det $(A) = 0$.

Example 14. Let $A = \begin{pmatrix} 2 & -1 & 2 \\ 3 & 2 & -2 \\ 1 & -4 & 6 \end{pmatrix}$ and $X = \begin{pmatrix} x \\ y \\ z \end{pmatrix}$. Then $AX = O$ is equivalent

to the system of linear equations

$$\left. \begin{array}{r} 2x - y + 2z = 0 \\ 3x + 2y - 2z = 0 \\ x - 4y + 6z = 0 \end{array} \right\}$$

Since

$$\det(A) = \begin{vmatrix} 2 & -1 & 2 \\ 3 & 2 & -2 \\ 1 & -4 & 6 \end{vmatrix} = 0$$

the system has non-trivial solutions. Two of the infinitely many possible non-trivial solutions are

$$X_1 = \begin{pmatrix} -2 \\ 10 \\ 7 \end{pmatrix}, \qquad X_2 = \begin{pmatrix} 4 \\ -20 \\ -14 \end{pmatrix}$$

Because of the above theorem we call a square matrix A *singular* if $\det(A) = 0$; otherwise, it is called *non-singular*.

A EXERCISES

1. Perform each of the following indicated operations so as to produce a single equivalent matrix.

(a) $3 \begin{pmatrix} 2 & -1 \\ 0 & 4 \end{pmatrix} - 4 \begin{pmatrix} 1 & 2 \\ 5 & -1 \end{pmatrix}.$

(b) $-2 \begin{pmatrix} 6 & 4 & -1 \\ -2 & 1 & 4 \\ 0 & 2 & -1 \end{pmatrix} + 5 \begin{pmatrix} -2 & 0 & 1 \\ 3 & -1 & 2 \\ 4 & -3 & 0 \end{pmatrix}.$

(c) $(2 \quad 1) \begin{pmatrix} -1 \\ 4 \end{pmatrix}.$

(d) $(3 \quad -2 \quad 2) \begin{pmatrix} 4 \\ -1 \\ -2 \end{pmatrix}.$

(e) $\begin{pmatrix} 3 & -1 \\ 2 & 4 \end{pmatrix} \begin{pmatrix} -2 & 1 \\ 0 & -1 \end{pmatrix}.$

(f) $\begin{pmatrix} 4 & 1 & -1 \\ 2 & -1 & 2 \\ -3 & 0 & -2 \end{pmatrix} \begin{pmatrix} 1 & 2 & -1 \\ 0 & -3 & 2 \\ -1 & 1 & 1 \end{pmatrix}.$

2. Find the values of a and b for which $(2a - b \quad 4a + 3) = (a + b \quad 3)\begin{pmatrix} -1 & 2 \\ 2 & 3 \end{pmatrix}$.

3. Given that $A = \begin{pmatrix} 2 & -1 \\ 1 & 0 \end{pmatrix}$, find $3A^2 - 4A + 6I$.

4. Given $A = \begin{pmatrix} 1 & -4 \\ 2 & 3 \end{pmatrix}$, $B = \begin{pmatrix} 2 & 3 \\ -1 & 0 \end{pmatrix}$, and $C = \begin{pmatrix} 3 & -5 \\ 2 & 1 \end{pmatrix}$. Show that:

(a) $A + (B + C) = (A + B) + C$.　(b) $A(B + C) = AB + AC$.　(c) $A(BC) = (AB)C$.

What laws do these results illustrate?

5. If A, B, C are the matrices given in Exercise 4, determine whether: (a) $AB = BA$. (b) $A(BC) = C(AB)$. What conclusion can you draw from these observations?

6. Express in matrix form:

(a) $\left.\begin{array}{l} 3x + 2y = 7 \\ 2x - 5y = -8 \end{array}\right\}$.　(b) $\left.\begin{array}{l} 2x - y + 3z = 9 \\ 4x + 3y - 2z = -3 \\ 3x - 2y + z = 7 \end{array}\right\}$.　(c) $\left.\begin{array}{l} x - z = 4 \\ y + 3z = 2 \\ x - 2y + z = 0 \end{array}\right\}$.

7. Given that $A = \begin{pmatrix} t^2 & e^{-t} \\ 3t - 2 & 4e^t \end{pmatrix}$, find $t\dfrac{d^2A}{dt^2} - 2\dfrac{dA}{dt} + 4t^2 I$.

8. Express the following systems of differential equations in matrix form:

(a) $\dfrac{dx}{dt} + 3x - 2y = e^{-t}$, 　$\dfrac{dy}{dt} - x + 4y = \sin 2t$

(b) $\dfrac{dx}{dt} - x + 2y - z = t^2$, 　$\dfrac{dy}{dt} + 3x - y + 4z = e^t$, 　$\dfrac{dz}{dt} - 2x + y - z = 0$

(c) $\dfrac{dx}{dt} + z = x$, 　$\dfrac{dy}{dt} - 2x = y + 3t$, 　$\dfrac{dz}{dt} + 4y = z - \cos t$

9. Find $\det(A)$ if: (a) $A = \begin{pmatrix} 4 & -1 \\ 2 & -3 \end{pmatrix}$. (b) $A = \begin{pmatrix} 5 & -2 & 1 \\ 3 & 1 & -2 \\ 0 & 4 & -1 \end{pmatrix}$.

10. Determine which if any of the following systems have non-trivial solutions, and determine a few of these solutions if they exist.

(a) $\left.\begin{array}{l} 2x - y + z = 0 \\ 4x + 3y - z = 0 \\ x - 3y + 2z = 0 \end{array}\right\}$.　(b) $\left.\begin{array}{l} 3x + 2y - z = 0 \\ x - 3y + 2z = 0 \\ 2x + 5y - 3z = 0 \end{array}\right\}$.

B EXERCISES

1. If $A, B,$ and C are any matrices having the same size, prove that (a) $A + B = B + A$. (b) $A + (B + C) = (A + B) + C$.

2. Prove that if $A, B,$ and C are conformable matrices then:
(a) $A(B + C) = AB + AC$.　(b) $A(BC) = (AB)C$.

3. If A, B are comformable matrices whose elements are differentiable functions of t, show that

$$\frac{d}{dt}(AB) = A\frac{dB}{dt} + \frac{dA}{dt}B$$

Illustrate by taking $\qquad A = \begin{pmatrix} t^2 & t \\ 1 & e^{-t} \end{pmatrix}, \qquad B = \begin{pmatrix} e^{2t} & -3t \\ 4 & e^t \end{pmatrix}$

Explain why the order indicated is essential.

4. If $AX = AB$, where A, B, X are assumed conformable, does it necessarily follow that $X = B$? Explain.

5. If $A = \begin{pmatrix} 2 & -1 \\ -3 & 4 \end{pmatrix}$, determine if possible the elements in the matrix $X = \begin{pmatrix} x_1 & x_2 \\ x_3 & x_4 \end{pmatrix}$ so that $AX = XA$.

6. Show that if I is the unit matrix (10), page 506, and A is a matrix, then if the products IA or AI exist, the result is A.

C EXERCISES

1. Given $A = \begin{pmatrix} 2 & -5 \\ 3 & -4 \end{pmatrix}$, $B = \begin{pmatrix} -1 & 2 \\ 3 & -2 \end{pmatrix}$ show that even though $AB \neq BA$ we still have

$\det(AB) = \det(BA)$.

2. Prove the result indicated in Exercise 1 if A and B are any (a) 2×2 matrices (b) 3×3 matrices. The result can be generalized to $n \times n$ matrices where $n > 3$.

3. Show that if A and B are any 2×2 determinants we have $\det(AB) = \det(A)\det(B)$ and illustrate by using the determinants in Exercise 1. (b) Extend the result of (a) to 3×3 determinants and illustrate by an example (the result is true for any $n \times n$ determinants). (c) Discuss the connection between the results of (a) and (b) and the results of Exercises 1 and 2.

4. A matrix X such that $AX = I$ where I is the identity or unit matrix is called the *inverse* of A and is denoted by A^{-1}. Find A^{-1} if: (a) $A = \begin{pmatrix} 2 & -1 \\ 3 & 4 \end{pmatrix}$. (b) $A = \begin{pmatrix} 3 & -2 & 5 \\ 2 & 1 & -1 \\ 4 & -3 & 2 \end{pmatrix}$.

5. Show that the inverse of a square matrix A cannot exist if A is a singular matrix.

6. Let $A = (a_{jk})$ be a square matrix, $C = (c_{jk})$ a square matrix of the same size whose elements c_{jk} denote the cofactors of the elements, and C^T the square matrix obtained from C by interchanging columns and rows (often called the *transpose* of C). (a) Using the matrices in Exercise 4 illustrate the fact that

$$AC^T = C^TA = I\Delta \quad \text{or} \quad A\left(\frac{C^T}{\Delta}\right) = \left(\frac{C^T}{\Delta}\right)A = I \quad \text{where } \Delta = \det(A) \neq 0$$

(b) Use (a) to show that the inverse of a non-singular matrix A is given by $A^{-1} = \dfrac{C^T}{\Delta}$ where $\Delta \neq 0$ and use this result to work Exercise 4. These results hold for $n \times n$ matrices.

7. Solve the system of equations (1), page 502, by using inverse matrices. [*Hint:* Use Illustrative Example 1, page 508, to write the system as $AX = B$ and then multiply on the left by A^{-1}.)

2 Matrix Differential Equations

Let us now turn our attention to showing how matrices can be used to solve systems of linear differential equations with constant coefficients by the complementary and particular solution method. We parallel this discussion to that on page 491 so that the student may see the relationships between both methods. As already mentioned, matrices serve to simplify the labor involved where the number of equations is large.

It was pointed out in Chapter Ten, page 443, that every system of linear differential equations can be written as a system of first-order linear differential equations if we are willing to use enough unknowns. In the case where, for example, we have three unknown functions of t, say x, y, z, such a system can be written in the form

$$\left.\begin{array}{l} \dfrac{dx}{dt} + a_{11}x + a_{12}y + a_{13}z = F_1 \\[2mm] \dfrac{dy}{dt} + a_{21}x + a_{22}y + a_{23}z = F_2 \\[2mm] \dfrac{dz}{dt} + a_{31}x + a_{32}y + a_{33}z = F_3 \end{array}\right\} \tag{1}$$

where the a's are given constants and the right-hand sides F_1, F_2, F_3 are given functions of t. The system (1) can be written in matrix form as

$$\frac{du}{dt} + Au = F \tag{2}$$

where
$$u = \begin{pmatrix} x \\ y \\ z \end{pmatrix}, \quad A = \begin{pmatrix} a_{11} & a_{12} & a_{13} \\ a_{21} & a_{22} & a_{23} \\ a_{31} & a_{32} & a_{33} \end{pmatrix}, \quad F = \begin{pmatrix} F_1 \\ F_2 \\ F_3 \end{pmatrix} \tag{3}$$

as we see by using a procedure similar to that of Illustrative Example 3, page 508. If (1) is replaced by a system of n linear equations in n unknowns, where $n \geq 2$, the corresponding matrix equation is still given by (2), but u and F are then n-dimensional vectors and A is an $n \times n$ matrix.

By a *solution* of (2), we of course mean any column vector u which when substituted into (2) yields an identity. To solve any linear system such as (1), we must thus learn how to solve (2).

For obvious reasons, we call (2) a *first-order linear matrix differential equation with constant coefficients.** The equation obtained from (2) on replacing the right side F by O, the zero or null matrix, is by analogy with previous chapters called the

---

* We could also call (2) a *vector differential equation* since u is a column vector.

complementary equation and is given by

$$\frac{du}{dt} + Au = O \qquad (4)$$

The *general solution* of (4), which involves a number of arbitrary constants equal to the dimension of u, is called the *complementary solution* of (2), and as in previous chapters we have the following fundamental theorems.

Theorem 1. Let u_c be the general solution of the complementary equation (4), i.e., the complementary solution of (2). Let u_p be any particular solution of (2). Then the general solution of (2) is given by

$$u = u_c + u_p \qquad (5)$$

Theorem 2 (Principle of Superposition). Let $u_1, u_2, \ldots$ be any number of solutions of the complementary equation (4), and let $c_1, c_2, \ldots$ be arbitrary constants. Then

$$u = c_1 u_1 + c_2 u_2 + \cdots \qquad (6)$$

is also a solution of (4).

Proofs of these theorems are left to the exercises.

Remark. It should be noticed that if we want the general solution of the complementary equation (4) where u is an n dimensional vector we require n independent solutions in (6). As in Chapter Four, this concept of independence requires a discussion of *Wronskians*, which we provide on page 522.

In view of the above theorems it is clear that to obtain the required general solution of (2) we must find the complementary solution of (2), or general solution of (4), and also a particular solution of (2). We now show how these can be found.

3 The Complementary Solution

To find the general solution of (4) or the complementary solution of (2), let us make the substitution in (4) given by

$$u = e^{\lambda t} v \qquad (7)$$

where λ is a constant and v is a constant n-dimensional column vector. Note that for $n = 2$, (7) amounts to making the substitution $x = a e^{\lambda t}$, $y = b e^{\lambda t}$ on page 493. Then (4) becomes

$$\lambda e^{\lambda t} v + A e^{\lambda t} v = O \qquad (8)$$

or on division by $e^{\lambda t} \neq 0$, $\qquad (\lambda I + A) v = O \qquad (9)$

where I is the identity or unit matrix. Now from the theorem on page 509, it follows that (9) will have non-trivial solutions, i.e., $v \neq O$, if and only if

$$\det (\lambda I + A) = 0 \qquad (10)$$

For the matrix (4) on page 503 with $m = n$ this yields the nth-order determinant equation

$$\begin{vmatrix} \lambda + a_{11} & a_{12} & \cdots & a_{1n} \\ a_{21} & \lambda + a_{22} & \cdots & a_{2n} \\ & & \vdots & \\ a_{n1} & a_{n2} & \cdots & \lambda + a_{nn} \end{vmatrix} = 0 \qquad (11)$$

3.1 EIGENVALUES AND EIGENVECTORS

The determinant equation (11) is equivalent to an nth-degree polynomial equation in λ. This equation, which is analogous to the *auxiliary equation* of Chapter Four, has n not necessarily distinct roots, say $\lambda_1, \lambda_2, \ldots, \lambda_n$. Corresponding to these roots there will be values of the column vector v obtained from (9), and from these values of v solutions of the matrix equation (2) are obtained. Before illustrating the procedure, let us examine the significance of the equation (9) which led to (10) or (11).

We can write (9) in the form

$$Av = -\lambda v \qquad (12)$$

On the left side of this equation A is an $n \times n$ matrix and v an n-dimensional column vector. A question naturally arises as to the significance of the product Av. Clearly, Av is another n-dimensional vector, so that we can think of A as an *operator* or *transformation* which when acting on v produces a new n-dimensional vector. This transformation in n-dimensional space can be visualized better if we consider $n = 3$, for example. In such case we can think of the 3×3 matrix A as taking the three-dimensional vector v and transforming it in some way into another three-dimensional vector. In three-dimensional space we can think of such a transformation as consisting of a *rotation* of the vector, a *stretching or contracting* of the vector, or possibly both. For the general case, where $n > 3$, we can think of A as a rotation and possible stretching or contraction of a vector in n-dimensional space, although such visualization is not as easy as for $n = 3$.

In the case of a three-dimensional vector, multiplication by a number or scalar $-\lambda$ represents a vector in the same or opposite direction, depending on whether the number $-\lambda$ is positive or negative. In any case, we can think of $-\lambda v$ as a vector *parallel* to v. This same idea also can apply to n-dimensional space.

With these interpretations the problem of finding values of λ and their corresponding vectors v amounts to asking the following

Question. Which vectors v in an n-dimensional space are such that when they undergo a transformation A (rotation and stretching or contraction) they result in a new vector (i.e., $-\lambda v$) parallel to the original vector and having magnitude $|\lambda|$ times that of the original vector?

In answering the above question we are of course looking for non-zero or non-trivial vectors which can be thought of as *proper vectors*. Now because much work on matrix theory was done in German in which the word for proper is *eigen*, the values of λ and the corresponding vectors v have come to be known by the hybrid words *eigenvalues* and *eigenvectors*, respectively, and we shall use this terminology. The determinant equation (10) or (11) used to find the values of λ is called the *eigenvalue equation.** There are several remarks which should be made.

Remark 1. The fact that a minus sign appears in (12) arises simply because we let $u = e^{\lambda t}v$ in the matrix differential equation (4) on page 513. If we had let $u = e^{-\lambda t}v$ instead, (12) would have been replaced by $Av = \lambda v$, which may appear to be more esthetic.† We could in such case call the λ's eigenvalues, but they would be the negatives of those given above. There are some advantages to using the equation (12) because of analogy with *Sturm–Liouville theory* of Chapter Eight, as we shall discuss on page 532. It should be noted that regardless of the sign for the eigenvalues there are no differences in corresponding eigenvectors, because any constant (or scalar) multiple of an eigenvector is also an eigenvector.

Remark 2. The problem of determining λ and v such that $Av = -\lambda v$ (or $Av = \lambda v$) can be thought of as a purely algebraic problem (i.e., apart from its connection with solutions of matrix differential equations). For this reason, we refer to the λ's and v's as the eigenvalues and eigenvectors of the matrix A. The matrix eigenvalue problem has many interesting theoretical ramifications, some of which are presented in the exercises and at the end of this chapter for those who are interested in them. The student who wishes to pursue further study is referred to books on matrix theory.‡

3.2 THE CASE OF REAL DISTINCT EIGENVALUES

To illustrate the above ideas, let us consider the following

PROBLEM FOR DISCUSSION

Solve the system
$$\begin{aligned} \frac{dx}{dt} - 2x - 6y &= 0 \\ \frac{dy}{dt} + 2x + 5y &= 0 \end{aligned} \right\} \tag{13}$$

The given system is equivalent to the matrix differential equation

$$\frac{du}{dt} + Au = 0, \quad \text{where } u = \begin{pmatrix} x \\ y \end{pmatrix}, A = \begin{pmatrix} -2 & -6 \\ 2 & 5 \end{pmatrix} \tag{14}$$

Putting $u = e^{\lambda t}v$ in the given differential equation or using (10), we have

$$(\lambda I + A)v = 0 \quad \text{or} \quad \begin{pmatrix} \lambda - 2 & -6 \\ 2 & \lambda + 5 \end{pmatrix} v = 0 \tag{15}$$

---

* The student who has read Chapter Eight will note the similarity in terminology with that given here. This similarity will be explored further on pages 530–535.

† Some writers express the matrix differential equation in the form $du/dt = Au + F$ in order to achieve this goal. However, this is not the form already used in writing first-order linear differential equations (see page 53).

‡ See [11], for example.

Now in order to have non-trivial solutions $v \neq O$, we must have

$$\begin{vmatrix} \lambda - 2 & -6 \\ 2 & \lambda + 5 \end{vmatrix} = 0, \quad \text{i.e.,} \quad \lambda^2 + 3\lambda + 2 = 0 \quad \text{or} \quad \lambda = -1, -2$$

Thus there are two real and distinct eigenvalues $\lambda = -1$ and $\lambda = -2$.

Case 1, $\lambda = -1$. Putting $\lambda = -1$ in (15) and using $v = \begin{pmatrix} a \\ b \end{pmatrix}$, where a and b are unknown constants, we have

$$\begin{pmatrix} -3 & -6 \\ 2 & 4 \end{pmatrix}\begin{pmatrix} a \\ b \end{pmatrix} = O \quad \text{or} \quad -3a - 6b = 0, \quad 2a + 4b = 0$$

From either of the last two equations we find $a = -2b$. In particular, if $b = 1$, then $a = -2$. Thus a solution of (15) corresponding to $\lambda = -1$ is

$$v = v_1 = \begin{pmatrix} -2 \\ 1 \end{pmatrix} \tag{16}$$

Using $u = e^{\lambda t}v$, this yields a corresponding solution to (14) given by

$$u = u_1 = \begin{pmatrix} -2 \\ 1 \end{pmatrix} e^{-t} \tag{17}$$

Case 2, $\lambda = -2$. Putting $\lambda = -2$ in (15) as in Case 1, we find

$$\begin{pmatrix} -4 & -6 \\ 2 & 3 \end{pmatrix}\begin{pmatrix} a \\ b \end{pmatrix} = O \quad \text{or} \quad -4a - 6b = 0, \quad 2a + 3b = 0$$

Thus $a = -\frac{3}{2}b$. In particular, if $b = 2$, we have $a = -3$, so that

$$v = v_2 = \begin{pmatrix} -3 \\ 2 \end{pmatrix}$$

and a solution to the differential equation (14) is

$$u = u_2 = \begin{pmatrix} -3 \\ 2 \end{pmatrix} e^{-2t} \tag{18}$$

Using the superposition principle, we see from the solutions (17) and (18) that

$$u = c_1 \begin{pmatrix} -2 \\ 1 \end{pmatrix} e^{-t} + c_2 \begin{pmatrix} -3 \\ 2 \end{pmatrix} e^{-2t} \tag{19}$$

is also a solution. Since this contains the necessary number (two) of arbitrary constants, it is the required general solution. We can write (19) as

$$x = -2c_1 e^{-t} - 3c_2 e^{-2t}, \qquad y = c_1 e^{-t} + 2c_2 e^{-2t} \tag{20}$$

Remark 3. In obtaining the above general solution, we have assumed the linear independence of the solutions u_1, u_2 given by (17) and (18), respectively. This is verified on page 523, Example 1.

Remark 4. The system (13) is equivalent to the system (9) on page 493, and the same general solution is obtained. We can use matrix methods to solve (9), page 493, directly as shown in B Exercise 6(a).

3.3 THE CASE OF REPEATED EIGENVALUES

Let us consider the following

PROBLEM FOR DISCUSSION

Solve the system
$$\left.\begin{array}{c} \dfrac{dx}{dt} + 3x + 4y = 0 \\[2mm] \dfrac{dy}{dt} - x - y = 0 \end{array}\right\} \tag{21}$$

This system is equivalent to the matrix differential equation

$$\frac{du}{dt} + Au = 0, \quad \text{where } u = \begin{pmatrix} x \\ y \end{pmatrix}, \; A = \begin{pmatrix} 3 & 4 \\ -1 & -1 \end{pmatrix} \tag{22}$$

The same technique of letting $u = e^{\lambda t}v$ used above yields

$$(\lambda I + A)v = 0 \quad \text{or} \quad \begin{pmatrix} \lambda + 3 & 4 \\ -1 & \lambda - 1 \end{pmatrix} v = 0 \tag{23}$$

Thus, to have non-trivial solutions $v \neq 0$, we must have

$$\begin{vmatrix} \lambda + 3 & 4 \\ -1 & \lambda - 1 \end{vmatrix} = 0, \quad \text{i.e.,} \quad \lambda^2 + 2\lambda + 1 = 0 \quad \text{or} \quad \lambda = -1, -1$$

or repeated eigenvalues.

Case 1, $\lambda = -1$. Putting $\lambda = -1$ in (23) yields

$$\begin{pmatrix} 2 & 4 \\ -1 & -2 \end{pmatrix} \begin{pmatrix} a \\ b \end{pmatrix} = 0 \quad \text{or} \quad 2a + 4b = 0, \quad -a - 2b = 0$$

so that $a = -2b$. In particular, if $b = 1$, then $a = -2$ so that

$$v = v_1 = \begin{pmatrix} -2 \\ 1 \end{pmatrix}$$

and a solution to the differential equation is

$$u = u_1 = \begin{pmatrix} -2 \\ 1 \end{pmatrix} e^{-t} \tag{24}$$

Obviously, it does no good to take Case 2, $\lambda = -1$, since this is just Case 1 all over again. We might be led from our experience with repeated roots to try as a possible solution

$$u = \begin{pmatrix} -2 \\ 1 \end{pmatrix} te^{-t} \tag{25}$$

obtained from (24) on multiplying by t. However, on substituting (25) in the differential equation of (22), we find

$$\begin{pmatrix} -2 \\ 1 \end{pmatrix} e^{-t} = 0 \tag{26}$$

which shows that (25) is not a solution. The fact that a term involving

$$\begin{pmatrix} -2 \\ 1 \end{pmatrix} e^{-t}$$

is left over in (26) leads us to assume instead of (25) the possible solution

$$u = u_2 = \begin{pmatrix} -2 \\ 1 \end{pmatrix} te^{-t} + \begin{pmatrix} K_1 \\ K_2 \end{pmatrix} e^{-t} \tag{27}$$

where K_1, K_2 are constants to be determined. Substituting (27) in (22) gives

$$\begin{pmatrix} -2 \\ 1 \end{pmatrix} e^{-t} + \begin{pmatrix} 2K_1 + 4K_2 \\ -K_1 - 2K_2 \end{pmatrix} e^{-t} = 0 \quad \text{or} \quad 2K_1 + 4K_2 = 2, \qquad K_1 + 2K_2 = 1$$

Taking $K_2 = 0$, so that $K_1 = 1$, we see from (27) that a solution is given by

$$u = u_2 = \begin{pmatrix} -2 \\ 1 \end{pmatrix} te^{-t} + \begin{pmatrix} 1 \\ 0 \end{pmatrix} e^{-t} \tag{28}$$

From the two solutions (24) and (28) and the superposition principle we obtain

$$u = c_1 \begin{pmatrix} -2 \\ 1 \end{pmatrix} e^{-t} + c_2 \left[\begin{pmatrix} -2 \\ 1 \end{pmatrix} te^{-t} + \begin{pmatrix} 1 \\ 0 \end{pmatrix} e^{-t} \right] \tag{29}$$

as the required general solution.

Remark 5. We can show that the solutions (24) and (28) are linearly independent by the method on pages 522–523.

Remark 6. The system (21) is equivalent to that of (14) on page 494, and the two general solutions obtained are the same.

3.4 THE CASE OF IMAGINARY EIGENVALUES

Let us consider the following

PROBLEM FOR DISCUSSION

Solve the system

$$\left. \begin{aligned} \frac{dx}{dt} + \tfrac{3}{2}x + \tfrac{5}{2}y &= 0 \\[2mm] \frac{dy}{dt} - \tfrac{5}{2}x - \tfrac{3}{2}y &= 0 \end{aligned} \right\} \tag{30}$$

The system is equivalent to the matrix differential equation

$$\frac{du}{dt} + Au = 0, \qquad \text{where } u = \begin{pmatrix} x \\ y \end{pmatrix}, \ A = \begin{pmatrix} \tfrac{3}{2} & \tfrac{5}{2} \\ -\tfrac{5}{2} & -\tfrac{3}{2} \end{pmatrix} \tag{31}$$

From this we find
$$\begin{pmatrix} \lambda + \frac{3}{2} & \frac{5}{2} \\ -\frac{5}{2} & \lambda - \frac{3}{2} \end{pmatrix} v = O \tag{32}$$

so that
$$\begin{vmatrix} \lambda + \frac{3}{2} & \frac{5}{2} \\ -\frac{5}{2} & \lambda - \frac{3}{2} \end{vmatrix} = 0, \quad \text{i.e.,} \quad \lambda^2 + 4 = 0 \quad \text{or} \quad \lambda = \pm 2i$$

Thus we have imaginary eigenvalues.

Case 1, $\lambda = 2i$. Putting $\lambda = 2i$ in (32), we have
$$\begin{pmatrix} 2i + \frac{3}{2} & \frac{5}{2} \\ -\frac{5}{2} & 2i - \frac{3}{2} \end{pmatrix} \begin{pmatrix} a \\ b \end{pmatrix} = O \quad \text{or} \quad (2i + \tfrac{3}{2})a + \tfrac{5}{2}b = 0, \quad -\tfrac{5}{2}a + (2i - \tfrac{3}{2})b = 0$$

Then $a = (4i - 3)b/5$. In particular if $b = 5$, we have $a = -3 + 4i$. Thus
$$v = v_1 = \begin{pmatrix} -3 + 4i \\ 5 \end{pmatrix} \tag{33}$$

which leads to a solution of the differential equation (31) given by
$$u = u_1 = \begin{pmatrix} -3 + 4i \\ 5 \end{pmatrix} e^{2it} \tag{34}$$

Case 2, $\lambda = -2i$. We could put $\lambda = -2i$ in (32) and proceed as in Case 1. It is easier to note that we can obtain the required solution for this case by simply replacing i by $-i$ in (34). This leads to a solution
$$u = u_2 = \begin{pmatrix} -3 - 4i \\ 5 \end{pmatrix} e^{-2it} \tag{35}$$

From (34) and (35) we could find the general solution by using the superposition principle. However, this solution is given in terms of imaginaries and we would like the solution to be real. We can easily prove that the real part of u_1 and the imaginary part of u_1 are solutions (see A Exercise 4). From (34) we have, on making use of Euler's identity, page 178,
$$\begin{pmatrix} -3 + 4i \\ 5 \end{pmatrix} e^{2it} = \left[\begin{pmatrix} -3 \\ 5 \end{pmatrix} + i \begin{pmatrix} 4 \\ 0 \end{pmatrix} \right] [\cos 2t + i \sin 2t]$$
$$= \begin{pmatrix} -3 \\ 5 \end{pmatrix} \cos 2t - \begin{pmatrix} 4 \\ 0 \end{pmatrix} \sin 2t + i \left[\begin{pmatrix} 4 \\ 0 \end{pmatrix} \cos 2t + \begin{pmatrix} -3 \\ 5 \end{pmatrix} \sin 2t \right] \tag{36}$$

This leads to the solutions
$$\begin{pmatrix} -3 \\ 5 \end{pmatrix} \cos 2t - \begin{pmatrix} 4 \\ 0 \end{pmatrix} \sin 2t, \quad \begin{pmatrix} 4 \\ 0 \end{pmatrix} \cos 2t + \begin{pmatrix} -3 \\ 5 \end{pmatrix} \sin 2t \tag{37}$$

so that by the superposition principle the general solution is
$$u = c_1 \left[\begin{pmatrix} -3 \\ 5 \end{pmatrix} \cos 2t - \begin{pmatrix} 4 \\ 0 \end{pmatrix} \sin 2t \right] + c_2 \left[\begin{pmatrix} 4 \\ 0 \end{pmatrix} \cos 2t + \begin{pmatrix} -3 \\ 5 \end{pmatrix} \sin 2t \right] \tag{38}$$

Remark 7. We can show that the solutions in (38) are linearly independent by the method on page 522, so that (38) gives the general solution.

Remark 8. The system (30) is equivalent to that in (22) on page 495, and the general solution obtained there is equivalent to (38).

3.5 A SLIGHTLY MORE COMPLICATED PROBLEM

The above ideas can of course be extended to systems involving $n \times n$ matrices where $n > 2$. The methods are essentially the same, and the only difficulty which can arise is the labor involved in the evaluation of the higher-order determinants. In order to show the procedure in such cases, let us consider the following

PROBLEM FOR DISCUSSION

Solve the system
$$
\left.
\begin{aligned}
\frac{d^2x}{dt^2} - \frac{dy}{dt} - x + 3y &= 0 \\[2mm]
\frac{dx}{dt} + \frac{dy}{dt} + x - y &= 0
\end{aligned}
\right\}
\tag{39}
$$

In order to reduce this to a first-order system, we make the substitution $\dfrac{dx}{dt} = z$.

Then the given system (39) can be written

$$
\frac{dx}{dt} - z = 0, \quad \frac{dy}{dt} + x - y + z = 0, \quad \frac{dz}{dt} + 2y + z = 0
\tag{40}
$$

This can be written as a first-order matrix differential equation

$$
\frac{du}{dt} + Au = O
\tag{41}
$$

where
$$
u = \begin{pmatrix} x \\ y \\ z \end{pmatrix}, \qquad
A = \begin{pmatrix} 0 & 0 & -1 \\ 1 & -1 & 1 \\ 0 & 2 & 1 \end{pmatrix}
\tag{42}
$$

If we now put $u = e^{\lambda t}v$ in (41), where v is a constant three-dimensional column vector, it becomes

$$
(\lambda I + A)v = O \quad \text{or} \quad
\begin{pmatrix} \lambda & 0 & -1 \\ 1 & \lambda - 1 & 1 \\ 0 & 2 & \lambda + 1 \end{pmatrix} v = O
\tag{43}
$$

Thus, in order to get non-trivial solutions $v \neq O$, we must have

$$
\begin{vmatrix} \lambda & 0 & -1 \\ 1 & \lambda - 1 & 1 \\ 0 & 2 & \lambda + 1 \end{vmatrix} = \lambda^3 - 3\lambda - 2 = 0, \qquad \text{i.e., } \lambda = 2, -1, -1
$$

Case 1, $\lambda = 2$. Putting $\lambda = 2$ in (43) yields

$$
\begin{pmatrix} 2 & 0 & -1 \\ 1 & 1 & 1 \\ 0 & 2 & 3 \end{pmatrix}
\begin{pmatrix} a_1 \\ a_2 \\ a_3 \end{pmatrix} = O \quad \text{or} \quad
2a_1 - a_3 = 0, \quad a_1 + a_2 + a_3 = 0, \quad 2a_2 + 3a_3 = 0
$$

from which $a_1 = \frac{1}{2}a_3$, $a_2 = -\frac{3}{2}a_3$. In particular, if $a_3 = 2$, then $a_1 = 1$, $a_2 = -3$.

Thus we have
$$v = v_1 = \begin{pmatrix} 1 \\ -3 \\ 2 \end{pmatrix} \tag{44}$$

and a corresponding solution is
$$u = u_1 = \begin{pmatrix} 1 \\ -3 \\ 2 \end{pmatrix} e^{2t} \tag{45}$$

Case 2, $\lambda = -1$. Putting $\lambda = -1$ in (43) yields

$$\begin{pmatrix} -1 & 0 & -1 \\ 1 & -2 & 1 \\ 0 & 2 & 0 \end{pmatrix} \begin{pmatrix} a_1 \\ a_2 \\ a_3 \end{pmatrix} = O \quad \text{or} \quad -a_1 - a_3 = 0, \quad a_1 - 2a_2 + a_3 = 0, \quad 2a_2 = 0$$

from which $a_1 = -a_3$, $a_2 = 0$. In particular if $a_3 = 1$, then $a_1 = -1$, $a_2 = 0$.

Thus we have
$$v = v_2 = \begin{pmatrix} -1 \\ 0 \\ 1 \end{pmatrix} \tag{46}$$

and a corresponding solution is
$$u = u_2 = \begin{pmatrix} -1 \\ 0 \\ 1 \end{pmatrix} e^{-t} \tag{47}$$

Since Case 3, $\lambda = -1$ would yield the same solution as Case 2, we must use the method for repeated roots given on page 517. For this purpose we take as solution

$$u = u_3 = \begin{pmatrix} -1 \\ 0 \\ 1 \end{pmatrix} te^{-t} + \begin{pmatrix} K_1 \\ K_2 \\ K_3 \end{pmatrix} e^{-t} \tag{48}$$

and seek to determine K_1, K_2, K_3. Substituting (48) into (41) yields

$$\frac{du}{dt} + Au = \begin{pmatrix} -K_1 - K_3 - 1 \\ K_1 - 2K_2 + K_3 \\ 2K_2 + 1 \end{pmatrix} e^{-t} = O$$

or $\quad K_1 + K_3 + 1 = 0, \quad K_1 - 2K_2 + K_3 = 0, \quad 2K_2 + 1 = 0$

from which $K_1 = -1 - K_3$, $K_2 = -\frac{1}{2}$. Since K_3 is arbitrary, we may choose $K_3 = 0$ so that $K_1 = -1$. Thus (48) gives the solution

$$u = u_3 = \begin{pmatrix} -1 \\ 0 \\ 1 \end{pmatrix} te^{-t} + \begin{pmatrix} -1 \\ -\frac{1}{2} \\ 0 \end{pmatrix} e^{-t} \tag{49}$$

Using the principle of superposition, we obtain from (45), (47), and (49) the required general solution:

$$u = c_1 \begin{pmatrix} 1 \\ -3 \\ 2 \end{pmatrix} e^{2t} + c_2 \begin{pmatrix} -1 \\ 0 \\ 1 \end{pmatrix} e^{-t} + c_3 \left[\begin{pmatrix} -1 \\ 0 \\ 1 \end{pmatrix} te^{-t} + \begin{pmatrix} -1 \\ -\frac{1}{2} \\ 0 \end{pmatrix} e^{-t} \right] \qquad (50)$$

This is equivalent to

$$x = c_1 e^{2t} - c_2 e^{-t} - c_3 te^{-t} - c_3 e^{-t},$$
$$y = -3c_1 e^{2t} - \tfrac{1}{2}c_3 e^{-t}, \qquad z = 2c_1 e^{2t} + c_2 e^{-t} + c_3 te^{-t} \qquad (51)$$

It should be noted that although (51) is the general solution for the system (40), the general solution for the system (39) is given by the values for x and y in (51), while z, which equals dx/dt, is extraneous. In spite of this, however, (50) or its equivalent (51) can be useful when initial conditions are given. For example, suppose that we must solve (39) subject to the initial conditions

$$x = 5, \qquad y = 2, \qquad \frac{dx}{dt} = 0, \qquad \text{at } t = 0 \qquad (52)$$

This can be used to write (50) as

$$c_1 \begin{pmatrix} 1 \\ -3 \\ 2 \end{pmatrix} + c_2 \begin{pmatrix} -1 \\ 0 \\ 1 \end{pmatrix} + c_3 \begin{pmatrix} -1 \\ -\frac{1}{2} \\ 0 \end{pmatrix} = \begin{pmatrix} 5 \\ 2 \\ 0 \end{pmatrix} \qquad (53)$$

from which we can find c_1, c_2, c_3.

3.6 LINEAR INDEPENDENCE AND WRONSKIANS

The use of the principle of superposition in preceding problems to find general solutions can be justified by employing concepts of linear independence and Wronskians as in Chapter Four. In this connection, we have the following fundamental definitions.

Definition 1. Let $u_1, u_2, \ldots, u_k$ be a set of n-dimensional column vectors, which we shall assume are functions of t defined in some interval denoted by J. The set is said to be *linearly independent* in J if for all t in J

$$c_1 u_1 + c_2 u_2 + \cdots + c_k u_k = O \qquad (54)$$

implies that $c_1 = c_2 = \cdots = c_k = 0$. Otherwise the set is *linearly dependent* in J.

Definition 2. Let $u_1, u_2, \ldots, u_n$ be a set of n-dimensional column vectors defined as in Definition 1. Then the *Wronskian* of this set is the determinant of the matrix obtained from these column vectors and is denoted by

$$W(u_1, u_2, \ldots, u_n) = \det(u_1, u_2, \ldots, u_n) \tag{55}$$

The following fundamental theorem can then be proved.

Theorem 3.   Let $u_1, u_2, \ldots, u_n$ be solutions of the matrix differential equation

$$\frac{du}{dt} + Au = 0 \tag{56}$$

for all t in J, where A is an $n \times n$ matrix and u is an n-dimensional column vector. Then
(a) the set is linearly independent in J if and only if the Wronskian $W \neq 0$ for some value, say t_1 in J.
(b) All solutions of (56) have the form

$$u = c_1 u_1 + c_2 u_2 + \cdots + c_n u_n \tag{57}$$

i.e., there is no singular solution.

Example 1.   In the problem on page 515 we found two solutions of the matrix equation (14) given by

$$u_1 = \begin{pmatrix} -2 \\ 1 \end{pmatrix} e^{-t} = \begin{pmatrix} -2e^{-t} \\ e^{-t} \end{pmatrix}, \qquad u_2 = \begin{pmatrix} -3 \\ 2 \end{pmatrix} e^{-2t} = \begin{pmatrix} -3e^{-2t} \\ 2e^{-2t} \end{pmatrix}$$

The Wronskian of this set of solutions is given by

$$W(u_1, u_2) = \begin{vmatrix} -2e^{-t} & -3e^{-2t} \\ e^{-t} & 2e^{-2t} \end{vmatrix} = -e^{-3t}$$

and since $W \neq 0$, we see by Theorem 3 that the set is linearly independent and

$$u = c_1 \begin{pmatrix} -2 \\ 1 \end{pmatrix} e^{-t} + c_2 \begin{pmatrix} -3 \\ 2 \end{pmatrix} e^{-2t} \text{ is the required general solution.}$$

Example 2.   In the problem on page 520 we found three solutions of the matrix equation (41), which can be written as

$$u_1 = \begin{pmatrix} e^{2t} \\ -3e^{2t} \\ 2e^{2t} \end{pmatrix}, \qquad u_2 = \begin{pmatrix} -e^{-t} \\ 0 \\ e^{-t} \end{pmatrix}, \qquad u_3 = \begin{pmatrix} -(1+t)e^{-t} \\ -\frac{1}{2}e^{-t} \\ te^{-t} \end{pmatrix}$$

The Wronskian of this set of solutions is given by

$$W(u_1, u_2, u_3) = \begin{vmatrix} e^{2t} & -e^{-t} & -(1+t)e^{-t} \\ -3e^{2t} & 0 & -\frac{1}{2}e^{-t} \\ 2e^{2t} & e^{-t} & te^{-t} \end{vmatrix} = \frac{9}{2}$$

Since $W \neq 0$, we see by Theorem 3 that the set is linearly independent and $u = c_1 u_1 + c_2 u_2 + c_3 u_3$ is the required general solution.

4 The Particular Solution

Now that we can find the complementary solution, we turn to methods for obtaining a particular solution of the matrix equation

$$\frac{du}{dt} + Au = F \tag{1}$$

As in Chapter Four, we can use the method of undetermined coefficients, the method of variation of parameters, or special operator methods. Of these the method of variation of parameters works best, since the others apply only to special types of functions as we have seen. We thus confine our attention here to this method. To illustrate the procedure, let us consider the following

PROBLEM FOR DISCUSSION

Solve the system

$$\left.\begin{array}{l} \dfrac{dx}{dt} - 2x - 6y = 2t - 2 - e^{-t} \\[2ex] \dfrac{dy}{dt} + 2x + 5y = 2e^{-t} - 3t + 3 \end{array}\right\} \tag{2}$$

This system can be written in matrix form as

$$\frac{du}{dt} + Au = F, \quad \text{where } u = \begin{pmatrix} x \\ y \end{pmatrix}, A = \begin{pmatrix} -2 & -6 \\ 2 & 5 \end{pmatrix}, F = \begin{pmatrix} 2t - 2 - e^{-t} \\ 2e^{-t} - 3t + 3 \end{pmatrix} \tag{3}$$

Now we have already found the complementary solution of the differential equation in (3) to be

$$u = c_1 \begin{pmatrix} -2 \\ 1 \end{pmatrix} e^{-t} + c_2 \begin{pmatrix} -3 \\ 2 \end{pmatrix} e^{-2t} \tag{4}$$

To find a particular solution, let us replace the constants c_1 and c_2 in (4) by functions of t, which we shall denote by γ_1 and γ_2, respectively, so that

$$u = \gamma_1 \begin{pmatrix} -2 \\ 1 \end{pmatrix} e^{-t} + \gamma_2 \begin{pmatrix} -3 \\ 2 \end{pmatrix} e^{-2t} \tag{5}$$

is a solution of the differential equation in (3). Substitution yields

$$\gamma_1' \begin{pmatrix} -2 \\ 1 \end{pmatrix} e^{-t} + \gamma_2' \begin{pmatrix} -3 \\ 2 \end{pmatrix} e^{-2t} = \begin{pmatrix} 2t - 2 - e^{-t} \\ 2e^{-t} - 3t + 3 \end{pmatrix} \tag{6}$$

where it is noted that only terms involving the derivatives γ_1', γ_2' are present, since all other terms satisfy the differential equation in (3) with F replaced by O. We can write (6) as two equations in γ_1', γ_2' namely

$$-2\gamma_1' e^{-t} - 3\gamma_2' e^{-2t} = 2t - 2 - e^{-t}, \qquad \gamma_1' e^{-t} + 2\gamma_2' e^{-2t} = 2e^{-t} - 3t + 3$$

which we can solve either by elimination or by determinants (Cramer's rule, in the Appendix). By either of these methods we find

$$\gamma_1' = 5te^t - 5e^t - 4, \qquad \gamma_2' = 3e^t - 4te^{2t} + 4e^{2t}$$

Integration of these, omitting the constants of integration, yields

$$\gamma_1 = 5te^t - 10e^t - 4t, \qquad \gamma_2 = 3e^t - 2te^{2t} + 3e^{2t} \tag{7}$$

Using these in (5) gives a particular solution, and so the required general solution is

$$u = c_1 \begin{pmatrix} -2 \\ 1 \end{pmatrix} e^{-t} + c_2 \begin{pmatrix} -3 \\ 2 \end{pmatrix} e^{-2t} + \begin{pmatrix} -2 \\ 1 \end{pmatrix} (5t - 10 - 4te^{-t}) + \begin{pmatrix} -3 \\ 2 \end{pmatrix} (3e^{-t} - 2t + 3) \tag{8}$$

Note that if we were to add the arbitrary constants c_1, c_2 to (7) and use the results in (5) we would also obtain (8).

In terms of x and y the solution is given by

$$\left. \begin{array}{l} x = -2c_1 e^{-t} - 3c_2 e^{-2t} + 8te^{-t} - 9e^{-t} - 4t + 11 \\ y = c_1 e^{-t} + 2c_2 e^{-2t} - 4te^{-t} + 6e^{-t} + t - 4 \end{array} \right\} \tag{9}$$

Remark. The system (2) is equivalent to the system given on page 497, and the general solutions are the same.

5 Summary of Procedure

Let us summarize the procedure used above. It is assumed that the given system to be solved is first written in the form (2), page 512.

I. *Write the complementary equation.* This is done by replacing the right-hand side of the matrix differential equation by the null or zero matrix, assuming of course that this is not already the case.

II. *Find the complementary solution.* To do this assume a solution of the form $u = ve^{\lambda t}$, where u is a column vector of the dependent variables, v is a constant column vector, and λ is a constant (scalar). In the case of the equation (4), page 513, this leads to $(\lambda I + A)v = 0$. From this we have $\det(\lambda I + A) = 0$, which leads to the eigenvalues λ. From these eigenvalues we can determine the corresponding eigenvectors v and from these the solutions. There are three situations which can arise.

(a) **Eigenvalues are real and distinct.** In this case the complementary solution can be written as a linear combination of the solutions (i.e., a sum of solutions each multiplied by a different constant).

(b) **Eigenvalues are repeated.** If there are two eigenvalues each equal to λ_1, for example, then one solution is given by $u = v_1 e^{\lambda_1 t}$, where v_1 is the eigenvector corresponding to λ_1. A second and independent solution is obtained by assuming

$$u = v_1 t e^{\lambda_1 t} + K e^{\lambda_1 t}, \quad K = \begin{pmatrix} K_1 \\ K_2 \end{pmatrix} \tag{10}$$

and determining the constants K_1, K_2 so that (10) satisfies the complementary equation. A linear combination of these two solutions will also be a solution. For three or more equal eigenvalues the procedure can be extended.

(c) *Eigenvalues are imaginary.* In this case we use the fact that the real and imaginary parts of solutions are also solutions.

III. *Find a particular solution.* Use the method of variation of parameters, i.e., replace the constants in the complementary solution by functions of t, which must then be determined.

IV. *Add the complementary and particular solutions.* The result is the required general solution of the given system.

Remark. It is assumed in the above procedure that the given system of equations is expressed as a linear first-order matrix differential equation with constant coefficients. If, for example, we have a system involving two differential equations, which may involve derivatives up to the second order, for both dependent variables the above procedure could involve four-dimensional vectors and 4×4 matrices. It is, however, possible to generalize the above procedure to *second-order* matrix differential equations involving only two-dimensional vectors and 2×2 matrices (see C Exercise 2).

6 Applications Using Matrices

We can use matrices to solve systems of differential equations formulated from problems arising in science and engineering. A typical illustration is that given in the following

ILLUSTRATIVE EXAMPLE

Find the currents at any time t in the electrical network shown in Fig. 11.1, assuming that they are zero at time $t = 0$.

Solution Using Kirchhoff's laws (see page 469), we find the equations

$$370 \sin t - 2I_1 - 3(I_1 - I_2) - \frac{dI_1}{dt} = 0, \qquad -2\frac{dI_2}{dt} + 3(I_1 - I_2) = 0 \quad (1)$$

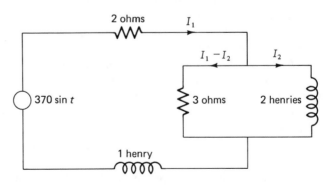

Figure 11.1

or
$$\frac{du}{dt} + \begin{pmatrix} 5 & -3 \\ -\frac{3}{2} & \frac{3}{2} \end{pmatrix} u = \begin{pmatrix} 370 \sin t \\ 0 \end{pmatrix}, \quad \text{where } u = \begin{pmatrix} I_1 \\ I_2 \end{pmatrix} \tag{2}$$

Thus the eigenvalue equation is given by

$$\begin{vmatrix} \lambda + 5 & -3 \\ -\frac{3}{2} & \lambda + \frac{3}{2} \end{vmatrix} = 0 \quad \text{or} \quad \lambda^2 + \tfrac{13}{2}\lambda + 3 = 0, \quad \text{i.e.,} \quad \lambda = -6, -\tfrac{1}{2}$$

Case 1, $\lambda = -6$. From
$$\begin{pmatrix} \lambda + 5 & -3 \\ -\frac{3}{2} & \lambda + \frac{3}{2} \end{pmatrix} \begin{pmatrix} a_1 \\ a_2 \end{pmatrix} = O \tag{3}$$

we have on putting $\lambda = -6$, $-a_1 - 3a_2 = 0$, $-\frac{3}{2}a_1 - \frac{9}{2}a_2 = 0$ or $a_1 = -3a_2$.

Taking $a_2 = 1$, we have $a_1 = -3$. Thus a solution of the complementary equation is

$$u_1 = \begin{pmatrix} -3 \\ 1 \end{pmatrix} e^{-6t} \tag{4}$$

Case 2, $\lambda = -1/2$. Putting $\lambda = -\frac{1}{2}$ in (3) leads to
$$\tfrac{9}{2}a_1 - 3a_2 = 0, \quad -\tfrac{3}{2}a_1 + a_2 = 0 \quad \text{or} \quad a_2 = \tfrac{3}{2}a_1$$

Taking $a_1 = 2$, we have $a_2 = 3$. Thus a second solution of the complementary equation is

$$u_2 = \begin{pmatrix} 2 \\ 3 \end{pmatrix} e^{-t/2} \tag{5}$$

From (4) and (5) we see that the complementary solution of (2) is

$$u_c = c_1 \begin{pmatrix} -3 \\ 1 \end{pmatrix} e^{-6t} + c_2 \begin{pmatrix} 2 \\ 3 \end{pmatrix} e^{-t/2} \tag{6}$$

To find a particular solution we use the method of variation of parameters. In accordance with this method we replace c_1 and c_2 in (6) by the functions of t given by γ_1 and γ_2 respectively to obtain

$$u = \gamma_1 \begin{pmatrix} -3 \\ 1 \end{pmatrix} e^{-6t} + \gamma_2 \begin{pmatrix} 2 \\ 3 \end{pmatrix} e^{-t/2} \tag{7}$$

which must be a solution of (2). Substitution leads to

$$\gamma_1' \begin{pmatrix} -3 \\ 1 \end{pmatrix} e^{-6t} + \gamma_2' \begin{pmatrix} 2 \\ 3 \end{pmatrix} e^{-t/2} = \begin{pmatrix} 370 \sin t \\ 0 \end{pmatrix} \tag{8}$$

where we use the fact that terms involving γ_1 and γ_2 disappear since the sum of such terms is a solution to (2) with the right side replaced by O. We have from (8)

$$-3\gamma_1' e^{-6t} + 2\gamma_2' e^{-t/2} = 370 \sin t, \qquad \gamma_1' e^{-6t} + 3\gamma_2' e^{-t/2} = 0$$

or
$$\gamma_1' = -\tfrac{1110}{11}e^{6t} \sin t, \qquad \gamma_2' = \tfrac{370}{11}e^{t/2} \sin t$$

Integration of these omitting the constants of integration leads to

$$\gamma_1 = -\tfrac{30}{11}e^{6t}(6 \sin t - \cos t), \qquad \gamma_2 = \tfrac{148}{11}e^{t/2}(\sin t - 2 \cos t)$$

Using these in (7) yields the particular solution

$$u_p - \tfrac{30}{11}\begin{pmatrix} -3 \\ 1 \end{pmatrix}(6\sin t - \cos t) + \tfrac{148}{11}\begin{pmatrix} 2 \\ 3 \end{pmatrix}(\sin t - 2\cos t) = \begin{pmatrix} -62 \\ -78 \end{pmatrix}\cos t + \begin{pmatrix} 76 \\ 24 \end{pmatrix}\sin t$$

The general solution is thus given by

$$u = c_1\begin{pmatrix} -3 \\ 1 \end{pmatrix}e^{-6t} + c_2\begin{pmatrix} 2 \\ 3 \end{pmatrix}e^{-t/2} + \begin{pmatrix} -62 \\ -78 \end{pmatrix}\cos t + \begin{pmatrix} 76 \\ 24 \end{pmatrix}\sin t \qquad (9)$$

Using the initial conditions $u = O$ at $t = 0$ in (9) yields

$$c_1 = -\tfrac{30}{11}, \qquad c_2 = \tfrac{296}{11} \qquad (10)$$

so that the required solution is

$$u = -\tfrac{30}{11}\begin{pmatrix} -3 \\ 1 \end{pmatrix}e^{-6t} + \tfrac{296}{11}\begin{pmatrix} 2 \\ 3 \end{pmatrix}e^{-t/2} + \begin{pmatrix} -62 \\ -78 \end{pmatrix}\cos t + \begin{pmatrix} 76 \\ 24 \end{pmatrix}\sin t \qquad (11)$$

or

$$\left.\begin{aligned} I_1 &= \tfrac{90}{11}e^{-6t} + \tfrac{592}{11}e^{-t/2} - 62\cos t + 76\sin t \\ I_2 &= -\tfrac{30}{11}e^{-6t} + \tfrac{888}{11}e^{-t/2} - 78\cos t + 24\sin t \end{aligned}\right\} \qquad (12)$$

A EXERCISES

1. Solve each of the following systems using matrices and determine that solution satisfying any given conditions.

(a) $\dfrac{dx}{dt} + 5x - 4y = 0,\ \dfrac{dy}{dt} - x + 2y = 0;\ x(0) = 3,\ y(0) = -2.$

(b) $\dfrac{dx}{dt} + x - 5y = 0,\ \dfrac{dy}{dt} + 4x + 5y = 0.$ (c) $\dfrac{dx}{dt} + 3y - 2x = 0,\ \dfrac{dy}{dt} - 2x + 3y = 0.$

(d) $\dfrac{dx}{dt} + 3x - 6y = 0,\ \dfrac{dy}{dt} = x - 3y;\ x(0) = 0,\ y(0) = 2.$

(e) $\dfrac{dx}{dt} = x + 8y,\ \dfrac{dy}{dt} = -2x - 7y.$ (f) $\dfrac{dx}{dt} = -12x - 7y,\ \dfrac{dy}{dt} = 19x + 11y.$

2. Solve each of the following systems using matrices and determine that solution satisfying any given conditions.

(a) $\dfrac{dx}{dt} - y = t,\ \dfrac{dy}{dt} + x = t^2;\ x(0) = 2,\ y(0) = -1.$

(b) $\dfrac{dx}{dt} + 3x + 4y = 8e^t,\ \dfrac{dy}{dt} - x - y = 0.$ (c) $\dfrac{dx}{dt} - 2x + y = e^{-t},\ \dfrac{dy}{dt} - 3x + 2y = t.$

(d) $\dfrac{dx}{dt} + 2x - y = 100\sin t,\ \dfrac{dy}{dt} - 4x - y = 36t;\ x(0) = -8,\ y(0) = -21.$

(e) $\dfrac{dx}{dt} - 3x - 6y = 9(1 - t),\ \dfrac{dy}{dt} + 3x + 3y = 9te^{-3t}.$

(f) $\dfrac{dx}{dt} = 2x - 3y + te^{-t},\ \dfrac{dy}{dt} = e^{-t} + 2x - 3y.$

3. Use matrices to solve the systems of (a) A Exercise 2(a), page 499, (b) A Exercise 2(b), page 499, (c) A Exercise 2(f), page 499.

4. Prove that if A is any 2×2 matrix and u_1 is a solution of $\qquad \dfrac{du}{dt} + Au = 0$

then the real and imaginary parts of u_1 denoted by Re (u_1), Im (u_1) respectively are also solutions (see page 519). Can you generalize this?

B EXERCISES

1. Use matrices to find the solution of the system

$$\frac{dx}{dt} + 4x + 2y - z = 12e^t, \qquad \frac{dy}{dt} - 2x - 5y + 3z = 0, \qquad \frac{dz}{dt} + 4x + z = 30e^{-t}$$

2. Solve C Exercise 1, page 500 by matrices.

3. Show how the method of undetermined coefficients can be used with matrices to find a particular solution by solving the system of A Exercise 2(c).

4. Solve by matrices and the method of undetermined coefficients (a) A Exercise 2(d), (b) A Exercise 2(e), (c) A Exercise 2(f), (d) Exercise 1.

5. Work the Problem for Discussion on page 524 by using the method of undetermined coefficients.

6. Show how to solve directly by matrices the systems (a) (9), page 493, (b) (14), page 494, (c) (22), page 495. (*Hint*: Write the systems in matrix form and let $u = e^{\lambda t}v$ as usual.)

7. Show that the solutions to (31), page 518, are linearly independent.

C EXERCISES

1. (a) Show that the system of differential equations for the vibrating masses (see page 463) can be written in matrix form as

$$\frac{du}{dt} + Au = 0$$

where $x_3 = dx_1/dt$, $x_4 = dx_3/dt$, and $\quad u = \begin{pmatrix} x_1 \\ x_2 \\ x_3 \\ x_4 \end{pmatrix}, \quad A = \begin{pmatrix} 0 & 0 & -1 & 0 \\ 0 & 0 & 0 & -1 \\ 2\omega^2 & -\omega^2 & 0 & 0 \\ -\omega^2 & 2\omega^2 & 0 & 0 \end{pmatrix}$

(b) Use matrix methods to solve the system in (a) subject to the conditions given on page 464. Compare with results already obtained.

2. (a) Show that the system of equations for the vibrating masses (page 463) can be written as a second-order matrix differential equation

$$\frac{d^2u}{dt^2} + Bu = 0 \quad \text{where} \quad u = \begin{pmatrix} x_1 \\ x_2 \end{pmatrix}, \quad B = \begin{pmatrix} 2\omega^2 & -\omega^2 \\ -\omega^2 & 2\omega^2 \end{pmatrix}$$

(b) Show how matrix methods can be used to solve the equation in (a) directly by letting $u = e^{\lambda t}v$? What advantages would this procedure have over the method given in Exercise 1?

3. Use the method of Exercises 1 and 2 to work A Exercise 5, page 472.

Some Special Topics

7.1 ORTHOGONALITY

Suppose that we have two n-dimensional column vectors given by

$$A = \begin{pmatrix} a_1 \\ a_2 \\ \vdots \\ a_n \end{pmatrix}, \quad B = \begin{pmatrix} b_1 \\ b_2 \\ \vdots \\ b_n \end{pmatrix} \tag{1}$$

The scalar product of the vectors was defined on page 504 as the sum of the products of corresponding components of the vectors. Equivalently, this is obtained by changing one of the vectors (1) into a row vector and using it to multiply the other vector.

Given any matrix A we can obtain a new matrix by changing the columns into rows (or rows into columns) so that the jth column becomes the jth row. This new matrix is called the *transpose* of A and is denoted by A^T. Thus we are led to the following

Definition 1. The *transpose* of any matrix A, denoted by A^T, is the matrix obtained from A by changing its rows into columns (or columns into rows), the jth row becoming the the jth column. In symbols, if $A = (a_{jk})$, then $A^T = (a_{kj})$.

Example 1. If $A = \begin{pmatrix} 1 \\ -3 \\ 2 \end{pmatrix}$, then $A^T = (1 \quad -3 \quad 2)$.

Example 2. If $A = \begin{pmatrix} 2 & -1 \\ 4 & 0 \end{pmatrix}$, then $A^T = \begin{pmatrix} 2 & 4 \\ -1 & 0 \end{pmatrix}$.

By using the transpose, we could define the scalar product as follows.

Definition 2. Let A and B be two column vectors of the same dimension. Then the *scalar product* of A and B is given by

$$A^T B = B^T A \tag{2}$$

Example 3. If $A = \begin{pmatrix} 1 \\ -3 \\ 2 \end{pmatrix}$, $B = \begin{pmatrix} 2 \\ 1 \\ -1 \end{pmatrix}$, then

$$A^T B = B^T A = (1 \quad -3 \quad 2) \begin{pmatrix} 2 \\ 1 \\ -1 \end{pmatrix} = (2 \quad 1 \quad -1) \begin{pmatrix} 1 \\ -3 \\ 2 \end{pmatrix} = -3$$

By analogy with orthogonality (perpendicularity) of vectors in three dimensions, we make the following

Definition 3. Two column vectors A and B having the same dimension are said to be *orthogonal* if their scalar product is zero, i.e., if $A^T B = B^T A = 0$.

Whereas orthogonality means that the vectors are perpendicular in three dimensions, such visualization can only be imagined for dimensions larger than three.

ILLUSTRATIVE EXAMPLE 1

Find the value of K so that the vectors A and B are orthogonal where

$$A = \begin{pmatrix} -2 \\ 1 \\ K \\ 0 \end{pmatrix}, \quad B = \begin{pmatrix} 1 \\ 2K \\ -3 \\ 4 \end{pmatrix}$$

Solution We have scalar product $= A^T B = B^T A = -2 + 2K - 3K + 0 = -2 - K$ and this is zero for $K = -2$.

We have the following theorem.

Theorem 1. The transpose of the product of matrices is equal to the product of their transposes in reverse order, assuming of course that the products are defined. For the case of products involving two and three matrices, this can be written

(a) $(AB)^T = B^T A^T$.
(b) $(ABC)^T = C^T B^T A^T$.

Example 4. If $A = \begin{pmatrix} 3 & -2 \\ 4 & 1 \end{pmatrix}$ and $B = \begin{pmatrix} -2 \\ 5 \end{pmatrix}$, then

$$AB = \begin{pmatrix} 3 & -2 \\ 4 & 1 \end{pmatrix}\begin{pmatrix} -2 \\ 5 \end{pmatrix} = \begin{pmatrix} -16 \\ -3 \end{pmatrix}, \quad B^T A^T = (-2 \quad 5)\begin{pmatrix} 3 & 4 \\ -2 & 1 \end{pmatrix} = (-16 \quad -3)$$

Thus $(AB)^T = (-16 \quad -3) = B^T A^T$.

A proof of this theorem is left to B Exercise 1.

7.2 LENGTH OF A VECTOR

If we have a three-dimensional vector with components a_1, a_2, a_3 assumed to be real numbers, we know from elementary calculus that the length of the vector found by using the Pythagorean theorem is given by

$$l = \sqrt{a_1^2 + a_2^2 + a_3^2} \tag{3}$$

If the vector is given by

$$v = \begin{pmatrix} a_1 \\ a_2 \\ a_3 \end{pmatrix} \tag{4}$$

the result (3) can be expressed in terms of the scalar product of v with itself as

$$l = \sqrt{v^T v} \tag{5}$$

This leads us to define the length of a vector in n-dimensional space as follows:

Definition 4. Let v be any column vector with real components. Then the length of v is given by

$$l = \sqrt{v^T v}$$

In particular, if $l = 1$, we call v a *unit vector*.

Given a vector v the *process of multiplying* v by a scalar c so that cv is a unit vector is called *normalizing* the vector v. This can always be accomplished by dividing v by its length, assuming of course that this length is not zero.

ILLUSTRATIVE EXAMPLE 2

A four-dimensional column vector has components $2, -1, -3, 4$. (a) Determine the length of the vector and (b) normalize the vector.

Solution (a) The length of the vector is given by

$$l = \sqrt{v^T v} = \sqrt{(2)^2 + (-1)^2 + (-3)^2 + (4)^2} = \sqrt{30}$$

(b) Dividing v by $l = \sqrt{30}$, we obtain the normalized or unit vector

$$\begin{pmatrix} 2/\sqrt{30} \\ -1/\sqrt{30} \\ -3/\sqrt{30} \\ 4/\sqrt{30} \end{pmatrix}$$

We sometimes have a set of vectors, such as for example the eigenvectors, which correspond to the eigenvalues of a matrix. If we convert them all to unit or normalized vectors, the set is often referred to as a *normalized set*. If every pair of vectors in this set is in addition orthogonal, in which case we can say that the vectors are *mutually orthogonal*, we often refer to the set as an *orthonormal set*. The word *ortho-normal* is of course a combination of the words *orthogonal* and *normalized*.

Remark. Those who have studied Chapter Eight will notice the remarkable analogies which exist involving the concepts of orthogonality, the concept of eigenvalues, in particular the eigenvalue equation $Av = -\lambda v$, and the Sturm–Liouville differential equation of pages 357–358. Further analogies will appear in the next few pages and in the advanced exercises.

Suppose that we have a square matrix A whose elements are all real numbers called a *real matrix*. Then as we have seen the eigenvalues and eigenvectors corresponding to A are found by seeking non-trivial solutions, i.e., $v \neq 0$, of

$$Av = -\lambda v \tag{6}$$

Now, as we have seen, even though A is real the eigenvalues and eigenvectors need not be real. Thus, if we take the complex conjugate of both sides of (6), we obtain $\bar{A}\bar{v} = -\bar{\lambda}\bar{v}$, or since A is real so that $\bar{A} = A$,

$$A\bar{v} = -\bar{\lambda}\bar{v} \tag{7}$$

Let us multiply both sides of (6) by $\bar{v}^T$ to obtain

$$\bar{v}^T Av = -\lambda \bar{v}^T v \tag{8}$$

and multiply both sides of (7) by v^T to obtain

$$v^T A\bar{v} = -\bar{\lambda} v^T \bar{v} \tag{9}$$

Since $v^T \bar{v}$ and $\bar{v}^T v$ are transposes of each other, we are led to take the transpose of both sides of (8) and then use the theorem on page 531 to obtain

$$v^T A^T \bar{v} = -\lambda v^T \bar{v} \tag{10}$$

Now it can be seen that the left sides of (9) and (10) are equal if A is a matrix such that

$$A = A^T \tag{11}$$

Writing $A = (a_{jk})$, we see that (11) implies that

$$(a_{jk}) = (a_{kj}) \quad \text{or} \quad a_{jk} = a_{kj} \tag{12}$$

This means that the element in the jth row and kth column is the same as the element in the kth row and jth column, so that there is a symmetry of the elements in the matrix about the main diagonal. This leads us to call a real matrix having property (11) a *real symmetric matrix*. We state this in the following

Definition 5. A matrix A is called a *real symmetric matrix* if A is real and $A = A^T$ or $a_{jk} = a_{kj}$.

Example 5. Since $A = \begin{pmatrix} 2 & -3 & 0 \\ -3 & 1 & 5 \\ 0 & 5 & 4 \end{pmatrix}$ so that $A^T = \begin{pmatrix} 2 & -3 & 0 \\ -3 & 1 & 5 \\ 0 & 5 & 4 \end{pmatrix}$

we see that A is real, $A = A^T$, and so A is a real symmetric matrix.

Getting back to (9) and (10), we see that if A is symmetric we have $\bar{\lambda} v^T \bar{v} = \lambda v^T \bar{v}$ or

$$(\lambda - \bar{\lambda})v^T \bar{v} = 0 \tag{13}$$

From this it follows that either $\lambda = \bar{\lambda}$, in which case any eigenvalue must be real, or $v^T \bar{v} = 0$. However, it is easy to show that $v^T \bar{v} \neq 0$. To do this, let us assume that v is given by the column vector

$$v = \begin{pmatrix} a_1 \\ a_2 \\ \vdots \\ a_n \end{pmatrix}$$

where some or all of the elements may not be real. Then we have

$$v^T \bar{v} = a_1 \bar{a}_1 + a_2 \bar{a}_2 + \cdots + a_n \bar{a}_n = |a_1|^2 + |a_2|^2 + \cdots + |a_n|^2$$

Since at least one of the a's is not zero because $v \neq O$, we see that $v^T \bar{v} \neq 0$, which means that any eigenvalues are real. As a necessary consequence of this, we also see from (6) that corresponding eigenvectors can be chosen so as to be real. Because of this we have the following

Theorem 2. If A is a real symmetric matrix, then all its eigenvalues are real and corresponding eigenvectors can be chosen real.

ILLUSTRATIVE EXAMPLE 3

Verify the above theorem for the matrix $A = \begin{pmatrix} 3 & -2 \\ -2 & 6 \end{pmatrix}$.

Solution The eigenvalues and eigenvectors are obtained as usual from $Av = -\lambda v$ or $(\lambda I + A)v = O$; i.e.,

$$\begin{pmatrix} \lambda + 3 & -2 \\ -2 & \lambda + 6 \end{pmatrix} v = O \tag{14}$$

For non-trivial solutions $v \neq O$, we must have

$$\begin{vmatrix} \lambda + 3 & -2 \\ -2 & \lambda + 6 \end{vmatrix} = 0 \quad \text{or} \quad \lambda^2 + 9\lambda + 14 = 0, \quad \text{i.e.,} \quad \lambda = -2, -7$$

Case 1, $\lambda = -2$. Putting $\lambda = -2$ in (14) leads to

$$\begin{pmatrix} 1 & -2 \\ -2 & 4 \end{pmatrix} \begin{pmatrix} a_1 \\ a_2 \end{pmatrix} = O, \quad \text{i.e.,} \quad a_1 - 2a_2 = 0, \quad -2a_1 + 4a_2 = 0$$

so that $a_1 = 2a_2$. In particular, choosing $a_2 = 1$, we have $a_1 = 2$. Thus an eigenvector is

$$v_1 = \begin{pmatrix} 2 \\ 1 \end{pmatrix} \tag{15}$$

Case 2, $\lambda = -7$. Putting $\lambda = -7$ in (14) leads to

$$\begin{pmatrix} -4 & -2 \\ -2 & -1 \end{pmatrix} \begin{pmatrix} a_1 \\ a_2 \end{pmatrix} = O, \quad \text{i.e.,} \quad -4a_1 - 2a_2 = 0, \quad -2a_1 - a_2 = 0$$

so that $a_2 = -2a_1$. In particular, choosing $a_1 = 1$, we have $a_2 = -2$. Thus another eigenvector is

$$v_2 = \begin{pmatrix} 1 \\ -2 \end{pmatrix} \tag{16}$$

Since A is a real symmetric matrix and since we have found the eigenvalues to be real the above theorem has been verified for this case. It should also be noted that the corresponding eigenvectors are real.

Now that we know that the eigenvalues and eigenvectors for a real symmetric matrix are real, let us suppose that λ_1 and λ_2 are any two different eigenvalues with corresponding eigenvectors v_1 and v_2, respectively. Then by definition

$$Av_1 = -\lambda_1 v_1 \qquad Av_2 = -\lambda_2 v_2 \tag{17}$$

Let us now multiply each side of the first equation in (17) by v_2^T to obtain

$$v_2^T A v_1 = -\lambda_1 v_2^T v_1 \tag{18}$$

Similarly, let us multiply each side of the second equation in (17) by v_1^T to obtain

$$v_1^T A v_2 = -\lambda_2 v_1^T v_2 \tag{19}$$

If we take the transpose of both sides of (18), we find

$$v_1^T A^T v_2 = -\lambda_1 v_1^T v_2 \tag{20}$$

where we have made use of part (b) of Theorem 1, page 531, and the fact that the transpose of the transpose of a given matrix is clearly the given matrix. It follows that if A is symmetric, we obtain from (19) and (20), $\lambda_1 v_1^T v_2 = \lambda_2 v_1^T v_2$ or

$$(\lambda_1 - \lambda_2)v_1^T v_2 = 0 \tag{21}$$

and since $\lambda_1 \neq \lambda_2$

$$v_1^T v_2 = 0 \tag{22}$$

which states that v_1 and v_2 are orthogonal. We have thus proved the rather remarkable result given in the following

Theorem 3. If A is any square real symmetric matrix, then the eigenvectors belonging to any two different eigenvalues are orthogonal.

ILLUSTRATIVE EXAMPLE 4

Verify Theorem 3 for the matrix of Illustrative Example 3.

Solution From Illustrative Example 3 we have the two eigenvectors

$$v_1 = \begin{pmatrix} 2 \\ 1 \end{pmatrix}, \qquad v_2 = \begin{pmatrix} 1 \\ -2 \end{pmatrix}$$

Then $\qquad v_1^T v_2 = (2 \ \ 1)\begin{pmatrix} 1 \\ -2 \end{pmatrix} = (2)(1) + (1)(-2) = 0$

so that v_1 and v_2 are orthogonal and the theorem is verified.

1. Find the scalar product of (a) $A = \begin{pmatrix} 2 \\ -1 \\ 3 \end{pmatrix}$, $B = \begin{pmatrix} -3 \\ 2 \\ -4 \end{pmatrix}$. (b) $A = \begin{pmatrix} 2 \\ -1 \\ 1 \\ -2 \end{pmatrix}$, $B = \begin{pmatrix} -1 \\ -2 \\ 0 \\ 1 \end{pmatrix}$.

2. Find the lengths of the column vectors with components (a) $0, -2, 3, 6$. (b) $1, 2, -3, -2, 0$.

3. Normalize the vectors given in Exercise 2.

4. Given the column vectors A and B whose components are given by $3, -1, k + 2, -1$ and $-2, k, 0, -4$, respectively. (a) Find the value of k so that the vectors are orthogonal, and (b) determine the corresponding orthonormal vectors.

5. Given column vectors A and B with components $2, 1, -4$ and $4, -3, 2$, respectively. Find a normalized vector which is orthogonal to both A and B. Interpret the results geometrically.

6. Given $A = \begin{pmatrix} 2 & 1 \\ 3 & -2 \end{pmatrix}$, $B = \begin{pmatrix} -2 & 5 \\ 1 & 0 \end{pmatrix}$, $C = \begin{pmatrix} 1 & 0 \\ -3 & 4 \end{pmatrix}$, verify that

   (a) $(AB)^T = B^T A^T$. (b) $(ABC)^T = C^T B^T A^T$.

7. Verify Theorems 2 and 3 for the matrices (a) $A = \begin{pmatrix} 5 & 3 \\ 3 & -3 \end{pmatrix}$. (b) $A = \begin{pmatrix} 2 & -1 \\ -1 & 4 \end{pmatrix}$.

8. Find a set of mutually orthogonal and normalized eigenvectors belonging to the matrix

$$A = \begin{pmatrix} 7 & -3 & 0 \\ -3 & -1 & 0 \\ 0 & 0 & 4 \end{pmatrix}$$

1. If A, B, C are suitably conformable matrices, prove that (a) $(AB)^T = B^T A^T$. (b) $(ABC)^T = C^T B^T A^T$.

2. Prove that if A and B are matrices of the same size then $(A + B)^T = A^T + B^T$.

3. Given the matrix $A = \begin{pmatrix} 2 & -1 \\ -3 & 4 \end{pmatrix}$, find a matrix $B = \begin{pmatrix} b_1 & b_2 \\ b_3 & b_4 \end{pmatrix}$ such that $(AB)^T = A^T B^T$.

4. Show that the matrices A and B, assumed suitably conformable, commute if and only if $(AB)^T = A^T B^T$. Illustrate by use of Exercise 3.

5. Given the vectors A_1, A_2, A_3 where $A_1^T = (1 \quad -1 \quad 2)$, $A_2^T = (2 \quad 1 \quad 0)$, $A_3^T = (0 \quad 2 \quad -1)$, determine an orthonormal set of vectors $c_{11}A_1$, $c_{21}A_1 + c_{22}A_2$, $c_{31}A_1 + c_{32}A_2 + c_{33}A_3$, where the c's are constants (scalars). Discuss the geometrical significance. (This is the *Gram–Schmidt orthonormalization method* of page 410 for the special case of three-dimensional vectors.

6. Generalize the ideas of Exercise 5.

7. Let $A_1, A_2, \ldots, A_k$ be n-dimensional vectors. If there exist constant (scalars) $c_1, c_2, \ldots, c_k$ not all zero such that

$$c_1 A_1 + c_2 A_2 + \cdots + c_k A_k \equiv 0$$

the vectors are said to be *linearly dependent*; otherwise they are *linearly independent*. Determine whether the vectors (a) $(2 \quad -1 \quad 1), (3 \quad 2 \quad -2), (4 \quad 5 \quad -5), (b) (3 \quad 1 \quad 5), (-2 \quad 2 \quad 4),$ $(1 \quad 1 \quad 0)$ (c) $(-2 \quad 1 \quad 4 \quad 3), (1 \quad -2 \quad 3 \quad -1), (5 \quad -7 \quad 5 \quad -6)$ are linearly dependent or independent.

8. Discuss the geometrical significance of linearly dependent and independent vectors and their relationship with solutions of systems of linear differential equations considered in the last section.

9. Prove that a set of more than n different vectors in n-dimensional space must be linearly dependent. Discuss the relationship of this with the theory of systems of differential equations.

10. Discuss the relationship of the Gram–Schmidt orthonormalization method (see Exercise 5) with the concepts of linear dependence.

C EXERCISES

1. (a) Solve the matrix differential equation
$$\frac{du}{dt} + Au = O$$

where
$$A = \begin{pmatrix} 7 & -3 & 0 \\ -3 & -1 & 0 \\ 0 & 0 & 4 \end{pmatrix}, \qquad u = \begin{pmatrix} x \\ y \\ z \end{pmatrix}$$

(b) Show that there are three linearly independent and mutually orthogonal solutions u_1, u_2, u_3 to the system in (a), and explain their relationship to the general solution obtained in (a).

2. Let S be the matrix whose columns are the normalized eigenvectors of the matrix
$$A = \begin{pmatrix} 2 & -2 \\ -2 & 6 \end{pmatrix}$$

given in Illustrative Example 3, page 534. (a) Show that $S^T S = I$, where I is the 2×2 identity matrix, or equivalently $S^T = S^{-1}$. A matrix S having the property that $S^T = S^{-1}$ is often called an *orthogonal matrix*. (b) Show that the matrix $S^{-1}AS$ is a matrix with elements zero everywhere except in the main diagonal. (c) Show that the elements in the main diagonal of the matrix in (b) are the negatives of the eigenvalues of A, or equivalently that the eigenvalues of $S^{-1}AS$ are equal to the eigenvalues of A.

3. Demonstrate the results of Exercise 2 by using the matrix A of Exercise 1.

4. Prove directly from the definition $Av = -\lambda v$ the results indicated in Exercises 2(b) and (c).

5. Discuss the relationships of the matrix formulations of the vibrating masses problem given in C Exercises 1 and 2, page 529, with Theorems 2 and 3 on pages 534–535.

III

partial differential equations

twelve
partial differential equations in general

1 The Concept of a Partial Differential Equation

1.1 INTRODUCTION

In preceding chapters we were concerned with ordinary differential equations involving derivatives of one or more dependent variables with respect to a single independent variable. We learned how such differential equations arise, methods by which their solutions can be obtained, both exact and approximate, and have considered applications to various scientific fields.

By using ordinary differential equations to solve applied problems we are in effect greatly simplifying (and often seriously oversimplifying) the mathematical model of physical reality which leads to these problems. This is because in the mathematical formulations of such problems we are restricting ourselves to one independent variable on which all other pertinent variables in the problem depend. While this is often useful, as we have seen, it limits the kinds of problems of the real world which can be investigated, because in many such cases two or more independent variables are required.

Mathematical formulations of problems involving two or more independent variables lead to partial differential equations. As one might expect, the introduction of more independent variables makes the subject of partial differential equations more complicated than ordinary differential equations, and thus comparatively little is known concerning them. Nevertheless, the subject is so large that we shall be able to touch upon it only briefly in this book.

1.2 SOLUTIONS OF SOME SIMPLE PARTIAL DIFFERENTIAL EQUATIONS

In order to get some ideas concerning the nature of solutions of partial differential equations, let us consider the following

PROBLEM FOR DISCUSSION

Obtain solutions of the partial differential equation

$$\frac{\partial^2 U}{\partial x \, \partial y} = 6x + 12y^2 \tag{1}$$

Here the dependent variable U depends on the two independent variables x and y. To find solutions, we seek to determine U in terms of x and y, i.e., $U(x, y)$. If we write (1) as

$$\frac{\partial}{\partial x}\left(\frac{\partial U}{\partial y}\right) = 6x + 12y^2 \tag{2}$$

we can integrate with respect to x, keeping y constant, to find

$$\frac{\partial U}{\partial y} = 3x^2 + 12xy^2 + F(y) \tag{3}$$

where we have added the arbitrary "constant" of integration, which can depend on y and so is really an arbitrary function of y denoted by $F(y)$.

We now integrate (3) with respect to y keeping x constant to find

$$U = 3x^2y + 4xy^3 + \int F(y)dy + G(x) \tag{4}$$

this time adding an arbitrary function of x given by $G(x)$. Since the integral of an arbitrary function of y is another arbitrary function of y, we can write (4) as

$$U = 3x^2y + 4xy^3 + H(y) + G(x) \tag{5}$$

This can be checked by substituting it back into (1) and obtaining an identity. Since (1) is a second-order partial differential equation, while the solution (5) has two arbitrary functions, we are led by analogy with ordinary differential equations to call (5) the *general solution* of (1). Using the same analogy, we would of course call any solution obtained from the general solution (5) by particular choices of the arbitrary functions, such as for example $H(y) = y^3$, $G(x) = \sin 2x$, a *particular solution*. We are thus led to make the following

Definition. Given an nth-order partial differential equation, a solution containing n arbitrary functions is called the *general solution*, and any solution obtained from this general solution by particular choices of the arbitrary functions is called a *particular solution.**

As in the case of ordinary differential equations, we are often required to determine solutions of partial differential equations which satisfy given conditions. For example, suppose that we want to solve the differential equation (1) subject to the two conditions

$$U(1, y) = y^2 - 2y, \qquad U(x, 2) = 5x - 5 \tag{6}$$

Then from the general solution (5) and the first condition in (6) we are led to

$$U(1, y) = 3(1)^2y + 4(1)y^3 + H(y) + G(1) = y^2 - 2y$$

or

$$H(y) = y^2 - 5y - 4y^3 - G(1)$$

so that

$$U = 3x^2y + 4xy^3 + y^2 - 5y - 4y^3 - G(1) + G(x) \tag{7}$$

If we now use the second condition in (6), we are led to

$$U(x, 2) = 3x^2(2) + 4x(2)^3 + (2)^2 - 5(2) - 4(2)^3 - G(1) + G(x) = 5x - 5$$

from which

$$G(x) = 33 - 27x - 6x^2 + G(1)$$

Using this in (7), we obtain the required solution

$$U = 3x^2y + 4xy^3 + y^2 - 5y - 4y^3 - 27x - 6x^2 + 33 \tag{8}$$

We could use the same terminology of initial and boundary value problems for partial differential equations as was done for ordinary differential equations.

* As in the case of ordinary differential equations, it may happen that there are *singular solutions* which cannot be obtained from the general solution by any choice of the arbitrary functions.

However, because there is generally a combination of boundary and initial conditions, we usually refer to all such problems as *boundary value problems*.

As in the problem discussed on page 541, the form of a partial differential equation may suggest a method of solution. An especially simple type of partial differential equation is one which can be treated by methods of ordinary differential equations using one independent variable at a time. The problem on page 541 is an example. A slightly more complicated example is provided in the following

ILLUSTRATIVE EXAMPLE 1

Find a solution of the boundary value problem*

$$\frac{\partial^2 U}{\partial x \, \partial y} = \frac{\partial U}{\partial x} + 2, \qquad U(0, y) = 0, \qquad U_x(x, 0) = x^2$$

Solution Writing the equation as

$$\frac{\partial}{\partial x}\left(\frac{\partial U}{\partial y} - U\right) = 2$$

and integrating with respect to x yields

$$\frac{\partial U}{\partial y} - U = 2x + F(y)$$

which is a linear equation having integrating factor e^{-y}. Hence,

$$\frac{\partial}{\partial y}(e^{-y}U) = 2xe^{-y} + e^{-y}F(y)$$

or $\qquad U(x, y) = -2x + e^y \int e^{-y}F(y)dy + e^y G(x)$

where $G(x)$ is arbitrary. Writing $H(y) \equiv e^y \int e^{-y}F(y)dy$, we have

$$U(x, y) = -2x + H(y) + e^y G(x) \tag{9}$$

From $U(0, y) = 0$ we find $H(y) = -G(0)e^y$, so (9) becomes

$$U(x, y) = -2x - G(0)e^y + e^y G(x)$$

Differentiating with respect to x and placing $y = 0$, we find

$$U_x(x, 0) = -2 + G'(x) = x^2 \quad \text{or} \quad G(x) = \frac{x^3}{3} + 2x + c$$

Hence, $\qquad U(x, y) = -2x - G(0)e^y + e^y\left(\frac{x^3}{3} + 2x + c\right)$

Since $c = G(0)$, $\qquad U(x, y) = \frac{x^3 e^y}{3} + 2xe^y - 2x$

---

* Subscripts x and y are often used to denote partial derivatives. For example, U_x or $U_x(x, y)$ is the same as $\partial U/\partial x$ while U_y or $U_y(x, y)$ is the same as $\partial U/\partial y$. Similarly U_{xy} is the same as $\partial^2 U/\partial x \, \partial y$.

As in the case of ordinary differential equations, applied problems provide an important source of partial differential equations to be solved subject to associated conditions, which we have called *boundary value problems*. Given some problem of science or engineering, we proceed as usual to make a *mathematical model* which simplifies but hopefully closely approximates reality. We then formulate the problem mathematically, arriving at the boundary value problem. It should be pointed out that in practice formulations of partial differential equations and the associated conditions may sometimes be difficult and even impossible. Later in this chapter we shall derive some of the important partial differential equations arising in various fields.

If we are successful in formulating a boundary value problem, there still remains the task of finding a solution of this boundary value problem, i.e., finding a solution of the partial differential equation which satisfies the conditions. Sometimes it is easy to find a solution, in fact many solutions, of the partial differential equation, but difficult or even impossible to find that solution which satisfies the given conditions. As in ordinary differential equations there are, logically, three questions which we as scientists should *ask* even though we may not be able to *answer*.

1. *Does a solution to our problem exist?* If in any way we can show that a solution does not exist there is clearly no point in looking for it. Mathematicians have succeeded in proving that certain types of boundary value problems have solutions. Theorems guaranteeing the existence of solutions are called *existence theorems* and are very valuable.

2. *If a solution exists, is it unique?* If it is not unique, i.e., if we have two possible answers to a given physical problem it might prove utterly embarrassing. Theorems guaranteeing the uniqueness of solutions are called *uniqueness theorems*.

3. *If a solution does exist and is unique, what is this solution?*

In an elementary treatment we can naturally be concerned only with the last question, how to determine one solution which satisfies the equation and the conditions. This solution must of course be in agreement with experiment or observation; otherwise we would have to revise the equations.

1.3 GEOMETRIC SIGNIFICANCE OF GENERAL AND PARTICULAR SOLUTIONS

In the problem on page 541, we obtained the general solution

$$U = 3x^2y + 4xy^3 + H(y) + G(x) \tag{10}$$

Let us suppose now that we choose particular functions for $H(y)$ and $G(x)$, and replace U by z. Then (10) takes the form

$$z = f(x, y) \tag{11}$$

which can be interpreted geometrically as a surface S in a rectangular or *xyz* coordinate system such as indicated in Fig. 12.1. The surface is made up of points with coordinates (x, y, z) which satisfy (11).

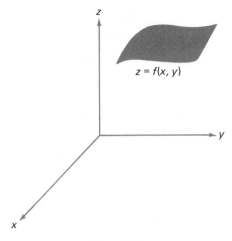

$z = f(x, y)$

Figure 12.1

For arbitrary functions $H(y)$ and $G(x)$, we obtain a *family of surfaces* each member of which corresponds to a particular choice of $H(y)$ and $G(x)$, i.e., a particular solution. The differential equation having this as a solution is then called the *differential equation of the family of surfaces*. The student will note the analogy with ordinary differential equations in which the general solution with arbitrary constants (rather than functions) represents a family of curves, each member of which corresponds to a particular solution, i.e., a particular choice of these arbitrary constants.

These ideas can be generalized to cases where there are more than two independent variables. Thus, for instance, in the case where U is a function of three independent variables, which we can denote by x_1, x_2, x_3, we could think of a particular solution of a partial differential equation with these variables as given by

$$U = f(x_1, x_2, x_3) \tag{12}$$

This could not be visualized geometrically as in Fig. 12.1. However, we can think of the quadruple of numbers (x_1, x_2, x_3, U) as representing a point in *four-dimensional space* and then refer to (12) as a *four-dimensional surface* or *hypersurface*. For example, just as $x^2 + y^2 + z^2 = c^2$ represents a *sphere* of radius c, in three-dimensional space, $x_1^2 + x_2^2 + x_3^2 + U^2 = c^2$ would represent a *hypersphere* of radius c in four-dimensional space.

1.4 PARTIAL DIFFERENTIAL EQUATIONS ARISING FROM ELIMINATION OF ARBITRARY FUNCTIONS

Since general solutions of partial differential equations involve arbitrary functions, it seems logical that we ought to obtain partial differential equations by the reverse process of eliminating such functions. This idea proves useful because it helps to build up our knowledge of how partial differential equations can be solved. Let us consider some examples.

Find a first-order partial differential equation which has as its general solution

$$U = y^2 F(x) - 3x + 4y \tag{13}$$

where $F(x)$ is an arbitrary function of x.

Solution If we differentiate (13) with respect to y, we get

$$\frac{\partial U}{\partial y} = 2yF(x) + 4 \tag{14}$$

Then eliminating $F(x)$ between (13) and (14), we find the required equation

$$y\frac{\partial U}{\partial y} - 2U = 6x - 4y \tag{15}$$

Check. $y\dfrac{\partial U}{\partial y} - 2U = y[2yF(x) + 4] - 2[y^2 F(x) - 3x + 4y] = 6x - 4y.$

ILLUSTRATIVE EXAMPLE 3

Find a first-order partial differential equation which has as its general solution

$$z = F(3x - 4y) \tag{16}$$

where F is an arbitrary function.

Solution Let $u = 3x - 4y$. Then (16) becomes

$$z = F(u) \tag{17}$$

Differentiating (17) with respect to x, we have

$$\frac{\partial z}{\partial x} = \frac{\partial z}{\partial u} \cdot \frac{\partial u}{\partial x} = F'(u)(3) = 3F'(u) \tag{18}$$

Differentiating (17) with respect to y, we have

$$\frac{\partial z}{\partial y} = \frac{\partial z}{\partial u} \cdot \frac{\partial u}{\partial y} = F'(u)(-4) = -4F'(u) \tag{19}$$

Eliminating $F'(u)$ between (18) and (19) yields the required equation:

$$4\frac{\partial z}{\partial x} + 3\frac{\partial z}{\partial y} = 0 \tag{20}$$

ILLUSTRATIVE EXAMPLE 4

Find a second-order partial differential equation which has as its general solution

$$U = xF(y) + yG(x) \tag{21}$$

where F and G are arbitrary functions.

Solution We can eliminate $F(y)$ in (21) by dividing both sides of (21) by x and differentiating the result with respect to x. Then we find

$$\frac{\partial}{\partial x}\left[\frac{U}{x}\right] = \frac{\partial}{\partial x}\left[F(y) + \frac{y}{x}G(x)\right] \quad \text{i.e.,} \quad x\frac{\partial U}{\partial x} - U = xyG'(x) - yG(x)$$

which can be written $\quad x\dfrac{\partial U}{\partial x} - U = y[xG'(x) - G(x)]$ ⠀⠀⠀⠀⠀(22)

If we now divide both sides of (22) by y and differentiate with respect to y, we find

$$\frac{\partial}{\partial y}\left[\frac{1}{y}\left(x\frac{\partial U}{\partial x} - U\right)\right] = 0 \quad \text{or} \quad xy\frac{\partial^2 U}{\partial y\,\partial x} - x\frac{\partial U}{\partial x} - y\frac{\partial U}{\partial y} + U = 0 \quad (23)$$

which gives the required second-order equation. Note that the second equation of (23) can also be written

$$xy\frac{\partial^2 U}{\partial x\,\partial y} - x\frac{\partial U}{\partial x} - y\frac{\partial U}{\partial y} + U = 0 \quad \text{since} \quad \frac{\partial^2 U}{\partial x\,\partial y} = \frac{\partial^2 U}{\partial y\,\partial x}$$

Several remarks should be made about the above results.

Remark 1. In differentiating the arbitrary functions we of course assume that they are differentiable. Otherwise, we have no right to differentiate.

Remark 2. The differential equation obtained in each example represents the differential equation of the family represented by the general solution.

Remark 3. If a solution has a given number n of arbitrary functions, it is often easy to write a differential equation of order greater than n having this solution. For example, it is easy to see that (21) is a solution of

$$\frac{\partial^4 U}{\partial x^2\,\partial y^2} = 0$$

but this is of order 4 and not 2. When finding the differential equation, we seek the one of *least order*.

A EXERCISES

1. Obtain solutions to the following boundary value problems:

(a) $\dfrac{\partial U}{\partial x} = \sin y$; $U(0, y) = 0$. ⠀⠀(b) $\dfrac{\partial^2 U}{\partial y^2} = x^2 \cos y$; $U(x, 0) = 0$, $U\left(x, \dfrac{\pi}{2}\right) = 0$.

(c) $\dfrac{\partial^2 V}{\partial x\,\partial y} = 0$; $V(0, y) = 3\sin y$, $V_x(x, 1) = x^2$.

(d) $\dfrac{\partial^2 U}{\partial x\,\partial y} = 4xy + e^x$; $U_y(0, y) = y$, $U(x, 0) = 2$.

(e) $\dfrac{\partial^2 Z}{\partial x\,\partial y} = 3\dfrac{\partial Z}{\partial y} + 2y$; $Z_y(0, y) = y^2 - 2y$, $Z(x, 0) = x + 3e^{-x}$.

2. Obtain partial differential equations (of least order) by eliminating the arbitrary functions in each given relation. In each case verify that the given relation in a solution of the equation obtained. Take z as dependent variable unless otherwise stated.

(a) $U = x^2F(y) + 3xy$.   (b) $z = e^{xy}G(x)$.   (c) $z = e^{-y}F(x) + e^xG(y)$.

(d) $U = \sqrt{x}F(y) + (\ln y)G(x)$.   (e) $z = F(x - 2y)$.   (f) $z = F(xy)$.

(g) $x = F(y/z)$(let x be dependent).   (h) $z = F(x^2 - y^2)$.   (i) $z = e^{3y}F(x - 2y)$.

(j) $z = F(x + 3y) + G(2x - y)$.

B EXERCISES

1. Considering $z = F(3x - 2y)$, it would seem that $3x - 2y = G(z)$, where F and G are inverse functions. Assuming this true, are the partial differential equations obtained from these two relations equivalent? Justify your conclusion.

2. Obtain a partial differential equation with $z = F(x^2y) + G(xy^2)$, where F and G are arbitrary differentiable functions, as solution.

3. If a function $F(x, y)$ can be written $x^nG(y/x)$ it is called homogeneous of degree n. Show that any homogeneous function which is differentiable satisfies the partial differential equation

$$x\frac{\partial F}{\partial x} + y\frac{\partial F}{\partial y} = nF$$

This is called *Euler's theorem on homogeneous functions.*

4. Show that if $F = \sqrt{x^2 + y^2}\,\tan^{-1}(y/x)$ then $x\frac{\partial F}{\partial x} + y\frac{\partial F}{\partial y} = F$.

5. Determine n so that $z = x^3\tan^{-1}\left(\dfrac{x^2 - xy + y^2}{x^2 + xy + y^2}\right)$ satisfies $x\frac{\partial z}{\partial x} + y\frac{\partial z}{\partial y} = nz$.

6. Show that the function $V(x, y, z) = (x^2 + y^2 + z^2)^{-1/2}$ satisfies Laplace's equation

$$\frac{\partial^2 V}{\partial x^2} + \frac{\partial^2 V}{\partial y^2} + \frac{\partial^2 V}{\partial z^2} = 0$$

7. Use Exercise 3 to solve the boundary value problem

$$x\frac{\partial U}{\partial x} + y\frac{\partial U}{\partial y} = 2U, \quad U(1, y) = 20\cos y$$

C EXERCISES

1. (a) If $z = F(y/x) + xG(y/x)$ show that $x^2\frac{\partial^2 z}{\partial x^2} + 2xy\frac{\partial^2 z}{\partial x\,\partial y} + y^2\frac{\partial^2 z}{\partial y^2} = 0$.

(b) Obtain a solution to the partial differential equation in (a) which satisfies the conditions $z = \cos y$ for $x = 1$ and $z = e^{-2y}$ for $x = \frac{1}{2}$.

2. (a) In the relation $F(u, v) = 0$, F is an arbitrary differentiable function of u and v, which are given differentiable functions of x, y, and z. By differentiation with respect to x and y, prove

prove that

$$\frac{\partial F}{\partial u}\left(\frac{\partial u}{\partial x}+\frac{\partial u}{\partial z}\frac{\partial z}{\partial x}\right)+\frac{\partial F}{\partial v}\left(\frac{\partial v}{\partial x}+\frac{\partial v}{\partial z}\frac{\partial z}{\partial x}\right)=0,\quad \frac{\partial F}{\partial u}\left(\frac{\partial u}{\partial y}+\frac{\partial u}{\partial z}\frac{\partial z}{\partial y}\right)+\frac{\partial F}{\partial v}\left(\frac{\partial v}{\partial y}+\frac{\partial v}{\partial z}\frac{\partial z}{\partial y}\right)=0$$

(b) By eliminating $\partial F/\partial u$ and $\partial F/\partial v$ from the equations of (a), show that the resulting partial differential equation has the form

$$P\frac{\partial z}{\partial x}+Q\frac{\partial z}{\partial y}=R$$

where P, Q, R are known functions of x, y, and z. Discuss any restrictions which must be imposed in performing this elimination. This result is the partial differential equation corresponding to $F(u, v) = 0$.

3. Using the method of Exercise 2, find partial differential equations corresponding to each of the following, where F is an arbitrary function.

(a) $F(2x + 3z, x - 2y) = 0$.
(c) $F(z \sin x, z \cos y) = 0$.

(b) $F(x^2 + y^2, yz) = 0$.
(d) $F(x - y - z, x^2 - 2xy) = 0$.

4. (a) Show that the differential equation

$$\frac{\partial^3 z}{\partial x^3}+3\frac{\partial^3 z}{\partial x^2\,\partial y}+3\frac{\partial^3 z}{\partial x\,\partial y^2}+\frac{\partial^3 z}{\partial y^3}=0$$

has solution $z = F(x - y) + xG(x - y) + x^2 H(x - y)$, where F, G, and H are arbitrary differentiable functions.

(b) Show that the differential equation of (a) also has solutions given by

$$z = F(x - y) + yG(x - y) + y^2 H(x - y),\ z = F(x - y) + yG(x - y) + xyH(x - y)$$

Are all these solutions related? Explain.

5. Suppose that in the partial differential equation

$$\frac{\partial^2 U}{\partial x^2}+\frac{\partial^2 U}{\partial y^2}=0$$

we make the change of variables from rectangular coordinates (x, y) to polar coordinates (r, ϕ) according to the transformation equations $x = r\cos\phi$, $y = r\sin\phi$.

(a) Show that

$$\frac{\partial^2 U}{\partial x^2}+\frac{\partial^2 U}{\partial y^2}=\frac{\partial^2 U}{\partial r^2}+\frac{1}{r}\frac{\partial U}{\partial r}+\frac{1}{r^2}\frac{\partial^2 U}{\partial \phi^2}$$

(b) Show that the given differential equation is given in polar coordinates by

$$\frac{\partial^2 U}{\partial r^2}+\frac{1}{r}\frac{\partial U}{\partial r}+\frac{1}{r^2}\frac{\partial^2 U}{\partial \phi^2}=0$$

6. Suppose that in the partial differential equation

$$\frac{\partial^2 U}{\partial x^2}+\frac{\partial^2 U}{\partial y^2}+\frac{\partial^2 U}{\partial z^2}=0$$

we change variables from (x, y, z) to (r, ϕ, θ) according to the transformation equations

$$x = r\sin\theta\cos\phi,\ y = r\sin\theta\sin\phi,\ z = r\cos\theta$$

(a) Show that $$\frac{\partial^2 U}{\partial x^2}+\frac{\partial^2 U}{\partial y^2}+\frac{\partial^2 U}{\partial z^2}=\frac{1}{r^2}\left[r\frac{\partial^2}{\partial r^2}(rU)+\frac{1}{\sin\theta}\frac{\partial}{\partial\theta}\left(\sin\theta\frac{\partial U}{\partial\theta}\right)+\frac{1}{\sin^2\theta}\frac{\partial^2 U}{\partial\phi^2}\right]$$

(b) Show that the given partial differential equation can be written as

$$\frac{\partial^2 U}{\partial r^2}+\frac{2}{r}\frac{\partial U}{\partial r}+\frac{1}{r^2}\frac{\partial^2 U}{\partial\theta^2}+\frac{\cot\theta}{r^2}\frac{\partial U}{\partial\theta}+\frac{1}{r^2\sin^2\theta}\frac{\partial^2 U}{\partial\phi^2}=0$$

2 The Method of Separation of Variables

As in ordinary differential equations, we can classify partial differential equations into two types, *linear* and *non-linear*. If we consider, for example, two independent variables x, y and the dependent variable U, a linear equation has the form

$$\phi(D_x, D_y)U = F(x, y) \tag{1}$$

where the operator $\phi(D_x, D_y)$ is a polynomial in the two operators

$$D_x \equiv \frac{\partial}{\partial x}, \qquad D_y \equiv \frac{\partial}{\partial y} \tag{2}$$

having coefficients which are functions of only the independent variables x and y. If these coefficients are constants, we call the equation a *linear equation with constant coefficients*; otherwise it is a *linear equation with variable coefficients*. A *non-linear* partial differential equation is one which is not linear.

Example 1. If $\phi = D_x^2 + 4D_xD_y - 2D_y^2 - 3D_x + 5$, $F(x, y) = x^3 - e^y$, then (1) becomes

$$\frac{\partial^2 U}{\partial x^2} + 4\frac{\partial^2 U}{\partial x\,\partial y} - 2\frac{\partial^2 U}{\partial y^2} - 3\frac{\partial U}{\partial x} + 5U = x^3 - e^y$$

which is a linear partial differential equation with constant coefficients.

Example 2. If $\phi = xD_x + yD_y$, $F(x, y) = 1$, then (1) becomes

$$x\frac{\partial U}{\partial x} + y\frac{\partial U}{\partial y} = 1$$

which is a linear partial differential equation with variable coefficients.

Example 3. The equation $\left(\dfrac{\partial U}{\partial x}\right)^2 + \left(\dfrac{\partial U}{\partial y}\right)^2 = 3x - 2y$

is a non-linear partial differential equation since it cannot be expressed in the form (1).

Extensions to more than two independent variables are easily made.

As might be expected, non-linear equations are in general difficult to handle, and we shall not discuss them in this book. We shall in fact discuss only those linear partial differential equations which are most useful in applied problems.

From the analogy with ordinary differential equations, the student might expect the following theorems, which are in fact correct and not difficult to prove (see **B** Exercise 9).

Theorem 1. Consider the linear partial differential equation

$$\phi(D_x, D_y, \ldots)U = F(x, y, \ldots) \tag{3}$$

where x, y, ... are independent variables and $\phi(D_x, D_y, \ldots)$ is a polynomial operator in $D_x, D_y, \ldots$. Then the general solution of (3) is the sum of the general solution U_c of the *complementary equation*

$$\phi(D_x, D_y, \ldots)U = 0 \tag{4}$$

and any particular solution U_p of (3), i.e.,

$$U = U_c + U_p \tag{5}$$

The general solution U_c of (4) is often called the *complementary solution* of (3).

Theorem 2. Let U_1, U_2, ... be solutions of the equation

$$\phi(D_x, D_y, \ldots)U = 0 \tag{6}$$

Then if a_1, a_2, ... are any constants

$$U = a_1 U_1 + a_2 U_2 + \cdots \tag{7}$$

is also a solution, This theorem is often referred to as the *principle of superposition.*

Let us consider how we might solve equation (4). When we had an ordinary differential equation

$$\phi(D)y = 0 \tag{8}$$

with constant coefficients, we used the substitution $y = e^{mx}$, which led to the *auxiliary equation* or *characteristic equation* for the determination of the constant m. For the case (6) with constant coefficients, we would thus be led by analogy to assume as a solution $U = e^{ax+by+\cdots}$ and try to determine the constants $a, b, \ldots$. While this leads to success in some cases (see B Exercises 2 and 3), a better approach is to assume a solution of the form

$$U = X(x)Y(y)\ldots \quad \text{or briefly} \quad U = XY \ldots \tag{9}$$

i.e., a function of x alone times a function of y alone, and so on, as suggested by writing $U = e^{ax+by+\cdots}$ as $U = e^{ax} \cdot e^{by} \cdots$. The method of solution using (9) is often called, for obvious reasons, the method of *separation of variables*. This method, which is useful in obtaining solutions of linear partial differential equations in cases of constant or variable coefficients, will be the primary method of solution in the remainder of the book.

The best way to illustrate the method of separation of variables is to present some examples of its use. Let us begin with the following

PROBLEM FOR DISCUSSION

Solve the boundary value problem

$$\frac{\partial U}{\partial x} + 3\frac{\partial U}{\partial y} = 0, \qquad U(0, y) = 4e^{-2y} - 3e^{-6y} \tag{10}$$

In order to work this problem let us assume that there are solutions having the form

$$U(x, y) = X(x)Y(y) \quad \text{or} \quad U = XY \tag{11}$$

i.e., U can be expressed as a function of x alone times a function of y alone in accordance with the method of separation of variables. Using (11) in the differential equation of (10), we have, if $X' = dX/dx$, $Y' = dY/dy$,

$$X'Y + 3XY' = 0 \quad \text{or} \quad \frac{X'}{3X} + \frac{Y'}{Y} = 0 \tag{12}$$

on dividing both sides by $3XY$ (assumed not zero). Suppose now that we write (12) in the form

$$\frac{X'}{3X} = -\frac{Y'}{Y} \tag{13}$$

Then we see that one side depends only on x, while the other side depends only on y. Since x and y are independent variables, they do not depend on each other, and thus (13) can be true if and only if each side of equation (13) is equal to the same constant, which we call c. From (13) we have therefore

$$X' - 3cX = 0, \qquad Y' + cY = 0 \tag{14}$$

These equations have solutions given, respectively, by

$$X = a_1 e^{3cx}, \qquad Y = a_2 e^{-cy} \tag{15}$$

Thus from (11) $\qquad U = XY = a_1 a_2 e^{c(3x-y)} = B e^{c(3x-y)} \tag{16}$

where $B = a_1 a_2$ is a constant. If we now use the condition in (10), we must have

$$Be^{-cy} = 4e^{-2y} - 3e^{-6y} \tag{17}$$

Unfortunately, (17) cannot be true for any choice of the constants B and c so it would seem as if the method fails. Of course, if we had only one of the terms on the right of (17), the method would work. Thus, if we had only $4e^{-2y}$, for example, we would have $Be^{-cy} = 4e^{-2y}$, which would be satisfied if $B = 4$, $c = 2$, and would lead to the required solution from (16) given by $U = 4e^{2(3x-y)}$.

The situation is saved, however, if we use Theorem 2, page 551, on the *superposition of solutions*. For we see from (16) that $U_1 = b_1 e^{c_1(3x-y)}$ and $U_2 = b_2 e^{c_2(3x-y)}$ are both solutions, and so we must also have as solution

$$U = b_1 e^{c_1(3x-y)} + b_2 e^{c_2(3x-y)} \tag{18}$$

The boundary condition of (10) now leads to

$$b_1 e^{-c_1 y} + b_2 e^{-c_2 y} = 4e^{-2y} - 3e^{-6y}$$

which is satisfied if we choose

$$b_1 = 4, \qquad c_1 = 2, \qquad b_2 = -3, \qquad c_2 = 6$$

This leads to the required solution (18) given by

$$U = 4e^{2(3x-y)} - 3e^{6(3x-y)} \tag{19}$$

Remark 1. The student may wonder why we did not work the above problem by first finding the general solution and then getting the particular solution. A reason is that except in very simple cases the general solution is often difficult to find, and even when it can be obtained, it may be difficult to determine the particular solution from it. However, experience shows that for more difficult problems which arise in practice the method of separation of variables combined with the principle of super-position proves to be successful. For this reason, we shall employ the separation of variables method, unless otherwise stated, in the remaining chapters of the book.

As a more complicated example of the method of separation of variables, let us consider the following

ILLUSTRATIVE EXAMPLE

Solve the boundary value problem

$$\frac{\partial U}{\partial t} = 2 \frac{\partial^2 U}{\partial x^2}$$

$$U(0, t) = 0, \qquad U(10, t) = 0, \qquad U(x, 0) = 50 \sin \frac{3\pi x}{2} + 20 \sin 2\pi x - 10 \sin 4\pi x$$

Solution   Here the independent variables are x and t, so we substitute $U = XT$ in the given differential equation, where X depends only on x and T depends only on t, to find

$$\frac{\partial}{\partial t}(XT) = 2 \frac{\partial^2}{\partial x^2}(XT) \quad \text{or} \quad XT' = 2X''T \tag{20}$$

where $X'' = d^2X/dx^2$, $T' = dT/dt$. Writing (20) as

$$\frac{T'}{2T} = \frac{X''}{X}$$

we see that each side must be a constant denoted by c, so that

$$T' - 2cT = 0, \qquad X'' - cX = 0 \tag{21}$$

Now to write the solution of the second equation in (21), we must know whether the constant c is positive, negative, or zero. We should thus consider three cases.

Case 1, $c = 0$. In this case the solutions to (21) are given by $T = c_1$ and $X = c_2 x + c_3$, where c_1, c_2, c_3 are constants, so that a solution of the given differential equation is

$$U = XT = c_1(c_2 x + c_3) \tag{22}$$

From the first and second boundary conditions we have

$$c_1 c_3 = 0, \qquad c_1(10c_2 + c_3) = 0 \tag{23}$$

These are satisfied if $c_1 = 0$, but in such case the solution is the trivial one $U = 0$, which cannot satisfy the third boundary condition. Thus $c_1 \neq 0$. However, in such case we see from (23) that $c_3 = 0$, and so $c_2 = 0$, again yielding $U = 0$. We thus see that c cannot be zero.

Case 2, $c > 0$. Here the solutions of (21) are given by

$$T = c_1 e^{2ct}, \qquad X = c_2 e^{\sqrt{c}x} + c_3 e^{-\sqrt{c}x}$$

which yields

$$U = XT = e^{2ct}(A e^{\sqrt{c}x} + B e^{-\sqrt{c}x})$$

where

$$A = c_1 c_2, \qquad B = c_1 c_3.$$

From the first boundary condition we have

$$U(0, t) = e^{2ct}(A + B) = 0 \quad \text{or} \quad A + B = 0 \quad \text{and} \quad A = -B$$

since e^{2ct} cannot be zero. The solution thus far is given by

$$U(x, t) = B e^{2ct}(e^{\sqrt{c}x} - e^{-\sqrt{c}x}) \tag{24}$$

From the second boundary condition we have

$$U(10, t) = B e^{2ct}(e^{10\sqrt{c}} - e^{-10\sqrt{c}}) = 0$$

Since $e^{2ct} \neq 0$, this leads to either $B = 0$ or $e^{10\sqrt{c}} - e^{-10\sqrt{c}} = 0$, i.e., $e^{20\sqrt{c}} = 1$. Now if $B = 0$, the solution (24) is the trivial one $U = 0$, which of course cannot satisfy the third boundary condition, which has not even been considered yet. Also it is impossible to have $e^{20\sqrt{c}} = 1$ for positive values of c, since for $c > 0$, $e^{20\sqrt{c}} > 1$. This state of affairs shows that we cannot have $c > 0$.

Case 3, $c < 0$. It is convenient in this case to write $c = -\lambda^2$ to show that c is negative. Then (21) becomes

$$T' + 2\lambda^2 T = 0, \qquad X'' + \lambda^2 X = 0$$

or

$$T = c_1 e^{-2\lambda^2 t}, \qquad X = c_2 \cos \lambda x + c_3 \sin \lambda x$$

This yields

$$U = XT = e^{-2\lambda^2 t}(A \cos \lambda x + B \sin \lambda x)$$

where $A = c_1 c_2$, $B = c_1 c_3$. From the first boundary condition we have

$$U(0, t) = A e^{-2\lambda^2 t} = 0 \quad \text{or} \quad A = 0$$

since $e^{-2\lambda^2 t}$ cannot be zero. The solution thus far is

$$U(x, t) = B e^{-2\lambda^2 t} \sin \lambda x \tag{25}$$

From the second boundary condition we have

$$U(10, t) = B e^{-2\lambda^2 t} \sin 10\lambda = 0$$

Since $B \neq 0$ (otherwise we have the trivial solution $U = 0$), we must have

$$\sin 10\lambda = 0, \quad \text{i.e.,} \quad 10\lambda = m\pi \quad \text{or} \quad \lambda = \frac{m\pi}{10}$$

where $m = 0, \pm 1, \pm 2, \ldots$, and (25) becomes

$$U(x, t) = B e^{-m^2\pi^2 t/50} \sin \frac{m\pi x}{10} \tag{26}$$

The last boundary condition yields

$$U(x, 0) = B \sin \frac{m\pi x}{10} = 50 \sin \frac{3\pi x}{2} + 20 \sin 2\pi x - 10 \sin 4\pi x$$

However, we cannot find a single pair of constants m and B which will satisfy this condition. Fortunately, the principle of superposition comes to the rescue, since we know that sums of solutions of the type (26) for different values of B and integers m will also be a solution. Since we need only three terms having the form (26), we consider the solution

$$U(x, t) = b_1 e^{-m_1^2 \pi^2 t/50} \sin \frac{m_1 \pi x}{10} + b_2 e^{-m_2^2 \pi^2 t/50} \sin \frac{m_2 \pi x}{10} + b_3 e^{-m_3^2 \pi^2 t/50} \sin \frac{m_3 \pi x}{10} \quad (27)$$

so that the last boundary condition gives

$$b_1 \sin \frac{m_1 \pi x}{10} + b_2 \sin \frac{m_2 \pi x}{10} + b_3 \sin \frac{m_3 \pi x}{10} = 50 \sin \frac{3\pi x}{2} + 20 \sin 2\pi x - 10 \sin 4\pi x$$

This can be satisfied if we choose

$$b_1 = 50, \quad \frac{m_1 \pi}{10} = \frac{3\pi}{2} \quad \text{or} \quad m_1 = 15; \qquad b_2 = 20, \quad \frac{m_2 \pi}{10} = 2\pi \quad \text{or} \quad m_2 = 20;$$

$$b_3 = -10, \quad \frac{m_3 \pi}{10} = 4 \quad \text{or} \quad m_3 = 40$$

Using these in (27) gives the required solution

$$U(x, t) = 50 e^{-9\pi^2 t/2} \sin \frac{3\pi x}{2} + 20 e^{-8\pi^2 t} \sin 2\pi x - 10 e^{-32\pi^2 t} \sin 4\pi x \quad (28)$$

Remark 2. For most problems we can anticipate the fact that we must make the choice $c = -\lambda^2$, since otherwise we do not get the sine terms present in the last boundary condition of Illustrative Example 1. Thus we need not bother with any choice of the constant other than $-\lambda^2$. This fact can also be deduced from a physical point of view if U represents some physical variable, such as temperature, and t denotes time. This is because if c were positive then as $t \to \infty$ we would have $U \to \infty$, or as we often say U would be *unbounded*, and this would contradict conservation principles in nature. Also, if c were zero, U would be independent of time, which is not the usual case.

A EXERCISES

Use the method of "separation of variables" to obtain solutions to each of the following boundary value problems.

1. $\dfrac{\partial U}{\partial x} = \dfrac{\partial U}{\partial y}$; $U(0, y) = e^{2y}$.

2. $\dfrac{\partial U}{\partial x} + U = \dfrac{\partial U}{\partial y}$; $U(x, 0) = 4e^{-3x}$.

3. $\dfrac{\partial U}{\partial x} + \dfrac{\partial U}{\partial y} = U$; $U(0, y) = 2e^{-y} + 3e^{-2y}$.

4. $4\dfrac{\partial Y}{\partial t} + \dfrac{\partial Y}{\partial x} = 3Y$; $Y(x, 0) = 4e^{-x} - e^{-5x}$.

5. $\dfrac{\partial U}{\partial t} = 4\dfrac{\partial^2 U}{\partial x^2}$; $U(0, t) = 0$, $U(10, t) = 0$, $U(x, 0) = 5 \sin 2\pi x$.

6. $2\dfrac{\partial U}{\partial t} = \dfrac{\partial^2 U}{\partial x^2}$; $U(0, t) = 0$, $U(\pi, t) = 0$, $U(x, 0) = 2 \sin 3x - 5 \sin 4x$.

7. $\dfrac{\partial^2 Y}{\partial t^2} = \dfrac{\partial^2 Y}{\partial x^2}$; $Y(0, t) = 0$, $Y(20, t) = 0$, $Y_t(x, 0) = 0$, $Y(x, 0) = 10 \sin \dfrac{\pi x}{2}$.

8. $\dfrac{\partial^2 Y}{\partial t^2} = 4\dfrac{\partial^2 Y}{\partial x^2}$; $Y(0, t) = 0$, $Y(10, t) = 0$, $Y_t(x, 0) = 0$, $Y(x, 0) = 3 \sin 2\pi x - 4 \sin \dfrac{5\pi x}{2}$.

9. $9\dfrac{\partial^2 Y}{\partial t^2} = \dfrac{\partial^2 Y}{\partial x^2}$; $Y(0, t) = 0$, $Y(\pi, t) = 0$, $Y_t(x, 0) = 2 \sin x - 3 \sin 2x$, $Y(x, 0) = 0$.

10. $\dfrac{\partial U}{\partial t} = \dfrac{\partial^2 U}{\partial x^2} + U$; $U(0, t) = 0$, $U(10, t) = 0$, $U(x, 0) = 5 \sin 2\pi x - \sin 4\pi x$.

B EXERCISES

1. Can the method of separation of variables be applied to

$$\frac{\partial^2 U}{\partial x^2} + 2\frac{\partial^2 U}{\partial x\,\partial y} - 3\frac{\partial^2 U}{\partial y^2} = 0$$

2. (a) Show that the partial differential equation $\quad A\dfrac{\partial^2 U}{\partial x^2} + B\dfrac{\partial^2 U}{\partial x\partial y} + C\dfrac{\partial^2 U}{\partial y^2} = 0$

where A, B, and C are constants, has solutions of the form e^{ax+by} where a and b are constants. Show that if $A \neq 0$, then $a = m_1 b$ and $a = m_2 b$ so that $e^{b(y+m_1 x)}$ and $e^{b(y+m_2 x)}$ are solutions, where b is arbitrary. (b) By actual substitution show that $F(y + m_1 x)$ and $G(y + m_2 x)$, where F and G are arbitrary functions, are also solutions. We can think of these as being obtained by superimposing solutions of the types $ce^{b(y+m_1 x)}$ and $ce^{b(y+m_2 x)}$, where b and c can assume any values. What modification if any should be made if $A = 0$? (c) Use (b) to show that the general solution of the equation in (a) is $U = F(y + m_1 x) + G(y + m_2 x)$. (d) Use the results of (c) to obtain the general solution of the equation of Exercise 1. (The student should compare the assumption e^{ax+by} to the assumption e^{mx} used in ordinary linear differential equations with constant coefficients.)

3. Show how Exercise 2 may be used to find general solutions of each of the following:

(a) $\dfrac{\partial^2 U}{\partial x^2} - 3\dfrac{\partial^2 U}{\partial x\,\partial y} + 2\dfrac{\partial^2 U}{\partial y^2} = 0.$

(b) $\dfrac{\partial^2 U}{\partial x^2} + \dfrac{\partial^2 U}{\partial y^2} = 0.$

(c) $\dfrac{\partial^2 Y}{\partial t^2} = a^2\dfrac{\partial^2 Y}{\partial x^2}.$

(d) $\dfrac{\partial^2 U}{\partial x^2} + 4\dfrac{\partial^2 U}{\partial x\,\partial y} = 0.$

4. Can the method of Exercise 2 be applied to the equation $x\dfrac{\partial U}{\partial x} + \dfrac{\partial U}{\partial y} = 0$?

Could the separation of variables method be applied here? To what types of partial differential equations do you think the methods can be applied?

5. If m_1 and m_2 of Exercise 2 are equal, then $F(y + m_1 x)$ is a solution. Show that $xG(y + m_1 x)$ or $yG(y + m_1 x)$ is also a solution, and hence that $F(y + m_1 x) + xG(y + m_1 x)$ is a general solution. Use this to obtain general solutions of each of the following:

(a) $\dfrac{\partial^2 U}{\partial x^2} - 4\dfrac{\partial^2 U}{\partial x\,\partial y} + 4\dfrac{\partial^2 U}{\partial y^2} = 0.$ (b) $4\dfrac{\partial^2 Y}{\partial t^2} - 12\dfrac{\partial^2 Y}{\partial x\,\partial t} + 9\dfrac{\partial^2 Y}{\partial x^2} = 0.$

(Compare with the case of repeated roots in the auxiliary equation in connection with linear ordinary differential equations.)

6. Use Theorem 1, page 550 to obtain the general solution of $\dfrac{\partial^2 Z}{\partial x^2} + 4\dfrac{\partial^2 Z}{\partial x\partial y} + 3\dfrac{\partial^2 Z}{\partial y^2} = 4e^{2x+3y}.$

(*Hint:* Assume a particular solution ae^{2x+3y} and determine a.)

7. Find the general solution of $\dfrac{\partial^2 U}{\partial x^2} + 2\dfrac{\partial^2 U}{\partial x\,\partial y} + \dfrac{\partial^2 U}{\partial y^2} = 3\sin(x - 4y) + x^2.$

8. Find the general solution of $\dfrac{\partial^2 U}{\partial x^2} + 2\dfrac{\partial^2 U}{\partial x\,\partial y} - 3\dfrac{\partial^2 U}{\partial y^2} = 6e^{y-3x} + \cos(x - 2y).$

(*Hint:* Corresponding to the term $6e^{y-3x}$, assume particular solution axe^{y-3x} or aye^{y-3x} and determine a.)

9. Prove (a) Theorem 1 and (b) Theorem 2 on pages 550–551.

C EXERCISES

1. (a) If the system of equations $\dfrac{dx}{P} = \dfrac{dy}{Q} = \dfrac{dz}{R}$

where P, Q, and R are functions of x, y, and z, has solution $u(x, y, z) = c_1$, $v(x, y, z) = c_2$, show that the partial differential equation

$$P\dfrac{\partial z}{\partial x} + Q\dfrac{\partial z}{\partial y} = R$$

has solution $F(u, v) = 0$, where F is an arbitrary differentiable function. Compare with C Exercise 2, page 548.

(b) Use the result of (a) to find a solution of $x\dfrac{\partial z}{\partial x} + (x - y)\dfrac{\partial z}{\partial y} = y.$

2. (a) Given the function $U(x, t)$. Assuming suitable restrictions, find the Laplace transform of $\partial U/\partial t$ with respect to the variable t, showing that

$$\mathscr{L}\left\{\dfrac{\partial U}{\partial t}\right\} = su - U(x, 0) \text{ where } u = u(x, s) = \mathscr{L}\{U(x, t)\}$$

(b) Show that $\mathscr{L}\left\{\dfrac{\partial^2 U}{\partial t^2}\right\} = s^2 u - sU(x, 0) - U_t(x, 0).$

(c) Show that $\mathscr{L}\left\{\dfrac{\partial U}{\partial x}\right\} = \dfrac{du}{dx}, \mathscr{L}\left\{\dfrac{\partial^2 U}{\partial x^2}\right\} = \dfrac{d^2 u}{dx^2}.$

(d) Show how Laplace transforms can be used to solve the boundary value problem

$$\dfrac{\partial U}{\partial t} = \dfrac{\partial^2 U}{\partial x^2}, \qquad U(0, t) = 0,\ U(1, t) = 0,\ U(x, 0) = 8\sin 2\pi x$$

(*Hint:* Take Laplace transforms of the first two boundary conditions.)

3. Use the method of Laplace transforms as in Exercise 2 to obtain solutions to (a) A Exercise 7; (b) A Exercise 8; (c) A Exercise 10.

4. Can you use Laplace transform methods to solve the boundary value problem

$$\frac{\partial U}{\partial t} = \frac{\partial^2 U}{\partial x^2}, \qquad U(0, t) = 0, \, U(1, t) = 0, \, U(x, 0) = x(1 - x)$$

3

Some Important Partial Differential Equations Arising from Physical Problems

We have already seen how some simple boundary value problems involving partial differential equations can be solved. We now turn to the derivations of several important partial differential equations which arise in various applied problems. There are three important types of problems which we shall consider.

1. Problems involving vibrations or oscillations.
2. Problems involving heat conduction or diffusion.
3. Problems involving electrical or gravitational potential.

3.1 PROBLEMS INVOLVING VIBRATIONS OR OSCILLATIONS. THE VIBRATING STRING

One of the simplest problems in vibrations or oscillations leading to boundary value problems involving partial differential equations is the problem of a vibrating string, such as a violin or piano string. Suppose that such a string is tightly stretched

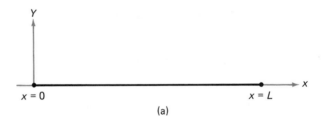

(a)

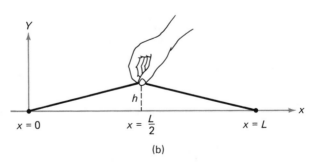

(b)

Figure 12.2

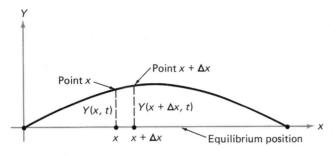

Figure 12.3

between two fixed points $x = 0$ and $x = L$ on the x axis of Fig. 12.2(a). At time $t = 0$ the string is picked up at the middle [Fig. 12.2(b)], to a distance h. Then the string is released. The problem is to describe the motion which takes place.

Clearly many things could happen. The string could be so tightly stretched that when we lifted the middle a height h the string would break. This case is simple and we shall not consider it. It is more natural to assume that the string is perfectly flexible and elastic. Also, to simplify the problem, we assume that h is small compared with L. Other assumptions will be made as we proceed.

Mathematical Formulation. Let us suppose that at some instant t, the string has a shape as shown in Fig. 12.3. We shall call $Y(x, t)$ the displacement of point x on the string (measured from the equilibrium position which we take as the x axis) at the time t. The displacement, at time t, of the neighboring point $x + \Delta x$ will then be given by $Y(x + \Delta x, t)$.

In order to describe the motion which ensues, we consider the forces acting on the small element of string of length Δs between x and $x + \Delta x$, shown considerably enlarged in Fig. 12.4. There will be two forces acting on the element, the tension $\tau(x)$ due to the portion of the string to the left, and the tension $\tau(x + \Delta x)$ due to the portion to the right. Note that we have for the moment assumed that the tension

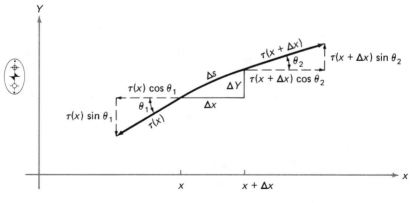

Figure 12.4

depends on position. Resolving these forces into components gives

$$\left.\begin{array}{l} \text{net vertical force (upward)} = \tau(x + \Delta x) \sin \theta_2 - \tau(x) \sin \theta_1 \\ \text{net horizontal force (to right)} = \tau(x + \Delta x) \cos \theta_2 - \tau(x) \cos \theta_1 \end{array}\right\} \tag{1}$$

We now assume that there is no right and left motion of the string, i.e., to a high degree of approximation the net horizontal force is zero. This agrees with the physical situation.* The net vertical force in (1) produces an acceleration of the element. Assuming the string has density (mass per unit length) ρ, the mass of the element is $\rho \, \Delta s$. The vertical acceleration of the string is given approximately by $\partial^2 Y/\partial t^2$.† Hence by Newton's law,

$$\tau(x + \Delta x) \sin \theta_2 - \tau(x) \sin \theta_1 = \rho \, \Delta s \, \frac{\partial^2 Y}{\partial t^2} \tag{2}$$

to a high degree of accuracy. If θ is the angle which the tangent at any point of the element makes with the positive x axis, then θ is a function of position and we write $\theta_1 = \theta(x)$, $\theta_2 = \theta(x + \Delta x)$. Substitution into (2) and dividing by Δx yields

$$\frac{\tau(x + \Delta x) \sin \theta(x + \Delta x) - \tau(x) \sin \theta(x)}{\Delta x} = \rho \, \frac{\Delta s}{\Delta x} \cdot \frac{\partial^2 Y}{\partial t^2} \tag{3}$$

Now the slope of the tangent at any point on the string is given by

$$\tan \theta(x) = \frac{\partial Y}{\partial x}$$

so that

$$\sin \theta(x) = \frac{\partial Y/\partial x}{\sqrt{1 + (\partial Y/\partial x)^2}} \tag{4}$$

Thus if we assume the slope to be small compared with 1, we can neglect $(\partial Y/\partial x)^2$ in the denominator of (4) which amounts to the approximation‡

$$\sin \theta(x) = \tan \theta(x) = \frac{\partial Y}{\partial x}$$

Using this in (3) and taking the limit as $\Delta x \to 0$, (3) becomes

$$\frac{\partial}{\partial x}[\tau(x) \tan \theta(x)] = \rho \, \frac{\partial^2 Y}{\partial t^2} \quad \text{or} \quad \frac{\partial}{\partial x}\left[\tau(x) \frac{\partial Y}{\partial x}\right] = \rho \, \frac{\partial^2 Y}{\partial t^2} \tag{5}$$

which is called the *vibrating string equation*. If $\tau(x) = \tau$, a constant,

$$\tau \frac{\partial^2 Y}{\partial x^2} = \rho \, \frac{\partial^2 Y}{\partial t^2} \quad \text{or} \quad \frac{\partial^2 Y}{\partial t^2} = a^2 \frac{\partial^2 Y}{\partial x^2} \tag{6}$$

where $a^2 \equiv \tau/\rho$. We shall take the tension as constant unless otherwise specified.

* Actually this consequence follows from some of the other assumptions which we make.
† More exactly it is $(\partial^2 Y/\partial t^2) + \epsilon$ where $\epsilon \to 0$ as $\Delta x \to 0$.
‡ The degree of smallness depends, of course, on accuracy desired.

Let us now see what the boundary conditions are. Since the string is fixed at points $x = 0$ and $x = L$, we have

$$Y(0, t) = 0, \qquad Y(L, t) = 0, \quad \text{for} \quad t \geq 0 \tag{7}$$

These state that the displacements at the ends of the string are always zero. Referring to Fig. 12.2(b) it is seen that

$$Y(x, 0) = \begin{cases} \dfrac{2hx}{L}, & 0 \leq x \leq \dfrac{L}{2} \\[3mm] \dfrac{2h}{L}(L - x), & \dfrac{L}{2} \leq x \leq L \end{cases} \tag{8}$$

This merely gives the equations of the two straight line portions of that figure. $Y(x, 0)$ denotes the displacement of any point x at $t = 0$. Since the string is released from rest, its initial velocity everywhere is zero. Denoting by Y_t the velocity $\partial Y / \partial t$ we may write

$$Y_t(x, 0) = 0 \tag{9}$$

which says that the velocity at any place x at time $t = 0$ is zero.

There are many other boundary value problems which can be formulated using the same partial differential equation (6). For example, the string could be plucked at another point besides the midpoint or even at two or more points. We could also have a string or rope of which one end is fixed while the other is moved up and down according to some law of motion.

It is also possible to generalize the vibrating string equation (6). For example, suppose that we have a membrane or drumhead in the form of a square in the xy plane whose boundary is fixed (see Fig. 12.5). If we set it into vibration, such as occurs when beating a drum, each point (x, y) of the square is set into motion in a

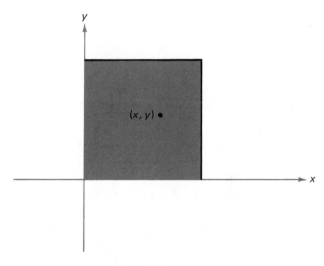

Figure 12.5

direction perpendicular to the plane. If we let Z denote the displacement of point (x, y) from the plane, which is the equilibrium position, at any time t, then the partial differential equation for the vibration is given by

$$\frac{\partial^2 Z}{\partial t^2} = a^2 \left(\frac{\partial^2 Z}{\partial x^2} + \frac{\partial^2 Z}{\partial y^2} \right), \qquad a^2 = \frac{\tau}{\rho} \tag{10}$$

where τ is the tension per unit length along any curve in the surface of the drumhead which is assumed constant, and ρ is the density (mass per unit area). Here Z is a function of x, y, and t, and can be denoted by $Z(x, y, t)$.

We can even generalize (6) and (10) to three dimensions. This generalization would be

$$\frac{\partial^2 U}{\partial t^2} = a^2 \left(\frac{\partial^2 U}{\partial x^2} + \frac{\partial^2 U}{\partial y^2} + \frac{\partial^2 U}{\partial z^2} \right) \tag{11}$$

and we could think of this as having applications to the vibrations of a spherical or other shaped surface. The equation (11) also turns up in electromagnetic theory in connection with the propagation of waves such as radio or television waves. For this reason we often call (11), or any of the special cases (6) or (10), the *wave equation*. When necessary to distinguish the different equations from each other, we refer to (6), (10), and (11) as the *one-*, *two-*, and *three-dimensional* wave equations respectively.

If we introduce the partial derivative operator

$$\nabla^2 \equiv \frac{\partial^2}{\partial x^2} + \frac{\partial^2}{\partial y^2} + \frac{\partial^2}{\partial z^2} \tag{12}$$

equation (11) can be written

$$\nabla^2 U = \frac{1}{a^2} \frac{\partial^2 U}{\partial t^2} \tag{13}$$

In case U does not depend on t, this equation becomes

$$\nabla^2 U = 0 \quad \text{or} \quad \frac{\partial^2 U}{\partial x^2} + \frac{\partial^2 U}{\partial y^2} + \frac{\partial^2 U}{\partial z^2} = 0 \tag{14}$$

We often call (14) *Laplace's equation* and ∇^2 the *Laplacian* after the mathematician *Laplace*, who investigated many of its important and interesting properties.

3.2 PROBLEMS INVOLVING HEAT CONDUCTION OR DIFFUSION

Suppose that a thin metal bar of length L is placed on the x axis of an xy coordinate system (Fig. 12.6). Suppose that the bar is immersed in boiling water so that it is at temperature $100°$C. Then it is removed and the ends $x = 0$ and $x = L$ are kept in ice so that the ends are at temperature $0°$C. We shall suppose that no heat escapes from the surface of the bar, i.e., the surface is insulated. We shall not be concerned here with how this can be accomplished physically. Let us inquire as to what the temperature will be at any place in the bar at any time. Denoting the temperature of the bar by U it is easily seen that U depends on the position x of the bar, as well as the time t (measured from time zero when the bar is at $100°$C) of

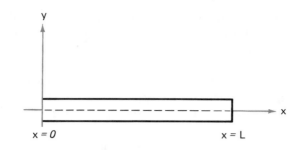

Figure 12.6

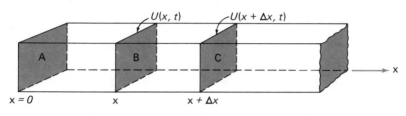

Figure 12.7

observation. We denote this dependence by $U(x, t)$. Thus we have the dependent variable U depending on the two independent variables x and t. Let us try to formulate this problem mathematically.

Mathematical Formulation. Suppose that we have a bar of constant cross section A (as shown in Fig. 12.7, where the cross section is rectangular, although it could be any shape such as that of a cylinder). Consider the element of volume of the bar included between two neighboring planes, parallel to A and denoted by B and C at distances x and $x + \Delta x$, respectively, from A. Denote the temperature in plane B at time t by $U(x, t)$; the temperature in plane C at time t will be then given by $U(x + \Delta x, t)$. To proceed further we shall need the following two physical laws concerning heat transfer.*

   1. *The amount of heat necessary to raise the temperature of an object of mass m by an amount ΔU is ms ΔU, where s is a constant dependent on the material used, and is called the specific heat.*
   2. *The amount of heat flowing across an area (such as B or C) per unit time is proportional to the rate of change of temperature with respect to distance perpendicular to the area (i.e., normal distance x).*

Taking the direction from left to right in Fig. 12.7 as positive, we may write the second law as

$$Q = -KA \, \Delta t \, \frac{\partial U}{\partial x} \tag{15}$$

* A discussion of heat transfer is given in Chapter Three, page 101.

where Q is the quantity of heat flowing to the right, Δt is the length of time during which the flow takes place, and K is the constant of proportionality called the *thermal conductivity*, which depends on the material. The minus sign in (15) shows that Q is positive (i.e., flow is to the right) when $\partial U/\partial x$ is negative (i.e., when the temperature is decreasing as we go to the right). Similarly Q is negative when $\partial U/\partial x$ is positive. This is in agreement with the physical facts. Using (15) we may say that the amount of heat flowing from left to right across plane B (Fig. 12.7) is

$$-KA\,\Delta t\,\frac{\partial U}{\partial x}\bigg|_{x}$$

Similarly, the amount which flows from left to right across plane C is

$$-KA\,\Delta t\,\frac{\partial U}{\partial x}\bigg|_{x+\Delta x}$$

The net amount of heat which accumulates in the volume between B and C is the amount that flows in at B minus the amount that flows out at C, i.e.,

$$\left[-KA\,\Delta t\,\frac{\partial U}{\partial x}\bigg|_{x}\right] - \left[-KA\,\Delta t\,\frac{\partial U}{\partial x}\bigg|_{x+\Delta x}\right] = KA\,\Delta t\left[\frac{\partial U}{\partial x}\bigg|_{x+\Delta x} - \frac{\partial U}{\partial x}\bigg|_{x}\right] \quad (16)$$

This amount of accumulated heat raises or lowers the temperature of the volume element depending on whether (16) is $+$ or $-$. By law 1, then

$$KA\,\Delta t\left(\frac{\partial U}{\partial x}\bigg|_{x+\Delta x} - \frac{\partial U}{\partial x}\bigg|_{x}\right) = ms\,\Delta U = \rho A\,\Delta x\,s\,\Delta U \quad (17)$$

since the mass of the volume element is the density ρ times the volume $A\,\Delta x$. It should be stated that (17) is only approximately true, the degree of the approximation being better, the smaller the values of Δx, ΔU, and Δt. Dividing both sides of (17) by $A\,\Delta x\,\Delta t$ and letting Δx and Δt approach zero, we obtain

$$K\,\frac{\partial^2 U}{\partial x^2} = \rho s\,\frac{\partial U}{\partial t} \quad \text{or} \quad \frac{\partial U}{\partial t} = \kappa\,\frac{\partial^2 U}{\partial x^2} \quad (18)$$

where $\kappa \equiv K/\rho s$ is called the *diffusivity* of the material. This equation is called the *one-dimensional heat flow or heat conduction equation*.

It should be noted that if the surface were not insulated we should have had to consider an extra term in (17), namely, the amount of heat escaping from (or flowing into) the element (see B Exercise 1).

Taking the special case where the ends are kept at 0°C and where the initial temperature of the bar is 100°C, the following boundary conditions result:

$$U(0, t) = 0, \qquad U(L, t) = 0 \quad \text{for } t > 0 \left.\vphantom{\begin{matrix}1\\1\end{matrix}}\right\}$$
$$U(x, 0) = 100, \qquad \text{for } 0 < x < L \qquad\qquad (19)$$

The first two express the fact that the temperatures at $x = 0$ and $x = L$ are zero for all time. The third expresses the fact that the temperature at any place x between 0 and L at time zero is 100°C. Thus the required boundary value problem is that of determining the solution of the partial differential equation (18) which satisfies the conditions (19).

Remark. The same boundary value problem may have different interpretations, i.e., it can arise from apparently different physical situations. For example, suppose that we have an infinite slab of conducting material initially at 100°C where the bounding planes $x = 0$ and $x = L$ are kept at 0°C. Then the temperature $U(x, t)$ at any location x of the slab at any time $t > 0$ is given by the boundary value problem defined by (18) and (19) above.

It is easy to generalize equation (18) to the case where heat may flow in more than one direction. For example, if we have heat conduction in three dimensions the equation is

$$\frac{\partial U}{\partial t} = \kappa \left(\frac{\partial^2 U}{\partial x^2} + \frac{\partial^2 U}{\partial y^2} + \frac{\partial^2 U}{\partial z^2} \right) \tag{20}$$

where the constant κ has the same meaning given above. We call (20) the *three-dimensional heat conduction equation*. The temperature U in this case is a function of x, y, z, and t, written $U(x, y, z, t)$. In case U for some reason, such as symmetry for example, depends only on x, y, and t but not on z, then (20) reduces to

$$\frac{\partial U}{\partial t} = \kappa \left(\frac{\partial^2 U}{\partial x^2} + \frac{\partial^2 U}{\partial y^2} \right) \tag{21}$$

which is called the *two-dimensional heat conduction equation*.

The equation (20) can be written in terms of the *Laplacian operator* [see (12), page 562] as

$$\frac{\partial U}{\partial t} = \kappa \nabla^2 U \tag{22}$$

In case U does not depend on time t we call U the *steady-state temperature*, and this reduces to

$$\nabla^2 U = 0 \quad \text{or} \quad \frac{\partial^2 U}{\partial x^2} + \frac{\partial^2 U}{\partial y^2} + \frac{\partial^2 U}{\partial z^2} = 0 \tag{23}$$

which is *Laplace's equation* (compare (14), page 562).

We can think of heat conduction as due to the random motion of particles of matter such as atoms or molecules; the greater the speeds the higher will be the temperature. The flow of heat from places of higher to lower temperature is due to the spreading or *diffusion* of such particles from places where their concentration or density is high to places where it is low. Such diffusion problems occur in chemistry and biology. For example, in biology we can have the diffusion of drugs from the blood stream into cells or organs. The same equations derived above for heat conduction can be used for diffusion. The only essential difference is that U denotes the *density* or *concentration* of the moving particles. The diffusion interpretation of heat conduction has in fact been hinted above when we referred to the constant κ as the *diffusivity*.

3.3 PROBLEMS INVOLVING ELECTRICAL OR GRAVITATIONAL POTENTIAL

Consider a three-dimensional region $\mathscr{R}$ as in Fig. 12.8, which may represent a continuous distribution of electric charges (or a continuous distribution of mass). Let ρ, which may vary from point to point, denote the charge per unit volume (or mass per unit volume). The quantity ρ is thus the charge density (or mass density). The electrical potential at P due to charge q at Q (or gravitational potential at P due to mass m at Q) is defined to be q/r (or m/r), where r is the distance from P to Q given by

$$r = \sqrt{(x - X)^2 + (y - Y)^2 + (z - Z)^2}$$

If we let dV be the potential, electrical or gravitational, due to the charge or mass given by $\rho \, dX \, dY \, dZ$, then we have

$$dV = \frac{\rho \, dX \, dY \, dZ}{r} = \frac{\rho \, dX \, dY \, dZ}{\sqrt{(x - X)^2 + (y - Y)^2 + (z - Z)^2}} \tag{24}$$

From this it follows that the total potential V due to the entire charge or mass distribution in the region $\mathscr{R}$ is found by integrating (24) over the region $\mathscr{R}$ to obtain

$$V = \iiint\limits_{\mathscr{R}} \frac{\rho \, dX \, dY \, dZ}{\sqrt{(x - X)^2 + (y - Y)^2 + (z - Z)^2}} \tag{25}$$

In working potential problems it is convenient to find a partial differential equation which is satisfied by V. To obtain such a differential equation, let us take partial derivatives of V with respect to x, y, and z and see if we can eliminate the integral in (25). It turns out on taking second derivatives with respect to x, y, z, respectively, and then adding them that we find

$$\frac{\partial^2 V}{\partial x^2} + \frac{\partial^2 V}{\partial y^2} + \frac{\partial^2 V}{\partial z^2} = 0 \quad \text{or} \quad \nabla^2 V = 0 \tag{26}$$

which is again Laplace's equation, already seen twice before (pages 562 and 565). The result (26) is not difficult to establish. To do so we take ∇^2 of both sides of (25). By interchanging the order of differentiation and integration (which can be justified), the result amounts to showing that the Laplacian of $1/r$ is zero. But by ordinary

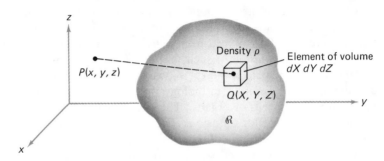

Figure 12.8

differentiation we easily find

$$\frac{\partial^2}{\partial x^2}\left(\frac{1}{r}\right) = \frac{2(x-X)^2 - (y-Y)^2 - (z-Z)^2}{[(x-X)^2 + (y-Y)^2 + (z-Z)^2]^{5/2}}$$

$$\frac{\partial^2}{\partial y^2}\left(\frac{1}{r}\right) = \frac{2(y-Y)^2 - (z-Z)^2 - (x-X)^2}{[(x-X)^2 + (y-Y)^2 + (z-Z)^2]^{5/2}}$$

$$\frac{\partial^2}{\partial z^2}\left(\frac{1}{r}\right) = \frac{2(z-Z)^2 - (x-X)^2 - (y-Y)^2}{[(x-X)^2 + (y-Y)^2 + (z-Z)^2]^{5/2}}$$

where to save labor we can obtain the last two results from the first on using symmetry. Since these add up to zero, the required result follows.*

In arriving at Laplace's equation (26), it was assumed that the potential is to be found at points not occupied by matter or electric charge. In case we wish to find the potential at points occupied by matter or charge, it turns out that the equation is given by

$$\nabla^2 V = -4\pi\rho$$

which is called *Poisson's equation*. The special case $\rho = 0$ yields (26).

3.4 REMARKS ON THE DERIVATION OF PARTIAL DIFFERENTIAL EQUATIONS

It may be remarked that we have been somewhat sloppy from a mathematical viewpoint in deriving the above partial differential equations. For example, in taking derivatives under the integral sign in (25) we have not stated appropriate conditions under which operations may be justified. On further consideration of this matter, it makes little sense to be rigorous in deriving an equation from some postulated mathematical model of physical reality when we do not know whether the model can provide even a reasonably accurate description of such reality. It makes much more sense, however, to use plausible reasoning, intuition, ingenuity, etc., to obtain such equations and then simply *postulate* the equations. Then we can say that the equations are either correct or incorrect according as results obtained from them agree or do not agree with observed or experimental results. The most rigorously derived equation is useless if it leads to results which disagree with the physical facts it was intended to describe.

A EXERCISES

1. A string vibrates in a vertical plane. Show that the differential equation describing the small vibrations of the string if gravitation is considered is

$$\frac{\partial^2 Y}{\partial t^2} = a^2 \frac{\partial^2 Y}{\partial x^2} - g$$

---

* Although we have used the concept of gravitational or electrical potential to arrive at Laplace's equation, it turns out that this equation also occurs in other fields such as, for example, in the motion of an incompressible fluid of *aerodynamics* or *hydrodynamics*. In such cases V is a *velocity potential*.

2. The string of Exercise 1 vibrates in a viscous medium, and a damping force proportional to the instantaneous velocity is assumed to act. Write a partial differential equation describing the motion.

3. Show that the quantity a in the wave equation (6) has dimensions of velocity.

4. If the ends of the bar in the heat flow problem of the text are insulated instead of being kept at 0°C, express mathematically the new boundary conditions. Can you think of the solution to the boundary value problem by physical reasoning?

B EXERCISES

1. The surface of a thin metal bar is not insulated but, instead, radiation can take place into the surroundings. Assuming Newton's law of cooling (page 107) is applicable, show that the heat equation becomes

$$\frac{\partial U}{\partial t} = \kappa \frac{\partial^2 U}{\partial x^2} - c(U - U_0)$$

where c is a constant and U_0 is the temperature of the surroundings.

2. Show that on making the change of variable $U - U_0 = Ve^{\alpha t}$, where α is a suitably chosen constant, the equation of Exercise 1 is transformed into the heat equation (18), page 564, in which the surface is insulated. Discuss the significance of this.

3. A string has one endpoint fixed at $x = L$ while the other end at $x = 0$ is moved in the xy plane so that the displacement from the x axis varies sinusoidally with time. Set up a boundary value problem describing the vibrations of the string, assuming no damping and an appropriate initial shape and velocity distribution. Describe physically how you would expect the string to move.

4. Work Exercise 3 if the endpoint at $x = L$ is free instead of being fixed.

C EXERCISES

1. Suppose that it is required to derive the heat conduction equation in three dimensions. To do this consider an element of volume within a region $\mathscr{R}$ (Fig. 12.9). This element is taken

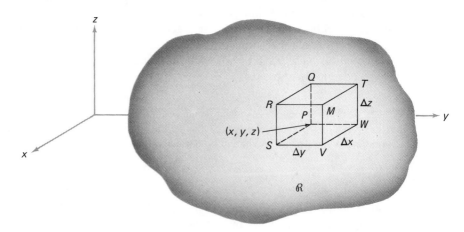

Figure 12.9

as a parallellepiped and is shown considerably enlarged in the figure. Show that the heat equation is

$$\frac{\partial U}{\partial t} = \kappa \left(\frac{\partial^2 U}{\partial x^2} + \frac{\partial^2 U}{\partial y^2} + \frac{\partial^2 U}{\partial z^2} \right)$$

where the various symbols have the same meaning as in the text. (*Hint:* The net amount of heat which accumulates in the volume element is the sum of the amounts of heat entering through the faces $PQRS$, $PQTW$, $PSVW$ minus the sum of the amounts which leave through faces $WTMV$, $SRMV$, $QRMT$.)

2. Since $V = 1/r$, where $r = \sqrt{x^2 + y^2 + z^2}$, is a solution of Laplace's equation in three dimensions, i.e., equation (26) on page 566, one might expect that $V = 1/r$, where $r = \sqrt{x^2 + y^2}$, is a solution of Laplace's equation in two dimensions, i.e.,

$$\frac{\partial^2 V}{\partial x^2} + \frac{\partial^2 V}{\partial y^2} = 0$$

Show that this is in fact not correct and explain why.

3. Show that a solution to the two-dimensional Laplace equation of Exercise 2 is given by $V = \ln r$, where $r = \sqrt{x^2 + y^2}$.

4. Derive the vibrating string equation if it is assumed that the displacement need not be small.

5. Give physical interpretations to the partial differential equations

(a) $\dfrac{\partial U}{\partial t} = \kappa \dfrac{\partial^2 U}{\partial x^2} + F(x, t)$.   (b) $\dfrac{\partial^2 Y}{\partial t^2} = a^2 \dfrac{\partial^2 Y}{\partial x^2} + F(x, t)$.

6. Show that $V = \displaystyle\int_{-\pi}^{\pi} f(z + ix \cos \theta + iy \sin \theta)\, d\theta$ satisfies Laplace's equation.

7. Show that if the thermal conductivity K is not constant but depends on x then the heat equation (18), page 564, is replaced by

$$\frac{\partial}{\partial x}\left[K(x)\, \frac{\partial U}{\partial x} \right] = \rho s \frac{\partial U}{\partial t}$$

8. Use the method of separation of variables in equation (5), page 560, to show that its solution depends on the solutions of the *Sturm-Liouville differential equation*

$$\frac{d}{dx}\left[\tau(x)\, \frac{dX}{dx} \right] + \lambda^2 \rho X = 0$$

From your knowledge of the results of Chapter Eight what conclusions might you draw?

9. Work Exercise 8 for the heat equation in Exercise 7.

thirteen

solutions of boundary value problems using Fourier series

1 Boundary Value Problems Involving Heat Conduction

1.1 FOURIER'S PROBLEM

In the last chapter, page 564, we arrived at the boundary value problem

$$\frac{\partial U}{\partial t} = \kappa \frac{\partial^2 U}{\partial x^2} \tag{1}$$

$$U(0, t) = 0, \qquad U(L, t) = 0, \qquad U(x, 0) = 100 \tag{2}$$

by a consideration of heat conduction in an insulated bar of length L which was initially at 100°C and had its ends kept at temperature 0°C. Fourier, a French scientist and mathematician of the early 19th century, was led to a boundary value problem very similar to this in his researches on heat, and we shall therefore refer to the problem as *Fourier's problem*.

In attempting to solve this problem, Fourier used the method of separation of variables on the equation (1) as we have already done in Chapter Twelve; i.e., he assumed a solution of the form $U = XT$, where X depends only on x and T depends only on t. Substituting into (1) and separating the variables, he thus obtained

$$\frac{T'}{\kappa T} = \frac{X''}{X} \tag{3}$$

From this point Fourier used the same reasoning given in Chapter Twelve. Since one side of (3) depends on x, while the other depends on t, it follows that each side is constant. Calling this constant c and considering the cases $c = 0, c > 0, c < 0$, we can show, as in the Illustrative Example on page 553, that only $c < 0$ yields anything.* Hence, we assume that $c = -\lambda^2$ and obtain from (3),

$$T' + \kappa\lambda^2 T = 0, \qquad X'' + \lambda^2 X = 0$$

or
$$T = c_1 e^{-\kappa\lambda^2 t}, \qquad X = c_2 \cos \lambda x + c_3 \sin \lambda x$$

As before, a solution is

$$U(x, t) = e^{-\kappa\lambda^2 t}(A \cos \lambda x + B \sin \lambda x)$$

To satisfy the first of conditions (2), we have $A = 0$, and the solution thus far is

$$U(x, t) = Be^{-\kappa\lambda^2 t} \sin \lambda x$$

To satisfy the second of conditions (2), we must have $\sin \lambda L = 0$, $\lambda L = m\pi$, where m is any integer, so that the solution thus far is

$$U(x, t) = Be^{-\kappa m^2 \pi^2 t/L^2} \sin \frac{m\pi x}{L} \tag{4}$$

---

* As already pointed out on page 555, this can be seen physically by observing that as $t \to \infty$ the temperature becomes unbounded if $c > 0$, thus violating a fundamental physical fact. Also, if $c = 0$ the temperature is independent of time.

We now consider the last boundary condition of (2). This yields

$$U(x, 0) = B \sin \frac{m\pi x}{L} = 100$$

Here we are apparently stuck, and probably so was Fourier. In Chapter Twelve, page 552, we saw that the *principle of superposition* helped us out of our difficulty in the Problem for Discussion. Can it help us out of the difficulty here too? Fourier attempted this. He reasoned that the sum of solutions of the type (4) satisfied the given differential equation and the first and second boundary conditions of (2). Since a finite number of terms in this solution still did not appear to help in the satisfaction of the last condition of (2), he reasoned that perhaps an *infinite number* of terms would help, i.e., that the solution is given by

$$U(x, t) = b_1 e^{-\kappa\pi^2 t/L^2} \sin \frac{\pi x}{L} + b_2 e^{-4\kappa\pi^2 t/L^2} \sin \frac{2\pi x}{L} + \cdots \tag{5}$$

However, when dealing with an infinite number of terms we must naturally worry about things like convergence. Assuming that such questions are put aside, the assumption leads to

$$U(x, 0) = 100 = b_1 \sin \frac{\pi x}{L} + b_2 \sin \frac{2\pi x}{L} + \cdots + b_n \sin \frac{n\pi x}{L} + \cdots \tag{6}$$

The more we look at the requirement (6), the stranger it appears. It states that an infinite number of sinusoidal terms must be so combined as to give a constant for *all* values of x within the range $0 \leqq x \leqq L$. We see at once, however, that when $x = 0$ and $x = L$ the right-hand side is zero, and so (6) cannot possibly be true at the end points. If it is true at all, it can only hold for $0 < x < L$.

The problem Fourier faced was the determination of the constants $b_1, b_2, \ldots$ so that (6) would be true. That he succeeded in solving the problem and opened up whole new fields in mathematics and applied science is now a matter of history. Suffice it to say that when he did publish his results many mathematicians and scientists thought it pure nonsense, for it was not at that time placed on a rigorous basis. Now mathematicians have developed the theory of *Fourier series* [one example of which is the right-hand side of (6)] to such an extent that whole volumes have been written concerning it.

Remark 1. The student who has read the optional Chapter Eight will recognize the values $\lambda = m\pi/L$, $m = 1, 2, \ldots$, as *eigenvalues* and $b_m \sin m\pi x/L$, $m = 1, 2, \ldots$, as corresponding *eigenfunctions*. The problem of determining the coefficients b_1, $b_2, \ldots$ in the series (6) would for such a student not be new, since as we have seen these problems arise in connection with Sturm–Liouville differential equations and orthogonal functions. Since these topics actually arose much later in the 19th century, and were most likely even inspired by the work of Fourier and others, Fourier of course had to devise his own scheme for arriving at the coefficients. The student who has not had occasion to read Chapter Eight is thus in a situation much as Fourier was, and it is interesting to speculate how he arrived at the technique.

First, since sines and cosines have much in common and since cosine series arose in several variations of Fourier's problem, Fourier was led to consider the more general problem of expansion of functions into series having the form

$$A + a_1 \cos \frac{\pi x}{L} + a_2 \cos \frac{2\pi x}{L} + \cdots + b_1 \sin \frac{\pi x}{L} + b_2 \sin \frac{2\pi x}{L} + \cdots$$

which we shall often denote briefly using the summation notation as

$$A + \sum_{n=1}^{\infty} \left(a_n \cos \frac{n\pi x}{L} + b_n \sin \frac{n\pi x}{L} \right) \tag{7}$$

In this more general problem we seek values of the constants A, a_n, b_n so that the series (7) will equal a given function, say $f(x)$. Since sines and cosines are periodic functions, Fourier concluded that $f(x)$ should also be periodic. To determine the period the following reasoning can be employed. The functions $\sin(\pi x/L)$ and $\cos(\pi x/L)$ have periods $2\pi/(\pi/L) = 2L$, or $4L$, $6L$, ... Similarly, $\sin(2\pi x/L)$ and $\cos(2\pi x/L)$ have periods $2\pi/(2\pi/L) = L$, or $2L$, $3L$, ... In general, the functions $\sin(n\pi x/L)$ and $\cos(n\pi x/L)$ each have periods equal to $2\pi/(n\pi/L) = 2L/n$, or $4L/n$, $6L/n$, ..., $2nL/n = 2L$, ..., It is thus seen that *all* the terms have a common period $2L$. This is actually the *least* period for all the terms. We may say, then, that if the infinite series (7) is equal to $f(x)$ where x lies in an interval of length $2L$, it will hold in any other interval, provided $f(x)$ has period $2L$. We shall restrict ourselves often to the interval $-L < x < L$, although our results will be capable of extension to any other interval of length $2L$.

Fourier arrived at an ingenious way of determining the constants in (7) for which that series is supposed to equal $f(x)$. How he arrived at the method is, as they say, a long story and we cannot enter into it here.* The final method is, however, very simple. It consists of the following steps:

1. Assume that
$$A + \sum_{n=1}^{\infty} \left(a_n \cos \frac{n\pi x}{L} + b_n \sin \frac{n\pi x}{L} \right) = f(x) \tag{8}$$

To find A integrate both sides of (8) from $-L$ to L to obtain

$$\int_{-L}^{L} A\,dx + \sum_{n=1}^{\infty} \left(a_n \int_{-L}^{L} \cos \frac{n\pi x}{L}\,dx + b_n \int_{-L}^{L} \sin \frac{n\pi x}{L}\,dx \right) = \int_{-L}^{L} f(x)\,dx$$

Then $\qquad \int_{-L}^{L} A\,dx = 2LA = \int_{-L}^{L} f(x)\,dx \quad$ or $\quad A = \dfrac{1}{2L} \int_{-L}^{L} f(x)\,dx \qquad (9)$

since all the other integrals are zero.

2. To find a_n, $n = 1, 2, \ldots$, multiply both sides of (8) by $\cos(k\pi x/L)$ and then integrate from $-L$ to L to obtain

$$\int_{-L}^{L} A \cos \frac{k\pi x}{L}\,dx + \sum_{n=1}^{\infty} \left(a_n \int_{-L}^{L} \cos \frac{k\pi x}{L} \cos \frac{n\pi x}{L}\,dx + b_n \int_{-L}^{L} \cos \frac{k\pi x}{L} \sin \frac{n\pi x}{L}\,dx \right)$$

$$= \int_{-L}^{L} f(x) \cos \frac{k\pi x}{L}\,dx \tag{10}$$

* For an interesting account see [17]. Also see [10].

Using
$$\int_{-L}^{L} \cos \frac{k\pi x}{L} \cos \frac{n\pi x}{L} \, dx = \begin{cases} 0, & n \neq k \\ L, & n = k \end{cases} \tag{11}$$

and
$$\int_{-L}^{L} \cos \frac{k\pi x}{L} \sin \frac{n\pi x}{L} \, dx = 0 \tag{12}$$

we see that all the terms on the left of (10) with one exception, i.e., $n = k$, are zero.*
Thus we are led to

$$a_n = \frac{1}{L} \int_{-L}^{L} f(x) \cos \frac{n\pi x}{L} \, dx \tag{13}$$

3. To find b_n, $n = 1, 2, \ldots,$ multiply both sides of (8) by $\sin(k\pi x/L)$ and then integrate from $-L$ to L. Then we obtain

$$b_n = \frac{1}{L} \int_{-L}^{L} f(x) \sin \frac{n\pi x}{L} \, dx \tag{14}$$

on using (12) and the analogue of (11) for sines given by

$$\int_{-L}^{L} \sin \frac{k\pi x}{L} \sin \frac{n\pi x}{L} \, dx = \begin{cases} 0, & n \neq k \\ L, & n = k \end{cases} \tag{15}$$

Remark 2. The student who has read Chapter Eight will observe that (11), (12), and (15) are simply statements of the orthogonality of the functions in the interval $x = -L$ to $x = L$.

It is interesting to observe that if we formally put $n = 0$ in (13) we obtain

$$a_0 = \frac{1}{L} \int_{-L}^{L} f(x) dx \tag{16}$$

But we found in (9) that
$$A = \frac{1}{2L} \int_{-L}^{L} f(x) dx \tag{17}$$

Comparison of (16) and (17),
$$A = \frac{a_0}{2} \tag{18}$$

An advantage of this observation is that we do not need to have *three* separate formulas for A, a_n, b_n, but instead need only two, (13) and (14), *provided* we remember that $A = a_0/2$. It is also evident that (13) and (14) are alike except that one has a cosine while the other has a sine.

These results are of such importance in the solution of boundary value problems by Fourier series that we shall state them in the form of a

Summary. If it is required to obtain the expansion of a given function $f(x)$, $-L < x < L$, in a Fourier series so that

$$f(x) = \frac{a_0}{2} + \sum_{n=1}^{\infty} \left(a_n \cos \frac{n\pi x}{L} + b_n \sin \frac{n\pi x}{L} \right) \tag{19}$$

* To show this use the trigonometric formulas

$\sin A \cos B = \frac{1}{2}[\sin(A+B) + \sin(A-B)]$, $\cos A \cos B = \frac{1}{2}[\cos(A+B) + \cos(A-B)]$

then the Fourier coefficients a_n, b_n are given by

$$a_n = \frac{1}{L}\int_{-L}^{L} f(x)\cos\frac{n\pi x}{L}\,dx, \qquad b_n = \frac{1}{L}\int_{-L}^{L} f(x)\sin\frac{n\pi x}{L}\,dx \qquad (20)$$

Since $f(x)$ has period $2L$, these coefficients can more generally be written as

$$a_n = \frac{1}{L}\int_{c}^{c+2L} f(x)\cos\frac{n\pi x}{L}\,dx, \qquad b_n = \frac{1}{L}\int_{c}^{c+2L} f(x)\sin\frac{n\pi x}{L}\,dx \qquad (21)$$

where c is any real number. The special choice $c = -L$ then yields (20). Another case which can arise in practice is $c = 0$. A simplification results in case the series involves only sine terms or only cosine terms, as follows:

Sine Terms. In this case $f(x)$ is an *odd function*, i.e., $f(-x) = -f(x)$, and we can easily show from (20) that

$$a_n = 0, \qquad b_n = \frac{2}{L}\int_{0}^{L} f(x)\sin\frac{n\pi x}{L}\,dx \qquad (22)$$

Cosine Terms. In this case $f(x)$ is an *even function*, i.e., $f(-x) = f(x)$, and we can easily show from (20) that

$$a_n = \frac{2}{L}\int_{0}^{L} f(x)\cos\frac{n\pi x}{L}\,dx, \qquad b_n = 0 \qquad (23)$$

The series (19) with coefficients (22) and (23) are known, respectively, as *Fourier sine* and *Fourier cosine series* for $f(x)$. As is clear from the formulas (22) and (23), we need only know the function $f(x)$ for $0 < x < L$, the corresponding values for $-L < x < 0$ being *automatically* determined once we know whether the function is odd or even.

To illustrate the use of these results, let us complete the solution to Fourier's problem. As indicated on page 572, to do this we must find b_n so as to obtain the expansion

$$100 = b_1\sin\frac{\pi x}{L} + b_2\sin\frac{2\pi x}{L} + \cdots = \sum_{n=1}^{\infty} b_n\sin\frac{n\pi x}{L}$$

in the interval $0 < x < L$. Since this only involves sine terms, we use (22) to obtain

$$b_n = \frac{2}{L}\int_{0}^{L} 100\sin\frac{n\pi x}{L}\,dx = \frac{200}{n\pi}(1 - \cos n\pi)$$

or $b_1 = \dfrac{400}{\pi}$, $b_2 = 0$, $b_3 = \dfrac{400}{3\pi}$, $b_4 = 0$, $b_5 = \dfrac{400}{5\pi}$, $b_6 = 0, \ldots$

Thus (5) becomes

$$U(x, t) = \frac{400}{\pi}\left(e^{-\kappa\pi^2 t/L^2}\sin\frac{\pi x}{L} + \tfrac{1}{3}e^{-9\kappa\pi^2 t/L^2}\sin\frac{3\pi x}{L} + \tfrac{1}{5}e^{-25\kappa\pi^2 t/L^2}\sin\frac{5\pi x}{L} + \cdots\right)$$

$$(24)$$

This is only a formal solution, since we still have to show that it satisfies the given differential equation and boundary conditions, and also we have to show that if it is a solution it is unique. All these can be shown, but the details are difficult and are omitted here.*

As another variation of Fourier's problem let us consider the following

ILLUSTRATIVE EXAMPLE 1

A metal bar 100 cm long has ends $x = 0$ and $x = 100$ kept at 0°C. Initially, half of the bar is at 60°C, while the other half is at 40°C. Assuming a diffusivity of 0.16 cgs units and that the surface of the bar is insulated, find the temperature everywhere in the bar at time t.

Mathematical Formulation. The heat conduction equation is

$$\frac{\partial U}{\partial t} = 0.16 \frac{\partial^2 U}{\partial x^2} \tag{25}$$

where $U(x, t)$ is the temperature at place x and time t. The boundary conditions are

$$U(0, t) = 0, \qquad U(100, t) = 0, \qquad U(x, 0) = \begin{cases} 60, & 0 < x < 50 \\ 40, & 50 < x < 100 \end{cases} \tag{26}$$

Solution Assuming a solution $U = XT$, in (25) we find

$$XT' = 0.16X''T \quad \text{or} \quad \frac{T'}{0.16T} = \frac{X''}{X}$$

Setting these equal to a constant which, as our previous experience indicated was negative, and which we therefore denote by $-\lambda^2$, we find

$$T' + 0.16\lambda^2 T = 0, \qquad X'' + \lambda^2 X = 0$$

and are thus led to the solution $U(x, t) = e^{-0.16\lambda^2 t}(A \cos \lambda x + B \sin \lambda x)$.

The first two of conditions (26) show that $A = 0$, $\lambda = n\pi/100$. To satisfy the last condition we use the superposition of solutions to obtain

$$U(x, t) = b_1 e^{-16 \cdot 10^{-6}\pi^2 t} \sin \frac{\pi x}{100} + b_2 e^{-64 \cdot 10^{-6}\pi^2 t} \sin \frac{2\pi x}{100} + \cdots \tag{27}$$

For $t = 0$,

$$b_1 \sin \frac{\pi x}{100} + b_2 \sin \frac{2\pi x}{100} + \cdots = U(x, 0)$$

Thus, we have

$$b_n = \frac{2}{100} \int_0^{100} U(x, 0) \sin \frac{n\pi x}{100} \, dx$$

$$= \frac{2}{100} \int_0^{50} (60) \sin \frac{n\pi x}{100} \, dx + \frac{2}{100} \int_{50}^{100} (40) \sin \frac{n\pi x}{100} \, dx$$

$$= \frac{120}{n\pi} \left(1 - \cos \frac{n\pi}{2} \right) + \frac{80}{n\pi} \left(\cos \frac{n\pi}{2} - \cos n\pi \right)$$

* See [5] for example.

Thus $b_1 = 200/\pi$, $b_2 = 40/\pi$, ..., and (27) becomes

$$U(x, t) = \frac{200}{\pi} e^{-16\cdot10^{-6}\pi^2 t} \sin \frac{\pi x}{100} + \frac{40}{\pi} e^{-64\cdot10^{-6}\pi^2 t} \sin \frac{2\pi x}{100} + \cdots$$

which can be shown to be a unique solution.

1.2 PROBLEMS INVOLVING INSULATED BOUNDARIES

In Fourier's problem, both ends of the metal bar were kept at $0°C$. However, other end conditions are also possible. For example, the ends could be insulated so that heat could neither enter nor escape through the ends. However, this leads to a rather trivial problem, since if the bar initially has a temperature of $100°C$ and if heat cannot escape or enter through the ends or surface, then the temperature will clearly remain at $100°C$ for all time. To make a non-trivial problem, we would have to assume that the initial temperature is not constant. Let us therefore assume that the bar is insulated at both ends as well as the surface, but that the initial temperature distribution is specified by some given function $f(x)$ where $0 < x < L$.

Mathematical Formulation. The differential equation for this problem is, as in Fourier's problem, the one-dimensional heat conduction equation

$$\frac{\partial U}{\partial t} = \kappa \frac{\partial^2 U}{\partial x^2} \tag{28}$$

We must now express mathematically the conditions of insulation at the ends $x = 0$ and $x = L$ of the bar. To do so let us recall from equation (15), page 563, that the quantity of heat which crosses the end $x = 0$ is proportional to the partial derivative of the temperature U with respect to x at $x = 0$, i.e., $U_x(0, t)$. Thus, if the end $x = 0$ is insulated, it means that no heat crosses this end, so that $U_x(0, t) = 0$. Similarly, the mathematical formulation of the condition that $x = L$ be insulated is $U_x(L, t) = 0$. Combining these with the condition that the initial temperature is given by $f(x)$, we have the following boundary conditions

$$U_x(0, t) = 0, \qquad U_x(L, t) = 0, \qquad U(x, 0) = f(x) \tag{29}$$

Solution The boundary value problem consists in the determination of that solution of (28) which satisfies the conditions (29). Using the method of separation of variables as on page 571, we arrive at the solution of (28) given by

$$U(x, t) = e^{-\kappa\lambda^2 t}(A \cos \lambda x + B \sin \lambda x) \tag{30}$$

To satisfy the first condition in (29), differentiate (30) with respect to x and obtain

$$U_x(x, t) = e^{-\kappa\lambda^2 t}(-A\lambda \sin \lambda x + B\lambda \cos \lambda x)$$

Then we have

$$U_x(0, t) = B\lambda e^{-\kappa\lambda^2 t} = 0$$

so that $B = 0$ and

$$U(x, t) = Ae^{-\kappa\lambda^2 t} \cos \lambda x \tag{31}$$

To satisfy the second condition in (29), differentiate (31) with respect to x to obtain

$$U_x(x, t) = -A\lambda e^{-\kappa\lambda^2 t} \sin \lambda x$$

so that
$$U_x(L, t) = -A\lambda e^{-\kappa\lambda^2 t} \sin \lambda L = 0$$

or $\quad \sin \lambda L = 0, \quad \lambda L = n\pi, \quad \lambda = \dfrac{n\pi}{L}, \quad n = 0, 1, 2, 3, \ldots$

and so
$$U(x, t) = Ae^{-\kappa n^2\pi^2 t/L^2} \cos \frac{n\pi x}{L} \tag{32}$$

To satisfy the last condition in (29), we must use the principle of superposition to arrive at the possible solution

$$U(x, t) = \frac{a_0}{2} + \sum_{n=1}^{\infty} a_n e^{-\kappa n^2\pi^2 t/L^2} \cos \frac{n\pi x}{L} \tag{33}$$

The last condition then yields

$$f(x) = \frac{a_0}{2} + \sum_{n=1}^{\infty} a_n \cos \frac{n\pi x}{L} \tag{34}$$

from which we can conclude that [see (23), page 575]

$$a_n = \frac{2}{L} \int_0^L f(x) \cos \frac{n\pi x}{L} dx \tag{35}$$

Using these in (33) yields the required solution. As an example of this type, let us consider the following

ILLUSTRATIVE EXAMPLE 2

Work Illustrative Example 1 if the ends $x = 0$ and $x = L$ are insulated instead of being kept at $0°$C.

Solution We could proceed as above, i.e., use separation of variables, satisfying the various conditions, etc., but since we already have the required solution in (33) with coefficients (35), we need only adapt that solution by using

$$L = 100, \quad \kappa = 0.16, \quad f(x) = \begin{cases} 60, & 0 < x < 50 \\ 40, & 50 < x < 100 \end{cases}$$

In such case, if $n = 1, 2, 3, \ldots,$

$$a_n = \frac{2}{100} \int_0^{50} 60 \cos \frac{n\pi x}{100} dx + \frac{2}{100} \int_{50}^{100} 40 \cos \frac{n\pi x}{100} dx = \frac{40}{n\pi} \sin \frac{n\pi}{2}$$

and
$$a_0 = \frac{2}{100} \int_0^{50} 60 \, dx + \frac{2}{100} \int_{50}^{100} 40 \, dx = 100$$

Thus $\quad U(x, t) = 50 + \sum_{n=1}^{\infty} \left(\frac{40}{n\pi} \sin \frac{n\pi}{2}\right) e^{-16 \cdot 10^{-6} n^2\pi^2 t} \cos \frac{n\pi x}{100} \tag{36}$

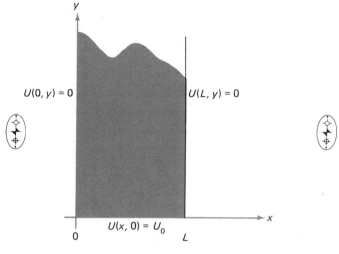

Figure 13.1

1.3 STEADY-STATE TEMPERATURE IN A SEMI-INFINITE PLATE

Consider a thin rectangular metal plate as shown in Fig. 13.1 whose width is L and whose length is so large compared to its width that for all practical purposes it can be considered infinite. Let us assume that the infinite sides are maintained at $0°C$ and that the base of the plate (on the x axis) is maintained at a constant temperature U_0 °C. We shall also assume that the faces of the plate are insulated so that heat can neither enter nor escape from the faces. We seek to determine the *steady-state temperature* of the plate, i.e., the temperature which is independent of time.

Mathematical Formulation. Since this is a problem in two-dimensional heat conduction where the temperature depends only on x and y and not on the time t, the required partial differential equation is given by

$$\frac{\partial^2 U}{\partial x^2} + \frac{\partial^2 U}{\partial y^2} = 0 \tag{37}$$

The boundary conditions are given by

$$U(0, y) = 0, \qquad U(L, y) = 0, \qquad U(x, 0) = U_0 \tag{38}$$

the first two being the conditions on the two infinite sides and the last being the condition on the base. Since it is physically impossible for the temperature to become infinite anywhere in the plate, we shall also assume that $U(x, y)$ remains *bounded*, i.e., there is some constant, say M, which is independent of x and y for which

$$|U(x, y)| < M \tag{39}$$

for all x and y in the plate. Such boundedness conditions are generally implicitly assumed in applied problems, even if they are not always explicitly stated.

Solution In order to find a solution of (37), let us try the method of separation of variables, i.e., assume that $U = XY$, where X depends only on x while Y depends only on y. Then (37) becomes

$$X''Y + XY'' = 0 \quad \text{or} \quad \frac{X''}{X} = -\frac{Y''}{Y} \tag{40}$$

on dividing by XY. Setting each side of the second equation in (40) equal to $-\lambda^2$,

$$X'' + \lambda^2 X = 0, \qquad Y'' - \lambda^2 Y = 0 \tag{41}$$

or $\qquad X = c_1 \cos \lambda x + c_2 \sin \lambda x, \qquad Y = c_3 e^{\lambda y} + c_4 e^{-\lambda y}$

Then $\qquad U(x, y) = (c_1 \cos \lambda x + c_2 \sin \lambda x)(c_3 e^{\lambda y} + c_4 e^{-\lambda y})$

Assuming that $\lambda > 0$, this solution becomes unbounded as $y \to \infty$ and thus violates condition (39). To avoid this "catastrophe," we choose $c_3 = 0$ so that the solution becomes

$$U(x, y) = e^{-\lambda y}(A \cos \lambda x + B \sin \lambda x)$$

on placing $A = c_1 c_4$, $B = c_2 c_4$.

From the first boundary condition in (38), we have

$$U(0, y) = A e^{-\lambda y} = 0$$

which gives $A = 0$, and $\qquad U(x, y) = B e^{-\lambda y} \sin \lambda x \tag{42}$

From the second boundary condition in (38), we have

$$U(L, y) = B e^{-\lambda y} \sin \lambda L = 0 \quad \text{or} \quad \sin \lambda L = 0, \quad \text{and} \quad \lambda = \frac{n\pi}{L}, n = 1, 2, \ldots$$

yielding the solution $\qquad U(x, y) = B e^{-n\pi y/L} \sin \frac{n\pi x}{L}$

From this we see by the principle of superposition that

$$U(x, y) = \sum_{n=1}^{\infty} b_n e^{-n\pi y/L} \sin \frac{n\pi x}{L} \tag{43}$$

is also a solution. Using the third boundary condition of (38), we see that

$$U(x, 0) = U_0 = \sum_{n=1}^{\infty} b_n \sin \frac{n\pi x}{L}$$

from which by the usual technique of Fourier series we have

$$b_n = \frac{2}{L} \int_0^L U_0 \sin \frac{n\pi x}{L} \, dx = \frac{2U_0(1 - \cos n\pi)}{n\pi}$$

Thus $\qquad U(x, y) = \frac{2U_0}{\pi} \sum_{n=1}^{\infty} \frac{1 - \cos n\pi}{n} e^{-n\pi y/L} \sin \frac{n\pi x}{L} \tag{44}$

or, writing out the first few terms,

$$U(x, y) = \frac{4U_0}{\pi}\left(e^{-\pi y/L}\sin\frac{\pi x}{L} + \tfrac{1}{3}e^{-3\pi y/L}\sin\frac{3\pi x}{L} + \tfrac{1}{5}e^{-5\pi y/L}\sin\frac{5\pi x}{L} + \cdots\right) \quad (45)$$

It is interesting that this series can be summed in closed form by a long and tedious procedure outlined in C Exercise 1, the final result being

$$U(x, y) = \frac{2U_0}{\pi}\tan^{-1}\left[\frac{\sin(\pi x/L)}{\sinh(\pi y/L)}\right] \quad (46)$$

We can verify directly that this is a solution (see C Exercise 2).

Interpretation. By using this solution we can determine the steady-state temperature at any point of the plate. Instead of this, however, it is of interest to obtain the curves of equal temperature, or *isothermal curves* as they are often called. Setting (46) equal to a constant, we see that these are given by

$$\sin\frac{\pi x}{L} = \alpha \sinh\frac{\pi y}{L} \quad (47)$$

where α is a parameter, so that (47) is a one-parameter family of curves. The orthogonal trajectories of this family are given by (see C Exercise 3)

$$\cos\frac{\pi x}{L} = \beta \cosh\frac{\pi y}{L} \quad (48)$$

where β is a parameter. These are the *flux curves* and indicate the directions in which heat flows in order to maintain the steady temperatures.

1.4 DIFFUSION INTERPRETATION OF HEAT CONDUCTION

As we have already mentioned in Chapter Twelve, problems involving diffusion of a chemical substance, drug, etc., can be formulated in a manner similar to one involving heat conduction since the same partial differential equations occur. The only difference is that the temperature U is interpreted as a density or concentration of the chemical substance, drug, etc. Thus we can solve problems of diffusion in chemistry, biology, etc., in the same way as we have solved problems in heat conduction, the only question being one of interpretation of the variables.

To illustrate the solution of such a diffusion problem, let us consider a cylinder situated as shown in Fig. 13.2, which is made up of some porous material through which a chemical can pass. Suppose that half of the cylinder contains the chemical of constant concentration or density C_0 kg/m^3, while the other half has the same chemical but with constant concentration C_1 kg/m^3. As time goes by the chemical will of course diffuse from places of high concentration to places of lower concentration, and ultimately some equilibrium or steady-state condition should be reached. The question which naturally arises is, what is the concentration of the chemical in any part of the cylinder at any time? We shall suppose for simplicity that the

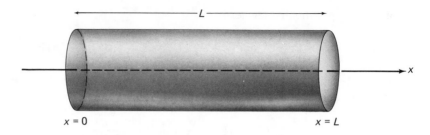

Figure 13.2

entire surface of the cylinder, including the ends, is coated with a substance through which the chemical cannot pass. The corresponding heat conduction interpretation of this is of course that the entire cylinder is insulated so that heat cannot escape or enter.

Mathematical Formulation. Since as we have seen the laws governing diffusion are very much analogous to those which govern heat conduction, the diffusion equation can be written

$$\frac{\partial C}{\partial t} = D\left(\frac{\partial^2 C}{\partial x^2} + \frac{\partial^2 C}{\partial y^2} + \frac{\partial^2 C}{\partial z^2}\right) = D\nabla^2 C \tag{49}$$

where C or more specifically $C(x, y, z, t)$ denotes the concentration of the diffusing substance at the position (x, y, z) at any time t and D is a constant called the *diffusion constant.**

For the present problem let us assume that diffusion takes place only in the x direction so that there is no y or z dependence. Then (49) becomes

$$\frac{\partial C}{\partial t} = D\frac{\partial^2 C}{\partial x^2} \tag{50}$$

where $C = C(x, t)$. Since the concentration of the chemical is given by C_0 in half the cylinder and C_1 in the other half, we arrive at the boundary condition

$$C(x, 0) = \begin{cases} C_0, & 0 < x < L/2 \\ C_1, & L/2 < x < L \end{cases} \tag{51}$$

The fact that the chemical cannot diffuse through the convex portion of the cylindrical surface has already been taken into account in equation (50). To account for the fact that the chemical cannot diffuse through the ends $x = 0$ and $x = L$, we impose the conditions

$$C_x(0, t) = 0 \qquad C_x(L, t) = 0 \tag{52}$$

This is because, as in the case of heat flow, the quantity $-D\, \partial C/\partial x$ represents the flux of the diffusing substance across a plane perpendicular to the x axis at the position x, and this is zero when there is no diffusing substance crossing the plane.

---

* We use D rather than κ, reserving the latter for heat conduction.

Our boundary value problem thus consists of the partial differential equation (50) to be solved subject to the conditions (51) and (52) and of course the boundedness condition, which is always understood and which we did not bother to state explicitly.

Solution As we have already seen several times before, a solution of (50) obtained by the method of separation of variables is given by

$$C(x, t) = e^{-D\lambda^2 t}(A \cos \lambda x + B \sin \lambda x)$$

Since

$$C_x(x, t) = e^{-D\lambda^2 t}(-A\lambda \sin \lambda x + B\lambda \cos \lambda x)$$

the boundary conditions (52) require that

$$C_x(0, t) = B\lambda e^{-D\lambda^2 T} = 0, \qquad C_x(L, t) = e^{-D\lambda^2 T}(-A\lambda \sin \lambda L + B\lambda \cos \lambda L) = 0$$

from which

$$B = 0 \quad \text{and} \quad \sin \lambda L = 0$$

i.e.,

$$\lambda = \frac{n\pi}{L}, \qquad n = 0, 1, 2, 3, \ldots$$

Thus

$$C(x, t) = Ae^{-n^2\pi^2 Dt/L^2} \cos \frac{n\pi x}{L} \tag{53}$$

Superposition of solutions having the form (53) leads to the solution

$$C(x, t) = \frac{a_0}{2} + \sum_{n=1}^{\infty} a_n e^{-n^2\pi^2 Dt/L^2} \cos \frac{n\pi x}{L} \tag{54}$$

The boundary condition (51) then leads to the requirement

$$C(x, 0) = \frac{a_0}{2} + \sum_{n=1}^{\infty} a_n \cos \frac{n\pi x}{L}$$

By Fourier series methods we thus have

$$a_0 = \frac{2}{L} \int_0^L C(x, 0)dx = \frac{2}{L} \int_0^{L/2} C_0 \, dx + \frac{2}{L} \int_{L/2}^L C_1 \, dx = C_0 + C_1$$

$$a_n = \frac{2}{L} \int_0^L C(x, 0) \cos \frac{n\pi x}{L} dx = \frac{2}{L} \int_0^{L/2} C_0 \cos \frac{n\pi x}{L} dx + \frac{2}{L} \int_{L/2}^L C_1 \cos \frac{n\pi x}{L} dx$$

$$= \frac{2(C_0 - C_1)}{n\pi} \sin \frac{n\pi}{2}$$

for $n = 1, 2, \ldots$. Substituting these into (54) yields the solution

$$C(x, t) = \tfrac{1}{2}(C_0 + C_1) + \frac{2(C_0 - C_1)}{\pi} \sum_{n=1}^{\infty} \frac{\sin(n\pi/2)}{n} e^{-n^2\pi^2 Dt/L^2} \cos \frac{n\pi x}{L}$$

$$= \tfrac{1}{2}(C_0 + C_1) + \frac{2(C_0 - C_1)}{\pi} \left[e^{-\pi^2 Dt/L^2} \cos \frac{\pi x}{L} - \tfrac{1}{3}e^{-9\pi^2 Dt/L^2} \cos \frac{3\pi x}{L} + \cdots \right]$$

Interpretation. As t increases the concentration at any location approaches the mean $\tfrac{1}{2}(C_0 + C_1)$ of the initial concentrations as we might have expected. The time needed to reach a specified value close to this equilibrium or steady-state value depends on the diffusion constant. The smaller the diffusion constant, the more time is needed.

A EXERCISES

1. A metal bar 100 cm long has its ends $x = 0$ and $x = 100$ kept at $0°C$. Initially, the right half of the bar is at $0°C$, while the other half is at $80°C$. Assuming a diffusivity of 0.20 cgs units and an insulated surface, find the temperature at any position of the bar at any time.

2. If the bar of the previous exercise has initial temperature $f(x)$ given by

$$f(x) = \begin{cases} 0, & 0 < x < 40 \\ 100, & 40 < x < 60 \\ 0, & 60 < x < 100 \end{cases}$$

calculate the temperature of the bar at any time.

3. A metal bar 40 cm long with a diffusivity of 0.20 cgs units has its surface insulated. If the ends are kept at $20°C$ and the initial temperature is $100°C$, find the temperature of the bar at any time. (*Hint:* Lower all temperature readings by $20°C$.)

4. Give a three-dimensional analogue of Fourier's problem, page 571.

5. Work the problem of page 579 if the semi-infinite edges are insulated instead of being kept at $0°C$.

B EXERCISES

1. (a) A bar of diffusivity κ whose surface is insulated and whose ends are located at $x = 0$ and $x = L$ has an initial temperature distribution $f(x)$. Assuming that the ends of the bar are insulated, determine the temperature of the bar at any time. (b) Find the temperature of the bar if

$$f(x) = \begin{cases} \dfrac{2U_0 x}{L}, & 0 < x \leq \dfrac{L}{2} \\ \dfrac{2U_0}{L}(L - x), & \dfrac{L}{2} \leq x < L \end{cases}$$

2. Work Exercise 1 if one end is insulated, the other at constant temperature U_0.

3. If the surface of a metal bar is not insulated but instead radiates heat into a medium of constant temperature U_0 according to Newton's law of cooling, the differential equation for heat flow becomes

$$\frac{\partial U}{\partial t} = \kappa \frac{\partial^2 U}{\partial x^2} - c(U - U_0)$$

Assuming the ends of the bar of length L are kept at $0°C$ and the initial temperature distribution is $f(x)$, while $U_0 = 0°C$, find $U(x, t)$.

4. Give a diffusion interpretation to (a) A Exercise 1; (b) A Exercise 5; (c) the problem of the semi-infinite plate on page 579.

5. A rectangular plate of dimensions a and b, as shown in Fig. 13.3, has its plane faces insulated. Three of its edges are kept at temperature zero while the fourth is kept at constant temperature

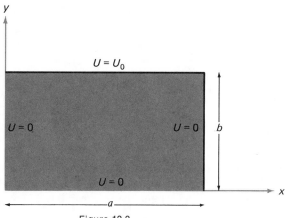

Figure 13.3

U_0, as indicated in the figure. Show that the steady-state temperature is given by

$$U(x, y) = \frac{2U_0}{\pi} \sum_{k=1}^{\infty} \frac{(1 - \cos k\pi) \sin (k\pi x/a) \sinh (k\pi y/a)}{k \sinh (k\pi b/a)}$$

6. Work Exercise 5 if U_0 is replaced by the temperature distribution $f(x)$.

C EXERCISES

1. Derive the result (46) on page 581 from (45). [*Hint:* Write the series of (45) as the imaginary part of

$$ue^{i\phi} + \tfrac{1}{3}(ue^{i\phi})^3 + \tfrac{1}{5}(ue^{i\phi})^5 + \cdots$$

where $u = e^{-\pi y/L}$, $\phi = \pi x/L$ using Euler's formula (14) on page 178. Then use the fact that for $|v| < 1$

$$v + \frac{v^3}{3} + \frac{v^5}{5} + \cdots = \frac{1}{2} \ln \left(\frac{1 + v}{1 - v} \right)$$

putting $v = ue^{i\phi}$ and taking the imaginary part.]

2. Show that (46) is a solution of the boundary value problem given by (37) and (38), page 579. Thus show that (45) is a solution. Could you show directly that (45) is a solution? Discuss what difficulties are involved if any.

3. Obtain the family of flux curves (48) from the isothermal curves (47). Sketch some curves of each family and discuss the physical interpretation.

2 Boundary Value Problems Involving Vibratory Motion

2.1 THE VIBRATING STRING PROBLEM

If a tightly stretched flexible string with fixed ends is given some initial displacement $f(x)$ and then released, the boundary value problem for the displacement

$Y(x, t)$ of the string from its equilibrium position on the x axis is (see page 560)

$$\frac{\partial^2 Y}{\partial t^2} = a^2 \frac{\partial^2 Y}{\partial x^2} \tag{1}$$

$$Y(0, t) = 0, \qquad Y(L, t) = 0, \qquad Y(x, 0) = f(x), \qquad Y_t(x, 0) = 0 \tag{2}$$

where L is the length of the string. Assuming a solution of the form $Y = XT$ where X depends only on x and T depends only on t, (1) becomes

$$XT'' = a^2 X''T \quad \text{or} \quad \frac{X''}{X} = \frac{T''}{a^2 T} \tag{3}$$

which shows that the method of separation of variables applies. Setting each side of the second equation in (3) equal to $-\lambda^2$, we have

$$X'' + \lambda^2 X = 0, \qquad T'' + \lambda^2 a^2 T = 0$$

or $\qquad X = c_1 \cos \lambda x + c_2 \sin \lambda x, \qquad T = c_3 \cos \lambda at + c_4 \sin \lambda at$

Thus $\qquad Y(x, t) = XT = (c_1 \cos \lambda x + c_2 \sin \lambda x)(c_3 \cos \lambda at + c_4 \sin \lambda at) \tag{4}$

Using the first boundary condition in (2), we see that $c_1 = 0$ so that (4) becomes

$$Y(x, t) = (c_2 \sin \lambda x)(c_3 \cos \lambda at + c_4 \sin \lambda at)$$

From the second boundary condition we have

$$\sin \lambda L = 0, \quad \text{i.e.,} \quad \lambda L = n\pi \quad \text{or} \quad \lambda = \frac{n\pi}{L}, \qquad \text{for } n = 0, 1, 2, 3, \ldots$$

so that $\qquad Y(x, t) = \left(c_2 \sin \frac{n\pi x}{L}\right)\left(c_3 \cos \frac{n\pi at}{L} + c_4 \sin \frac{n\pi at}{L}\right) \tag{5}$

Since the third boundary condition in (2) is somewhat complicated, we skip to the simpler fourth boundary condition. This yields $c_4 = 0$. Thus the solution of (1) which satisfies the first, second, and fourth conditions in (2) is given by

$$Y(x, t) = B \sin \frac{n\pi x}{L} \cos \frac{n\pi at}{L} \tag{6}$$

where we have let $B = c_2 c_3$. To satisfy the third condition in (2), we first superimpose solutions of the form (6) to obtain the solution

$$Y(x, t) = \sum_{n=1}^{\infty} b_n \sin \frac{n\pi x}{L} \cos \frac{n\pi at}{L} \tag{7}$$

Then the third boundary condition in (2) requires that

$$f(x) = \sum_{n=1}^{\infty} b_n \sin \frac{n\pi x}{L} \tag{8}$$

From this we have by the method of Fourier series

$$b_n = \frac{2}{L} \int_0^L f(x) \sin \frac{n\pi x}{L} \, dx \tag{9}$$

and using this in (7) we obtain the required solution:

$$Y(x, t) = \frac{2}{L} \sum_{n=1}^{\infty} \left(\int_0^L f(x) \sin \frac{n\pi x}{L} \, dx \right) \sin \frac{n\pi x}{L} \cos \frac{n\pi a t}{L} \tag{10}$$

The precise form of the series depends of course on the particular initial displacement $f(x)$ of the string. Regardless of this initial displacement, however, it is possible to give an interesting interpretation to the various terms in the series (10).

Let us consider the first term in (10) corresponding to $n = 1$. Apart from a constant, this term has the form

$$\sin \frac{\pi x}{L} \cos \frac{\pi a t}{L}$$

If we suppose that $f(x)$ is such that only this term is present in the series, i.e., if $f(x)$ has the form $\sin (\pi x/L)$ apart from some multiplicative constant, then initially the string has the shape shown in Fig. 13.4, where the vertical scale has been enlarged. As t varies the string tends to vibrate as a whole about the equilibrium position with frequency determined from $\cos (\pi a t/L)$ and given by

$$f_1 = \frac{a}{2L} = \frac{1}{2L} \sqrt{\frac{\tau}{\rho}} \tag{11}$$

where τ is the tension and ρ is the mass per unit length (see page 560). This type of vibration is called the *first mode* or *fundamental mode of vibration,* and the corresponding frequency (11) is called the *fundamental frequency* or *first harmonic.*

If $f(x)$ is such that only the second term $n = 2$ in (10) is present, i.e., if $f(x)$ has the form $\sin (2\pi x/L)$ apart from a multiplicative constant, the string appears initially as in Fig. 13.5. As t varies the string vibrates so that the part above the x axis initially moves below the axis while at the same time the part below the x axis moves above it, the point N at $x = L/2$, called a *node* or *nodal point,* being fixed. This type of vibration is called the *second mode of vibration,* and the corresponding frequency of vibration is given by

$$f_2 = \frac{a}{L} = \frac{1}{L} \sqrt{\frac{\tau}{\rho}} \tag{12}$$

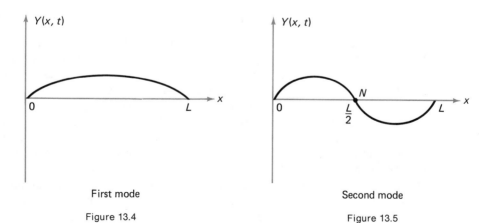

First mode

Figure 13.4

Second mode

Figure 13.5

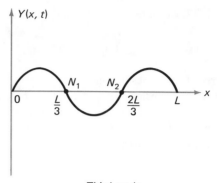

Third mode

Figure 13.6

and is called the *second harmonic* or *first overtone*. Note that the frequency is twice the fundamental frequency (11).

Similarly, if $f(x)$ is such that only the third term is present in (10), the initial shape of the string is as in Fig. 13.6, and the string vibrates in three sections, the points N_1 or $L/3$ and N_2 or $2L/3$ representing the *nodes* or *nodal points*, which are fixed. This type of vibration is called the *third mode of vibration*, and the corresponding frequency is given by

$$f_3 = \frac{3a}{2L} = \frac{3}{2L}\sqrt{\frac{\tau}{\rho}} \tag{13}$$

called the *third harmonic* or *second overtone*. Note that this frequency is three times the fundamental frequency. Carrying this further, the nth term in (10) represents the *n*th mode of vibration in which the string vibrates in n sections with $n - 1$ fixed or nodal points. The frequency of this vibration is called the *n*th *harmonic* or $(n - 1)$st *overtone* and is given by

$$f_n = \frac{na}{2L} = \frac{n}{2L}\sqrt{\frac{\tau}{\rho}} \tag{14}$$

which is n times the fundamental frequency.

Remark 1. In the terminology of Chapter Eight, the frequencies (14) with $n = 1, 2, \ldots$ represent the *eigenvalues* (and are sometimes even called *eigenfrequencies*), while the various modes of vibration described in the above figures represent the *eigenfunctions*. These modes are also called *standing waves*.

For an arbitrary initial displacement $f(x)$ of the string, the motion is more complicated since it represents a combination in general of all the modes of vibration and thus all the frequencies. However, all the harmonic frequencies or overtones are integer multiples of the fundamental frequency. Whenever such a situation arises in a vibrating system, as in a violin or piano string, we say that we have *music*, provided of course that sounds are produced, i.e., that the frequencies are in the audible range. If a vibrating system produces sounds where frequencies are not integer multiples of some fundamental frequency, we say that we have *noise*. As we shall see later, such a situation prevails when for example we have a circular drumhead which is set into vibration by beating it.

Work the problem of the plucked string given on page 558.

Solution As seen on pages 560–561, the boundary value problem is given by

$$\frac{\partial^2 Y}{\partial t^2} = a^2 \frac{\partial^2 Y}{\partial x^2} \tag{15}$$

$$Y(0, t) = 0, \ Y(L, t) = 0, \ Y(x, 0) = \begin{cases} \dfrac{2hx}{L}, & 0 \le x \le \dfrac{L}{2} \\[2mm] \dfrac{2h(L - x)}{L}, & \dfrac{L}{2} \le x \le L \end{cases}, \quad Y_t(x, 0) = 0 \tag{16}$$

We could use the method of separation of variables on equation (15) and proceed to satisfy the boundary conditions (16). However, since this has already been done on page 587 for initial displacement $f(x)$, we need only use

$$f(x) = \begin{cases} \dfrac{2hx}{L} & 0 \le x \le \dfrac{L}{2} \\[2mm] \dfrac{2h(L - x)}{L}, & \dfrac{L}{2} \le x \le L \end{cases} \tag{17}$$

In such case we have from (9) and (10), pages 586–587,

$$b_n = \frac{2}{L} \int_0^L f(x) \sin \frac{n\pi x}{L} \, dx = \frac{2}{L} \int_0^{L/2} \frac{2hx}{L} \sin \frac{n\pi x}{L} \, dx + \frac{2}{L} \int_{L/2}^L \frac{2h(L - x)}{L} \sin \frac{n\pi x}{L} \, dx$$

$$= \frac{8h}{n^2\pi^2} \sin \frac{n\pi}{2}$$

so that

$$Y(x, t) = \frac{8h}{\pi^2} \sum_{n=1}^{\infty} \frac{\sin (n\pi/2)}{n^2} \sin \frac{n\pi x}{L} \cos \frac{n\pi at}{L} \tag{18}$$

or $\quad Y(x, t) = \dfrac{8h}{\pi^2} \left[\sin \dfrac{\pi x}{L} \cos \dfrac{\pi at}{L} + \dfrac{1}{3^2} \sin \dfrac{3\pi x}{L} \cos \dfrac{3\pi at}{L} + \dfrac{1}{5^2} \sin \dfrac{5\pi x}{L} \cos \dfrac{5\pi at}{L} + \cdots \right]$

This result shows that for a string plucked in the center only the odd modes of vibration and associated odd harmonics are present.

2.2 THE VIBRATING STRING WITH DAMPING

From a realistic point of view the mathematical model leading to the vibrating string solution just obtained fails because the vibration continues for all time, whereas in actuality it should gradually cease. The situation is somewhat analogous to the vibrating spring of Chapter Five and leads us to introduce a damping term proportional to the instantaneous velocity of the string. If this is done, the differential equation for the motion becomes

$$\frac{\partial^2 Y}{\partial t^2} + \beta \frac{\partial Y}{\partial t} = a^2 \frac{\partial^2 Y}{\partial x^2} \tag{19}$$

where we call β the *damping constant* as in the case of the spring. If we use the method of separation of variables in (19), i.e., let $Y = XT$, we find

$$XT'' + \beta XT' = a^2 X''T \quad \text{or} \quad \frac{X''}{X} = \frac{T''}{a^2 T} + \frac{\beta T'}{a^2 T} \tag{20}$$

Setting each side of the last equation equal to $-\lambda^2$ yields

$$X'' + \lambda^2 X = 0, \qquad T'' + \beta T' + \lambda^2 a^2 T = 0$$

or $\qquad X = c_1 \cos \lambda x + c_2 \sin \lambda x, \qquad T = e^{-\beta t/2}(c_3 \cos \omega t + c_4 \sin \omega t)$

where $\qquad\qquad\qquad\qquad \omega = \sqrt{\lambda^2 a^2 - \beta^2/4} \tag{21}$

Then a solution of (19) is given by

$$Y(x, t) = XT = e^{-\beta t/2}(c_1 \cos \lambda x + c_2 \sin \lambda x)(c_3 \cos \omega t + c_4 \sin \omega t) \tag{22}$$

Let us now seek to find that solution which satisfies the same boundary conditions (16) for the undamped plucked string. Using the first boundary condition in (16) leads to $c_1 = 0$ and

$$Y(x, t) = c_2 e^{-\beta t/2} \sin \lambda x (c_3 \cos \omega t + c_4 \sin \omega t) \tag{23}$$

Using the second boundary condition in (16) leads to

$$\sin \lambda L = 0 \quad \text{or} \quad \lambda L = n\pi, \qquad \text{i.e.,} \quad \lambda = \frac{n\pi}{L}, \quad n = 0, 1, 2, \ldots$$

and $\qquad\qquad Y(x, t) = c_2 e^{-\beta t/2} \sin \frac{n\pi x}{L} (c_3 \cos \omega t + c_4 \sin \omega t) \tag{24}$

The last boundary condition in (16) leads to $c_4 = \beta c_3/2\omega$ and

$$Y(x, t) = Be^{-\beta t/2} \sin \frac{n\pi x}{L} \left[\cos \sqrt{\frac{n^2 \pi^2 a^2}{L^2} - \frac{\beta^2}{4}}\, t + \frac{\beta}{2\omega} \sin \sqrt{\frac{n^2 \pi^2 a^2}{L^2} - \frac{\beta^2}{4}}\, t \right] \tag{25}$$

where we have let $B = c_2 c_3$ and used the value of ω given in (21) with $\lambda = n\pi/L$. In order to satisfy the third and final boundary condition in (16), we superimpose the solutions (25) to find

$$Y(x, t) = e^{-\beta t/2} \sum_{n=1}^{\infty} b_n \sin \frac{n\pi x}{L} \left[\cos \sqrt{\frac{n^2 \pi^2 a^2}{L^2} - \frac{\beta^2}{4}}\, t + \frac{\beta}{2\omega} \sin \sqrt{\frac{n^2 \pi^2 a^2}{L^2} - \frac{\beta^2}{4}}\, t \right] \tag{26}$$

so that the third boundary condition leads to

$$f(x) = \sum_{n=1}^{\infty} b_n \sin \frac{n\pi x}{L}$$

from which
$$b_n = \frac{2}{L} \int_0^L f(x) \sin \frac{n\pi x}{L} dx$$

which is identical with (9) for the undamped case. As a check we can note that, for $\beta = 0$, (26) reduces to (10) as it should.

There are several important interpretations which we can give to (26). First, because of the damping factor $e^{-\beta t/2}$ the successive oscillations of the string tend to diminish with time, the degree of rapidity depending on the magnitude of β, the larger the value of β, the more rapid the decrease. Second, the frequency of the nth harmonic is a function of the damping constant and is given by

$$f_n = \frac{1}{2\pi} \sqrt{\frac{n^2\pi^2 a^2}{L^2} - \frac{\beta^2}{4}} = \frac{1}{2\pi} \sqrt{\frac{n^2\pi^2 \tau}{L^2 \rho} - \frac{\beta^2}{4}} \tag{27}$$

It should be noted that these frequencies are smaller than the corresponding frequencies in the undamped case. The normal modes as described by the figures on pages 587–588 still exist, but their frequencies are smaller and their amplitudes decrease with time. It should also be noted that in order for the above analysis to be valid the fundamental frequency obtained by putting $n = 1$ in (27) must be real, which implies that

$$\beta < \frac{2\pi}{L} \sqrt{\frac{\tau}{\rho}}$$

This means that the damping must not be too large if the fundamental frequency is to be present.

2.3 VIBRATIONS OF A BEAM

Suppose that we have a thin beam located on the x axis with its ends at $x = 0$ and $x = L$. If we set the beam into vibration in the direction of the x axis by hitting the end $x = 0$ with a hammer, for example, we say that the beam *vibrates longitudinally* or undergoes *longitudinal vibrations*. If we denote the *longitudinal displacement* of any cross section at time t from its equilibrium position at x by $Y(x, t)$, it is not difficult to show (see C Exercise 9) that the differential equation for the motion, assuming small vibrations is given by the same equation as that for the vibrating string, i.e.,

$$\frac{\partial^2 Y}{\partial t^2} = a^2 \frac{\partial^2 Y}{\partial x^2} \tag{28}$$

In this case a is a constant given by

$$a^2 = \frac{E}{\rho} \tag{29}$$

where E is Young's modulus and ρ is the density or mass per unit volume. By solving (28) subject to various boundary conditions, many problems involving the longitudinal vibrations of a beam can be worked, the procedure being identical with that for the vibrating string. Some problems of this type are given in the exercises.

If the beam is set into vibration in a direction perpendicular to the x axis, such as for example by hitting the side with a hammer instead of an end, we say that the beam *vibrates transversely* or undergoes *transverse vibrations*.

Remark 2. We could have referred to the vibrations of a string as *transverse vibrations* since they occur in a direction perpendicular to the x axis, and this in fact is the terminology used by some writers. We have avoided such terminology in connection with the vibrating string because there was no need to consider longitudinal vibrations.

The differential equation of motion for the transverse vibrations of a beam can be derived by using the ideas of C Exercise 8, page 147. If we let $Y(x, t)$ denote the transverse displacement from the equilibrium position of cross section x at time t, we then find, again assuming small vibrations, that

$$\frac{\partial^2 Y}{\partial t^2} + a^2 \frac{\partial^4 Y}{\partial x^4} = 0 \tag{30}$$

In this case a is a constant given by

$$a^2 = \frac{EI}{A\rho} \tag{31}$$

where EI is the flexural rigidity (see page 138), A is the cross-sectional area of the beam assumed constant, and ρ is the density. The student should observe the similarities and differences of the two equations (28) and (30). Various applications involving (30) are also considered in the exercises.

It is possible to generalize equations (28) and (30) to the vibrations of plates and shells. Such problems are considered in the *theory of elasticity** and have important applications to mechanical, aeronautical, and civil engineering.

A EXERCISES

1. A tightly stretched string has its endpoints fixed at $x = 0$ and $x = L$. If it is given an initial displacement $f(x) = \alpha x(L - x)$ from the equilibrium position, where α is a constant and then released, find the displacement at any later time t. Discuss the modes of vibration.

2. A string of length 2 ft weighs 4 oz and is stretched until the tension throughout is 1 lb force. The center of the string is picked up $\frac{1}{4}$ in. above the equilibrium position and then released. Find the subsequent displacement of the string as a function of time, and describe the modes.

3. A string of length 4 ft weighs 2 oz and is stretched until the tension throughout is 4 lb force. The string is assumed to lie along the x axis with one end fixed at $x = 0$ and the other at $x = 4$. If at $t = 0$ the string is given a shape $f(x)$ and then released, determine the subsequent displacement of the string as a function of time for the cases (a) $f(x) = 0.25 \sin \frac{\pi x}{4}$. (b) $f(x) =$ 0.1 $\sin \pi x - 0.02 \sin 3\pi x$. (c) $f(x) = \begin{cases} 0.02x, & 0 \le x \le 2 \\ 0.02(4 - x), & 2 \le x \le 4. \end{cases}$

4. Show that the constant a in the wave equation has the dimensions of a velocity.

* See [27] for example.

5. Suppose that the string of the text is plucked at the point $x = b$, a small distance h from the equilibrium position, and then released. Show that the subsequent displacement is given by

$$Y(x, t) = \frac{2hL^2}{\pi^2 b(L - b)} \sum_{n=1}^{\infty} \frac{1}{n^2} \sin \frac{n\pi b}{L} \sin \frac{n\pi x}{L} \cos \frac{n\pi at}{L}$$

6. Work Exercise 1 if there is a damping force proportional to the instantaneous velocity.

7. Work Exercise 2 if there is damping numerically equal to twice the instantaneous velocity.

B EXERCISES

1. A tightly stretched string with endpoints fixed at $x = 0$ and $x = L$ is initially in equilibrium. At $t = 0$ it is set into vibration by giving each of its points a velocity distribution defined by $g(x)$. (a) Set up the boundary value problem, and (b) show that the displacement of any point of the string at any time $t > 0$ is given by

$$Y(x, t) = \sum_{n=1}^{\infty} b_n \sin \frac{n\pi x}{L} \sin \frac{n\pi at}{L} \qquad \text{where} \qquad b_n = \frac{2}{n\pi a} \int_0^L g(x) \sin \frac{n\pi x}{L} \, dx$$

2. Work Exercise 1 if $g(x) = \alpha x(L - x)$.

3. A flexible string is stretched tightly between $x = 0$ and $x = L$. At $t = 0$ it is struck at the position $x = b$, where $0 < b < L$, in such a way that the initial velocity v is given by

$$v = \begin{cases} v_0/2\epsilon, & |x - b| < \epsilon \\ 0, & |x - b| \geq \epsilon \end{cases}$$

where $\epsilon > 0$ is assumed small. Find the resulting displacement of any point x of the string at time t. Discuss the motion and examine the case where $\epsilon \to 0$.

4. Show that the solution to the vibrating string problem of the text given by (15) on page 589 can be written in the form

$$Y(x, t) = \tfrac{1}{2}\{f(x + at) + f(x - at)\}$$

where $f(x)$ satisfies certain "end conditions." Show how this solution (often called *D'Alembert's solution* after the mathematician who first discovered it) can be arrived at by use of B Exercise 2, page 556.

5. (a) Show that $f(x + at)$ and $f(x - at)$ represent "waves" traveling to the left and right, respectively, with velocity a. (b) Using this interpretation and Exercise 4 show how the shape of the plucked string can be described at various times. (*Hint:* Because of the "end conditions" of Exercise 4, the waves are *reflected* at the ends.)

C EXERCISES

1. If a thin beam is fixed at both ends $x = 0$ and $x = L$, show that the natural frequencies of its longitudinal vibrations are given by

$$f_n = \frac{n}{2L} \sqrt{\frac{E}{\rho}}, \qquad n = 1, 2, 3, \ldots$$

2. Show that the normal modes for the beam of Exercise 1 are given by $c_n \sin \frac{n\pi x}{L}$, where c_n is a suitable function of t and find this function. What would be the longitudinal displacement of any point of the beam at any time?

3. Suppose that a beam is fixed at both ends $x = 0$ and $x = L$. If a constant axial force (i.e., a force in the direction of the beam) is applied at the middle of the beam and then released, the initial displacement is given by

$$f(x) = \begin{cases} \alpha x, & 0 \le x \le L/2 \\ \alpha(L - x), & L/2 \le x \le L \end{cases}$$

where the constant α depends on the magnitude of the force. Show that the subsequent longitudinal displacement is given by

$$Y(x, t) = \frac{4\alpha L}{\pi^2} \sum_{n=1}^{\infty} \frac{(-1)^{n-1}}{(2n-1)^2} \sin \frac{(2n-1)\pi x}{L} \cos \frac{(2n-1)\pi a t}{L}$$

where $a = \sqrt{E/\rho}$. Does this result contradict Exercise 1? Explain.

4. Show that if the axial force of Exercise 3 has magnitude F_0 then $\alpha = F_0/2AE$, where A is the cross-sectional area of the beam, and E is Young's modulus.

5. A beam of length L moves to the left along the x axis with constant speed v_0. It is suddenly stopped at $x = 0$ at time $t = 0$. Show that the resultant longitudinal vibrations set up in the beam are given by

$$Y(x, t) = \frac{8v_0 L}{\pi^2 a} \sum_{n=1}^{\infty} \frac{1}{(2n-1)^2} \sin \frac{(2n-1)\pi x}{2L} \sin \frac{(2n-1)\pi a t}{2L} \qquad \text{where } a = \sqrt{E/\rho}$$

6. Work Exercise 1 if (a) one end is fixed and the other is free; (b) both ends are free.

7. A beam has its ends fixed or built in concrete. (a) Show that if the beam is set into transverse vibrations the natural frequencies are given by

$$f_n = \frac{\beta_n^2}{2\pi} \sqrt{\frac{EI}{A\rho}}, \qquad n = 1, 2, 3, \ldots$$

where $\beta = \beta_n$ are the roots of the equation $\cos \beta L \cosh \beta L = 1$. (b) Show that there are infinitely many positive roots. (*Hint*: At the ends the displacement Y and the slope $\partial Y/\partial x$ must be zero. For the second part obtain the graphs of $y = \cos x$ and $y = 1/\cosh x$ and obtain their intersections.)

8. A beam is simply supported at both ends $x = 0$ and $x = L$, which means that the displacement Y and the bending moment $-EI \, \partial^2 Y/\partial x^2$ are zero at both ends. (a) Show that the natural frequencies for the transverse vibrations of the beam are given by

$$f_n = \frac{\pi n^2}{2L^2} \sqrt{\frac{EI}{A\rho}}, \qquad n = 1, 2, 3, \ldots$$

(b) If a constant transverse force of magnitude F_0 is applied to the beam at its midpoint and is then released, show that the resulting transverse displacement is given by

$$Y(x, t) = \frac{2F_0 L^3}{\pi^4 EI} \sum_{n=1}^{\infty} \frac{(-1)^{n-1}}{(2n-1)^4} \sin \frac{(2n-1)\pi x}{L} \cos \frac{(2n-1)^2 \pi^2 a t}{L^2} \qquad \text{where } a = \sqrt{EI/A\rho}$$

9. Derive equation (28), page 591, for the longitudinal vibrations of a beam as follows. Consider a beam of cross-sectional area A and an element of volume located between x and $x + \Delta x$ (such as indicated in Fig. 12.7, page 563). Let $Y(x, t)$, $Y(x + \Delta x, t)$ denote respectively the longitudinal displacements of the ends x and $x + \Delta x$ from their equilibrium positions at any time t and $P(x, t)$, $P(x + \Delta x, t)$ the forces per unit area (stresses) acting at these ends in the longitudinal direction. (a) Show that the strain at location x is given by $\partial Y/\partial x$ and thus that

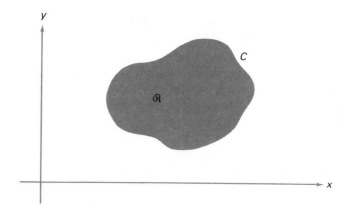

Figure 13.7

the stress is given by $P = E\, \partial Y/\partial x$ (see page 144). (b) Use Newton's law to show that

$$[P(x + \Delta x, t) - P(x, t)]A = \rho A\, \Delta x \frac{\partial^2 Y}{\partial t^2}$$

approximately where ρ is the density. (c) Divide both sides of the result in (b) by $A\, \Delta x$, take the limit as $\Delta x \to 0$ and then use the result in (a).

3 Boundary Value Problems Involving Laplace's Equation

As derived in Chapter Twelve page 566, the electrical or gravitational potential due to an electric charge or mass distribution satisfies *Laplace's equation*

$$\nabla^2 V = \frac{\partial^2 V}{\partial x^2} + \frac{\partial^2 V}{\partial y^2} + \frac{\partial^2 V}{\partial z^2} = 0 \tag{1}$$

This equation involving three independent variables is the three-dimensional case. However, we can also have Laplace's equation in less than three dimensions as in the two-dimensional case

$$\frac{\partial^2 V}{\partial x^2} + \frac{\partial^2 V}{\partial y^2} = 0 \tag{2}$$

or as a generalization in more than three dimensions, although geometric visualizations in such cases are lost. Laplace's equation (1) or (2) can also be interpreted as a heat conduction equation for the determination of steady-state temperature, i.e., temperature independent of time. Thus we have in effect already worked a problem involving Laplace's equation on page 579, which conversely can be given an electrical or gravitational potential interpretation.

Further research on the theory of Laplace's equation or, as it is sometimes called, *potential theory* reveals the following interesting and important

Theorem 1. Let $\mathscr{R}$ be a region in the xy plane as in Fig. 13.7 bounded by a closed curve C which does not intersect itself anywhere, often called a *simple closed curve*. Then there exists a unique solution V of Laplace's equation in the region which takes prescribed values on the boundary C, i.e., V is some specified function on the curve C. The theorem can be generalized to regions bounded by closed surfaces.

The boundary value problem which seeks to determine the solution V described in this theorem is often referred to as a *Dirichlet problem*. A function which is a solution of Laplace's equation is often called a *harmonic function*.

From a practical point of view it is easy to understand why Theorem 1 should be true. For suppose that the region $\mathscr{R}$ represents a metal plate of negligible thickness whose faces are insulated so that heat cannot enter or escape through them. Specification of V on the boundary C amounts to maintaining this boundary at some prescribed temperature distribution, and we would expect that each point of the plate would ultimately reach some unique equilibrium or steady-state temperature and remain at this temperature as long as the conditions are maintained.

Theorem 1 can be extended to unbounded regions by appropriate limiting procedures. The steady-state temperature problem solved on page 579 provides an example. It is also possible to arrive at theorems similar to Theorem 1 involving different prescribed conditions on the boundary C (such as specification of the derivative), or involving Laplace's equation in higher dimensions.

We have already on many occasions solved Laplace's equation in rectangular coordinates in connection with steady-state heat conduction problems in squares, rectangles, etc., and have thus illustrated Theorem 1 for these cases. To present an illustration of a different type we take this opportunity to solve Laplace's equation for a circle. In particular, let us consider the following

PROBLEM FOR DISCUSSION

Solve Laplace's equation in a region $\mathscr{R}$ bounded by a circle C of unit radius (see Fig. 13.8) if V is a specified function on the boundary C. As seen from Theorem 1 this is a Dirichlet problem for which a unique solution exists.

Mathematical Formulation. As we might very well expect, rectangular coordinates are not suitable for solving problems involving circles, and for this reason we are led to use *polar coordinates*. Since the relationship between rectangular coordinates (x, y) and polar coordinates (r, ϕ) is expressed by the transformation equations

$$x = r \cos \phi, \qquad y = r \sin \phi \tag{3}$$

we can by using methods learned in calculus transform (2) into the equation

$$\frac{\partial^2 V}{\partial r^2} + \frac{1}{r} \frac{\partial V}{\partial r} + \frac{1}{r^2} \frac{\partial^2 V}{\partial \phi^2} = 0 \tag{4}$$

(See C Exercise 5, page 549.) In this equation V is of course a function of r and ϕ written $V(r, \phi)$.

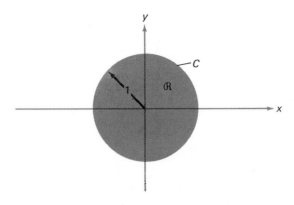

Figure 13.8

To complete the boundary value problem, we need boundary conditions. One condition involves specification of V on the circle C. Since the unit circle is represented by $r = 1$, the value of V on C is given by $V(1, \phi)$, where the angle ϕ is such that $0 \leq \phi < 2\pi$. Assuming that this is some prescribed function of ϕ, say $f(\phi)$, we have the boundary condition

$$V(1, \phi) = f(\phi) \qquad (5)$$

In addition to this, since we would want V to be bounded at all points of $\mathcal{R}$, we have

$$|V(r, \phi)| < M \qquad (6)$$

for all r and ϕ in $\mathcal{R}$, where M is some constant independent of r and ϕ.

Solution Let us seek to find solutions of (4) which have variables separable. For this purpose we let $V = R\Phi$, where it is assumed that R is a function of r alone and Φ is a function of ϕ alone. Using this in (4) it becomes

$$R''\Phi + \frac{1}{r}R'\Phi + \frac{1}{r^2}R\Phi'' = 0 \quad \text{or} \quad \frac{R''}{R} + \frac{1}{r}\frac{R'}{R} + \frac{1}{r^2}\frac{\Phi''}{\Phi} = 0 \qquad (7)$$

on dividing by $R\Phi$. This last equation can be written with R on one side and Φ on the other in the form

$$\frac{R''}{R} + \frac{1}{r}\frac{R'}{R} = -\frac{1}{r^2}\frac{\Phi''}{\Phi} \qquad (8)$$

However, although the left side depends only on r, the right side depends on both r and ϕ, so that we have not separated the variables. The situation is fortunately easily remedied by multiplying both sides of (8) by r^2 to obtain

$$r^2\frac{R''}{R} + r\frac{R'}{R} = -\frac{\Phi''}{\Phi} \qquad (9)$$

If we now set each side of (9) equal to λ^2, we arrive at the two equations

$$r^2R'' + rR' - \lambda^2 R = 0, \qquad \Phi'' + \lambda^2\Phi = 0 \qquad (10)$$

The first equation in (10) is an Euler equation which can be solved as on page 212. As a result the solutions of (10) are given, respectively, by

$$R = c_1 r^\lambda + c_2 r^{-\lambda}, \qquad \Phi = c_3 \cos \lambda\phi + c_4 \sin \lambda\phi$$

if $\lambda \neq 0$, and

$$R = c_5 + c_6 \ln r, \qquad \Phi = c_7 + c_8\phi$$

if $\lambda = 0$. We thus arrive at the possible solutions

$$(c_1 r^\lambda + c_2 r^{-\lambda})(c_3 \cos \lambda\phi + c_4 \sin \lambda\phi) \quad \text{and} \quad (c_5 + c_6 \ln r)(c_7 + c_8\phi) \quad (11)$$

Now it is evident that a solution must be bounded at $r = 0$. Taking $\lambda \geq 0$, this requires the choices $c_2 = 0$ and $c_6 = 0$ in (11). It is also clear that if we change ϕ to $\phi + 2\pi$ in any solution, the solution should not change since the points (r, ϕ) and $(r, \phi + 2\pi)$ are the same. This periodicity in ϕ requires us to choose $c_8 = 0$ and λ to be an integer, i.e.,

$$\lambda = n, \qquad \text{where } n = 0, 1, 2, 3, \ldots \qquad (12)$$

We can thus restrict ourselves to solutions having the form

$$V(r, \phi) = r^n(A \cos n\phi + B \sin n\phi) \qquad (13)$$

where A and B are arbitrary constants.

To satisfy the boundary condition (5), we first superimpose solutions of the form (13) corresponding to $n = 0, 1, 2, \ldots$ to obtain the solution

$$V(r, \phi) = \frac{a_0}{2} + \sum_{n=1}^{\infty} r^n(a_n \cos n\phi + b_n \sin n\phi) \qquad (14)$$

The boundary condition (5) then requires that

$$f(\phi) = \frac{a_0}{2} + \sum_{n=1}^{\infty} (a_n \cos n\phi + b_n \sin n\phi) \qquad (15)$$

By the method of Fourier series we see that this yields the Fourier coefficients

$$a_n = \frac{1}{\pi} \int_0^{2\pi} f(\phi) \cos n\phi \, d\phi, \qquad b_n = \frac{1}{\pi} \int_0^{2\pi} f(\phi) \sin n\phi \, d\phi \qquad (16)$$

Thus (14) with the coefficients (16) gives the required solution. As an illustration, let us consider the following

ILLUSTRATIVE EXAMPLE 1

(a) Find that solution of Laplace's equation inside the unit circle $r = 1$ which has values on the boundary given by

$$f(\phi) = \begin{cases} V_0, & 0 < \phi < \pi \\ 0, & \pi < \phi < 2\pi \end{cases} \qquad (17)$$

where V_0 is a constant. (b) Give two physical interpretations.

Solution   (a) Using (16) and (17), we have if $n = 1, 2, 3, \ldots,$

$$a_n = \frac{1}{\pi} \int_0^\pi V_0 \cos n\phi \, d\phi + \frac{1}{\pi} \int_\pi^{2\pi} (0) \cos n\phi \, d\phi = 0$$

If $n = 0$, we have $\qquad\qquad a_0 = \frac{1}{\pi} \int_0^\pi V_0 \, d\phi + \frac{1}{\pi} \int_\pi^{2\pi} (0) d\phi = V_0$

Similarly, for $n = 1, 2, 3, \ldots,$

$$b_n = \frac{1}{\pi} \int_0^\pi V_0 \sin n\phi \, d\phi + \frac{1}{\pi} \int_\pi^{2\pi} (0) \sin n\phi \, d\phi = \frac{V_0(1 - \cos n\pi)}{n\pi}$$

Then from (14) we have the required solution

$$V(r, \phi) = \frac{V_0}{2} + \frac{V_0}{\pi} \sum_{n=1}^\infty \left(\frac{1 - \cos n\pi}{n} \right) r^n \sin n\phi$$

or $\qquad V(r, \phi) = \frac{V_0}{2} + \frac{2V_0}{\pi} \left[r \sin \phi + \frac{r^3 \sin 3\phi}{3} + \frac{r^5 \sin 5\phi}{5} + \cdots \right]$ $\qquad$ (18)

(b) The problem has the following two interpretations:

(i) *Steady-state heat flow interpretation.* In this case (18) represents the steady-state temperature at any point (r, ϕ) of the circular plate indicated in Fig. 13.8, which has negligible thickness, insulated faces, and where the top half of the boundary is maintained at temperature V_0 while the bottom half is maintained at temperature zero.

(ii) *Electrical or gravitational potential interpretation.* In this case (18) represents the electrical or gravitational potential at any point (r, ϕ) inside the unit circle due to a system of electric charges or masses so distributed that the top half of the boundary is maintained at potential V_0, while the bottom half is maintained at potential zero (sometimes referred to as *ground potential*).

It is of interest and importance that the series (14) with coefficients (16) can be summed in closed form. The result, known as *Poisson's integral formula*, is given by

$$V(r, \phi) = \frac{1}{2\pi} \int_0^{2\pi} \frac{1 - r^2}{1 - 2r \cos (w - \phi) + r^2} f(w) dw \qquad (19)$$

The method of summation as well as related problems are given in the exercises.

In Illustrative Example 1 we found the potential inside the unit circle, given the potential on the circle. It is natural to ask whether we can also find the potential outside this unit circle. This can be done, as we see in the following

ILLUSTRATIVE EXAMPLE 2

Find the potential outside the unit circle $r = 1$, given that the potential on the circle is that of Illustrative Example 1.

Solution In this case we do not require boundedness at $r = 0$ since this is not in the given region. However, we do require boundedness for infinite r. If $\lambda \geq 0$, we see from (11) that we must have $c_1 = 0$, $c_6 = 0$. Also, as before, we must have the condition (12) to insure the periodicity of the solution as a function of ϕ. Thus we arrive at the possible solutions

$$V(r, \phi) = r^{-n}(A \cos n\phi + B \sin n\phi)$$

where A and B are arbitrary constants. To satisfy the given potential on the circle, we use the superposition principle to obtain

$$V(r, \phi) = \frac{a_0}{2} + \sum_{n=1}^{\infty} r^{-n}(a_n \cos n\phi + b_n \sin n\phi)$$

The required boundary condition then leads to

$$f(\phi) = \frac{a_0}{2} + \sum_{n=1}^{\infty} (a_n \cos n\phi + b_n \sin n\phi)$$

which is the same as in (15) of Illustrative Example 1. Thus the coefficients a_n and b_n are the same as already obtained, which leads to the required solution

$$V(r, \phi) = \frac{V_0}{2} + \frac{2V_0}{\pi} \left[\frac{\sin \phi}{r} + \frac{\sin 3\phi}{3r^3} + \frac{\sin 5\phi}{5r^5} + \cdots \right] \tag{20}$$

It should be noted that this is the same as (18) with r replaced by $1/r$.

As in Illustrative Example 1, page 599, we can give a steady-state temperature interpretation to (20). To supply such interpretation, we consider an infinite (or practically speaking, very large) conducting plate represented by the xy plane with its faces insulated. If a portion of this plate represented by the interior of the unit circle is removed from the plate, and if a temperature distribution given by $f(\phi)$ is applied to this boundary, then the steady-state temperature in the plate is given by (20).

Remark 1. The result arrived at in Illustrative Example 2 can be considered as a special case of Theorem 1, page 596, in which C is the boundary of the region represented by $r > 1$, and so we have a Dirichlet problem in this case as well.

Remark 2. The fact that (20) is obtained from (18) on replacing r by $1/r$ suggests that Poisson's integral formula (19) also holds for $r > 1$ if r is replaced by $1/r$. That this is true is seen in B Exercise 6.

Remark 3. The result (18) can be expressed in closed form by

$$\frac{V_0}{2} + \frac{V_0}{\pi} \tan^{-1} \left(\frac{2r \sin \phi}{1 - r^2} \right) \tag{21}$$

See B Exercise 2. The result (20) is obtained from (21) on replacing r by $1/r$.

A EXERCISES

1. Find the potential (a) inside and (b) outside the unit circle $r = 1$ if the potential on the circle is given by $f(\phi) = V_0, 0 < \phi < \pi/2; -V_0, \pi/2 < \phi < \pi; 0, \pi < \phi < 2\pi$. Give a temperature interpretation to this problem.

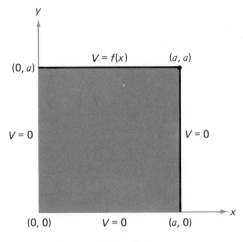

Figure 13.9

2. A square of side a (see Fig. 13.9) has three edges at potential zero and the fourth at potential given by $f(x) = V_0 x/a$, $0 < x < a/2$; $V_0(a - x)/a$, $a/2 < x < a$. Show that the potential at any point inside is given by

$$V(x, y) = \frac{4V_0}{\pi} \sum_{n=0}^{\infty} \frac{\sin (2n + 1)\pi x/a \sinh (2n + 1)\pi y/a}{(2n + 1)^2 \sinh (2n + 1)\pi}$$

3. Verify that (21) is a solution to Illustrative Example 1, page 598, for the case where $0 < r < 1$.

4. Verify that (21) with r replaced by $1/r$ is a solution to Illustrative Example 2, page 599, for the case where $r > 1$.

5. Work Illustrative Examples 1 and 2 if the circle has radius c instead of 1. Explain why the result can be obtained simply by replacing r by r/c.

6. Show that the steady-state temperature at the center of a circular plate is the average (mean) temperature of the boundary. Give an interpretation involving potential. Illustrate by using the results of (a) Illustrative Example 1, page 598; (b) Exercise 1. [*Hint:* Use (19) with $r = 0$.]

7. Verify the result of Exercise 6 if the boundary temperature is

(a) $f(\phi) = 100 \sin^3 \phi$, $0 \leq \phi < 2\pi$,   (b) $f(\phi) = \begin{cases} 50 \cos \phi, & 0 < \phi < \pi \\ 0, & \pi < \phi < 2\pi \end{cases}$

B EXERCISES

1. Show that $\dfrac{1}{2\pi} \displaystyle\int_0^{2\pi} \dfrac{1 - r^2}{1 - 2r \cos (w - \phi) + r^2}\, dw = 1$ and interpret physically.

2. Derive (21) directly from Poisson's integral formula (19).

3. A plate has the form of an annular region bounded by two concentric circles of radius r_1 and r_2, respectively, as indicated in Fig. 13.10. Taking $r_1 = 1$ and $r_2 = 2$, and assuming that the boundary temperatures are given, respectively, by

$$U_1 = 75 \sin \phi, \qquad U_2 = 60 \cos \phi, \qquad 0 < \phi < 2\pi$$

find the steady-state temperature at each point of the plate. Interpret the results in terms of potential theory.

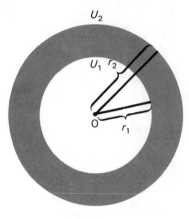

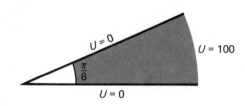

Figure 13.10   Figure 13.11

4. (a) Find the steady-state temperature in a plate in the shape of a circular sector of unit radius and angle $\pi/6$, as indicated in Fig. 13.11, if the temperatures of the sides are kept at 0, the temperature of the circular arc is kept at 100, and the plane faces are insulated. (b) Describe a three-dimensional problem having the same solution as (a). (c) What is the solution if the sector has radius c?

5. Work Exercise 4 if the sector has angle $\pi/2$.

6. Verify Remark 2.

C EXERCISES

1. Use C Exercise 1, page 585, to sum (18) and (20) in closed form.

2. Derive Poisson's integral formula (19) from the results (14) and (16) on page 598. [*Hint*: Change the dummy variable in (16) from ϕ to w, substitute the results into (14), and write as the integral of a series which can be summed by use of C Exercise 1, page 585.]

3. Suppose that the temperature (or potential) $U(1, \phi)$ on the unit circle is such that $|U(1, \phi)| \leq M$, where M is some constant. (a) Prove that $|U(r, \phi)| \leq M$ for all points inside the circle. (b) Deduce from (a) that the maximum or minimum temperature (or potential) cannot occur inside the circle unless $U(r, \phi)$ is constant everywhere.

4. Work B Exercise 3 if the second boundary condition is replaced by the condition that the boundary $r = 2$ is insulated.

5. The result in A Exercise 6 that the temperature (or potential) at the center of a circle is equal to the mean temperature of the boundary can be extended to other cases such as a rectangle. Using this and B Exercise 5, page 585, with $a = b = 1$, obtain the interesting series

$$\frac{1}{\cosh (\pi/2)} - \frac{1}{3 \cosh (3\pi/2)} + \frac{1}{5 \cosh (5\pi/2)} - \cdots = \frac{\pi}{8}$$

6. Arrive at a series similar to that of Exercise 5 by using A Exercise 2, page 601.

4 Miscellaneous Problems

In the last three sections we solved some important types of boundary value problems requiring the use of Fourier series. In this section we extend the techniques or, as we can say, the "tricks of the trade" to work various problems which are some-what more complicated by the nature of either the boundary conditions or the partial differential equations which are involved. It should be emphasized that the methods are capable of being applied to different fields, so that, for instance, just because we use a problem in heat conduction to illustrate a particular procedure does not mean that it cannot also be applied to some problem in vibrations, potential theory, etc.

4.1 THE VIBRATING STRING UNDER GRAVITY

In deriving the equation of a vibrating string the effects of gravity on the string were neglected. Let us now show how the effects can be taken into account. For this purpose we consider the same string as considered in Chapter Twelve and suppose that it is taken horizontal along the x axis with its ends fixed as before at $x = 0$ and $x = L$, respectively. We shall also assume as before that the string is given some initial shape or displacement, for example by plucking it, and then released.

If g is the acceleration due to gravity, we can then show (A Exercise 1, page 567) that the partial differential equation for the vibrating string is

$$\frac{\partial^2 Y}{\partial t^2} = a^2 \frac{\partial^2 Y}{\partial x^2} - g \tag{1}$$

where $Y = Y(x, t)$ is as usual the displacement of point x of the string from the equili-brium position (x axis) at any time t. We can choose as the boundary conditions the same ones used on page 586,

$$Y(0, t) = 0, \qquad Y(L, t) = 0, \qquad Y(x, 0) = f(x), \qquad Y_t(x, 0) = 0 \tag{2}$$

If we assume according to the method of separation of variables that $Y = XT$ in (1), it is simple to find that the method fails. It is not difficult to see that the cause of this failure is the presence of g, and it would be nice if in some way we could remove it. After some thought as to how this might be done, we may be led to the idea of changing or transforming the dependent variable Y to some other variable. One possibility is to let

$$Y(x, t) = W(x, t) + \psi(x) \tag{3}$$

where $W(x, t)$ is a new dependent variable and $\psi(x)$ is an unknown function of x which we would like to determine so as to remove g in (1). If we substitute (3) in (1), we obtain

$$\frac{\partial^2 W}{\partial t^2} = a^2 \frac{\partial^2 W}{\partial x^2} + a^2 \psi'' - g \tag{4}$$

Our goal of removing g is thus achieved if we choose ψ in (4) so that

$$\psi'' = \frac{g}{a^2} \quad \text{or} \quad \psi = \frac{gx^2}{2a^2} + \alpha x + \beta \tag{5}$$

where α and β are arbitrary constants. In such case (4) becomes

$$\frac{\partial^2 W}{\partial t^2} = a^2 \frac{\partial^2 W}{\partial x^2} \tag{6}$$

The boundary conditions (2) in terms of W and ψ become

$$W(0, t) = -\psi(0), \qquad W(L, t) = -\psi(L), \qquad W(x, 0) = f(x) - \psi(x), \qquad W_t(x, 0) = 0 \quad (7)$$

Equation (6) has the same form now as the vibrating string equation *without* gravity, which of course is separable. The only trouble now is that the first two conditions in (7) are complicated by the fact that the right sides are not zero. This offers no difficulty, however, because we can make the right sides zero through the choices

$$\psi(0) = 0, \qquad \psi(L) = 0 \tag{8}$$

In fact these two conditions are just what we need to determine the two constants α and β in (5). Using conditions (8) in (5) leads to

$$\beta = 0, \quad \frac{gL^2}{2a^2} + \alpha L = 0 \quad \text{i.e.,} \quad \alpha = \frac{-gL}{2a^2}, \beta = 0 \quad \text{so that} \quad \psi(x) = -\frac{gx(L - x)}{2a^2} \tag{9}$$

Our boundary conditions (7) thus become

$$W(0, t) = 0, \qquad W(L, t) = 0, \qquad W(x, 0) = f(x) - \psi(x), \qquad W_t(x, 0) = 0 \tag{10}$$

Now the boundary value problem consisting of (6) and (10) is exactly the same as that on page 586, except that we have W instead of Y and $f(x) - \psi(x)$ instead of $f(x)$, so that from (10), page 587, we see that the required solution is

$$W(x, t) = \frac{2}{L} \sum_{n=1}^{\infty} \left(\int_0^L [f(x) - \psi(x)] \sin \frac{n\pi x}{L} \, dx \right) \sin \frac{n\pi x}{L} \cos \frac{n\pi at}{L} \tag{11}$$

from which we obtain the required displacement $Y(x, t)$ by using (3). We could of of course obtain (11) by going through the method of separation of variables and satisfying the boundary conditions as before.

4.2 HEAT CONDUCTION IN A BAR WITH NON-ZERO END CONDITIONS

In the statement of Fourier's problem on page 571, the ends of the bar at $x = 0$ and $x = L$ were both maintained at $0°C$. From a *physical* point of view there is no reason why we could not have chosen two other conditions, such as for example the end $x = 0$ kept at $20°C$ and the end $x = L$ kept at $60°C$. The revised boundary value problem in this case would be

$$\frac{\partial U}{\partial t} = \kappa \frac{\partial^2 U}{\partial x^2} \tag{12}$$

$$U(0, t) = 20, \qquad U(L, t) = 60, \qquad U(x, 0) = 100 \tag{13}$$

As before, separation of variables in equation (12) yields the solution

$$U(x, t) = e^{-\kappa \lambda^2 t}(A \cos \lambda x + B \sin \lambda x) \tag{14}$$

but we are frustrated *mathematically* in our attempts to satisfy even the first condition in (13) because no conclusion can be drawn unless the right side in this condition is zero. Our first thought might be that since temperature is a relative concept we could introduce a new temperature scale in which all centigrade temperatures are reduced by 20°C. This would help us proceed in the first condition, but the second condition in (13) would only serve to frustrate us again.

However, having succeeded in the problem of the vibrating string under gravity, we might be led to try a transformation of the dependent variable from $U(x, t)$ to $W(x, t)$, given by

$$U(x, t) = W(x, t) + \psi(x) \tag{15}$$

where we would like to determine $\psi(x)$ so as to suit our needs. Using (15), the above boundary value problem becomes

$$\frac{\partial W}{\partial t} = \kappa \frac{\partial^2 W}{\partial x^2} + \kappa \psi'' \tag{16}$$

$$W(0, t) = 20 - \psi(0), \qquad W(L, t) = 60 - \psi(L), \qquad W(x, 0) = 100 - \psi(x) \tag{17}$$

Since we would like (16) to be one for which the method of separation of variables is applicable, it would be nice to choose ψ so that

$$\psi'' = 0 \quad \text{or} \quad \psi = \alpha x + \beta \tag{18}$$

In such case (16) becomes
$$\frac{\partial W}{\partial t} = \kappa \frac{\partial^2 W}{\partial x^2} \tag{19}$$

It would also be nice if the right sides in the first two conditions of (17) were both zero. This would be accomplished if we could choose ψ so that

$$\psi(0) = 20, \qquad \psi(L) = 60 \tag{20}$$

Fortunately, the two arbitrary constants α and β are just enough to enable us to satisfy the two conditions in (20); using these conditions in (18) gives

$$\beta = 20, 60 = \alpha L + 20 \quad \text{i.e.,} \quad \alpha = \frac{40}{L}, \beta = 20$$

so that
$$\psi(x) = \frac{40x}{L} + 20 \tag{21}$$

Thus the boundary conditions (17) can be written

$$W(0, t) = 0, \qquad W(L, t) = 0, \qquad W(x, 0) = 100 - \psi(x) \tag{22}$$

To solve the boundary value problem given by (19) and (22), we use the same procedure as in Fourier's problem. Thus from (19) we find by the usual method of separation of variables the solution

$$W(x, t) = e^{-\kappa \lambda^2 t}(A \cos \lambda x + B \sin \lambda x) \tag{23}$$

The first two conditions in (22) then lead to

$$A = 0, \qquad \lambda = \frac{n\pi}{L}, \qquad n = 1, 2, 3, \ldots$$

so that
$$W(x, t) = Be^{-\kappa n^2 \pi^2 t / L^2} \sin \frac{n\pi x}{L} \tag{24}$$

To satisfy the final condition in (22) we first use the superposition principle on (24) to obtain the solution

$$W(x, t) = \sum_{n=1}^{\infty} b_n e^{-\kappa n^2 \pi^2 t / L^2} \sin \frac{n\pi x}{L} \tag{25}$$

Then from the last condition in (22) we have

$$100 - \psi(x) = \sum_{n=1}^{\infty} b_n \sin \frac{n\pi x}{L}$$

so that
$$b_n = \frac{2}{L} \int_0^L [100 - \psi(x)] \sin \frac{n\pi x}{L} dx \tag{26}$$

Use of these values in (25) gives us $W(x, t)$, and the required solution $U(x, t)$ is then obtained from (15).

It is interesting to seek a physical interpretation of the function ψ. This can be accomplished by noting that after a long time has elapsed, i.e., as $t \to \infty$, $W(x, t)$ approaches zero so that $U(x, t)$ approaches $\psi(x)$. This means that $\psi(x)$ is the *steady-state temperature*. This is also clear from a mathematical point of view, since for a temperature independent of time t, (12) reduces to (18), which leads to (21).

4.3 THE VIBRATING STRING WITH NON-ZERO INITIAL VELOCITY

In the problem on the vibrating string, solved on page 586, the string was given some initial shape and then released so that the initial velocity was taken as zero. A question arises as to how the solution is modified in case the initial velocity is not zero, whereas the other conditions remain the same.

Mathematical Formulation. In such case the differential equation for the motion is, as before,

$$\frac{\partial^2 Y}{\partial t^2} = a^2 \frac{\partial^2 Y}{\partial x^2} \tag{27}$$

and the boundary conditions are

$$Y(0, t) = 0, \qquad Y(L, t) = 0, \qquad Y(x, 0) = f(x), \qquad Y_t(x, 0) = g(x) \tag{28}$$

besides the usual one on boundedness.

Solution As before a solution satisfying the first two boundary conditions in (28) is

$$Y(x, t) = \sin \frac{n\pi x}{L} \left(A \sin \frac{n\pi a t}{L} + B \cos \frac{n\pi a t}{L} \right) \tag{29}$$

Superposition of these leads to the solution

$$Y(x, t) = \sum_{n=1}^{\infty} \sin \frac{n\pi x}{L} \left(a_n \sin \frac{n\pi a t}{L} + b_n \cos \frac{n\pi a t}{L} \right) \tag{30}$$

The last two boundary conditions then lead to the requirements

$$f(x) = \sum_{n=1}^{\infty} b_n \sin \frac{n\pi x}{L} \tag{31}$$

$$g(x) = \sum_{n=1}^{\infty} \frac{n\pi a}{L} a_n \sin \frac{n\pi x}{L} \tag{32}$$

These are two Fourier sine series for the determination of a_n and b_n, respectively. We find

$$b_n = \frac{2}{L} \int_0^L f(x) \sin \frac{n\pi x}{L} \, dx \tag{33}$$

and $\quad \dfrac{n\pi a}{L} a_n = \dfrac{2}{L} \displaystyle\int_0^L g(x) \sin \dfrac{n\pi x}{L} \, dx \quad$ or $\quad a_n = \dfrac{2}{n\pi a} \displaystyle\int_0^L g(x) \sin \dfrac{n\pi x}{L} \, dx \tag{34}$

Using these values of a_n and b_n in (30) yields the required solution.

Another Method. On page 586 we solved the boundary value problem defined by (27) and (28) with $g(x) = 0$. The result is given by (30) with b_n given by (33) and $a_n = 0$. It is also easy to solve the boundary value problem defined by (27) and (28) with $f(x) = 0$. The result, which we leave to the student to verify (see A Exercise 7), is given by (30) with a_n given by (34) and $b_n = 0$. If we denote these two solutions, respectively, by $Y_1(x, t)$ and $Y_2(x, t)$, we see that the solution $Y(x, t)$ for the case $f(x) \neq 0, g(x) \neq 0$ is equal to the sum of the solutions $Y_1(x, t)$ [for the case $f(x) \neq 0, g(x) = 0$] and $Y_2(x, t)$ [for the case $f(x) = 0, g(x) \neq 0$]. A proof of this is easily supplied. For suppose that $Y = Y(x, t)$ is a solution of the boundary value problem given by (27) and (28). Then letting $Y = Y_1 + Y_2$ and using the subscript t to denote partial derivatives with respect to t, so that $Y_{1t} = \partial Y_1/\partial t$, etc., the problem can be written

$$\frac{\partial^2 Y_1}{\partial t^2} + \frac{\partial^2 Y_2}{\partial t^2} = a^2 \left(\frac{\partial^2 Y_1}{\partial x^2} + \frac{\partial^2 Y_2}{\partial x^2} \right) \tag{35}$$

$$\begin{aligned} Y_1(0, t) + Y_2(0, t) = 0, \qquad & Y_1(L, t) + Y_2(L, t) = 0, \\ Y_1(x, 0) + Y_2(x, 0) = f(x), \qquad & Y_{1t}(x, 0) + Y_{2t}(x, 0) = g(x) \end{aligned} \tag{36}$$

Let us now choose Y_1 so that it is a solution of the boundary value problem

$$\frac{\partial^2 Y_1}{\partial t^2} = a^2 \frac{\partial^2 Y_1}{\partial x^2} \tag{37}$$

$$Y_1(0, t) = 0, \qquad Y_1(L, t) = 0, \qquad Y_1(x, 0) = f(x), \qquad Y_{1t}(x, 0) = 0 \tag{38}$$

Then it follows from (35) and (36) that Y_2 must be a solution of the boundary value problem

$$\frac{\partial^2 Y_2}{\partial t^2} = a^2 \frac{\partial^2 Y_2}{\partial x^2} \tag{39}$$

$$Y_2(0, t) = 0, \qquad Y_2(L, t) = 0, \qquad Y_2(x, 0) = 0, \qquad Y_{2t}(x, 0) = g(x) \tag{40}$$

and this supplies the required proof.

Remark. This technique of "breaking" a given problem into a number of different problems using zeros in appropriate places and then adding the various solutions to these simpler problems is often also referred to as a *method of superposition* of solutions. The technique can also be used in other problems such as heat conduction, potential theory, etc., and is of course only applicable in cases involving linear differential equations and conditions.

4.4 VIBRATIONS OF A SQUARE DRUMHEAD: A PROBLEM INVOLVING DOUBLE FOURIER SERIES

On page 561 we described a problem concerning the vibrations of a square drumhead. The equation for these vibrations is given by

$$\frac{\partial^2 Z}{\partial t^2} = a^2 \left(\frac{\partial^2 Z}{\partial x^2} + \frac{\partial^2 Z}{\partial y^2} \right) \tag{41}$$

where $Z(x, y, t)$ is the displacement of any point (x, y) of the drumhead from its equilibrium position (in the xy plane) at any time t, and a^2 is a constant dependent on the tension and the density of the drumhead. We shall assume that the drumhead is situated as in Fig. 12.5, page 561, and that the edges are fixed and of unit length. Let us also assume that the drumhead is set into vibration by being given some initial shape, as is described for example by the surface with equation $Z = f(x, y)$ and then released.

Mathematical Formulation. The partial differential equation for the motion is given by (41). The fact that the edges are fixed provides us with the four boundary conditions

$$Z(0, y, t) = 0, \quad Z(1, y, t) = 0, \quad Z(x, 0, t) = 0, \quad Z(x, 1, t) = 0 \tag{42}$$

The fact that the drumhead is given some prescribed initial shape leads to

$$Z(x, y, 0) = f(x, y) \tag{43}$$

Finally, the fact that the drumhead is released after having been given this shape tells us that

$$Z_t(x, y, 0) = 0 \tag{44}$$

These conditions hold of course for $0 < x < 1, 0 < y < 1, t > 0$. In addition it is realized that Z must be bounded.

Solution Assume that (41) has a separable solution of the form $Z = XYT$, where X, Y, T are functions of x, y, t, respectively. Then substitution into (41) yields

$$XYT'' = a^2(X''YT + XY''T) \quad \text{or} \quad \frac{T''}{a^2 T} = \frac{X''}{X} + \frac{Y''}{Y} \tag{45}$$

after division by $a^2 XYT$. Since one side is a function of t alone, while the other side is a function of x and y, it follows that each side must be a constant, which we denote by $-\lambda^2$, i.e.,

$$\frac{T''}{a^2 T} = -\lambda^2, \qquad \frac{X''}{X} + \frac{Y''}{Y} = -\lambda^2 \tag{46}$$

If we now write the second equation of (46) in the form

$$\frac{X''}{X} = -\left(\frac{Y''}{Y} + \lambda^2\right) \tag{47}$$

we see that each side must be a constant, which we can take to be $-\mu^2$. Then

$$\frac{X''}{X} = -\mu^2, \qquad \frac{Y''}{Y} = -v^2 \tag{48}$$

where for brevity we have written $v^2 = \lambda^2 - \mu^2$ so that $\lambda^2 = \mu^2 + v^2$. From equations (48) and the first equation in (46) we then have

$$X'' + \mu^2 X = 0, \qquad Y'' + v^2 Y = 0, \qquad T'' + a^2(\mu^2 + v^2)T = 0$$

so that $\quad X = c_1 \cos \mu x + c_2 \sin \mu x, \qquad Y = c_3 \cos vy + c_4 \sin vy,$

$$T = c_5 \cos a\sqrt{\mu^2 + v^2}\, t + c_6 \sin a\sqrt{\mu^2 + v^2}\, t \tag{49}$$

Thus a solution to (41) is $Z = XYT$ where X, Y, and T are given by (49).

From the first and third conditions in (42) we find $c_1 = 0, c_3 = 0$ so that

$$Z = c_2 c_4 \sin \mu x \sin vy (c_5 \cos a\sqrt{\mu^2 + v^2}\, t + c_6 \sin a\sqrt{\mu^2 + v^2}\, t) \tag{50}$$

From the second and fourth conditions in (42) we find

$$\mu = m\pi, \qquad v = n\pi, \qquad \text{where } m, n = 1, 2, 3, 4, \ldots$$

so that $\quad Z = c_2 c_4 \sin m\pi x \sin n\pi y (c_5 \cos a\sqrt{m^2 + n^2}\, \pi t + c_6 \sin a\sqrt{m^2 + n^2}\, \pi t)$

The condition (44) leads to $c_6 = 0$ so that

$$Z = B \sin m\pi x \sin n\pi y \cos a\sqrt{m^2 + n^2}\, \pi t \tag{51}$$

where we have written $B = c_2 c_4 c_5$.

To satisfy (43) we shall have to use the principle of superposition, i.e., summation of solutions of the form (51) over all integer values of m and n. This leads to the *double series* solution

$$Z = \sum_{m=1}^{\infty} \sum_{n=1}^{\infty} B_{mn} \sin m\pi x \sin n\pi y \cos a\sqrt{m^2 + n^2}\, \pi t \tag{52}$$

where we have replaced B in (51) by B_{mn} since each solution can have a different coefficient. This will satisfy condition (43) if

$$f(x, y) = \sum_{m=1}^{\infty} \sum_{n=1}^{\infty} B_{mn} \sin m\pi x \sin n\pi y \tag{53}$$

We can write the double series (53) in the form

$$f(x, y) = \sum_{m=1}^{\infty} \left(\sum_{n=1}^{\infty} B_{mn} \sin n\pi y \right) \sin m\pi x \tag{54}$$

i.e.,

$$f(x, y) = \sum_{m=1}^{\infty} b_m \sin m\pi x \tag{55}$$

where

$$b_m = \sum_{n=1}^{\infty} B_{mn} \sin n\pi y \tag{56}$$

Since (55) represents a Fourier sine series in x, we have by the usual methods

$$b_m = \frac{2}{1} \int_0^1 f(x, y) \sin m\pi x \, dx \tag{57}$$

Since (56) also represents a Fourier sine series in y, we have similarly

$$B_{mn} = \frac{2}{1} \int_0^1 b_m \sin n\pi y \, dy \tag{58}$$

If we substitute (57) into (58), we find

$$B_{mn} = 4 \int_0^1 \int_0^1 f(x, y) \sin m\pi x \sin n\pi y \, dx \, dy \tag{59}$$

Putting these coefficients back into (52) then yields the required solution.

For obvious reasons the series (53) with coefficients (59) is called a *double Fourier sine series* corresponding to $f(x, y)$. We could have similarly double Fourier cosine series or double Fourier series involving both sines and cosines. As might be expected, it is also possible to have triple, quadruple, etc., Fourier series. Some applications involving such series are given in the exercises.

   Interpretation. It is of interest to examine the possible physical significance of the various terms in the series (52). To do this, let us assume that $f(x, y)$ is such that all terms of the series (52) are zero except the one for which $m = 3$ and $n = 3$. Then this term is

$$B_{33} \sin 3\pi x \sin 3\pi y \cos \pi a \sqrt{3^2 + 3^2} t \tag{60}$$

As in the case of the string, we call this a *mode of vibration*. It will be noticed that, for all values of time t, the displacement (60) will be zero along the lines $x = \frac{1}{3}, \frac{2}{3}$ and $y = \frac{1}{3}, \frac{2}{3}$, as indicated in Fig. 13.12. Because all points on these lines never move, we refer to the lines as *nodal lines*. If we should take a motion picture of the vibration taking place over an interval of time, we would observe that portions of the drumhead within the small squares bounded by these nodal lines each vibrate, sometimes in one direction and then in the other. For example, in Fig. 13.12 we have indicated this motion by using shading to indicate motion in one direction and no shading for motion in the opposite direction where the motions take place simultaneously. The fact that two adjacent squares are vibrating in opposite directions at a particular time is often indicated by saying that the vibrations in these

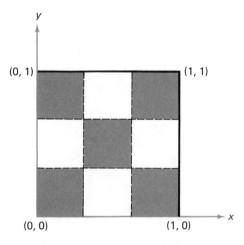

Figure 13.12

regions are *out of phase* with each other. What we have shown in Fig. 13.12 is of course a picture at one particular time. At another time the picture could change so that the directions of vibration are reversed.

By looking at the time factor in (60) it is evident that there is a periodicity in the pictures, i.e., if we find the drumhead in one particular state of motion at one instant of time, then it will be in exactly the same state at some minimum time later called the *period*. The corresponding frequency which is the reciprocal of this period is given for the case (60) by

$$f_{33} = \frac{\pi a \sqrt{3^2 + 3^2}}{2\pi} = \frac{3}{2}\sqrt{\frac{2\tau}{\rho}} \tag{61}$$

In a similar manner, each term in (52) corresponding to a pair of values of m and n represents a particular mode of vibration having a characteristic frequency given by

$$f_{mn} = \frac{\sqrt{m^2 + n^2}}{2}\sqrt{\frac{\tau}{\rho}} \tag{62}$$

For brevity we can speak of the (m, n) *mode* or the *frequency of the (m, n) mode*.

It is possible for two or more different modes to have the same frequency. For example, if we consider only the two terms in (52) corresponding to the modes (1, 2) and (2, 1), we find that the sum of these terms is

$$(B_{12} \sin \pi x \sin 2\pi y + B_{21} \sin 2\pi x \sin \pi y) \cos \sqrt{5}\pi a t \tag{63}$$

It is at once clear that for all choices of the coefficients B_{12} and B_{21} the frequency is given by

$$f_{12} = \frac{\sqrt{5}\pi a}{2\pi} = \frac{1}{2}\sqrt{\frac{5\tau}{\rho}}$$

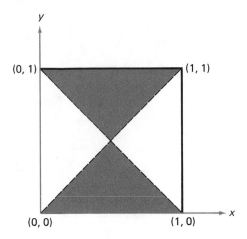

Figure 13.13

If in particular $B_{12} = B_{21}$, then the displacement for this case is given by

$$Y(x, t) = B_{12} \cos \sqrt{5}\pi a t (\sin \pi x \sin 2\pi y + \sin 2\pi x \sin \pi y)$$
$$= B_{12} \cos \sqrt{5}\pi a t \{\sin \pi x (2 \sin \pi y \cos \pi y) + (2 \sin \pi x \cos \pi x) \sin \pi y\}$$
$$= 2B_{12} \cos \sqrt{5}\pi a t \sin \pi x \sin \pi y (\cos \pi x + \cos \pi y)$$

from which we can conclude that the nodal lines are $x \pm y = 1$ (for which $\cos \pi x +$ $\cos \pi y = 0$) as indicated by the dashed lines in Fig. 13.13. Also indicated in the figure by shading or lack of shading are the directions of vibrations of the triangular shaped regions bounded by these lines at one particular instant of time, these being reversed half a period later.

We sometimes refer to two or more different modes which have the same frequency as *degenerate modes*. As seen above, all modes corresponding to the cases $m \neq n$ are degenerate.

The series (52) shows that the general displacement of a drumhead can be considered as a sum or superposition of displacements corresponding to the various modes. One interesting question which arises by analogy with the vibrating string is whether a square drumhead can create a musical tone. Let us recall that in the case of the vibrating string we have a smallest or fundamental frequency, and that we have defined *music* to be the state where all higher frequencies or harmonics are integer multiples of this fundamental frequency. If we use the same definition here, it turns out that we have a mixture of music and noise (i.e., non-music), since there are frequencies which are integer multiples of a fundamental frequency as well as frequencies which are not. An important consideration in this connection are the magnitudes of the coefficients B_{mn}, which serve to indicate those modes and corresponding frequencies of prominent importance.

4.5 HEAT CONDUCTION WITH RADIATION

In Fourier's problem involving a metal bar of length L, the boundary conditions at the ends $x = 0$ and $x = L$ treated thus far involved the cases where the ends were maintained at certain temperatures or were insulated, or a combination of these. An interesting possibility which can arise is the case where one of the ends, say $x = 0$, is maintained at some temperature, say $0°C$, while the other end $x = L$ radiates into the surrounding medium, which is assumed to be also at temperature $0°C$. We can assume as before that the initial temperature distribution is specified by $f(x)$.

Mathematical Formulation. As in Fourier's problem we assume that the convex surface of the bar is insulated so that the heat equation as before is given by

$$\frac{\partial U}{\partial t} = \kappa \frac{\partial^2 U}{\partial x^2} \tag{64}$$

The only question which we may have concerning the boundary conditions is that of the mathematical formulation of the radiation condition at the end $x = L$. To accomplish this formulation, we first recall that the flux of heat across the end $x = L$ is given by $-KU_x(L, t)$, where K is the thermal conductivity which is assumed constant. Using a Newton's law of cooling type of radiation, (i.e. where the flux is proportional to the difference in temperature between the end temperature $U(L, t)$ of the bar and the temperature of the surrounding medium taken as zero) we obtain the required boundary condition

$$U_x(L, t) = -hU(L, t) \tag{65}$$

where h is a positive constant. The remaining boundary conditions are given by

$$U(0, t) = 0, \qquad U(x, 0) = f(x) \tag{66}$$

and the obvious condition that U be bounded.

Solution By the usual method of separation of variables, a solution of (64) is found to be

$$U(x, t) = e^{-\kappa \lambda^2 t}(A \cos \lambda x + B \sin \lambda x) \tag{67}$$

To satisfy the first boundary condition in (66), we require $A = 0$ so that

$$U(x, t) = Be^{-\kappa \lambda^2 t} \sin \lambda x$$

From the boundary condition (65) we see that

$$B\lambda e^{-\kappa \lambda^2 t} \cos \lambda L = -hBe^{-\kappa \lambda^2 t} \sin \lambda L, \quad \text{i.e.,} \quad \tan \lambda L = -\lambda/h$$

To determine the values of λ which satisfy this last equation, let us write $\alpha = \lambda L$ and note that the equation can be written

$$\tan \alpha = -\frac{\alpha}{hL} \tag{68}$$

The required roots of equation (68) can be obtained as the intersection of the graphs of the equations

$$y = \tan \alpha \quad \text{and} \quad y = -\frac{\alpha}{hL} \tag{69}$$

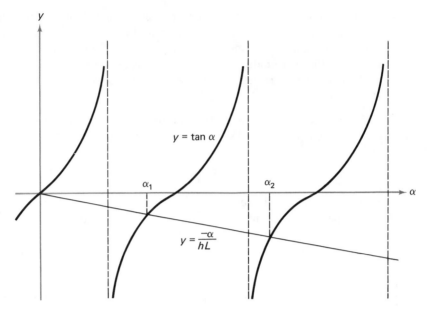

Figure 13.14

sketched in Fig. 13.14. As seen, there are infinitely many such intersections, revealing that there are infinitely many positive roots $\alpha_1, \alpha_2, \ldots$ of equation (68), and thus infinitely many corresponding positive values of λ denoted by $\lambda_1, \lambda_2, \ldots$. From this it follows that there are infinitely many solutions given by

$$U(x, t) = Be^{-\kappa\lambda_n^2 t} \sin \lambda_n x \tag{70}$$

where $n = 1, 2, \ldots$. To satisfy the second boundary condition in (66), we are led to superimpose the solutions (70) leading to

$$U(x, t) = \sum_{n=1}^{\infty} b_n e^{-\kappa\lambda_n^2 t} \sin \lambda_n x \tag{71}$$

The second condition in (66) thus leads to the requirement

$$f(x) = \sum_{n=1}^{\infty} b_n \sin \lambda_n x \tag{72}$$

which as in Fourier's problem requires the expansion of $f(x)$ into a series of trigonometric functions. The series (72) bears much resemblance to the Fourier sine series except for the fact that, whereas the values of λ_n are not equally spaced (as is evidenced from Fig. 13.14), the corresponding values $\lambda_n = n\pi/L$ in the Fourier series case are equally spaced.

The same method used in the Fourier series for finding the coefficients will also work in this case if it is true that

$$\int_0^L \sin \lambda_m x \sin \lambda_n x \, dx = 0, \qquad \lambda_m \neq \lambda_n \tag{73}$$

The student familiar with the concept of orthogonality of functions discussed in Chapter Eight will of course recognize that the statement (73) is the same as the statement that the functions $\sin \lambda_n x$, $n = 1, 2, \ldots$, are *mutually orthogonal* in the interval $0 \leq x \leq L$. This can be proved by direct integration of (73), as indicated in C Exercise 5, or by the method using the theory of Sturm–Liouville differential equations presented in Chapter Eight. No matter which way this is done, the end result amounts to the fact that, if we multiply both sides of (72) by $\sin \lambda_m x$ and then integrate from $x = 0$ to $x = L$, we find

$$\int_0^L f(x) \sin \lambda_m x \, dx = \sum_{n=1}^{\infty} b_n \int_0^L \sin \lambda_m x \sin \lambda_n x \, dx$$

or on using (73)
$$b_n = \frac{\int_0^L f(x) \sin \lambda_n x \, dx}{\int_0^L \sin^2 \lambda_n x \, dx} \tag{74}$$

The denominator in (74) is easily evaluated as

$$\int_0^L \sin^2 \lambda_n x \, dx = \frac{Lh + \cos^2 \lambda_n L}{2h} \tag{75}$$

so that
$$b_n = \frac{2h}{Lh + \cos^2 \lambda_n L} \int_0^L f(x) \sin \lambda_n x \, dx \tag{76}$$

Substitution of (76) into (71) yields the required solution.

A EXERCISES

1. Suppose that in the problem of the text concerning the vibrating string under gravity (page 603) we have $f(x) = 0$. (a) Show that

$$Y(x, t) = \frac{4gL^2}{\pi^3 a^2} \left(\frac{1}{1^3} \sin \frac{\pi x}{L} \cos \frac{\pi a t}{L} + \frac{1}{3^3} \sin \frac{3\pi x}{L} \cos \frac{3\pi a t}{L} + \cdots \right) - \frac{gx(L - x)}{2a^2}$$

(b) Discuss the physical significance.

2. A metal bar 50 cm long whose surface is insulated is at temperature 60°C. At $t = 0$ a temperature of 30°C is applied at one end and a temperature of 80°C to the other end, and these temperatures are maintained. Determine the temperature of the bar at any time assuming $\kappa = 0.15$ cgs unit.

3. A slab of material having diffusivity κ is bounded by the planes $x = 0$ and $x = L$. If the plane faces are kept at constant temperatures U_1 and U_2, respectively, while the initial temperature is U_0, find the temperature at any point at any later time t.

4. Work Exercise 3 if the initial temperature is given by $U_0(1 - x/L)$.

5. A string has its ends fixed at $x = 0$ and $x = L$. Suppose that initially the string has a parabolic shape and each point is moving in the same direction with the same speed. Find the displacement of any point of the string at any time.

6. Solve and interpret physically the boundary value problem

$$\frac{\partial U}{\partial t} = 1 + \frac{\partial^2 U}{\partial x^2}, \qquad 0 < x < 1, t > 0$$

$$U(0, t) = 0, \qquad U(1, t) = 0, \qquad U(x, 0) = \sin \pi x$$

7. Complete the second method given on page 607.

B EXERCISES

1. If a string with endpoints fixed at $x = 0$ and $x = L$ is under the influence of gravity but does not vibrate, show that its shape is given by that portion of the parabola

$$Y = -\frac{gx(L - x)}{2a^2}$$

between $x = 0$ and $x = L$. Does this conflict with the results obtained on pages 111–114 for the hanging cable? Explain.

2. Solve and give a physical interpretation to the boundary value problem

$$\frac{\partial^2 Y}{\partial t^2} = a^2 \frac{\partial^2 Y}{\partial x^2}, \qquad 0 < x < L, t > 0$$

$$Y(0, t) = 0, \qquad Y_x(L, t) = K, \qquad Y(x, 0) = 0, \qquad Y_t(x, 0) = 0$$

3. Solve and interpret physically the boundary value problem

$$Y_{tt} = 1 + Y_{xx}, \qquad 0 < x < \frac{\pi}{2}, t > 0$$

$$Y(0, t) = 1, \qquad Y(x, 0) = 1 + \tfrac{1}{2}x^2 - \tfrac{1}{2}\pi x, \qquad Y_x\left(\frac{\pi}{2}, t\right) = -\frac{\pi}{2}, \qquad Y_t(x, 0) = 0$$

4. Find the steady-state temperature in a square plate of unit side as indicated in Fig. 13.15 if the plane faces are insulated and the sides are kept at constant temperatures U_1, U_2, U_3, U_4,

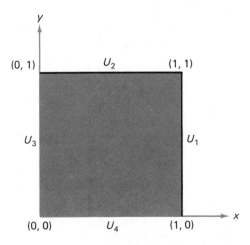

Figure 13.15

respectively. (*Hint*: Solve the problem for the case where three of the sides are kept at temperature zero and the fourth kept at a constant temperature different from zero. Then apply the method of superposition to the four cases.)

5. Work the problem of the vibrating string under gravity if damping proportional to the velocity is also taken into account.

6. A thin bar with ends at $x = 0$ and $x = L$ has constant diffusivity κ and initial temperature U_0. Due to nuclear phenomena taking place in the bar, heat is generated at a constant rate α. Assuming that the ends are kept at temperatures U_1 and U_2, respectively, and that the surface is insulated, find the temperature everywhere in the bar at any time.

7. Work Exercise 6 if $U_0 = U_1 = 0$ and the end $x = L$ is insulated.

8. Work Exercise 6 if $U_0 = U_1 = U_2 = 0$ and the rate of heat generation in the bar is proportional to the distance from the end $x = 0$.

9. A vibrating string has ends fixed at $x = 0$ and $x = L$. At $t = 0$ the shape is given by $f(x)$ and the velocity distribution by $g(x)$. Show that

$$Y(x, t) = \tfrac{1}{2}[f(x - at) + f(x + at)] + \frac{1}{2a} \int_{x-at}^{x+at} g(u)\,du$$

10. Interpret the boundary value problem defined by (1) and (2), page 586 as a problem on beams.

11. Work Exercise 4 for a solid cube having unit side.

12. Interpret Exercises 4 and 11 as problems involving potential.

C EXERCISES

1. Suppose that the square drumhead of the text is given an initial displacement $f(x, y) = xy(1 - x)(1 - y)$ and then released. Describe the subsequent vibrations which take place by finding the displacement of each point of the drumhead from the equilibrium position.

2. Work the problem of the vibrating drumhead if it is given both an initial shape and velocity distribution.

3. Describe the modes of vibration for the square drumhead in the cases
   (a) $m = 2, n = 2$.   (b) $m = 1, n = 3$.

4. If the square drumhead of the text is "plucked" by lifting the center a small distance h and then released, investigate the subsequent vibrations.

5. Verify by direct integration the results (73) and (75), pages 614 and 615.

6. Show that the successive roots of equation (68), page 613, are closely approximated by the values $(2n - 1)\pi/2$.

7. Discuss the conducting bar with radiation of the text for the cases (a) $h = 0$; (b) $h = \infty$.

8. Work the problem of the conducting bar with radiation assuming that the end $x = 0$ is insulated.

9. A sphere of unit radius and constant diffusivity κ has an initial temperature which depends only on the distance r from its center, which is chosen as the origin of a three-dimensional coordinate system. The surface of the sphere is kept at temperature zero. (a) If $U(r, t)$ is the temperature at any point of the sphere at time t, show that the boundary value problem

is given by

$$\frac{\partial U}{\partial t} = \kappa \left(\frac{\partial^2 U}{\partial r^2} + \frac{2}{r} \frac{\partial U}{\partial r} \right)$$

$$U(1, t) = 0, \qquad U(r, 0) = f(r), \qquad |U(r, t)| < M$$

(b) Solve the boundary value problem in (a) showing that

$$U(r, t) = \frac{2}{r} \sum_{k=1}^{\infty} e^{-k^2 \pi^2 \kappa t} \sin k\pi r \int_0^1 vf(v) \sin k\pi v \, dv$$

[*Hint:* Make the transformation $V(r, t) = rU(r, t)$ or use A Exercise 3, page 329.]

10. Work Exercise 9 if (a) $f(r) = U_0$; (b) $f(r) = \begin{cases} U_0, & 0 < r < \frac{1}{2} \\ 0, & \frac{1}{2} < r < 1 \end{cases}$, and interpret physically.

11. A region of diffusivity κ is bounded by the spheres $r = 1$ and $r = 2$ whose surfaces are kept at temperature zero. If the initial temperature of the region is U_0, show that the temperature at any point distant r from the common center is given at any time $t > 0$ by

$$U(r, t) = \frac{2U_0}{\pi r} \sum_{k=1}^{\infty} (-1)^k \left(\frac{1 - 2 \cos k\pi}{k} \right) e^{-k^2 \pi^2 \kappa t} \sin k\pi r, \qquad 1 < r < 2$$

12. Solve and interpret physically the boundary value problem

$$\frac{\partial^2 U}{\partial t^2} = \frac{\partial^2 U}{\partial r^2} + \frac{2}{r} \frac{\partial U}{\partial r}$$

$$U(1, t) = 0, \qquad U(r, 0) = f(r), \qquad U_t(r, 0) = 0, \qquad |U(r, t)| < M$$

fourteen

◆ solutions of boundary value problems using Bessel and Legendre functions

1 Introduction

In the last chapter we solved various boundary value problems in heat conduction, vibrations, and potential theory by using Fourier or other trigonometric series. Fundamental in the solution of such problems was the idea of the expansion of a function in a series having the form

$$f(x) = c_1 u_1(x) + c_2 u_2(x) + \cdots = \sum_{n=1}^{\infty} c_n u_n(x) \tag{1}$$

where the functions $u_n(x)$, $n = 1, 2, \ldots$, have the property

$$\int_a^b u_m(x) u_n(x) dx = 0, \qquad m \neq n \tag{2}$$

Remark. A set of functions having the property (2) is called (as the student who has read Chapter Eight will recall) an *orthogonal set* in the interval $a \leq x \leq b$, and the functions are said to be *mutually orthogonal* in this interval. The series (1) is then often called an *orthogonal series.**

From the fact that most problems of the last chapter involved rectangular coordinates and Fourier series, we might suspect that a choice of other types of coordinates, such as cylindrical or spherical coordinates, would lead to orthogonal series of the form (1) involving functions other than trigonometric functions. We shall find in this chapter that this suspicion proves to be correct. In fact, the orthogonal functions which are used in such case are Bessel and Legendre functions already familiar to us from Chapter Seven. In the first part of this chapter we shall deal with problems like those of the last chapter, i.e., involving heat conduction, vibration, and potential theory, formulated in cylindrical coordinates and leading to Bessel functions. In the second part we do the same thing with spherical coordinates and Legendre functions.

2 Boundary Value Problems Leading to Bessel Functions

2.1 THE LAPLACIAN IN CYLINDRICAL COORDINATES

In order to deal with boundary value problems involving cylinders having circular cross section, it is natural for us to use cylindrical coordinates which are made up of polar coordinates (r, ϕ) in the xy plane together with the additional z coordinate to give (r, ϕ, z). These coordinates are used to represent any point P in three dimensions as shown in Fig. 14.1. The values of r, ϕ, and z are such that

$$r \geq 0, \qquad 0 \leq \phi < 2\pi, \qquad -\infty < z < \infty \tag{1}$$

---

* Since we shall be making increasing use of the results in Chapter Eight, the student who wishes to derive the greatest benefit from the present chapter is advised to read Chapter Eight.

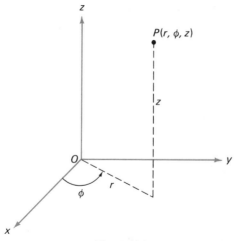

Figure 14.1

although any other interval of length 2π, such as $-\pi < \phi \leq \pi$, can also be chosen for ϕ. The transformation equations between cylindrical and rectangular coordinates are given by

$$x = r \cos \phi, \qquad y = r \sin \phi, \qquad z = z \tag{2}$$

Note that since z remains unchanged the transformation is the same as that involving polar coordinates in the xy plane. Now as we have seen in problems of the last two chapters, the partial differential equations often involve the Laplacian operator defined in rectangular coordinates by

$$\nabla^2 U = \frac{\partial^2 U}{\partial x^2} + \frac{\partial^2 U}{\partial y^2} + \frac{\partial^2 U}{\partial z^2} \tag{3}$$

It is thus natural for us to seek a corresponding expression for this in cylindrical coordinates. Since we have already found this for the first two terms in working the problem on page 596, we see that the required expression is

$$\nabla^2 U = \frac{\partial^2 U}{\partial r^2} + \frac{1}{r} \frac{\partial U}{\partial r} + \frac{1}{r^2} \frac{\partial^2 U}{\partial \phi^2} + \frac{\partial^2 U}{\partial z^2} \tag{4}$$

2.2 HEAT CONDUCTION IN A CIRCULAR CYLINDER

As a first illustration, let us consider a metal pipe in the form of a circular cylinder of unit radius, which is so long that for all practical purposes it can be considered to be of infinite length. We shall assume that initially, i.e., $t = 0$, the temperature of the pipe at a distance r from the axis is given by $f(r)$, and that its surface is kept at $0°C$ for $t > 0$. The problem consists of finding the temperature of the pipe at any point at any time.

Mathematical Formulation. Let us choose the cylinder so that its cross section in the xy plane is the circle of unit radius shown in Fig. 14.2. Since the pipe is considered as infinite, the temperature U should not involve any z dependence. Also since the boundary is maintained at $0°C$, we see by the symmetry that there is no ϕ dependence.

Since the heat conduction equation is

$$\frac{\partial U}{\partial t} = \kappa \nabla^2 U \tag{5}$$

and since there is no z or ϕ dependence, (5) becomes

$$\frac{\partial U}{\partial t} = \kappa \left(\frac{\partial^2 U}{\partial r^2} + \frac{1}{r} \frac{\partial U}{\partial r} \right) \tag{6}$$

where $U = U(r, t)$, i.e., the temperature U depends only on r and t. With U equal to zero at $r = 1$ for all $t > 0$, we have the boundary condition

$$U(1, t) = 0 \tag{7}$$

Also since U is equal to $f(r)$ initially, i.e., for $t = 0$, we have the boundary condition

$$U(r, 0) = f(r) \tag{8}$$

In addition to these we also have the usual condition that U must be bounded at all points in the cylinder, which is of course implied from physical reality. The required boundary value problem thus consists of determining that bounded solution of (6) which satisfies conditions (7) and (8).

Remark. It should be pointed out that the same boundary value problem is arrived at by considering a circular metal plate of negligible thickness whose faces are insulated and whose initial temperature is $f(r)$ while the boundary temperature is kept at $0°C$.

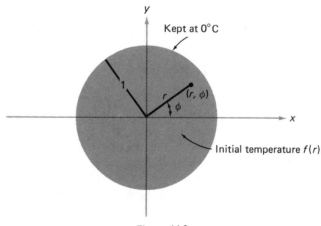

Figure 14.2

Solution Assuming, according to the method of separation of variables, that $U = RT$, where R depends only on r and T depends only on t, (6) becomes

$$RT' = \kappa\left(R''T + \frac{1}{r}R'T\right) \quad \text{or} \quad \frac{T'}{\kappa T} = \frac{R''}{R} + \frac{1}{r}\frac{R'}{R} \tag{9}$$

on division by κRT. Setting each side of the second equation in (9) equal to the constant $-\lambda^2$, we have

$$T' + \kappa\lambda^2 T = 0, \qquad R'' + \frac{1}{r}R' + \lambda^2 R = 0 \tag{10}$$

The first equation in (10) has solution $\qquad T = c_1 e^{-\kappa\lambda^2 t} \tag{11}$

To solve the second equation in (10) let $\lambda r = u$. Then the equation becomes

$$\frac{d^2 R}{du^2} + \frac{1}{u}\frac{dR}{du} + R = 0 \tag{12}$$

This is recognized to be Bessel's differential equation of order zero whose solution is

$$R = c_2 J_0(u) + c_3 Y_0(u) \tag{13}$$

(see page 337). Thus the second equation in (10) has solution

$$R = c_2 J_0(\lambda r) + c_3 Y_0(\lambda r) \tag{14}$$

From (11) and (14) we see that a solution of (6) is

$$U(r, t) = RT = c_1 e^{-\kappa\lambda^2 t}[c_2 J_0(\lambda r) + c_3 Y_0(\lambda r)] \tag{15}$$

Now the solution (15) must be bounded inside the cylinder and in particular for $r = 0$. However, for $r = 0$ the function $Y_0(\lambda r)$ is infinite. To avoid this catastrophe, we choose $c_3 = 0$ so that the solution (15) becomes

$$U(r, t) = Ae^{-\kappa\lambda^2 t}J_0(\lambda r) \tag{16}$$

where we have taken $A = c_1 c_2$. Using the boundary condition (7) in (16), we obtain

$$J_0(\lambda) = 0 \tag{17}$$

Fortunately, we know already from our discussion on page 338 that there are infinitely many positive roots of equation (17) which we shall denote by*

$$\lambda = \lambda_1, \lambda_2, \lambda_3, \ldots, \quad \text{i.e.,} \quad \lambda = \lambda_n, \quad \text{where } n = 1, 2, 3, \ldots \tag{18}$$

Thus $\qquad U(r, t) = Ae^{-\kappa\lambda_n^2 t}J_0(\lambda_n r) \tag{19}$

To satisfy the final boundary condition (8), we first use the superposition principle by adding solutions of the form (19) with different coefficients. This leads to

$$U(r, t) = \sum_{n=1}^{\infty} a_n e^{-\kappa\lambda_n^2 t}J_0(\lambda_n r) \tag{20}$$

* The roots are denoted in this chapter by $\lambda_1, \lambda_2, \lambda_3, \ldots$ rather than the equivalent ones $r_1, r_2, r_3, \ldots$ of Chapter Eight.

and from (8)
$$f(r) = \sum_{n=1}^{\infty} a_n J_0(\lambda_n r) \tag{21}$$

This problem is similar to that of Fourier except that, instead of requiring the expansion of a function in a trigonometric series, we require it in terms of a series of Bessel functions often called a *Bessel series* or *Fourier–Bessel series* because of the analogy to Fourier series.

The student familiar with Chapter Eight (see page 397) realizes that the coefficients a_n in (21) can be determined by multiplying both sides of (21) by $rJ_0(\lambda_m r)$ and then integrating from $r = 0$ to $r = 1$, obtaining

$$\int_0^1 rf(r)J_0(\lambda_m r)dr = \sum_{n=1}^{\infty} a_n \int_0^1 rJ_0(\lambda_m r)J_0(\lambda_n r)dr \tag{22}$$

Using the fact that
$$\int_0^1 rJ_0(\lambda_m r)J_0(\lambda_n r)dr = 0, \qquad m \neq n \tag{23}$$

i.e., that the functions $\sqrt{r}J_0(\lambda_n r)$, $n = 1, 2, 3, \ldots$, are mutually orthogonal in the interval $0 \leq r \leq 1$, all the terms in the series on the right of (22) except the one where $m = n$ are zero so that

$$\int_0^1 rf(r)J_0(\lambda_n r)dr = a_n \int_0^1 r[J_0(\lambda_n r)]^2 \, dr \quad \text{or} \quad a_n = \frac{\int_0^1 rf(r)J_0(\lambda_n r)dr}{\int_0^1 r[J_0(\lambda_n r)]^2 \, dr} \tag{24}$$

This can also be written according to Table 8.1, page 370,

$$a_n = \frac{2}{[J_1(\lambda_n)]^2} \int_0^1 rf(r)J_0(\lambda_n r)dr \tag{25}$$

Using (25) in (20) yields the required solution.

ILLUSTRATIVE EXAMPLE

Find the temperature at any point of the cylindrical pipe if the initial temperature is 100°C.

Solution   In this case we have $f(r) = 100$, so that the coefficients (25) are given by

$$a_n = \frac{2}{[J_1(\lambda_n)]^2} \int_0^1 r(100)J_0(\lambda_n r)dr = \frac{200}{\lambda_n^2[J_1(\lambda_n)]^2} \int_0^{\lambda_n} uJ_0(u)du = \frac{200}{\lambda_n J(\lambda_n)}$$

where we have made the substitution $u = \lambda_n r$ and used the result A Exercise 2(b), page 341. Substituting these values of a_n in (20) gives the required result

$$U(r, t) = \sum_{n=1}^{\infty} \frac{200e^{-\kappa \lambda_n^2 t}}{\lambda_n J_1(\lambda_n)} J_0(\lambda_n r) \tag{26}$$

2.3 HEAT CONDUCTION IN A RADIATING CYLINDER

Suppose that instead of having the surface of the cylindrical pipe kept at 0°C there is radiation of heat from the surface into the surrounding medium, which is assumed to be at 0°C. If we assume a Newton's law of cooling type of radiation, then the boundary condition at the surface $r = 1$ is given by

$$U_r(1, t) = -hU(1, t) \tag{27}$$

which expresses the fact that the flux across the surface is proportional to the difference in temperature between the surface temperature $U(1, t)$ and the surrounding temperature zero. Using the solution (16), page 623, condition (27) gives

$$\lambda J_0'(\lambda) = -hJ_0(\lambda) \tag{28}$$

As pointed out in Chapter Eight, page 370, equation (28) has infinitely many positive roots, which we shall denote by $\lambda_1, \lambda_2, \ldots$ or λ_n, $n = 1, 2, 3, \ldots$. Thus as in (19),

$$U(r, t) = Ae^{-\kappa\lambda_n^2 t}J_0(\lambda_n r) \tag{29}$$

where λ_n are the positive roots of (28). To prepare for the final boundary condition (8). we first use superposition to obtain the solution

$$U(r, t) = \sum_{n=1}^{\infty} a_n e^{-\kappa\lambda_n^2 t}J_0(\lambda_n r) \tag{30}$$

from which

$$f(r) = \sum_{n=1}^{\infty} a_n J_0(\lambda_n r) \tag{31}$$

Multiplying both sides of (31) as before by $rJ_0(\lambda_m r)$ and integrating from $r = 0$ to $r = 1$ yields

$$\int_0^1 rf(r)J_0(\lambda_m r)dr = \sum_{n=1}^{\infty} a_n \int_0^1 rJ_0(\lambda_m r)J_0(\lambda_n r)dr \tag{32}$$

Since

$$\int_0^1 rJ_0(\lambda_m r)J_0(\lambda_n r)dr = 0, \qquad m \neq n$$

as seen on page 370 of Chapter Eight, we find from (32)

$$a_n = \frac{\int_0^1 rf(r)J_0(\lambda_n r)dr}{\int_0^1 r[J_0(\lambda_n r)]^2 \, dr} \quad \text{or} \quad a_n = \frac{2\lambda_n^2}{(\lambda_n^2 + h^2)[J_0(\lambda_n)]^2} \int_0^1 rf(r)J_0(\lambda_n r)dr \tag{33}$$

the last result obtained by making use of the table on page 370.

2.4 VIBRATIONS OF A CIRCULAR DRUMHEAD

In the last chapter we analyzed the vibrations of a drumhead in the shape of a square. We now examine the vibrations of a more conventional drumhead, which has circular shape. Let us assume that the drumhead is of unit radius and is fixed at the boundary, so that it can be indicated by the region shown shaded in Fig. 14.2.

Mathematical Formulation. The partial differential equation is given by

$$\frac{\partial^2 Z}{\partial t^2} = a^2 \nabla^2 Z \tag{34}$$

or in polar coordinates

$$\frac{\partial^2 Z}{\partial t^2} = a^2 \left(\frac{\partial^2 Z}{\partial r^2} + \frac{1}{r}\frac{\partial Z}{\partial r} + \frac{1}{r^2}\frac{\partial^2 Z}{\partial \phi^2} \right) \tag{35}$$

where $Z = Z(r, \phi, t)$ is the displacement of any point (r, ϕ) of the drumhead at any time t from the equilibrium position in the xy plane. The boundary conditions are

$$Z(1, \phi, t) = 0, \qquad Z(r, \phi, 0) = f(r, \phi), \qquad Z_t(r, \phi, 0) = 0 \qquad (36)$$

the first expressing the fact that the boundary of the drumhead is fixed, the second that the drumhead is given some prescribed shape initially, i.e., at $t = 0$, dependent on both r and ϕ, and the third that the drumhead is then released so that its initial velocity is zero. An additional requirement which we shall not state explicitly is the obvious one that Z is bounded.

Solution Assuming a separable solution of the form $Z = R\Phi T$, where R, Φ, T depend only on r, ϕ, t, respectively, (35) becomes

$$R\Phi T'' = a^2 \left(R''\Phi T + \frac{1}{r} R'\Phi T + \frac{1}{r^2} R\Phi''T \right) \quad \text{or} \quad \frac{T''}{a^2 T} = \frac{R''}{R} + \frac{1}{r}\frac{R'}{R} + \frac{1}{r^2}\frac{\Phi''}{\Phi} \quad (37)$$

on dividing by $a^2 R\Phi T$. Setting each side of the second equation in (37) equal to $-\lambda^2$, we have

$$T'' + \lambda^2 a^2 T = 0, \qquad \frac{R''}{R} + \frac{1}{r}\frac{R'}{R} + \frac{1}{r^2}\frac{\Phi''}{\Phi} = -\lambda^2 \qquad (38)$$

The first equation in (38) has solution $\qquad T = c_1 \cos \lambda at + c_2 \sin \lambda at \qquad (39)$

The second equation in (38) can be written

$$\frac{R''}{R} + \frac{1}{r}\frac{R'}{R} + \lambda^2 = -\frac{1}{r^2}\frac{\Phi''}{\Phi} \qquad (40)$$

and we can make one side of (40) a function of r alone and the other side a function of ϕ alone if we multiply both sides by r^2. In such case (40) becomes

$$r^2 \frac{R''}{R} + r\frac{R'}{R} + \lambda^2 r^2 = -\frac{\Phi''}{\Phi} \qquad (41)$$

Setting each side equal to μ^2 leads to the equations

$$r^2 R'' + rR' + (\lambda^2 r^2 - \mu^2)R = 0, \qquad \Phi'' + \mu^2\Phi = 0 \qquad (42)$$

Letting $\lambda r = u$ in the first equation of (42), it becomes

$$u^2 R'' + uR + (u^2 - \mu^2)R = 0 \qquad (43)$$

which is Bessel's equation of order μ having solution (see page 337)

$$R = c_3 J_\mu(u) + c_4 Y_\mu(u)$$

Thus the solution of the first equation of (42) is $R = c_3 J_\mu(\lambda r) + c_4 Y_\mu(\lambda r)$.

The solution of the second equation in (42) is $\Phi = c_5 \cos \mu\phi + c_6 \sin \mu\phi$.

From these results we see that a solution of (35) is

$$Z = R\Phi T = [c_3 J_\mu(\lambda r) + c_4 Y_\mu(\lambda r)][c_5 \cos \mu\phi + c_6 \sin \mu\phi][c_1 \cos \lambda at + c_2 \sin \lambda at] \quad (44)$$

and we shall now try to satisfy the boundary conditions. Since Z must be bounded and since $Y_\mu(\lambda r)$ is infinite for $r = 0$, which is the center of the drumhead, we must take $c_4 = 0$ so that the solution becomes

$$Z = c_3 J_\mu(\lambda r)[c_5 \cos \mu\phi + c_6 \sin \mu\phi][c_1 \cos \lambda at + c_2 \sin \lambda at] \tag{45}$$

From the first boundary condition in (36) we can then see that

$$J_\mu(\lambda) = 0$$

As discussed in Chapter Seven, this equation has infinitely many positive roots, which can be denoted by $\lambda_1, \lambda_2, \dots$. However, since the roots are in general different for different orders μ, we shall add the additional subscript μ and write these roots as $\lambda_{1\mu}, \lambda_{2\mu}, \dots$, or $\lambda_{m\mu}$, $m = 1, 2, 3, \dots$. The solution (45) then becomes

$$Z = c_3 J_\mu(\lambda_{m\mu} r)[c_5 \cos \mu\phi + c_6 \sin \mu\phi][c_1 \cos \lambda_{m\mu} at + c_2 \sin \lambda_{m\mu} at] \tag{46}$$

The third condition in (36) is satisfied if we choose $c_2 = 0$, and thus we have

$$Z = J_\mu(\lambda_{m\mu} r)[A \cos \mu\phi + B \sin \mu\phi] \cos \lambda_{m\mu} at \tag{47}$$

where we have taken $A = c_1 c_3 c_5$, $B = c_1 c_3 c_6$.

Since any point (r, ϕ) of the drumhead can also be expressed as $(r, \phi + 2\pi)$ so that Z is a periodic function of ϕ, the order μ must be an integer denoted by n, where $n = 0, 1, 2, 3, \dots$. This leads us to write (47) as

$$Z = J_n(\lambda_{mn} r)[A \cos n\phi + B \sin n\phi] \cos \lambda_{mn} at \tag{48}$$

To prepare for the final boundary condition in (36) we must use the principle of superposition on the solutions given by (48). Since the coefficients A and B can depend on m and n and since m can assume the values $1, 2, 3, \dots$, while n can assume the values $0, 1, 2, 3, \dots$, this solution can be written

$$Z = \sum_{m=1}^{\infty} \sum_{n=0}^{\infty} J_n(\lambda_{mn} r)[a_{mn} \cos n\phi + b_{mn} \sin n\phi] \cos \lambda_{mn} at \tag{49}$$

Thus the boundary condition leads to

$$f(r, \phi) = \sum_{n=0}^{\infty} \left[\left\{ \sum_{m=1}^{\infty} a_{mn} J_n(\lambda_{mn} r) \right\} \cos n\phi + \left\{ \sum_{m=1}^{\infty} b_{mn} J_n(\lambda_{mn} r) \right\} \sin n\phi \right] \tag{50}$$

We can think of the summation over n in (50) as a Fourier series in the variable ϕ. According to the usual methods for finding Fourier coefficients, we see that the coefficient of $\cos n\phi$, $n = 1, 2, 3, \dots$ is given by

$$\sum_{m=1}^{\infty} a_{mn} J_n(\lambda_{mn} r) = \frac{1}{\pi} \int_0^{2\pi} f(r, \phi) \cos n\phi \, d\phi \tag{51}$$

the corresponding coefficient for $n = 0$ being given by

$$\sum_{n=1}^{\infty} a_{m0} J_0(\lambda_{m0} r) = \frac{1}{2\pi} \int_0^{2\pi} f(r, \phi) d\phi \tag{52}$$

Similarly, the coefficient of $\sin n\phi$, $n = 1, 2, 3, \ldots$, is given by

$$\sum_{m=1}^{\infty} b_{mn} J_n(\lambda_{mn}r) = \frac{1}{\pi} \int_0^{2\pi} f(r, \phi) \sin n\phi \, d\phi \qquad (53)$$

To find the coefficients a_{mn} for $m = 1, 2, 3, \ldots$, we use the fact that (51) is a Bessel series, and by the usual procedure obtain

$$a_{mn} = \frac{2}{\pi[J_{n+1}(\lambda_{mn})]^2} \int_0^1 rJ_n(\lambda_{mn}r)\left[\int_0^{2\pi} f(r, \phi) \cos n\phi \, d\phi\right] dr \qquad (54)$$

Similarly, from (52) we have

$$a_{m0} = \frac{1}{\pi[J_1(\lambda_{m0})]^2} \int_0^1 rJ_0(\lambda_{m0}r)\left[\int_0^{2\pi} f(r, \phi) \, d\phi\right] dr \qquad (55)$$

In the same way (53) leads to

$$b_{mn} = \frac{2}{\pi[J_{n+1}(\lambda_{mn})]^2} \int_0^1 rJ_n(\lambda_{mn}r)\left[\int_0^{2\pi} f(r, \phi) \sin n\phi \, d\phi\right] dr \qquad (56)$$

The required solution is thus given by (49) using the coefficients just found.

Interpretation. As in the case of the square drumhead, each term in the double series solution (49) corresponding to a given pair of values of m and n represents a particular mode of vibration. To each mode there is also a corresponding characteristic frequency. For example, the term corresponding to the (m, n) mode is

$$J_n(\lambda_{mn}r)[a_{mn} \cos n\phi + b_{mn} \sin n\phi] \cos \lambda_{mn}at \qquad (57)$$

and the frequency corresponding to this mode is

$$f_{mn} = \frac{\lambda_{mn}a}{2\pi} = \frac{\lambda_{mn}}{2\pi}\sqrt{\frac{\tau}{\rho}} \qquad (58)$$

The frequencies are the *eigenvalues* and the modes represent the *eigenfunctions*. The lowest frequency, called the *fundamental frequency*, corresponds to the mode for which $m = 1$ and $n = 0$. This mode is given, apart from a constant, by

$$J_0(\lambda_{10}r) \cos \lambda_{10}at \qquad (59)$$

The fundamental frequency, using $\lambda_{10} = 2.405$ as the least positive root $J_0(x) = 0$, is given by

$$f_{10} = \frac{\lambda_{10}a}{2\pi}\sqrt{\frac{\tau}{\rho}} = \frac{2.405}{2\pi}\sqrt{\frac{\tau}{\rho}} \qquad (60)$$

Since it can be shown that the higher frequencies are not integer multiples of this fundamental frequency, we would expect *noise* and not *music* from the vibrations of a drumhead, which is of course confirmed in reality.

It is of interest to examine the vibrations for various modes. In the case $n = 0$ the modes are independent of ϕ and, as seen from (57), are given by

$$J_0(\lambda_{m0}r) \cos \lambda_{m0}at$$

Mode $m = 1, n = 0$	Mode $m = 2, n = 0$	Mode $m = 2, n = 1$
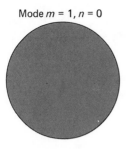	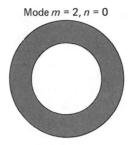	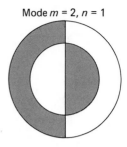
Figure 14.3	Figure 14.4	Figure 14.5

In the case $m = 1$, this reduces to (59) and corresponds to the situation indicated in Fig. 14.3 where the drumhead vibrates back and forth as a whole with frequency given by (60). For $m = 2$ the mode is given by

$$J_0(\lambda_{20}r) \cos \lambda_{20}at \tag{61}$$

and the corresponding frequency is

$$f_{20} = \frac{\lambda_{20}a}{2\pi} = \frac{5.520}{2\pi} \sqrt{\frac{\tau}{\rho}} \tag{62}$$

since the second positive root of $J_0(x) = 0$ is 5.520. There is only one root between $r = 0$ and $r = 1$ for which (61) is zero, i.e., $\lambda_{20}r = \lambda_{10}$ or $r = 2.405/5.520 = 0.436$. This is represented in Fig 14.4 by the circle lying inside the drumhead. It is called a *nodal circle* because points on it do not move. The vibrations for this mode are such that the shaded and unshaded portions of the membrane are moving in opposite directions at any given instant of time. In general, if $n = 0$, there will be $m - 1$ nodal circles in the drumhead.

In the case of ϕ dependence, the modes of vibration are more complicated. As an example, consider the mode corresponding to $m = 2, n = 1$ given by

$$J_1(\lambda_{21}r)[a_{21} \cos \phi + b_{21} \sin \phi] \cos \lambda_{21}at \tag{63}$$

the frequency being given by

$$f_{21} = \frac{\lambda_{21}a}{2\pi} = \frac{7.016}{2\pi} \sqrt{\frac{\tau}{\rho}} \tag{64}$$

If we take the particular case $a_{21} = 1$, $b_{21} = 0$, this can be written

$$J_1(\lambda_{21} r) \cos \phi \cos \lambda_{21} at \tag{65}$$

Now $J_1(\lambda_{21}r) = 0$ for only one value of r between 0 and 1, i.e., for $\lambda_{21}r = \lambda_{11}$, or, $r = \lambda_{11}/\lambda_{21} = 3.832/7.016 = 0.546$. This represents a nodal circle as indicated in Fig. 14.5. Also $\cos \phi = 0$ for $\phi = \pi/2, 3\pi/2$, which represent two nodal lines as indicated in Fig. 14.5. The vibration for this mode is thus as indicated in Fig. 14.5 where the shaded portions of the drumhead are vibrating in one direction and the remaining portions in the opposite direction, both with frequency (64).

A EXERCISES

1. Find the temperature at any point of the cylindrical pipe of page 621 if the initial temperature is given by (a) $f(r) = 50J_0(\lambda_1 r)$; (b) $f(r) = 20J_0(\lambda_1 r) - 10J_0(\lambda_2 r)$ where λ_1 and λ_2 are the two smallest positive roots of $J_0(\lambda) = 0$.

2. A circular plate of unit radius and diffusivity κ has initial temperature given by

$$f(r) = \begin{cases} 100, & 0 < r < \frac{1}{2} \\ 0, & \frac{1}{2} < r < 1 \end{cases}$$

where r is the distance of any point from the center. Assuming that the temperature of the boundary is kept at zero and that the plane faces are insulated, find the temperature at any point of the plate at any time. Give an interpretation involving a pipe.

3. Work Exercise 2 if the initial temperature is

(a) $f(r) = 100(1 - r^2)$,   (b) $f(r) = \begin{cases} 90 & 0 < r < 1/3 \\ 60 & 1/3 < r < 2/3 \\ 30 & 2/3 < r < 1 \end{cases}$

4. Work the Illustrative Example of page 624 if the boundary is kept at 25°C.

5. Work the Illustrative Example of page 624 if the radius is c.

6. A semi-circular plate of unit radius has its faces and its bounding diameter insulated. If the initial temperature of the plate is 60°C and the temperature of the semi-circular boundary is kept at 0°C, find the temperature of the plate. Give an interpretation involving a solid.

7. Work Exercise 6 if the semi-circular-boundary is kept at a temperature of 100°C.

8. Work the problem of the text on page 621 if the surface of the pipe is insulated instead of being kept at 0°C.

9. Work Exercise 2 if the boundary is insulated.

10. Find the temperature at any point of the radiating cylinder of page 624 if the initial temperature is (a) $f(r) = U_0$; (b) $f(r) = U_0(1 - r^2)$.

11. Work Exercise 2 if the temperature condition at the boundary is replaced by the radiation condition at the boundary given by $U_r(1, t) = -2U(1, t)$.

12. Discuss the cases (a) $h = 0$, (b) $h = \infty$ in the problem of the radiating cylinder on page 624.

B EXERCISES

1. A solid is in the form of an infinitely long quarter cylinder of unit radius and diffusivity κ. The flat sides are insulated while the curved portion is kept at 100°C. Assuming that the temperature initially varies as the fourth power of the distance from the axis, find the temperature at any point at any time.

2. Discuss the modes of the circular drumhead defined by (a) $m = 2, n = 2$; (b) $m = 3, n = 1$.

3. (a) Suppose that the circular drumhead of the text is given a displacement $f(r)$ and then released. Find the displacement at any later time. (b) What is the displacement at any time if $f(r) = \alpha(1 - r^2)$?

4. Work the problem of the circular drumhead if (a) it is initially in equilibrium and is given an initial velocity distribution $g(r, \phi)$; (b) it is given an initial shape and velocity distribution defined by $f(r, \phi)$ and $g(r, \phi)$, respectively.

5. A cylinder of unit radius and length L (see Fig. 14.6) has its convex surface and base in the xy plane kept at temperature zero while the top end is kept at temperature $f(r)$. (a) Write the boundary value problem for the steady-state temperature at any point of the cylinder.

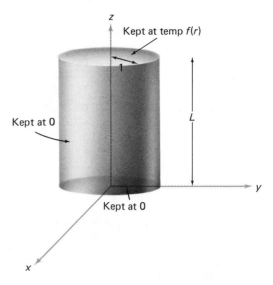

Figure 14.6

(b) Show that the steady-state temperature at any point of the cylinder is given by

$$U(r, z) = \sum_{n=1}^{\infty} a_n \sinh \lambda_n z J_0(\lambda_n r) \quad \text{where} \quad a_n = \frac{2}{[J_1(\lambda_n)]^2 \sinh \lambda_n L} \int_0^1 r f(r) J_0(\lambda_n r) dr$$

and λ_n, $n = 1, 2, \ldots$, are the positive roots of $J_0(x) = 0$. (c) Interpret the results as a problem in potential theory.

6. A cylinder of unit height and radius has its circular ends $z = 0$ and $z = 1$ kept at temperatures 0 and 1 respectively while its convex surface is kept at temperature 1. Find the steady-state temperature everywhere inside the cylinder.

7. Work Exercise 6 if the convex surface is insulated but the other conditions are the same.

8. Work the problem of the circular drumhead if it is "plucked" at its center a distance h and then released.

9. Work the problem of the circular drumhead if the radius is c.

C EXERCISES

1. A circular plate of unit radius and diffusivity κ has its faces insulated and its boundary kept at temperature zero. If the initial temperature is given by $f(r, \phi)$ show that the temperature of any point of the plate at a later time t is given by

$$U(r, \phi, t) = \sum_{m=1}^{\infty} \sum_{n=0}^{\infty} (a_{mn} \cos n\phi + b_{mn} \sin n\phi) e^{-\kappa \lambda_{mn}^2 t} J_n(\lambda_{mn} r)$$

where

$$a_{m0} = \frac{1}{\pi [J_1(\lambda_{m0})]^2} \int_{r=0}^{1} \int_{\phi=0}^{2\pi} r f(r, \phi) J_0(\lambda_{m0} r) dr \, d\phi$$

$$\begin{Bmatrix} a_{mn} \\ b_{mn} \end{Bmatrix} = \frac{2}{\pi [J_{n+1}(\lambda_{mn})]^2} \int_{r=0}^{1} \int_{\phi=0}^{2\pi} r f(r, \phi) J_n(\lambda_{mn} r) \begin{Bmatrix} \cos n\phi \\ \sin n\phi \end{Bmatrix} dr \, d\phi$$

$n = 1, 2, 3, \ldots$, and where λ_{mn} is the mth positive root of $J_n(x) = 0$.

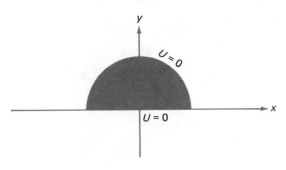

Figure 14.7

2. Work Exercise 1 if the initial temperature is given by $U_0 r \sin \phi$ where U_0 is a constant.

3. A semi-circular plate of unit radius has its boundaries kept at temperature zero and its faces insulated (see Fig. 14.7). If the initial temperature is given by $f(r, \phi) = 100r^2 \cos 2\phi$, find the temperature $U(r, \phi, t)$ at any point inside the region at any time.

4. Solve and interpret physically the boundary value problem

$$\frac{\partial U}{\partial t} = \frac{\partial^2 U}{\partial r^2} + \frac{1}{r}\frac{\partial U}{\partial r} + \alpha r \qquad 0 < r < 1, t > 0$$

$$U(1, t) = 0, \qquad U(r, 0) = 0, \qquad |U(r, t)| < M$$

5. A semi-infinite circular cylinder $z \geq 0, 0 \leq r \leq 1$ has its convex surface $r = 1$ kept at potential zero while its base $z = 0$ is kept at potential $f(r)$. Show that the potential everywhere inside the cylinder is given by

$$V(r, \phi, z) = \sum_{m=1}^{\infty} \sum_{n=1}^{\infty} a_{mn} J_n(\lambda_{mn} r) \sin n\phi \sinh (\lambda_{mn} z)$$

where λ_{mn} is the mth root of the equation $J_n(x) = 0$ and

$$a_{mn} = \frac{4}{\pi(\sinh \lambda_{mn} L)J_{n+1}^2(\lambda_{mn})} \int_{r=0}^{1} \int_{\phi=0}^{\pi} rf(r, \phi) J_n(\lambda_{mn} r) \sin n\phi \, dr$$

6. A solid in the shape of half a circular cylinder of unit radius and length L has its base in the plane as indicated in Fig. 14.8. If the surface $z = L$ is kept at potential $f(r, \phi)$, $0 < r < 1$, $0 < \phi < \pi$ while the remaining surfaces are kept at potential zero show that the potential everywhere inside the cylinder is given by

$$V(r, z) = \sum_{n=1}^{\infty} a_n e^{-\lambda_n z} J_0(\lambda_n r), \qquad a_n = \frac{2}{[J_1(\lambda_n)]^2} \int_0^1 rf(r)J_0(\lambda_n r)dr$$

where λ_n is the nth positive root of $J_0(x) = 0$.

7. A solid circular cylinder is bounded by the surfaces $r = 1$, $z = 0$, and $z = 1$. If $r = 1$ is kept at temperature $f(z)$ while $z = 0$ and $z = 1$ are kept at temperature zero, show that the steady-state temperature is given by

$$U(r, z) = 2 \sum_{n=1}^{\infty} \left(\int_0^1 f(u) \sin n\pi u \, du \right) \frac{I_0(n\pi r)}{I_0(n\pi)} \sin n\pi z$$

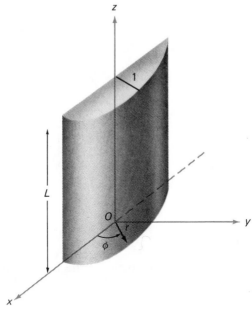

Figure 14.8

where I_0 is the modified Bessel function of order zero given by

$$I_0(x) = J_0(ix) = 1 + \frac{x^2}{2^2} + \frac{x^4}{2^2 4^2} + \frac{x^6}{2^2 4^2 6^2} + \cdots$$

3 Boundary Value Problems Leading to Legendre Functions

3.1 THE LAPLACIAN IN SPHERICAL COORDINATES

In order to deal with boundary value problems involving spheres, it is natural for us to use *spherical coordinates*. These coordinates for the location of any point P are given by (r, ϕ, θ), as indicated in Fig. 14.9.* The values are such that

$$r \geq 0, \qquad 0 \leq \phi < 2\pi, \qquad 0 \leq \theta \leq \pi \tag{1}$$

* There is no standard notation for spherical (or cylindrical) coordinates and as a result various notations are in use. A majority of writers seem to favor the spherical coordinates used in Fig. 14.9 although the coordinates of P are sometimes written (r, θ, ϕ). Others interchange θ and ϕ or use ρ in place of r. Going along with the majority who use Fig. 14.9 it follows that for the special case where point P is in the xy plane, we have $\theta = \pi/2$ and then (r, ϕ) are the plane polar coordinates of P. This is consistent with the notation already used and is also in agreement with the usage (r, ϕ, z) for cylindrical coordinates given on page 620. In the latter the special case $z = 0$ also yields the polar coordinates (r, ϕ) in the xy plane.

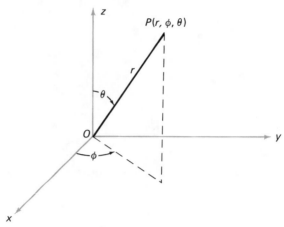

Figure 14.9

The transformation equations between the rectangular and spherical coordinates are given by (see A Exercise 9)

$$x = r \sin \theta \cos \phi, \; y = r \sin \theta \sin \phi, \; z = r \cos \theta \qquad (2)$$

Using these we find that the Laplacian in spherical coordinates is given by (see C Exercise 6, page 549)

$$\nabla^2 U = \frac{1}{r^2}\left[r \frac{\partial^2}{\partial r^2}(rU) + \frac{1}{\sin\theta}\frac{\partial}{\partial\theta}\left(\sin\theta \frac{\partial U}{\partial\theta}\right) + \frac{1}{\sin^2\theta}\frac{\partial^2 U}{\partial\phi^2} \right] \qquad (3)$$

3.2 HEAT CONDUCTION IN A SPHERE

As a first illustration, let us consider a solid metal sphere of unit radius situated as shown in Fig. 14.10. Let us assume that the surface of the top half of the sphere (for which $0 \le \theta < \pi/2$) is kept at some constant temperature $U_0 \ne 0$, while the bottom half (for which $\pi/2 < \theta \le \pi$) is kept at temperature zero. To make this physically possible we can suppose for example that the top and bottom halves of the surface are separated slightly or insulated from each other. As a result of heat conduction the sphere will ultimately reach some steady-state temperature distribution which will depend, because of the way in which we have chosen the sphere's orientation in Fig. 14.10, on the variables r and θ but not on ϕ and t. We would like to determine this steady-state temperature.

Mathematical Formulation. The partial differential equation which we must solve is Laplace's equation $\nabla^2 U = 0$. Using (3) and the fact that U depends only on r and θ, i.e., $U = U(r, \theta)$, this equation is

$$r \frac{\partial^2}{\partial r^2}(rU) + \frac{1}{\sin\theta}\frac{\partial}{\partial\theta}\left(\sin\theta \frac{\partial U}{\partial\theta}\right) = 0 \qquad (4)$$

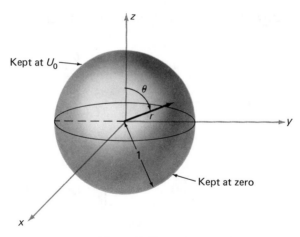

Kept at U_0

θ

r

z

y

Kept at zero

x

Figure 14.10

The boundary conditions are
$$U(1, \theta) = \begin{cases} U_0, & 0 \leq \theta < \pi/2 \\ 0, & \pi/2 \leq \theta \leq \pi \end{cases} \tag{5}$$

together with the obvious one that U is bounded inside the sphere.

Solution Assume the separable solution to (4) given by $U = R\Theta$, where R depends only on r and Θ depends only on θ. Then (4) becomes

$$r\Theta \frac{d^2}{dr^2}(rR) + \frac{R}{\sin\theta}\frac{d}{d\theta}\left(\sin\theta\frac{d\Theta}{d\theta}\right) = 0 \quad \text{or} \quad \frac{r}{R}\frac{d^2}{dr^2}(rR) = -\frac{1}{\Theta\sin\theta}\frac{d}{d\theta}\left(\sin\theta\frac{d\Theta}{d\theta}\right) \tag{6}$$

on dividing by $R\Theta$. Since the second equation in (6) has one side dependent only on r and the other only on θ, each side must be a constant which we take as $-\lambda^2$. We are thus led to the equations

$$r\frac{d^2}{dr^2}(rR) + \lambda^2 R = 0, \qquad \frac{d}{d\theta}\left(\sin\theta\frac{d\Theta}{d\theta}\right) - \lambda^2\Theta\sin\theta = 0 \tag{7}$$

The first equation in (7) can be written as

$$r^2 R'' + 2rR' + \lambda^2 R = 0 \tag{8}$$

and is recognized as an Euler equation having solution

$$R = c_1 r^n + \frac{c_2}{r^{n+1}}, \qquad \text{where } \lambda^2 = -n(n + 1) \tag{9}$$

The second equation in (7) is more complicated. Because of the presence of $\sin\theta$ in the equation we might be led to try the transformation from the independent variable θ to another independent variable x given by $x = \sin\theta$. However this does not lead to anything useful. A better choice is provided by the transformation

$$x = \cos\theta \tag{10}$$

In such case we have from the chain rule for differentiation

$$\frac{d\Theta}{d\theta} = \frac{d\Theta}{dx} \cdot \frac{dx}{d\theta} = -\sin\theta \frac{d\Theta}{dx} \quad \text{so that} \quad \sin\theta \frac{d\Theta}{d\theta} = -\sin^2\theta \frac{d\Theta}{dx} = (x^2 - 1)\frac{d\Theta}{dx}$$

Then $\quad \dfrac{d}{d\theta}\left(\sin\theta \dfrac{d\Theta}{d\theta}\right) = \dfrac{d}{dx}\left[(x^2 - 1)\dfrac{d\Theta}{dx}\right]\dfrac{dx}{d\theta} = \sin\theta \dfrac{d}{dx}\left[(1 - x^2)\dfrac{d\Theta}{dx}\right]$

Using this last result in the second of equations (7) together with $\lambda^2 = -n(n + 1)$, it can be written

$$\frac{d}{dx}\left[(1 - x^2)\frac{d\Theta}{dx}\right] + n(n + 1)\Theta = 0 \tag{11}$$

This is *Legendre's differential equation*, which we have already investigated in Chapter Seven. Its solution is given by

$$\Theta = c_3 P_n(x) + c_4 Q_n(x) \tag{12}$$

where $P_n(x)$ and $Q_n(x)$ are Legendre functions. From (9) and (12) we see that

$$U(r, \theta) = \left[c_1 r^n + \frac{c_2}{r^{n+1}}\right][c_3 P_n(x) + c_4 Q_n(x)] \tag{13}$$

Since this solution becomes infinite for $r = 0$, i.e., at the center of the sphere, we must choose $c_2 = 0$. However, this is not enough to satisfy the boundedness condition for U, because the Legendre functions also become infinite for $x = \pm 1$ or equivalently for $\theta = 0, \pi$. We can obtain boundedness if we take n as a non-negative integer, i.e., $n = 0, 1, 2, \ldots$, in which case the functions $P_n(x)$ are the Legendre polynomials, and choose $c_4 = 0$ to avoid the functions $Q_n(x)$ which are unbounded for $x = \pm 1$ or $\theta = 0, \pi$. Thus (13) gives

$$U(r, \theta) = A r^n P_n(x) \tag{14}$$

on taking $A = c_1 c_3$. By superposition of solutions having the form (14), we have

$$U(r, \theta) = \sum_{n=0}^{\infty} a_n r^n P_n(x) \tag{15}$$

We can now determine the coefficients a_n from the boundary condition (5). To do this let us first note that in terms of $x = \cos\theta$ the boundary condition (5) is

$$U(1, \theta) = f(x) = \begin{cases} U_0, & 0 < x \leq 1 \\ 0, & -1 \leq x < 0 \end{cases} \tag{16}$$

or using (15) $$f(x) = \sum_{n=0}^{\infty} a_n P_n(x) \tag{17}$$

This, however, requires the expansion of $f(x)$ into a series of Legendre polynomials or *Legendre series*, already done in Chapter Eight, page 402. As in the case of Fourier and Bessel series, we multiply both sides of (17) by a suitable function, in this case

$P_m(x)$, and integrate both sides from $x = -1$ to 1 to obtain

$$\int_{-1}^{1} f(x)P_m(x)dx = \sum_{n=0}^{\infty} a_n \int_{-1}^{1} P_m(x)P_n(x)dx \tag{18}$$

Since

$$\int_{-1}^{1} P_m(x)P_n(x)dx = 0, \qquad m \neq n \tag{19}$$

the series on the right of (18) reduces to only one term, where $m = n$, i.e.,

$$\int_{-1}^{1} f(x)P_n(x)dx = a_n \int_{-1}^{1} [P_n(x)]^2 \, dx \quad \text{or} \quad a_n = \frac{\int_{-1}^{1} f(x)P_n(x)dx}{\int_{-1}^{1} [P_n(x)]^2 \, dx} \tag{20}$$

Since

$$\int_{-1}^{1} [P_n(x)]^2 \, dx = \frac{2}{2n + 1}$$

(20) reduces to

$$a_n = \frac{2n + 1}{2} \int_{-1}^{1} f(x)P_n(x)dx \tag{21}$$

Using (16) this gives

$$a_n = \frac{2n + 1}{2} U_0 \int_{0}^{1} P_n(x)dx$$

Putting $n = 0, 1, 2, \ldots$, using the first few Legendre polynomials,

$$P_0(x) = 1, \qquad P_1(x) = x, \qquad P_2(x) = \frac{3x^2 - 1}{2}, \qquad P_3(x) = \frac{5x^3 - 3x}{2}, \ldots$$

we find

$$a_0 = \frac{U_0}{2}, \qquad a_1 = \frac{3U_0}{4}, \qquad a_2 = 0, \qquad a_3 = -\frac{7U_0}{16}, \ldots$$

Thus

$$U(r, \theta) = \frac{U_0}{2} \left[1 + \tfrac{3}{2}rP_1(x) - \tfrac{7}{8}r^3 P_3(x) + \cdots \right] \tag{22}$$

Remark 1. The general term of this series can be found by making use of the result (93) obtained on page 404. Using this we find

$$U(r, \theta) = \frac{U_0}{2} + \tfrac{3}{4}U_0 rP_1(x) + \frac{U_0}{2} \sum_{n=3,5,7,\ldots} (-1)^{(n-1)/2} \frac{1 \cdot 3 \cdots (n - 2)}{2 \cdot 4 \cdots (n + 1)} r^n P_n(x) \tag{23}$$

where only terms corresponding to $n = 3, 5, 7, \ldots$ occur in the last summation.

3.3 ELECTRICAL OR GRAVITATIONAL POTENTIAL DUE TO A SPHERE

The problem just solved can be interpreted as one in potential theory in which there is an electric charge or mass so distributed as to give the top and bottom halves of the spherical surface potentials U_0 and 0, respectively. The potential inside the sphere is then given by (23).

In such case it is also natural to ask for the potential outside the sphere. To determine this we must solve Laplace's equation (4) for the region $r > 1, 0 \leq \theta \leq \pi$ where $U(r, \theta)$ reduces to (5) for $r = 1$. To satisfy the boundedness in the region $r > 1$, we must choose $c_1 = 0$ in (13), otherwise the solution would become infinite as $r \to \infty$. Similarly, since we must have boundedness for $\theta = 0, \pi$, i.e., $x = \pm 1$, we must choose

$c_4 = 0$ as before. This leads to the solution

$$U(r, \theta) = \frac{A}{r^{n+1}} P_n(x)$$

or by superposition $$U(r, \theta) = \sum_{n=0}^{\infty} \frac{a_n}{r^{n+1}} P_n(x) \tag{24}$$

We must now determine the coefficients a_n so that (24) reduces to (16) for $r = 1$. This leads to (17). Thus we get the same values of a_n as before, i.e., those given by (21). Using these in (24), we see that the required potential outside the sphere is given by

$$U(r, \theta) = \frac{U_0}{2r} + \frac{3}{4} \frac{U_0}{r^2} P_1(x) + \frac{U_0}{2} \sum_{n=3,5,7,\dots} (-1)^{(n-1)/2} \frac{1 \cdot 3 \cdots (n-2)}{2 \cdot 4 \cdots (n+1)} \frac{P_n(x)}{r^{n+1}} \tag{25}$$

Remark 2. The result (25) also has a heat flow interpretation. It represents the steady-state temperature in an infinite conducting region having a "spherical hole" of unit radius, where the boundary of the hole is kept at temperature given by (16).

A EXERCISES

1. The surface of a sphere of radius c with center at the origin is charged so that its potential is constant and equal to V_0. (a) Find the potential inside the sphere. (b) Find the potential outside the sphere. (c) What charge q on the surface is needed to produce the potential in (b)?

2. Suppose that the top half of the spherical surface shown in Fig. 14.10, page 635, is charged to potential V_0 while the bottom half is charged to potential $-V_0$, where V_0 is a constant. Find the potential (a) inside and (b) outside the sphere.

3. Show that at points very far from the spherical surface of Exercise 2 the potential is to a high degree of approximation given by $V = 3V_0 \cos \theta / 2r^2$.

4. Find the steady-state temperature inside a uniform spherical conducting solid of unit radius situated as in Fig. 14.10, page 635, if the surface temperature is given by $40 \cos \theta - 20 \cos^3 \theta$.

5. Find the potential (a) inside and (b) outside the unit sphere of Fig. 14.10, page 635, if the potential on the surface is given by $V_0 \cos 2\theta$. Give an interpretation involving heat flow.

6. Find the potential (a) inside and (b) outside of a spherical surface of unit radius whose upper half is kept at potential $V_0 \cos \theta$ and whose lower half is kept at potential zero.

7. A solid hemisphere of unit radius has its base in the xy plane and center at the origin. If the base is insulated and the convex surface is kept at temperature $60 \cos \theta$, find the temperature everywhere in the solid.

8. Derive the transformation equations (2), page 634, for spherical coordinates.

B EXERCISES

1. A spherical shell has an inner radius r_1, and outer radius r_2. The temperatures of the inner and outer surfaces are given respectively, by $F_1(x)$ and $F_2(x)$, where $x = \cos \theta$. Show that the temperature at any point of the shell is given by

$$U(r, \theta) = \sum_{n=0}^{\infty} \left(a_n r^n + \frac{b_n}{r^{n+1}} \right) P_n(x)$$

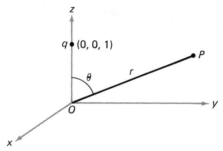

Figure 14.11

where a_n and b_n are determined by solving simultaneously the equations

$$a_n r_1^n + \frac{b_n}{r_1^{n+1}} + \frac{2n+1}{2} \int_{-1}^{1} F_1(x)P_n(x)dx, \quad a_n r_2^n + \frac{b_n}{r_2^{n+1}} = \frac{2n+1}{2} \int_{-1}^{1} F_2(x)P_n(x)dx$$

2. Work Exercise 1 if $r_1 = 1$, $r_2 = 2$, $F_1(x) = 100 - 20x$, and $F_2(x) = 50 + 30x$.

3. A charge q is located at the point $(0, 0, 1)$ of a rectangular coordinate system (see Fig. 14.11). Show that the potential due to this charge at any point P distant r from the origin is given by

$$V = \frac{q}{\sqrt{1 - 2r\cos\theta + r^2}} = q \sum_{n=0}^{\infty} r^n P_n(\cos\theta)$$

Discuss the relationship with the generating function for Legendre polynomials (page 345).

4. Two charges q and $-q$ (see Fig. 14.12) are located at points A and B, respectively, at a distance d from each other. If $OP = r$ is the distance of any point P from the midpoint O of AB, and θ is the angle between OP and OA, show that the potential at P is given very nearly by

$$V = \frac{p\cos\theta}{r^2}$$

where $p = qd$, The combination of two opposite charges q and $-q$ is often called a *dipole* and the product $p = qd$ is called the *dipole moment*.

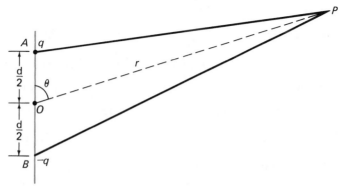

Figure 14.12

5. Use Exercise 4 to show that the spherical charge distribution of A Exercise 2 is equivalent to a dipole having dipole moment $3V_0/2$.

6. Show that the force on a unit charge placed at point P of the dipole of Fig. 14.12 has magnitude

$$\frac{p}{r^3}\sqrt{1 + 3\cos^2 \theta}$$

Use the result to find the magnitude of the force due to the charge distribution of A Exercise 2. (*Hint*: The relationship between force F and potential V is given by $F = -\nabla V$, i.e., the force is the negative of the gradient of the potential.)

7. Describe the surfaces of equal potential or *equipotential surfaces* at distances which are far from the charge distribution of A Exercise 2. What are the orthogonal trajectories corresponding to these surfaces and explain their physical significance?

C EXERCISES

1. If Laplace's equation is written in spherical coordinates, the result is

$$r\frac{\partial^2}{\partial r^2}(rU) + \frac{1}{\sin\theta}\frac{\partial}{\partial\theta}\left(\sin\theta\frac{\partial U}{\partial\theta}\right) + \frac{1}{\sin^2\theta}\frac{\partial^2 U}{\partial\phi^2} = 0$$

using (3) on page 634. (a) Making the assumption that the variables are separable so that $U = R\Theta\Phi$, where R, Θ, Φ are functions of only r, θ, ϕ, respectively, show that

$$r^2 R'' + 2rR' - n(n + 1)R, \qquad \Phi'' + m^2\Phi = 0$$

$$\sin^2\theta\Theta'' + \sin\theta\cos\theta\Theta' + [n(n + 1)\sin^2\theta - m^2]\Theta = 0$$

(b) Show that the third equation in (a) becomes

$$(1 - x^2)\frac{d^2\Theta}{dx^2} - 2x\frac{d\Theta}{dx} + \left[n(n + 1) - \frac{m^2}{1 - x^2}\right]\Theta = 0$$

on letting $x = \cos\theta$. This equation which reduces to Legendre's equation for the case $m = 0$ is often called *Legendre's associated equation.*

2. (a) Show that if $m = 0, 1, 2, \ldots$ a solution to Legendre's associated equation in Exercise 1(b) is given by

$$P_n^m(x) = (1 - x^2)^{m/2}\frac{d^m}{dx^m}P_n(x)$$

which are called *associated Legendre functions*. What are these if (i) $m = 0$; (ii) $m > n$?

(b) Show that for any given integer $n \geq 0$, $\quad \int_{-1}^{1} P_n^m(x)P_k^m(x)dx = 0, \qquad m \neq k$

i.e., the functions $P_n^m(x)$, $m = 0, 1, 2, \ldots$, are mutually orthogonal in $-1 \leq x \leq 1$.

(c) Verify for some special cases the result $\quad \int_{-1}^{1} [P_n^m(x)]^2\, dx = \frac{2}{2n + 1}\frac{(n + m)!}{(n - m)!}$

Can you prove this in general?

(d) Show by using the preceding results how to expand a given function in a series of associated Legendre functions and illustrate by an example.

3. Use Exercises 1 and 2 to show that a solution of Laplace's equation in spherical coordinates which is bounded for $r = 0$ and $-1 \leq x \leq 1$ is

$$U = \sum_{m=0}^{\infty} \sum_{n=0}^{\infty} r^n (a_{mn} \cos m\phi + b_{mn} \sin m\phi) P_n^m(\cos \theta)$$

4. Show how the orthogonality of the associated Legendre functions can be used to solve various problems involving spheres in which there may be ϕ dependence as well as r and θ dependence. Illustrate by making up an exercise and solving it.

5. How would you modify the solution of Exercise 2 in the cases where (a) U must be bounded at $r = \infty$; (b) the region for which we seek U is given by $r_1 < r < r_2$?

6. Make up and solve exercises illustrating the cases in Exercise 5.

4 Miscellaneous Problems

There are many kinds of problems arising in science and engineering which require use of Bessel or Legendre functions and sometimes a combination of these. The problems need not necessarily involve the geometry of cylinders or spheres associated with such problems. Also, in many cases the differential equations which arise from their mathematical formulations are not quite the standard ones given in Chapter Twelve. We devote this section to three examples of such problems, the first two being rather harmless ones involving vibrations of a chain and the potential due to a circular wire, and the third an atomic bomb problem.

4.1 THE PROBLEM OF THE VIBRATING CHAIN

Let us suppose, as indicated in Fig. 14.13, that we have a uniform chain of length L and mass M suspended vertically from a fixed point O. Suppose that each point of the chain is given a small horizontal displacement at time $t = 0$, and then

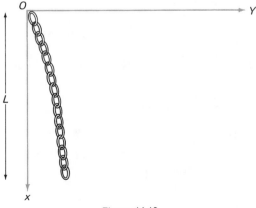

Figure 14.13

the chain is released. As a result the chain will undergo vibrations, and we would, as in the cases of the vibrating string or vibrating drumhead, like to investigate these vibrations. Since this problem is at least as complicated as the problem of a vibrating simple pendulum, for which difficulties arise when vibrations are not small, we shall assume small vibrations for the chain.

Mathematical Formulation. It is convenient to choose a rectangular coordinate system having origin at O and with positive x and y axes as indicated in Fig. 14.13. The equilibrium position of the chain is coincident with the x axis. The transverse or horizontal displacement of each point x of the chain at time t can be denoted by $Y(x, t)$ or simply Y. It is this function which we wish to determine.

Since the chain has mass, there will be a tension at each point x which is equal to the weight of that portion of the chain below x. If we denote by $\rho = M/L$ the mass per unit length of the chain, then since the length of that portion of the chain below x is $L - x$, the mass of this portion is $\rho(L - x)$. Thus the weight or tension is $\tau = \rho g(L - x)$. Now in Chapter Twelve, page 560, when we formulated the equation for a vibrating string, we allowed for the possibility that the string have variable tension. That same equation of motion will apply to our chain problem, provided that we assume conditions of small displacement and slope, which were assumed in that derivation. In this case the equation of motion is

$$\frac{\partial}{\partial x}\left\{[\rho g(L - x)]\frac{\partial Y}{\partial x}\right\} = \rho\frac{\partial^2 Y}{\partial t^2} \quad \text{or} \quad g\frac{\partial}{\partial x}\left\{(L - x)\frac{\partial Y}{\partial x}\right\} = \frac{\partial^2 Y}{\partial t^2} \tag{1}$$

Since end $x = 0$ is fixed, $\qquad Y(0, t) = 0, \qquad t > 0$ (2)

If we assume that initially each point x of the chain is given a displacement $f(x)$, we have the condition

$$Y(x, 0) = f(x), \qquad 0 < x < L \tag{3}$$

Also, the fact that the chain is released so that each point has velocity zero leads to

$$Y_t(x, 0) = 0, \qquad 0 < x < L \tag{4}$$

In addition we must also impose the usual requirement of boundedness.

Solution   Letting $Y(x, t) = X(x)T(t)$ in (1), and separating the variables, we obtain

$$\frac{1}{X}\frac{d}{dx}\left\{(L - x)\frac{dX}{dx}\right\} = \frac{1}{gT}\frac{d^2 T}{dt^2} \tag{5}$$

Setting each side equal to the constant $-\lambda^2$, we obtain the equations

$$\frac{d}{dx}\left\{(L - x)\frac{dX}{dx}\right\} + \lambda^2 X = 0, \qquad \frac{d^2 T}{dt^2} + \lambda^2 gT = 0 \tag{6}$$

The second equation is easy to solve, and we find

$$T = c_1 \cos \lambda\sqrt{g}\,t + c_2 \sin \lambda\sqrt{g}\,t \tag{7}$$

The first equation can be written in the form

$$(L - x)\frac{d^2 X}{dx^2} - \frac{dX}{dx} + \lambda^2 X = 0 \tag{8}$$

but it is certainly not obvious what its solution is. The presence of $L - x$ in the coefficient of d^2X/dx^2 suggests the transformation $L - x = u$. This yields

$$u \frac{d^2X}{du^2} + \frac{dX}{du} + \lambda^2 X = 0 \qquad (9)$$

We can solve this equation by using the method of Frobenius (see Chapter Seven), but we shall instead employ another method involving transformation of variables which is often useful in solving various types of differential equations. According to this method, we change the independent variable from u to v by using the transformation $u = cv^\alpha$, where at present we do not know what values to choose for the constants c and α, but hope to choose them judiciously so as to obtain a differential equation which we may recognize. With this transformation we find

$$\frac{dX}{du} = \frac{dX}{dv} \frac{dv}{du} = \frac{dX/dv}{du/dv} = \frac{1}{c\alpha v^{\alpha-1}} \frac{dX}{dv}$$

$$\frac{d^2X}{du^2} = \frac{d}{du}\left(\frac{dX}{du}\right) = \frac{d}{du}\left(\frac{1}{c\alpha v^{\alpha-1}} \frac{dX}{dv}\right) = \frac{\frac{d}{dv}\left(\frac{1}{c\alpha v^{\alpha-1}} \frac{dX}{dv}\right)}{du/dv} = \frac{d}{dv}\left(\frac{1}{c\alpha v^{\alpha-1}} \frac{dX}{dv}\right)\left(\frac{1}{c\alpha v^{\alpha-1}}\right)$$

$$= \frac{1}{c^2\alpha^2}\left\{v^{2-2\alpha} \frac{d^2X}{dv^2} + (1-\alpha)v^{1-2\alpha} \frac{dX}{dv}\right\}$$

Substituting these into equation (9) and simplifying, we obtain

$$\frac{v^{2-\alpha}}{c\alpha^2} \frac{d^2X}{dv^2} + \frac{v^{1-\alpha}}{c\alpha^2} \frac{dX}{dv} + \lambda^2 X = 0$$

or

$$\frac{d^2X}{dv^2} + \frac{1}{v} \frac{dX}{dv} + \frac{c\alpha^2\lambda^2}{v^{2-\alpha}} X = 0 \qquad (10)$$

If we now make the judicious choice $\alpha = 2$ and $c\alpha^2\lambda^2 = 1$ or $c = \dfrac{1}{4\lambda^2}$

(10) becomes

$$\frac{d^2X}{dv^2} + \frac{1}{v} \frac{dX}{dv} + X = 0 \quad \text{or} \quad v \frac{d^2X}{dv^2} + \frac{dX}{dv} + vX = 0 \qquad (11)$$

which we recognize as Bessel's equation of zero order having general solution

$$X = c_3 J_0(v) + c_4 Y_0(v) \qquad (12)$$

Recalling now that $\quad u = cv^\alpha = \dfrac{v^2}{4\lambda^2} \quad$ or $\quad v = 2\lambda\sqrt{u}$

and that $u = L - x$, we find $v = 2\lambda\sqrt{L-x}$, and so (12) can be written

$$X = c_3 J_0(2\lambda\sqrt{L-x}) + c_4 Y_0(2\lambda\sqrt{L-x}) \qquad (13)$$

which is the general solution of the first of equations (6).

Thus a possible solution to our boundary value problem has the form

$$Y(x, t) = \{c_3 J_0(2\lambda\sqrt{L-x}) + c_4 Y_0(2\lambda\sqrt{L-x})\}\{c_1 \cos(\lambda\sqrt{g}t) + c_2 \sin(\lambda\sqrt{g}t)\} \qquad (14)$$

Since the solution must be bounded, in particular at $x = L$, we must choose $c_4 = 0$ in (14). Also by condition (4) we must choose $c_2 = 0$. Thus the solution so far is

$$Y(x, t) = AJ_0(2\lambda\sqrt{L - x}) \cos(\lambda\sqrt{g}t) \tag{15}$$

where we have written $A = c_3c_1$.

The condition (2) yields $\qquad J_0(2\lambda\sqrt{L}) = 0 \tag{16}$

Denoting the positive roots of this equation by $\lambda_1, \lambda_2, \ldots$, we have

$$2\lambda\sqrt{L} = \lambda_1, \lambda_2, \ldots, \quad \text{or} \quad \lambda = \frac{\lambda_1}{2\sqrt{L}}, \frac{\lambda_2}{2\sqrt{L}}, \ldots \tag{17}$$

This leads to the solution $\qquad a_nJ_0\left(\lambda_n\sqrt{1 - \dfrac{x}{L}}\right) \cos\left(\dfrac{\lambda_n}{2}\sqrt{\dfrac{g}{L}}\,t\right) \tag{18}$

Superimposing these solutions in preparation for the condition (3), we find the possible solution

$$Y(x, t) = \sum_{n=1}^{\infty} a_nJ_0\left(\lambda_n\sqrt{1 - \frac{x}{L}}\right) \cos\left(\frac{\lambda_n}{2}\sqrt{\frac{g}{L}}\,t\right) \tag{19}$$

Condition (3) requires that $\qquad f(x) = \sum_{n=1}^{\infty} a_nJ_0\left(\lambda_n\sqrt{1 - \dfrac{x}{L}}\right) \tag{20}$

To find the coefficients a_n in (20), it is convenient to let $\sqrt{1 - x/L} = u$ so that $x = L[1 - u^2]$. Then (20) becomes

$$f(L[1 - u^2]) = F(u) = \sum_{n=1}^{\infty} a_nJ_0(\lambda_nu) \tag{21}$$

Hence $\qquad a_n = \dfrac{\displaystyle\int_0^1 uF(u)J_0(\lambda_nu)\,du}{\displaystyle\int_0^1 u[J_0(\lambda_nu)]^2\,du} = \dfrac{2}{[J_1(\lambda_n)]^2} \int_0^1 uF(u)J_0(\lambda_nu)\,du \tag{22}$

Using these coefficients in (19), we arrive at the required solution to our problem.

Interpretation. The terms in (19) corresponding to $n = 1, 2, \ldots$ represent the various modes of vibration of the chain. If the chain is vibrating in the mode corresponding to $n = 1$, the coefficient a_1 in (19) is not zero, while all the other coefficients $a_2, a_3, \ldots$ are zero. In this case the shape of the chain at $t = 0$ is given by $a_1J_0(\lambda_1\sqrt{1 - x/L})$. Then as time goes by the chain vibrates as a whole about the equilibrium position. The frequency of this lowest mode, sometimes called the *fundamental frequency* by analogy with the vibrating string, is given by

$$f_1 = \frac{\lambda_1}{4\pi}\sqrt{\frac{g}{L}} = \frac{2.405}{4\pi}\sqrt{\frac{g}{L}} \tag{23}$$

Similarly, the frequency corresponding to the nth mode is given by

$$f_n = \frac{\lambda_n}{4\pi}\sqrt{\frac{g}{L}}, \quad n = 1, 2, 3, \ldots \tag{24}$$

The frequencies corresponding to $n = 2, 3, \ldots$, sometimes called the *second, third, ...* *harmonics*, are given approximately by

$$f_2 = \frac{5.520}{4\pi} \sqrt{\frac{g}{L}}, \qquad f_3 = \frac{8.654}{4\pi} \sqrt{\frac{g}{L}}, \ldots \tag{25}$$

It is possible to show that these higher frequencies are not integer multiples of the fundamental frequency, so that in a general motion of the chain, which corresponds to a superposition of the various modes, we would not expect to hear "music."

It is of interest to compare the period of a vibrating chain in its lowest mode with that of a simple pendulum. In the case of a simple pendulum of length L, the period is given by

$$T_p = 2\pi \sqrt{\frac{L}{g}} \tag{26}$$

In the case of the chain, this is given by

$$T_c = \frac{4\pi}{2.405} \sqrt{\frac{L}{g}} \tag{27}$$

Hence the period for the chain is slightly smaller than that for the simple pendulum.

4.2 ELECTRICAL POTENTIAL DUE TO A UNIFORMLY CHARGED CIRCULAR WIRE

Consider a circular wire of unit radius which is given a constant electric charge density ρ. We wish to determine the electrical potential due to this charge distribution. Clearly, the problem could also be interpreted as one involving gravitational potential, in which case ρ is a mass density rather than a charge density.

Mathematical Formulation. Let us choose the wire in the xy plane, as shown in Fig. 14.14, so that there will be no ϕ dependence. The potential V at any point P must then satisfy Laplace's equation with no ϕ dependence, as given by equation (4) on page 634. There is, however, no boundary condition evident in this problem other than the usual one involving boundedness, and it might thus appear that we cannot proceed further. We reason nevertheless that, if somehow we can determine the potential for some set of points, this might serve as a boundary condition. We choose as the set of points a line perpendicular to the plane of the circular wire (the xy plane) and passing through the center of this circle, i.e., the z axis.

Since ρ is the charge per unit length of wire, the charge on an arc of the circular wire of length ds is $\rho \, ds$. The potential at point P due to this charge is

$$\frac{\rho \, ds}{\sqrt{z^2 + 1}} = \frac{\rho(1 \, d\phi)}{\sqrt{z^2 + 1}} = \frac{\rho \, d\phi}{\sqrt{z^2 + 1}}$$

Hence, by integration, the potential at P due to the entire circular wire is

$$\int_{\phi=0}^{2\pi} \frac{\rho \, d\phi}{\sqrt{z^2 + 1}} = \frac{2\pi\rho}{\sqrt{z^2 + 1}}$$

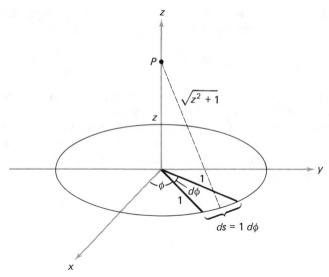

Figure 14.14

Since V depends only on r and θ, we can take as our boundary condition for $V(r, \theta)$ the condition that for $r = z$ and $\theta = 0$

$$V(z, 0) = \frac{2\pi\rho}{\sqrt{z^2 + 1}} \tag{28}$$

We shall consider $z > 0$. The case $z < 0$ is obtained by symmetry.

Solution Two cases must be considered.

Case 1, $0 < r < 1$. In this case, as we have already noted, a possible solution is given by [see (15) on page 636],

$$V(r, \theta) = \sum_{n=0}^{\infty} a_n r^n P_n(\cos \theta)$$

Then using the boundary condition (28) and the fact that $P_n(1) = 1$, we are led to the requirement

$$V(z, 0) = \sum_{n=0}^{\infty} a_n z^n = \frac{2\pi\rho}{\sqrt{z^2 + 1}} \tag{29}$$

To find a_n we must expand the function on the right into a power series in z. This can be done either by using the method of Taylor series or more easily by the binomial theorem. Using the latter approach, we find

$$\frac{2\pi\rho}{\sqrt{1 + z^2}} = 2\pi\rho(1 + z^2)^{-1/2} = 2\pi\rho \left\{ 1 + (-\tfrac{1}{2})z^2 + \frac{(-\tfrac{1}{2})(-\tfrac{3}{2})}{2!} z^4 + \cdots \right\}$$

$$= 2\pi\rho \left\{ 1 - \tfrac{1}{2}z^2 + \frac{1 \cdot 3}{2 \cdot 4} z^4 - \frac{1 \cdot 3 \cdot 5}{2 \cdot 4 \cdot 6} z^6 + \cdots \right\} \tag{30}$$

Comparing (29) and (30), we see that

$$a_0 = 2\pi\rho, \qquad a_1 = 0, \qquad a_2 = -\pi\rho, \qquad a_3 = 0, \qquad a_4 = \tfrac{3}{4}\pi\rho, \ldots$$

or

$$a_{2k} = (-1)^k 2\pi\rho \left[\frac{1 \cdot 3 \cdots (2k-1)}{2 \cdot 4 \cdots 2k}\right], \qquad a_{2k+1} = 0, \qquad k = 1, 2, \ldots \quad (31)$$

Hence

$$V(r, \theta) = 2\pi\rho\, \{1 - \tfrac{1}{2}r^2 P_2(\cos\theta) + \tfrac{3}{8}r^4 P_4(\cos\theta) - \cdots\}$$

or

$$V(r, \theta) = 2\pi\rho \left\{1 + \sum_{k=1}^{\infty} (-1)^k \left[\frac{1 \cdot 3 \cdots (2k-1)}{2 \cdot 4 \cdots 2k}\right] r^{2k} P_{2k}(\cos\theta)\right\} \quad (32)$$

Case 2, $r > 1$. In this case we must take $c_1 = 0$ in the solution (13) on page 636 because the solution must be bounded as $r \to \infty$. Also, as before, we choose $c_4 = 0$. Then we have

$$V(r, \theta) = \sum_{n=0}^{\infty} \frac{b_n}{r^{n+1}} P_n(\cos\theta) \quad (33)$$

The boundary condition (28) yields the requirement

$$V(z, 0) = \sum_{n=0}^{\infty} \frac{b_n}{z^{n+1}} = \frac{2\pi\rho}{\sqrt{z^2 + 1}} \quad (34)$$

To find b_n we must expand the function on the right into a power series in $1/z$. Using the binomial theorem once again, we have

$$\frac{2\pi\rho}{\sqrt{z^2 + 1}} = \frac{2\pi\rho}{z}\left(1 + \frac{1}{z^2}\right)^{-1/2}$$

$$= \frac{2\pi\rho}{z}\left\{1 - \frac{1}{2}\left(\frac{1}{z^2}\right) + \frac{1 \cdot 3}{2 \cdot 4}\left(\frac{1}{z^4}\right) - \frac{1 \cdot 3 \cdot 5}{2 \cdot 4 \cdot 6}\left(\frac{1}{z^6}\right) + \cdots\right\} \quad (35)$$

Then comparison of (34) and (35) shows that

$$b_0 = 2\pi\rho, \qquad b_1 = 0, \qquad b_2 = -\pi\rho, \qquad b_3 = 0, \qquad b_4 = \tfrac{3}{4}\pi\rho, \ldots \quad (36)$$

and by substitution in (33) the required solution is seen to be

$$V(r, \theta) = \frac{2\pi\rho}{r}\left\{1 - \frac{1}{2}\frac{P_2(\cos\theta)}{r^2} + \frac{3}{8}\frac{P_4(\cos\theta)}{r^4} - \cdots\right\} \quad (37)$$

or

$$V(r, \theta) = \frac{2\pi\rho}{r}\left\{1 + \sum_{k=1}^{\infty} (-1)^k \left[\frac{1 \cdot 3 \cdots (2k-1)}{2 \cdot 4 \cdots 2k}\right] \frac{P_{2k}(\cos\theta)}{r^{2k}}\right\} \quad (38)$$

Remark 3. From (38) we see that a very good approximation to the potential far from the wire is given by

$$V(r, \theta) = \frac{2\pi\rho}{r} - \frac{\pi\rho P_2(\cos\theta)}{r^3} = \frac{2\pi\rho}{r} - \frac{\pi\rho(3\cos^2\theta - 1)}{2r^3} \quad (39)$$

the second term supplying a correction indicating the angular dependence. If the distance is large enough, this angular dependence is negligible, and the result is for all practical purposes given by q/r, where $q = 2\pi\rho$ is the total charge on the wire and agrees with our expectation.

4.3 THE ATOMIC BOMB PROBLEM

If the nucleus of a uranium atom is bombarded with neutrons, it may happen that the nucleus "captures" a neutron. When this occurs, the nucleus splits into two parts, resulting in the release of neutrons already present in the nucleus and also liberating a relatively large amount of energy. This process is called *nuclear fission*. The neutrons which are released in this fashion can in turn bombard nuclei of other uranium atoms, resulting in further nuclear fission. This results in a *chain reaction* in which larger and larger amounts of energy can be released.

If the process can be made to take place in a short enough interval of time, there will be a tremendous burst of energy and an explosion will take place.

This, in brief, is the basis of the atomic bomb. However, just as we have already observed that the principle of resonance can be used for either constructive or destructive purposes, so it turns out that nature has given us the choice of employing nuclear fission constructively or destructively. For by "properly controlling" the chain reaction in a *nuclear reactor*, large amounts of energy can be obtained for long periods of time. This energy can provide us with electrical power so vital to our civilization.

Although it is obvious that with the above information we are not able to rush right out and build ourselves an atomic bomb or a nuclear reactor, it is of interest that with a mathematical formulation of the simple facts presented above we can arrive at some important ideas.

To begin somewhere, let us suppose that by some means we have succeeded in introducing neutrons into a mass of fissionable material which occupies some type of container, such as a cylinder, sphere, etc. This source of neutrons produces nuclear fission, which in turn produces more neutrons and more nuclear fission, etc. We would like to formulate this mathematically.

Mathematical Formulation. Let us suppose for our present purposes that the container is a cylinder* having height h and radius a situated in a rectangular co-ordinate system so that its base is in the xy plane and its axis coincides with the z axis, as shown in Fig. 14.15. Now the fact that neutrons are in the process of traveling in this cylinder is analogous to diffusion of a chemical in a porous medium, which we discussed earlier in this book. Instead of talking about the concentration of a diffusing chemical substance, we can talk about the density of neutrons, i.e., the number of neutrons per unit volume. Letting N be this neutron density, we would have, by analogy with the usual diffusion equation, an equation for N given by

$$\frac{\partial N}{\partial t} = D\nabla^2 N \tag{40}$$

* Other types of containers are considered in the exercises.

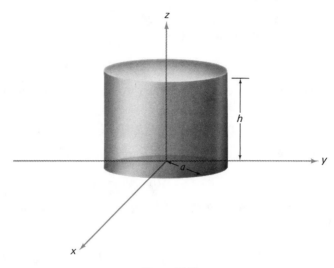

Figure 14.15

The coefficient D, which depends on various physical factors and which we shall assume is a constant, will be called the *neutron diffusivity*. The equation (40), however, does not account for the fact that a new supply of neutrons is constantly being generated within the cylinder by the fission process. We must therefore modify the equation by including a source term due to this extra supply of neutrons. A simple assumption to make is that the extra supply of neutrons per unit volume per unit time is proportional to the instantaneous neutron density; i.e., we take as the source term γN, where γ is a constant of proportionality. The resulting equation is

$$\frac{\partial N}{\partial t} = D\nabla^2 N + \gamma N \tag{41}$$

If we assume that the instantaneous distribution of neutrons depends only on the cylindrical coordinates r and z but not ϕ, then we can denote the neutron density by $N(r, z, t)$, and equation (41) becomes

$$\frac{\partial N}{\partial t} = D\left(\frac{\partial^2 N}{\partial r^2} + \frac{1}{r}\frac{\partial N}{\partial r} + \frac{\partial^2 N}{\partial z^2}\right) + \gamma N \tag{42}$$

We must now investigate some possible boundary conditions. One condition which we have always assumed in previous formulations of boundary value problems was the condition of boundedness of a physical variable both in space and in time. We ask the question whether we want such boundedness in the present problem. To answer this we first ask whether we want the neutron density to increase with time. The answer is that whether we wish to make an atomic bomb or a controlled nuclear reactor it would be desirable that the neutron density increase with time; certainly we would not want it to decrease with time. We thus have the condition

$$N(r, z, t) \quad \text{must not decrease as time } t \text{ increases} \tag{43}$$

As a function of r and z for fixed t, it would actually be physically impossible for the neutron density to be unbounded. Hence we must require the condition

$$N(r, z, t) \quad \text{is bounded for} \quad 0 \leq r \leq a, 0 \leq z \leq h \text{ and any fixed } t \qquad (44)$$

Further conditions are obtained when we realize that the neutron density must be zero on the surface of the cylinder. Thus

$$N(r, 0, t) = 0, \qquad N(a, z, t) = 0, \qquad N(r, h, t) = 0 \qquad (45)$$

Also if we assume that there is some initial distribution of neutrons, we have

$$N(r, z, 0) = f(r, z) \qquad (46)$$

Solution Letting $N = RZT$, where R, Z, T are, respectively, functions of r, z, t, we find from (42)

$$RZT' = D\left(R''ZT + \frac{1}{r}R'ZT + RZ''T\right) + \gamma RZT$$

Dividing by $DRZT$,
$$\frac{T'}{DT} - \frac{\gamma}{D} = \frac{R''}{R} + \frac{1}{r}\frac{R'}{R} + \frac{Z''}{Z} \qquad (47)$$

Hence each side is a constant, which we denote by $-\lambda^2$, and we obtain

$$\frac{T'}{DT} - \frac{\gamma}{D} = -\lambda^2 \qquad (48)$$

$$\frac{R''}{R} + \frac{1}{r}\frac{R'}{R} + \frac{Z''}{Z} = -\lambda^2 \qquad (49)$$

From (49),
$$\frac{R''}{R} + \frac{1}{r}\frac{R'}{R} = -\frac{Z''}{Z} - \lambda^2 \qquad (50)$$

so that each side is a constant which we can call $-\mu^2$, Thus

$$\frac{R''}{R} + \frac{1}{r}\frac{R'}{R} = -\mu^2, \qquad -\frac{Z''}{Z} - \lambda^2 = -\mu^2 \qquad (51)$$

If we define $\lambda^2 = \mu^2 + \nu^2$, equations (48) and (51) become

$$T' - [\gamma - D(\mu^2 + \nu^2)]T = 0, \qquad R'' + \frac{1}{r}R' + \mu^2 R = 0, \qquad Z'' + \nu^2 Z = 0 \quad (52)$$

These have the solutions

$$T = c_1 e^{[\gamma - D(\mu^2 + \nu^2)]t}, \qquad R = c_2 J_0(\mu r) + c_3 Y_0(\mu r), \qquad Z = c_4 \cos \nu z + c_5 \sin \nu z$$

Thus $N(r, z, t) = c_1 e^{[\gamma - D(\mu^2 + \nu^2)]t}[c_2 J_0(\mu r) + c_3 Y_0(\mu r)][c_4 \cos \nu z + c_5 \sin \nu z]$ (53)

From condition (44), i.e., the boundedness at $r = 0$, we must choose $c_3 = 0$. Also from the first of conditions (45) we find $c_4 = 0$. Calling $c_1 c_2 c_5 = A$, we have

$$N(r, z, t) = A e^{[\gamma - D(\mu^2 + \nu^2)]t} J_0(\mu r) \sin \nu z \qquad (54)$$

The second condition in (45) requires that

$$J_0(\mu a) = 0 \quad \text{or} \quad \mu = \frac{\lambda_m}{a}, m = 1, 2, \ldots \tag{55}$$

where $\lambda_1, \lambda_2, \ldots$ are the successive positive roots of $J_0(x) = 0$. Similarly the third condition in (45) requires that

$$\sin vh = 0 \quad \text{or} \quad v = \frac{n\pi}{h}, \quad n = 1, 2, \ldots \tag{56}$$

Hence

$$N(r, z, t) = a_{mn}e^{[\gamma - D(\mu_m^2 + v_n^2)]t} J_0\left(\frac{\lambda_m r}{a}\right) \sin \frac{n\pi z}{h} \tag{57}$$

where

$$\mu_m = \frac{\lambda_m}{a} \quad \text{and} \quad v_n = \frac{n\pi}{h} \quad \text{for } m, n = 1, 2, 3, \ldots$$

Thus by the principle of superposition we have the solution

$$N(r, z, t) = \sum_{m=1}^{\infty} \sum_{n=1}^{\infty} a_{mn}e^{[\gamma - D(\mu_m^2 + v_n^2)]t} J_0\left(\frac{\lambda_m r}{a}\right) \sin \frac{n\pi z}{h} \tag{58}$$

From (46)

$$f(r, z) = \sum_{m=1}^{\infty} \sum_{n=1}^{\infty} a_{mn} J_0\left(\frac{\lambda_m r}{a}\right) \sin \frac{n\pi z}{h} \tag{59}$$

so that

$$a_{mn} = \frac{\int_0^a \int_0^h rf(r, z)J_0(\lambda_m r/a) \sin (n\pi z/h)dr\, dz}{\int_0^a \int_0^h r[J_0(\lambda_m r/a)]^2 \sin^2 (n\pi z/h)dr\, dz} \tag{60}$$

Substituting these into (58) yields the required solution.

Interpretation. In order to have the neutron density increase with t, let us look at the time factor in the solution (58), i.e.,

$$e^{[\gamma - D(\mu_m^2 + v_n^2)]t} \tag{61}$$

The coefficient of t in this factor is

$$\gamma - D\left(\frac{\lambda_m^2}{a^2} + \frac{n^2\pi^2}{h^2}\right) \tag{62}$$

and it is at once clear that this will be negative for large enough integers m and n. However, we are not interested in the negative values since in this case the neutron density decreases, i.e., we have *attenuation*. Of great interest for our purposes is the case where the coefficient (62) is positive. The largest possible value occurs when $m = 1$ and $n = 1$, in which case the coefficient is

$$\gamma - D\left(\frac{\lambda_1^2}{a^2} + \frac{\pi^2}{h^2}\right) \tag{63}$$

where λ_1, the smallest positive root of $J_0(x) = 0$ is 2.405 approximately. If we want (63) to be non-negative, we must choose the radius a and height h of the cylinder so that

$$\frac{\lambda_1^2}{a^2} + \frac{\pi^2}{h^2} \leq \frac{\gamma}{D} \tag{64}$$

Now the mass M of fissionable material is given by

$$M = \pi \rho a^2 h \tag{65}$$

where ρ is the density. If a and h are chosen so that

$$\frac{\lambda_1^2}{a^2} + \frac{\pi^2}{h^2} = \frac{\gamma}{D} \tag{66}$$

then for these values we refer to the mass M given by (65) as the *critical mass*, because any increase in the dimensions a and h, and thus M, could result in a chain reaction which, if violent enough, could cause an explosion as in the atomic bomb.

It should be pointed out that the same conclusions could have been seen directly from equation (53), so that it was actually not necessary to obtain the complete solution. The same conclusion thus would hold even if we assumed ϕ dependence. Of course, these solutions are important if we should wish to determine the actual variations of neutron density in the cylinder.

A EXERCISES

1. Obtain the displacement of the chain of the text at any time if it is given an initial displacement $f(x) = \alpha x$, where α is a constant and then released.

2. Work Exercise 1 if (a) $f(x) = \beta x^2$; (b) $f(x) = \alpha x + \beta x^2$ where α and β are constants.

3. Determine the displacement of the chain at any time if (a) the initial displacement is $f(x) = 0$ and the initial velocity is $g(x) \neq 0$; (b) the initial displacement is $f(x) \neq 0$ and the initial velocity is $g(x) \neq 0$.

4. Work the problem of the vibrating chain if the initial displacement is zero and the chain is given a constant initial velocity v_0 in a horizontal direction.

5. Discuss by use of diagrams the various modes of the vibrating chain.

6. Obtain the general solution of the differential equation (9), page 643, by using the method of Frobenius.

7. Derive the coefficients (60), page 651, needed to determine the neutron density in a cylindrical atomic bomb or reactor.

8. Discuss the density of neutrons in the cylinder of page 648 for the lowest mode corresponding to $m = 1$, $n = 1$.

9. Work Exercise 8 for the higher modes and discuss the differences.

B EXERCISES

1. Work the vibrating chain problem if damping proportional to velocity is taken into account.

2. Suppose that initially the chain has the lower endpoint fixed at $x = L$ while the midpoint is displaced horizontally a distance h. If it is then released from this position find the displacement at any later time.

3. Suppose that the circular wire of the text is replaced by a circular plate of unit radius and constant charge density ρ. (a) Show that the potential at any point on the z axis (assumed perpendicular to the plate and passing through its center) is given by $2\pi \rho(\sqrt{z^2 + 1} - z)$.

(b) Use (a) to show that the potential at any point due to the charged plate is given by

$$2\pi\rho \left[1 - rP_1(x) + \tfrac{1}{2}r^2 P_2(x) - \frac{1}{2\cdot 4} r^4 P_4(x) + \frac{1\cdot 3}{2\cdot 4\cdot 6} r^6 P_6(x) - \cdots \right], 0 \le r < 1$$

$$2\pi\rho \left[\frac{1}{2r} - \frac{1}{2\cdot 4} \frac{P_2(x)}{r^3} + \frac{1\cdot 3}{2\cdot 4\cdot 6} \frac{P_4(x)}{r^5} - \frac{1\cdot 3\cdot 5}{2\cdot 4\cdot 6\cdot 8} \frac{P_6(x)}{r^7} + \cdots \right], r > 1$$

4. Show that the most economic proportions for a cylindrical atomic bomb or reactor are obtained if the radius a and height h are given by

$$a = 2.405 \sqrt{\frac{3D}{\gamma}}, \qquad h = \pi\sqrt{\frac{3D}{\gamma}}$$

C EXERCISES

1. Suppose that the potential at all points on a line, taken as the z axis, in an infinite region is given by $f(z) = \sum_{n=0}^{\infty} a_n z^n$ where the series converges for all z. (a) Show that the potential at any other point of the region is given by

$$V(r, \theta) = \sum_{n=0}^{\infty} a_n r^n P_n(\cos\theta)$$

(b) Interpret the result in (a) in terms of a temperature problem.

2. Work Exercise 1 for the case $f(z) = e^{-|z|}$.

3. An infinite cone $\theta = \alpha$, where $0 < \alpha < \pi/2$, has its surface kept at potential

$$f(r) = \sum_{n=0}^{\infty} a_n r^n, \qquad r > 0$$

where r, θ are the usual spherical coordinates. (a) Show that the potential at all points inside the cone is given by

$$V(r, \theta) = \sum_{n=0}^{\infty} \frac{a_n r^n P_n(\cos\theta)}{P_n(\cos\alpha)}$$

(b) Interpret the result in (a) in terms of a temperature problem.

4. The end $x = 0$ of a chain of length L is attached to a vertical rod which rotates with constant angular velocity ω as indicated in Fig. 14.16. If ω is large enough, the chain remains in a horizontal plane and the effects of gravity are negligible. (a) If the chain is subjected to small transverse displacements, show that the equation of motion is given by

$$\frac{\partial^2 Y}{\partial t^2} = \tfrac{1}{2}\omega^2 \frac{\partial}{\partial x}\left[(L^2 - x^2)\frac{\partial Y}{\partial x} \right]$$

(b) Show that the frequency of the nth mode is given by $f_n = \frac{\omega}{2\pi}\sqrt{\frac{n(n+1)}{2}}$.

(c) If the chain is given an initial displacement $f(x)$ and then released, find the displacement at any later time.

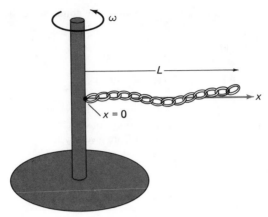

Figure 14.16

5. Work the atomic bomb or nuclear reactor problem for the case of a rectangular parallelepiped with sides a, b, c showing that the critical mass is given by

$$M = \rho abc \qquad \text{where} \quad \frac{1}{a^2} + \frac{1}{b^2} + \frac{1}{c^2} = \frac{\gamma}{\pi^2 D}$$

and that the most economic proportions occur for a cube.

6. (a) Show that the equation for the neutron density N in a spherical container assuming only radial dependence is given by

$$\frac{\partial N}{\partial t} = D \left(\frac{\partial^2 N}{\partial r^2} + \frac{2}{r} \frac{\partial N}{\partial r} \right) + \gamma N$$

(b) Show that the critical mass is given by $M = \dfrac{4}{3} \rho \pi^4 \left(\dfrac{D}{\gamma} \right)^{3/2}$ where ρ is the density.

7. By assuming some initial neutron density distribution in Exercise 6 depending only on r, show how the neutron density can be found after time t.

8. Work Exercises 6 and 7 if the neutron density depends on (a) only r and θ (b) all the spherical coordinates.

determinants

Consider the linear system of equations with two unknowns x and y given by

$$\left.\begin{array}{l} a_{11}x + a_{12}y = b_1 \\ a_{21}x + a_{22}y = b_2 \end{array}\right\} \tag{1}$$

Solving for x and y by the method of elimination yields

$$x = \frac{b_1 a_{22} - a_{12} b_2}{a_{11} a_{22} - a_{12} a_{21}}, \qquad y = \frac{a_{11} b_2 - b_1 a_{21}}{a_{11} a_{22} - a_{12} a_{21}} \tag{2}$$

In an effort to write these in a form which is easily remembered we note that the denominators in (2) are the same and can be obtained by writing the four coefficients of x and y of equations (1) in a form called a 2 by 2, 2×2, or second-order *determinant* given by

$$\begin{vmatrix} a_{11} & a_{12} \\ a_{21} & a_{22} \end{vmatrix} = a_{11} a_{22} - a_{12} a_{21} \tag{3}$$

where we multiply the upper left and lower right numbers and then subtract the product of the upper right and lower left numbers as indicated by the arrows. Using this same procedure for the numerators in (2) the required solution can then be written as

$$x = \frac{\begin{vmatrix} b_1 & a_{12} \\ b_2 & a_{22} \end{vmatrix}}{\begin{vmatrix} a_{11} & a_{12} \\ a_{21} & a_{22} \end{vmatrix}}, \qquad y = \frac{\begin{vmatrix} a_{11} & b_1 \\ a_{21} & b_2 \end{vmatrix}}{\begin{vmatrix} a_{11} & a_{12} \\ a_{21} & a_{22} \end{vmatrix}} \tag{4}$$

This can easily be remembered if we note the following rules called *Cramer's rules*.

CRAMER'S RULES

(i) The equal denominators in (4) are written as (3) which is called the *determinant of the coefficients* of x and y in equations (1).

(ii) The numerator for x in (4) is obtained on replacing the entries in the *first* column of the denominator by the constants b_1, b_2 on the right of equations (1) and leaving the entries in the *second* column unchanged.

(iii) The numerator for y in (4) is obtained on replacing the entries in the *second* column of the denominator by the constants b_1, b_2 and leaving the entries in the *first* column unchanged.

Example 1. To solve

$$3x - 4y = 30 \atop 5x + 7y = -32 \Big\}$$

we write

$$x = \frac{\begin{vmatrix} 30 & -4 \\ -32 & 7 \end{vmatrix}}{\begin{vmatrix} 3 & -4 \\ 5 & 7 \end{vmatrix}} = \frac{(30)(7) - (-4)(-32)}{(3)(7) - (-4)(5)}, \qquad y = \frac{\begin{vmatrix} 3 & 30 \\ 5 & -32 \end{vmatrix}}{\begin{vmatrix} 3 & -4 \\ 5 & 7 \end{vmatrix}} = \frac{(3)(-32) - (30)(5)}{(3)(7) - (-4)(5)}$$

i.e.,

$$x = 82/41 = 2, \; y = -246/41 = -6,$$

which can be checked by substitution.

Success in solving systems of two equations in two unknowns naturally inspires one to see whether the same procedure works for systems of 3 equations in 3 unknowns, namely

$$a_{11}x + a_{12}y + a_{12}z = b_1 \atop a_{21}x + a_{22}y + a_{23}z = b_2 \atop a_{31}x + a_{32}y + a_{33}z = b_3 \Bigg\} \tag{5}$$

In such case we would by analogy with Cramer's rule for the case of 2 equations, write the solution in terms of 3 by 3, 3×3, or third-order determinants as

$$x = \frac{\begin{vmatrix} b_1 & a_{12} & a_{13} \\ b_2 & a_{22} & a_{23} \\ b_3 & a_{32} & a_{33} \end{vmatrix}}{\Delta}, \qquad y = \frac{\begin{vmatrix} a_{11} & b_1 & a_{13} \\ a_{21} & b_2 & a_{23} \\ a_{31} & b_3 & a_{33} \end{vmatrix}}{\Delta}, \qquad z = \frac{\begin{vmatrix} a_{11} & a_{12} & b_1 \\ a_{21} & a_{22} & b_2 \\ a_{31} & a_{32} & b_3 \end{vmatrix}}{\Delta} \tag{6}$$

where the determinant of the coefficients in (5) is given by

$$\Delta = \begin{vmatrix} a_{11} & a_{12} & a_{13} \\ a_{21} & a_{22} & a_{23} \\ a_{31} & a_{32} & a_{33} \end{vmatrix} \tag{7}$$

The scheme for evaluating the determinants in (6) can be arrived at by actually solving the system (5) by the method of elimination. If this is done we find that agreement occurs provided we use the following schematic diagram for the evaluation.

$$\begin{vmatrix} a_{11} & a_{12} & a_{13} \\ a_{21} & a_{22} & a_{23} \\ a_{31} & a_{32} & a_{33} \end{vmatrix} \begin{matrix} a_{11} & a_{12} \\ a_{21} & a_{22} \\ a_{31} & a_{32} \end{matrix} = a_{11}a_{22}a_{33} + a_{12}a_{23}a_{31} + a_{13}a_{21}a_{32} \\ - a_{13}a_{22}a_{31} - a_{11}a_{23}a_{32} - a_{12}a_{21}a_{33}$$

$$(-) \quad (-) \quad (-) \qquad (+) \qquad (+) \quad (+)$$

(8)

In this diagram we have repeated the first two columns of the determinant (7) as shown. We then form the products indicated by following the arrows in the diagram using plus signs for arrows pointing to the right and minus signs for arrows pointing to the left and add the results.

Example 2. To solve

$$\left. \begin{matrix} 3x - 2y + 5z = 7 \\ 2x + y - z = -6 \\ 4x - 3y + 2z = -5 \end{matrix} \right\}$$

use Cramer's rules to write

$$x = \frac{\begin{vmatrix} 7 & -2 & 5 \\ -6 & 1 & -1 \\ -5 & -3 & 2 \end{vmatrix}}{\Delta}, \quad y = \frac{\begin{vmatrix} 3 & 7 & 5 \\ 2 & -6 & -1 \\ 4 & -5 & 2 \end{vmatrix}}{\Delta}, \quad z = \frac{\begin{vmatrix} 3 & -2 & 7 \\ 2 & 1 & -6 \\ 4 & -3 & -5 \end{vmatrix}}{\Delta}$$

where

$$\Delta = \begin{vmatrix} 3 & -2 & 5 \\ 2 & 1 & -1 \\ 4 & -3 & 2 \end{vmatrix}$$

(9)

Evaluating all these determinants we find

$$x = \frac{74}{-37} = -2, \quad y = \frac{-37}{-37} = 1, \quad z = \frac{-111}{-37} = 3$$

which can be checked in the original equations to be the required solution.

These ideas can be extended to linear systems of n equations in n unknowns in which case we have the n by n, $n \times n$, or nth-order determinant

$$\Delta = \begin{vmatrix} a_{11} & a_{12} & \cdots & a_{1n} \\ a_{21} & a_{22} & \cdots & a_{2n} \\ \vdots & \vdots & & \vdots \\ a_{n1} & a_{n2} & \cdots & a_{nn} \end{vmatrix}$$

(10)

Since evaluation becomes more difficult for $n > 3$ (e.g., a 4×4 or fourth-order determinant has $4! = 24$ terms) various definitions and theorems are used for simplifying procedures. We illustrate these below with third-order determinants, but they can be used for determinants of any order n.

Definition 1. Let a_{jk} be the element in the jth row and kth column of the determinant (14). Then the *minor* of a_{jk} is the determinant M_{jk} obtained by crossing out the row and column in which a_{jk} appears.

Example 3. In the determinant (9), given below, the minor of the element -3 in the third row and second column is obtained by crossing out the third row and second column and then writing down all the remaining elements, i.e.,

$$\begin{vmatrix} 3 & -2 & 5 \\ 2 & 1 & -1 \\ 4 & -3 & 2 \end{vmatrix} \text{ or } \begin{vmatrix} 3 & 5 \\ 2 & -1 \end{vmatrix}$$

Definition 2. The cofactor of a_{jk} is obtained on multiplying the minor M_{jk} by $(-1)^{j+k}$, i.e., $+1$ if the sum of the row and column in which a_{jk} appears is even, and -1 if the sum is odd. Denoting the cofactor a_{jk} by A_{jk}, we have

$$A_{jk} = (-1)^{j+k} M_{jk}$$

Example 4. In the determinant of Example 3 the cofactor of the element -3 in the third row and second column is

$$A_{32} = (-1)^{3+2} \begin{vmatrix} 3 & 5 \\ 2 & -1 \end{vmatrix}$$

Definition 3. The determinant (10) is obtained by finding the sum of the products of the elements in any row (or column) by its corresponding cofactor. In symbols if we choose elements in the jth row, for example, then

$$\Delta = a_{j1}A_{j1} + a_{j2}A_{j2} + \cdots + a_{jn}A_{jn} = \sum_{r=1}^{n} a_{jr}A_{jr} \tag{11}$$

Example 5. Using the elements in the third row the determinant of Example 3 can be evaluated as

$$(4)(-1)^{3+1} \begin{vmatrix} -2 & 5 \\ 1 & -1 \end{vmatrix} + (-3)(-1)^{3+2} \begin{vmatrix} 3 & 5 \\ 2 & -1 \end{vmatrix} + (2)(-1)^{3+3} \begin{vmatrix} 3 & -2 \\ 2 & 1 \end{vmatrix} = -37$$

Note that simplifications occur if elements in a row (or column) are zero since we do not have to bother with such terms. Theorem 5 below enables us to produce zero elements in a row (or column).

Theorem 1. The value of a determinant remains the same if rows and columns are interchanged.

Example 6.

$$\begin{vmatrix} 3 & -2 & 5 \\ 2 & 1 & -1 \\ 4 & -3 & 2 \end{vmatrix} = \begin{vmatrix} 3 & 2 & 4 \\ -2 & 1 & -3 \\ 5 & -1 & 2 \end{vmatrix} = -37$$

Theorem 2. If we multiply all the elements in any row (or column) of a given determinant by some constant c then the new determinant has a value which is c times that of the given determinant. This theorem also enables us to take out a common factor when $c \neq 0$ from any row (or column).

Example 7.

$$-2 \begin{vmatrix} 3 & -2 & 5 \\ 2 & 1 & -1 \\ 4 & -3 & 2 \end{vmatrix} = \begin{vmatrix} 3 & (-2)(-2) & 5 \\ 2 & (-2)(1) & -1 \\ 4 & (-2)(-3) & 2 \end{vmatrix} = \begin{vmatrix} 3 & 4 & 5 \\ 2 & -2 & -1 \\ 4 & 6 & 2 \end{vmatrix}$$

Example 8.

$$\begin{vmatrix} 3 & -2 & 4 \\ 5 & 1 & -8 \\ -7 & 5 & -4 \end{vmatrix} = \begin{vmatrix} 3 & -2 & (4)(1) \\ 5 & 1 & (4)(-2) \\ -7 & 5 & (4)(-1) \end{vmatrix} = 4 \begin{vmatrix} 3 & -2 & 1 \\ 5 & 1 & -2 \\ -7 & 5 & -1 \end{vmatrix}$$

Theorem 3. If we interchange any two rows (or columns) of a given determinant the new determinant has a value which is -1 times that of the given determinant.

Example 9. Interchanging the first and second columns gives

$$\begin{vmatrix} 3 & -2 & 5 \\ 2 & 1 & -1 \\ 4 & -3 & 2 \end{vmatrix} = - \begin{vmatrix} -2 & 3 & 5 \\ 1 & 2 & -1 \\ -3 & 4 & 2 \end{vmatrix}$$

Theorem 4. If any two rows (or columns) of a determinant have the same elements, then the value of the determinant is zero.

Example 10. Since the first and third columns have the same elements

$$\begin{vmatrix} -4 & 2 & -4 \\ 3 & -5 & 3 \\ -4 & 1 & -4 \end{vmatrix} = 0$$

Theorem 5. If the elements of any row (or column) are changed by adding to these elements some constant times the corresponding elements of any other row (or column) the value of the determinant is unchanged. By strategic choice of the constant this theorem enables us to produce zeros in a row (or column) and simplifies the evaluation.

Example 11. In the determinant (9) multiply the elements of the second row by 2 and add to the corresponding elements of the first row. Also multiply the elements of the second row by 3 and add to the corresponding elements of the third row. Then we obtain

$$\begin{vmatrix} 3 & -2 & 5 \\ 2 & 1 & -1 \\ 4 & -3 & 2 \end{vmatrix} = \begin{vmatrix} (2)(2)+3 & (2)(1)-2 & (2)(-1)+5 \\ 2 & 1 & -1 \\ (3)(2)+4 & (3)(1)-3 & (3)(-1)+2 \end{vmatrix} = \begin{vmatrix} 7 & 0 & 3 \\ 2 & 1 & -1 \\ 10 & 0 & -1 \end{vmatrix}$$

$$= 1(-1)^{2+2} \begin{vmatrix} 7 & 3 \\ 10 & -1 \end{vmatrix} = -37$$

where the last 3×3 determinant is expanded according to Definition 3 using the elements in the second column.

Theorem 6. Suppose that we are given a system of n linear equations in n unknowns with right sides all equal to zero, for example

$$\begin{aligned} a_{11}x + a_{12}y + a_{13}z &= 0 \\ a_{21}x + a_{22}y + a_{23}z &= 0 \\ a_{31}x + a_{32}y + a_{33}z &= 0 \end{aligned}$$

Let Δ denote the determinant of the coefficients. Then

1. All unknowns equal zero (the *trivial solution*) if and only if $\Delta \neq 0$.
2. Not all unknowns equal zero (*non-trivial solutions*) if and only if $\Delta = 0$.

Example 12. Given the system

$$\begin{aligned} 2x + y - z &= 0 \\ x - 2y + 3z &= 0 \\ 4x - 3y + 5z &= 0 \end{aligned}$$

we have

$$\Delta = \begin{vmatrix} 2 & 1 & -1 \\ 1 & -2 & 3 \\ 4 & -3 & 5 \end{vmatrix} = 0$$

Then the system has non-trivial solutions. To find these let z take on different values i.e., $z = 1, \frac{1}{2}, -1$, etc., and find corresponding values of x and y. We obtain solutions $(x, y, z) = (-\frac{1}{5}, \frac{7}{5}, 1), (-\frac{1}{10}, \frac{7}{10}, \frac{1}{2}), (\frac{1}{5}, -\frac{7}{5}, -1)$, etc. There are infinitely many non-trivial solutions.

answers to exercises

Chapter One

A EXERCISES, p. 12

1.

Differential equation	Ordinary or partial	Order	Independent variables	Dependent variables
(a)	Ordinary	1	x	y
(b)	Ordinary	2	x	y
(c)	Partial	2	x, y, t	U
(d)	Ordinary	3	t	s
(e)	Ordinary	1	ϕ	r
(f)	Ordinary	2	y	x
(g)	Partial	2	x, y	V
(h)	Ordinary	1	x or y	y or x
(i)	Ordinary	2	x	y
(j)	Partial	2	x, y, z	T

3. (a), (b), (f) Linear; (d), (e), (h), (i) Non-linear.

6. (b) $x = 12t - t^3 - 16.$ (c) $x = 0, x = -7.$

(d) $t = 2, t = -4$ (assuming the law of motion holds for $t < 0$).

7. (b) $x = 5t^2 - t^4 - 4.$ (c) $x = 0, v = 6.$

8. (b) $x = 5t^2 - t^4 - 6t + 3.$ (c) $x = 1, v = 0.$

9. (b) $y = 4x - x^2.$ **10.** (b) $y = 2 - 2e^{-2x}.$

11. (a) $y = -3 \cos x - 4$.   (b) $x = 7 - 2t - 4e^{-t}$.   (c) $x = \dfrac{t^4}{12} - \dfrac{2t^3}{3} + 4t^2 - 3t + 1$.

     (d) $s = 6u^{3/2} - 32$.   (e) $y = 8x^3 - x^4 - 14x + 7$.

12. (a) $y = \dfrac{4 + cx}{x}; y = \dfrac{4 - 2x}{x}$.

     (b) $y = \dfrac{x^2}{2} + \cos x + c_1 x + c_2; y = \dfrac{x^2}{2} + \cos x + 2x - 1$.

     (c) $y = \dfrac{(2x + 1)^{5/2}}{15} + c_1 x + c_2; y = \dfrac{(2x + 1)^{5/2}}{15} - \dfrac{181x}{30} + \dfrac{74}{15}$.

B EXERCISES, p. 14

2. (a) $m = 2$.   (b) $m = 1, -4$.   (c) $m = 1, 2, 3$.

4. (b) $y = c_1 e^x + c_2 e^{-4x}$.   (c) $y = \frac{3}{5}(4e^x + e^{-4x})$.

8. (a) $y = \dfrac{192}{\pi^3} \sin \dfrac{\pi x}{2} + c_1 x^2 + c_2 x + c_3$.

     (b) $y = \dfrac{192}{\pi^3} \sin \dfrac{\pi x}{2} + \left(2 + \dfrac{192}{\pi^3}\right) x^2 + \left(2 - \dfrac{384}{\pi^3}\right) x - 4$.

C EXERCISES, p. 14

3. (a) $y = e^x$.   (b) $y = -\ln(1 - x)$.   (c) $y = \sin^{-1} x$.

4. (b) $y = c_1 x + c_2 - \sin x$.

A EXERCISES, p. 21

1. Particular solutions are (a) $y = -4 \cos x$.   (b) $y = 3e^x - e^{-x} - 4x$.

     (c) $x^2 - y^2 + 2xy = -17$, or explicitly $y = x + \sqrt{2x^2 + 17}$.   (d) $y = 1 - 2x + 3x^2$.

     (e) $xy^2 = 13x - 36$, or explicitly $y = \sqrt{\dfrac{13x - 36}{x}}$.

2. (a) $y' + 2y = 6x(x + 1)$.   (b) $y' = 1 + (y - x) \cot x$.

     (c) $(x^2 - 1)y' = xy$.   (d) $xy' \ln x = y(\ln x - 1)$.

     (e) $y'' + 16y = 16x$.   (f) $\dfrac{d^2 I}{dt^2} + 2\dfrac{dI}{dt} + I = 12 \cos 3t - 16 \sin 3t$.

     (g) $2xyy' = y^2 - x^2$.   (h) $x^3 y''' - 3x^2 y'' + 6xy' - 6y = 0$.

     (i) $y''' - 3y'' - 10y' + 24y = 0$.   (j) $\phi^2(1 - \ln \phi)\dfrac{d^3 r}{d\phi^3} + \phi\dfrac{d^2 r}{d\phi^2} - \dfrac{dr}{d\phi} = 0$.

3. (a) $yy' + x = 0$.   (b) $2xy' = y$.   (c) Let (c, c) be any point on $y = x$. Then the required circle is $(x - c)^2 + (y - c)^2 = c^2$. The required differential equation is $(x - y)^2[1 + (y')^2] = (x + yy')^2$.   (d) $xyy'' - yy' + x(y')^2 = 0$.

B EXERCISES, p. 22

3. $A = B = 0$ gives the required particular solution.   **5.** $x = -\frac{1}{2}y^{-2} + c$.

2. (a) $y'' - y' - 6y = 0.$ (b) $(y''' + y')(x^3 + 12x) = (y'' + y)(4x^2 + 24).$

7. $xy' \ln y - xy' \ln y' + yy' - y^2 = 0.$

Chapter Two

A EXERCISES, p. 37

1. (a) $x^2 + y^2 = 5.$ (b) $xy = 3.$ (c) $(x^2 + 2)^3(y^2 + 1) = c.$

     (d) $2e^{3x} + 3 \ln y = c.$ (e) $x^2 - 4 \ln (1 + y^2) = 1.$ (f) $\ln r - \tan^{-1} \phi = c.$

     (g) $\tan x \tan y = 1.$ (h) $\sqrt{1 + x^2} - \sqrt{1 + y^2} = c.$ (i) $y^3 \sin^2 x = 8.$

     (j) $y = ce^{4x^2 + 3x}.$ (k) $I = 2(1 - e^{-5t}).$ (l) $x^{-2} - y^2 - 2 \ln y = c.$

2. $(x^2 + 2)(y^2 + 3) = 24.$

B EXERCISES, p. 37

1. $(x - 1)(y + 3)^5 = c(y - 1)(x + 3)^5.$ **2.** $2 \tan^{-1} (e^r) + \tan^{-1} (\cos \phi) = \dfrac{\pi}{2}.$

3. $25(3x^2 - 1)e^{3x^2} + 9(5y^2 + 1)e^{-5y^2} = c.$

4. $2\sqrt{U} + \ln (U + 1) - 2 \tan^{-1} \sqrt{U} = 2\sqrt{s} + c.$

C EXERCISES, p. 37

1. $x = 32/3.$ **2.** $x^4 + 4y^2 = cx^2.$ **3.** $x^3 - 3 \cos xy = c.$

A EXERCISES, p. 40

1. $y = x \ln x + cx.$ **2.** $x = y - y \ln x.$ **3.** $y = cx^3 - x.$

4. $x^3 - 3xy^2 = c.$ **5.** $x^2 + 4xy + y^2 = c.$ **6.** $1 + \ln x = \tan \dfrac{y}{x}.$

7. $y + \sqrt{x^2 + y^2} = c.$ **8.** $x + 3y = cy^2.$ **9.** $y^3 = 3x^3 \ln x.$

10. $x^2 - y^2 = cx.$ **11.** $\dfrac{2y}{x} + \sin \dfrac{2y}{x} = 4 \ln x + c.$

12. $(x + y)^5 = c(x - 2y)^2.$

B EXERCISES, p. 40

1. $y^2 + y\sqrt{x^2 + y^2} + x^2 \ln (y + \sqrt{x^2 + y^2}) - 3x^2 \ln x = cx^2.$

2. $(x + y)^3 = c(2x + y)^4.$ **3.** $(y - x)(y - 3x)^9 = c(y - 2x)^{12}.$

4. $x - \tan^{-1} (x + y) = c.$

5. $6\sqrt{2x + 3y} - 4 \ln (3\sqrt{2x + 3y} + 2) - 9x = c.$

6. $\ln \left[(x - 1)^2 + (y + 1)^2\right] - 3 \tan^{-1} \left(\dfrac{y + 1}{x - 1}\right) = c.$

7. $y - 2x + 7 = c(x + y + 1)^4$.

8. $\ln(2x + 3y + 2) = 2y - x + c$.

9. $2x + y - 3\ln(x + y + 2) = c$.

10. $x^2\left(\sin\dfrac{y}{x} + \tan\dfrac{y}{x}\right) = c$.

C EXERCISES, p. 41

1. (a) $x + \sqrt{x^2 - y^2} = c$.　(b) $x - y - 2\sqrt{x - y} = 2x + c$; singular solution $y = x$.

2. $x^2 \cos\dfrac{y}{x^2} + y\sin\dfrac{y}{x^2} = cx^3$.

3. $\frac{1}{2}\ln(x^6 + y^4) + \tan^{-1}\dfrac{x^3}{y^2} = c$.

4. Choose $n = \frac{3}{4}$, $2 + 5xy^2 = cx^{5/2}$.

5. (a) $(Y - \alpha X)^{2 + \sqrt{2}}(Y - \beta X)^{2 - \sqrt{2}} = c(Y - X)^2$ where $\alpha = 3 + 2\sqrt{2}$, $\beta = 3 - 2\sqrt{2}$, $X = x - 2$, $Y = y + 1$.

(b) $v^2 + 2v - v\sqrt{v^2 - 1} + \ln(v + \sqrt{v^2 - 1}) = 4x + c$ where $v = x + y$.

6. $xy + \ln(x/y) = c$.

A EXERCISES, p. 47

1. (a) $3x^2 + 4y^2 = c$.　(b) $x^2 - 2xy - y^2 = c$.　(c) $3xy^2 - x^3 = c$.

(d) Not exact.　(e) $x^2 - y^2 - 2y\sin x = c$.　(f) $r^2\cos\phi - r = c$.

(g) $ye^{-x} - \cos x + y^2 = c$.　(h) $x^3 + 3y\ln x + 3y^2 = c$.　(i) $ye^x - xy^2 = c$.

(j) Not exact.

2. (a) $x^2 - xy + y^2 = 3$.　(b) $x^2y + y + 6 = 0$.　(c) $x^2 - x\sin y = 4$.

(d) $2x\tan y + \cos 2x = 2\pi + 1$.　(e) $x^2y + y^2e^{2x} = 1$.

3. (a) $x^2y^2 + x^4 = c$.　(b) $6xy^4 - y^6 = c$.　(c) $y\cos^2 x + 3\sin x = c$.

(d) $\frac{1}{2}\ln(x^2 + y^2) + \tan^{-1}\dfrac{y}{x} = c$.

B EXERCISES, p. 48

1. $x + y^2 - x^2 = c(x + y)$.

4. $N(x, y) = -\cos x + \dfrac{x^3}{3} - \dfrac{x^2}{2}\sec y\tan y + F(y)$,

$\frac{1}{3}x^3y - y\cos x - \frac{1}{2}x^2\sec y + \int F(y)\,dy = c$.

C EXERCISES, p. 48

2. $y^4(3 - xy)^5 = cx^2$.

A EXERCISES, p. 52

1. (a) $x^3 + x^2y^2 = c$.　(b) $y = 2x - x^3$.　(c) $y^2 - x = y(c - \sin x)$.

(d) $x^2(c + \cos 2y) = 2y$.　(e) $x = \left(\dfrac{\pi}{2} - y\right)\sin y$.　(f) $y = (x + c)\cos^2 x$.

(g) $6x^4y - x^6 = c$.　(h) $6xy^3 - y^6 = c$.　(i) $I = 1 - \dfrac{1}{\sqrt{t^2 + 1}}$.

(j) $y^3e^x + ye^{2x} = c$.

2. $y = x \ln \left(\dfrac{x}{3} \right)$.

1. $y^2 + \sin^2 x = c \sin^3 x$.

2. $x^2 + y^2 + 1 = ce^{y^2}$.

3. $e^{x^3}(1 + xy) = c$.

4. $(x^2 + y^2)\sqrt{1 + y^2} = c$.

2. $e^{xy}(x + y) = c$.

3. $x^2y^4 + x^4y^3 = c$.

1. (a) $x^2 - 2xy = c$.

(b) $x^5 - 5x^3y = c$.

(c) $xy = y^2 + \ln y + c$.

(d) $x^2 \cos 3x + 3y = cx^2$.

(e) $I = e^{-2t} + 4e^{-3t}$.

(f) $2y \sin x - \sin^2 x = c$.

(g) $x - 3y - 3 = ce^y$.

(h) $7r = 3\phi^2 + 4\phi^{-1/3}$.

2. $I = 10te^{-2t}$.

3. $y = (1 - x + ce^{-x})^{-1}$.

4. $2x^2 - 3y = cx^2y^3$.

5. $y = c_1x^4 - \frac{4}{3}x^3 + c_2$.

7. $y = \left[\dfrac{\beta}{\alpha} + ce^{(1-n)ax} \right]^{1/(1-n)}$

2. $\ln y = 2x^2 + cx$.

4. $y = \ln (x - 8x^{-3})$.

1. $xy^2 - y = cx$.

2. $y^3 + 2x = cy$.

3. $x^2 - 2 \tan^{-1} \dfrac{y}{x} = c$.

4. $\sqrt{x^2 + y^2} = x + y + c$.

5. $x^3 + xy^2 - 2y = cx$.

6. $x + y - \tan^{-1} \dfrac{y}{x} = c$.

7. $\ln (x^2 + y^2) + 2y - 2x = c$.

8. $\ln (x^2 + y^2) = 2xy + c$.

1. $(x^2 + y^2)^{3/2} - 3xy = c$.

2. $3x^2 + 3y^2 + 2(xy)^{-3} = c$.

3. $x^2 - y^2 + \ln \left| \dfrac{x - y}{x + y} \right| = c$.

1. $\ln (x^2 + 2y^2) = x^2 + y^2 + c$.

2. $x^3 - x^2y - y = cx^2$.

3. $x^2 - 2 \ln (y^2 + \sin^2 x) = c$.

4. $(y - x)^{-1} + \ln \cos (y/x) = c$.

1. $y = \dfrac{x^3}{3} + 10x.$

2. $y = \dfrac{x^5}{360} + c_1 x^3 + c_2 x^2 + c_3 x + c_4.$

3. $y = 3 \cos x + \dfrac{x^2}{2} - 2.$

4. $y = \dfrac{e^x - e^{-x}}{2} - \dfrac{x^3}{6} - x.$

5. $I = \dfrac{t^4}{12} + \dfrac{t^2}{2} + 3t + 2.$

6. $y = \dfrac{x^2}{2} - \ln x + \dfrac{1}{2}.$

7. $y = \frac{1}{2} \ln x + \frac{8}{3} x^{1/2} + c_1 x^2 + c_2 x + c_3.$

8. $y = \dfrac{(2x + 1)^{3/2}}{3} + \dfrac{14}{3}.$

9. $y = 3 \cos 2x + \sin 2x.$

10. $y = \dfrac{c_1}{x} + c_2.$

11. $y = c_1 e^x + c_2 e^{-x}.$

12. $x = \int dy/(\ln y + c_1) + c_2, \; y = c_3.$

13. $y = \ln \cosh (x + c_1) + c_2.$

14. $y + 1 = c_1 \tan \frac{1}{2}(x + c_2).$

15. $y = x + c_1 - c_2 \int e^{-x^2/2} \, dx.$

1. $y = \dfrac{x^4}{24} \ln x - \dfrac{25x^4}{288} + \dfrac{x^3}{6} - \dfrac{x^2}{8} + \dfrac{x}{18} - \dfrac{1}{96}.$

2. $y = \dfrac{x^5}{240} - \dfrac{x^4}{96} + \dfrac{x^3}{48} - \dfrac{x^2}{32} + \dfrac{x}{32} - \dfrac{1}{64} + \dfrac{e^{-2x}}{64}.$

3. $y = \dfrac{x^2}{2} + c_1 x \ln x + c_2 x + c_3.$

4. $y = -4 \ln (x + c_1) + c_2 x + c_3.$

5. $y = c_1 e^x + c_2 e^{-x} + c_3.$

6. $(x + c_1)^2 + y^2 = c_2^2.$

7. $y = (x + c_1) \ln x + c_2 x + c_3.$

1. $\pm \sqrt{15}.$

2. $(x - A)^2 + (y - B)^2 = 1.$

3. $(x - c_1)(y - c_2) = 4.$

4. $y = \ln \sec x.$

5. $y = \displaystyle\int_0^x \dfrac{u^2 \, du}{\sqrt{1 - u^4}}.$

1. $y = cx - c^2, \; y = x^2/4.$

2. $y = cx + 1 + 4c^2, \; y = 1 - x^2/16.$

3. $y = cx - \tan c; \; x = \sec^2 v, \; y = c \sec^2 v - \tan v$ are parametric equations.

4. $y = cx + \sqrt{1 + c^2}; \; x = -\dfrac{v}{\sqrt{1 + v^2}}, \; y = \dfrac{1}{\sqrt{1 + v^2}}$ or $x^2 + y^2 = 1.$

6. $y = \frac{1}{4}(x + c)^2, \; y = 0.$

C EXERCISES, p. 64

1. $y = c \sin x - c^2$, $y = \frac{1}{4} \sin^2 x$.

2. $y = cx^2 + 1/4c$, $y = \pm x$.

MISCELLANEOUS EXERCISES, CHAPTER TWO

A EXERCISES, p. 65

1. $x - x^{-1} = \frac{1}{3} \ln (y^3 - 1) + c$.

2. $x^2 y + y^2 x = c$.

3. $x^2 + xy + y^2 = c(x + y)$.

4. $5x^2 y = x^5 + c$.

5. $(3 - y)^2 = 4x$.

6. $y = 2x - x^2 + c$.

7. $2s^3 + 3t^2 + 24 \ln t = c$.

8. $x^3 + 3xy^2 = c$.

9. $3ye^x - 2x^3 = c$.

10. $xy = x + \ln x + c$.

11. $\ln x = \int \dfrac{dv}{\tan^{-1} v} + c$, where $v = \dfrac{y}{x}$.

12. $x + y + 1 = ce^x$.

13. $y = x^2 - 2 + ce^{-x^2/2}$.

14. $\frac{1}{2}x^2 y^2 + 4x - 3y = c$.

15. $r^2 \cos \phi + 10r = c$.

16. $y = ce^{2x} - \dfrac{x^2}{2} - \dfrac{x}{2} - \dfrac{1}{4}$.

17. $x^2 y^{-3} - x = c$.

18. $x^2 + y^2 - 2x + 2 = 6e^{-x}$.

19. $3xy^2 + x^3 - 9y = c$.

20. $y^2 \ln y - x = cy^2$.

21. $u^3 v^{-3} - 3 \ln v = c$.

22. $x \cot y - \sin x = c$.

23. $x^2 - y^2 + 4xy = c$.

24. $y = (x + c) \sin x$.

25. $x^2 + y^2 = cx$.

26. $2x^3 - 3ye^x = c$.

27. $y - \ln (x + y + 1) = c$.

28. $x^3 + 3x^2 y = c$.

29. $y^2 - 2x \sin y = c$.

30. $e^{-y/x} + \ln x = c$.

31. $\cos x \cos y = c$.

32. $y = x^3 + cx^2$.

33. $x^3 y^2 + x^2 = c$.

34. $xy^{-1} + 2y = c$.

35. $y = c_1 e^x - x^2 - 2x + c_2$.

36. $2y^{1/2} + \frac{2}{3}y^{3/2} = \dfrac{x^2}{2} + c$.

37. $\sec x(\csc y - \cot y)^3 = c$.

38. $\sec \dfrac{y}{x} + \tan \dfrac{y}{x} = cx$.

39. $(1 - s)^{3/2} - (1 - t)^{3/2} = 1$.

40. $x^2 y + x^3 = c$.

41. $y^3(1 + x^3) = c$.

42. $2xy + 2 \cos y + 2x^2 = \pi + 2$.

43. $N = N_0 e^{-\alpha t}$.

44. $xye^{x/y} = c$.

45. $I = ce^{-t} + \frac{1}{2}e^t$.

46. $3xy = x^3 + 5$.

47. $xy^2 - 2y = cx$.

48. $e^{p^2} - e^{q^2} = c$.

49. $y^3 \cos x + y^2 + 48 = 0$.

50. $y + \sin y = cx$.

51. $y = ce^{-2x} - \frac{3}{2}x - \frac{3}{4}$.

52. $y^2 = x(c - y)$.

53. $2 \ln r + (\ln r)^2 = 2 \ln \phi + (\ln \phi)^2 - 5$.

54. $U = 100t - \dfrac{100}{a}(1 - e^{-at})$.

55. $(u - 1)v = cu^2$.

56. $I = 3 \sin t - \cos t + e^{-3t}$.

57. $s - \ln (s + t + 2) = c$.

58. $y^2 = c_1 x + c_2$.

59. $\sqrt{1 - x^2} + \sqrt{1 - y^2} = c$.

60. $2y \sin x + \cos^2 x = c$.

61. $(y + 3)^{-1} = (4x)^{-1} + c$.

62. $y = cx^3 - x^3 e^{-x}$.

63. $\ln \sin y + \cos x = c$.

64. $y^2 = 2x^2 \ln x + cx^2$.

65. $2xy^3 - 3y = cx$.

66. $\ln y = cx + \ln x$.

67. $xy - x^2 = c$.

68. $y = x + c_1 \ln x + c_2$.

69. $t^3 I^{-3} + 3 \ln I = c$.

70. $x = ye^y - 3 + ce^y$.

71. $r = \frac{1}{4}e^\phi + \frac{3}{4}e^{-3\phi}$.

72. $y = ae^{bx}$.

73. $6xy = c_1 x^3 + c_2 x^2 + c_3 x + 1$.

74. $\ln (x^2 + 2xy + 3y^2) = 2\sqrt{2} \tan^{-1}\left(\dfrac{x + 3y}{x\sqrt{2}}\right) + c$.

75. $y(\sec x - \tan x) = 2 \ln (1 + \sin x) - 2 \sin x + c$.

76. $e^{4x} + 2e^{2y} = c$.

77. $r^4 = a^4 \sin (4\phi + c)$.

78. $2x^3 - 3ye^x = c$.

79. $x^2 y = (x^2 - 2) \sin x + 2x \cos x + c$.

80. $2\sqrt{1 + x^3} = 3 \ln (y + 1) + c$.

81. $x^3 y^2 + x^4 y = c$.

82. $y^{-1} = -e^{-x^2/2} \int e^{x^2/2} \, dx + ce^{-x^2/2}$.

83. $y = e^{x^2/2} \int x^2 e^{-x^2/2} \, dx + ce^{x^2/2}$.

84. $U = 12 + 4r - r^2 - 2 \ln r$.

85. $(y - x)(x + 1) = 1$.

86. $(y - 1)e^y - e^x = c$.

B EXERCISES, p. 67

1. $x^2 y^2 = 2 \sin x - 2x \cos x + c$.

4. $x + 2y + 1 = ce^{y/2}$.

5. $\sqrt{y + \sin x} = \dfrac{x}{2} + c$.

6. $\sin (x + y) + \cos (x + y) = ce^{x-y}$.

7. $3e^{4x} + 4e^{3x - 3y} = c$.

8. $y = \frac{3}{4}e^{2x} + \dfrac{1}{4} - \dfrac{3x}{2} - \dfrac{3x^2}{2} - x^3 - \dfrac{x^4}{2}$.

9. $xy + 2 \ln (x/y) - (xy)^{-1} = c$.

10. $(y - ce^{-3x})(y - x^2 - c) = 0$.

11. $y = \sqrt{2e^x - x - 1}$.

12. $(x + 2\sqrt{y})(x - \sqrt{y})^2 = c$.

C EXERCISES, p. 68

1. $\ln |x| = \ln |u^2 + 1| - \frac{16}{31} \ln |2u - 1| - \frac{23}{31} \ln |3u^2 + 4u + 5|$

$$-\dfrac{68}{31\sqrt{11}} \tan^{-1}\left(\dfrac{3u + 2}{\sqrt{11}}\right), \text{ where } u = \sqrt{\dfrac{5x - 6y}{5x + 6y}}.$$

4. $5x^{12}y^{12} + 12x^{10}y^{15} = c$.

5. $x^2 + cxy + y^2 = 1$.

13. $y = 2 + e^{2x^2 - 2x}[c - \int xe^{2x^2 - 2x} \, dx]^{-1}$.

14. $y = x + e^{-x^2}[c + \int e^{-x^2} \, dx]^{-1}$.

15. $e^x(y - 1) = c(y - x)$.

Chapter Three

A EXERCISES, p. 78

1. (b) 4410 cm, 2940 cm/sec.   (c) 3430 cm, 4410 cm.

2. (a) 2940 cm, 490 cm/sec upward after 2 sec.
        1960 cm, 1470 cm/sec downward after 4 sec.

(b) 3062.5 cm above starting point (assuming $g = 980$ cm/sec² and answers accurate to at least five significant figures), 2.5 sec.

(c) 2940 cm, 4165 cm.

3. 9.08 sec, 290.7 ft/sec.

4. (a) $v = 49(1 - e^{-20t})$, $x = 49t + 2.45e^{-20t} - 2.45$.　　(b) 49 cm/sec.

5. (a) 25 ft/sec.　　(b) $v = 25(1 - e^{-2t/15})$, $x = 25t + \frac{375}{2}e^{-2t/15} - \frac{375}{2}$.

6. (a) 50 ft/sec.　　(b) $v = 50 - 10e^{-0.64t}$, $x = 50t + \frac{125}{8}e^{-0.64t} - \frac{125}{8}$.

7. (a) $v = 16(1 - e^{-2}) = 13.8$ ft/sec.　　(b) $t = \frac{1}{2} \ln 16 = 2 \ln 2 = 1.39$ sec.

8. (a) $v = 16 \left(\dfrac{e^4 - 1}{e^4 + 1} \right) = 15.4$ ft/sec.　　(b) $t = \frac{1}{4} \ln 31 = 0.86$ sec.

9. 2 ft/sec.

B EXERCISES, p. 79

4. (b) $v = p \left[\dfrac{(p + v_0)e^{\alpha t} - (p - v_0)}{(p + v_0)e^{\alpha t} + (p - v_0)} \right]$ where $p = V \sqrt{\dfrac{W}{F}}$, $\alpha = \dfrac{2g}{V} \sqrt{\dfrac{F}{W}}$, limiting velocity $=$

$p = V \sqrt{\dfrac{W}{F}}$. If $v_0 = 0$, $v = p \left[\dfrac{e^{\alpha t} - 1}{e^{\alpha t} + 1} \right] = p \tanh \dfrac{\alpha t}{2}$ using hyperbolic functions.

(c) $v = \sqrt{p^2 - (p^2 - v_0^2)e^{-2kx}}$ where $p = V\sqrt{W/F}$, $k = Fg/WV^2$.

5. 37.5 ft, 8 ft/sec.

6. (a) $v = 100\sqrt{1 - e^{-0.0064x}}$.　　(b) $v = 100 \left(\dfrac{e^{0.64t} - 1}{e^{0.64t} + 1} \right) = 100 \tanh (0.32t)$.

(c) 97.9 ft/sec.　　(d) 100 ft/sec.　　(e) 784 ft.

10. $\sqrt{\dfrac{L}{g}} \ln \left(\dfrac{L + \sqrt{L^2 - a^2}}{a} \right)$.

C EXERCISES, p. 80

1. (b) 200 ft, 80 ft/sec, 16 ft/sec².　　　　**5.** 130.8 ft, 52.3 ft/sec, 10.5 ft/sec².

7. $v = \dfrac{cFt}{\sqrt{F^2t^2 + m_0^2 c^2}}$, $x = \dfrac{c}{F}\sqrt{F^2 t^2 + m_0^2 c^2} - \dfrac{m_0 c^2}{F}$.

8. (a) $x = \frac{5}{2}t^2$.　　(b) $x = \dfrac{c^2}{5}(\sqrt{1 + 25t^2/c^2} - 1)$.

15. $\sqrt{\dfrac{L}{g(1 + \sin \alpha)}} \ln \left\{ \dfrac{L + \sqrt{L^2 - [a + (a - L)\sin \alpha]^2}}{a + (a - L)\sin \alpha} \right\}$; $\sin \alpha < \dfrac{a}{L - a}$.

A EXERCISES, p. 88

1. $I = \frac{1}{2}(1 - e^{-20t})$.　　　　　　　　**2.** $I = 2 \sin 10t - \cos 10t + e^{-20t}$.

3. $I = 20e^{-4t}$.　　　　　　　　　　　　　**4.** $Q = \frac{1}{4}(1 - e^{-8t})$, $I = 2e^{-8t}$.

5. $Q = 0.16 \cos 6t + 0.12 \sin 6t - 0.16e^{-8t}$,　　**6.** $Q = 0.05e^{-10t}$, $I = -0.5e^{-10t}$.
$I = 0.72 \cos 6t - 0.96 \sin 6t + 1.28e^{-8t}$.

7. $I(t) = 2e^{-4t} \sin 50t$.

3. (a) $Q = 10 - 5(1 + 0.01t)^{-1000}$, $I = 50(1 + 0.01t)^{-1001}$.   (b) 10 coulombs.

4. (a) $I(t) = 4 - 4(1 + 0.02t)^{-10,000}$.   (b) 4 amp.

5. $Q = 0.99CEe^{-t/RC}$.

1. $I(t) = \begin{cases} 1 - e^{-100t}, & 0 \leq t \leq 5 \\ (1 - e^{-500})e^{-100(t-5)}, & t \geq 5. \end{cases}$

2. $Q = CE_0(e^{-T/RC} - e^{-2T/RC} + e^{-3T/RC} - e^{-4T/RC})$.

1. (a) $2xy' = y$.   (b) $yy' = -2x$, $2x^2 + y^2 = c_1$.

2. $3x^2 + 2y^2 = c_1$.

3. (a) $e^{x^2+y^2} = c_1 x^2$; $x^2 - 3y^2 = 1$, $4e^{x^2+y^2-5} = x^2$.

(b) $x^3 + 3xy^2 = c_1$; $x^2 + 8y = y^2$, $x^3 + 3xy^2 = 36$.

(c) $4y^2 - 8y + \sin^2 2x = c_1$; $y = 1 - \tan 2x$, $8y^2 - 16y + 2\sin^2 2x = 1$.

(d) $9x - 3y + 5 = c_1 e^{-6y}$; $y = 3e^{-2x} + 3x$, $9x - 3y + 5 = -4e^{6(3-y)}$.

(e) $e^{-x^2-y^2} = c_1 x^2$; $y^2 = 5(1 + x^2)$, $4e^{29-x^2-y^2} = x^2$.

1. $a = \frac{1}{3}$.

4. (a) $y^2 = \dfrac{2x^{2-p}}{2 - p} - x^2 + c_1$ if $p \neq 2$; $e^{x^2+y^2} = c_1 x^2$ if $p = 2$.

(b) $x^2 - y^2 = c_1 e^{x^2+y^2}$.

1. $2\tan^{-1}(y/x) + \ln(x^2 + y^2) = c$, or in polar coordinates $r = ae^{-\phi}$.

2. $\ln(x^2 + y^2) \pm 2\sqrt{3}\tan^{-1}(y/x) = \pm 2\pi\sqrt{3}/3$ or $r = e^{\pm\sqrt{3}(\phi - \pi/3)}$.

3. $\pm y\tan\alpha + \sec^2\alpha \ln|y \mp \tan\alpha| = x + c_1$.

5. $r = c_1 \sin\phi$.   **6.** $(\ln r)^2 + \phi^2 = c_1$.   **7.** $r = c_1(1 + \cos\phi)$.

1. (b) $x = 24 - 22e^{-t/2}$.   (c) 3 lb/gal approx.   (d) 24 lb.

2. (a) $x = 40(1 - e^{-t/20})$.   (b) 13.9 min.

3. 20.8 min.   **4.** 34.7 min, 184.5 min.

5. (a) 5.1 lb, 6.6 lb, 7.6 lb.   (b) 3 hr 53 min.

6. (a) $x = \dfrac{300[(18/17)^t - 1]}{3(18/17)^t - 2}$.   (b) 26.6 lb.   (c) 100 lb.

1. (a) $x = 1.5(10 - t) - 0.0013(10 - t)^4, 0 \leq t \leq 10$.   (b) 1.5 lb/gal.
2. (a) $3[1 - (1 - t/120)^4], 0 \leq t \leq 120$.   (b) 2.81 lb/gal.   (c) None.   (d) 96.3 lb.
3. (a) 2.33 lb.   (b) 1.12 lb/gal.   (c) 10.3 hr.

4. (a) 1.44×10^{-2} liters moles^{-1} min^{-1}.   (b) 127 g sodium acetate, 71 g ethyl alcohol.

1. (a) $75 - \frac{1}{4}x$.   (c) 0.0375 cal.
2. (a) $U_0 + \left(\dfrac{U_1 - U_0}{L}\right)x$.   (c) $\dfrac{K(U_0 - U_1)}{L}$.
3. (a) $741 \ln r - 2120$.   (b) 265°C.   (c) 4.2×10^7 cal/min.
4. (a) $548 - 216 \ln r$ or $200 - 216 \ln (r/5), 5 \leq r \leq 10$.
   (b) 112°C.   (c) 1.96×10^7 cal/min.
5. (a) $0.08x + 25$.   (b) 144 cal/hr.   6. (a) $C_0(1 - x/L)$.

1. (a) 65.3°C.   (b) 52 min, 139 min.   2. (a) 34.1°C, 37.4°C.   (b) 8.5 min.
3. 98%, 96%, 92%.   4. About 24 years.   5. 64.5 days.
6. 79 min.   7. 1.35 hr, 1.98 hr.   8. 9 days.
9. 10 g.   10. 3.6 days.   11. 324°F.

3. $1492, 4.92%.

2. 22,400 years.

1. Taking x axis horizontal, with origin at P, $y = x^2/500, -100 \leq x \leq 100$.
2. $y = x^2/625, -250 \leq x \leq 250$; absolute value of slope $= 0.8$.
3. $y = 2.0 \times 10^{-3}x^2 + 8.3 \times 10^{-10}x^4, -100 \leq x \leq 100$, taking x axis horizontal with origin at P.

1. (a) 15.6.   (b) $y = 31.2[\cosh (x/31.2) - 1]$, choosing the minimum point at (0, 0).

2. (a) 66 ft.   (b) 34 lb.
3. (a) 55 ft.   (b) 16.4 lb.   (c) 21.4 lb.

1. 1.25 mi/sec.   3. 18 hr, 50 hr.   5. 216,000 mi.

3. At 50 ft/sec, heights are 39 ft; at 6 mi/sec, heights are 2970 miles, assuming constant gravitational acceleration and 11,500 miles otherwise.

1. $V = b \ln \left(\dfrac{M_0}{M_0 - at} \right) - gt, \; 0 \leq t < M_0/a.$

2. $x = bt - \dfrac{b}{a}(M_0 - at) \ln \left(\dfrac{M_0}{M_0 - at} \right) - \frac{1}{2}gt^2, \; 0 \leq t < M_0/a.$

1. (a) 220 meters/sec, 448 meters/sec, 350 meters/sec.   (b) 445 meters.

1. (a) 18 hr.   (b) 30 hr.   2. (a) 23.8 min.   (b) 2.16 hr.

3. 37.7 min, 62.8 min.   4. $T = \dfrac{16A}{15ac} \sqrt{\dfrac{H}{2g}};$ 100 min, 168 min.

1. $y = x(1 + \ln x).$ 2. $y = ce^{x/2} - 4x - 8.$ 3. $x^2 + y^2 - 4y = 9.$

4. $y = \frac{1}{3}(x + 1).$

6. $\ln (x^2 + y^2) + 2 \tan^{-1} (x/y) = \ln 2 + \pi/2.$

1. $x(\sqrt{x^2 + y^2} + x)/y^2 + \ln |(\sqrt{x^2 + y^2} + x)/y^3| = \frac{3}{2} - 3 \ln 2$

3. $\sqrt{y^2 + 1} + \ln (\sqrt{y^2 + 1} - 1) = \pm x + c$

5. The straight line $y = a\sqrt{2m} - mx$ or the hyperbola $xy = a^2/2.$

6. The straight line $y = am/\sqrt{1 + m^2} - mx$ or the hypocycloid $x^{2/3} + y^{2/3} = a^{2/3}.$

7. (a) $y^2 = 2cx + c^2, \; y^2 = 16 - 8x.$

2. $r = \pm a/(\phi + c)$ where a is the constant length.

3. $r = a \sin (\phi + c)$, where a is the constant length.

4. $r = ce^{\alpha \phi}$ where $\alpha = \pm \sqrt{\frac{1}{2}(\sqrt{5} - 1)}.$

5. $r = \sqrt{3}(\csc \phi - \cot \phi)$ or $r = (\sqrt{3}/3)(\csc \phi + \cot \phi).$

6. (b) $x \cos c + y \sin c = a$ and the envelope $x^2 + y^2 = a^2.$

1. $y = \dfrac{S}{6EI}(3Lx^2 - x^3), 0 \le x \le L, \dfrac{SL^3}{3EI}.$

2. $y = \dfrac{S}{48EI}(3L^2x - 4x^3), 0 \le x \le L/2, \dfrac{SL^3}{48EI}, \dfrac{SL^2}{16EI}$ (y for $L/2 \le x \le L$ is obtained by sym-

metry).

3. (a) $y = \dfrac{S}{6EI}(3Lx^2 - x^3) + \dfrac{w}{24EI}(x^4 - 4Lx^3 + 6L^2x^2).$ **(b)** $\dfrac{SL^3}{3EI} + \dfrac{wL^4}{8EI}.$

4. (a) $y = \dfrac{w}{24EI}(x^4 - 2Lx^3 + L^3x) + \dfrac{S}{48EI}(3L^2x - 4x^3), 0 \le x \le L/2$ (y for $L/2 \le x \le L$

is obtained by symmetry).

(b) $\dfrac{5wL^4}{384EI} + \dfrac{SL^3}{48EI}.$

5. $y = \begin{cases} \dfrac{S}{12EI}(3Lx^2 - 2x^3), 0 \le x \le L/2. \\[2mm] \dfrac{SL^2}{48EI}(6x - L), L/2 \le x \le L. \end{cases}$ $\quad \dfrac{5SL^2}{48EI}.$

6. $y = \begin{cases} \dfrac{w}{24EI}(x^4 + 4Lx^3 + 6L^2x^2) + \dfrac{S}{12EI}(3Lx^2 - 2x^3), 0 \le x \le \dfrac{L}{2}. \\[2mm] \dfrac{w}{24EI}(x^4 - 4Lx^3 + 6L^2x^2) + \dfrac{SL^2}{48EI}(6x - L), \dfrac{L}{2} \le x \le L. \end{cases}$

1. $y = \dfrac{w}{24EI}(x^4 - 2Lx^3 + L^2x^2), 0 \le x \le L/2, \dfrac{wL^4}{384EI}$ (y for $L/2 \le x \le L$ is obtained by

symmetry).

2. $y = \begin{cases} \dfrac{1}{48EI}[(3SL + 2wL^2)x^2 - 4(wL + S)x^3 + 2wx^4], 0 \le x \le L/2. \\[2mm] \dfrac{1}{48EI}[(3SL + 2wL^2)(L - x)^2 - 4(wL + S)(L - x)^3 + 2w(L - x)^4], L/2 \le x \le L. \end{cases}$

Maximum deflection $= \dfrac{L^3}{384EI}(2S + wL)$. (Note that this is valid only for $S \ge 0$; for $S < 0$

the maximum deflection need not occur at $x = L/2$.)

3. (a) $y = \dfrac{w}{48EI}(2x^4 - 5Lx^3 + 3L^2x^2).$

4. $y = \dfrac{w}{384EI}(16x^4 - 12Lx^3 + L^3x), 0 \le x \le L/2$ (y for $L/2 \le x \le L$ is obtained by sym-

metry).

6. (a) $y = \dfrac{k}{360EI}(3x^5 - 10L^2x^3 + 7L^4x), 0 \le x \le L.$

(b) $y = \dfrac{k}{360EI}(x^6 - 5L^3x^3 + 4L^5x), 0 \le x \le L.$

(k is the constant of proportionality in each case.)

1. $y = \dfrac{66.9}{1 + 2.45e^{-1.31t}}$ or $y = \dfrac{66.9}{1 + 2.45(0.270)^t}.$

3. (a) $y = \dfrac{38.47}{1 + 31.06e^{1.165x}}$ or $y = \dfrac{38.47}{1 + 31.06(0.03061)^{x/3}}.$

(b) $y = 1.20, 3.59, 9.52, 19.7, 29.7, 35.2, 37.4$ corresponding to $x = 0, 1, \dots, 6$ respectively.

(c) 38.5 cm^2.

4. (a) 61.3 cm. (b) $y = \dfrac{61.27}{1 + (1.220)(0.3269)^t}$ or $y = \dfrac{61.27}{1 + 1.220e^{-1.118t}}$ where $t = 0, 1, 2$ corresponds to $16,32,48$ days respectively.

(c) 19.1 cm, 36.6 cm, 58.8 cm.

2. $37.2\%, 66.5\%, 86.9\%, 95.7\%.$

1. In another 23 days.

1. (a) $0.08(1 - e^{-1/2}), 0.08(1 - e^{-2}).$ (b) $0.08.$

2. (a) $60 \ln 2.$ (b) $120 \ln 2.$

3. (a) $0.08 + 0.12e^{-1/2}, 0.08 + 0.12e^{-2}.$ (b) $0.08.$

1. (a) $p = 25e^{3t/2} - 20.$ (b) No price stability, no equilibrium price.

2. (a) $p = 15 + 20e^{-2t}.$ (b) There is price stability, equilibrium price $= 20.$

3. (a) $p = 30t + 20.$ (b) No price stability or equilibrium price.

4. $p = 14 + 6e^{-15t}.$

5. $2440 + 360e^{-15t}.$

1. (a) $p = 8 + 4e^{-2t} + te^{-2t}.$ (b) There is price stability, equilibrium price $= 8.$

2. $p = 18 - 3e^{-2t} - 6e^{-t/3}.$

Chapter Four

1. (a) $(D^2 + 3D + 2)y = x^3$. (b) $(3D^4 - 5D^3 + 1)y = e^{-x} + \sin x$.

(c) $(D^2 + \beta D + \omega^2)s = 0, D \equiv d/dt$. (d) $(x^2D^2 - 2xD - 1)y = 1$.

2. (a) $x^3 + 6x^2 - 12x - 6 - 2e^{-x}$. (b) $48 \sin 2x - 16 \cos 2x$.

(c) $6x^2 - 6x - 6 - 2e^{-x} - 8 \cos 2x - 16 \sin 2x$.

(d) $26x^3 - 36x^2 + (4x^2 - 12x - 8)e^{-x} + (12x^2 + 54x + 6) \sin 2x + (36x^2 - 18x + 18) \cos 2x$.

1. $(D - 1)(x^3 + 2x) = -x^3 + 3x^2 - 2x + 2$.

$(D - 2)(D - 1)(x^3 + 2x) = 2x^3 - 9x^2 + 10x - 6$.

$(D^2 - 3D + 2)(x^3 + 2x) = 2x^3 - 9x^2 + 10x - 6$.

1. $(D - x)(2x^3 - 3x^2) = -2x^4 + 3x^3 + 6x^2 - 6x$.

$(D + x)(D - x)(2x^3 - 3x^2) = -2x^5 + 3x^4 - 2x^3 + 3x^2 + 12x - 6$.

$(D^2 - x^2)(2x^3 - 3x^2) = -2x^5 + 3x^4 + 12x - 6$.

$(D - x)(D + x)(2x^3 - 3x^2) = -2x^5 + 3x^4 + 2x^3 - 3x^2 + 12x - 6$.

3. $(D - 1)(D - 2)(D - 3)$; no. **5.** $y = c_1 e^x + c_2 e^{2x} + \dfrac{x}{2} + \dfrac{3}{4}$.

6. (a) $y = c_1 e^{-2x} + c_2 e^x - \frac{1}{2}e^{-x}$. (b) $y = c_1 e^{3x} + c_2 e^{-2x} + c_3 e^x - \frac{1}{2}xe^x - \frac{1}{3}$.

(c) $y = c_1 e^x + c_2 x e^x + 1$. (d) $y = c_1 e^{-x} + c_2 e^{-x} \int e^{(x^2/2) + x} \, dx - 1$.

1. (a) $y = c_1 e^x + c_2 e^{-5x}$. (b) $y = c_1 e^{5x/2} + c_2 e^{-5x/2}$. (c) $y = c_1 e^{2x} + c_2 e^{-2x}$.

(d) $y = c_1 e^{2x} + c_2 e^{x/2} + c_3$. (e) $I = c_1 e^{(2+\sqrt{2})t} + c_2 e^{(2-\sqrt{2})t}$.

(f) $y = c_1 e^{-x} + c_2 e^{-3x} + c_3 e^{2x}$.

2. (a) $y = -\frac{1}{2}e^x + \frac{5}{2}e^{-x}$. (b) $y = e^{2x} - 2e^x$. (c) $y = \frac{1}{2}(e^{4x} + e^{-4x}) - 1$.

1. $y = c_1 e^{-3x} + c_2 e^{(\sqrt{5} - 1)x} + c_3 e^{-(\sqrt{5} + 1)x}$.

1. $y = c_1 e^{(2 - \sqrt{6})x} + c_2 e^{(2 + \sqrt{6})x} + c_3 e^{-(2 - \sqrt{6})x} + c_4 e^{-(2 + \sqrt{6})x}$.

2. $y = c_1 e^{(3 + \sqrt{3})x} + c_2 e^{(3 - \sqrt{3})x} + c_3 e^{-(2 + \sqrt{2})x} + c_4 e^{-(2 - \sqrt{2})x}$.

1. (a) $y = c_1 e^{2x} + c_2 x e^{2x}$. (b) $y = c_1 e^{x/4} + c_2 x e^{x/4}$. (c) $I = c_1 e^{3t/2} + c_2 t e^{3t/2}$.

(d) $y = c_1 + c_2 x + c_3 x^2 + c_4 x^3 + c_5 e^{2x} + c_6 e^{-2x}$.

(e) $y = c_1 + c_2 x + c_3 e^x + c_4 x e^x$.

(f) $y = c_1 e^{\sqrt{5/2}x} + c_2 x e^{\sqrt{5/2}x} + c_3 e^{-\sqrt{5/2}x} + c_4 x e^{-\sqrt{5/2}x}$.

2. (a) $y = e^x - 3xe^x$. (b) $y = 1$. (c) $s = -4te^{-8t}$.

1. $y = c_1 x + c_2[x \int e^{x^2/2}\, dx - e^{x^2/2}]$. 2. $p = -1$, $y = c_1 x^{-1} + c_2 x^{-1} \ln x$.

3. $y = c_1 \left[\dfrac{x}{2} \ln \left(\dfrac{1+x}{1-x} \right) - 1 \right] + c_2 x$.

$y = c_1 \left[\dfrac{x}{2} \ln \left(\dfrac{1+x}{1-x} \right) - 1 \right] + c_2 x - \dfrac{x}{6} \ln (1 - x^2)$.

1. (a) $y = c_1 \sin 2x + c_2 \cos 2x$. (b) $y = e^{-2x}(c_1 \sin x + c_2 \cos x)$.

(c) $s = c_1 \sin \dfrac{3t}{2} + c_2 \cos \dfrac{3t}{2}$. (d) $y = e^x \left(c_1 \sin \dfrac{\sqrt{3}}{2} x + c_2 \cos \dfrac{\sqrt{3}}{2} x \right)$.

(e) $y = c_1 + c_2 x + c_3 \sin 4x + c_4 \cos 4x$. (f) $y = c_1 e^x + e^{-x}(c_2 \sin x + c_3 \cos x)$.

2. (a) $y = 4 \cos x$. (b) $U = \sin 4t$. (c) $I = e^{-t}(\sin 2t + 2 \cos 2t)$.

1. $y = c_1 e^{-2x} + c_2 e^{2x} + e^x(c_3 \sin \sqrt{3}x + c_4 \cos \sqrt{3}x) + e^{-x}(c_5 \sin \sqrt{3}x + c_6 \cos \sqrt{3}x)$.

3. $y = (c_1 \sin \sqrt{2}x + c_2 \cos \sqrt{2}x) + x(c_3 \sin \sqrt{2}x + c_4 \cos \sqrt{2}x)$.

1. $y = e^x(c_1 \sin x + c_2 \cos x) + e^{-x}(c_3 \sin x + c_4 \cos x)$.

2. $y = e^x(c_1 \sin 2x + c_2 \cos 2x) + e^{-x}(c_3 \sin 2x + c_4 \cos 2x)$.

MISCELLANEOUS REVIEW EXERCISES ON COMPLEMENTARY SOLUTIONS

1. (a) $y = c_1 e^{3x} + c_2 e^{-x}$. (b) $y = c_1 e^{4x} + c_2 + c_3 e^{-2x}$.

(c) $y = c_1 e^{2x} + c_2 x e^{2x} + c_3 x^2 e^{2x} + c_4 + c_5 x$.

(d) $y = c_1 e^{-x} + c_2 \sin x + c_3 \cos x + c_4 e^{-2x}$.

(e) $y = e^{2x}(c_1 \sin 3x + c_2 \cos 3x) + e^{-x}(c_3 \sin 2x + c_4 \cos 2x) + c_5 e^{5x} + c_6 e^{-x}$.

(f) $y = e^x(c_1 \sin \sqrt{3}x + c_2 \cos \sqrt{3}x) + c_3 e^{-2x} + c_4 x e^{-2x} + c_5 + c_6 e^{-x}$.

(g) $y = c_1 e^{-x} + c_2 x e^{-x} + c_3 e^x + c_4 x e^x + c_5 + c_6 e^{-2x} + e^{-x}(c_7 \sin 2x + c_8 \cos 2x)$.

(h) $y = e^x(c_1 \sin x + c_2 \cos x) + x e^x(c_3 \sin x + c_4 \cos x)$.

(i) $y = c_1 e^{\sqrt{3}x} + c_2 e^{-\sqrt{3}x} + c_3 e^{4x} + c_4 e^{-4x} + e^{x/2}(c_5 \sin 2x + c_6 \cos 2x)$
$\qquad\qquad\qquad\qquad\qquad\qquad + e^{-x}(c_7 \sin 3x + c_8 \cos 3x)$.

(j) $y = c_1 e^x + c_2 x e^x + c_3 x^2 e^x + c_4 + c_5 x + c_6 \sin x + c_7 \cos x + x(c_8 \sin x + c_9 \cos x)$.

2. (a) $y = e^{-x/2} \left(c_1 \sin \dfrac{\sqrt{3}}{2} x + c_2 \cos \dfrac{\sqrt{3}}{2} x \right)$.

(b) $y = c_1 e^x + c_2 e^{-x} + c_3 \sin x + c_4 \cos x$.

(c) $y = c_1 + c_2 x + c_3 \sin x + c_4 \cos x + x(c_5 \sin x + c_6 \cos x)$.

(d) $y = c_1 + c_2 e^{2x} + c_3 x e^{2x}.$ (e) $y = c_1 + c_2 x + c_3 e^x.$

(f) $S = c_1 \sin 2t + c_2 \cos 2t + c_3 e^{\sqrt{2}t} + c_4 e^{-\sqrt{2}t}.$

B EXERCISES, p. 181

1. $a = 5, b = 24, c = 20.$

2. $(D^6 - 2D^5 + 7D^4 + 4D^3 - D^2 + 30D + 25)y = 0,$
   $y = c_1 e^{-x} + c_2 x e^{-x} + e^x(c_3 \sin 2x + c_4 \cos 2x) + x e^x(c_5 \sin 2x + c_6 \cos 2x).$

C EXERCISES, p. 181

2. (a) $y = c_1 e^x + e^{-x/2}\left(c_2 \sin \dfrac{\sqrt{3}}{2} x + c_3 \cos \dfrac{\sqrt{3}}{2} x \right).$

   (b) $y = c_1 e^x + e^{x \cos 4\pi/5}[c_2 \sin (x \sin 4\pi/5) + c_3 \cos (x \sin 4\pi/5)]$
   $\qquad\qquad\qquad\qquad\qquad + e^{x \cos 2\pi/5}[c_4 \sin (x \sin 2\pi/5) + c_5 \cos (x \sin 2\pi/5)].$

3. $y = c_1 e^{\sqrt[3]{4}x} + e^{-\sqrt[3]{4}x/2}\left(c_2 \sin \dfrac{\sqrt{3}\sqrt[3]{4}}{2} x + c_3 \cos \dfrac{\sqrt{3}\sqrt[3]{4}}{2} x \right)$

A EXERCISES, p. 189

1. l.d. denotes *linearly dependent*; l.i. denotes *linearly independent*.

   (a) l.i.   (b) l.d.   (c) l.i.   (d) l.i.   (e) l.d.   (f) l.i.   (g) l.d.   (h) l.d.

3. (a) $y = c_1 e^x + c_2 e^{-3x}.$ (b) $y = e^x(c_1 \cos 2x + c_2 \sin 2x).$

   (c) $y = c_1 + c_2 x + c_3 e^{3x}.$ (d) $y = (c_1 + c_2 x)e^{2x} + (c_3 + c_4 x)e^{-2x}.$

B EXERCISES, p. 189

4. The third function is the sum of the first two so that the functions are linearly dependent.

A EXERCISES, p. 194

1. (a) $y = c_1 \sin x + c_2 \cos x + \frac{1}{5}e^{3x}.$ (b) $y = c_1 e^{-x} + c_2 x e^{-x} - \frac{12}{25} \sin 2x - \frac{16}{25} \cos 2x.$

   (c) $y = c_1 e^{2x} + c_2 e^{-2x} - 2x^2 - 1.$

   (d) $y = e^{-2x}(c_1 \sin x + c_2 \cos x) + \frac{1}{2}e^{-x} + 3x - \frac{12}{5}.$

   (e) $I = c_1 \sin \dfrac{t}{2} + c_2 \cos \dfrac{t}{2} + t^2 - 8 - \dfrac{2}{35} \cos 3t.$

   (f) $y = c_1 + c_2 \sin 2x + c_3 \cos 2x + \frac{1}{5}e^x - \frac{1}{3}\cos x.$

2. (a) $y = \frac{1}{3}\sin x - \frac{1}{12}\sin 4x.$ (b) $s = 8e^{2t} - 24e^t + 4t^2 + 12t + 14 + 2e^{-t}.$

B EXERCISES, p. 195

1. $y = 3 - 4 \sin x - 2 \cos x - \cos 2x.$

C EXERCISES, p. 195

1. $y = c_1 e^x + c_2 e^{2x} + \frac{27}{130} \cos 3x - \frac{21}{130} \sin 3x + \frac{79}{6970} \sin 9x - \frac{27}{6970} \cos 9x.$

2. $y = \begin{cases} x - \sin x, 0 \le x \le \pi. \\ -\pi \cos x - 2 \sin x, x \ge \pi. \end{cases}$

A EXERCISES, p. 197

1. (a) $y = c_1 e^x + c_2 e^{-3x} + \frac{1}{2} x e^x$.   (b) $y = c_1 \sin x + c_2 \cos x + x^2 - 2 - \frac{1}{2} x \cos x$.

(c) $y = c_1 + c_2 e^{-x} + \dfrac{x^3}{3} + \dfrac{x^2}{2} - x + \frac{1}{12} e^{3x}$.

(d) $y = c_1 e^x + c_2 x e^x + \frac{1}{2} x^2 e^x$.

(e) $y = c_1 \sin 2x + c_2 \cos 2x + 2x \sin 2x - x$.

(f) $y = c_1 + c_2 \sin x + c_3 \cos x + \dfrac{x^2}{2} - \dfrac{x}{2} (\sin x + \cos x)$.

2. (a) $I = 4 \cos 3t + 2t \sin 3t$.   (b) $s = 2 - 2e^{-t} + \dfrac{t^2}{2} - t - t e^{-t}$.

B EXERCISES, p. 197

1. $y = c_1 \sinh x + c_2 \cosh x + c_3 \sin x + c_4 \cos x + \dfrac{x}{4} \sinh x$.

2. $y = c_1 \sin x + c_2 \cos x + \frac{1}{4}(x \sin x - x^2 \cos x)$.

C EXERCISES, p. 197

3. $y = c_1 \sin 2x + c_2 \cos 2x + \frac{3}{32} - \frac{1}{96} \cos 4x - \dfrac{x}{8} \sin 2x$.

A EXERCISES, p. 199

1. (a) $c_1 \sin x + c_2 \cos x; (ax + b)e^{-x} + x(c \sin x + d \cos x)$.

(b) $c_1 e^{-x} + c_2 e^{3x}; (ax + b) \sin 2x + (cx + d) \cos 2x + (fx^4 + gx^3 + hx^2 + kx)e^{3x}$.

(c) $c_1 + c_2 x + c_3 \sin x + c_4 \cos x; ax^4 + bx^3 + cx^2 + de^x$.

(d) $c_1 e^x + c_2 x e^x; (ax^4 + bx^3 + cx^2)e^x$.

(e) $c_1 \sin x + c_2 \cos x; e^{-x}(a \sin x + b \cos x) + cx + d$.

(f) $c_1 e^x + c_2 e^{3x}; axe^x + (bx^3 + cx^2 + dx + f)e^{-x}$.

2. (a) $y = c_1 e^x + c_2 e^{-x} + \dfrac{e^x}{4}(x^2 - x)$.

(b) $y = c_1 \sin 2x + c_2 \cos 2x + \dfrac{x^2}{4} - \dfrac{1}{8} + \dfrac{3x^2}{8} \sin 2x + \dfrac{3x}{16} \cos 2x$.

(c) $y = c_1 e^{-x} + c_2 x e^{-x} + \frac{1}{6} x^3 e^{-x} - \frac{1}{50}(3 \cos 3x + 4 \sin 3x)$.

(d) $Q = c_1 \sin t + c_2 \cos t + \frac{3}{4} t \sin t - \frac{1}{4} t^2 \cos t$.

B EXERCISES, p. 200

1. $y = c_1 e^{-2x} + c_2 e^{4x} - \left(\dfrac{x^3}{15} + \dfrac{4x^2}{25} + \dfrac{42x}{125} + c_3 \right) e^{3x}$.

2. $y = \dfrac{t}{4\omega^2}(\sin \omega t + \cos \omega t) + \dfrac{t^2}{4\omega}(\sin \omega t - \cos \omega t) - \dfrac{1}{4\omega^3} \sin \omega t$.

3. $y = c_1 e^x + c_2 e^{2x} + \frac{1}{6} e^{-x} + \dfrac{e^{-x}}{52}(\cos 2x - 5 \sin 2x)$.

1. $y = c_1 \sin 2x + c_2 \cos 2x + \frac{1}{16} + \frac{1}{16}x \sin 2x - \frac{1}{48} \cos 4x - \frac{1}{128} \cos 6x.$

2. $y = c_1 e^{2x} + c_2 e^{-(3+\sqrt{3})x} + c_3 e^{-(3-\sqrt{3})x} + \dfrac{e^{4x}}{1472} - \dfrac{xe^{2x}}{88} - \dfrac{1}{32} - \dfrac{e^{-2x}}{32} + \dfrac{e^{-4x}}{192}.$

1. $y = c_1 \sin x + c_2 \cos x + \sin x \ln (\csc x - \cot x).$

2. $y = c_1 \sin x + c_2 \cos x + x \sin x + \cos x \ln \cos x.$

3. $y = c_1 \sin 2x + c_2 \cos 2x + \frac{1}{4} \sin 2x \ln \sin 2x - (x/2) \cos 2x.$

4. $y = c_1 e^x + c_2 e^{-x} + \frac{1}{2}xe^x.$ **5.** $y = c_1 e^{-x} + c_2 e^{-2x} + \dfrac{x}{2} - \dfrac{3}{4} - 3xe^{-2x}.$

6. $y = c_1 e^{-2x} + c_2 e^x - \dfrac{1}{2} \ln x + \dfrac{e^x}{3} \int \dfrac{e^{-x}}{x} dx + \dfrac{e^{-2x}}{6} \int \dfrac{e^{2x}}{x} dx.$

7. $y = c_1 e^{-x} + c_2 e^{-x/2} + \frac{1}{10}e^{-3x}.$

8. $y = c_1 e^x + c_2 e^{-x} + \dfrac{e^x}{48}(8x^3 - 12x^2 + 12x).$

9. $y = c_1 e^x + c_2 e^{-x} + 2e^x \int e^{-x-x^2} dx - 2e^{-x} \int e^{x-x^2} dx.$

10. $y = c_1 e^{2x} + c_2 xe^{2x} - e^{2x} \int x^{3/2}e^{-2x} dx + xe^{2x} \int x^{1/2}e^{-2x} dx.$

3. (b) $y = c_1 e^x + c_2 e^{-x} + c_3 e^{2x} - \frac{1}{2}xe^x.$

5. $y = c_1 x^2 + c_2 x - xe^{-x} - (x^2 + x) \int \dfrac{e^{-x}}{x} dx.$

1. $y = c_1 + c_2 e^x - x.$ **2.** $y = c_1 e^x + c_2 xe^x + \frac{1}{2}x^2 e^x.$

3. $y = c_1 e^{-x} + c_2 e^{-2x} + \frac{1}{6}e^x - xe^{-x}.$

4. $y = c_1 e^x + c_2 e^{-x} - 2x^4 - 24x^2 + 3x - 49.$

5. $y = c_1 + c_2 e^{-x} + x^4 - 4x^3 + 12x^2 - 24x - \frac{1}{3}e^{2x}.$

6. $y = c_1 e^{-x} + c_2 xe^{-x} + \frac{1}{12}x^4 e^{-x} + 1.$ **7.** $y = c_1 e^{2x} + c_2 xe^{2x} + \frac{1}{9}e^{2x} \sin 3x.$

8. $y = c_1 + c_2 e^x + c_3 e^{-x} - \frac{1}{6}x^6 - 5x^4 - 60x^2 - x.$

3. $\frac{1}{5}e^{4x}, \quad y = c_1 e^{3x} + c_2 e^{-x} + \frac{1}{5}e^{4x}.$

4. $\dfrac{e^{3x}}{15} + \dfrac{2e^{-5x}}{21}, \quad y = c_1 e^{2x} + c_2 e^{-2x} + c_3 e^{-3x} + \dfrac{e^{3x}}{15} + \dfrac{2e^{-5x}}{21}.$

9. (a) $-e^{2x}(x^3 + 6x).$ (b) $2e^{-2x}(x^2 + 4x + 6) + \frac{1}{3}e^{2x}.$

11. (a) $-\frac{1}{8} \sin 3x.$ (b) $\frac{1}{15}(\cos 2x - 2 \sin 2x).$

13. (a) $\frac{1}{130}(9 \cos 3x - 7 \sin 3x).$ (b) $\dfrac{8 \cos 2x - 19 \sin 2x}{85}.$

14. (a) $-\dfrac{e^{2x}}{50}(3 \cos 3x + \sin 3x).$ **(b)** $\dfrac{e^{-x}}{58}(2 \cos 2x - 5 \sin 2x).$

16. (a) $\dfrac{e^{3x}}{10}.$ **(b)** $-x^5 - 3x^4 + 2x^3 - 60x^2 - 72x + 12.$

(c) $\dfrac{e^{2x}}{125}(25x^2 - 40x + 22).$ **(d)** $-\tfrac{2}{17}(\cos 4x + 4 \sin 4x).$

(e) $\cos x.$ **(f)** $-\tfrac{1}{64}e^{3x} \sin 4x.$

(g) $\dfrac{e^x}{5}(4 \sin x + 7 \cos x).$ **(h)** $e^{-x}\left(\dfrac{x^6}{120} - \dfrac{x^5}{10} + \dfrac{3x^4}{4} - 4x^3 + \dfrac{1}{2} \sin x\right).$

(i) $-x^2 + 2x - 4 + \tfrac{3}{4}(\cos x - \sin x).$

(j) $e^x(x^2 - 4x + 4) - \tfrac{4}{5}x^5 - 4x^4 + 48x^2 + 96x.$

C EXERCISES, p. 210

1. $\dfrac{e^{2x}}{60}\left(x^4 - \dfrac{4}{5}x^3 + \dfrac{12}{25}x^2 - \dfrac{24}{125}x + \dfrac{24}{125}\right).$

4. (a) $\tfrac{1}{2}x^2 e^x.$ **(b)** $-2xe^x - xe^{2x}.$ **(c)** $\tfrac{1}{24}x^4 e^{-2x}.$

(d) $-\tfrac{1}{6}x \cos 3x.$ **(e)** $-\dfrac{x}{2}(\sin x + \cos x).$ **(f)** $-\dfrac{x}{4}\cos x - \dfrac{x^2}{8}\sin x.$

5. $y = \left(\dfrac{x^2}{4} - \dfrac{1}{8}\right)\cos x + \left(\dfrac{x^3}{6} - \dfrac{x}{4}\right)\sin x.$

A EXERCISES, p. 213

1. (a) $y = c_1 x^2 + c_2 x.$ **(b)** $y = \tfrac{1}{2}x^{1/2}(2 - \ln x).$

(c) $y = c_1 x^2 + \dfrac{c_2}{x} - \dfrac{x}{2}.$ **(d)** $y = x[c_1 \sin (\ln x) + c_2 \cos (\ln x)] + \tfrac{1}{2} \ln x + \tfrac{1}{2}.$

(e) $y = \dfrac{c_1}{x^2} + \dfrac{c_2 \ln x}{x^2} + \dfrac{x^2}{16} + 4(\ln x)^2 - 8 \ln x + 6.$

(f) $y = x^{1/2}\left[c_1 \cos \left(\dfrac{\sqrt{3}}{2} \ln x\right) + c_2 \sin \left(\dfrac{\sqrt{3}}{2} \ln x\right)\right] + 16 \cos (\ln x).$

(g) $I = \dfrac{c_1 + c_2 \ln t}{t} + \dfrac{t}{4}(\ln t - 1).$

(h) $y = 2x^{4/5} - 3x^{1/5} + x.$

(i) $y = c_1 x^3 + c_2 x^{-3} - \tfrac{4}{35}(x^{1/2} + x^{-1/2}).$

(j) $y = c_1 x^3 + c_2 \ln x + c_3 - \tfrac{5}{2}x \ln x + \tfrac{15}{4}x.$

2. $y = c_1 + \dfrac{c_2}{x} + c_3 x - \ln x + \tfrac{1}{2}x \ln x.$

3. (a) $y = c_1 x + c_2 x \ln x + c_3 x(\ln x)^2 + \dfrac{x(\ln x)^4}{24}$

(b) $y = c_1 x + \dfrac{c_2}{x} + c_3 \cos (\ln x) + c_4 \sin (\ln x) - 1.$

1. $-\frac{1}{2}(\ln x)^2 - \ln x + c_1 x + c_2$.

3. $m = -3, 1; y = c_1 x^{-3} + c_2 x$.

5. $y = \dfrac{c_1}{\sqrt{2x + 3}} + c_2(2x + 3) + \frac{3}{5}(2x + 3)^2 - 6(2x + 3) \ln (2x + 3) - 27$.

6. $y = c_1(x + 2)^{(1/2)(1 + \sqrt{5})} + c_2(x + 2)^{(1/2)(1 - \sqrt{5})} - 4$.

7. $R = c_1 r^n + \dfrac{c_2}{r^{n+1}}$.

1. $y = c_1 \sin (\sin x) + c_2 \cos (\sin x)$.

2. $y = c_1 e^{x^2} + c_2 e^{-x^2}$.

6. (a) $y = c_1 e^{\cos x} + c_2 e^{2 \cos x}$. (b) $y = c_1 \sin \left(\dfrac{1}{x}\right) + c_2 \cos \left(\dfrac{1}{x}\right) + \dfrac{1}{x^2} - 2$.

MISCELLANEOUS EXERCISES ON CHAPTER FOUR

1. $y = \dfrac{2}{\sqrt{3}} \sin \sqrt{3}x - \frac{1}{9} \cos \sqrt{3}x + \frac{1}{3}x^2 + \frac{1}{9}$.

2. $y = c_1 e^x + c_2 e^{2x} + \frac{1}{10}(\sin x + 3 \cos x)$.

3. $y = c_1 e^{-x} + c_2 x e^{-x} + \frac{1}{4}e^x + \frac{1}{2}x^2 e^{-x}$.

4. $y = c_1 e^{2x} + c_2 e^{-2x} + c_3 - \frac{1}{2}x^2 - \frac{1}{2}x + \frac{3}{8}xe^{-2x}$.

5. $I = e^{-t}(c_1 \cos 2t + c_2 \sin 2t) + 2(\cos 2t + 4 \sin 2t)$.

6. $x = c_1 e^t + c_2 e^{-t} + c_3 \cos t + c_4 \sin t - 2te^{-t}$.

7. $y = \dfrac{e^{2x}}{64}(8x^2 - 4x + 1) - \frac{1}{64}e^{-2x}$.

8. $y = \frac{4}{5}x^3 + \frac{6}{5}x^{-2}$.

9. $y = c_1 e^{2x} + c_2 x + c_3 - \frac{1}{4}x^2$.

10. $y = c_1 x + c_2 + c_3 \cos 4x + c_4 \sin 4x + \frac{1}{2}x \sin 4x$.

11. $y = c_1 \cos 2x + c_2 \sin 2x + \dfrac{x}{4} + \frac{1}{3}x \cos x + \frac{2}{9} \sin x$.

12. $r = c_1 e^{\sqrt{2}\phi} + c_2 e^{-\sqrt{2}\phi} - \frac{1}{2}e^{-2\phi}$.

13. $y = c_1 e^{2x} + c_2 x e^{2x} + c_3 + 3x^2 e^{2x} + 2x^3 + 6x^2 + 9x$.

14. $y = \cos x + 2 \sin x + \cos x \ln (\cos x) + x \sin x$.

15. $y = c_1 x^4 + c_2 x - 8x \ln x + 6$.

16. $s = c_1 e^t + c_2 t e^t + c_3 e^{-t} + c_4 t e^{-t} + \cos 3t$.

17. $y = c_1 e^{x/2} + c_2 x e^{x/2} - e^{x/2} \int xe^{-x/2} \ln x \, dx + xe^{x/2} \int e^{-x/2} \ln x \, dx$.

18. $y = c_1 + c_2 e^x + c_3 e^{-x} + c_4 e^{3x} + c_5 e^{-3x} + \dfrac{x^3}{27} - \dfrac{x^2}{18} + \dfrac{20x}{81} - \dfrac{xe^x}{16}$.

19. $I = c_1 \cos 3t + c_2 \sin 3t + c_3 t + c_4 + 2e^{-t}$.

20. $y = c_1 x^2 + c_2 x^2 \ln x + c_3 + \frac{1}{8}(\ln x)^2 + \frac{1}{4} \ln x$.

1. $y = c_1 e^{2x} + c_2 \cos 2x + c_3 \sin 2x + 4x \cos 2x - 4x \sin 2x.$

2. $y = 8x^2.$

3. $x = 2e^{-4t}(\cos 2t + \sin 2t) - 2e^{-2t}.$

4. $y = c_1 \cos x \int e^{-\sin x} \sec^2 x \, dx + c_2 \cos x.$

5. $y = \sin x + c_1 \cos x \int e^{-\sin x} \sec^2 x \, dx + c_2 \cos x.$

6. $y = c_1 x^4 + x^{-1/2} \left[c_1 \cos \left(\dfrac{\sqrt{23}}{2} \ln x \right) + c_2 \sin \left(\dfrac{\sqrt{23}}{2} \ln x \right) \right] - x.$

7. $y = c_1 e^x + c_2 e^{-x} + c_3 e^{-2x} + \dfrac{e^x}{6} \int \dfrac{e^{-x}}{x} \, dx - \dfrac{e^{-x}}{2} \int \dfrac{e^x}{x} \, dx + \dfrac{e^{-2x}}{3} \int \dfrac{e^{2x}}{x} \, dx.$

1. $y = e^{-x^2}(c_1 \cos x + c_2 \sin x).$

2. $y = \dfrac{c_1 \cos x + c_2 \sin x}{x}.$

3. Eigenvalues are given by $\lambda = n^2$, $n = 1, 2, 3, \ldots$. Eigenfunctions are $B_n \sin nx$, $n = 1, 2, 3, \ldots$, where B_n, $n = 1, 2, 3, \ldots$ are constants.

4. (b) $y = \dfrac{c \sin x - \cos x}{x(c \cos x + \sin x)}.$

5. $x = c_1 e^{-y} + c_2 y e^{-y} + 4y^2 - 16y + 24.$

6. (a) $Q = Q_0 \cos \sqrt{k} t + \dfrac{1}{\sqrt{k}} \int_0^t E(u) \sin \sqrt{k}(t - u) du.$

9. $y = c_1 - c_1 x e^{-x} \int \dfrac{e^x}{x} \, dx + c_2 x e^{-x}.$

10. $y = c_1 e^x + c_2 e^x \int \dfrac{e^{-x}}{x^2} \, dx.$

Chapter Five

A EXERCISES, p. 225

1. Take positive direction downward. Then

   (b) $x = \frac{1}{4} \cos 16t$ (ft), $v = -4 \sin 16t$ (ft/sec).

   (c) Amplitude $= \frac{1}{4}$ ft, period $= T = \pi/8$ sec, frequency $= f = 8/\pi$ cycles per sec.

   (d) $x = \sqrt{2}/8$ ft, $v = -2\sqrt{2}$ ft/sec, $a = -32\sqrt{2}$ ft/sec^2.

2. (a) $x = \frac{1}{4} \sin 8t$ (ft), $v = 2 \cos 8t$ (ft/sec).

   (b) Amplitude $= \frac{1}{4}$ ft, $T = \pi/4$ sec, $f = 4/\pi$ cycles per sec.

   (c) 1.89 ft/sec, 5.33 ft/sec^2.

3. $x = \frac{5}{12} \cos 4\sqrt{3} t$, amplitude $= \frac{5}{12}$ ft, $T = 2\pi/4\sqrt{3} = \pi\sqrt{3}/6$ sec, $f = 2\sqrt{3}/\pi$ cycles per sec.

4. $x = \dfrac{5}{12} \cos 4\sqrt{3} t + \dfrac{5\sqrt{3}}{12} \sin 4\sqrt{3} t = \dfrac{5}{6} \sin \left(4\sqrt{3} t + \dfrac{\pi}{6} \right).$

   Amplitude $= \frac{5}{6}$ ft, $T = \pi\sqrt{3}/6$ sec, $f = 2\sqrt{3}/\pi$ cycles per sec.

5. (a) Amplitude $= 5$ cm, $T = \pi/\sqrt{5}$ sec, $f = \sqrt{5}/\pi$ cycles per sec.

   (b) $\pi/6\sqrt{5}, 5\pi/6\sqrt{5}, 7\pi/6\sqrt{5}, 11\pi/6\sqrt{5}, \ldots$ seconds.

6. (a) $x = 1.5$ in. above equilibrium position, $v = 5\sqrt{3}/8$ ft/sec moving upward.

(b) Amplitude $= 3$ in., $T = 2\pi/5$ sec, $f = 5/2\pi$ cycles per sec.

(c) $2\pi/15, 8\pi/15, 14\pi/15, \ldots$ seconds.

7. (a) $x = 4\cos 2t - 3\sin 2t$, $v = -8\sin 2t - 6\cos 2t$.

(b) Amplitude $= 5$ cm, period $= \pi$ sec, frequency $= 1/\pi$ cycles per sec. [*Hint:* $4\cos 2t - 3\sin 2t = 5(\frac{4}{5}\cos 2t - \frac{3}{5}\sin 2t) = 5\cos(2t + \phi)$ where $\cos\phi = \frac{4}{5}$, $\sin\phi = -\frac{3}{5}$. Thus the maximum value of x, i.e., the amplitude, is 5. This can also be found by calculus.]

(c) Maximum velocity $= 10$ cm/sec, maximum acceleration $= 20$ cm/sec^2.

8. (a) $x = \pm 4\sin 5t$, $v = \pm 20\cos 5t$, $a = \mp 100\sin 5t$ (the choice of sign depends on the direction of motion of the particle at $t = 0$).

(b) Amplitude $= 4$ cm, period $= 2\pi/5$ sec, frequency $= 5/2\pi$ cycles per sec.

(c) Magnitude of force $= 100\sqrt{2}$ dynes.

9. $x = 10\cos 3t$, amplitude $= 10$ cm, period $= 2\pi/3$ sec. Frequency $= 3/2\pi$ cycles per sec.

10. (a) $2\sqrt{3}$ ft/sec toward or away from O.

(b) Amplitude $= 1$ ft, period $= \dfrac{\pi}{2}$ sec, frequency $= \dfrac{2}{\pi}$ cycles per second.

(c) $\sqrt{2}/2$ ft from O, $2\sqrt{2}$ ft/sec toward O, $8\sqrt{2}$ ft/sec^2 toward O.

11. (a) $20\sqrt{3}$ cm/sec toward or away from O, 40 cm/sec^2 toward O.

(b) Amplitude $= 20$ cm, period $= \pi$ sec, frequency $= 1/\pi$ cycles per sec.

(c) 10 cm from O, $20\sqrt{3}$ cm/sec away from O, 40 cm/sec^2 toward O.

(d) After $\dfrac{\pi}{4}$ sec, $\dfrac{3\pi}{4}$ sec, $\dfrac{5\pi}{4}$ sec, etc.

B EXERCISES, p. 227

3. 4.5 lb.

6. (a) 85 min.   (b) 4.9 mi/sec.

A EXERCISES, p. 234

1. (b) $x = \dfrac{e^{-8t}}{2}(\sin 8t + \cos 8t)$, taking downward as positive.

(c) $x = \dfrac{\sqrt{2}}{2}e^{-8t}\sin\left(8t + \dfrac{\pi}{4}\right)$, $A(t) = \dfrac{\sqrt{2}}{2}e^{-8t}$, $\omega = 8$, $\phi = \dfrac{\pi}{4}$,

quasi period $= 2\pi/8 = \pi/4$ sec.

2. (a) $x = -\frac{25}{32}e^{-4.8t}\sin 6.4t$, $v = e^{-4.8t}(3.75\sin 6.4t - 5\cos 6.4t)$, taking downward as positive.

(b) $x = \frac{25}{32}e^{-4.8t}\sin(6.4t + \pi)$.

3. $x = 1.5e^{-3t}(3\sin 4t + 4\cos 4t) = 7.5e^{-3t}\sin(4t + \phi)$ where $\cos\phi = \frac{3}{5}$, $\sin\phi = \frac{4}{5}$ or $\phi = 0.927$ radians $= 53°$. Here, x is in inches.

4. (a) $x = 5e^{-3t}(3\sin 4t + 4\cos 4t)$ taking downward as positive.

(b) $v = -125e^{-3t}\sin 4t$.

5. (a) $x = 5e^{-3t}(4 \cos 4t + 9 \sin 4t)$, $v = 5e^{-3t}(24 \cos 4t - 43 \sin 4t)$.

(b) $x = 5e^{-3t}(4 \cos 4t - 3 \sin 4t)$, $v = -5e^{-3t}(24 \cos 4t + 7 \sin 4t)$.

6. (a) $x = e^{-16t}(0.5 + 8t)$ (ft).   (b) Critically damped.

7. (a) $x = 10te^{-16t}$ (ft), $v = 10(1 - 16t)e^{-16t}$ (ft/sec).   (b) $5/8e = 0.23$ ft.

A EXERCISES, p. 237

1. (a) $x = 0.960e^{-1.56t} - 0.695e^{-6.45t} - 0.298 \sin 10t - 0.265 \cos 10t$.

(b) Steady-state part: $-0.298 \sin 10t - 0.265 \cos 10t = 0.397 \sin (10t + 3.87)$.

(c) Steady-state amplitude $= 0.397$ ft, period $= 2\pi/10 = \pi/5$ sec, frequency $= 5/\pi$ cycles per sec.

2. $x = \frac{2}{3} \cos 2t - 5 \sin 2t - \frac{2}{3} \cos 4t$, $v = \frac{8}{3} \sin 4t - \frac{4}{3} \sin 2t - 10 \cos 2t$.

3. (a) $x = 10te^{-5t} - 2 \cos 5t$.   (b) $x = -2 \cos 5t$.

B EXERCISES, p. 238

2. (a) $x = 5 \cos 10t + \cos 8t$.   (b) $x = \sin 10t - \cos 10t + \cos 8t$.

(c) $x = 5 \cos 10t + \sin 10t + \cos 8t$.

A EXERCISES, p. 241

1. (b) $x = 2 \sin 2t - 4t \cos 2t$, $v = 8t \sin 2t$.

2. $x = \frac{1}{2} \cos 2t - 4t \cos 2t$, $v = 8t \sin 2t - \sin 2t - 4 \cos 2t$; yes.

3. (a) $x = \dfrac{15 \cos \omega t}{100 - \omega^2}$.   (b) $\omega = 10$.

A EXERCISES, p. 244

1. (a) $Q = 4 - 2e^{-5t/2}(2 \cos 5t + \sin 5t)$, $I = 25e^{-5t/2} \sin 5t$.

(b) Transient terms of Q and I are $-2e^{-5t/2}(2 \cos 5t + \sin 5t)$ and $25e^{-5t/2} \sin 5t$ respectively. Steady-state term of Q is 4.

(c) $Q = 4, I = 0$.

2. (a) Period $= 2\pi/50 = \pi/25$ sec, frequency $= 25/\pi$ cycles per sec.   (b) 0.04 coulombs, 1 amp.

3. $Q = 2(\cos 40t - \cos 50t)$, $I = 20(5 \sin 50t - 4 \sin 40t)$.

4. $Q = 5.07e^{-5t} - 1.27e^{-20t} + 0.20$, $I = 25.4(e^{-20t} - e^{-5t})$ approximately.

A EXERCISES, p. 252

1. 3.26 ft, 13.04 ft.

2. (a) $\theta = 5 \cos 4t$ (degrees) or $(\pi/36) \cos 4t$ (radians).

(b) Frequency $= 2/\pi = 0.636$ cycles per sec.

(c) $2\pi/9$ ft.

(d) Velocity $= 2\pi/9$ ft/sec, acceleration $= 2\pi^2/81$ ft/sec^2.

3. Period $= \pi/5$ sec, frequency $= 5/\pi$ cycles per sec.

4. Frequency $= 1.6$ cycles per sec, period $= 0.625$ sec.

B EXERCISES, p. 253

2. $1/4$ density of water $= 15.6$ lb/ft^3.

5. (a) $P = P_0 + \dfrac{\beta}{k_1 k_2} + \dfrac{\alpha \sin \sqrt{k_1 k_2}t}{\sqrt{k_1 k_2}} - \dfrac{\beta \cos \sqrt{k_1 k_2}t}{k_1 k_2}$

$\quad S = S_0 + \dfrac{\alpha + \beta t}{k_1} - \dfrac{\beta \sin \sqrt{k_1 k_2}t}{k_1 \sqrt{k_1 k_2}} - \dfrac{\alpha \cos \sqrt{k_1 k_2}t}{k_1}.$

(b) $P = P_0 + \dfrac{K\omega(\cos \sqrt{k_1 k_2}t - \cos \omega t)}{\omega^2 - k_1 k_2}$

$\quad S = S_0 + \dfrac{K}{k_1} \sin \omega t + \dfrac{K\omega \sqrt{k_1 k_2} \sin \sqrt{k_1 k_2}t - K\omega^2 \sin \omega t}{k_1(\omega^2 - k_1 k_2)}.$

C EXERCISES, p. 253

7. $\dfrac{WEI}{LF^2}(u \tan u - \sec u - \tfrac{1}{2}u^2 + 1)$ where $u = L\sqrt{F/EI}$.

Chapter Six

A EXERCISES, p. 265

1. (a) $\dfrac{3}{s^2} - \dfrac{2}{s}, s > 0.$ (b) $\dfrac{12}{s^2 + 9}, s > 0.$ (c) $\dfrac{5s}{s^2 + 4}, s > 0.$

(d) $\dfrac{10}{s + 5}, s > -5.$ (e) $\dfrac{2}{s - 1} - \dfrac{3}{s + 1} + \dfrac{8}{s^3}, s > 1.$ (f) $\dfrac{15 - 4s}{s^2 + 25}, s > 0.$

(g) $\dfrac{6s}{s^2 - 9} - \dfrac{10}{s^2 - 25}, s > 5.$ (h) $\dfrac{1}{(s + 3)^2} - \dfrac{6}{s^4} + \dfrac{1}{s^2}, s > 0.$

3. (a) $\dfrac{2}{(s - 3)^3},$ (b) $\dfrac{6 - 2s}{s^2 + 4s + 8}.$ (c) $\dfrac{2s}{(s^2 + 1)^2} + \dfrac{1}{(s + 1)^2}.$

(d) $\dfrac{s^4 + 4s^2 + 24}{s^5}.$ (e) $\dfrac{s^2 + 4}{(s^2 - 4)^2} - \dfrac{4}{s^3}.$ (f) $\dfrac{4s}{s^2 - 36} - \dfrac{4}{s}.$

4. (a) $\dfrac{3\sqrt{\pi}}{4s^{5/2}}.$ (b) $\dfrac{\sqrt{\pi}}{2s^{3/2}} + \dfrac{2}{s} + \dfrac{\sqrt{\pi}}{s^{1/2}}.$ (c) $\dfrac{2\Gamma(\frac{2}{3})}{3s^{5/3}}.$ (d) $\dfrac{\sqrt{\pi}}{2(s - 1)^{3/2}}.$

5. (b) $\dfrac{1}{s^2} - \dfrac{e^{-4s}}{s^2} - \dfrac{4}{s}e^{-4s}.$

7. (a) $1 + H(t - 1).$ (b) $2t + (t^2 - 2t)H(t - 3).$

(c) $\cos t\{1 - H(t - 2\pi)\}.$

8. (a) $e^{-s}(s + 1)/s^2.$ (b) $\dfrac{e^{-2(s-1)}}{s - 1} - \dfrac{e^{-3(s+1)}}{s + 1}.$

2. $\dfrac{2s + 2}{(s^2 + 2s + 2)^2}.$

5. $\dfrac{1}{s}\left(\dfrac{e^{as} - 1}{e^{as} + 1}\right) = \dfrac{1}{s}\tanh\dfrac{as}{2}.$

7. $3\sin t + (t^2 - 3\sin t)H(t - \pi) + (t - t^2 - \cos t)H(t - 2\pi).$

4. (b) $\dfrac{\pi}{s^2 + \pi^2}\coth\dfrac{s}{2}.$

1. (a) $-1.$ **(b)** $\frac{1}{4}e^{-3/2}.$ **(c)** $1/27.$

   (d) $0.$ **(e)** $27/8.$ **(f)** $\sin(\pi^2/4).$

2. (a) $e^{-2(s-1)}.$ **(b)** $e^{-\pi(s+3)}.$ **(c)** $e^{-s}.$ **(d)** $0.$

2. $e^{-\pi s}\cos\pi^3.$

5. (a) $-1.$ **(b)** $\frac{1}{3}e.$

1. (a) $4e^{-2t}.$ **(b)** $3\cos 3t.$ **(c)** $3\sin 5t.$

   (d) $6\cos 2t - 5\sin 2t - 3e^{4t}.$ **(e)** $\dfrac{2}{\sqrt{5}}\sin\sqrt{5}t - \cos\sqrt{5}t.$ **(f)** $t^2 + 3t - 1.$

2. (a) $\frac{1}{2}t^2 e^{-2t}.$ **(b)** $\frac{1}{2}e^{2t}(4\cos 4t - 3\sin 4t).$ **(c)** $\frac{1}{3}e^{-t/2}\left(3\cos\dfrac{\sqrt{3}}{2}t + \sqrt{3}\sin\dfrac{\sqrt{3}}{2}t\right).$

   (d) $\frac{1}{6}e^t(t^3 + 3t^2).$ **(e)** $\frac{1}{2}e^{-t/2}(\cos t - \sin t).$

3. (a) $\frac{9}{2}e^t - \frac{7}{2}e^{-3t}.$ **(b)** $\frac{1}{2}e^{4t} + \frac{5}{2}e^{-4t}.$

   (c) $3e^{2t} - e^{-t} - 2e^{3t}.$ **(d)** $(2t - 3)e^{-t} + 5e^{2t}.$

   (e) $2 - e^t(2\cos 2t - \sin 2t).$ **(f)** $\frac{1}{6}(2\cos t - 2\cos 2t + 2\sin t - \sin 2t).$

4. (a) $\frac{1}{4}(e^{2t} - e^{-2t}) = \frac{1}{2}\sinh 2t.$ **(b)** $(t + 2)e^{-t} + t - 2.$ **(c)** $\frac{1}{2}(\sin t - t\cos t).$

5. (a) $Y = e^{3t} + 2e^t.$ **(b)** $Y = 3e^{-2t} + 2t - 2.$ **(c)** $Y = 2e^{-t} + \sin 3t - 2\cos 3t.$

   (d) $Y = 24 + 12t + e^t(9t - 20).$ **(e)** $Y = 4 - 2e^{-4t}(\cos 3t - 2\sin 3t).$

1. $\sin t - 2\cos t + 2e^t(\cos 2t + 2\sin 2t).$ **2.** $(2t - 1)e^{-t} + (2 - 3t)e^{-2t}.$

3. $Y = e^{-t}(1 + t - t^2 + 2t^3).$

4. $Y = \frac{1}{4}\cos t - \frac{1}{2}\sin t + \frac{1}{8}e^t + \frac{5}{8}e^{-t} - \frac{1}{4}t\sin t.$

5. (a) $\begin{cases} \frac{1}{2}(t - 2)^2, & t \geq 2 \\ 0, & t < 2 \end{cases}$ **(b)** $\begin{cases} 2(t - 1)^{1/2}e^{-(t-1)}/\sqrt{\pi}, & t \geq 1 \\ 0, & t < 1 \end{cases}.$

6. $Y = 4\sin t.$

8. (a) $Y = \frac{1}{3}(e^{3t} + 2)$.   (b) $Y = 1 - t$.

9. $\frac{1}{2}t \sinh t$.   **10.** $Y = \pm \sin t$.

C EXERCISES, p. 284

3. $Y = t$.

5. (a) $Y = 2e^{-2t} + 5e^{-2(t-1)}H(t-1) = \begin{cases} 2e^{-2t}, & t < 1 \\ 2e^{-2t} + 5e^{-2(t-1)}, & t > 1 \end{cases}$.

(b) $Y = 6 \cos t + 3 \sin (t - \pi)H(t - \pi) = \begin{cases} 6 \cos t, & t < \pi \\ 6 \cos t + 3 \sin t, & t > \pi \end{cases}$.

(c) $Y = 6(t - 2)e^{-2(t-2)}H(t - 2) = \begin{cases} 0, & t < 2 \\ 6(t - 2)e^{-2(t-2)}, & t > 2 \end{cases}$.

11. (c) $\dfrac{d^{1/2}}{dt^{1/2}}(t^2) = \dfrac{8}{3\sqrt{\pi}} t^{3/2}$, $\dfrac{d^{1/2}}{dt^{1/2}}\left(\dfrac{8}{3\sqrt{\pi}} t^{3/2}\right) = 2t$.

A EXERCISES, p. 296

1. $I = \dfrac{E}{R}(1 - e^{-Rt/L})$.   **2.** $X = v_0 t - \frac{1}{2}gt^2$, if $X = 0$ at $t = 0$.

3. (a) $\begin{cases} (E_0/R)(1 - e^{-Rt/L}), & t \leq T \\ (E_0/R)(e^{-R(t-T)/L} - e^{-Rt/L}), & t > T \end{cases}$.

(b) $\dfrac{E_0}{R^2 + \omega^2 L^2}(R \sin \omega t - \omega L \cos \omega t + \omega L e^{-Rt/L})$.

B EXERCISES, p. 296

11. (b) $x = \begin{cases} 0, & t < t_0 \\ \dfrac{P_0(t - t_0)}{m}, & t > t_0 \end{cases}$.

Chapter Seven

A EXERCISES, p. 316

1. (a) $y = 4\left(1 - x + \dfrac{x^2}{2!} - \dfrac{x^3}{3!} + \dfrac{x^4}{4!} - \cdots\right) = 4e^{-x}$, all x.

(b) $y = 5\left(1 + \dfrac{x^2}{2} + \dfrac{x^4}{2^2 2!} + \dfrac{x^6}{2^3 3!} + \dfrac{x^8}{2^4 4!} + \cdots\right) = 5e^{x^2/2}$, all x.

(c) $y = c - cx + (c + 2)\left(\dfrac{x^2}{2} - \dfrac{x^3}{3!} + \dfrac{x^4}{4!} - \cdots\right) = 2x - 2 + (c + 2)e^{-x}$, all x.

(d) $y = 1 + \dfrac{x^2}{2!} + \dfrac{x^4}{4!} + \dfrac{x^6}{6!} + \cdots = \cosh x = \frac{1}{2}(e^x + e^{-x})$, all x.

(e) $y = 2 + 3[(x - 1) - \frac{1}{2}(x - 1)^2 + \frac{1}{3}(x - 1)^3 - \frac{1}{4}(x - 1)^4 + \cdots]$, $|x - 1| < 1$, actual solution for all x is $y = 2 + 3 \ln x$.

(f) $y = c_1 \left[1 - \dfrac{(x-1)^2}{2!} + \dfrac{(x-1)^3}{3!} - \dfrac{(x-1)^4}{4!} + \dfrac{2(x-1)^5}{5!} - \cdots \right]$

$\qquad + c_2 \left[(x-1) - \dfrac{(x-1)^3}{3!} + \dfrac{2(x-1)^4}{4!} - \dfrac{5(x-1)^5}{5!} + \cdots \right]$,

$\qquad |x-1| < 1$.

(g) $y = c_1 \left(1 - \dfrac{x^2}{2} + \dfrac{x^4}{2 \cdot 4} - \dfrac{x^6}{2 \cdot 4 \cdot 6} + \cdots \right) + c_2 \left(x - \dfrac{x^3}{3} + \dfrac{x^5}{3 \cdot 5} - \dfrac{x^7}{3 \cdot 5 \cdot 7} + \cdots \right)$,

$\qquad$ all x.

(h) $y = c_1 x + c_2 \left[1 - \dfrac{(x-2)}{2} + \dfrac{(x-2)^2}{2^2} - \dfrac{(x-2)^3}{2^3} + \cdots \right]$, $|x-2| < 2$, actual solu-

$\qquad$ tion for all $x \ne 0$ is $y = c_1 x + 2c_2/x$.

(i) $y = 1 - \dfrac{x^2}{2!} - \dfrac{x^4}{4!} - \dfrac{11x^6}{6!} - \dfrac{311x^8}{8!} - \cdots$, $|x| < 1$.

(j) $y = c_1 + c_2 \left[1 - \dfrac{(x-1)}{2} + \dfrac{(x-1)^2}{2^2} - \dfrac{(x-1)^3}{2^3} + \cdots \right]$, $|x-1| < 2$, actual solution

$\qquad$ for all $x \ne -1$ is $y = c_1 + 2c_2/(x+1)$.

2. (a) Yes; all x. (b) Yes; $|x-2| < 2$. (c) Yes; $|x| < 1$.

$\quad$ (d) Yes; $|x - \frac{1}{3}| < \frac{1}{3}$. (e) Yes; $|x| < 2$. (f) Yes; $|x-1| < 1$.

4. $y = c_1 \left(1 - \dfrac{x^3}{3!} + \dfrac{1 \cdot 4}{6!} x^6 - \dfrac{1 \cdot 4 \cdot 7}{9!} x^9 + \cdots \right)$

$\qquad + c_2 \left(x - \dfrac{2}{4!} x^4 + \dfrac{2 \cdot 5}{7!} x^7 - \dfrac{2 \cdot 5 \cdot 8}{10!} x^{10} + \cdots \right)$,

$\quad$ convergence for all x.

B EXERCISES, p. 316

1. (a) $y_4 = 1 - 3x + \dfrac{(3x)^2}{2!} - \dfrac{(3x)^3}{3!}$. (b) $y_4 = 2 \left(\dfrac{x^3}{3!} - \dfrac{x^4}{4!} + \dfrac{x^5}{5!} \right)$.

$\quad$ (c) $y_4 = 3e^x - \dfrac{x^2}{2} - 2x - 3$.

$\quad$ (d) $y_4 = c \left(1 + \dfrac{x}{2} - \dfrac{x^2}{8} - \dfrac{5x^3}{48} + \dfrac{x^5}{64} - \dfrac{x^6}{384} \right)$.

2. $y_4 = 2 + 3(x-1) + 2(x-1)^2 + \dfrac{2(x-1)^3}{3} + \dfrac{(x-1)^4}{24}$.

4. $y = x^2 + 1$.

5. $y = c_1(1 - 2x^2) + c_2 \left(x - \dfrac{x^3}{3} - \dfrac{x^5}{30} - \dfrac{x^7}{210} - \dfrac{x^9}{1512} - \cdots \right)$.

8. $y = c_1 \left(1 - \dfrac{x^3}{6} + \dfrac{x^6}{180} - \cdots \right) + c_2 \left(x - \dfrac{x^4}{12} + \dfrac{x^7}{504} + \cdots \right)$

$\qquad + \dfrac{x^3}{6} - \dfrac{x^5}{120} - \dfrac{x^6}{180} + \dfrac{x^7}{5040} + \cdots$.

1. $y = 1 - \dfrac{x^2}{2!} - \dfrac{x^3}{3!} + \dfrac{3x^5}{5!} + \dfrac{9x^6}{6!} + \dfrac{16x^7}{7!} + \cdots$, all x.

1. (a) $y = c_1\left(x^{-1} - \dfrac{x}{2!} + \dfrac{x^3}{4!} - \dfrac{x^5}{6!} + \cdots\right) + c_2\left(1 - \dfrac{x^2}{3!} + \dfrac{x^4}{5!} - \dfrac{x^6}{7!} + \cdots\right)$

$\qquad = \dfrac{c_1 \cos x + c_2 \sin x}{x}$, all $x \neq 0$.

(b) $y = c_1\left(1 + \dfrac{x^2}{2 \cdot 3} + \dfrac{x^4}{(2 \cdot 4)(3 \cdot 7)} + \dfrac{x^6}{(2 \cdot 4 \cdot 6)(3 \cdot 7 \cdot 11)} + \cdots\right)$

$\qquad + c_2 x^{1/2}\left(1 + \dfrac{x^2}{2 \cdot 5} + \dfrac{x^4}{(2 \cdot 4)(5 \cdot 9)} + \dfrac{x^6}{(2 \cdot 4 \cdot 6)(5 \cdot 9 \cdot 13)} + \cdots\right)$, all x.

(c) $y = cx^{-1} + c_2\left(1 - x + \dfrac{x^2}{3}\right)$, all $x \neq 0$.

(d) $y = c_1(x - \tfrac{5}{3}x^3) + c_2(1 - 6x^2 + 3x^4 + \tfrac{4}{5}x^6 + \tfrac{3}{7}x^8 + \tfrac{2}{7}x^{10} + \cdots)$, last series converges
for $|x| < 1$.

(e) $y = c_1 U(x) + c_2 V(x)$, where

$\qquad U(x) = x - \dfrac{x^2}{1!2!} + \dfrac{x^3}{2!3!} - \dfrac{x^4}{3!4!} + \cdots$

$\qquad V(x) = U\displaystyle\int \dfrac{dx}{U^2} = U(x) \ln |x| + 1 - \dfrac{x}{2} + \tfrac{3}{4}x^2 - \tfrac{7}{48}x^3 + \dfrac{5x^4}{1728} + \cdots$, all $x \neq 0$.

(f) $y = c_1\left(x^{-1/2} - \dfrac{x^{3/2}}{2!} + \dfrac{x^{7/2}}{4!} - \cdots\right) + c_2\left(x^{1/2} - \dfrac{x^{5/2}}{3!} + \dfrac{x^{9/2}}{5!} - \cdots\right)$

$\qquad = x^{-1/2}(c_1 \cos x + c_2 \sin x)$, all $x \neq 0$.

(g) $y = c_1(1 - x) + c_2(1 - x)\displaystyle\int \dfrac{e^x \, dx}{x(1 - x)^2}$

$\qquad = c_1(1 - x) + c_2\left((1 - x) \ln |x| + 3x - \dfrac{x^2}{4} - \dfrac{x^3}{36} - \dfrac{x^4}{72} - \cdots\right)$, all $x \neq 0$.

(h) $y = c_1\left(1 - \dfrac{x^4}{2!} + \dfrac{x^8}{4!} - \dfrac{x^{12}}{6!} + \cdots\right) + c_2\left(x^2 - \dfrac{x^6}{3!} + \dfrac{x^{10}}{5!} - \dfrac{x^{14}}{7!} + \cdots\right)$

$\qquad = c_1 \cos x^2 + c_2 \sin x^2$, all x.

(i) $y = c_1\left(x^{1/2} + \dfrac{x^{3/2}}{2!} + \dfrac{x^{5/2}}{4!} + \cdots\right) + c_2\left(x^{3/2} + \dfrac{x^{7/2}}{3!} + \dfrac{x^{11/2}}{5!} + \cdots\right)$

$\qquad = c_1 x^{1/2}\left(\dfrac{e^x + e^{-x}}{2}\right) + c_2 x^{1/2}\left(\dfrac{e^x - e^{-x}}{2}\right)$

$\qquad = \sqrt{x}(c_1 \cosh x + c_2 \sinh x) = \sqrt{x}(Ae^x + Be^{-x})$, all x.

(j) $y = c_1 U(x) + c_2 V(x)$, where

$$U(x) = 1 - \frac{x}{(1!)^2} + \frac{x^2}{(2!)^2} - \frac{x^3}{(3!)^2} + \cdots,$$

$$V(x) = U \int \frac{dx}{xU^2}, \text{ all } x \neq 0.$$

(k) $y = c_1 U(x) + c_2 V(x)$, where

$$U(x) = 1 - \frac{x^2}{2^2} + \frac{x^4}{2^2 4^2} - \frac{x^6}{2^2 4^2 6^2} + \cdots,$$

$$V(x) = U \int \frac{dx}{xU^2}, \text{ all } x \neq 0.$$

(l) $y = c_1 U(x) + c_2 V(x)$, where

$$U(x) = x^2 \left(1 - \frac{x^2}{(2)(6)} + \frac{x^4}{(2 \cdot 4)(6 \cdot 8)} - \frac{x^6}{(2 \cdot 4 \cdot 6)(6 \cdot 8 \cdot 10)} + \cdots\right),$$

$$V(x) = U \int \frac{dx}{xU^2}, \text{ all } x \neq 0.$$

3. $U = \begin{cases} r^{-1}(A \sin \beta r + B \cos \beta r) & \text{if } \alpha = \beta^2 > 0 \\ r^{-1}(A \sinh \beta r + B \cosh \beta r) & \text{if } \alpha = -\beta^2 < 0 \\ A + Br^{-1} & \text{if } \alpha = 0. \end{cases}$

B EXERCISES, p. 330

2. $y = c_1 \left(1 + \frac{x^2}{2} + \frac{x^4}{2 \cdot 4} + \frac{x^6}{2 \cdot 4 \cdot 6} + \cdots\right) + c_2 \left(x + \frac{x^3}{1 \cdot 3} + \frac{x^5}{1 \cdot 3 \cdot 5} + \cdots\right)$

$\quad + \frac{4}{3}x^{5/2} \left(1 + \frac{2x^2}{9} + \frac{4x^4}{9 \cdot 13} + \frac{8x^6}{9 \cdot 13 \cdot 17} + \cdots\right).$

3. $y = c_1 \left(x^{-1} - \frac{x}{2!} + \frac{x^2}{4!} - \frac{x^6}{6!} + \cdots\right) + c_2 \left(1 - \frac{x^2}{3!} + \frac{x^4}{5!} - \cdots\right) + 2$

$\quad = \frac{c_1 \cos x + c_2 \sin x}{x} + 2.$

4. (a) $y = c_1 x^{-1} + c_2(1 + x + x^2 + x^3 + \cdots) = c_1 x^{-1} + c_2(1 - x)^{-1}.$

(b) $y = c_1 x^{-1} + c_2(1 - x)^{-1} + \frac{2}{3}x^5 - \frac{4}{3}x^4 + (1 - x)^{-1}(x^4 - \frac{8}{5}x^5 + \frac{2}{3}x^6).$

C EXERCISES, p. 330

2. (a) $y = AU(x) + B \left\{U(x) \ln x + 2x \left[1 - \frac{x}{1^2 \cdot 2^2}\left(1 + \frac{1}{2}\right)\right.\right.$

$\quad\quad\quad\quad\quad\quad\quad\quad\quad \left.\left. + \frac{x^2}{1^2 \cdot 2^2 \cdot 3^2}\left(1 + \frac{1}{2} + \frac{1}{3}\right) - \cdots\right]\right\},$

where $U(x) = 1 - \frac{x}{1^2} + \frac{x^2}{1^2 \cdot 2^2} - \frac{x^3}{1^2 \cdot 2^2 \cdot 3^2} + \cdots.$

(b) $y = Axe^{-x} + B\left\{1 - xe^{-x}\ln x - x^2\left[1 - \dfrac{x}{2!}\left(1 + \dfrac{1}{2}\right)\right.\right.$

$$\left.\left. + \dfrac{x^2}{3!}\left(1 + \dfrac{1}{2} + \dfrac{1}{3}\right) - \cdots\right]\right\}.$$

(c) $y = AV(x) + B\left\{V(x)\ln x + x^{-2}\left[1 + \dfrac{x^2}{2^2} + \dfrac{x^4}{2^2 4^2} - \dfrac{11x^6}{2^2 4^2 6^2} + \cdots\right]\right\}$,

where $V(x) = -\dfrac{1}{16}\left(x^2 - \dfrac{x^4}{(2)(6)} + \dfrac{x^6}{(2\cdot 4)(6\cdot 8)}\right.$

$$\left. - \dfrac{x^8}{(2\cdot 4\cdot 6)(6\cdot 8\cdot 10)} + \cdots\right).$$

3. (b) $y = A\left(1 - \dfrac{x^{-2}}{2!} + \dfrac{x^{-4}}{4!} + \cdots\right) + B\left(x^{-1} - \dfrac{x^{-3}}{3!} + \dfrac{x^{-5}}{5!} + \cdots\right)$

$$= A\cos(1/x) + B\sin(1/x).$$

4. (a) $y = A\left(1 + x^{-1} + \dfrac{x^{-2}}{2!} + \dfrac{x^{-3}}{3!} + \cdots\right) + B\left(x^{-1} + x^{-2} + \dfrac{x^{-3}}{2!} + \dfrac{x^{-4}}{3!} + \cdots\right)$

$$= (A + Bx^{-1})e^{1/x}.$$

(b) $y = A\left(x^{-1} - \dfrac{x^{-2}}{3!} + \dfrac{x^{-3}}{5!} - \cdots\right) + B\left(x^{-1/2} - \dfrac{x^{-3/2}}{2!} + \dfrac{x^{-5/2}}{4!} - \cdots\right)$

$$= x^{-1/2}[A\sin(x^{-1/2}) + B\cos(x^{-1/2})].$$

A EXERCISES, p. 341

5. (a) $y = c_1 J_3(x) + c_2 Y_3(x)$.   (b) $y = c_1 J_{2\sqrt 2}(x) + c_2 J_{-2\sqrt 2}(x)$.

8. (a) $y = c_1 J_2(\sqrt 3 x) + c_2 Y_2(\sqrt 3 x)$.

(b) $y = c_1 J_{1/2}(\sqrt 2 x/2) + c_2 J_{-1/2}(\sqrt 2 x/2)$

$$= \dfrac{A\cos(\sqrt 2 x/2) + B\sin(\sqrt 2 x/2)}{\sqrt x}.$$

B EXERCISES, p. 341

1. $5J_1(2x)/J_1(4)$.

5. (a) $xJ_1(x) + c$.   (b) $x^3 J_3(x) + c$.   (c) $-x^{-4}J_4(x) + c$.

(d) $-x^{1/2}J_{-1/2}(x) + c$ or $-\sqrt{2/\pi}\cos x + c$.

6. $(x^3 - 4x)J_1(x) + 2x^2 J_0(x) + c$.

7. (a) $-J_2(x) - \dfrac{2J_1(x)}{x} + c$.   (b) $-\tfrac13 J_1(x) - \dfrac{1}{3}\dfrac{J_2(x)}{x} + \dfrac{1}{3}\int J_0(x)dx$.

10. $(x^5 - 16x^3 + 64x)J_1(x) + (4x^4 - 32x^2)J_0(x) + c$.

A EXERCISES, p. 346

2. $y = 40 - 120x^2$.

4. (a) $\dfrac{1}{2}\ln\left|\dfrac{1+x}{1-x}\right|$, $y = c_1 + c_2\ln\left|\dfrac{1+x}{1-x}\right|$.

(b) $1 - \dfrac{x}{2} \ln \left| \dfrac{1+x}{1-x} \right|$, $y = c_1 x + c_2 \left(1 - \dfrac{x}{2} \ln \left| \dfrac{1+x}{1-x} \right| \right)$.

B EXERCISES, p. 347

1. $y = 4x + 3 - \dfrac{3x}{2} \ln \left| \dfrac{1+x}{1-x} \right|$.

C EXERCISES, p. 347

5. (a) $y = c_1 H_n(x) + c_2 H_n(x) \displaystyle\int \dfrac{e^{x^2}}{H_n^2(x)} \, dx$.

(b) $y = c_1 L_n(x) + c_2 L_n(x) \displaystyle\int \dfrac{e^x}{x L_n^2(x)} \, dx$.

6. $y = c_1 (1-x)^{-1} + c_2 (1-x)^{-1} \ln x$.

Chapter Eight

A EXERCISES, p. 354

1. (a) Finite.   (b) Non-countably infinite.   (c) Finite.

(d) Non-countably infinite.   (e) Countably infinite.

2. (a) $3, -6$.   (b) $3, 0, -3$.   (c) $4, -1$.

3. (a) 0.   (b) $1/6$.   (c) $12/5$.

4. (a) $\sqrt{30}$.   (b) $\sqrt{3}/3$.   (c) $2\sqrt{2}$.   (d) $\sqrt{34}$.

5. (a) $c = 4$.   (b) $c = 14/51$.   (c) $c = 0$ or $-14/9$.

6. (a) $\vec{u} = \left(-\dfrac{4}{\sqrt{54}}, \dfrac{3}{\sqrt{54}}, \dfrac{5}{\sqrt{54}}, \dfrac{-2}{\sqrt{54}} \right)$.   (b) $\phi(x) = \sqrt{\dfrac{3}{14}} (2x - 3), 0 \leq x \leq 2$.

(c) $\phi(x) = \sqrt{10}\, e^{-5x}, x \geq 0$.   (d) $\phi(x) = \begin{cases} x/5 & 0 \leq x < 3 \\ 4/5 & 3 \leq x \leq 4. \end{cases}$

7. (a) $c_1 = -6, c_2 = 6, c_3 = -6$.

(b) $\phi_1(x) = 2 - 3x$, $\phi_2(x) = \sqrt{5}(6x^2 - 6x + 1)$, $\phi_3(x) = \sqrt{3}(x - 1)$.

B EXERCISES, p. 355

3. (b) $\phi_1(x) = \sqrt{\dfrac{2}{\pi}} \sin x$, $\phi_2(x) = \sqrt{\dfrac{2}{\pi}} \sin 2x$, $\phi_3(x) = \sqrt{\dfrac{2}{\pi}} \sin 3x$.

4. $\phi_n(x) = \sqrt{\dfrac{2}{\pi}} \sin nx$, $n = 1, 2, 3, \ldots$.

A EXERCISES, p. 374

3. (a) $y = c_1 \cos x + c_2 \sin x - \tfrac{1}{2} x \cos x$.   (b) $\tfrac{1}{2}(\sin x - x \cos x)$.

5. (a) $\dfrac{d}{dx}\left(x^2 \dfrac{dy}{dx} \right) + (x^2 + \lambda x) y = 0$.   (b) $\dfrac{d}{dx}\left(e^{-x} \dfrac{dy}{dx} \right) + (e^{-2x} + \lambda e^{-x}) y = 0$.

(c) $y'' + (3 + \lambda \cos x)y = 0$.
(d) $\dfrac{d}{dx}\left(\cos x\,\dfrac{dy}{dx}\right) + (\cos x + \lambda \sin x)y = 0$.

6. (a) $\lambda_n = \dfrac{n^2\pi^2}{16}$; $y_n = B_n \sin \dfrac{n\pi x}{4}$; $\phi_n(x) = \dfrac{\sqrt{2}}{2}\sin \dfrac{n\pi x}{4}$ where $n = 1, 2, \dots$.

(b) $\lambda_n = \dfrac{(2n-1)^2}{4}$; $y_n = A_n \cos \dfrac{(2n-1)}{2}x$; $\phi_n(x) = \sqrt{\dfrac{2}{\pi}}\cos \dfrac{(2n-1)}{2}x$ where $n = 1, 2, \dots$.

(c) $\lambda_n = \dfrac{n^2\pi^2}{4}$; $y_n = A_n \cos \dfrac{n\pi x}{2}$; $\phi_n(x) = \cos \dfrac{n\pi x}{2}$ where $n = 1, 2, \dots$

B EXERCISES, p. 374

1. (b) $\lambda_n = n^2\pi^2$; $A_n \cos n\pi x + B_n \sin n\pi x$, where $n = 0, 1, 2, \dots$.

(d) $\phi_n(x) = (A_n \cos n\pi x + B_n \sin n\pi x)/\sqrt{A_n^2 + B_n^2}$.

2. (b) $\lambda_n = (2n-1)^2\pi^2$; $A_n \cos (2n-1)\pi x$ where $n = 1, 2, \dots$.

3. (b) $\lambda_n = \dfrac{(2n-1)^2\pi^2}{4}$; $A_n \cos \dfrac{(2n-1)\pi x}{2} + B_n \sin \dfrac{(2n-1)\pi x}{2}$, $n = 1, 2, \dots$.

5. (a) $\dfrac{\sqrt{2x}J_0(r_k x)}{J_1(r_k)}$ where r_k, $k = 1, 2, \dots$ are the roots of $J_0(x) = 0$.

(b) $\dfrac{\sqrt{2x}J_0(r_k x)}{J_0(r_k)}$ where r_k, $k = 1, 2, \dots$ are the roots of $J_0'(x) = 0$ $[$or $J_1(x) = 0]$.

(c) $\dfrac{\sqrt{2x}\,r_k J_0(r_k x)}{\sqrt{\frac{1}{4} + r_k^2}\,J_0(r_k)}$ where r_k, $k = 1, 2, \dots$ are the roots of $\frac{1}{2}J_0(x) + xJ_0'(x) = 0$.

6. (a) $\dfrac{\sqrt{2x}\,r_k J_2(r_k x)}{\sqrt{r_k^2 - 4}\,J_2(r_k)}$ where r_k, $k = 1, 2, \dots$ are the roots of $J_2'(x) = 0$.

(b) $\dfrac{\sqrt{2x}\,r_k J_5(r_k x)}{\sqrt{r_k^2 - 221/9}}$ where r_k, $k = 1, 2, \dots$ are the roots of $\frac{2}{3}J_5(x) + xJ_5'(x) = 0$.

7. (a) $k^2\pi^2$, $\sqrt{2}\,\dfrac{\sin k\pi x}{\sqrt{x}}$ orthogonal in $0 \le x \le 1$ with respect to weight function x, $k = 1, 2, \dots$.

(b) r_k^2, $k = 1, 2, 3, \dots$ where r_k are the roots of $\tan x = 2x$, $\dfrac{\sqrt{2r_k}\,\sin r_k x}{\sqrt{r_k^2 - \frac{1}{4}(\sin r_k)}\sqrt{x}}$ orthogonal in

$0 \le x \le 1$ with respect to weight function x.

(c) $\dfrac{(2k-1)^2\pi^2}{4}$, $\sqrt{\dfrac{2}{x}}\sin \dfrac{(2k-1)\pi x}{2}$ orthogonal in $0 \le x \le 1$ with respect to the weight function x.

C EXERCISES, p. 375

2. $k^2\pi^2$, $\dfrac{\sqrt{2}\sin(k\pi \ln x)}{x^2}$, $k = 1, 2, 3, \dots$, orthogonal with respect to weight function x^3.

1. (a) $f(x) = \sum\limits_{n=1}^{\infty} \dfrac{-4}{n\pi}(1 - \cos n\pi)\sin nx$

$f(x)$ is odd, points of discontinuity are $x = 0, \pm\pi, \pm 2\pi, \ldots$ at which series converges to 0.

(b) $f(x) = 1$. $f(x)$ is even and there are no points of discontinuity.

(c) $f(x) = 1 + \sum\limits_{n=1}^{\infty} \dfrac{2}{n\pi}(1 - \cos n\pi)\sin \dfrac{n\pi x}{3}$

$f(x) - 1$ is odd, points of discontinuity are $x = 0, \pm 3, \pm 6, \ldots$ where series converges to 1.

(d) $f(x) = \dfrac{1}{4} + \sum\limits_{n=1}^{\infty} \left\{ \left(\dfrac{\cos n\pi - 1}{n^2\pi^2} \right)\cos n\pi x - \dfrac{\cos n\pi}{n\pi}\sin n\pi x \right\}$

$f(x)$ is neither even nor odd, points of discontinuity are $x = \pm 1, \pm 3, \ldots$ at which series converges to $\tfrac{1}{2}$.

(e) $f(x) = \dfrac{\pi}{2} + \dfrac{2}{\pi}\sum\limits_{n=1}^{\infty} \left(\dfrac{\cos n\pi - 1}{n^2} \right)\cos nx$

$f(x)$ is even and there are no points of discontinuity.

(f) $f(x) = \sum\limits_{n=1}^{\infty} \dfrac{6}{n\pi}(1 - \cos n\pi)\sin \dfrac{n\pi x}{4}$

$f(x)$ is odd, points of discontinuity are $x = 0, \pm 4, \pm 8, \ldots$ at which series converges to 0.

(g) $f(x) = \dfrac{8}{3} - \sum\limits_{n=1}^{\infty} \dfrac{16}{n^2\pi^2}\cos \dfrac{n\pi x}{2}$

$f(x)$ is even and there are no points of discontinuity.

(h) $f(x) = \sum\limits_{n=1}^{\infty} \dfrac{1}{n\pi}\left(2 - \cos \dfrac{2n\pi}{3} - \cos \dfrac{4n\pi}{3} \right)\sin nx.$

$f(x)$ is odd. Points of discontinuity are $x = 0, \pm 2\pi, \pm 4\pi$ at which series converges to 0; $x = 2\pi/3 \pm 2m\pi$ $(m = 0, 1, 2, \ldots)$ where series converges to $\tfrac{1}{2}$ and $x = -4\pi/3 \pm 2m\pi$ $(m = 0, 1, 2, \ldots)$ where series converges to $-\tfrac{1}{2}$.

2. (a) $f(x) = \sum\limits_{n=1}^{\infty} \dfrac{10}{n\pi}(1 - \cos n\pi)\sin \dfrac{n\pi x}{3}$

Series converges to $f(x)$ except at $x = 0, \pm 3, \pm 6, \ldots$ where it converges to 0.

(b) $f(x) = \dfrac{1}{2} + \sum\limits_{n=1}^{\infty} \left(\dfrac{2}{n\pi}\sin \dfrac{n\pi}{2} \right)\cos nx$

Series converges to $f(x)$ except at $x = \pm\pi/2, \pm 3\pi/2, \ldots$ where it converges to $\tfrac{1}{2}$.

(c) $f(x) = \sum\limits_{n=1}^{\infty} \left(\dfrac{20}{n^2\pi^2}\sin \dfrac{n\pi}{2} - \dfrac{10}{n\pi}\cos \dfrac{n\pi}{2} \right)\sin \dfrac{n\pi x}{10}.$

Series converges to $f(x)$ except at $x = 5 \pm 20m$ $(m = 0, 1, 2, \ldots)$ where it converges to $5/2$ and $x = -5 \pm 20m$ $(m = 0, 1, 2, \ldots)$ where it converges to $-5/2$.

(d) $f(x) = 2$.

(e) $f(x) = \sum\limits_{n=1}^{\infty} \dfrac{64}{n^3\pi^3}(1 - \cos n\pi)\sin \dfrac{n\pi x}{4}$

Series converges to $f(x)$ for all x.

(f) $f(x) = \sum\limits_{n=1}^{\infty} -\dfrac{4}{n\pi} \sin \dfrac{n\pi}{2} \cos \dfrac{n\pi x}{4}$

Series converges to $f(x)$ except at $x = \pm 2, \pm 6, \pm 10, \ldots$ where it converges to 0.

3. (a) $8\pi = \dfrac{64}{\pi} \left(\dfrac{1}{1^2} + \dfrac{1}{3^2} + \dfrac{1}{5^2} + \cdots \right).$ (b) $\dfrac{2\pi^3}{3} = \dfrac{\pi^3}{2} + \dfrac{16}{\pi} \left(1 + \dfrac{1}{3^4} + \dfrac{1}{5^4} + \cdots \right).$

(c) $\dfrac{512}{15} = \dfrac{256}{9} + \dfrac{512}{\pi^4} \left(\dfrac{1}{1^4} + \dfrac{1}{2^4} + \dfrac{1}{3^4} + \cdots \right).$

B EXERCISES, p. 395

3. $f(x) = \sum\limits_{n=1}^{\infty} \dfrac{2L^2}{n^3\pi^3} [(2 - n^2\pi^2) \cos n\pi - 2] \sin \dfrac{n\pi x}{L}.$

A EXERCISES, p. 400

3. $x^2 = 8 \sum\limits_{k=1}^{\infty} \dfrac{J_2(r_k x)}{r_k^2 J_3(r_k)}.$

5. $1 = 2 \sum\limits_{k=1}^{\infty} \dfrac{r_k J_1(r_k)}{(1 + r_k^2) J_0^2(r_k)} J_0(r_k x).$

7. (a) $\dfrac{1}{64} = \sum\limits_{k=1}^{\infty} \dfrac{r_k^2 J_1^2(r_k) + (r_k^2 - 1)J_1^2(r_k)}{2r_k^6 J_2^2(r_k)}.$

(b) $\dfrac{1}{6} = 32 \sum\limits_{k=1}^{\infty} \dfrac{J_2'^2(r_k)}{r_k^4 J_3^2(r_k)}.$

A EXERCISES, p. 406

1. (a) $\frac{4}{3}P_0(x) + \frac{8}{3}P_2(x).$ (b) $\frac{17}{3}P_0(x) + \frac{1}{5}P_1(x) + \frac{10}{3}P_2(x) + \frac{4}{3}P_3(x).$

(c) $2P_0(x) + 3P_1(x) - \frac{7}{4}P_3(x) + \frac{11}{8}P_5(x) - \frac{75}{64}P_7(x) + \cdots.$

(d) $\frac{5}{2}P_0(x) + \frac{3}{4}P_1(x) - \frac{7}{16}P_3(x) + \frac{11}{32}P_5(x) - \frac{75}{256}P_7(x) + \cdots.$

2. $\frac{1}{4}P_0(x) + \frac{1}{2}P_1(x) + \frac{5}{16}P_2(x) - \frac{3}{32}P_4(x) + \cdots.$

4. (a) $\int_{-1}^{1} (4x^2)^2 \, dx = \dfrac{16}{9} \int_{-1}^{1} [P_0(x)]^2 \, dx + \dfrac{64}{9} \int_{-1}^{1} [P_2(x)]^2 \, dx$ or $\dfrac{32}{5} = \dfrac{16}{9}\left(\dfrac{2}{1}\right) + \dfrac{64}{9}\left(\dfrac{2}{5}\right).$

(b) $\int_{-1}^{1} (2x^3 + 5x^2 - x + 4)^2 \, dx = \dfrac{289}{9} \int_{-1}^{1} [P_0(x)]^2 \, dx + \dfrac{1}{25} \int_{-1}^{1} [P_1(x)]^2 \, dx$

$+ \dfrac{100}{9} \int_{-1}^{1} [P_2(x)]^2 \, dx + \dfrac{16}{9} \int_{-1}^{1} [P_3(x)]^2 \, dx.$

$= \dfrac{289}{9}\left(\dfrac{2}{1}\right) + \dfrac{1}{25}\left(\dfrac{2}{3}\right) + \dfrac{100}{9}\left(\dfrac{2}{5}\right) + \dfrac{16}{9}\left(\dfrac{2}{7}\right).$

5. $\dfrac{2}{3} = \dfrac{1}{4} + \sum\limits_{k=1}^{\infty} \dfrac{4k + 1}{2^{2k+1}[(k + 1)!]^2}.$

B EXERCISES, p. 407

1. (a) $\frac{3}{2}H_0(x) + \frac{1}{4}H_2(x).$ (b) $\frac{3}{4}H_1(x) + \frac{1}{8}H_3(x).$

2. (a) $4L_0(x) - 10L_1(x) + 3L_2(x).$ (b) $15L_0(x) - 40L_1(x) + 19L_2(x) - 2L_3(x).$

3. (a) $\int_{-\infty}^{\infty} e^{-x^2}(x^2 + 1)^2 \, dx = \dfrac{9}{4} \int_{-\infty}^{\infty} e^{-x^2}[H_0(x)]^2 \, dx + \dfrac{1}{16} \int_{-\infty}^{\infty} e^{-x^2}[H_2(x)]^2 \, dx.$

(b) $\int_0^\infty e^{-x}(3x^2 - 2x)^2\,dx = 16\int_0^\infty e^{-x}[L_0(x)]^2\,dx + 100\int_0^\infty e^{-x}[L_1(x)]^2\,dx$

$$+ 9\int_0^\infty e^{-x}[L_2(x)]^2\,dx.$$

A EXERCISES, p. 412

1. (a) $y = c_1 x^{-1} + c_2 x^{-2} + 2x.$ (b) $y = \frac{3}{4}x - \frac{3}{8} + c_1 e^{-2x} + c_2 e^{-2x}\int \frac{e^{2x}}{x}\,dx.$

    (c) $y = \frac{5}{2}x\csc^2 x - \frac{5}{2}\cot x + c_1\csc^2 x + c_2\csc x\cot x.$

2. (a) $y = 2x - 2 + c_1 e^{-x} + c_2 e^{-x}\int \frac{e^x}{x^2}\,dx.$

    (b) $y = (\sin x)\ln(\csc x - \cot x) + c_1\cos x + c_2\sin x.$

3. (b) $y = \dfrac{c_1\cos x + c_2\sin x}{x}.$

B EXERCISES, p. 412

2. (a) $\dfrac{d^2\mu}{dx^2} + x^2\dfrac{d\mu}{dx} + 3x\mu = 0.$ (b) $\mu = c_1 x e^{-x^3/3} + c_2 x e^{-x^3/3}\int \dfrac{e^{x^3/3}}{x^2}\,dx.$

4. $y = c_1 x e^{x\cos x - \sin x} + c_2 x e^{x\cos x - \sin x}\int \dfrac{e^{\sin x - x\cos x}}{x}\,dx.$

5. $y = c_1 x e^{-(8/3)x^{-3}} + c_2 x e^{-(8/3)x^{-3}}\int \dfrac{e^{(8/3)x^{-3}}\,dx}{x^4}.$

6. The Laguerre polynomials.   **7.** The Hermite polynomials.

C EXERCISES, p. 412

3. $y = c_1 e^{-x^2/2}\int e^{x^2/2}\,dx + c_2 e^{-x^2/2}\int \dfrac{e^{x^2/2}}{x}\,dx + c_3 e^{-x^2/2} + 4e^{-x^2/2}\int x^2 e^{x^2/2}\,dx.$

Chapter Nine

A EXERCISES, p. 426

1. Exact 1.95.   **2.** Exact 0.81.   **3.** Exact 4.60.

4. Exact 2.00.   **5.** Exact 0.80.   **6.** Exact 1.33.

7. Exact 1.96.

B EXERCISES, p. 426

1. $y(1) = e = 2.7183$ approx.   **2.** Exact 1.439

3. Exact 0.819.   **4.** Exact 0.197.   **5.** Exact 0.670.

7. 1.55.

2. $y(0.8) = 1.31, y'(0.8) = 0.95$.

1. 1.946.   **2.** 0.811.   **3.** 4.6.   **4.** 2.80.

Chapter Ten

1. (a) $x = \cos t, y = \sin t$.

(b) $u = c_1 e^{2x} + c_2 e^{-2x} - \frac{1}{2}, v = c_1 e^{2x} - c_2 e^{-2x} + \frac{1}{2}$.

(c) $x = c_1 e^{\sqrt{2}t} + c_2 e^{-\sqrt{2}t}, y = c_1(\sqrt{2} - 1)e^{\sqrt{2}t} - c_2(\sqrt{2} + 1)e^{-\sqrt{2}t}$.

(d) $x = c_1 \sin t + c_2 \cos t, y = c_3 e^t + c_4 e^{-t}$.

(e) $x = c_1 \sin t + c_2 \cos t + c_3 e^t + c_4 e^{-t} + 2$,

$\quad y = -c_1 \sin t - c_2 \cos t + c_3 e^t + c_4 e^{-t} - 2$.

(f) $x = 4e^{2t} - 2e^{-3t}, y = e^{2t} + 2e^{-3t}$.

(g) $x = c_1 e^{3t} + 2c_2 e^{-2t} + \frac{1}{3}t + \frac{4}{9}, y = -2c_1 e^{3t} + c_2 e^{-2t} - \frac{2}{3}t + \frac{1}{9}$.

(h) $x = \frac{3}{5}c_1 e^{-t/8} + \frac{2}{5} \sin t - \frac{1}{5} \cos t$,

$\quad y = c_1 e^{-t/8} + \sin t + \cos t$.

3. $x = 12 \cos 2t + 2e^{-2t} - 8e^{-t} + 4t, y = -12 \sin 2t + 6e^{-2t} - 36e^{-t} + 6$.

4. (a) Linear.   (b) Non-linear.   (c) Linear.   (d) Non-Linear.

6. $x = 1 + t - 2t^2 + \frac{2}{3}t^3 - \frac{1}{6}t^4 + e^{-t} - \sin t$,

$\quad y = -6 - 3t - 4t^2 - \frac{1}{6}t^4 + e^t + e^{-t} - \frac{1}{2} \sin t - \frac{1}{2} \cos t$.

1. $x = \left(1 + \dfrac{\sqrt{2}}{2}\right)e^{\sqrt{2}t} + \left(1 - \dfrac{\sqrt{2}}{2}\right)e^{-\sqrt{2}t}, y = \dfrac{\sqrt{2}}{2}\left(e^{\sqrt{2}t} - e^{-\sqrt{2}t}\right), z = e^{\sqrt{2}t} + e^{-\sqrt{2}t} - 2$.

2. (a) $x^2 - y^2 = c_1, y^2 - z^2 = c_2$.   (b) $x = c_1\sqrt{y^2 + 1}, \tan^{-1} y = \ln z + c_2$.

3. $y = \dfrac{1}{4}(\ln x)^2 + \dfrac{1}{2}\ln x - \dfrac{x}{2} + c_1, z = \dfrac{1}{2}\ln x + \dfrac{x}{2} - \dfrac{1}{2}$.

4. $x = c_1 e^t + c_2 e^{\omega t} + c_3 e^{\omega^2 t}$

$\quad y = c_1 e^t + \omega c_2 e^{\omega t} + \omega^2 c_3 e^{\omega^2 t}$

$\quad z = c_1 e^t + \omega^2 c_2 e^{\omega t} + \omega c_3 e^{\omega^2 t}$

where 1, ω, and ω^2 are the three cube roots of unity.

1. $P_n(t) = \dfrac{(\lambda t)^n e^{-\lambda t}}{n!}$ where $n = 0, 1, 2, 3, \ldots$.

2. $c_1(z - y) = c_2(x - z) = \sqrt{c_3/(x + y + z)}$.

1. (b) $x = 80t$, $y = 80\sqrt{3}t - 16t^2$.

(c) Range $= 400\sqrt{3} = 693$ ft, max. height $= 300$ ft, time of flight $= 8.66$ sec.

(d) Position after 2 sec (160, 213), position after 4 sec (320, 298), velocity after 2 sec has magnitude 109 ft/sec, velocity after 4 sec has magnitude 80.7 ft/sec.

2. Max. range $= 165$ miles, height $= 41\frac{1}{4}$ miles, time of flight $= 3$ min 53 sec.

3. (a) 4 sec.   (b) 200 ft.

1. 283 ft/sec, 15.8 sec.

6. (a) $2a/(1 - \epsilon)$.   (b) At the point (r, ϕ) where $r = 2a\epsilon/(1 - \epsilon)$, $\phi = \pi$.

8. (a) 7.7 km/sec.   (b) 2600 km.   (c) 6.3 km/sec.

4. 484 million miles.

5. (a) 36.0 million miles, 35.2 million miles.

(b) 28.4 million miles, 43.6 million miles.

6. 236,000 miles.

7. Force is inversely proportional to r^5.

1. (a) $x_1 = a \cos \omega t$, $x_2 = a \cos \omega t$.   (b) $x_1 = a \cos \sqrt{3}\omega t$, $x_2 = -a \cos \sqrt{3}\omega t$.

2. $x_1 = \dfrac{v_1 + v_2}{2\omega} \sin \omega t + \dfrac{v_1 - v_2}{2\sqrt{3}\omega} \sin \sqrt{3}\omega t$,

$x_2 = \dfrac{v_1 + v_2}{2\omega} \sin \omega t - \dfrac{v_1 - v_2}{2\sqrt{3}\omega} \sin \sqrt{3}\omega t$.

1. $I_1 = 3 - 2e^{-5t} - e^{-20t}$, $I_2 = 4e^{-5t} - e^{-20t} - 3$; $I_1 = 3$, $I_2 = -3$.

2. $I_1 = 2e^{-5t} + e^{-20t} + 3 \sin 10t - 3 \cos 10t$, $I_2 = 3 \cos 10t - 4e^{-5t} + e^{-20t}$; $I_1 = 3 \sin 10t - 3 \cos 10t$, $I_2 = 3 \cos 10t$.

3. Charge on capacitor $= 2 - e^{-t}$, current through capacitor $= e^{-t}$, current through inductor $= 20 - 20e^{-4t}$, current through battery $= 20 - 20e^{-4t} + e^{-t}$.

4. Charge on capacitor $= \sin 2t - 2 \cos 2t + 3e^{-t}$, current through capacitor $= 2 \cos 2t + 4 \sin 2t - 3e^{-t}$, current through inductor $= 40 \sin 2t - 20 \cos 2t + 20e^{-4t}$, current through battery $= 44 \sin 2t - 18 \cos 2t + 20e^{-4t} - 3e^{-t}$.

5. $I_1 = 500te^{-50t}$, $I_2 = 5e^{-50t} + 250te^{-50t} - 5\cos 50t$,
$I_3 = 5\cos 50t - 5e^{-50t} + 250te^{-50t}$,
$Q_3 = 0.1\sin 50t - 5te^{-50t}$.

A EXERCISES, p. 478

3. (a) $x_1 = \frac{10}{17} + \frac{24}{17}e^{-0.85t}$, $x_2 = \frac{24}{17}(1 - e^{-0.85t})$ milligrams.

(b) $\frac{10}{17}$ milligrams, $\frac{24}{17}$ milligrams.

A EXERCISES, p. 488

1. (a) 500 predators, 4000 prey. (b) 100π or 314 days.

(c) $\dfrac{(x - 4000)^2}{16c^2} + \dfrac{(y - 500)^2}{c^2} = 1$.

A EXERCISES, p. 491

1. (a) $X = 2\cos t - \sin t$, $Y = -2\sin t - \cos t$.

(b) $X = e^{-3t}(2\sin 4t - \cos 4t)$, $Y = 2e^{-3t}\cos 4t$.

(c) $X = -3te^{-t}$, $Y = \frac{1}{4}e^t + \frac{3}{4}e^{-t} + \frac{3}{2}te^{-t}$.

(d) $X = 3e^t - 9t^2 + 6t + 2$, $Y = -e^t - 6t$.

2. (a) $X = 10(1 - e^{-t})$, $Y = 5e^{-t}$.

(b) $X = 2\sin t - 3\cos t + e^{-t} + 2$, $Y = -2\sin t + 3\cos t + e^{-t} - 2$.

3. $X = 2 + \frac{1}{2}\sin t$, $Y = \frac{1}{2}\sin t$.

B EXERCISES, p. 491

2. $X = -2$, $Y = -1$, $Z = 3$.

5. $x = \sin t - \frac{1}{2}\cos t + e^{-2t} + \frac{1}{4}e^t - \frac{3}{4}e^{-t}$, $y = \frac{1}{2}\cos t - \sin t + \frac{1}{4}e^t - \frac{3}{4}e^{-t}$.

A EXERCISES, p. 499

1. (a) $\left.\begin{array}{l} x = c_1e^{2t} + 5c_2e^{-4t} \\ y = -c_1e^{2t} + c_2e^{-4t} \end{array}\right\}$; $\left.\begin{array}{l} x = 5e^{-4t} - 3e^{2t} \\ y = 3e^{2t} + e^{-4t} \end{array}\right\}$.

(b) $\left.\begin{array}{l} x = c_1e^{-2t} + c_2te^{-2t} \\ y = c_1e^{-2t} + c_2te^{-2t} - c_2e^{-2t} \end{array}\right\}$.

(c) $\left.\begin{array}{l} x = (2A - B)\cos t + (A + 2B)\sin t \\ y = A\cos t + B\sin t \end{array}\right\}$.

(d) $\left.\begin{array}{l} x = -8c_1e^{7t/2} + c_2e^{-2t} \\ y = 3c_1e^{7t/2} + c_2e^{-2t} \end{array}\right\}$.

(e) $\left.\begin{array}{l} x = (1 - \sqrt{7})c_1e^{(\sqrt{7}-2)t} + (1 + \sqrt{7})c_2e^{-(\sqrt{7}+2)t} \\ y = 3c_1e^{(\sqrt{7}-2)t} + 3c_2e^{-(\sqrt{7}+2)t} \end{array}\right\}$.

(f) $\left.\begin{array}{l} x = c_1 + 2c_2t \\ y = -c_1 + c_2 - 2c_2t \end{array}\right\}$; $\left.\begin{array}{l} x = 10t - 1 \\ y = 2t + 6 \end{array}\right\}$.

(g) $\left.\begin{array}{l} x = e^{-t}[(2A - B)\cos t + (A + 2B)\sin t] \\ y = e^{-t}(A\cos t + B\sin t) \end{array}\right\}$.

(h) $\begin{aligned} x &= e^{2t}[(A - 2B) \cos 2t + (2A + B) \sin 2t] \\ y &= 5e^{2t}(A \cos 2t + B \sin 2t) \end{aligned}\Big\}$; $\begin{aligned} x &= e^{2t}(2 \cos 2t - \sin 2t) \\ y &= -5e^{2t} \sin 2t \end{aligned}\Big\}$.

(i) $\begin{aligned} x &= c_1e^{-3t} + c_2te^{-3t} \\ y &= (c_1 - \tfrac{1}{2}c_2)e^{-3t} + c_2te^{-3t} \end{aligned}\Big\}$; $\begin{aligned} x &= 20te^{-3t} \\ y &= 20te^{-3t} - 10e^{-3t} \end{aligned}\Big\}$.

(j) $\begin{aligned} x &= (14B - 27A) \cos 4t - (14A + 27B) \sin 4t \\ y &= 25A \cos 4t + 25B \sin 4t \end{aligned}\Big\}$.

2. (a) $\begin{aligned} x &= c_1e^{3t} + 2c_2e^{-2t} + e^{-t} \\ y &= -2c_1e^{3t} + c_2e^{-2t} + 4 \end{aligned}\Big\}$.

(b) $\begin{aligned} x &= 2c_1e^{-t} - c_2e^{2t} + 2e^{2t} \\ y &= c_1e^{-t} + c_2e^{2t} \end{aligned}\Big\}$; $\begin{aligned} x &= 4e^{-t} + 3e^{2t} \\ y &= 2e^{-t} - e^{2t} \end{aligned}\Big\}$.

(c) $\begin{aligned} x &= c_1e^{-2t} + c_2te^{-2t} + 4t \\ y &= c_1e^{-2t} + c_2te^{-2t} - c_2e^{-2t} + t - 3 \end{aligned}\Big\}$; $\begin{aligned} x &= 4t \\ y &= t - 3 \end{aligned}\Big\}$.

(d) $\begin{aligned} x &= A \cos t + B \sin t - e^{-3t} \\ y &= (2A - B) \cos t + (A + 2B) \sin t \end{aligned}\Big\}$.

(e) $\begin{aligned} x &= e^{-t}(A \cos t + B \sin t) + 8 \sin t + 14 \cos t - 3 \\ y &= e^{-t}[(B - A) \cos t - (A + B) \sin t] - 14 \sin t - 12 \cos t + 6 \end{aligned}\Big\}$.

(f) $\begin{aligned} x &= c_1e^{t} + c_2e^{-t} - 8te^{-t} - 8t \\ y &= c_1e^{t} + 3c_2e^{-t} - 24te^{-t} + 24e^{-t} - 16t + 8 \end{aligned}\Big\}$.

(g) $\begin{aligned} x &= c_1 + 2c_2e^{-t} - 2t^2e^{-t} \\ y &= c_1 + 3c_2e^{-t} + 3te^{-t} - 3t^2e^{-t} \end{aligned}\Big\}$.

(h) $\begin{aligned} x &= e^{-t}[(2A - B) \cos t + (A + 2B) \sin t] + 35 \sin 2t - 10 \cos 2t \\ y &= e^{-t}(A \cos t + B \sin t) - 9 \sin 2t - 8 \cos 2t \end{aligned}\Big\}$.

C EXERCISES, p. 500

1. $x = c_1e^{t} + 2c_2e^{2t} + c_3e^{3t} + 2e^{-t}$,
$y = 2c_2e^{2t} + 2c_3e^{3t} - 4e^{-t}$,
$z = 2c_1e^{t} + c_2e^{2t} - 2c_3e^{3t} + 16e^{-t}$.

Chapter Eleven

A EXERCISES, p. 509

1. (a) $\begin{pmatrix} 2 & -11 \\ -20 & 16 \end{pmatrix}$.   (b) $\begin{pmatrix} -22 & -8 & 7 \\ 19 & -7 & 2 \\ 20 & -19 & 2 \end{pmatrix}$.   (c) 2.

(d) 10.   (e) $\begin{pmatrix} -6 & 4 \\ -4 & -2 \end{pmatrix}$.   (f) $\begin{pmatrix} 5 & 4 & -3 \\ 0 & 9 & -2 \\ -1 & -8 & 1 \end{pmatrix}$.

2. $a = 2, b = -1$.   **3.** $\begin{pmatrix} 1 & 2 \\ -2 & 5 \end{pmatrix}$.

6. (a) $\begin{pmatrix} 3 & 2 \\ 2 & -1 \end{pmatrix}\begin{pmatrix} x \\ y \end{pmatrix} = \begin{pmatrix} 7 \\ -8 \end{pmatrix}.$ (b) $\begin{pmatrix} 2 & -1 & 3 \\ 4 & 3 & -2 \\ 3 & -2 & 1 \end{pmatrix}\begin{pmatrix} x \\ y \\ z \end{pmatrix} = \begin{pmatrix} 9 \\ -3 \\ 7 \end{pmatrix}.$

(c) $\begin{pmatrix} 1 & 0 & -1 \\ 0 & 1 & 3 \\ 1 & -2 & 1 \end{pmatrix}\begin{pmatrix} x \\ y \\ z \end{pmatrix} = \begin{pmatrix} 4 \\ 2 \\ 0 \end{pmatrix}.$

7. $\begin{pmatrix} 4t^4 - 2t & 4t^2 e^{-t} + te^{-t} + 2e^{-t} \\ 12t^3 - 8t^2 - 6 & 16t^2 e^t + 4te^t - 8e^t \end{pmatrix}.$

8. (a) $\dfrac{d}{dt}\begin{pmatrix} x \\ y \end{pmatrix} + \begin{pmatrix} 3 & -2 \\ -1 & 4 \end{pmatrix}\begin{pmatrix} x \\ y \end{pmatrix} = \begin{pmatrix} e^{-t} \\ \sin 2t \end{pmatrix}.$

(b) $\dfrac{d}{dt}\begin{pmatrix} x \\ y \\ z \end{pmatrix} + \begin{pmatrix} -1 & 2 & -1 \\ 3 & -1 & 4 \\ -2 & 1 & -1 \end{pmatrix}\begin{pmatrix} x \\ y \\ z \end{pmatrix} = \begin{pmatrix} t^2 \\ e^t \\ 0 \end{pmatrix}.$

(c) $\dfrac{d}{dt}\begin{pmatrix} x \\ y \\ z \end{pmatrix} + \begin{pmatrix} -1 & 0 & 1 \\ -2 & -1 & 0 \\ 0 & 4 & -1 \end{pmatrix}\begin{pmatrix} x \\ y \\ z \end{pmatrix} = \begin{pmatrix} 0 \\ 3t \\ -\cos t \end{pmatrix}.$

9. (a) $-10.$ (b) $41.$

10. (a) Has only trivial solution $x = 0, y = 0, z = 0.$ (b) Has non-trivial solutions; e.g.,

$$\begin{pmatrix} x \\ y \\ z \end{pmatrix} = \begin{pmatrix} -1/11 \\ 7/11 \\ 1 \end{pmatrix}, \begin{pmatrix} 1 \\ -7 \\ -11 \end{pmatrix}, \text{ etc.}$$

B EXERCISES, p. 510

5. $X = x_1 \begin{pmatrix} 1 & 0 \\ 0 & 1 \end{pmatrix} + x_2 \begin{pmatrix} 0 & 1 \\ 3 & -2 \end{pmatrix}$ where x_1, x_2 can have any value.

C EXERCISES, p. 511

4. (a) $A^{-1} = \begin{pmatrix} \dfrac{4}{11} & \dfrac{1}{11} \\ -\dfrac{3}{11} & \dfrac{2}{11} \end{pmatrix}.$ (b) $A^{-1} = \begin{pmatrix} 1/37 & 11/37 & 3/37 \\ 8/37 & 14/37 & -13/37 \\ 10/37 & -1/37 & -7/37 \end{pmatrix}.$

A EXERCISES, p. 528

1. (a) $\begin{pmatrix} x \\ y \end{pmatrix} = c_1 \begin{pmatrix} 1 \\ 1 \end{pmatrix} e^{-t} + c_2 \begin{pmatrix} 4 \\ -1 \end{pmatrix} e^{-6t}; \begin{pmatrix} x \\ y \end{pmatrix} = \begin{pmatrix} -1 \\ -1 \end{pmatrix} e^{-t} + \begin{pmatrix} 1 \\ -4 \end{pmatrix} e^{-6t}$

or $x = c_1 e^{-t} + 4c_2 e^{-6t}, y = c_1 e^{-t} - c_2 e^{-6t}; x = e^{-6t} - e^{-t}, y = -e^{-t} - 4e^{-6t}.$

(b) $x = 5e^{-3t}(A \cos 4t + B \sin 4t), y = 2e^{-3t}[(2B - A) \cos 4t - (2A + B) \sin 4t].$

(c) $x = 3c_1 + c_2 e^{-t}, y = 2c_1 + c_2 e^{-t}.$

(d) $\begin{aligned} x &= \sqrt{6}e^{-3t}(c_1 e^{\sqrt{6}t} - c_2 e^{-\sqrt{6}t}) \\ y &= e^{-3t}(c_1 e^{\sqrt{6}t} + c_2 e^{-\sqrt{6}t}) \end{aligned}; \begin{aligned} x &= \sqrt{6}e^{-3t}(e^{\sqrt{6}t} - e^{-\sqrt{6}t}) \\ y &= e^{-3t}(e^{\sqrt{6}t} + e^{-\sqrt{6}t}) \end{aligned}.$

(e) $x = 8(c_1 e^{-3t} + c_2 te^{-3t}), y = (c_2 - 4c_1)e^{-3t} - 4c_2 te^{-3t}.$

(f)
$$x = 14e^{-t/2}\left(A\cos\frac{\sqrt{3}}{2}t + B\sin\frac{\sqrt{3}}{2}t\right)$$
$$y = e^{-t/2}\left[(A\sqrt{3} - 23B)\sin\frac{\sqrt{3}}{2}t - (23A + B\sqrt{3})\cos\frac{\sqrt{3}}{2}t\right]\}.$$

2. (a) $x = A\cos t + B\sin t + t^2 - 1;\; x = 3\cos t - \sin t + t^2 - 1\}$
$\qquad y = -A\sin t + B\cos t + t \quad\}; \quad y = -3\sin t - \cos t + t \quad\}.$

(b) $x = -2c_1e^{-t} - 2c_2te^{-t} + c_2e^{-t}\}$
$\qquad y = c_1e^{-t} + c_2te^{-t} + 2e^t \quad\}.$

(c) $x = c_1e^t + c_2e^{-t} - \frac{1}{2}te^{-t} + t \qquad\}$
$\qquad y = c_1e^t + 3c_2e^{-t} - \frac{3}{2}te^{-t} + \frac{3}{2}e^{-t} + 2t - 1\}.$

(d) $x = c_1e^{-3t} + c_2e^{2t} + 16\sin t - 12\cos t - 6t - 1 \quad\}$
$\qquad y = -c_1e^{-3t} + 4c_2e^{2t} - 8\cos t - 56\sin t - 12t - 8\};$

$\qquad x = 5e^{-3t} + 16\sin t - 12\cos t - 6t - 1 \quad\}$
$\qquad y = -5e^{-3t} - 8\cos t - 56\sin t - 12t - 8 \;\}.$

(e) $x = 2A\cos 3t + 2B\sin 3t + 3te^{-3t} + e^{-3t} - 3t + 2 \qquad\}$
$\qquad y = (B - A)\cos 3t - (A + B)\sin 3t - 3te^{-3t} - \frac{1}{2}e^{-3t} + 3t - 3\}.$

(f) $x = 3c_1 + c_2e^{-t} - t^2e^{-t},\; y = 2c_1 + c_2e^{-t} + te^{-t} - t^2e^{-t}.$

B EXERCISES, p. 529

1. $x = 5c_1e^{4t} + 2c_2e^{-3t} - 3e^{-t} + 4e^t \qquad\}$
$\quad\; y = -22c_1e^{4t} + c_2e^{-3t} + c_3e^{-t} - 8e^t + 21te^{-t} \quad\}.$
$\quad\; z = -4c_1e^{4t} + 4c_2e^{-3t} + 2c_3e^{-t} - 9e^{-t} - 8e^t + 42te^{-t}\}$

A EXERCISES, p. 536

1. (a) -20.   (b) -2.   **2.** (a) 7.   (b) $3\sqrt{2}$.

3. (a) Column vector with components $0, -2/7, 3/7, 6/7$.

(b) Column vector with components $\sqrt{2}/6, 2\sqrt{2}/3, -\sqrt{3}/2, -2\sqrt{2}/3, 0$.

4. (a) -2.

(b) Column vectors with components $\dfrac{3}{\sqrt{11}}, \dfrac{-1}{\sqrt{11}}, 0, \dfrac{-1}{\sqrt{11}}$ and $\dfrac{-1}{\sqrt{6}}, \dfrac{-1}{\sqrt{6}}, 0, \dfrac{-2}{\sqrt{6}}$ respectively.

5. $\begin{pmatrix} 1/\sqrt{6} \\ 2/\sqrt{6} \\ 1/\sqrt{6} \end{pmatrix}$.

8. $\begin{pmatrix} 1/\sqrt{10} \\ 3/\sqrt{10} \\ 0 \end{pmatrix}, \begin{pmatrix} 0 \\ 0 \\ 1 \end{pmatrix}, \begin{pmatrix} -3/\sqrt{10} \\ 1/\sqrt{10} \\ 0 \end{pmatrix}$.

B EXERCISES, p. 536

3. $x_1\begin{pmatrix} 1 & 0 \\ 0 & 1 \end{pmatrix} + x_2\begin{pmatrix} 0 & 1 \\ 3 & -2 \end{pmatrix}$ where x_1 and x_2 can have any value.

5. $\begin{pmatrix} 1/\sqrt{6} \\ -1/\sqrt{6} \\ 2/\sqrt{6} \end{pmatrix}, \begin{pmatrix} -11/\sqrt{174} \\ -7/\sqrt{174} \\ 2/\sqrt{174} \end{pmatrix}, \begin{pmatrix} 2/\sqrt{29} \\ -4/\sqrt{29} \\ -3/\sqrt{29} \end{pmatrix}$.

7. (a) Linearly dependent.   (b) Linearly independent.   (c) Linearly dependent.

1. (a) $u = c_1 \begin{pmatrix} 1 \\ 3 \\ 0 \end{pmatrix} e^{2t} + c_2 \begin{pmatrix} -3 \\ 1 \\ 0 \end{pmatrix} e^{-8t} + c_3 \begin{pmatrix} 0 \\ 0 \\ 1 \end{pmatrix} e^{-4t}.$

(b) $u_1 = \begin{pmatrix} 1 \\ 3 \\ 0 \end{pmatrix} e^{2t}, \; u_2 = \begin{pmatrix} -3 \\ 1 \\ 0 \end{pmatrix} e^{-8t}, \; u_3 = \begin{pmatrix} 0 \\ 0 \\ 1 \end{pmatrix} e^{-4t}.$

Chapter Twelve

A EXERCISES, p. 547

1. (a) $U = x \sin y.$ **(b)** $U = x^2(1 - \cos y) - 2x^2 y/\pi.$

(c) $V = \frac{1}{3}x^3 + 3 \sin y.$ **(d)** $U = x^2 y^2 + ye^x + \frac{1}{2}y^2 - y + 2.$

(e) $Z = \frac{1}{6}e^{3x}(y^3 - 2y^2) - \frac{1}{3}y^2 + x + 3e^{-x}.$

2. (a) $x\dfrac{\partial U}{\partial x} - 2U + 3xy = 0.$ **(b)** $\dfrac{\partial z}{\partial y} = xz.$

(c) $\dfrac{\partial^2 z}{\partial x\, \partial y} + \dfrac{\partial z}{\partial x} - \dfrac{\partial z}{\partial y} - z = 0.$ **(d)** $2xy(\ln y)\dfrac{\partial^2 U}{\partial x\, \partial y} - 2x\dfrac{\partial U}{\partial x} - y(\ln y)\dfrac{\partial U}{\partial y} + U = 0.$

(e) $2\dfrac{\partial z}{\partial x} + \dfrac{\partial z}{\partial y} = 0.$ **(f)** $x\dfrac{\partial z}{\partial x} = y\dfrac{\partial z}{\partial y}.$

(g) $y\dfrac{\partial x}{\partial y} + z\dfrac{\partial x}{\partial z} = 0.$ **(h)** $y\dfrac{\partial z}{\partial x} + x\dfrac{\partial z}{\partial y} = 0.$

(i) $2\dfrac{\partial z}{\partial y} + \dfrac{\partial z}{\partial y} = 3z.$ **(j)** $3\dfrac{\partial^2 z}{\partial x^2} + 5\dfrac{\partial^2 z}{\partial x\, \partial y} - 2\dfrac{\partial^2 z}{\partial y^2} = 0.$

B EXERCISES, p. 548

2. $2x^2 \dfrac{\partial^2 z}{\partial x^2} - 5xy\dfrac{\partial^2 z}{\partial x\, \partial y} + 2y^2 \dfrac{\partial^2 z}{\partial y^2} + 2x\dfrac{\partial z}{\partial x} + 2y\dfrac{\partial z}{\partial y} = 0.$

5. $n = 3.$

7. $U = 20x^2 \cos (y/x).$

C EXERCISES, p. 548

1. (b) $z = (2x - 1) \cos \dfrac{y}{x} + 2(1 - x)e^{-y/x}.$

3. (a) $6\dfrac{\partial z}{\partial x} + 3\dfrac{\partial z}{\partial y} = -4.$ **(b)** $y^2 \dfrac{\partial z}{\partial x} - xy\dfrac{\partial z}{\partial y} = xz.$

(c) $(\tan x)\dfrac{\partial z}{\partial x} - (\cot y)\dfrac{\partial z}{\partial y} + z = 0.$

(d) $x\dfrac{\partial z}{\partial x} + (x - y)\dfrac{\partial z}{\partial y} = y.$

1. $U = e^{2x+2y}$.

2. $U = 4e^{-3x-2y}$.

3. $U = 2e^{2x-y} + 3e^{3x-2y}$.

4. $Y = 4e^{-x+t} - e^{-5x+2t}$.

5. $U = 5e^{-16\pi^2 t} \sin 2\pi x$.

6. $U = 2e^{-4.5t} \sin 3x - 5e^{-8t} \sin 4x$.

7. $Y = 10 \sin \dfrac{\pi x}{2} \cos \dfrac{\pi t}{2}$.

8. $Y = 3 \sin 2\pi x \cos 4\pi t - 4 \sin \dfrac{5\pi x}{2} \cos 5\pi t$.

9. $Y = 6 \sin x \sin (t/3) - 4.5 \sin 2x \sin (2t/3)$.

10. $U = 5e^{(1-4\pi^2)t} \sin 2\pi x - e^{(1-16\pi^2)t} \sin 4\pi x$.

3. (a) $U = F(y + x) + G(y + 2x)$. (b) $U = F(x + iy) + G(x - iy)$.

   (c) $Y = F(x - at) + G(x + at)$. (d) $U = F(y - 4x) + G(y)$.

5. (a) $U = F(y + 2x) + xG(y + 2x)$. (b) $Y = F(2x + 3t) + xG(2x + 3t)$.

6. $Z = F(y - x) + G(y - 3x) + \frac{4}{55}e^{2x+3y}$.

7. $U = F(x - y) + xG(x - y) - \frac{1}{3}\sin(x - 4y) + \frac{1}{12}x^4$.

8. $U = F(x + y) + G(y - 2x) - \frac{3}{2}xe^{y-3x} + \frac{1}{15}\cos(x - 2y)$.

1. (b) $F(x - y - z, x^2 - 2xy) = 0$.

2. (d) $U = 8c^{-4\pi^2 t} \sin 2\pi x$.

2. $\dfrac{\partial^2 Y}{\partial t^2} + \beta \dfrac{\partial Y}{\partial t} = a^2 \dfrac{\partial^2 Y}{\partial x^2} - g$.

4. $U_x(0, t) = 0$, $U_x(L, t) = 0$, $U(x, 0) = 100$; $U(x, t) = 100$.

Chapter Thirteen

1. $U(x, t) = \displaystyle\sum_{n=1}^{\infty} \dfrac{160}{n\pi}\left(1 - \cos \dfrac{n\pi}{2}\right) e^{-0.20(n\pi/100)^2 t} \sin \dfrac{n\pi x}{100}$.

2. $U(x, t) = \displaystyle\sum_{n=1}^{\infty} \dfrac{200}{n\pi}\left(\cos \dfrac{2n\pi}{5} - \cos \dfrac{3n\pi}{5}\right) e^{-0.20(n\pi/100)^2 t} \sin \dfrac{n\pi x}{100}$.

3. $U(x, t) = 20 + \displaystyle\sum_{n=1}^{\infty} \dfrac{160}{n\pi}(1 - \cos n\pi)e^{-0.20(n\pi/40)^2 t} \sin \dfrac{n\pi x}{40}$.

5. $U(x, y) = U_0$.

1. (a) $U(x, t) = \dfrac{1}{L} \displaystyle\int_0^L f(x)\,dx + \sum_{n=1}^{\infty} \left\{ \dfrac{2}{L} \int_0^L f(x) \cos \dfrac{n\pi x}{L}\,dx \right\} e^{-\kappa n^2 \pi^2 t/L^2} \cos \dfrac{n\pi x}{L}.$

(b) $U(x, t) = \dfrac{U_0}{2} + \displaystyle\sum_{n=1}^{\infty} \left\{ \dfrac{4U_0}{n^2\pi^2} \left(2 \cos \dfrac{n\pi}{2} - 1 - \cos n\pi \right) \right\} e^{-\kappa n^2 \pi^2 t/L^2} \cos \dfrac{n\pi x}{L}.$

2. $U(x, t) = U_0 + \displaystyle\sum_{n=1}^{\infty} \left\{ \dfrac{2}{L} \int_0^L [f(x) - U_0] \cos \dfrac{(2n-1)\pi x}{L}\,dx \right\} e^{-\kappa (2n-1)^2 \pi^2 t/L^2} \cos \dfrac{(2n-1)\pi x}{L}.$

3. $U(x, t) = e^{-ct} \displaystyle\sum_{n=1}^{\infty} \left\{ \dfrac{2}{L} \int_0^L f(x) \sin \dfrac{n\pi x}{L}\,dx \right\} e^{-\kappa n^2 \pi^2 t/L^2} \dfrac{\sin n\pi x}{L}.$

1. $Y(x, t) = \dfrac{8\alpha L^2}{\pi^3} \left(\sin \dfrac{\pi x}{L} \cos \dfrac{\pi a t}{L} + \dfrac{1}{3^3} \sin \dfrac{3\pi x}{L} \cos \dfrac{3\pi a t}{L} + \dfrac{1}{5^3} \sin \dfrac{5\pi x}{L} \cos \dfrac{5\pi a t}{L} + \cdots \right).$

2. $Y(x, t) = \dfrac{1}{6\pi^2} \left(\sin \dfrac{\pi x}{2} \cos 8\pi t - \dfrac{1}{3^2} \sin \dfrac{3\pi x}{2} \cos 24\pi t + \dfrac{1}{5^2} \sin \dfrac{5\pi x}{2} \cos 40\pi t - \cdots \right).$

3. (a) $Y(x, t) = 0.25 \sin \dfrac{\pi x}{4} \cos 16\pi t.$

(b) $Y(x, t) = 0.1 \sin \pi x \cos 64\pi t - 0.02 \sin 3\pi x \cos 192\,\pi t.$

(c) $Y(x, t) = \displaystyle\sum_{n=1}^{\infty} \dfrac{0.32}{n^2\pi^2} \sin \dfrac{n\pi}{2} \sin \dfrac{n\pi x}{4} \cos 16 n\pi t.$

6. $Y(x, t) = \dfrac{8\alpha L^2 e^{-\beta t/2}}{\pi^3} \left(\sin \dfrac{\pi x}{L} \cos \sqrt{\dfrac{\pi^2 a^2}{L^2} - \dfrac{\beta^2}{4}}\, t + \dfrac{1}{3^3} \sin \dfrac{3\pi x}{L} \cos \sqrt{\dfrac{9\pi^2 a^2}{L^2} - \dfrac{\beta^2}{4}}\, t \right.$

$\left. + \dfrac{1}{5^3} \sin \dfrac{5\pi x}{L} \cos \sqrt{\dfrac{25\pi^2 a^2}{L^2} - \dfrac{\beta^2}{4}}\, t + \cdots \right)$

7. $Y(x, t) = \dfrac{e^{-t}}{6\pi^2} \left(\sin \dfrac{\pi x}{2} \cos 2\sqrt{16\pi^2 - 1}\, t - \dfrac{1}{3^2} \sin \dfrac{3\pi x}{2} \cos 2\sqrt{144\pi^2 - 1}\, t \right.$

$\left. + \dfrac{1}{5^2} \sin \dfrac{5\pi x}{2} \cos 2\sqrt{400\pi^2 - 1}\, t - \cdots \right)$

2. $Y(x, t) = \dfrac{4\alpha L^3}{a\pi^4} \displaystyle\sum_{n=1}^{\infty} \dfrac{1 - \cos n\pi}{n^4} \sin \dfrac{n\pi x}{L} \sin \dfrac{n\pi a t}{L}.$

3. (a) $Y(x, t) = \dfrac{2v_0 L}{\pi^2 a\epsilon} \displaystyle\sum_{n=1}^{\infty} \dfrac{1}{n} \sin \dfrac{n\pi\epsilon}{L} \sin \dfrac{n\pi b}{L} \sin \dfrac{n\pi x}{L} \sin \dfrac{n\pi a t}{L}.$

1. (a) $V(r, \phi) = \dfrac{V_0}{\pi} \displaystyle\sum_{n=1}^{\infty} r^n \left[\dfrac{2 \sin (n\pi/2)}{n} \cos n\phi + \dfrac{1 - 2 \cos (n\pi/2) + \cos n\pi}{n} \sin n\phi \right].$

(b) $V(r, \phi) = \dfrac{V_0}{\pi} \displaystyle\sum_{n=1}^{\infty} r^{-n} \left[\dfrac{2 \sin (n\pi/2)}{n} \cos n\phi + \dfrac{1 - 2 \cos (n\pi/2) + \cos n\pi}{n} \sin n\phi \right].$

3. $U(r, \phi) = 40\left(r - \dfrac{1}{r}\right)\cos\phi - 25\left(r - \dfrac{4}{r}\right)\sin\phi.$

4. (a) $U(r, \phi) = \dfrac{400}{\pi}\left(r^6\sin 6\phi + \dfrac{r^{12}\sin 12\phi}{3} + \dfrac{r^{18}\sin 18\phi}{5} + \cdots\right).$

5. (a) $U(r, \phi) = \dfrac{400}{\pi}\left(r^2\sin 2\phi + \dfrac{r^4\sin 4\phi}{3} + \dfrac{r^6\sin 6\phi}{5} + \cdots\right).$

4. $U(r, \phi) = 15\left(r + \dfrac{4}{r}\right)\sin\phi.$

2. $U(x, t) = 30 + x + \displaystyle\sum_{n=1}^{\infty}\left(\dfrac{60 + 40\cos n\pi}{n\pi}\right)e^{-0.15(n\pi/50)^2 t}\sin\dfrac{n\pi x}{50}.$

3. $U(x, t) = U_1 + \dfrac{U_2 - U_1}{L}x + \displaystyle\sum_{n=1}^{\infty}b_n e^{-\kappa n^2\pi^2 t/L^2}\sin\dfrac{n\pi x}{L}$

where $b_n = \dfrac{2}{n\pi}\left[U_0 - U_1 + (U_2 - U_0)\cos n\pi\right].$

4. $U(x, t) = U_1 + \dfrac{U_2 - U_1}{L}x + \displaystyle\sum_{n=1}^{\infty}b_n e^{-\kappa n^2\pi^2 t/L^2}\sin\dfrac{n\pi x}{L}$

where $b_{n'} = \dfrac{2}{n\pi}\left[U_0 - U_1 + U_2\cos n\pi\right].$

5. $Y(x, t) = \displaystyle\sum_{n=1}^{\infty}\sin\dfrac{n\pi x}{L}\left(a_n\sin\dfrac{n\pi at}{L} + b_n\cos\dfrac{n\pi at}{L}\right)$

where $a_n = \dfrac{2v_0 L}{n^2\pi^2 a}(1 - \cos n\pi),\ b_n = \dfrac{4\alpha L^3}{n^3\pi^3}(1 - \cos n\pi)$

$v_0 = $ speed, $\alpha = $ a constant.

6. $U = \tfrac{1}{2}x(1 - x) + \displaystyle\sum_{n=1}^{\infty}\left[1 - \dfrac{2(1 - \cos n\pi)}{n^3\pi^3}\right]e^{-n^2\pi^2 t}\sin n\pi x.$

2. $Y = Kx + \dfrac{8KL}{\pi^2}\displaystyle\sum_{n=1}^{\infty}\dfrac{(-1)^n}{(2n - 1)^2}\sin\dfrac{(2n - 1)\pi x}{2L}\cos\dfrac{(2n - 1)\pi at}{2L}.$

3. $Y = 1 - \dfrac{x^2}{2} - \dfrac{4}{\pi}\displaystyle\sum_{n=1}^{\infty}\left[\dfrac{(2n - 1)\pi\cos n\pi + 2}{(2n - 1)^3}\right]\sin(2n - 1)x\cos(2n - 1)t.$

6. $U = U_1 + \left(\dfrac{U_2 - U_1}{L} + \dfrac{\alpha L}{2\kappa}\right)x - \dfrac{\alpha x^2}{2\kappa}$

$\quad + \displaystyle\sum_{n=1}^{\infty}\left[\dfrac{\{2\kappa n^2\pi^2(U_2 - U_0) + 2\alpha\}\cos n\pi + 2\kappa n^2\pi^2(U_0 - U_1) - 2\alpha}{\kappa L n^3\pi^3}\right]e^{-\kappa n^2\pi^2 t/L^2}\sin\dfrac{n\pi x}{L}.$

7. $U = \dfrac{\alpha L x}{\kappa} - \dfrac{\alpha x^2}{2\kappa} - \dfrac{16\alpha L^2}{\kappa\pi^3} \displaystyle\sum_{n=1}^{\infty} \dfrac{e^{-(2n-1)^2\pi^2\kappa t/4L^2}}{(2n-1)^3} \sin\dfrac{(2n-1)\pi x}{2L}.$

8. $U = \dfrac{\beta x(L^2 - x^2)}{6\kappa} + \dfrac{2\beta L^3}{\kappa\pi^3} \displaystyle\sum_{n=1}^{\infty} \dfrac{\cos n\pi}{n^3} e^{-n^2\pi^2\kappa t/L^2} \sin\dfrac{n\pi x}{L}$

where β is the constant of proportionality.

1. $Z = \dfrac{16}{\pi^6} \displaystyle\sum_{m=1}^{\infty}\sum_{n=1}^{\infty} \dfrac{(1 - \cos m\pi)(1 - \cos n\pi)}{m^3 n^3} \sin m\pi x \sin n\pi y \cos a\sqrt{m^2 + n^2}\pi t.$

8. $U(x, t) = \displaystyle\sum_{n=1}^{\infty} a_n e^{-\kappa\lambda_n^2 t} \cos\lambda_n x$ where λ_n are the positive roots of $\cot\lambda L = \lambda/h$ and

$$a_n = \dfrac{2h}{Lh + \sin^2\lambda_n L} \int_0^L f(x)\cos\lambda_n x\, dx.$$

10. (a) $U(r, t) = \dfrac{2U_0}{r} \displaystyle\sum_{n=1}^{\infty} \left(\dfrac{1 - \cos n\pi}{n\pi}\right) e^{-n^2\pi^2\kappa t} \sin n\pi r.$

(b) $U(r, t) = \dfrac{2U_0}{r} \displaystyle\sum_{n=1}^{\infty} \left(\dfrac{\sin n\pi/2}{n^2\pi^2} - \dfrac{\cos n\pi/2}{2n\pi}\right) e^{-n^2\pi^2\kappa t} \sin n\pi r.$

12. $U = \dfrac{2}{r} \displaystyle\sum_{n=1}^{\infty} \left(\int_0^1 vf(v)\sin n\pi v\, dv\right) \sin n\pi r \cos n\pi t.$

Chapter Fourteen

1. (a) $U(r, t) = 50e^{-\kappa\lambda_1^2 t} J_0(\lambda_1 r).$ **(b)** $U(r, t) = 20e^{-\kappa\lambda_1^2 t} J_0(\lambda_1 r) - 10e^{-\kappa\lambda_2^2 t} J_0(\lambda_2 r).$

2. $U(r, t) = 100 \displaystyle\sum_{n=1}^{\infty} \dfrac{e^{-\kappa\lambda_n^2 t} J_1(\lambda_n/2)}{\lambda_n[J_1(\lambda_n)]^2} J_0(\lambda_n r)$ where $J_0(\lambda_n) = 0.$

3. (a) $U(r, t) = 800 \displaystyle\sum_{n=1}^{\infty} \dfrac{e^{-\kappa\lambda_n^2 t}}{\lambda_n^3 J_1(\lambda_n)} J_0(\lambda_n r)$ where $J_0(\lambda_n) = 0.$

(b) $U(r, t) = 20 \displaystyle\sum_{n=1}^{\infty} \dfrac{e^{-\kappa\lambda_n^2 t}[J_1(\lambda_n/3) + 2J_1(2\lambda_n/3) + 3J_1(\lambda_n)]}{\lambda_n[J_1(\lambda_n)]^2} J_0(\lambda_n r)$ where $J_0(\lambda_n) = 0.$

4. $U(r, t) = 25 + 150 \displaystyle\sum_{n=1}^{\infty} \dfrac{e^{-\kappa\lambda_n^2 t}}{\lambda_n J_1(\lambda_n)} J_0(\lambda_n r).$

5. $U(r, t) = 100 \displaystyle\sum_{n=1}^{\infty} \dfrac{e^{-\kappa\lambda_n^2 t/c^2} J_1(\lambda_n/2)}{\lambda_n[J_1(\lambda_n)]^2} J_0\left(\dfrac{\lambda_n r}{c}\right).$

6. $U(r, t) = 120 \displaystyle\sum_{n=1}^{\infty} \dfrac{e^{-\kappa\lambda_n^2 t}}{\lambda_n J_1(\lambda_n)} J_0(\lambda_n r)$ where $J_0(\lambda_n) = 0.$

7. $U(r, t) = 100 - 80 \displaystyle\sum_{n=1}^{\infty} \dfrac{e^{-\kappa\lambda_n^2 t}}{\lambda_n J_1(\lambda_n)} J_0(\lambda_n r)$ where $J_0(\lambda_n) = 0.$

8. $U(r, t) = \displaystyle\sum_{n=1}^{\infty} a_n e^{-\kappa\lambda_n^2 t} J_0(\lambda_n r)$ where $J_1(\lambda_n) = 0$ and $a_n = \dfrac{2}{[J_0(\lambda_n)]^2} \int_0^1 rf(r)J_0(\lambda_n r)dr.$

9. $U(r, t) = 100 \sum\limits_{n=1}^{\infty} \dfrac{e^{-\kappa\lambda_n^2 t} J_1(\lambda_n/2)}{\lambda_n [J_0(\lambda_n)]^2} J_0(\lambda_n r)$ where $J_1(\lambda_n) = 0$.

10. (a) $U(r, t) = \sum\limits_{n=1}^{\infty} a_n e^{-\kappa\lambda_n^2 t} J_0(\lambda_n r)$ where $a_n = \dfrac{2U_0 J_1(\lambda_n)}{(\lambda_n^2 + h^2)[J_0(\lambda_n)]^2}$.

(b) Same as in (a) except that $a_n = \dfrac{4U_0[2J_1(\lambda_n) - \lambda_n J_0(\lambda_n)]}{\lambda_n(\lambda_n^2 + h^2)[J_0(\lambda_n)]^2}$.

11. $U(r, t) = \sum\limits_{n=1}^{\infty} a_n e^{-\kappa\lambda_n^2 t} J_0(\lambda_n r)$ where $\lambda_n J_0'(\lambda_n) + 2J_0(\lambda_n) = 0$ and $a_n = \dfrac{100\lambda_n J_1(\lambda_n/2)}{(\lambda_n^2 + 4)[J_0(\lambda_n)]^2}$.

B EXERCISES, p. 630

1. $U(r, t) = 100 - 200\alpha \sum\limits_{n=1}^{\infty} \dfrac{(\lambda_n^3 - 16\lambda_n^2 + 64)}{\lambda_n^5 J_1(\lambda_n)} e^{-\kappa\lambda_n^2 t} J_0(\lambda_n r)$ where $J_0(\lambda_n) = 0$ and α is a proportionality constant.

3. (a) $Z(r, t) = \sum\limits_{n=1}^{\infty} a_n J_0(\lambda_n r) \cos \lambda_n at$ where $J_0(\lambda_n) = 0$ and $a_n = \dfrac{2}{[J_1(\lambda_n)]^2} \int\limits_0^1 rf(r) J_0(\lambda_n r) dr$.

(b) Same as in (a) where $a_n = \dfrac{8\alpha}{\lambda_n^3 J_1(\lambda_n)}$.

6. $U(r, z) = 1 - 2 \sum\limits_{n=1}^{\infty} \dfrac{\sinh \lambda_n z J_0(\lambda_n r)}{\lambda_n \sinh \lambda_n J_1(\lambda_n)}$ where $J_0(\lambda_n) = 0$.

7. $U(r, z) = \sum\limits_{n=1}^{\infty} a_n \sinh \lambda_n z J_0(\lambda_n r)$ where $J_1(\lambda_n) = 0$ and

$a_n = \dfrac{2}{[J_0(\lambda_n)]^2 \sinh \lambda_n} \int\limits_0^1 rf(r) J_0(\lambda_n r) dr$.

C EXERCISES, p. 631

2. $U(r, \phi, t) = 2U_0 \sin \phi \sum\limits_{m=1}^{\infty} \dfrac{e^{-\kappa\lambda_m^2 t} J_1(\lambda_m r)}{\lambda_m J_2(\lambda_m)}$ where λ_m is the mth positive root of $J_1(\lambda) = 0$.

3. $U(r, \phi, t) = 200 \cos 2\phi \sum\limits_{m=1}^{\infty} \dfrac{e^{-\kappa\lambda_m^2 t} J_2(\lambda_m r)}{\lambda_m J_3(\lambda_m)}$ where λ_m is the mth positive root of $J_2(\lambda) = 0$.

4. $U(r, t) = \dfrac{\alpha}{4}(1 - r^2) - 2\alpha \sum\limits_{n=1}^{\infty} \dfrac{e^{-\lambda_n^2 t} J_0(\lambda_n r)}{\lambda_n^3 J_1(\lambda_n)}$ where $J_0(\lambda_n) = 0$.

A EXERCISES, p. 638

1. (a) V_0.   (b) cV_0/r.   (c) $q = cV_0$.

2. (a) $\tfrac{3}{2}V_0 r P_1(x) + V_0 \sum\limits_{n=3,5,\ldots} (-1)^{(n-1)/2} \dfrac{1 \cdot 3 \cdots (n-2)}{2 \cdot 4 \cdots (n+1)} r^n P_n(x)$.

(b) $\dfrac{3V_0}{2r^2} P_1(x) + V_0 \sum\limits_{n=3,5,\ldots} (-1)^{(n-1)/2} \dfrac{1 \cdot 3 \cdots (n-2)}{2 \cdot 4 \cdots (n+1)} \dfrac{P_n(x)}{r^{n+1}}$.

4. $U(r, \theta) = 28r P_1(\cos \theta) - 8r^3 P_3(\cos \theta)$.

5. (a) $V(r, \theta) = \tfrac{4}{3}V_0 r^2 P_2(\cos \theta) - \dfrac{V_0}{3} P_0(\cos \theta)$.   (b) $V(r, \theta) = \dfrac{4V_0}{3r^3} P_2(\cos \theta) - \dfrac{V_0}{3r} P_0(\cos \theta)$.

6. (a) $V_0 \left[\dfrac{1}{4} + \dfrac{r}{2} P_1(\cos \theta) + \dfrac{5r^2}{16} P_2(\cos \theta) - \dfrac{3r^4}{32} P_4(\cos \theta) + \cdots \right].$

(b) $V_0 \left[\dfrac{1}{4r} + \dfrac{P_1(\cos \theta)}{2r^2} + \dfrac{5P_2(\cos \theta)}{16r^3} - \dfrac{3P_4(\cos \theta)}{32r^5} + \cdots \right].$

7. $30 \left[1 + \frac{5}{4} r^2 P_2(\cos \theta) - \frac{3}{16} r^4 P_4(\cos \theta) + \cdots \right].$

B EXERCISES, p. 638

2. $\dfrac{100}{r} + \left(20r - \dfrac{40}{r^2} \right) P_1(\cos \theta).$

A EXERCISES, p. 652

1. $Y(x, t) = 8\alpha L \displaystyle\sum_{n=1}^{\infty} \dfrac{J_0(\lambda_n \sqrt{1 - x/L})}{\lambda_n^3 J_1(\lambda_n)} \cos \dfrac{\lambda_n}{2} \sqrt{\dfrac{g}{L}} \, t$ where $J_0(\lambda_n) = 0.$

2. (a) $Y(x, t) = 16\beta L^2 \displaystyle\sum_{n=1}^{\infty} \dfrac{(8 - \lambda_n^2) J_0(\lambda_n \sqrt{1 - x/L})}{\lambda_n^5 J_1(\lambda_n)} \cos \dfrac{\lambda_n}{2} \sqrt{\dfrac{g}{L}} \, t.$

(b) The sum of the results in (a) and Exercise 1.

3. (a) $Y(x, t) = \displaystyle\sum_{n=1}^{\infty} b_n J_0(\lambda_n \sqrt{1 - x/L}) \sin \left(\dfrac{\lambda_n}{2} \sqrt{\dfrac{g}{L}} \, t \right)$

where $b_n = \dfrac{4\sqrt{L/g}}{\lambda_n [J_1(\lambda_n)]^2} \displaystyle\int_0^1 u G(u) J_0(\lambda_n u) \, du$

$G(u) = g(L[1 - u^2])$ and $J_0(\lambda_n) = 0.$

(b) The sum of the result in (a) and equation (19) page 644 with coefficients (22).

4. $Y(x, t) = 4v_0 \sqrt{\dfrac{L}{g}} \displaystyle\sum_{n=1}^{\infty} \dfrac{J_0(\lambda_n \sqrt{1 - x/L})}{\lambda_n^2 J_1(\lambda_n)} \sin \left(\dfrac{\lambda_n}{2} \sqrt{\dfrac{g}{L}} \, t \right).$

B EXERCISES, p. 652

2. $Y(x, t) = \displaystyle\sum_{n=1}^{\infty} a_n J_0(\lambda_n \sqrt{1 - x/L}) \cos \left(\dfrac{\lambda_n}{2} \sqrt{\dfrac{g}{L}} \, t \right)$

where $a_n = \dfrac{8h}{\lambda_n^3 [J_1(\lambda_n)]^2} \left[\lambda_n J_0 \left(\dfrac{\sqrt{2}}{2} \lambda_n \right) - 2\sqrt{2} J_1 \left(\dfrac{\sqrt{2}}{2} \lambda_n \right) + 2J_1(\lambda_n) \right].$

C EXERCISES, p. 653

2. $V(r, \theta) = \displaystyle\sum_{n=0}^{\infty} \dfrac{(-1)^n}{n!} r^n P_n(\cos \theta).$

4. (c) $Y(x, t) = \displaystyle\sum_{n=0}^{\infty} a_n P_n(x/L) \cos \left(\dfrac{\omega \sqrt{n(n + 1)} \, t}{2} \right)$ where $a_n = \dfrac{2n + 1}{2L} \displaystyle\int_0^L f(x) P_n(x/L) \, dx.$

bibliography

[1] Akhiezer, N. I., *The Calculus of Variations*, Blaisdell, 1962.

[2] Allen, R. G. D., *Mathematical Economics*, Macmillan, 1960.

[3] Andronow, A. A., and C. E. Chaikin, *Theory of Oscillations*, Princeton University Press, 1953.

[4] Carslaw, H. S., and J. C. Jaeger, *Conduction of Heat in Solids*, 2nd edition, Oxford University Press, 1959.

[5] Churchill, R. V., and J. V. Brown, *Fourier Series and Boundary Value Problems*, 3rd edition, McGraw-Hill, 1978.

[6] Churchill, R. V., *Operational Mathematics*, 3rd edition, McGraw-Hill, 1972.

[7] Davis, H. T., *Introduction to Non-Linear Differential and Integral Equations*, Dover, 1962.

[8] Dirac, P. A. M., *Principles of Quantum Mechanics*, 4th edition, Oxford University Press, 1958.

[9] Forsyth, A. R., *Calculus of Variations*, Dover, 1960.

[10] Fourier, J., *The Analytic Theory of Heat*, Dover, 1955.

[11] Frazer, R. A., W. J. Duncan, and A. R. Collar, *Elementary Matrices*, Cambridge University Press, 1947.

[12] Hildebrand, F. B., *Introduction to Numerical Analysis*, 2nd edition, McGraw-Hill, 1974.

[13] Ince, E. L., *Ordinary Differential Equations*, 4th edition, Dover, 1956.

[14] Jackson, D., *Fourier Series and Orthogonal Polynomials*, Carus Monograph No. 6, Mathematical Association of America, 1941.

[15] Kooros, A., *Elements of Mathematical Economics*, Houghton Mifflin, 1965.

[16] Lancaster, K., *Mathematical Economics*, Macmillan, 1968.

[17] Langer, R. E., Fourier's Series, *The Genesis and Evolution of a Theory*, Slaught Memorial Paper No. 1, Mathematical Association of America, 1947.

[18] Lotka, A. J., *Elements of Mathematical Biology*, Dover, 1956.

[19]   Mikusinski, J., *Operational Calculus*, Pergamon, 1959.

[20]   Minorsky, N., *Non-Linear Oscillation*, Krieger, 1974.

[21]   Oldham, K. B., and J. Spanier, *Fraction Calculus, Theory and Application*, Academic Press, 1974.

[22]   Pogorzelski, W., *Integral Equations and Their Applications*, Pergamon Press, 1966.

[23]   Rashevsky, N., *Mathematical Biophysics*, 3rd edition, Dover, 1960.

[24]   Stoker, J. J., *Non-Linear Vibrations in Mechanical and Electrical Systems*, Interscience. 1950.

[25]   Szego, G., *Orthogonal Polynomials*, 2nd edition, American Mathematical Society, 1959.

[26]   Taylor, A. E., *Advanced Calculus*, Ginn, 1955.

[27]   Timoshenko, S., and J. N. Goodier, *Theory of Elasticity*, McGraw-Hill, 1951.

[28]   Watson, G. N., *Theory of Bessel Functions*, 2nd edition, Cambridge University Press, 1962.

[29]   Whittaker, E. T., and G. N. Watson, *Modern Analysis*, 4th edition, Cambridge University Press, 1958.

[30]   Widder, D. V., *The Laplace Transform*, Princeton University Press, 1946.

index

Fourier series (*cont.*)
  double, 608-612
Fourier sine series, 387
Fourier's problem, 571, 576, 584, 613
FPS system, 72
Frequency
  for simple pendulum, 247
  for vibrating beam, 593, 594
  for vibrating chain, 644, 645
  for vibrating spring, 223, 225
  for vibrating string, 587, 588
Friction, coefficient of, 81
Frobenius, method of, 317-332
  examples using, 321-332
Frobenius-type solution, 318
Functions
  elimination of arbitrary, 545-548
  as vectors, 349-351
Fundamental frequency, 587, 628, 644

Gamma function, 261, 262
  recurrence formula for, 262
  use of in Bessel functions, 270
Gauss's differential equation, 346, 347
Geiger counter, 445
Generalized functions, 270
General solution, 15-20, 169, 542
Generating function, 340, 345, 346, 403
Generator, 82
Geometry, problems involving, 123-136
Gradient, 102
Gram-Schmidt orthonormalization method, 410-412, 536, 537
  for generating Legendre polynomials, 411, 412
Gravitation, Newton's law of universal, 116, 451, 454
Gravitational potential, 566, 595, 599, 633, 637, 638, 645-648
  deduced from Kepler's laws, 403
Gravity, acceleration due to, 72
Green's function, 218
Ground potential, 599
Grouping of terms, solution by, 46, 47
Growth and decay, applications involving, 106-111, 148-153

Half derivative, 169, 285, 298
Half-life, 109
Half range Fourier sine and cosine series, 389
Halley's comet, 457
Hanging cable or chain, 111-116
Harmonic frequencies, 587, 588, 596, 611, 612
Harmonic function, 596
Heat conduction, boundary value problems involving, 562-565, 571-584, 604-606, 621-625, 634-637
Heat flow, 91 (*see also* Heat conduction)
Heaviside, 204, 256
  expansion formula, 284
  unit step function, 264-268
Henry, 84
Hermite polynomials, 345, 346, 373
Hermite's differential equation, 345, 346

Hermite series, 405
Higher-order differential equations, 57-60
Hilbert space, 354
Homogeneous differential equations, 38, 39
  functions, 548
Hooke's law, 220, 221
  generalized, 144
Hosts, 488
Huygens, 292
Hydrodynamics, 339, 567
Hyperbolic orbits, 455, 457
Hypergeometric differential equation, 346
Hypergeometric functions, 346
Hypersphere, 545
Hypersurface, 545

ICBM, 460
Identity matrix, 506
Imaginary axis, 309, 311
Imaginary eigenvalues, 518-520
Imaginary roots, 178
Immunity, 154
Impedance, 245
Implicit functions, 6
Impulse of a force, 268
Impulse functions, 268-273
  unit, 270
Inclined plane, 80-82, 449
Independence (*see* Linear independence)
Independent variable, 3
Indicial equation, 322
Indicial roots, 322, 328
Inductance, 83
Inductors, 83
Inflation, 250
Initial condition, 8
Initial value problem, 8, 434
Input function, 295
Inspection, method of, 46, 50, 56, 57
Insulated boundaries, problems involving, 577, 578
Integral equations, 280, 289-293, 298
Integrating factors, 42, 48-52, 56, 57, 409
Intercontinental ballistic missile, 460
Interest, compound, 110
Inventory, 162-165
Inverse Laplace transform, 274-285
Irregular singular point, 319-321
Isobars, 91
Isoclines, method of, 28
Isothermal curves and surfaces, 91, 101, 581
Isotope, radioactive, 110

Jupiter, period of revolution for, 461

Kepler's laws, 453, 454
Kinetic energy, 80, 124, 290
Kirchhoff's laws, 82, 84, 242, 469

Lagrange, 200
Laguerre polynomials, 346, 373

Laguerre's differential equation, 346
Laguerre series, 405
Laplace's equation, 548, 562, 565-567, 569
  boundary value problems involving, 595-602, 637, 638
Laplace transform operator, 257
  linear property of, 259
Laplace transforms, 204, 255-298, 489-491
  existence of, 262-264
  for partial differential equations, 557, 558
Laplace transform tables, use of, 275, 276
Laplacian operator, 562, 565, 566
  in cylindrical coordinates, 620, 621
  in spherical coordinates, 633, 634
Law of exponential growth and decay, 106
Law of mass action, 97, 99
Left hand limit, 263
Legendre functions, 344 (*see also* Legendre polynomials)
  associated, 640, 641
Legendre polynomials, 344, 345
  boundary value problems using, 633-641
  generating function for, 345, 403, 639
  orthogonality of, 366, 371-373
  recurrence formula for, 344, 345
  Rodrigues' formula for, 344
Legendre's associated equation, 640
Legendre's differential equation, 333, 343, 345, 636
Legendre series, 402-405, 636
Leibniz, 3
Leibniz's rule for differentiating an integral, 260
Length of a vector, 352, 353, 531, 532
Lerch's theorem, 283
L'Hôpital's rule, 337, 368
Liapunov, 482
Light, absorption of, 110, 111
Limit cycles, 489
Limiting velocity, 78
Limit in mean, 379
Linear combination, 182
Linear dependence (*see* Linear independence)
Linear differential equations, 4, 53, 54, 167-218, 220-254, 438, 550
Linear independence, 181-190, 356, 522, 523
Linearization, method of, 441, 442
Linear operator, 169
Linear partial differential equation, 550
Linear systems of differential equations, 438
Lines of force, 90
Logarithmic decrement, 235
Logistic curve, 152, 155
Longitudinal vibrations of a beam, 591
Lotka-Volterra equations, 481, 488, 489
Lower limit, in a summation, 303

Maclaurin series, 302, 309, 404
Magnet, 90, 91

SOME USEFUL INTEGRALS
(The constant of integration is omitted in each indefinite integral)

1. $\int u^n \, du = \dfrac{u^{n+1}}{n+1}, n \neq -1$

2. $\int \dfrac{du}{u} = \ln |u| = \begin{cases} \ln u & \text{if } u > 0 \\ \ln(-u) & \text{if } u < 0 \end{cases}$

3. $\int \sin u \, du = -\cos u$

4. $\int \cos u \, du = \sin u$

5. $\int \tan u \, du = \ln |\sec u|$

6. $\int \cot u \, du = \ln |\sin u|$

7. $\int \sec u \, du = \ln |\sec u + \tan u|$

8. $\int \csc u \, du = \ln |\csc u - \cot u|$

9. $\int \sin^2 u \, du = \dfrac{u}{2} - \dfrac{\sin 2u}{4}$

10. $\int \cos^2 u \, du = \dfrac{u}{2} + \dfrac{\sin 2u}{4}$

11. $\int \tan^2 u \, du = \tan u - u$

12. $\int \cot^2 u \, du = -\cot u - u$

13. $\int \sec^2 u \, du = \tan u$

14. $\int \csc^2 u \, du = -\cot u$

15. $\int e^{au} \, du = \dfrac{e^{au}}{a}, a \neq 0$

16. $\int \ln u \, du = u \ln u - u$

17. $\int e^{au} \sin bu \, du = \dfrac{e^{au}(a \sin bu - b \cos bu)}{a^2 + b^2}$

18. $\int e^{au} \cos bu \, du = \dfrac{e^{au}(a \cos bu + b \sin bu)}{a^2 + b^2}$

19. $\int \dfrac{du}{u^2 + a^2} = \dfrac{1}{a} \tan^{-1} \dfrac{u}{a}$

20. $\int \dfrac{du}{\sqrt{a^2 - u^2}} = \sin^{-1} \dfrac{u}{a}$

21. $\int \dfrac{du}{\sqrt{u^2 + a^2}} = \ln |u + \sqrt{u^2 + a^2}| = \sinh^{-1} \dfrac{u}{a}$

22. $\int \dfrac{du}{\sqrt{u^2 - a^2}} = \ln |u + \sqrt{u^2 - a^2}| = \cosh^{-1} \dfrac{u}{a}$

23. $\int \dfrac{du}{a^2 - u^2} = \dfrac{1}{2a} \ln \left| \dfrac{a+u}{a-u} \right| = \begin{cases} (1/a) \tanh^{-1}(u/a) & \text{if } |u| < |a| \\ (1/a) \coth^{-1}(u/a) & \text{if } |u| > |a| \end{cases}$

24. $\displaystyle\int_0^{\pi/2} \sin^{2n} u \, du = \int_0^{\pi/2} \cos^{2n} u \, du = \dfrac{1 \cdot 3 \cdot 5 \cdots (2n-1)}{2 \cdot 4 \cdot 6 \cdots (2n)} \dfrac{\pi}{2}, n = 1, 2, \ldots$

25. $\displaystyle\int_0^{\pi/2} \sin^{2n+1} u \, du = \int_0^{\pi/2} \cos^{2n+1} u \, du = \dfrac{2 \cdot 4 \cdot 6 \cdots (2n)}{1 \cdot 3 \cdot 5 \cdots (2n+1)}, n = 1, 2, \ldots$